Climate Change Management

Series editor

Walter Leal Filho, Hamburg, Germany

More information about this series at http://www.springer.com/series/8740

Walter Leal Filho · Belay Simane
Jokasha Kalangu · Menas Wuta
Pantaleo Munishi · Kumbirai Musiyiwa
Editors

Climate Change Adaptation in Africa

Fostering Resilience and Capacity to Adapt

Springer

Editors
Walter Leal Filho
FTZ-NK
HAW Hamburg
Hamburg
Germany

Belay Simane
Environmental and Development Studies
Addis Ababa University
Addis Ababa
Ethiopia

Jokasha Kalangu
School of Water Resources
South Eastern Kenya University
Nairobi
Kenya

Menas Wuta
Soil Science and Agricultural Engineering
University of Zimbabwe
Harare
Zimbabwe

Pantaleo Munishi
Department Ecosystems and Conservation, College of Forestry, Wildlife and Tourism
Sokoine University of Agriculture
Morogoro
Tanzania

Kumbirai Musiyiwa
Agricultural Sciences and Technology
Chinhoyi University of Technology
Chinhoyi
Zimbabwe

ISSN 1610-2010 ISSN 1610-2002 (electronic)
Climate Change Management
ISBN 978-3-319-49519-4 ISBN 978-3-319-49520-0 (eBook)
DOI 10.1007/978-3-319-49520-0

Library of Congress Control Number: 2016960722

Printed on acid-free paper

This Springer imprint is published by Springer Nature
The registered company is Springer International Publishing AG
The registered company address is: Gewerbestrasse 11, 6330 Cham, Switzerland

Preface

It is widely known that Africa is one of the most vulnerable continents to climate change. As the Fifth Assessment Report (AR5) produced by the Intergovernmental Panel on Climate Change (IPCC) has shown, climate change is expected to have widespread impacts on African societies and Africans' interaction with the natural environment. There are also signs that the impacts of climate change are already being felt, not only in terms of increases in temperature, but also in respect of agriculture (with lower crop yields) and the availability of water resources, among others. The links between climate change and the incidence of diseases such as malaria are also becoming clearer.

The above state of affairs illustrates the need for a better understanding of how climate change affects African countries, and for the identification of processes, methods and tools which may help African nations to adapt. There is also a perceived need to showcase successful examples of how to cope with the social, economic and political problems posed by climate change in Africa.

It is against this background that the "Symposium on Climate Change Adaptation in Africa" was organised by Manchester Metropolitan University (UK), Addis Ababa University, the Research and Transfer Centre "Applications of Life Sciences" of the Hamburg University of Applied Sciences (Germany) and the International Climate Change Information Programme (ICCIP). The Symposium, held in Addis Ababa in February 2016, was a truly interdisciplinary event, mobilising African and non-African scholars undertaking research and/or executing climate change projects in the African continent.

This book, which contains a set of papers presented at the Symposium, focuses on "Fostering African Resilience and Capacity to Adapt", meaning that it will serve the purpose of showcasing experiences from research, field projects and best practice in climate change adaptation in African countries, which may be useful or implemented in other countries in the continent.

Consistent with the need for more cross-sectoral interactions among the various stakeholders working in the field of climate change adaptation in the African continent, this book aims to:

i. provide research institutions, universities, NGOs and enterprises from Africa and those working in Africa with an opportunity to display and present their works in the field of climate change adaptation;
ii. foster the exchange of information, ideas and experiences acquired in the execution of climate change adaptation projects, especially successful initiatives and good practice across the African continent;
iii. introduce methodological approaches and experiences deriving from case studies and projects, which aim to show how climate change adaptation may be implemented in practice; and
iv. to network African and non-African experts, and provide a platform so they can explore possibilities for cooperation.

Last but not least, a further aim of this book is to document and disseminate the wealth of experiences available today.

This book is divided into two parts:

- Part 1 contains papers that describe the adaptation methods and approaches.
- Part 2 entails institutional experiences on adaptation, as well as case studies, examples of projects and of good practice

We thank the authors for their willingness to share their knowledge, know-how and experiences, as well as the many peer reviewers, which have helped us to ensure the quality of the manuscripts. Thanks are also due to Magdalena Salewski for her valuable support for the manuscripts.

Enjoy your reading!

Hamburg, Germany — Walter Leal Filho
Addis Ababa, Ethiopia — Belay Simane
Nairobi, Kenya — Jokasha Kalangu
Harare, Zimbabwe — Menas Wuta
Morogoro, Tanzania — Pantaleo Munishi
Chinhoyi, Zimbabwe — Kumbirai Musiyiwa
Winter/Spring 2017

Contents

Part I
Adaptation Methods and Approaches

Convenient Solution for Convenient Truth: Adoption of Soil and Water Conservation Measures for Climate Change and Variability in Kuyu District, Ethiopia

Abayineh Amare and Belay Simane

1 Introduction

The Ethiopian economy has largely remained dependent on agriculture, which provides about 44% of the GDP, over 80% of the export revenue and employment for about 80% of the population (CSA 2012). At present, Ethiopia is facing greater land degradation problem to the development of agriculture and food security (FAO 1986; Merrey and Gebreselassie 2011). The study by Yesuf et al. (2008) found out soil erosion in Ethiopia is estimated to cause a damage of about one billion tons of topsoil annually. According to Berry (2009) the loss of soil and essential nutrients due to unsustainable agricultural practices is costing $139 million or 3–4% of its agricultural GDP. The problem of soil erosion in Ethiopia is attributed to erratic and erosive rainfall, steep terrain, deforestation, inappropriate land use, land fragmentation, overgrazing and farmers' management practices (Osman and Sauerborn 2001). Soil erosion problem is further exacerbated by intense and continuous cultivation on sloping land, without supplementary use of soil amendments and conservation technologies (Bekele and Holden 1998). Additionally, it has been reported that climate change can increase potential erosion rates, which can lower agricultural productivity by 10–20%, and more in extreme cases (Jorge et al. 2011). The Intergovernmental Panel on Climate Change report revealed that rainfall intensities will increase in many parts of the world, increasing the potential for soil erosion (IPCC 2007).

Give soil erosion problem posed by climate change, soil and water conservation technologies have been suggested as a key adaptation strategy for developing

A. Amare (✉) · B. Simane
College of Development Studies, Addis Ababa University, Addis Ababa, Ethiopia
e-mail: abayineh.amare@aau.edu.et

B. Simane
e-mail: belay.simane@aau.edu.et

W. Leal Filho et al. (eds.), *Climate Change Adaptation in Africa*,
Climate Change Management, DOI 10.1007/978-3-319-49520-0_1

countries. The study on the adoption and continued use of stone terraces by farmers for soil and water conservation by Amsalu and de Graaff (2006) revealed that these soil and water conservation technologies reduced run-off and soil loss through improvement in soil structure, increasing infiltration and soil resistance to detachment due to increased soil cover.

The problem of land degradation due to soil erosion received great attention in Ethiopia following the 1973/74 famine (Lundgren 1993). Since then, several soil and water conservation and land reclamation projects were initiated with the support of donor agencies and efforts have been put in place in order to rehabilitate degraded areas. For these purpose various SWC measures were introduced (Dejene 2003; Amsalu 2006). A variety of conservation structures, applicable to different soil types, rainfall conditions and topography such as soil bunds, stone bunds, *Fanya juu* (to throw up') were developed (Hurni 1993). Furthermore, planting trees on hillsides and catchments areas, water harvesting in drier areas, stream development, construction of earth dams, pond, gully plugging, traces, diversion of drains, and check dam are SWC practices developed in different part of Ethiopia to which Kuyu district is not an exception (Asrat et al. 2004). However, earlier studies on adoption of soil and water conservation to combat land degradation problem in Ethiopia showed that farmers have not been changed markedly nor adopt most of the recommended conservation measures (Shiferaw and Holden 1998; Gebremedhin and Swinton 2003; Beshah 2003; Merrey and Gebreselassie 2011).

The causes for failures and low adoption of introduced soil conservation practices were attributed mainly to the approach in the development and transfer of the conservation practices (Shiferaw and Holden 1998). Among others, failure to recognize differences in agro-ecological and socio-economic settings in which farmers operate has been considered as contributing to low adoption of soil and water conservation measures (Bekele and Drake 2003; Pender et al. 2001). For instance Pender et al. (2001) showed that farmers' decision to adopt soil and water conservation measures depends up on specific characteristics of farm plots and the importance of the plot to the household economy. Furthermore, adoption of soil and water conservation measures are constrained by a combination of unfavourable physical environment, population pressure, institutional set up, and short term household benefit (Pender et al. 2001; Tizale 2007). Hence due to these complex set of factors influenced farmer's incentive to soil and water conservation, adoption of *these practices remains below expectations* in different parts of Ethiopia including kuyu district.

In different parts of Ethiopia, factors influencing adoption and management of SWC have been investigated (Graaff et al. 2008; Kassie et al. 2008; Kato et al. 2011; Teshome et al. 2012). However, due to heterogeneity in agro ecology, socio-economic characteristics and institutional arrangements in different part of the country, it is difficult to extrapolate results from other area. Therefore, this study is aimed to analyse determinants of adoption of soil and water conservation measures by smallholder farmers in Kuyu district.

1.1 *Objectives of the Study*

The study is set to ascertain the use of soil and water conservation measures as climate change adaptation techniques by the smallholder farmers in the study area. Specifically, the study aimed at: (1) to explore soil and water conservation measures that smallholder farmers in Kuyu district employed in response to climate change and variability; and (2) to examine factors that constrain and/or facilitate the adoption of soil and water conservation measures in the district.

2 Methods and Materials

2.1 *Description of the Study Site*

This research is undertaken in Kuyu district located in North Shoa zone of Oromiya regional state. The area lies approximately 160 km to the northwest of Addis Ababa, the capital city, on the way to Gojam. At present, this area is facing severe soil degradation. The total population of the district is 126,546 of which the rural population is 103,065 (CSA 2008). The average household's size is 6.5. It has three livelihood zones via: Ambo Selale Ginde-Beret Teff and Wheat Livelihood Zone; Muger-Abay-Jema Sorghum and Teff Belt Livelihood Zone and Selale-Ambo Highland Barley, Wheat and Horse bean Belt Livelihood Zone.

This site is located at about 9°36′34″–9°56′56″ N latitude and 38°05′00″–38°34′ 13″ E longitude. The total area of Kuyu district is 974 km^2. It receives its maximum rainfall during summer season-June, July and August (Mesfin 1984). The predominant economic activity and land use is mixed agriculture; having land use systems of agricultural land (mainly rain fed), grazing land, and forest/bush. The main crops grown includes: *teff (Eragrostis tef)*, wheat, barley, and *nug* (*Guizotia abyssinica*) major rainy season in the area; and Sorghum in the next rainy season. Given the problem of soil erosion in the area, it is the most intervened area with various soil and water conservation measures.

2.2 *Research Design*

The study followed a cross-sectional research design in which data from households were collected from June to August 2015. In view of the diverse impact of climate change and variability on smallholder farmers and the nature of the information needed on various aspects of this research, employing a single method of data collection method is impossible to satisfy data requirements. Therefore, this demands a multi-methods of data collection approach to generate adequate and reliable data that will be enhanced through triangulation. Many authors advocate this approach in such a way that it makes possible to develop an integrated system in which the first method sequentially informs the second method, contradictions and fresh perspectives appear, and different facets of the phenomena emerge in order to keep the data both comprehensive and authentic (Mathison 1988; Greene et al. 1989; Swanson 1992).

In similar vein, this research employed mixed methods of data collection both to collect data from primary and secondary sources. Primary data were collected using a pre-tested semi structured questionnaire, key informant interview and focussed group discussions. Moreover, secondary data were gathered through reviewing documents, reports and records maintained at district rural development and agricultural office. A total of 100 households were selected and surveyed using systematic random sampling technique. Lists of households were obtained from the sampled kebele offices.

To analyse the data, descriptive statistics such as frequencies, percentages, and means were used. In order to make a decision on whether or not a significant relationship existed between adoption of SWC technologies in Kuyu district and the variables investigated, a chi-square test and t-test were performed for dummy and continuous variables, respectively (Table 1).

Table 1 Descriptions, definition and values of variables used in the independent t-test and chi-square test

Variables	Definition	Values
Adoption	Adopted stone bunds	0 = not adopted (no stone bund structures on his/her farmland; 1 = adopted (presence of stone bund on his/her farmland)
Farm size	Landholding of the family	Total landholding in hectares (continuous)
Family size	Number of people in the family	Refers to the number of members who are currently living within the family
Age	Age of family head	It is a continuous variable measured in years
Livestock	Number of livestock owned in TLU	Continuous
Distance of the plot	Distance from their residence	It is a continuous variable measured in hours and refers to distance of the plot from the farmer's house
Education	Education level of the family head	It is a categorical variable representing illiterate, read and write, grade 1–4, grade 5–8, and above grade 8 of the household heads
Perception of soil erosion	Household head perception of soil erosion problem	0 = not perceive the problem; 1 = otherwise
Slope of the plot	Perceived slope of the plot	0 = plain; 1 = steep
Training	Household access to training on soil bunds	0 = not access to training; 1 = access to training

(continued)

Table 1 (continued)

Variables	Definition	Values
Access to credit	Access to credit	It is a dummy variable, which takes the value 1 if the farm household access to credit and 0 otherwise
Gender	Sex of the household head	Refers to sex of the head of the household having a binary value. If the household head is male, it takes a value of 1; 0 otherwise
Number of dependents in the household	Family member <15 and ≥65 years old	Total number of family members in this age range

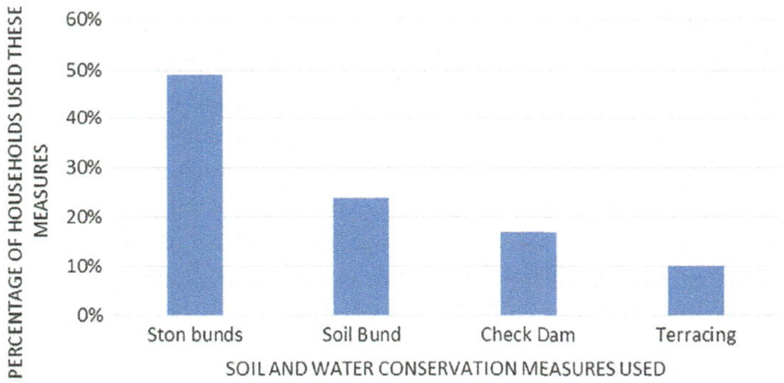

Fig. 1 Respondents use of soil conservation techniques as climate change adaptation strategies. *Source* Field Survey, 2015

3 Results and Discussions

3.1 Soil and Water Conservation Measures in the Study Area: An Overview

This study indicated that different types of soil and water conservation practices were undertaken in Kuyu district. Stone bund is the most widely and most intensively used soil conservation structures in the area. The study revealed that from 100 households interviewed, 49% were used stone bund on their farm plots as climate change adaptation strategy. The discussions with key informants revealed that it is due to high potential of stone in the area that farmers mostly used stone bund in their farm plots. Soil bund were used by 24% of farmers, next to stone bund. Moreover, Fig. 1 presents that check dam were used by smallholder farmers (17%) and followed by terracing (10%).

3.2 Determinants of Use of Soil and Water Conservation Measures

In this analysis, we used only stone bunds that the farmers used as climate change adaptation strategy in their farm plots. This is mainly because stone bund were used by many farmers and we are interested to examine the factors that influence adoption of soil bund. First, surveyed households were classified into adopters and non-adopters. Adopter households were those that had practiced conservation measures on their farm plots. The non-adopter households were those that had not practiced soil and water conservation measures on their farm plot (Bekele 2003).

The result indicated that among many explanatory variables, nine variables were found to significantly affect use of stone bund. The result of independent t-test showed that farm size, family size, age of the household head, number of livestock and distance of the plots from their residence were determining farmers use SWC practices (Table 2). Moreover, chi-square test result revealed that education of the household head, perception of soil erosion problem, slope of the plot, and training on soil and water conservation were significant determinant factors in adopting soil and water conservation measures in the study area. Interpretation of independent t-test and chi-square test results is discussed in detail as follow:

Farm Size. The effect of farm size is also found to be positive and significant, suggesting that farmers who hold large farms are more likely to invest in conservation. The positive influence might be explained by the propensity of retaining conservation structures increases with increasing availability of land resources. An increase in landholding size in the study area encouraged management of the land resource. This is true because farmers having larger farm size can allocate some part of the land to stone bund than those farmers who have small farms. Moreover, large farm operators will have more opportunity to use new practices on a trial basis and

Table 2 Differences of continuous explanatory variables between adopters and non-adopters

Variables	Adopters		Non-adopters		t-value
	Mean	St. deviation	Mean	St. dev	
Farm Size in hectare	1.82	0.232	1.12	0.33	2.8**
Family Size in number	6.048	1.678	3.234	2.123	0.224**
Age of HH in years	43.322	10.542	60.435	5.456	−7.14***
Number of dependents in the household	1.745	1.211	2.512	2.421	−0.135**
Number of Livestock in Tropical Livestock Unit	1.211	0.833	2.736	0.689	−9.074**
Distance of the plots in walking hours	22.571	10.567	30.492	11.688	−1.549***

Source Own survey, 2015
*** Significant at $p < 0.001$
** Significant at $p < 0.05$

more ability to deal with risk. This is consistent with the findings of earlier studies in different part of Ethiopia that report a positive and significant effect of farm size on the decision to use conservation measures (Amsalu and de Graaff 2007; Kassa et al. 2013). Contrary results were obtained from Toni sub-watershed, indicating that farmers with large farms have alternative land to plough, and can allow for a fallow period; hence, they may neglect the maintenance of SWC structures (Kebede and Mesele 2014).

Family Size. This is a major determinant in SWC, especially with respect to poor resource farmers who depend solely on family labour to maintain their farms. This might have a link with the high rate of adoption of SWC technologies in the district. Keil (2001) noted that household size influences the decision of farmers to undertake the conservation measures given household labour is the whole supplier of the required labour for undertaking the farming and soil conservation operation. This study proved that adoption of SWC is influenced significantly ($p < 0.05$) by household size. However, the result by Aklilu and de Graaff (2006) revealed that an increase in family size demands more food. Thus, family members may become involved in off-farm work to generate income for securing a consistent food supply that indeed reduce participating in the maintenance of SWC structures.

Age of the Household Head. The age of the household head was negatively and significantly ($p < 0.001$) related to the adoption of SWC in the study area. This may be explained by the fact that older farmers resisted the adoption of new technology. Another reason for the expected negative relationship between age and conservation effort is an assumed longer planning horizon for younger farmers relative to older ones. This finding is inconsistent with the result which showed the likelihood of adoption of conservation practices is more among older farmers than the younger ones, perhaps due to the experience of older farmers to perceive erosion problems and their limited participation in off-farm activities. With experience, farmers notice and recognize erosion problems, learn how to conserve soil and develop capacity and strategy to cope with erosion problems (Kebede and Mesele 2014; Kidane et al. 2012).

Number of Livestock Owned. Livestock constitutes an important component of farming system in the study area. However, the result showed that the effect of livestock size on adoption of stone bund was found to be negatively significant. This may be explained by wealthy farmers have other resource options besides farmland and less concerned about adopting SWC technologies for improving productivity. This indicates that farmers with large livestock have more tendency to focus more on livestock than on crop production. Moreover, farmers who have large numbers of cattle may ignore structure maintenance, expecting frequent damage by cattle. This result is consistent with a study conducted by Amsalu and de Graaff (2007) in central highlands of Ethiopia which showed that livestock ownership has negative influence to adopt stone terrace. On the contrary, many empirical studies have shown positively and significantly effect of number of livestock on adoption of SWC measures (Damena 2012; Kassa et al. 2013).

Distance of the Plots. The effect of distance of the plots from farmers' residence on adoption of stone bund is found to be significantly negative. This analysis revealed that the tendency of retention of conservation structures decreases with increasing distance of plot from the residential area. Farmers whose plots are near to their residence use soil conservation measures than farmers whose plots are far from their residence because time and energy spent is relatively lesser than distance plots. This is perhaps due to plots far away from home take more time and energy to construct soil conservation structures as well as other farming practices. The cost of soil conservation includes not only cash costs, but also transaction costs of travel to plots distant from the homestead. Similarly, many empirical studies conducted in different period likewise found that distance of plots from homestead discouraged investment in soil conservation (Shiferaw and Holden 1998; Bekele and Drake 2003; Regasa 2005).

Education Status of the Household Head. As expected, level of education is positively and significantly related with adoption of conservation structures in Kuyu district. This is perhaps due to literate farmers are in a better position to get information and use it in such a way that it contributes in their farming practices. This could be attributed to the fact that household heads with relatively better formal education are more likely to use appropriate SWC practices and they also able anticipate the consequences of soil erosion than non-educated farmers. This result is in line with evidence from different parts of the country where adoption is high among farmers with higher education (Anley et al. 2007; Tizale 2007).

Perception of Soil Erosion Problem. Majority of farmers interviewed (58%) attested to the fact that there exists soil erosion problems in the study area. The chi-square test revealed that there exists a significant relationship between adoption of SWC technologies and perception of soil erosion problem($X^2 = 24.042$, df = 1, p = 0.000), and that the tendency to adopt the technology was correspondingly high. However, this result contradicts with the finding by Awdenegest and Holden (2007) in Southern Ethiopia, where farmers' own initiatives were minimal, even under serious, advanced erosion.

Perceived Slope of the Plot. The slope condition of cultivated plots is an important determinant of farmer's investment in soil and water conservation measures. As expected, the influence of slope of the plot on farmer's decision to invest in soil bund is significantly positive ($p < 0.001$). In most SWC adoption studies, it has been shown that adoption of SWC measures are positively related to slope of the plot (Asrat et al. 2004; Rgasa 2005; Amsalu and de Graaff 2007). This might be because farmers cultivating sloping fields perceive the threat of soil loss better than farmers who cultivate gentle or level sloping fields that indeed motivate farmers more likely to adopt SWC technologies in their more steep farms than those cultivating less steep lands. Further, the slope of a plot influences the decision of SWC practices positively for the reason that erosion is more serious on steeper plot than flat plots. This suggests that targeting the stone bund on a steep plots might induces adoption of the measure.

Training on Soil and Water Conservation. Training on soil and water conservation significantly and positively influenced farmer's decision to adopt stone bund in the study area. This agrees with the argument that information obtained and the knowledge and skills gained through training accelerates farmer's decision on

Table 3 Differences of continuous explanatory variables between adopters and non-adopters

Variables	Chi-square values
Education status	13.48**
Perception of Soil Erosion Problem	24.042***
Access to credit	11.734
Slope of the plot	12.672***
Access to training	39.235***
Off-farm activities	0.044
Sex of the HH	2.13

Source Own survey, 2015
*** Significant at $p < 0.001$
** Significant at $p < 0.05$

conservation practices (Shiferaw and Holden 1998; Sidibe 2004). The possible explanation for this is that farmer who got training on SWC from development agent could be more encouraged to use SWC practices on their farm plots. This suggests that conservation efforts should encompass continuous training in order to encourage farmers to adopt. The chi-square result also showed that access to credit, off-farm activities, and sex of the household head no significant influence on adoption of stone bund (Table 3).

3.3 Summary

Soil erosion is one of the most serious environmental problems in Ethiopia. Climate change aggravates this problem. A number of soil and water conservation methods were introduced to combat soil erosion but adoption of these practices remains below expectations. Thus, this paper explored major adaptation strategies smallholder farmers used to combat soil erosion problem caused by climate change and variability. A special emphasis was given to investigate determinants of farmers' adoption of stone bund for adapting to climate change and variability.

The common climate adaptation strategies used to mitigate the effect of flooding among farmers in the study area were stone bud, soil bund check dam, and hillside terracing which have direct effect on soil erosion. The result depicted that adoption of stone bund is conditioned by different factors at different levels of significance. Results of independent t-test and chi-square test showed that adoption of stone bund was positively and significantly determined by size of landholding, family size, education of the household head, perception of soil erosion problem, slope of the plot, and access to training on soil and water conservation. Furthermore, adoption of stone bund was negatively and significantly determined by age of the household head, number of dependents, number of livestock, and distance of the plot.

3.4 Conclusions and Recommendations

The study showed that stone bund, soil bund, check dam, and hillside terracing were major soil and water conservation measures in the study area. In Kuyu district, a rage of factors influenced farmers' decisions to invest in soil and water conservation measures. Result of independent t-test and chi-square test showed that adoption of soil bund was positively and significantly influenced by farm size, family size, education, perception of soil erosion, slope of the plot and access to training on stone bund, on one hand. On the other hand, adoption of soil bund is negatively and significantly determined by increase in age of the household head, increase in number of livestock, and increase in distance of the plot from their residence.

On the basis of the survey results, the following recommendations were made: conservation interventions that failed to account for inter household variation (age, education, family size) and inter-plot variation (slope of the plot) are unlikely to be effective. Hence, any development intervention intended to enhance agricultural productivity through promoting SWC practices in the study area need to consider those differences in program design and implementation. Moreover, it is also imperative to give a due attention to the importance of farmers' perception on soil erosion problem for adoption of soil and water conservation measures. The study suggested strengthening agricultural extension services to make farmers more informed and knowledgeable about climate change impact on soil erosion and the adaptation strategies to use.

Acknowledgements We wish to express our profound gratitude to Addis Ababa University; the German Academic Exchange Service (DAAD), Germany; African Climate change Fellowship Program; and International Development Research Centre (IDRC) for their financial support in accomplishing this paper.

References

Amsalu, A. (2006). *Caring for the land best practices in soil and water conservation in Beressa Watershed, Highland of Ethiopia*. Tropical Resource Management Paper No. 76: Wageningen University.

Amsalu, A., & de Graaff, J. (2006). Farmers' views of soil erosion problems and their conservation knowledge at Beressa watershed, central highlands of Ethiopia. *Agriculture and Human Values, 23*(1), 99–108.

Amsalu, A., & de Graaff, J. (2007). Determinants of adoption and continued use of stone terraces for soil and water conservation in an Ethiopian highland watershed. *Ecological economics, 61*, 294–302.

Anley, Y., Bogale, A., & Haile-Gabriel, A. (2007). Adoption decision and use intensity of soil and water conservation measures by smallholder subsistence farmers in Dedo district, Western Ethiopia. *Land Degradation and Development, 18,* 289–302.

Asrat, P., Kassa, B., & Hamito, D. (2004). *Determinants of farmers' willing to pay for soil conservation practices in the southern highlands of Ethiopia: Land degradation and development, 15,* 423–438.

Awdenegest, M., & Holden, N. M. (2007). Farmers' perceptions of soil erosion and soil fertility loss in Southern Ethiopia. *Land Degradation & Development, 18*(5), 543–554

Bekele, S., & Holden, S. T. (1998). Resource degradation and adoption of land conservation technologies in the Ethiopian Highlands: A case study in Andit Tid, North Shewa. *Agricultural Economics, 18,* 233–247.

Bekele, W. (2003). *Economics of soil and water conservation: Theory and empirical application to subsistence farming in the eastern Ethiopian highlands.* PhD. Dissertation, Uppsala.

Bekele, W., & Drake, L. (2003). Soil and water conservation decision behavior of subsistence farmer s in the Eastern Highlands of Ethiopia: a case study of the Hunde-Lafto area. *Ecological Economics, 46*(3), 437–451.

Berry, L. (2009). *Land degradation in Ethiopia: Its extent and impact.* Global Mechanism and World Bank.

Beshah, T. (2003). *Understanding farmers: Explaining soil and water conservation in Konso, Wolaita and Wello, Ethiopia.* Tropical Resource Management Papers, vol. 41. The Netherland, Wageningen University.

Central Statistical Authority (CSA). (2008). Statistical Abstract, CSA, Addis Ababa.

CSA (Central statistics Agency). (2012). "*[Ethiopia] Agricultural sample survey 2011/2012 (2004 E.C.). (September–December, 2011), Volume IV, report on land utilization (Private Peasant Holdings, Meher Season),*" Statistical Bulletin 532. Addis Ababa: Federal Democratic Republic of Ethiopia, Central Statistical Agency.

Damena, D. (2012). Determinants of Farmers' Land Management Practices: The Case of Tole District, South West Shewa Zone, Oromia National Regional State, Ethiopia. *Journal of Sustainable Development in Africa, 14*(1), 76–96.

Dejene, A. (2003). *Integrated national resources management to enhance food security: The case of community based approaches in Ethiopia.* Rome: FAO.

FAO. (1986). *Ethiopian highlands reclamation study, Ethiopia. Final Report.* Rome: FAO.

Gebremedhin, B., & Swinton, S. M. (2003). Investment in soil conservation in northern Ethiopia: the role of land tenure security and public programs. *Agricultural Economics, 29,* 69–84.

de Graaff, J., Amsalu, A., Bodna, F., Kessler, A., Posthumus, H., & Tenge, A. (2008). Factors influencing adoption and continued use of long-term soil and water conservation measures in five developing countries. *Applied Geography, 28,* 271–280.

Greene, W. H., et al. (1989). Towards a conceptual framework for mixed methods evaluation design. *Educational Evaluation and policy Analysis, 11*(30), 255–274.

Hurni, H. (1993). Land degradation, famine and land resources scenario in Ethiopia. In D. Pimentel (ed.), *World soil erosion and conservation* (pp. 27–62). Cambridge: Cambridge university press. http://dx.doi.org/10.1017/CBO9780511735394.004

IPCC (Intergovernmental Panel on Climate Change). (2007). Summary for Policymakers. In Climate Change 2007: The Physical Science smallholder maize in central Malawi. *Experimental Agriculture, 50*(2), 229–249.

Jorge, A., Peter, M., Mark, A., Tom, G., Don, R., Rattan, L., et al. (2011). Conservation practices to mitigate and adapt to climate change. *Journal of Soil and Water Conservation, 66*(4), 118–129.

Kassie, M., Pender, J., Yesuf, M., Kohlin, G., Bluffstone, R., & Mulugeta, E. (2008). Estimating returns to soil conservation adoption in the northern Ethiopian highlands. *Agricultural Economics, 38,* 213–232.

Kassa, Y., Beyene F., Haji, J., & Legesse, B. (2013). Impact of integrated soil and water conservation program on crop production and income in West Harerghe Zone, Ethiopia. *International Journal of Environmental Monitoring and Analysis, 1*(4), 111–120.

Kato, E., Ringler, C., Yesuf, M., & Bryan, E. (2011). Soil and water conservation technologies: a buffer against production risk in the face of climate change? Insights from the Nile basin in Ethiopia. *Agricultural Economics, 42,* 593–604.

Wolka, K., & Negash, M. (2014). Farmers' adoption of soil and water conservation technology: a case study of the Bokole and Toni sub-watersheds, southern Ethiopia. *Journal of Science & Development, 2*(1), 35–48.
Keil, A. (2001). Adoption of leguminous tree fallows in Zambia. Discussion Paper No. 33, Universität Göttingen, Göttingen, Germany.
Tafa, Kidane, Beshah, Tesfaye, & Amsalu, Aklilu. (2012). Determinants of physical soil and water conservation practices in Ethiopia's semi-arid tropics: The case of Bati District. *Social and Basic Science Review Research, 2*(12), 525–541.
Lundgren, L. (1993). *Twenty years of soil and water conservation in eastern Africa*. SIDA, Nairobi: RSCU.
Mathison, S. (1988). Why Triangulate? *Educational Researcher, 17*(2), 13–27.
Merrey, D. J., & Gebreselassie, T. (2011). *Promoting improved rainwater and land management in the Blue Nile (Abay) basin of Ethiopia*. NBDC Technical Report 1. Nairobi, Kenya, ILRI.
Mesfin, W.-M. (1984). *Rural vulnerability to famine in Ethiopia 1958–1977*. Addis Ababa: Addis Ababa University Press.
Osman, M., & Sauerborn, P. (2001). Soil and water conservation in Ethiopia: Experiences and lessons. *Journal of Soils and Sediments, 1*(2), 117–123.
Pender, J., Gebremedhin, B., Benin, S., & Ehui, S. (2001). Strategies for sustainable agricultural development in the Ethiopian Highlands. *American Journal of Agricultural Economics, 83*(5), 1231–1240.
Regasa, S. (2005). Determinants of choice of land management practice: The Case of Ankober District. *Ethiopian Journal of Agricultural Science, 18,* 53–71.
Shiferaw, B., & Holden, S. (1998). Resource degradation and adoption of land conservation techno logies in the Ethiopian highlands: A case study in Andit Tid, North Shewa. *Agricultural Economics, 18,* 233–247.
Sidibe, A. (2004). Farm-level adoption of soil and water conservation techniques in Northern Burkina Faso. *Agricultural Water Management, 71*(3), 211–224.
Swanson, S. (1992). *Mixed-method triangulation: Theory and practice compared*. Paper presented at annual meeting of the American Educational Research Association, San Francisco.
Teshome, A., Rolker, D., & Jde, Graaff. (2012). Financial viability of soil and water conservation technologies in northwestern Ethiopian highlands. *Applied Geography, 37,* 139–149.
Tizale, C. Y. (2007). *The dynamics of soil degradation and incentives for optimal management in Central Highlands of Ethiopia*. PhD. Dissertation, South Africa: University of Pretoria.
Yesuf, M., et al. (2008). *The impact of climate change and adaptation on food production in low-income countries: evidence from the Nile Basin, Ethiopia*. International Food Policy Research Institute.

Between Climate Reliance and Climate Resilience: Empirical Analysis of Climate Variability and Impact on Nigerian Agricultural Production

Olawale Emmanuel Olayide and Isaac Kow Tetteh

1 Introduction

Climate change is now a global phenomenon that portends significant developmental challenges. The agricultural sector is no exception to the impact of climate change (Choptiany et al. 2015). The potential and predicted impacts of climate change are resulting in increased frequency and intensity of rainfall, floods and droughts (IPCC 2015). Rain-fed agricultural production system is vulnerable to seasonal variability which affects the livelihood outcomes of farmers and landless laborers who depend on such system of agricultural production. (Choptiany et al. 2015; Vermeulen et al. 2012). Climate change affect agriculture through rainfall variability (IPCC 2015). This situation, therefore, makes climate change an important consideration for sustainable agricultural production (Easterling et al. 2007). In the events of erratic rainfall, irrigated land area is insurance to rain-fed agriculture and a predictor of resilience of agriculture to rainfall-induced vagaries (including, droughts and heat waves) and impact of climate change (Cassman and Grassini 2013). Hence, the need for the empirical analysis of the impacts of rainfall and irrigation on agricultural production in Nigeria.

The agricultural sector is increasingly showing high level of vulnerability and impact. Climate change across Africa is exacerbated by low level of adaptation and mitigation (IPCC 2015; Montpellier Panel Report 2015). Further, literature suggests that farmers are now adapting to climate change and building resilience to vagaries of climate change (Choptiany et al. 2015; Wood et al. 2014; Kristjanson et al. 2012).

O.E. Olayide (✉)
Centre for Sustainable Development, University of Ibadan, Ibadan, Nigeria
e-mail: waleolayide@yahoo.com; oe.olayide@ui.edu.ng; oeolayide@knust.edu.gh

O.E. Olayide · I.K. Tetteh
Department of Theoretical and Applied Biology, Kwame Nkrumah University of Science and Technology, Kumasi, Ghana

W. Leal Filho et al. (eds.), *Climate Change Adaptation in Africa*,
Climate Change Management, DOI 10.1007/978-3-319-49520-0_2

The agricultural production risk imposed by rainfall variability may be a motivation or hindrance to investment in improved agricultural technology and climate resilient agriculture. Besides, farmers who are unable to adapt to changing climate may find alternative livelihoods or remain impoverished. Others may become resilient by developing alternative systems of production that help them to cope with changing climate. There is, therefore, pseudo choice-making process that is constrained by initial endowment or capacity to innovate so as to overcome vulnerability by becoming climate-resilient through appropriate adaptation and mitigation strategies (Montpellier Panel Report 2015; Wood et al. 2014). It has been noted that any strategy to adapt agriculture and food systems to a changing climate must, therefore, exploit the diversified means of climate resilient strategies. (Vermeulen et al. 2012).

Variability and extreme rainfall events have the potential to transform agricultural production system (rain-fed or irrigated) and diversifications of agricultural production (Liverman and Kapadia 2010; Nelson et al. 2009). The ability to circumvent the negative impact of climate and weather variability in agricultural production is an important consideration for climate adaptation and resilient agriculture for maximizing its benefits agricultural livelihoods and economic development.

The paper builds on emerging literature on the impact of climate variability on agricultural production (Ajetomobi et al. 2015; Craparo et al. 2015; Gourdji et al. 2015). It reveals the reliance and/or resilience of agricultural production to climate change and variability (Schlenker and Lobell 2010; Schlenker and Roberts 2009; Guiteras 2009; Kurukulasuriya et al. 2006). The study also underscores the contexts of vulnerability, impact and adaptation to climate change (Metternicht et al. 2014), including productivity, food security and livelihoods (Carandang et al. 2015; Arumugam et al. 2015).

2 Materials and Methods

2.1 Type, Measurement and Sources of Data

Time series data were extracted from harmonized databases of the Central Bank of Nigeria (CBN), National Bureau of Statistics (NBS), Nigerian Meteorological Agency (NIMET) and the Food and Agriculture Organization (FAO) of the United Nations in the Statistical Bulletin of the NBS (NBS 2013). Supplementary data on occurrence of flooding of the magnitude of national emergency situation were obtained from various publications. The specific data extracted included: agricultural production index, incidences or occurrence of flooding in a specific year, mean annual rainfall in millilitres, and value of agricultural (food) imports in million US dollars. The index of agricultural production is the relative level of the aggregate/composite volume of agricultural production (base year = 1990) (http://faostat.fao.org/site/362/DesktopDefault.aspx?PageID=362).

Since the impact of climate change is considered over a long period time (usually more than 30 years), this paper analysed times series dataset that spanned a period of 43 years (1970–2012) to estimate the impact of rainfall and irrigation on aggregate agricultural production in Nigeria. This criterion of sufficient time period also satisfies the econometric properties of large sample size (or observations) of the generalised methods of moment (GMM) estimation (Craparo et al. 2015; Gourdji et al. 2015; Hansen 2012), and consequently the estimation of the impact of rainfall and irrigation on agricultural production.

2.2 Analytical Methods

Descriptive analyses (means and standard deviations) were used to analyse the dataset to elaborate the variables. The GMM econometric technique was employed in estimating the impact of rainfall and irrigation on agricultural production. The choice of GMM was informed because the ordinary least squares estimation technique (regression) might result in biased estimation which is particularly linked to spurious regression and endogeneity problems (Fan et al. 2008). The issue that may cause spurious regressions is the possible existence of unit roots or non-stationarity of variables in the time series data analysis. This problem was handled by differencing while the problem of endogeneity of correlated independent variable (Fan et al. 2008) was resolved with the use of instrumental variables in the GMM estimation procedures.

Following Fan et al. (2008), and Arellano and Bond (1991), a GMM estimator as an estimation method was stated as:

$$\Delta y_{it} = \sum_{e=1}^{m} a_e \Delta y_{it-e} + \sum_{e=1}^{n} \beta_e \Delta x_{it-e} + \Delta \eta_{it} + \Delta u_{it} \quad (1)$$

where y is the dependent variable; x is a set of independent variables, i = 1, …, N; m and n are the lag (Δ) lengths sufficient to ensure that u_{it} is a stochastic error and η_i are instrumental variables. Blundell and Bond (1998) suggest that if the simple autoregressive AR(1) model is mean-stationary, the first differences Δy_{it} will be uncorrelated with individual effects.

The procedure for examining the nature of dataset for stationarity is to establish whether or not there exists a long-run relationship between the dependent variables and the independent variables. According to Engel and Granger (1987), homogenous non-stationary time series, which can be transformed to a stationary time series by differencing d times, is said to be integrated of order d. Thus, Y, (a time series variable) is integrated of order d [$Y \sim I(d)$] if differencing d times induces stationarity in Y_t. If $Y_t \sim I(0)$, then no differencing is required as Y is stationary (Jefferis and Okeahalam 2000). The test proposed by Dickey-Fuller to determine the stationarity properties of a time series is called the Unit Root test denoted by DF. The regression equation for the DF class of unit root test is:

$$\Delta Y_t = \varphi Y_{t-1} + \varepsilon_t;\ \varepsilon_t \sim N(0, \sigma^2), Y_0 = 0 \quad (2)$$

The unit root test above is valid only if the series is an autoregressive, AR(1) process. The Augmented Dickey-Fuller (ADF) tests use a difference method to control for higher-order serial correction in the time series. Another alternative test for stationarity is the Phillips-Perron (PP) test. The PP test allows for individual unit root process so that the autoregressive coefficient can vary across units (Olayide and Ikpi 2013; Ajetomobi 2008). The stationarity tests make a parametric correction for higher-order correlation by assuming that the Y series follows an AR(p) process and adjusting the test methodology. The ADF is identical to the standard DF regression, but augmented by k lags of the first difference of the series as follows:

$$\Delta Y_t = \alpha Y_{t-1} + \sum_{i=1}^{k} \omega i \Delta Y_{t-1} + \varepsilon_t \quad (3)$$

where the lag k is set so as to ensure that any autocorrelation in Y_t is absorbed and that a reasonable degree of freedom is preserved, while the error term is white noise or stationary.

The GMM is widely preferred and used in applied econometric research for empirical impact analysis. Zhang and Fan (2004) applied a GMM method to empirically test the causal relationship between productivity growth and infrastructure development using India district-level data, while Fan et al. (2008) assessed the impact of public expenditure in developing countries.

2.3 *Variables Used for the Estimation of the GMM*

In estimating the GMM model, aggregate agricultural production index was specified as the dependent variables while annual mean rainfall (in millilitres) and proportion of arable land under irrigation were the independent variables. The instrumental variables were incidence of flooding and annual total value of agricultural (food) imports (in million US dollars) (Quian and Schmidt 1999). These variables are predicted to have impact on aggregate agricultural production in Nigeria. The estimations were carried with E-Views 7 econometric computer software package.

3 Results and Discussion

3.1 *Description of Variables*

Results in Table 1 show the description of variable used in the analysis. The results reveal that the index of aggregate agricultural production was above the average for the base year (1990 = 100). This result indicates that agricultural production

Table 1 Description of variables used in estimating the generalized method of moment model

Variable and measurement	Mean	Std. deviation
Index of aggregate agricultural production	119.48	67.87
Mean rainfall in mm	355.39	64.24
Proportion of arable land under irrigation	0.80	0.10
Flood occurrence (dummy)	0.42	0.50
Total agricultural (food) imports in million US dollars	2236.47	1971.98

increased above the base year period. The mean rainfall for the study period was 355.39 (± 64.24) mm. The average proportion of arable land under irrigation was less than one percent (0.80 ± 0.10). Flooding incidence of national catastrophe magnitude was recorded for average of 42% of the study period. The total agricultural (food) imports in million US dollars were worth 2236.47 (±1971.98). The implications of the results in the context of the Nigeria agricultural policy call for concern. In that, the national agricultural policy agenda seek to promote food self-sufficiency by gradual reduction in the share of food imports that have comparative and competitive advantages. However, the country still spends a lot of foreign exchange on food imports (Olayide et al. 2011), and therefore, not self-sufficient in food production. For the country to move progressively towards self-sufficient in food production and food security (availability, access and stability), it should ensure increased food production under climatic changes which would have implications for rainfall and irrigation under the current agricultural production system.

3.2 *Results of the Stationarity Tests*

As a necessarily steps for estimating times series econometric models, we examined the variables used for the GMM model for stationarity or unit roots using comparable standard test statistic recommended in literature (Breitung 2002). The natural logarithms of the variables (except incidence of flooding which is a dummy variable) were tested for stationarity/unit root using comparable test methodologies of Augmented Dickey-Fuller and the Philips-Perron. Both tests yielded similar results (see Table 2). Only average annual rainfall and value of total agricultural (food) imports were stationary (white-noised) at level. All the variables (including, average annual rainfall and value of total agricultural (food) imports) were, however, stationary at first difference which suggests that they were auto-regressive of order I (ARI) variables (Breitung 2002), and they are co-integrated with their past values. This result also informed the estimation of the GMM by suggesting the incorporation of appropriate lag length (first difference) in the model estimation (Fan et al. 2008).

Further, the stationarity tests of the variables suggest that the interdependence with one-year lag or past values. For instance, this result has implications for

Table 2 Results of unit roots tests

Variable	At level (test statistic)		At first difference (test statistic)	
	Augmented Dickey-Fuller	Philips-Perron	Augmented Dickey-Fuller	Philips-Perron
Index of aggregate production	−1.0242 (0.7359)	−1.2427 (0.6469)	−5.9026*** (0.0000)	−5.9026*** (0.0000)
Mean annual rainfall	−4.9653*** (0.0002)	−4.9871*** (0.0002)	−8.9894*** (0.0000)	−23.0233*** (0.0001)
Proportion of arable land under irrigation	−1.1104 (0.7030)	−1.3873 (0.5794)	−5.9871*** (0.0000)	−5.9955*** (0.0000)
Value of total agricultural (food) imports	−3.5210** (0.0122)	−3.4116** (0.0161)	−6.5873*** (0.0000)	−7.8971*** (0.0000)

*** indicates 1% level of significance while
** indicates 5% level of significance

availability or retention rainfall from past year in the current year, all things being equal. But we know that soil water retention/availability is affected by many factors (Brooksbank et al. 2011), which are not captured in the present study. However, it is instructive that rainfall is required in current agricultural production system in Nigeria, almost on year-to-year basis. Again, the result supports the fact that the Nigerian agricultural production system is predominantly rain-fed. This result has implications for climate adaptation and resilient agriculture in Nigeria because with the climate change predictions of variability in rainfall, drought in previous year would have negative impact on agricultural production and food security (availability, access and stability) in Nigeria.

4 Impacts of Rainfall and Irrigation on Agricultural Production

4.1 Evidence for Promoting Climate Change Adaptation and Climate Resilient Agriculture in Nigeria

The results in Table 3 revealed that irrigation had positive and significant impact on aggregate agricultural production. The diagnostic statistic (R-squared) of the aggregate agricultural production model showed that the independent variables explained 74% of the variation in agriculture production. The impact of irrigation on aggregate agricultural production showed that a unit change in arable land under would lead to 4.3% in aggregate agricultural production. This result also revealed that rainfall has positive but insignificant impact on aggregate agricultural production. The results implies that irrigation system of production, rather than the current reliance on rain-fed agriculture, holds the key to sustainable agricultural production in Nigeria, especially the face of climate change and variability.

Table 3 Impacts of rainfall and irrigation on agricultural production

Variable	Coefficient	*t*-value	Probability	R-squared
Constant	1.0265	0.1951	0.8463	0.7455
Rainfall	0.7843	0.8877	0.3802	
Proportion of arable land under irrigation	4.3260***	9.6783	0.0000	

*** indicates 1% level of significance

Irrigation agriculture is a climate change adaptation strategy for managing extreme rainfall events (including, drought) and for promoting climate resilient agriculture. Irrigation agriculture could also ensure all-year round and sustainable agricultural production, afforestation and development of ranches and pasturelands for livestock production, including aquaculture. The pasturelands could also contribute to reduced greenhouse gas emission and enhance carbon sequestration in grazing lands through grazing intensity, increased productivity, nutrient management, fire management, and species introduction (O'Mara 2012).

It should be noted that irrigation agriculture is expensive to maintain, and a long-term investment (Mulangu and Kraybill 2015). However, the cost of irrigation system could be recouped through the production of high agricultural value commodities, especially in the dry season farming. The irrigation technology could also involve the combination of drainages and channeling of flood water to reservoirs to mitigate flooding. Irrigation practices should be encouraged at both small-holder famers' level and community level. Overall, the social and economic costs of irrigation agriculture (Connor et al. 2008) are often lower than the benefits that could accrue in terms of increased agricultural productivity and diversification, economic profitability, environmental amelioration and reduction of agricultural production risks.

5 Conclusion

This paper assessed the impacts of rainfall and irrigation on agricultural Production in Nigeria with a view to determining the level of reliance or resilience in the face of climate change. We found evidence for the impact of irrigation as a tool for promoting climate change adaptation and resilient agriculture in Nigeria. Irrigation had positive and significant impact on aggregate agricultural production. The findings suggest the need for the minimization of the impact of climate-induced production risks through the climate resilient agriculture in order to bring about the much-desired sustainable agricultural production in Nigeria. Expanding the area under irrigation in Nigeria would ensure sustainable agricultural production and food security in the country. It would also enhance the development of all subsectors of agriculture. It would not only be beneficial to the crop production subsector only (Ajetomobi et al. 2015), but it would inclusive and broad-based development of the agricultural economy (World Bank 2008) in Nigeria.

Therefore, there is the need for policy shift from the predominantly rain-fed agricultural production to irrigation system of agricultural production. Such policy shift would involve progressive expansion of arable areas under irrigation in Nigeria. The irrigation system of agricultural production is consistent with climate change adaptation and resilient agriculture. The irrigation system of agricultural production would enhance long-term sustainability of food production and food security. It is important to consider the cost on investment in irrigation and sustainability dimensions (social, economic and environmental) of irrigation agriculture. Although the issue of investments on irrigation agriculture was not considered by the present study, past studies (Mulangu and Kraybill 2015) have established that the initial costs of irrigation system of agricultural production (either at farm-level or landscape level), are huge but recoverable over a long-term investment period. Notwithstanding, the overall social, economic and environmental beneficial impacts of irrigation agriculture provide the guarantee for food security and sustainable agricultural production in the face of climate change and variability.

Acknowledgements This research is supported by funding from the Department for International Development (DFID) under the Climate Impact Research Capacity and Leadership Enhancement (CIRCLE) programme. We are grateful to our respective institutions for providing conducive environment for research collaboration and training. The contributions of Professor Labode Popoola of the Centre for Sustainable Development, University of Ibadan, Nigeria and Professor John Roy Porter of the Climate and Food Security Unit, University of Copenhagen, Denmark, are deeply appreciated. We are also thankful to anonymous reviewers for useful comments and suggestions.

References

Ajetomobi, J. O. (2008). *Total factor productivity of agricultural commodities in Economic Community of West African States (ECOWAS): 1961–2005* (p. 30). African Economic Research Consortium, Kenya (AERC), 2008.

Ajetomobi, J., Ajakaiye, O., & Gbadegesin, A. (2015). *The potential impact of climate change on Nigerian agriculture* (p. 44). AGRODEP Working Paper 0016. International Food Policy Research Institute (IFPRI).

Arellano, M., & Bond, S. (1991). Some tests of specification for panel data; Monte Carlo evidence and an application to employment equations. *Review of Economic Studies, 58,* 277–297.

Arumugam, S., Ashok, K. R., Kulshreshtha, S. N., Vellangany, I., & Govindasamy, R. (2015). Yield variability in rainfed crops as influenced by climate variables: A micro level investigation into agro-climatic zones of Tamil Nadu, India. *International Journal of Climate Change Strategies and Management, 7*(4), 442–459.

Blundell, R., & Bond, S. (1998). Initial conditions and moments restrictions in dynamic panel data models. *Journal of Econometric, 87,* 115–143.

Breitung, J. (2002). Nonparametric tests for unit roots and cointegration. *Journal of Econometrics, 108,* 343–363.

Brooksbank, K., Veneklaas, E. J., White, D. A., & Carter, J. L. (2011). Water availability determines hydrological impact of tree belts in dryland cropping systems. *Agricultural Water Management, 100,* 76–83.

Carandang, J. L., Banaguas, G. S., Flores, M. J. C., & Carandang VI, J. S. R. (2015). Modelling climate change risks for food security in the Philippines. *International Journal of Climate Change Strategies and Management, 7*(4), 499–515.

Cassman, K. G., & Grassini, P. (2013). Can there be a green revolution in Sub-Saharan Africa without large expansion of irrigated crop production? *Global Food Security, 2*(2013), 203–209.

Choptiany, J., Graub, B., Phillips, S., Colozza, D., & Dixon, J. (2015). *Biodiversity and ecosystem services in agricultural production systems: Self-evaluation and holistic assessment of climate resilience of farmers and pastoralists*. Food and Agricultural Organisation of the United Nations, Rome. 166 p.

Connor, D., Comas, J., Gomez-Macpherson, H., & Mateos, L. (2008). Impact of small-holder irrigation on the agricultural production, food supply and economic prosperity of a representative village beside the Senegal River, Mauritania. *Agricultural Systems, 96*(2008), 1–15.

Craparo, A. C. W., Van Asten, P. J. A., Läderach, P., Jassogne, L. T. P., & Grab, S. W. (2015). *Coffea arabica* yields decline in Tanzania due to climate change: Global implications. *Agricultural and Forest Meteorology, 207*(2015), 1–10.

Easterling, W. E., Aggarwal, P. K., Batima, P., Brander, K. M., Erda, L., Howden, S. M., et al. (2007). Food, fibre and forest products. In M. L. Parry, O. F. Canziani, J. P. Palutikof, P. J. van der Linden, & C. E. Hanson (Eds.), *Climate change 2007: Impacts, adaptation and vulnerability* (pp. 273–313). Contribution of Working Group II to the Fourth Assessment Report of the Intergovernmental Panel on Climate Change. Cambridge, UK: Cambridge University Press.

Engel, R. F., & Granger, C. W. (1987). Cointegration and error correction: Representation, estimation and testing. *Econometrica, 55*(2), 251–276.

Fan, S., Yu, B., & Saurkar, A. (2008). Public spending in developing countries: Trends, determination, and impact. In S. Fan (Ed.), *Public expenditures, growth, and poverty. Lessons from developing countries*. 249 p.

Gourdji, S., Läderach, P., Valle, A. M., Martinez, C. Z., & Lobell, D. B. (2015). Historical climate trends, deforestation, and maize and bean yields in Nicaragua. *Agricultural and Forest Meteorology, 200*(2015), 270–281.

Guiteras, R. (2009). *The impact of climate change on Indian agriculture*. College Park, MD: Department of Economics, University of Maryland.

Hansen, L. P. (2012). Proofs for large sample properties of generalized method of moments estimators. *Journal of Econometrics, 170,* 325–330.

http://faostat.fao.org/site/362/DesktopDefault.aspx?PageID=362. Cite visited May 10, 2015.

Intergovernmental Panel on Climate Change (IPCC). (2015). 2015: Climate change: Synthesis report. Contribution of Working Groups I, II and III to the Fifth Assessment Report of the Intergovernmental Panel on Climate Change [Core Writing Team, R. K. Pachauri & L. A. Meyer (Eds.)]. IPCC, Geneva, Switzerland, 151 p.

Jefferis, K. R., & Okeahalam, C. C. (2000). The impact of economic fundamentals on stock markets in southern Africa. *Development Southern Africa, 17,* 23–51.

Kristjanson, P., Neufeldt, H., Gassner, A., Mango, J., Kyazze, F. B., Desta, S., et al. (2012). Are food insecure smallholder households making changes in their farming practices? Evidence from East Africa. *Food Security*.

Kurukulasuriya, P., Mendelsohn, R., Hassan, R., Benhin, J., Deressa, T., Diop, M., et al. (2006). Will African agriculture survive climate change? *The World Bank Economic Review, 20,* 367–388.

Liverman, D. M., & Kapadia, K. (2010). Food systems and the global environment: An overview. In J. S. I. Ingram, P. J. Ericksen, & D. Liverman (Eds.), *Food security and global environmental change*. London: Earthscan.

Metternicht, G., Sabelli, A., & Spensley, J. (2014). Climate change vulnerability, impact and adaptation assessment. *International Journal of Climate Change Strategies and Management, 6*(4), 442–476. doi:10.1108/IJCCSM-06-2013-0076

Montpellier Panel Report. (2015). *The farms of change: African smallholders responding to an uncertain climate future*. A Montpellier Panel Report, 44 p.

Mulangu, F., & Kraybill, D. (2015). A cost-benefit analysis of improved irrigation when faced with risks of climate change on Mount Kilimanjaro. *Water Resources and Economics, 10,* 31–44.

National Bureau of Statistics (NBS). (2013). *Statistical bulletin* (p. 445). Abuja, Nigeria: Federal Republic of Nigeria.

Nelson, G. C., Rosegrant, M. W., Koo, J., Robertson, R., Sulser, T., Zhu, T., et al. (2009). Climate change: Impact on agriculture and costs of adaptation. Food Policy Report #19. IFPRI, Washington, DC.

Olayide, O. E., & Ikpi, A. E. (2013). *Agricultural production and rural welfare in Nigeria: Assessing agricultural production and rural welfare* (p. 173). Germany: Scholars' Press. AV Akademikerverlag GmbH & Co. KG. ISBN 978-3-639-51239-7.

Olayide, O. E., Ikpi, A. E., Okoruwa, V. O., & Akinyosoye, V. O. (2011). Agricultural trade balance and food self-sufficiency: Implications for sustainable development in Nigeria. *World Rural Observations, 3*(4), 59–64.

O'Mara, F. P. (2012). The role of grasslands in food security and climate change. *Annals of Botany, 110,* 1263–1270.

Quian, H., & Schmidt, P. (1999). Improved instrumental variables and generalized method of moments estimators. *Journal of Econometrics, 91,* 145–169.

Schlenker, W., & Roberts, M. J. (2009). Nonlinear temperature effects indicate severe damages to US crop yields under climate change. *Proceedings of the National Academy of Sciences of the United States of America, 106,* 15594–15598.

Schlenker, W., & Lobell, D. B. (2010). Robust negative impacts of climate change on African agriculture. *Environmental Research Letters, 5,* 014010.

Vermeulen, S. J., Aggarwal, P. K., Ainslie, A., Angelone, C., Campbell, B. M., Challinor, A. J., et al. (2012). Options for support to agriculture and food security under climate change. *Environmental Science & Policy, 15*(2012), 136–144.

Wood, S. A., Jina, A. S., Jain, M., Kristjanson, P., & DeFries, R. S. (2014). Smallholder farmer cropping decisions related to climate variability across multiple regions. *Global Environmental Change*. http://dx.doi.org/10.1016/j.gloenvcha.2013.12.011

World Bank. (2008). *World Development Report (WDR) 2008. Agriculture for development* (p. 386). Washington, DC: The World Bank.

Zhang, X., & Fan, S. (2004). How productive is infrastructure? New approach and evidence from rural India. *American Journal of Agricultural Economics, 86*(2), 492–501.

Characterization of Present Day Climate Trend Over Ethiopia for Impact Study

Tamene Mekonnen Adgeh

1 Introduction

Understanding climate variability and its change over Africa is an issue of increasingly importance. The society and its economy are strongly dependent on agriculture where substantial proportion of this agriculture is rain-fed. Agriculture accounts for 30% of African GDP (IPCC 2007), while in rural Africa, over 70% of per capita income derives from agriculture (Jayne et al. 2003). Most of African crop production depends upon sufficient rainfall throughout the growing season; almost 90% for cereals (Cooper 2004). Relatively small changes in rainfall have large socioeconomic impacts. Changes in the spatial and seasonal distribution and year-to-year variability drastically alters rainfed crop yields (Challinor et al. 2004, 2005). Similarly crop pests and disease transmission are sensitive to both temperature and rainfall, eg. Malaria (Morse et al. 2005).

Climate change and its variabilities are the key challenges in developing countries where their impact at regional/sub-regional and ecosystem levels is likely to be uneven and unpredictable (Reda et al. 2013). As part of the African continent, Ethiopia faces frequent and devastating climate extreme events that can brought negative economic and social consequences. These extremes are mainly associated to a lack of rainfall over wide regions which often affect the livelihood of millions with a profound impact on rain-fed agriculture and pastoralism, water and food security and also on public health (Ummenhofer et al. 2009).

Improved quality in previous climate information and understanding the cycle, frequency and trends of past climate extremes have a potential to reduce the severity of climate related disasters in the future. Such climate information over a range of

---

T.M. Adgeh (✉)
Atmospheric and Oceanic Science Unit, Institute of Geophysics,
Space Sciences and Astronomy (IGSSA), Addis Ababa University,
PO Box 1176, Addis Ababa, Ethiopia
e-mail: tame.phys@gmail.com

W. Leal Filho et al. (eds.), *Climate Change Adaptation in Africa*,
Climate Change Management, DOI 10.1007/978-3-319-49520-0_3

time frames (from short-term to long-term) has a tremendous implication to align the scales (spatio-temporal and sectoral) and reliability of the information with the scale and nature of the decision (Easterling et al. 2007). It also plays a crucial role to make further seasonal climate forecasting and long-term climate projections at a required spatio-temporal scale which is particularly important to make informed decisions, implement adaptation strategies and effectively manage potential climate change risks. Both the past and future climate informations will serve as a bridge to link long-term options with the short ones at a minimum management and/or policy decisions implemented and strength the ability to cope with potentially larger impacts later in the century.

Various studies on climate change and its impact in Ethiopia are more of focusing on the analyses of potential impacts of climate change on crops production such as maize, changes in rainfall only at a water shade scale, links between rainfall and sea surface temperatures and comparison of model performances against observations (Araya et al. 2015; Cheung et al. 2008; Ummenhofer et al. 2009). However, the entire agriculture and food system are sensitive to changes in climate variables particularly changes in rainfall and temperature among other (Odjugo 2010). Hence, investigating the detail of previous rainfall and temperature informations both together over the region can provide accurate understanding of climate variation and recognize their effect on human lives and environment (Mndez-Lzaro et al. 2014). In this study, the spatial and temporal variabilities of rainfall and temperature are assessed for a period of 31 years (1983–2013) over Ethiopia. The primary objective of the study was to explore the spatio-temporal variabilities and detecting the trends in rainfall and temperature climatologies. The secondary objective was to assess the impact of climate variability on agricultural production and productivity particularly in identifying best adaptation options that has been implemented in selected regions of Ethiopia. The last, not the list, objective was benchmarking to undertake further future climate projection research and developing adaptation strategies.

2 Materials and Methods

2.1 Study Area

The study area encompasses the entire Ethiopia which is located over greater horn of Africa (see Fig. 1a). The topography of the area is characterized by a large area (6–15 N, 35–41E) of mountains >2000 m (see Fig. 1b) incised by river valleys. The terrain affects the geographical distribution of rainfall. The seasonal cycle also varies due to the proximity of the ITCZ which gives a bi-modal distribution in the southeast and a uni-modal distribution in the north (Sylla et al. 2009). Based on the distribution and magnitude of rainfall, the study area classified into nine homogeneous rainfall regime as shown in Fig. 8b (see Tsidu 2012). The importance of the timing and amount of rainfall over this climatologically diversified country is not

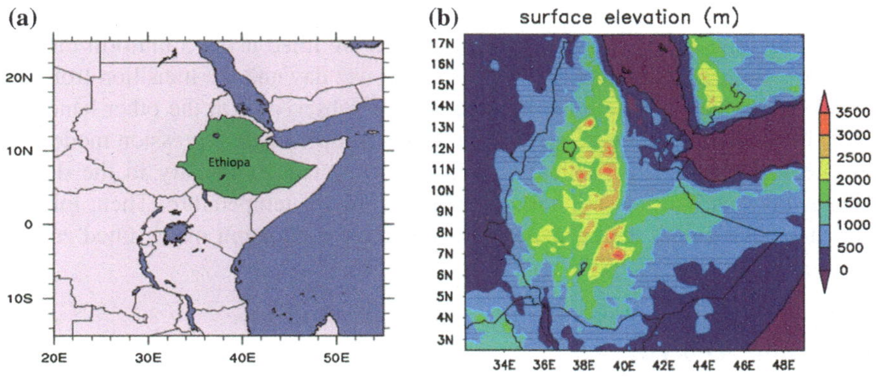

Fig. 1 Location (**a**) and surface elevation (**b**) of Ethiopian topography

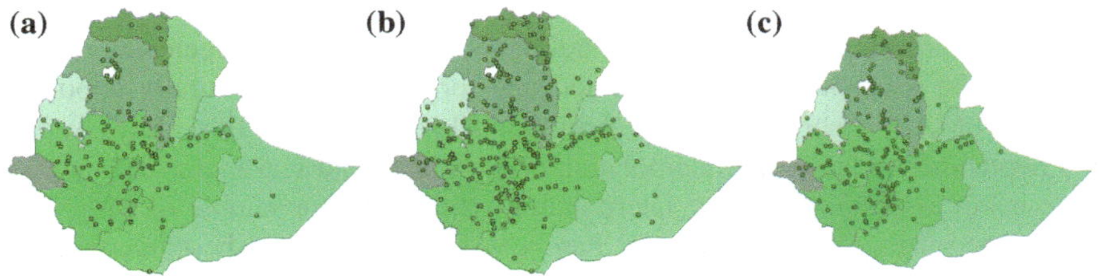

Fig. 2 Distribution of stations that measure rainfall (**a**), maximum temperature (**b**) and minimum temperature (**c**)

overstated (Cheung et al. 2008). The variation in seasonal rainfall regime with appraisal of variable onset, cessation and duration of rain is the most deterministic factor.

2.2 *Data Preprocessing*

Historical daily gauge data for rainfall, maximum and minimum temperature has been collected from Ethiopian National Meteorological Agency (NMA). The data were used from 115 stations for rainfall (Fig. 2a), 265 stations for maximum temperature (Fig. 2b) and 79 stations for minimum temperature, mean temperature and temperature range (Fig. 2c). The stations were selected based on the percentage of missing records less than 30% of the total available records. The mean temperature and temperature ranges were calculated by taking the averages and differences of the maximum and minimum temperatures respectively.

The missing records in rainfall data were filled using a stochastic weather generator. The persistence of wet day and transition from dry to wet are computed from available data on a monthly basis. The parameters of a mixed exponential

Probability Density Function (PDF) matching the available data are also computed at monthly time scale. Then the missing entries are filled using a uniform random generator compared to the critical values of the wet day and the transition from dry to wet for the month where the missing data are observed. On the other hand, the missed values in temperature were filled using multiple linear regression model that uses least squares. This method reduces the expected complexity in the dependencies between the missing and available records of temperature. Then, missing entries are filled with the estimates of the regression equation of the fitted curve.

2.3 Data Analysis

One of the distinguishing characteristics of analysis is the understanding of the outcomes. These outcomes can only be fully understood if their frequency and distribution are examined in terms of space and time. For this study, the annual and seasonal averages are calculated for individual station and homogeneous rainfall regions. The seasonal averages are calculated for the two agriculturally important seasons: FMAM (February, March, April and May) and JJAS (June, July, August and September). The spatial interpolation of the annual and seasonal avareges were done using a spherical kriging interpolation scheme with a sample of 12 points. This is because of that the method has been found to be the best interpolation technique without a repeated analysis of the structure of the variograms for irregularly spaced point climate data used. The interpolated spatial variabilities of rainfall, maximum and minimum temperature are compared with Climatic Research Unit (CRU) dataset (Tchotchou and Kamga 2009). Then, the linear trend was fitted to each time series using linear regression model at 95% significance level for the seasonal and annual averages. Here, linear trends, standardised residual (anomaly), coefficient of determination (R^2) and p Value were used as a measure of climate variability and detecting the changes in space and time (Lucio et al. 2011; Cheung et al. 2008).

Trend is one leg of change analysis that used for monitoring, forecasting, program evaluation, policy analysis and decision making and making future projections. Climate trend analysis used to detect and describe pattern, variability, anomaly and change in space and time that will enhance our insight into the possible monitoring and adaptation mechanisms for better preparedness against impacts (Reda et al. 2013). The linear regression equation is given by:

$$y_i = \beta_0 + \beta_1 x_i + \varepsilon_i \qquad (1)$$

where y_i is the ith scalar response, β_0 intercept, x_i the ith vector of input data, β_1 scalar coefficient (slope), ε_i is the ith scalar noise term which is independent random variable. The regression coefficients, β_0 and β_1 can be estimated as $\hat{\beta}_0$ and $\hat{\beta}_1$ using least squares estimation which is given by:

$$\hat{\beta}_1 = \frac{\sum_{i=1}^{n}(y_i - \bar{y})(x_i - \bar{x})}{\sum_{i=1}^{n}(x_i - \bar{x})^2} \tag{2}$$

and

$$\hat{\beta}_0 = \bar{y} - \hat{\beta}_1\bar{x} \tag{3}$$

where $\bar{y}$ and $\bar{x}$ are mean values (Everitt and Hothor 2010; Wilks 2006).

Residuals are used to detect outlying values of the variable and checking the linear regression assumptions with respect to the error term in the regression model. The standardised residual is defined as:

$$st_i = \frac{r_i}{\sqrt{\boldsymbol{MSE}(1 - h_{ii})}} \tag{4}$$

where $r_i = y_i - \bar{y}_i$ is raw residuals, MSE is the mean squared error and h_{ii} is the leverage value for observation i.

The coefficient of determination (R^2) measures how successful the fit is in explaining the variation of the data with time. The larger value in the R^2 tells the more variability of the dependent variable which is defined as:

$$R^2 = \frac{SSR}{SST} = 1 - \frac{SSE}{SST} \tag{5}$$

where SSE is the sum of squared error, SSR is the sum of squared regression, SST is the sum of squared total. P-value also used as a statistical measure to determine the result of the analysis within the normal range of values for the variables being observed. Usually, if the p-value of the dataset is below the pre-determined amount (say 0.05 which is the 95% significance level), the variabilities of the dataset had no meaningful effect on the result (Wilks 2006).

3 Results

This section examines the spatio-temporal variabilities of rainfall and temperature climatologies at various time scales (annual, seasonal and inter-annual). The Spatio-temporal analysis used to describe the distribution of climatological variables in space and time. The spatial patterns of these climatologies are compared with the CRU climatologies to build confidence on the interpolated climatologies from gauges. The results from the two datasets are in good agreement with few exceptions which are described in detail in the following sections.

3.1 *Mean Climatology*

3.1.1 Rainfall Climatology

Figure 3 shows the spatial distributions of average 1983–2013 rainfall amount (mm/month) from CRU (Fig. 3a) and gauges (Fig. 3b). Both CRU and gauge rainfall are in good agreement over most parts of the country though there are clear differences in magnitude. The differences might be resulted from (i) the coarser resolution in CRU and (ii) the representativeness of stations. Excluding Afar and Somali regions, the amount of rainfall over most part of the country is in the range between 50 and 187 mm/month which is large enough for rainfed agriculture and different socioeconomic consumptions.

The rainfall distribution over Ethiopia is mostly orographic and the seasonal cycle varies due to the proximity of the ITCZ. FMAM is the transition period for ITCZ from southern hemisphere towards northern hemisphere and during JJAS the ITCZ is confined over northern hemisphere at its northernmost position where the sun is almost over head over Ethiopia that bring high convections there.

Due to the dependent nature of seasonal shifts to ITCZ, the rainfall is mostly confined over southern and southwestern parts Ethiopia during FMAM (Fig. 4a) though it is sparsely distributed over many parts of the country with small extent in magnitude. This rainfall distribution is associated with moist southeasterly winds over the southern Indian Ocean and northeasterly winds over the northern Indian Ocean converging there following the ITCZ that generate intense convection. The FMAM rainfall distribution shows that the magnitude of the average rainfall amount is in the range between 62 and 184 mm/month mostly confined over southern regions.

In Ethiopia, the largest percentage of the total annual rainfall is contributed from JJAS rainfall and most of agricultural activities are practiced in JJAS following the availability of sufficient rainfall both in coverage and magnitude. This rainfall is mainly associated to the frequent occurrence of propagating meso-scale circulation

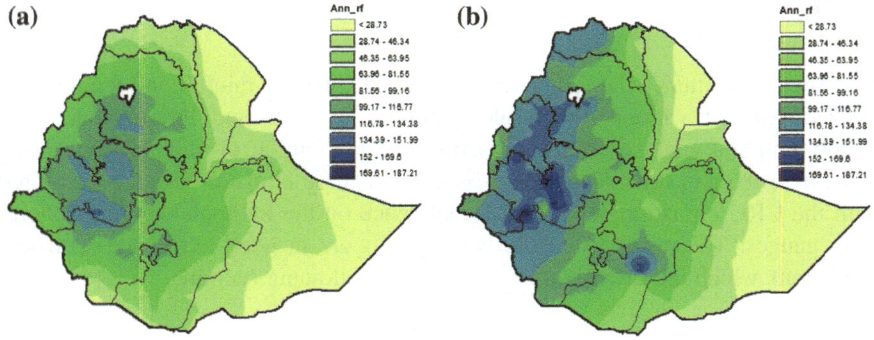

Fig. 3 Average 1983–2013 monthly total rainfall (mm/month) from CRU (**a**) and gauge (**b**)

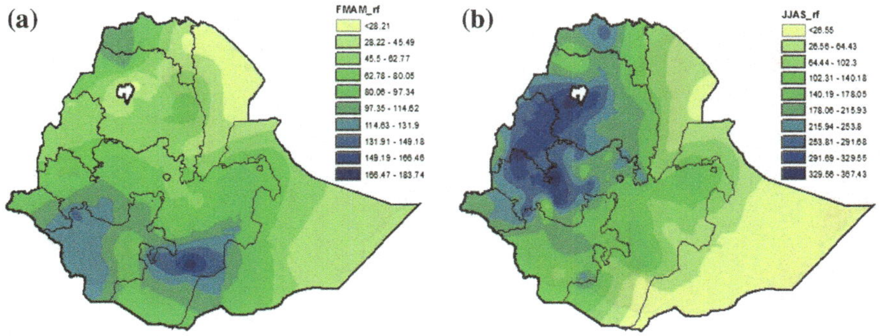

Fig. 4 Average 1983–2013 rainfall amount (mm/month) during FMAM (**a**) and JJAS (**b**) from gauge data

systems related to the dynamics of the African Easterly Jets (AEJ) and Tropical Easterly Jets (TEJ) as described by Sylla et al. (2009). The JJAS rainfall is mostly confined over northwestern, western, southwestern and central Ethiopia in large extent (Fig. 4b). However, the JJAS rainfall covers most parts of the country in the range 64–387 mm/month except the northeastern and southeastern regions which got rainfall amounts <28.55 mm/month.

Overall, the comparison between CRU and gauge rainfall shows some differences both in magnitude and spatial coverage. The differences might be related to coarse resolution and availability of gauge data during data reconstruction that CRU does not show heavy rainfall regions. Both CRU and gauge rainfall distributions clearly shows the seasonal oscillation as ITCZ does and the effect of orography and dynamic circulation systems. The mean annual rainfall reaches up to 187 mm/month while the FMAM and JJAS rainfall reach up to 183 and 367 mm/month respectively which is sufficient especially for rainfed agriculture.

3.2 Temperature Climatology

The mean annual maximum and minimum temperature from gauges and CRU are shown in Fig. 5. The peak values in maximum temperature are shown over lowland (>30 °C) while the lowest are over the highland regions (<26 °C) from both CRU (Fig. 5a) and gauge (Fig. 5b) datasets. Similarly, the the peak values in minimum temperature are over lowlands (>14 °C) and lowest over highlands (<10 °C) as shown in Figs. 5c, d for CRU and gauge respectively. The two datasets show good level of agreement in representing regions of maxima and minima for both maximum and minimum temperatures. However, the gauge temperature showed wider cold region following the complex mountainous areas of Ethiopia. The differences might be due to the relative coarser resolution in CRU to capture the fine scale features.

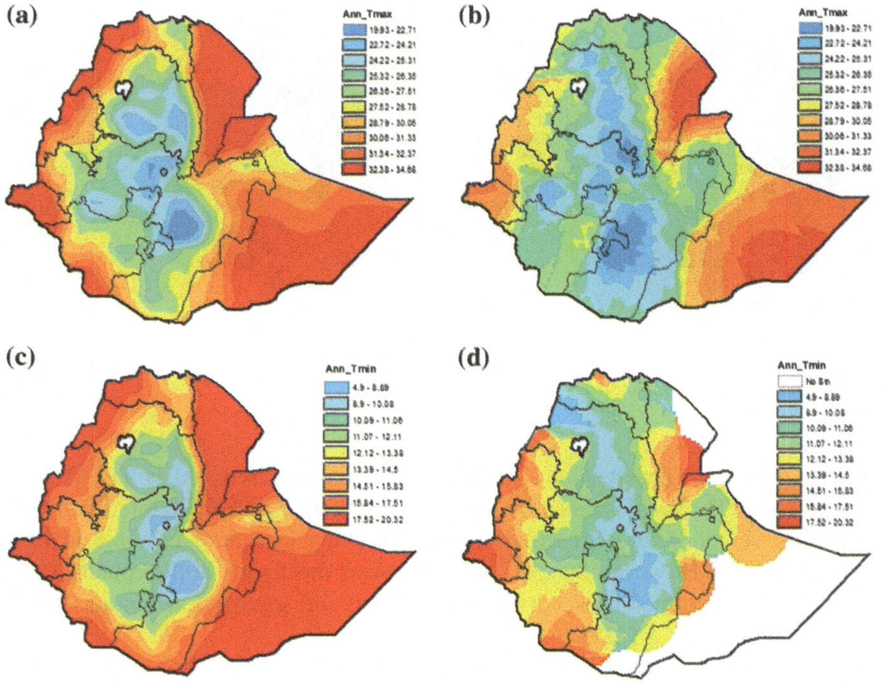

Fig. 5 Annual average (1983–2013) maximum temperature from CRU (*left panel*), gauges (*right panel*), for maximum temperature (*upper panel*) and minimum temperature (*lower panel*)

Figure 6 shows the mean FMAM and JJAS maximum temperature climatology. In FMAM (Fig. 6a) the lowest values of maximum temperature is shown over complex terrain highlands (above 2000 m) and thick forest parts of southwestern Ethiopia (21–27 °C). While the highest is shown over the low land areas (below 1000 m) especially over north eastern and south eastern regions which is mostly above 30 °C. In JJAS (Fig. 6b), the maximum temperature pattern show a shift in cold regions towards the west as compared to the pattern shown in FMAM maximum temperature. The maximum values of JJAS maximum temperature (>30 °C) is observed over northeastern and southeastern regions of the country. Similarly, the minimum/maximum temperature regions of FMAM minimum temperature are shown over highland/lowland regions (Fig. 6c). The JJAS (Fig. 6d) minimum temperature also show similar pattern in representing the cold/warm regions which is shown in FMAM. The coldest regions in JJAS minimum temperature cover larger area than FMAM.

Figure 7 shows the average mean temperature (upper) and temperature range (lower) during FMAM (left) and JJAS (right). The pattern in mean temperature for all time scales are similar to the one shown in maximum and minimum

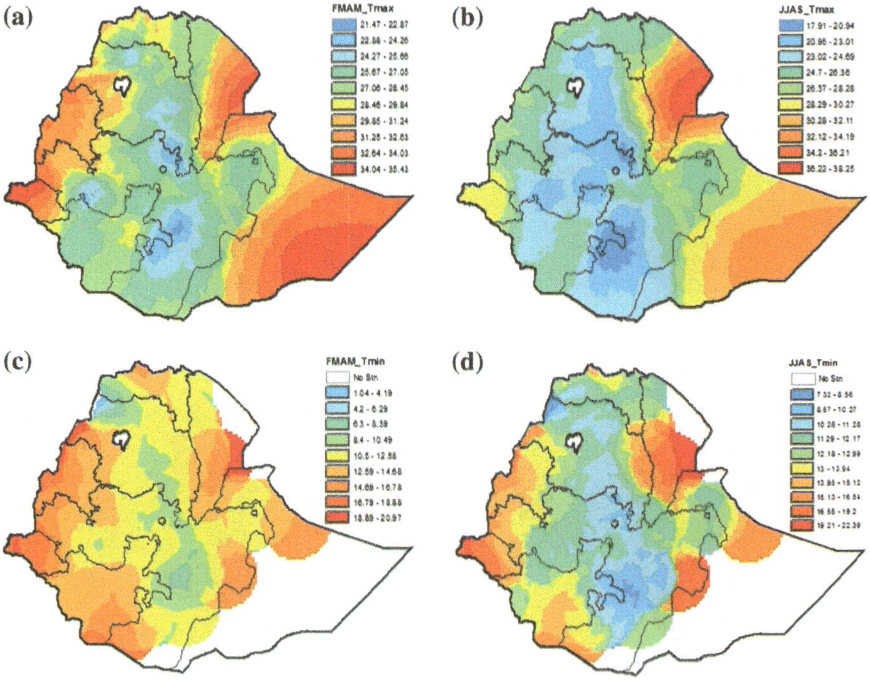

Fig. 6 Average (1983–2013) FMAM (*left*) and JJAS (*right*) maximum (*upper*) and minimum (*lower*) temperature

temperatures patterns except their magnitude. The mean temperature climatology are shown to be in the range between 5 and 30 °C in FMAM and 12 and 31 °C in JJAS. On the other hand, the FMAM (Fig. 7c) average temperature range showed higher differences over the highlands (>13.8 °C) than the difference shown over the lowlands and the situation is reversed in JJAS (Fig. 7d). The situation confirms that fluctuation in temperature are higher over the highlands than the lowland areas. The higher fluctuations are shown over central Amhara and Benishangul Gumz regions and also over some parts of southern Ethiopia. It is also shown that the temperature fluctuation is in small extent over northwestern part of Ethiopia. The JJAS temperature fluctuations are smaller than FMAM that might be due to thick cloud cover and moist wind fields from Indian Ocean.

In general, the comparison between CRU and gauge datasets show a good level agreement where both of them represent the regions of minima/maxima over highlands/lowlands very well. The lowest values in maximum temperature are confined over the highlands (>2000 m) while the highest are confined over the lowlands. The gauge maximum temperature represent low maximum temperature regions that are not shown in the CRU over regions below 1500 m. The differences

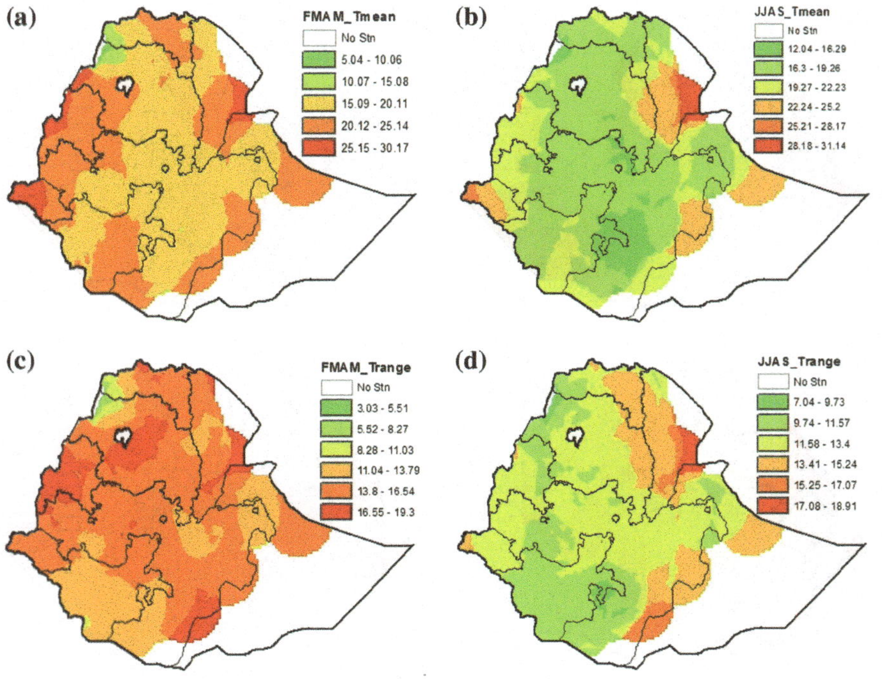

Fig. 7 Average (1983–2013) FMAM (*left*), JJAS (*right*) mean temperature (*upper*) and temperature range (*lower*)

between CRU and gauge in representing the regions peak temperature regions might be due to the differences in resolution. The temperature differences showed the highest temperature variation over mountainous areas in FMAM than JJAS and vice versa. It is clearly shown that the magnitude of maximum and minimum temperatures are higher over the lowlands where the magnitude of rainfall is very low. These contrasting phenomena may lead to severe drought over these regions in the near future.

3.3 Temporal Variability

Climatological trends on various time-scales were analysed to investigate the degree of climate variability and its change for impact study and adaptation purpose. The study was carried out for a 31 year time period (1983–2013) at monthly, seasonal, annual and interannual time scale. Temporal analysis used to reveal the temporal distribution of rainfall in space. The seasonal scale analysis was done for two agriculturally productive seasons (FMAM (February–May) and JJAS (June–September)) in Ethiopia. The degree of variability/changes and its significance is determined using a linear regression model at 95% Significance Level (SL).

3.3.1 Temperature Variability

This section describes the temporal variabilities of maximum and minimum temperatures during the entire year (12 months), FMAM and JJAS. The interannual and seasonal changes of maximum temperature is analysed for all 265 maximum temperature measuring stations. Among the 265 stations used, 99, 96 and 59 stations showed significant trend in the 12-month (annual), FMAM and JJAS time scale respectively. However, only 6, 3 and 5 stations showed significant decreasing trends in the three time scales respectively. While 41 stations showed significant increasing trend in maximum temperature for all time scales. From these stations, 21 of them are located over the highland areas >2000 m except Alemaya that showed significant decreasing trend in JJAS as shown in Table 1. The increasing

Table 1 Maximum temperature trend analysis (1983–2013) using linear regression model at 95% significance level (SL)

No.	Station name	Altitude	12 Months		FMAM		JJAS	
			Trend	SL (%)	Trend	SL (%)	Trend	SL (%)
1	Mehal Meda	3084	0.0295	100	0.04	99.95	0.0215	95.37
2	Ambagiorgis	2900	0.0618	100	0.06	99.91	0.1075	100
3	Hagre Selam	2840	0.0433	100	0.05	99.93	0.0435	99.81
4	Gohatsion	2507	0.0419	100	0.06	99.85	0.0523	99.96
5	Tikur Enchine	2467	0.053	100	0.07	99.97	0.0498	100
6	Shambu	2460	0.0318	100	0.05	99.71	0.0213	99.58
7	Debre Markos	2446	0.0353	100	0.05	99.95	0.0343	99.99
8	Kulubi	2436	0.0847	100	0.12	100	0.0749	99.97
9	Bulki(Mindre)	2430	0.0745	100	0.06	99.79	0.1096	100
10	Yetemen	2418	0.0487	100	0.09	99.86	0.0317	98.7
11	Gobessa III	2400	0.0343	100	0.05	99.86	0.0294	98.28
12	Alem Ketema	2280	0.04	99.98	0.1	100	0.03	100
13	Fincha	2248	0.0562	100	0.08	99.99	0.058	99.92
14	Hawzen	2242	0.0405	100	0.05	99.23	0.0363	99.03
15	Fiseha Genet	2240	0.1115	100	0.14	100	0.0918	100
16	Haisawita	2240	0.0276	99	0.04	99.54	0.0433	99.87
17	Dembecha	2117	0.0265	97	0.04	95.69	0.036	97.91
18	Ambo	2068	0.0321	100	0.05	99.99	0.0232	95.98
19	Amed Ber	2051	0.0388	100	0.05	99.94	0.0416	99.88
20	Boditi	2043	0.0549	100	0.07	99.99	0.0334	98.69
21	Alemaya	2020	0.0312	100	0.0287	99.68	−0.033	97.19

Table 2 Minimum temperature trend analysis (1983–2013) using linear regression model at 95% significance level (SL)

No.	Station name	Altitude	12 Months		FMAM		JJAS	
			Trend	SL (%)	Trend	SL (%)	Trend	SL (%)
1	Hagere Selam	2618	0.059	100	0.08	100	0.0553	100
2	Debre Work	2508	0.0743	98	0.07	95.13	0.0759	96.48
3	Gohatsion	2507	−0.1315	100	−0.14	99.07	−0.084	98.95
4	Lemi	2500	0.0934	100	0.11	97.38	0.1168	99.97
5	Yetemen	2418	0.0646	100	0.08	100	0.0647	99.93
6	Gobessa III	2400	0.157	100	0.18	99.84	0.1365	98.54
7	Haisawita	2240	0.0316	100	0.02	98.27	0.0345	100
8	Tulu Bolo	2190	0.0603	96	0.09	99.58	0.0847	98.42
9	Kimoye	2150	0.0512	100	0.07	99.95	0.0632	99.94
10	Dagaga	2067	0.0402	98	0.04	97.14	0.0476	98.44
11	Neshi	2060	0.0743	100	0.12	100	0.0798	99.99
12	Boditi	2043	0.1176	100	0.12	99.86	0.1247	99.8
13	Butajra	2000	−0.0758	99	−0.06	98.11	−0.0598	97.75

trends in maximum temperature are in the range between 0.0265 and 0.1115 °C per year during the 12 months, 0.0287 and 0.14 for FMAM, 0.0213 and 0.1096 for JJAS in °C per season respectively (see Table 1).

In the same manner, the temporal variabilities in minimum temperature were analysed for 79 stations that measure minimum temperature. From this stations, 13 stations show statistically significant trends which are located over the mountainous areas (>2000 m above sea level) as shown in Table 2. Only two of these stations (Butajra and Gohatsion) showed significant decreasing trends.

Overall, the changes in 12 month maximum temperature mostly in the range between 0.0265 and 0.1115 °C per year. Positive trends are mainly located over the mountainous highland areas while the changes minimum temperature are found to be in the range between −0.1315 and 0.157 °C per year but the positive trends are also over highlands. In FMAM, the increasing trends in maximum temperatures are relatively larger than from the 12-months but also over the highland areas which is in the range between 0.04 and 0.14 °C per season. Similarly, the changes in FMAM minimum temperature are in the range between −0.14 and 0.18 °C per season. The increasing trends in JJAS maximum temperature are also confined over highlands but in the range between −0.033 and 0.1096 °C per season where as for minimum temperature, it is between −0.084 and 0.1365 °C per season. It is clearly shown that increasing trends both in maximum and minimum temperatures are mostly confined over the mountainous highland areas of the country. If such trends continued at this or greater rate, there might be a severe drought followed by negative impact especially on crop production. To cope with such temperature changes, there should be integrated adaptation strategies such as changing the varieties of crops.

3.3.2 Rainfall Variability

This section examines the temporal variability of rainfall for individual stations and over each homogeneous rainfall regions shown in Fig. 8b at various time scales.

3.3.3 Mean Annual Cycle

The annual cycle of rainfall was examined only for six homogeneous rainfall regions (R-1, R-2, R-4, R-7, R-8, R-9) from the available nine regions (Fig. 8b) due to the scarcity of rainfall stations over the remaining three regions (R-3, R-5 and R-6) (see Fig. 2a). The values of the annual cycle in gauge rainfall are averaged for each homogeneous rainfall regions of over the whole analysis periods. AS shown in Fig. 8a, the rainfall overs each homogeneous regions are characterized by different rainfall cycles having different intensity, magnitude and length of rainy period.

The average annual cycle over region one (R-1 in Fig. 8b), which cover most of the south west part of the country, is characterized by a mono-modal and long rainy period which extends from March-November. The region is benefited from both FMAM and JJAS rainfall. The first peak rainfall shown in May while the second peak appear in September. Similarly, region nine (R-9 in Fig. 8b) show the same pattern but has greater rainfall in magnitude and the peak rainfall period appeared during July–August. Regions seven (R-7 in Fig. 8b) and eight (R-8 in Fig. 8b) are also characterized by the same pattern except their magnitude. Both regions got the highest rainfall peaks in July and the rainy period extend from April to October.

On the other hand, region two (R-2 in Fig. 8b), which cover the southern part of the country especially the Arsi and Bale highlands, is characterized by a quasi-bimodal rainfall pattern with long rainy season and moderate magnitude as compared to the other regions except region four. In this region, the primary rainfall

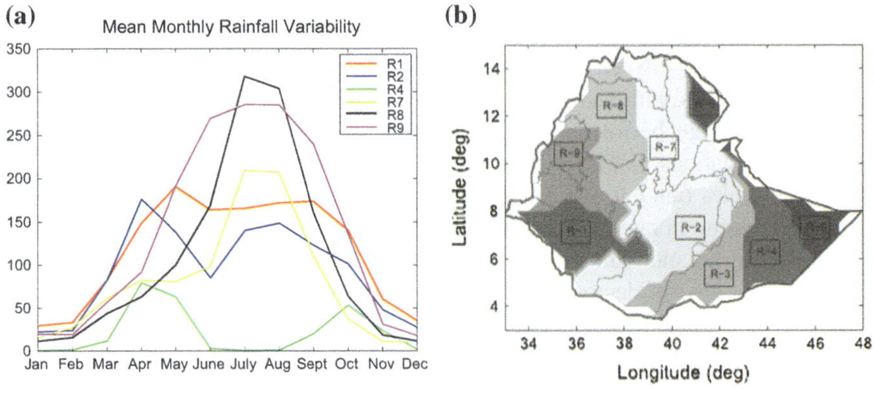

Fig. 8 **a** Average rainfall annual cycle over each homogeneous rainfall regions and **b** homogeneous rainfall regions (Tsidu 2012)

peak appear in April while the second is during July–August. It is also shown that, the region's highest peak rainfall of the year appeared in April. Similarly, region four (R-4 in Fig. 8b) which is located over Somali region is characterized by a bimodal rainfall pattern with short rainy season approximately two months during April–May and September–November (see Fig. 8a).

In general, a set of rainy seasons can be defined from mean annual rainfall cycle over each homogeneous rainfall regions. Based on this analysis, R-9, R-8, R-1 and R-7 considered as regions have single rainy season while R-2 and R-4 have double rainy seasons. The bimodal nature of rainfall over some parts of the county brought favorable condition for different socio-economic practices. Regions with bimodal rainfall pattern have a chance of agricultural production twice a year during FMAM (first growing period) and JJAS (second growing period)

3.3.4 Interannual and Seasonal Variability

Figure 9 shows the monthly rainfall time series over the six homogeneous rainfall regions which reveals the temporal distribution of rainfall in space. The distribution showed that the 31 years rainfall was distributed over time in a good manner for all homogeneous rainfall region.

Taking the Agricultural practice into consideration, the temporal variabilities/changes in rainfall was examined for the two growing seasons (FMAM and JJAS; see Figs. 10 and 11). The temporal distribution of rainfall showed remarkable variability/change over each subregions and individual stations. Among the six subregions, R-2 and R-7 show a decreasing trend in FMAM rainfall at about 3.4 mm/season (92.22% significant) over R-2 (Fig. 10a) and 2.5 mm/season (90.21% significant) over R-7 (Fig. 10b). In contrast, the JJAS rainfall distribution showed a statistically significant increasing trend over R-7 and R-8 at about 3 mm/season (96.46% significant) over R-7 (Fig. 11a) and 3.6 mm/season (98.99% Significant) over R-7 (Fig. 11b).

12-month, FMAM and JJAS rainfall respectively (see Table 3). From the 12 stations that show significant trend in FMAM rainfall, 5 of them have a negative trend in the range between −6.94 and −4.55 mm/season. On the other hand, only 5 of 15 (19) stations with significant trend in annual (JJAS) rainfall have a negative trend in the range between −6.1 (−6.3) and −19.06 (−15.8) mm/year (season). Some of the stations that have significant rainfall trend at-least two times from the three time scales (annual, FMAM and JJAS) are shown in Table 3. In general, the rate of decreasing in rainfall trend is higher at high altitudes or close to mountainous areas (eg. Dinsho and Gursum shown in Table 3). It is also showed that the rainfall trends are not always decreasing over mountainous areas.

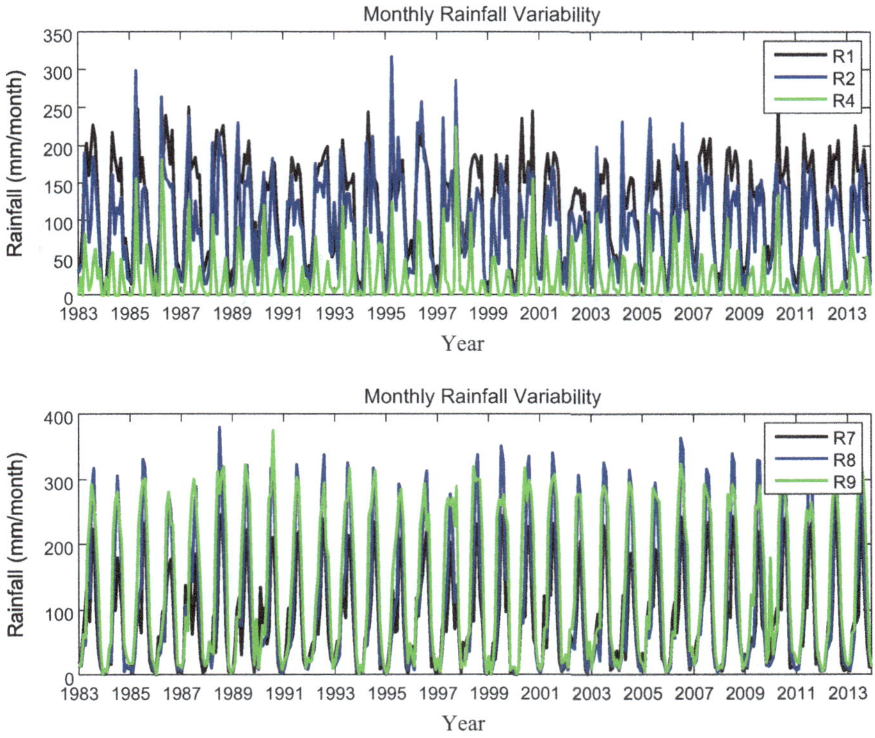

Fig. 9 Monthly time series of rainfall over each homogeneous rainfall region (mm/month)

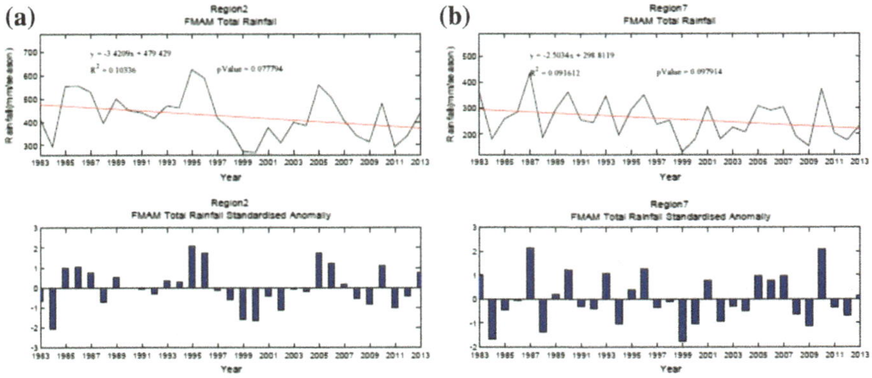

Fig. 10 FMAM seasonal time series of rainfall (mm/season)

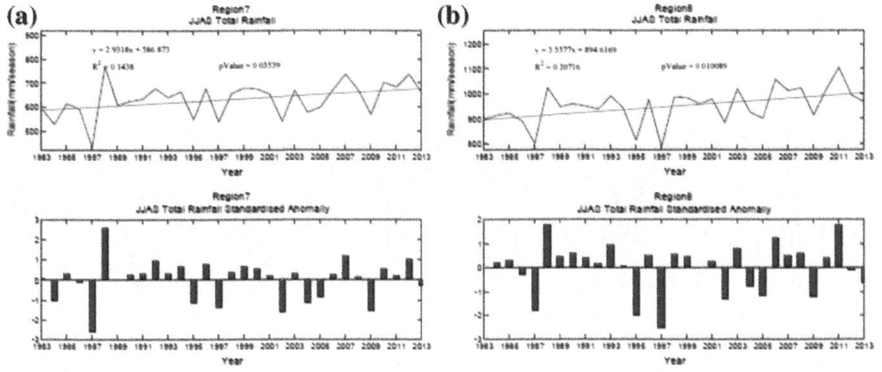

Fig. 11 JJAS seasonal time series of rainfall (mm/season)

Table 3 Rainfall trend analysis (1983–2013) using linear regression model at 95% significance level (SL)

No.	Station name	Altitude	12 Months		FMAM		JJAS	
			Trend	SL (%)	Trend	SL (%)	Trend	SL (%)
1	Metu	1711	−19.0642	100	−1.97	59.51	−15.8118	99.99
2	Dinsho	3072	−17.6828	98	−6.94	96.31	−9.3984	98.22
3	Derba	2385	−16.177	96	−1.79	34.83	−13.9322	98.68
4	Gursum	1900	−13.7861	99	−5.86	96.78	−7.1553	97.68
5	Gore	2033	−9.2699	94	−1.19	40.08	−6.2961	97.49
6	Combolcha	1857	2.9699	61	−5.64	99.39	8.1393	99.05
7	Debre Berhan	2750	4.2569	96	−1.82	86.27	5.6777	99.08
8	Haik	1985	5.231	75	−6.47	98.05	10.3943	99.17
9	Mehal Meda	3084	5.5714	95	−1.45	68	6.1828	97.33
10	Fiche	2784	6.4677	95	−4.55	98.6	9.9306	99.72
11	Melkasa (IAR)	1540	8.0174	100	−0.11	4.61	5.8543	98.92
12	Mojo	1763	11.2906	99	1.85	51.14	7.555	97.56
13	Tikur Enchine	2467	11.6048	96	1.47	45.23	7.7271	96.95
14	Ambagiorgis	2900	13.1059	100	−0.72	39.92	13.2856	99.88
15	Adaba	2420	13.4119	87	−1.15	29.84	10.9708	98.2
16	Debre Work	2508	16.0563	96	5.37	90.27	10.8516	96.32
17	Addis Zemen	1940	18.3645	98	0.64	37.28	17.1616	98.42

4 Summary and Conclusions

This paper evaluated the variabilities of rainfall and temperature climatologies and their trend in space and time. The spatial variabilities of gauges climatologies are compared with CRU. The comparison show a good level of agreement for both variables though there are some differences especially over lowlands. The bias between the two datasets might be due to the coarse resolution in the CRU dataset and also the sparse distribution of station used.

The study observed that the trends in maximum temperature showed statistically significant increasing trends approximately in the range between 0.027 and 0.112 °C/year mostly over highland regions (>2000 m). The changes shown in minimum temperature are not as drastic as the maximum temperature. However, 12 minimum temperature measuring stations which are located over the mountainous areas show statistically significant increasing trends. On the other hand, the trends in rainfall climatology are in the range −17.68 to 18.36 mm per year though the rate of changes are slower as compared to the total amount of rainfall in a year. The decreasing trends in rainfall are mostly confined to the high mountain area where as the increasing trends in rainfall are shown in some places both over lowland and highland areas. The mean annual cycle of rainfall over homogeneous rainfall regions showed a mono modal to bimodal pattern. The seasonal trend in rainfall show a decreasing tendency during FMAM over R-2 and R-7 and an increasing trend in JJAS over R-7 and R-8. The total annual rainfall, spatial and temporal distributions indicates that the amount and intensity of rainfall is large enough for agriculture and different socio-economic practices if it is supported by supplemental irrigation during a season in scarce rainfall such as in El Nino years.

Finally, the fast increasing trend in temperature and slow decreasing (tendency to decrease) trends in rainfall amount were observed in different parts of Ethiopia particularly over the highlands. The implication is that the warmer in temperatures and more variable rainfall followed by higher incidence of extreme events all will magnify the stresses on humane lives and the environment. These contrasting climatic phenomena might also negatively impact on crops grown in the region that will force the choice of crops to be adapted or altered. This study will provides additional insight into the connection between past climatological phenomena and future climate change projections at various spatio-temporal scale required. This task points toward the possibility of development and making available of more improved projected climate and weather informations as well as access to improved information sources that increase the ability of making informed decisions which is a vital component in water resource management in-light of projected climate change.

Furthermore, one of the limitation of this study is related to the station data used with large number of missing records that could cause spurious signals in the temperature and rainfall spatial interpolation and trends quantification. The other problem associated with the climate data was lack of high resolution country wide gridded historical climate records to investigate the changes in local scale climates. With such limitations, the spatio-temporal climate analysis such as the one discussed in this study cannot be accurately captured over complex topography and mountainous areas like Ethiopia. As the demand for local scale climate variability and climate change informations increase, additional climate monitoring stations needed to be deployed to capture climatological signals particularly for precipitation and temperature patterns among others. Future climate research should also include the development of high resolution historical and future climate data using various statistical and dynamical regional climate models to investigate the direct relationship between environmental factors and climate patterns.

Acknowledgements I wish to express my sincere gratitude to the Climate and Geospatial Research Directorate (CGRD) for providing financial support to present this manuscript as part of the symposium on climate change adaptation in Africa. Thanks to all colleagues for their insightful comments and technical supports.

References

Araya, A., Girma, A., & Getachew, F. (2015). Exploring impacts of climate change on maize yield in two contrasting agro-ecologies of Ethiopia. *Asian Journal of Applied Science and Engineering, 4*(1), 27–37.

Challinor, A. J., Wheeler, T. R., Slingo, J. M., Craufurd, P. Q., & Grimes, D. I. F. (2004). Design and optimisation of a large-area process-based model for annual crops. *Agricultural and Forest Meteorology, 124,* 99–120.

Challinor, A. J., Wheeler, T. R., Slingo, J. M., Craufurd, P. Q., & Grimes, D. I. F. (2005). Simulation of crop yields using the ERA40 re-analysis: Limits to skill and non-stationarity in weather-yield relationships. *Journal of Applied Meteorology, 44*(4), 516–531.

Cheung, W. H., Senay, G. B., & Singh, A. (2008). Trends and spatial distribution of annual and seasonal rainfall in Ethiopia. *International Journal of Climatology, 28*(13), 1723–1734.

Cooper, P. (2004). *Coping with climatic variability and adapting to climate change: Rural water management in dry-land areas*. London: International Development Research Centre.

Easterling, W., Aggarwal, P., Batima, P., Brander, K., Erda, L., Howden, M., et al. (2007). In M. L. Parry, O. F. Canziani, J. P. Palutikof, P. J. van der Linden, & C. E. Hanson (Eds.), *Climate change 2007: Impacts, adaptation and vulnerability* (pp. 273–313). Cambridge, UK: Cambridge University Press.

Everitt, B. S., & Hothor, I. (2010). *A handbook of statistical analyses using R edition* (2nd ed.). United States of America: Taylor and Francis Group, LLC.

IPCC. (2007). Summary for policymakers. In M. L. Parry, O. F. Canziani, J. P. Palutikof, P. J. van der Linden, & C. E. Hanson, (Eds.), *Climate change 2007: Impacts, Adaptation and vulnerability* (pp. 7–22). Contribution of Working Group II to the fourth assessment report of the intergovernmental panel on climate change. Cambridge: Cambridge University Press.

Jayne, T., Yamano, T., Weber, M. T., Tschirley, D., Benfica, R., Chapoto, A., et al. (2003). Smallholder income and land distribution in Africa: Implications for poverty reduction strategies. *Food Policy, 28*, 253–275. doi: 10.1016/S0306-9192(03)00046-0

Lucio, D. M., Daniela, V., Costanza, C. (2011). Climatic characterization and response of water resources to climate change in limestone areas: Some considerations on the importance of geological setting. In *Drought Research Initiative Network: 2011 Symposium on Data-Driven Approaches to Droughts*. Paper 41.

Mndez-Lzaro, P. A., Nieves-Santiango, A., & Miranda-Bermdez, J. (2014). Trends in total rainfall, heavy rain events, and number of dry days in sanjuan, puertorico, 1955–2009. *Ecology and Society, 19*(2), 50. http://dx.doi.org/10.5751/ES–06464–190250

Morse, A., Doblas-Reyes, F., Hoshen, M., Hagedorn, R., & Palmer, T. (2005). A forecast quality assessment of an end-to-end probabilistic multi-model seasonal forecast system using a malaria model. *Journal of Applied Meteorology, 57,* 464–475.

Odjugo, P. A. A. O. (2010). Adaptation to climate change in the agricultural sector in the semi-arid region of Nigeria. In *2nd International Conference: Climate, Sustainability and Development in Semi-arid Regions*, 16–20 August, 2010, Fortaleza—Cear, Brazil.

Reda, A. G., Tripathi, N. K., Soni, P., Tipdecho, T., & Phalke, A. (2013). Temporal climate trend of ping basin of Thailand and implications for Mekong region. *Earth Science and Climatic Change, 4*, 146. doi:10.4172/2157-7617.1000146

Sylla, M. B., Coppola, E., Mariotti, L., Giorgi, F., Ruti, P. M., Dell'Aquila, A., et al. (2009). Multiyear simulation of the African climate using a regional climate model (RegCM3) with the high resolution era-interim reanalysis. *Climate Dynamics.* doi:10.1007/s00382-009-0613-9

Tchotchou, L. A. D., & Kamga, F. M. (2009). Sensitivity of the simulated African monsoon of summers 1993 and 1999 to convective parameterization schemes in RegCM3.

Tsidu, G. M. (2012). High-resolution monthly rainfall database for Ethiopia: Homogenization, reconstruction, and gridding. *Journal of Climate, 25*, 8422–8443, doi:10.1175/JCLI-D-12-00027.1

Ummenhofer, C. C., Gupta, A. S., England, M. H., & Reason, C. J. C. (2009). Contributions of Indian Ocean Sea surface temperatures to enhanced east African rainfall. *Journal of Climate, 22*(4), 993–1013.

Wilks, D. (2006). *Statistical Methods in the atmospheric sciences* (2nd ed.), Elsevier Inc., Department of Earth and Atmospheric Sciences, Cornell University.

Adaptation Benefits of Climate-Smart Agricultural Practices in the Blue Nile Basin: Empirical Evidence from North-West Ethiopia

Paulos Asrat and Belay Simane

1 Introduction

The impact of climate change is detrimental in low income Tropical African countries including Ethiopia that depend on agriculture as the main livelihood (IAC 2004). Climate change events such as higher temperature, reduced rainfall and increased rainfall variability are mostly reflected in the form of reducing crop yield and threatening food security. In Ethiopia, the major factors responsible for the low productivity include climatic factors such as drought, flood and soil degradation; reliance on traditional farming techniques; and poor complementary services such as extension, credit, and marketing (Yirga 2007). These factors reduce the adaptive capacity or increase the vulnerability of farmers to climate change, which in turn affects the performance of the already weak agriculture. Studies indicated that farmers perceive climate change and also adapt to reduce the negative impacts (Mertz et al. 2009; Deressa et al. 2011). Further, the perception of climate change and taking adaptive measures are influenced by different socio-economic and environmental factors (Semenza et al. 2008; Deressa et al. 2011).

Climate-smart agricultural practices have been shown to be effective in adaptation to climate change in moisture stress areas. Empirical evidence has also shown synergistic relationships exists among climate-smart agricultural practices. Holding all else constant, a household that uses more than one practice is likely to have

P. Asrat (✉) · B. Simane
Center for Environment and Development Studies, Collegae of Development Studies, Addis Ababa University, Addis Ababa, Ethiopia
e-mail: paulosasrat63@gmail.com

W. Leal Filho et al. (eds.), *Climate Change Adaptation in Africa*,
Climate Change Management, DOI 10.1007/978-3-319-49520-0_4

better adaptation than a household using a single practice (Nkonya et al. 2011). Climate-smart agricultural practices like use of improved crop varieties (such as early maturing, drought, pests and diseases resistant); organic and inorganic soil fertility management practices; and agronomic practices such as changing the time of planting to reflect the new climatic patterns enhance adaptation to climate change and increases crop productivity (Lobell et al. 2008; Nkonya et al. 2011).

Adaptation to climate change is a two-step process; the first step requires farmers to perceive a change in climate and the second step requires them to act through adaptation (Deressa et al. 2011). So far various attempts have been made to analyses how farmers adapt to climate change in Ethiopia (Deressa et al. 2011). However, most of these studies failed to explicitly address how farmers perceive and what adaptation methods they employ at local level given the diverse agro-ecological setting of the country.

The results of these studies are highly aggregated and are of little help in addressing local agro-ecology specific perceptions and adaptations to climate change. Moreover, these studies have also paid little attention to the analysis of climate-smart agricultural practices as adaptation strategy and the likelihood impact of the practices on farm productivity and production risk management. Since adaptation is a local response to climate stimuli, addressing agro-ecology specific perception and adaptation decisions as well as measuring the impact of the decisions on rural livelihood is an important research gap that needs to be addressed. Therefore, the present study is aimed at filling these gaps. In doing so, the study first employs the Heckman sample selection model to analyze the two-step process of adaptation to climate change (perception and response) and then it employs nearest-neighbor matching techniques to measures the impact of adopting climate-smart agricultural practices on value of agricultural production.

2 Materials and Methods

2.1 Study Area

Dabus sub basin has an area of 21,030 km^2. The altitude in the sub basin ranges between 485 and 3150 masl. The sub basin has an annual rainfall ranging between 970 and 1985 mm. The annual maximum and minimum temperature in the sub basin varies between 20–35 and 8.5–20 °C respectively. The sub-basin is characterized by hot to warm moist and sub humid lowlands. Considerable part of the sub basin is cultivated and is characterized by Maize-sorghum and maize-sorghum-perennial complex.

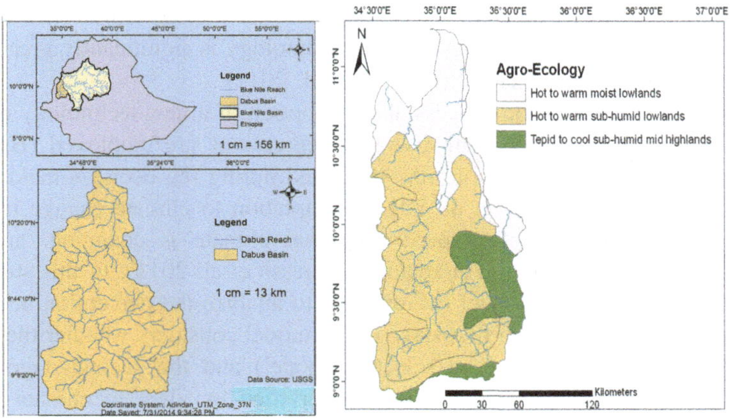

Local map of the research area and agro-ecological zones

2.2 Data Source

The study is based on a cross-sectional household survey data of 734 mixed farmers enumerated during July and August 2015 from the Dabus sub-baisn of the Blue Nile River in the North-west part of Ethiopia. The survey was conducted in four districts, spatially distributed throughout the sub-basin. The districts were purposefully drawn from two local agro-ecologies in the area namely, Wet Kola (wet lowland) and Dry Kola (dry lowland) to represent different aspects of the agricultural activity in the sub-basin. Following this, farm households were drawn randomly from each of the districts following probability proportional to size (PPS) sampling procedure.

2.3 Data Analysis

The study used descriptive analysis to reveal household perceptions of climate shocks and longer-term changes, the effects of the shocks and changes, responses implemented to address those shocks and changes, the impacts of those responses, and the constraints in implementing them.

The study also employed the Heckman's sample selection model to analyze the perception and adaptation to climate change. There are plausible methodological similarities among agricultural technology adoption and climate change adaptation methods as both involve decisions on whether or not to adopt a given course of action. The models are based on farmers' utility or profit-maximizing behaviors and the assumption here is that farmers adopt a new technology only when the perceived utility or profit from using this new technology is significantly greater than the traditional or the old method (Deressa et al. 2011).

When a farmer's decision process about adoption of a new technology requires more than one step, models with two-step regressions are employed (Heckman 1976) to correct for the selection bias generated during the decision-making processes (Yirga 2007; Deressa et al. 2011). Adaptation to climate change is also a two-step process that involves perceiving that climate is changing, and then responding to changes through adaptation (Deressa et al. 2011). In this study, the Heckman probit selection model is employed to analyze the perception and adaptation to climate change. The first stage of the model considers whether the farmer perceived a climate change (the selection model) and the second-stage model (outcome model) looks at whether the farmer tried to adapt to climate change conditional on the first stage (a perceived change in climate). The probit model for sample selection assumes that an underlying relationship exists and the latent equation is given by:

$$y_j = x_j\beta + u_{1j} \tag{1}$$

Such that we observe only the binary outcome given by the probit model as

$$y_i^{probit} = (y > 0) \tag{2}$$

The dependent variable is observed only if j is observed in the selection equation

$$y_i^{select} = z_j\delta + u_{2j} > 0 \tag{3}$$

$$\begin{aligned} \mathrm{u}_1 &\sim \mathrm{N}(0, 1) \\ \mathrm{u}_2 &\sim \mathrm{N}(0, 1) \\ corr(u_1, u_2) &= \rho \end{aligned}$$

where x iş a k-vector of regressors, z is an m vector of regressors; u_1 and u_2 are error terms.

when $\rho \neq 0$, standard probit techniques applied to Eq. (6) yield biased results. Thus, the Heckman probit (heckprob) provides consistent, asymptotically efficient estimates for all parameters in such models.

Similarly, the Nearest Neighbor Matching Technique is employed to explore the impact that climate-smart agricultural practices have on the value of agricultural output when used as an adaptation strategy. Value was used because many plots had more than one crop and there need to be some basis for aggregation. In doing so, we first use a probit regression technique to assess the type of household that is more likely to use and maintain the adaptation measures on private land. Then, we estimate the average treatment effect on the treated (ATT), using the nearest-neighbor matching method (NNM) which matches users and non-users/control households based on observable characteristics and calculates the mean difference in outcomes across the two groups. Thus, the control group is matched on the probability (propensity score) of using the adaptation practices given a set of observable characteristics from the probit regression.

Following (Quisumbing et al. 2011) we consider households that implemented and continue to maintain climate-smart agricultural practices on at least 33% of their total cultivated land holdings. Using this definition, we estimate a propensity score that is based on a probit regression of the probability of using the practices given observed household characteristics. The sample is then balanced by calculating and verifying that the means of the observed characteristics included in the probit model are similar for user households as compared to non-users. Individual user households are then paired with non-user households when their respective observable characteristics are similar, as determined by a weighted average of the distance between values of the observed characteristics. We then compare average outcomes of the user households with the matched non-user/comparison households.

Each user household is matched to a non-user household with its closest propensity score (allowing for five nearest neighbors in terms of absolute difference in propensity scores). Thus, for each household i, there are two potential outcomes: using the adaptation practice or no using. We denote users as *Ai(1)* and non-users as *Ai(0)*, whereby the impact of using the climate-smart agricultural practices is the difference in outcome between user and non-user. After obtaining the predicted probability values conditional on the observable covariates (the propensity scores) from the binary estimation, matching was done using a matching algorithm selected based on the data at hand. Then the effect of household's decision to use the practices on a given outcome (Y) is specified as:

$$\tau_i = Y_i(D_i = 1) - Y_i(D_i = 0) \tag{4}$$

where τ_i is treatment effect (effect due to the practices), Y_i is the outcome on household *i*, D_i is whether household *i* has got the treatment or not. In this particular case, variables that determine household's decision to use climate-smart agricultural practices could affect value of production at household level. Therefore, the outcomes of individuals from treatment and comparison group would differ even in the absence of treatment leading to a self-selection bias. By rearranging, and subtracting $E[Y(0)|D = 0]$ from both sides, one can get the following specification for ATT.

$$E[Y(1)|D = 1] - E[Y(0)|D = 0] = \tau_{ATT} + E[Y(0)|D = 1] - E[Y(0)|D = 0] \tag{5}$$

Both terms in the left hand side are observables and ATT can be identified, if and only if $E[Y(0)|D = 1] - E[Y(0)|D = 0] = 0$. i.e., when there is no self-selection bias.

Given Conditional Independence and Common support $[0 < P(D = 1|X) < 1]$ assumptions, the PSM estimator of ATT can be written as:

$$\tau_{ATT}^{PSM} = E_{P(X)/D=1}\{E[Y(1)|D=1, P(X)] - E[Y(0)|D=0, P(X)]\} \quad (6)$$

P(X) is the propensity score computed on the covariates X. This is explains that that the PSM estimator is the mean difference in outcomes over the common support, appropriately weighted by the propensity score distribution of users of the adaptation practice.

3 Results and Discussion

3.1 Perception and Adaptation to Climate Change

About 72% of the surveyed farmers perceived increasing temperatures, and 81% perceived decreasing precipitations over the past 20 years. About 78% of the farmers who claimed to have observed changes in climate over the past 20 years indicated that they have implemented at least one climate-smart agricultural practice as adaptation measure. Responses to the climate shock differ by agro-ecology and farmers have implemented various adaptation measures in their respective localities. Soil and water management practices are identified to be the most successful interventions in the face of changing climate and variability in both agro-ecologies. Particularly, use of manure as adaptation strategy is reported to be increasing from time to substitute fertilizer and enhance crop production in the face of declining soil productivity attributed to climate factors. Crop rotation, intercropping and use of irrigation as adaptation strategy are also practiced by considerable proportion of farmers in both agro-ecologies (Table 1).

Table 1 Farmers perception of climate change and commonly used climate smart agricultural practices for adaptation

Perceived change in temperature	Percent (N = 734)	Perceived change in precipitation	Percent (N = 734)
Increased	72	Increased	8
No change	20	No change	10
Decreased	8	Decreased	82
Adaptation strategy		Local agro-ecology (% of farmers)	
		Wet kola (N = 367)	Dry kola (N = 367)
No change/no adaptation		15	30
Early maturing verities		32	17
Change crop type		23	10
Adjust planting dates		48	19
Soil and water management		56	24
Crop rotation		30	13
Intercropping		28	11
Irrigation		12	10

Table 2 Reason for no response to climate change

Reasons for no adaptation	Percent of farmers (N = 165)
Lack of financial capital	63
Lack of input	51
Lack of appropriate information	38
Labor shortage	25
Lack of water	22
Lack of access to credit	12
Land shortage	10

Adjusting planting date is the second most common adaptation strategy in both agro-ecologies in response to perception of changing onset of rainfall. Planting early-maturing varieties as adaptation strategy was reported by 32% of the respondents in wet kola agro-ecology. This strategy is also justified by the fact that most households interviewed have experienced crop failure due to severe terminal moisture stress in the past 7 years. This is in line with the findings of Lobell et al. (2008) which indicate early maturing varieties as one of the key technologies for addressing climate change in areas where rainfall is expected to be more erratic.

Farm households are switching some of their land to production of high-value horticultural crops as a strategy to intensify and maximize returns on the increasingly scarce water and land resources. Partly, this strategy is driven by improved access to market (22%) declining farm productivity (15%) and to a growing experience of using irrigation. Some farmers (23%) in wet kola and 10% in dry kola have also preferred to diversify their livelihoods by planting more crop types the major reason being decreasing livestock population due to prolonged drought and disease prevalence. These findings are contrary to Jones and Thornton (2009), who predicted that climate change would induce a shift from crop production to livestock production but it is in line with the findings of Nkonya et al. (2011).

The proportion of farmers that reported no adaptation to climate change is much higher in dry kola agro-ecology (30%) as compared to 15% in the wet kola. These farmers also indicated major reasons for not responding to climate change (Table 2). Lack of financial capital was the major reason cited followed by lack of access to inputs for not adapting to climate change. Lack of information on appropriate adaptation strategies was the third most common reason for failing to adapt to climate change.

3.2 Results from the Heckman Probit Selection Model

The Heckman probit model was run and tested for its appropriateness over the standard probit model (i.e. a probit model that does not account for selection). The results indicated the presence of sample selection problem (dependence of the error terms from the outcome and selection models) justifying the use of Heckman probit model with rho significantly different from zero (Wald χ^2 = 10.43, with P = 0.001). Moreover, the likelihood function of the Heckman probit model was significant (Wald χ^2 = 81.24 with $P < 0.001$), showing its strong explanatory power (Tables 3 and 4).

Results from both the selection and outcome models show that most of the explanatory variables and their marginal values are significant at $P < 0.05$ and generally in the directions that would be expected (Table 5). The calculated

Table 3 Description of model variables for the Heckman probit outcome model

Dependent variable description	Farmers who adapt (%)	Farmers who did not adapt (%)
Adaptation to climate change (adapted = 1)	55	45
Independent variables	*Mean*	*SD*
Education of HH head (years of formal education)	1.3	1.05
Household size (number)	5.2	2.4
HH head sex (male = 1)	0.9	0.2
HH head age (years)	43.6	14.3
Farm income (Ethiopian currency)	5372	6897
Non-farm income (Ethiopian currency)	689	843
Livestock (owned = 1)	0.9	0.2
Extension advice (yes = 1))	0.4	0.3
Farm size (hectares)	2.3	1.4
Distance to market (Km)	9.5	5.2
Temperature (°C)	23	1.2
Precipitation (mm)	140.4	33.2
Plots with steep slope (%)	0.5	0.5
Plots with mixed slope (%)	0.5	0.5
Semi-fertile plots (%)	0.4	0.3
Non-fertile plots (%)	0.5	0.5

Table 4 Description of model variables for Heckman probit selection model

Dependent variable description	Farmers who perceive (%)	Farmers who did not perceive (%)
Perception of climate change (perceived = 1)	81	19
Independent variables	*Mean*	*SD*
Education of HH head (years of formal education)	1.3	1.05
HH head age (years)	43.6	14.3
Non-farm income (Ethiopian currency)	689	843
Climate change information (yes = 1)	0.5	0.6
Farmer to farmer extension (yes = 1)	0.5	0.5
Agro-ecology (wet kola = 1)	0.5	0.5
Agro-ecology (dry kola = 1)	0.5	0.5

Table 5 Results of the Heckman selection model

Explanatory variables	Outcome model				Selection model			
	Regression		Marginal effect		Regression		Marginal effect	
	Coefficients	*P*-values	Coefficients	*P*-values	Coefficients	*P*-values	Coefficients	*P*-values
Education	0.082	0.022	0.016	0.012	0.033	0.032	0.013	0.022
Household size	0.044	0.012	0.014	0.043				
Gender	0.580	0.010	0.177	0.012				
Age	0.138	0.012	0.000	0.031	0.015	0.000	0.008	0.000
Farm income	0.001	0.142	0.00	0.531				
Non-farm income	0.126	0.023	0.0211	0.044	0.034	0.131	0.031	0.113
Livestock ownership	1.012	0.003	0.309	0.004				
Extension advice	1.024	0.000	0.303	0.000				
Farm size	−0.565	0.034	−0.009	0.024				
Distance to market	0.025	0.310	0.012	0.310				
Temperature	0.122	0.011	0.033	0.001				
Precipitation	−0.013	0.021	−0.011	0.012				
Wet kola					1.418	0.000	0.278	0.000
Dry kola					0.934	0.054	0.169	0.047
Climate information					0.255	0.021	0.074	0.023
Farmer to farmer extension					0.83	0.001	0.212	0.000
Percentage of plots with steep slope	2.62	0.054	0.263	0.001				
Percentage of plots with mixed slope	0.056	0.113	0.012	0.110				
Percentage of semi-fertile plots	1.21	0.022	0.066	0.001				
Percentage of non-fertile plots	0.54	0.071	0.149	0.056				
Constant	−5.945	0.003			−1.245	0.000		
Total observations	444							

(continued)

Table 5 (continued)

Explanatory variables	Outcome model				Selection model			
	Regression		Marginal effect		Regression		Marginal effect	
	Coefficients	*P*-values	Coefficients	*P*-values	Coefficients	*P*-values	Coefficients	*P*-values
Censored	92							
Uncensored	353							
Wald chi square (zero slopes)	81.24 ($P < 0.001$)							
Wald chi square (independent equations)	10.43 ($P < 0.001$)							

marginal effects measure the expected changes in the probability of both perception of climate change and adaptation with respect to a unit change in an independent variable. The results from the selection model indicate that education, age of the head of the household (a proxy variable for farm experience); information on climate change; farmer to farm extension and agro-ecological settings affect the perception of climate change positively. The model result has mirrored that making a living and operating in wet kola agro-ecology appears strongly increase the likelihood of perceptions on climate ($P < 0.001$) as compared to dry kola, which positively but less strongly related to perception of climate change ($P < 0.1$). Unlike the prior expectations, farmers living in wet kola perceived more change in climate than farmers in dry kola (dry lowland). This could be associated with various environmental changes that reduced water availability or it might be linked to various problems which reduces yield like soil erosion.

The results from the outcome model indicated that most of the explanatory variables except the farm size affected the probability of adaptation as expected ($P < 0.05$). The fact that increasing household size increases the likelihood of adaptation is probably because large family size is normally associated with a higher labor endowment, which would enable a household to accomplish the adaptation measures. Gender of the household head is significantly and positively associated with adaptation indicating that male-headed households often have a higher probability of adopting agricultural technologies (Buyinza and Wambede 2008; Deressa et al. 2011) and here adapted better to climate change.

The model result also shows that the incidence of adaptation to climate change increased with temperature and decreases with precipitation. The negative relationship between average annual precipitation and adaptation probably reflects that increasing precipitation relaxes the constraint imposed by increasing temperature on crop growth. Plot size is significantly negatively associated with adaptation and may reveal that adaptation is plot-specific and it is the specific characteristics of the farm that dictates the need for a specific adaptation method to climate change rather than the size. These results are in line with the findings of Kurukulasuriya and Mendelsohn (2006) and Deressa et al. (2011).

Education and farm experience affects adaptation positively. Income from non-agricultural sources also positively affected adaptation to climate change. This could be attributed to the fact that the income from this source may provide farmers with additional capacity to finance adaptation measures. Likewise, livestock ownership, extension advice received; proportion of plot with steep slope, percentage of semi fertile plots and infertile plots affects adaptation positively as expected.

3.3 Impact of Climate-Smart Adaptation Practices

Households that applied adaptation to climate change are defined as those that that used the practice at least on 1/3 of their cultivated land since 2007 and maintain until August 2015. Given this condition, 32% of the households are considered as users of the adaptation practice. We first assess overall effects by matching all user households with non-use households to identify determinants of adaptation practices from the probit model estimations and then evaluate if any impact exists due to the practice at household level.

The probit model reveals that the share of land on steep slope, size of non-fertile and semi fertile land is significantly different between users and non-users of the adaptation measures. Thus these land characteristics are highly correlated with use of the adaptation practices. Moreover, those who use the practice have past experience of crop failure due to terminal moisture stress and depletion of fertile land as compared to non-users. Distance from market reveled significant negative correlation with probability of using the adaptation practices. In addition, we include share of plots received manure for productivity enhancement as a matching variable and we find that the share of plots on which manure is applied significantly differs between users and non-users suggesting willingness of farmers to invest labor on climate-smart agricultural practices as adaptation measures (Table 6).

Table 6 Probit results of determinants of using climate-smart adaptation practices

Variable	dy/dx	Std. Err.
HH head age (years)	0.035	(0.021)
HH head sex (male = 1)	0.003	(0.021)
Land size in hectares	0.024**	(0.011)
Household experienced erosion (yes = 1)	0.023*	(0.041)
Household experienced drought (yes = 1)	0.011**	(0.025)
Household size	0.021	(0.003)
Percentage of plots with steep slope	0.211***	(0.032)
Percentage of plots with mixed slope	0.018	(0.028)
Percentage of plots received manure	0.053***	(0.061)
Education of HH head (literate = 1)	0.031	(0.017)
Percentage of semi-fertile plots	0.071**	(0.037)
percentage of non-fertile plots	0.039**	(0.062)
Extension advice (yes = 1)	0.053	(0.024)
Wet Kola agro-ecology (1 = yes)	0.241***	(0.028)
Dry Kola agro-ecology (1 = yes)	0.067**	(0.046)
Distance from market (Km)	−0.031**	(0.011)
Assosa Woreda	0.261***	(0.053)
Bambasi Woreda	0.217***	(0.101)
Sherkole Woreda	0.042*	(0.006)
Mengie Woreda	0.135*	(0.015)
Number of observations = 506		
LR chi2(21) = 218.21		
Prob > chi^2 = 0		
Pseudo R^2 = 0.2323		

Note *, ** and *** are significance level at 10, 5 and 1%

Household level impact of climate-smart adaptation practices. The estimated propensity scores vary between 0.076 and 0.997 with mean of 0.684 for treated households and between 0.004 and 0.943 with mean of 0.316 for the control households. The common support region would then lie between 0.076 and 0.943. But for ATT it is sufficient to ensure that for each participant a close non-participant can be found (Caliendo and Kopeinig 2008). Therefore, households whose estimated propensity scores are less than 0.004 and larger than 0.943 are not considered for the matching exercise. Hence a total 11 observations have been dropped from users. This condition is presented in the common support graph of the propensity scores (Fig. 1).

The ATT (Average treatment effect on the threated) estimation result in Table 7 provides a significant difference between users and non-users in terms of value of agricultural production (the outcome variable). Value of production is significant at ($P < 0.001$), indicating that the use of the adaptation practices significantly increase productivity for users. In effect, households who implement and maintained the climate-smart agricultural practices since 2007 experience a 22.2% higher value of production in 2015.

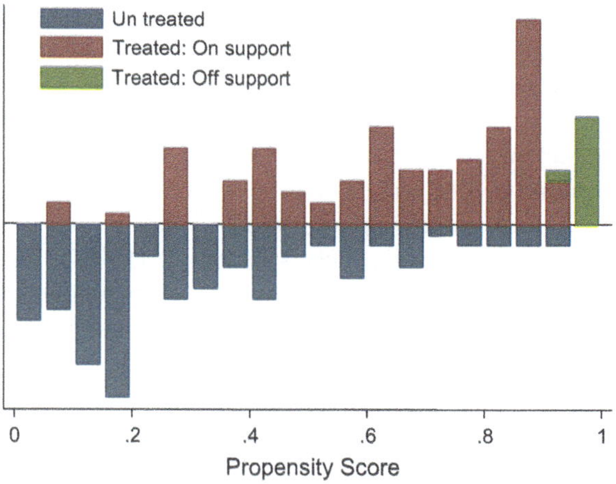

Fig. 1 Distribution and common support of the propensity scores

Table 7 Average household level impact of climate-smart adaptation practices

Outcome variable	Treated	Control	Difference	
			Value	Percent
Value of production	33,494	27,409	6085	(22.2%)
P-value			0.001*	
BSE	0.502			
Observation	497			

Note *Significant at ($P < 0.000$), *BSE* bootstrapped standard error

Table 8 Rosenbaum bound sensitivity analysis test for hidden bias

Gamma (e^{γ})	*p*-Critical
$e^{\gamma} = 1$	0
$e^{\gamma} = 1.25$	0
$e^{\gamma} = 1.5$	0
$e^{\gamma} = 1.75$	0
$e^{\gamma} = 2$	0
$e^{\gamma} = 2.25$	2.10e−15
$e^{\gamma} = 2.5$	4.90e−14
$e^{\gamma} = 2.75$	6.40e−13
$e^{\gamma} = 3$	5.40e−12

3.4 Sensitivity Aanalysis

Table 8 presents the critical level of e^{γ} (gamma) at which the causal inference of significant impact of practicing adaptation measures on value of production has to be questioned. The treatment effect is significant at ($P < 0.001$) and shows that the inference for the effect of practicing adaptation measures on value of production is not changing though the users and non-users has been allowed to differ in their odds of being treated up to 200% ($e^{\gamma} = 3$) in terms of unobserved covariates. That means for the outcome variable estimated, at various level of critical value of e^{γ}, the *p*-critical values are significant which further indicated that we have considered important covariates that affected both adaptation and outcome variable. We couldn't get the critical value e^{γ} where the estimated ATT is questioned even if we have set e^{γ} largely up to 3 which is larger value as compared to the value set in different literatures which is usually 2 (100%). Thus, we can conclude that our impact estimates (ATT) are insensitive to unobserved selection bias and are a pure effect of the use of adaptation practices on value of production.

4 Conclusion and Recommendation

The present study employed the Heckman sample selection model to analyze the two step process of adaptation to climate change, which initially requires farmers' perception that climate is changing prior to responding to changes through adaptation. Farmers' perception of climate change was significantly related to education, age, information on climate change; farmer to farm extension and agro-ecological settings. Adaptation to climate change through climate-smart agricultural practices was significantly influenced by family size, gender, local climatic conditions, farm size, education, farm experience, non-farm income, livestock ownership, extension advice, and farm characteristics. Moreover, results of the impact estimation from the Nearest Neighbor Matching Technique reveal that households that adopted climate-smart agricultural practices as adaptation strategy experienced higher value of production by 22.2% over non-users owing to reduced climate related risks that lead to yield variability. These practices are knowledge and resource intensive by their very nature and many not be implemented easily given the awareness level and resource endowments of farmers. Therefore, scaling up these adaptation benefits calls public investment to raise awareness and to provide technological support.

References

Buyinza, M., & Wambede, N. (2008). Extension of agroforestry technology adoption: Mixed intercropping of Crotalaria (*Crotaloria grahamiana*) and Maize (Zea mays L.) in Kabale district, Uganda. *Environmental Research Journal, 2,* 131–137.

Caliendo, M., & Kopeinig, S. (2008). Some practical guidance for the implementation of propensity score matching. *Journal of Economic Surveys, 22*(1), 31–72.

Deressa, T. T., Hassan, R. M., & Ringler, C. (2011). Perception of and adaptation to climate change by farmers in the Nile basin of Ethiopia. *Journal of Agricultural Science, 149,* 23–31.

Heckman, J. J. (1976). Common structure of statistical models of truncation, sample selection and limited dependent variables and a simple estimator for such models. *Annals of Economic and Social Measurement, 5,* 475–492.

Jones, P., & Thornton, P. (2009). Croppers to livestock keepers: Livelihood transitions to 2050 in Africa due to climate change. *Environmental Science and Policy, 12,* 427–437.

Kurukulasuriya, P., & Mendelsohn, R. (2006). *Crop selection: Adapting to climate change in Africa*. Pretoria, South Africa: Centre for Environmental Economics and Policy in Africa (Discussion Paper No. 26).

Lobell, D., Burke, M., Tebaldi, C., Mastrandrea, M., Falcon, W., & Naylor, R. (2008). Prioritizing climate change adaptation needs for food security in 2030. *Science, 319*(5863), 607–610.

Mertz, O., Mbow, C., Reenberg, A., & Diouf, A. (2009). Farmers' perceptions of climate change and agricultural adaptation strategies in rural Sahel. *Environmental Management, 43,* 804–816.

Nkonya., E, Place, F., Frank, P. John, P., Majaliwa M., Appollonia, O., et al. (2011). Climate risk management through sustainable land management in Sub-Saharan Africa. IFPRI discussion paper 01126.

Quisumbing, A. R., B. Baulch, & Kumar, N. (2011). *Evaluating the long-term impact of antipoverty interventions in Bangladesh: An overview* (IFPRI discussion paper 1077). Washington, DC: International Food Policy Research Institute (IFPRI).

Semenza, J. C., Hall, D. E., Wilson, D. J., Bontempo, B. D., Sailor, D. J., & George, L. A. (2008). Public perception of climate change voluntary mitigation and barriers to behaviour change. *American Journal of Preventive Medicine, 35,* 479–487.

Yirga, C. T. (2007). *The Dynamics of soil degradation and incentives for optimal management in central highlands of Ethiopia*. PhD thesis, Department of Agricultural Economics, Extension and Rural Development, University of Pretoria, South Africa.

Towards Mainstreaming Climate Change Adaptation into Urban Land Use Planning and Management: The Case of Ambo Town, Ethiopia

Gemechu Shale Ogato, Ketema Abebe, Amare Bantider and Davide Geneletti

1 Introduction

Adaptation to climate change and mainstreaming adaptation to climate change into urban land use planning and management underpin this paper as the two major concepts in climate change adaptation research. Adaptation to climate change may refer to "the process through which focus is given on building adaptive capacity of the most vulnerable people, reducing exposure or sensitivity to climate impacts, and ensuring that development initiatives don't inadvertently increase vulnerability" (Adger et al. 2002; Brooks et al. 2004; Füssel 2007; Huxtable and Yen 2009; UNPEI 2011). Mainstreaming climate change adaptation in this paper may refer to "the process of integrating considerations of climate change adaptation into

G.S. Ogato (✉)
Environmental Planning Programme, Ethiopian Institute of Architecture, Building Construction, and City Development, Addis Ababa University, P.O. Box 518, Addis Ababa, Ethiopia
e-mail: gemechushale2005@yahoo.com

K. Abebe
Urban Sociology, Ethiopian Institute of Architecture, Building Construction, and City Development, Addis Ababa University, P.O. Box 518, Addis Ababa, Ethiopia
e-mail: ketema_asrat@yahoo.com

A. Bantider
Center for Food Security Studies, College of Development Studies, Addis Ababa University, Addis Ababa, Ethiopia
e-mail: amare.bentider@gmail.com

D. Geneletti
Department of Civil, Environmental and Mechanical Engineering, University of Trento, Via Mesiano, 77, 38123, Trento, Italy
e-mail: davide.geneletti@unitn.it

W. Leal Filho et al. (eds.), *Climate Change Adaptation in Africa*, Climate Change Management, DOI 10.1007/978-3-319-49520-0_5

policy-making, budgeting, implementation and monitoring processes at urban land use planning and management level" (Lebel et al. 2012; Oates et al. 2011; UNPEI 2011).

Cities and towns in Least Developed Countries (LDCs) like Ethiopia are vulnerable to climate change and climate extremes in part because they concentrate many activities, people and wealth in limited areas (Sanderson 2000; Hallegatte et al. 2011). More importantly, urban poor can be very vulnerable to climate shocks, whether direct or indirect (Ziervogel et al. 2008; Watson 2009; Parikh et al. 2014; Stein and Moser 2014). However, traditional urban planning in cities and towns of least developed countries (LDCs) including those of Ethiopia has served to exclude the poor (Sanchez-Rodriguez 2009; Watson 2009; Nielsen and Reenberg 2010).

Adaptation to climate change and climate variability in cities demands a better understanding of the poor's adaptive capacity and of their autonomous coping strategies (Gill et al. 2007; Kazmierczak and Carter 2010; Lwasa 2010; UN-Habitat 2011; Chappin and van der Lei 2014; Taylor and Peter 2014). This is because the urban poor are affected by the 'double vulnerability' of climate change and poverty, which means that they are disproportionately affected in terms of both their exposure to climate related risks and the limited resources at their disposal to respond to such risks (Sanderson 2000; Van Aalst et al. 2008; Bartlett et al. 2012; Jabeen et al. 2012; Birkmann et al. 2010). Finding ways of integrating development interventions with the emergent climate adaptation responses is necessary in order not to duplicate efforts (Ziervogel et al. 2008; Hurlimann and March 2012; Loret and Ioannilli 2012; Jabareen 2013). Moreover, it is important to contextualize a response to climate change within the existing socio-political urban context (Bartlett et al. 2012; Jabeen et al. 2012; Fünfgeld 2012; Parikh et al. 2014). Adem and Bewket (2011) affirm that addressing current and future climate vulnerabilities in development planning and programming should be an immediate priority for Ethiopia as development-as-usual without consideration of climate risks and opportunities, will lead to maladaptive practices weakening national resilience to climate change.

According to UN-HABITAT (2006), high level of solid waste; water and air pollution; lack of adequate water and sanitation; and lack of legal access to land are some of the urban challenges for Ethiopian cities and towns. Sanderson (2000) contends that much urban legislation in LDCs including Ethiopia results in increased vulnerability of the poor. For instance, the prevention of permanent services to illegal settlements can increase ill-health, while the withholding of tenure inhibits consolidation of buildings, resulting in poorly-built shelters that easily collapse, catch fire or harbour disease (Sanderson 2000; Parikh et al. 2014; Stein and Moser 2014). Moreover, income inequality; poverty; growing job and residential informality; and high urbanization rates characterize cities and towns of Least Developed Countries (LDCs) including those of Ethiopia (Watson 2009; Sowers et al. 2011). Despite the aforementioned major urban challenges in those cities and towns, the capacity of local government is claimed to be weak and cities and towns have been shaped by national economic development policies and rampant market forces (Sanderson 2000; Ziervogel et al. 2008; Belinda and Kong 2012; Posey 2009; Watson 2009; McCarney 2012). Furthermore, their urban planning system is characterized by frequent corruption and clientelism. As a result, the ability to manage growth and deliver services equitably is lacking (Van Aalst et al. 2008; Watson 2009; Measham et al. 2011; Fünfgeld 2012).

Previous studies confirmed that urban areas play a key role in social and economic development as well as in change at global, regional, national, and local scales. Despite their key importance, far less attention has been given to the question on how to adapt cities and urban governance and planning systems to address climate change adaptation (Denton 2002; Birkmann et al. 2010; Measham et al. 2011; Dede et al. 2012; Jabareen 2013; Uittenbroek et al. 2013; Mashila 2014; Reid and Huq 2014; Vasileiadou et al. 2014; van den Brink et al. 2014; EEA 2015; Kelman et al. 2015; Yang et al. 2015). More importantly, studies on local socio-economic vulnerability to climate change and the quest of mainstreaming climate change adaptation into urban land use planning and management in Ethiopia are scant (Adem and Bewket 2011; Adem and Guta 2011; Ogato 2013b).

Mainstreaming adaptation to climate change into urban land use planning and management is currently being promoted as effective adaptation approach in cities and towns of least developed countries (LDCs) (Biesbroek 2009; Baloye et al. 2010; Hurlimann and March 2012; de Bruin et al. 2014; Kareem and Lwasa 2014; Mashila 2014; Reid et al. 2015; Sussams et al. 2015) and this necessitates careful understanding of vulnerabilities and adaptive capacities of vulnerable communities in small towns of Ethiopia like Ambo. Ambo town was taken as a case study town in this study in order to properly understand perceived climate change related disaster risks and local climate change adaptation strategies in Ambo town and its watershed. In a nutshell, this study intends to forward feasible recommendations for mainstreaming climate change adaptation into urban land use planning and management in Ambo town.

2 Research Methodology

This section deals with description of the study area, sampling procedures and methods, methods employed for data collection and methods employed for data analysis.

2.1 Study Area

The geographical location of Ambo town is approximately between 8°56′30″N–8° 59′30″N latitude and between 37°47′30″E–37°55′15″E longitude. It is located in the Western Shoa Zone of the Oromiya region (See Fig. 1). Relatively Ambo town is located 114 km far away West of Addis Ababa, 60 km North West of Weliso town

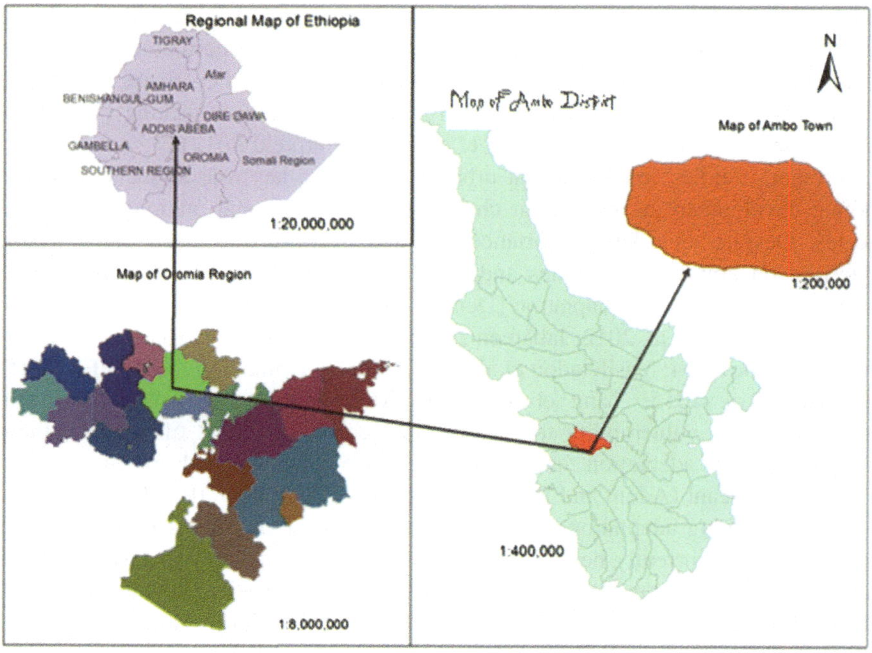

Fig. 1 Geographic location map of Ambo Town

and 12 km East of Guder town (UN-HABITAT 2008; Shanmugham and Tekele 2011; Ambo Town Administration Office 2013).

Most of the existing built up areas of the town is with gentle slope and undulating topography while some hilly slope and mountains are also seen in some parts of the town. Along the course of rivers and streams, steep slope and gullies are also observed. The town's altitude ranges from 1924 m above sea level (m.a.s) to 2384 m above sea level (m.a.s) (see Fig. 2). The slope classification of Ambo is largely dominated by terrain with flat to undulating and steep slopes. Slopes with 21–60% cover small area in the town whereas slopes with 2–20% cover the majority areas of the town (Ambo Town Administration Office 2013; Ogato 2013b).

The town is drained by Perennial and seasonal rivers and streams. The town is found within the Abay drainage basin, and it is particularly drained by major rivers (Huluka, Debis and Taltale); minor seasonal rivers (Aleltu, Awaro, Boji, Dobi, Kerise, Chafe Jara, Jalina, Maja, solbe, Jabdu and Sankale; and a number of intermittent or seasonal streams within the catchment area. The rivers and streams drain to the major Huluka river in the surrounding area of Ambo and Huluka river drains Westward to Guder river and finally Guder river drains to Abay drainage basin. The discharges of the streams are relatively small or absent during dry seasons, whereas the volume of these rivers/streams drastically increases during wet

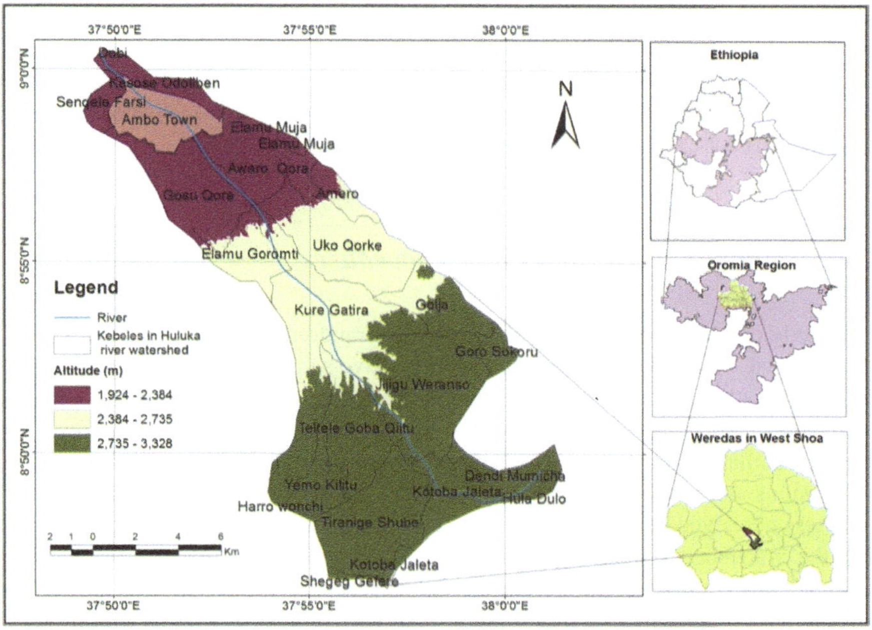

Fig. 2 Map of Huluka watershed where Ambo town is situated

season (June–October) and inundates the low gradient areas close to their banks (Prabu et al. 2010; Ambo Town Administration Office 2013; Ogato 2013b).

The mean annual temperature of the town over 30 years (1981–2010) is 18.64 °C while the mean annual rainfall of the town over 30 years (1981–2010) is 968.7 mm. The highest rainfall concentration occurs from June to September. The mean monthly relative humidity of the town varies from 64.6% in August to 35.8% in December. The prevailing winds of Autumn (Locally Meher or Birra), Winter (Locally Bega or Bona) and Spring (Locally Belg or Arfasaa) seasons are Easterly and South Easterly Winds while that of Summer (Locally Kremt or Ganna) season is Westerly and South Westerly Winds. Generally the most dominant prevailing wind of Ambo town is Easterly Wind (Ambo Town Administration Office 2013; Ogato 2013b).

The human population of Ambo town has been growing rapidly over the past few years. According to CSA (2007), the population of the town was 50,267 with the growth rate of 2.5%. The population of Ambo town is expected to reach more than 80,000 by 2016 with an average growth rate of 5% (UN-HABITAT 2008). The poor quality of housing and inability of the administration to increase supply could be taken as key indicators that a wide reform is necessary for Ambo town (UN-HABITAT 2008; Shanmugham and Tekele 2011).

Ambo is one of the oldest towns in Ethiopia (Established in 1889). It is among a few privileged towns of its time to have its own municipal administration since

1931, and a master plan since 1983 (UN-HABITAT 2008). It is governed through the Oromiya region municipal establishment proclamation no. 65/95 and has two tiers of administration. The highest level is the municipal council, which is responsible for service delivery, administering funds and management of the city (UN-HABITAT 2008; Ambo Town Administration Office 2013).

2.2 *Sampling Techniques and Procedures*

The study employed both purposive and simple random sampling techniques. First, Ambo town was selected by employing purposive sampling technique as the main purpose of the study is to assess the need for mainstreaming climate change adaptation into urban land use planning and management in Ambo town. To identify climate change related hazards in Ambo town, participants for focus group discussions and key informant interview were also purposively selected. Secondly, to assess perceived changes in climatic parameters (rainfall, temperature, and wind), Huluka watershed (where Ambo town is located) was stratified on the basis of altitude: upstream (2735–3328 m.a.s), midstream (2384–2735 m.a.s), and downstream (1924–2384 m.a.s). The sample size determination formula employed to determine sample size for households from Houlka Watershed was (Yemane 1967):

$$n = N/1 + N(e)^2$$

where

n = designates the sample size the study uses;
N = designates total number of households;
e = designates maximum variability or margin of error 5%;
1 = designates the probability of the event occurring.

Entering data of the study area into the aforementioned formula,

$$n = 3488/1 + 3488(0.05)^2 = 349$$

After determining the sample size for the study area, the sample for each altitude (stream) was distributed proportional to the size of the total households in each stream. Finally, 100 households from each stream were randomly selected based on the purpose of the study and resource limitation (see Table 1).

2.3 *Methods of Data Collection*

Both secondary and primary data sources were used as sources of data. The secondary sources were reports of urban development sectors in Ambo town, national

Table 1 Household sample design for households' survey questionnaire in Huluka Watershed

Sr. No.	Altitude	Target population (households)	Sample households	Actual sample
1	Upstream (2735–3328 m.a.s)	476	48	100[a]
2	Midstream (2384–2735 m.a.s)	1594	159	100[a]
3	Downstream (1924–2384 m.a.s)	1418	142	100[a]
Total		3488	349	300

[a]Decided based on the purpose of the study and limitation of financial resources

and regional climate change adaptation policies and strategies, and robust published materials (books, journal articles, reports of national and international organizations, and internet sources) on the issues under investigation. The primary sources were urban planners of Ambo town, urban households in Ambo town, and households in Huluka Watershed. The methods employed to collect primary data are presented hereunder.

2.3.1 Semi-structured Questionnaire

A semi-structured questionnaire was designed, pre-tested and administered through interview schedule for 300 households in Huluka Watershed where Ambo town is located to understand households' perceptions on trends of change in climatic parameters (rainfall, temperature and wind intensity) in Huluka watershed.

2.3.2 Personal Observation

Personal observation was also employed to visually observe and document indicators of urban vulnerabilities to climate change and environmental problems in Ambo town and Huluka watershed. The personal observation was guided by a semi-structured checklist and knowledgeable local person in each study village in the watershed.

2.3.3 Focus Group Discussion

Focus group discussion was also employed to collect data from members of urban local communities' development associations. Accordingly, a total of six focus group discussions (60 participants) were administered in three urban villages of Ambo town. Participants of the focus group discussion were members of the urban local communities' development associations who lived in Ambo town for more than 20 years and knowledgeable about negative impacts of climate change and climate variability. Each focus group discussion participated 10 persons and the discussion in each group took about 60 min. Each focus group discussion was started and closed with blessings of local elders as per the norm of Oromo culture in each urban village of Ambo town.

2.3.4 Key Informant Interview

Key informant interview was also employed to collect primary qualitative data. Accordingly, in-depth interview was made with 12 key informants from three urban villages (Village 1, village 2, and village 3 of Ambo town) and 15 key informants from pre-urban and rural villages (Awaro Kora, Sankile Farisi, Uko Korke, Kure Gatira, and Dendi Mumicha) in Huluka Watershed on their personal experiences with climate change and their adaptation measures in Ambo town and Huluka Watershed. The key informants were men and women who lived in Ambo town and Huluka watershed for more than 20 years and older than or equal to 50 years of age at the time of interview.

2.4 Methods of Data Analysis

Both qualitative and quantitative methods of data analysis were employed to analyze the collected data. The qualitative data captured through focus group discussion, personal observation, and key informant interview were analyzed in the form of narrations and descriptions. Simple descriptive statistics like mean, standard deviation, frequencies and percentages were employed to analyze data collected through household survey questionnaire and quantitative secondary data. SPSS software (SPSS-Version 20) was used to help the quantitative analysis of the study.

3 Results

3.1 Demographic and Socio-economic Profile of Respondents

96.3% (n = 289) of the respondents were males while 3.7% (n = 11) were females. 43.7% (n = 131) of the respondents were above 50 years of age. 94.7% (n = 284)

Table 2 Demographic and socio-economic profile of respondents

Variables	Frequency (n = 300)	Percentage (%)
Sex of respondent		
Male	289	96.3
Female	11	3.7
Age of respondent		
18–30	55	18.3
31–50	114	38.0
>50	131	43.7
Marital status		
Single	8	2.7
Married	284	94.7
Divorced	3	1.0
Widow	3	1.0
Widower	2	0.7
Educational level		
No education	43	14.3
Non-formal	70	23.3
Primary	150	50.0
Secondary	35	11.7
College	2	0.7
Agricultural land size		
Landless	77	25.7
0.25–1 ha	129	43.0
1.1–4 ha	89	29.7
4.1–10 ha	5	1.7
Main source of livelihood		
Agriculture	288	96
Off-farm activities	1	0.3
Employed	1	0.3
Labourer	2	0.7
Stone mining	8	2.7

of the respondents were with married marital status. 50% of the respondents were with primary education level. 43% of the respondents were within the agricultural land size category of 0.25–1 ha. 96% of the respondents pursue agriculture as the main source of livelihoods at the time of the interview (see Table 2).

3.2 *Perceptions, Negative Effects of Change in Climatic Parameters on Livelihoods and Adaptation Strategies of Households in Huluka Watershed*

3.2.1 Perceptions of Households in Huluka Watershed on Change in Climatic Parameters

Respondents were asked how they perceive changes in climatic parameters in their area. 99.3, 99.0, and 97.7% of the respondents perceived decreasing trend in annual rainfall, rainfall during rainy period, and rainfall during dry period respectively in their area. On the other hand, 98% of the respondents perceived increasing trend of rainfall intensity during rainy period in their area. 99.3 and 97.7% of the respondents perceived increasing trend in temperature of hot period and cold period respectively in their area. Similarly, 97% of the respondents perceived increasing trend of heat intensity during hot period in their area. 65.3 and 58.7% of the respondents perceived increasing trend of wind intensity during dry period and rainy period respectively in their area (see Table 3).

3.3 *Negative Effects of Change in Climatic Parameters on Livelihoods of Households in Huluka Watershed*

Key informants were asked to identify negative effects of change in climatic parameters on their livelihoods. Accordingly, the following negative economic, environmental, and social effects of change in climatic parameters were identified as analyzed hereunder.

Table 3 Perception of households in Huluka Watershed on change in climatic parameters

Climatic parameter	Attribute	Increase		No change		Decrease	
Rainfall		Freq.	%	Freq.	%	Freq.	%
	Annual	1	0.3	1	0.3	298	99.3
	Rainy period	1	0.3	2	0.7	297	99.0
	Dry period	–	–	7	2.3	293	97.7
	Rain intensity	294	98.0	2	0.7	4	1.3
Temperature	Hot period	298	99.3	1	0.3	1	0.3
	Cold period	293	97.7	3	1.0	4	1.3
	Heat intensity	291	97.0	7	2.3	2	0.7
Wind intensity	Dry period	196	65.3	87	29.0	17	5.7
	Rainy period	176	58.7	95	31.7	29	9.7

Similar answers were received across the three streams of the watershed (upstream, midstream, and downstream) on the negative economic effects. The negative economic effects reported were: Low crop yield associated with variability and change in climatic parameters, high cost of labour at the time of pick harvesting or planting, disease infestation resulting from shortage/excess of rainfall, stunted growth and eventual die back of trees and annual crops during prolong dry-season, destruction of trees and crops due to wind throw and flooding, and reducing production of crops to one cycle due to delay in rainfall.

No similar answers were received across the three streams of the watershed (upstream, midstream, and downstream) on the negative social effects. The negative social effects reported were: outbreak of diseases like malaria and water related diseases. These diseases were reported to exist in downstream of Huluka watershed where Ambo town is situated. For instance, the key informants from downstream of Huluka watershed reported prevalence of malaria in their residential area linked to change in climatic parameters (temperature and rainfall). On the other hand, key informants from upstream and midstream of Huluka watershed did not report any outbreak of diseases associated with change in climatic parameters in their area.

Similar answers were received across the three streams of the watershed (upstream, midstream, and downstream) on the negative environmental effects. The negative environmental effects reported were: Flooding in rainy season, too much heat in dry season, outbreak of pests associated with change and variability in climatic parameters, and loss of biodiversity resulting from rapid expansion of agricultural activities to forest areas in Huluka watershed.

3.4 Adaptation Strategies of Households in Huluka Watershed and Challenges for Adaptation to Change in Climatic Parameters

Key informants were asked to identify their adaptation strategies to negative effects of change in climatic parameters in their residential areas. Similar answers were received across the three streams of the watershed (upstream, midstream, and downstream). The following adaptation strategies were reported: Planting different crops, planting of short duration varieties, adjustment in planting period, and Prayer.

Key informants were also asked to identify challenges for adaptation to change in climatic parameters in their area. Similar answers were received across the three streams of the watershed (upstream, midstream, and downstream). The following challenges were reported: Lack of information on weather forecast, inadequate supply of improved varieties, limited access to water for irrigation, inadequate information on modern adaptation techniques, lack of financial capital, and lack of modern equipment.

3.5 Climate Change Related Disaster Risks and Urban Households Adaptation Strategies in Ambo Town

3.5.1 Climate Change Related Disaster Risks in Ambo Town

Participants of the focus group discussion were asked to discuss on climate change related disaster risks in Ambo town. They reported urban flooding disaster risk (see Fig. 3), water stress/water shortage, urban heat island effect/increased urban heat, wind storms, and dust storms as the five critical climate change related disaster risks in Ambo town (See Table 4).

The local criteria used in the ranking process were: duration of occurrence, current negative effect of the disaster risk, future negative effect of the disaster risk, severity of the disaster risk, and possibility of planned adaptation to the negative effects of climate change related disaster risk. They also indicated in which months of the year these disaster risks occur in Ambo town (see Table 5).

Fig. 3 Some of the negative effects of urban flooding disaster risk in Ambo town. *Source* Field Observation, July, 2015

Table 4 Community-based pair-wise ranking chart for climate change related disaster risks in Ambo town

	UF	Water Stress	Wind storms	UHIE	Dust Storms	Score	Rank
Urban Flooding		UF	UF	UF	UF	4	1
Water Stress			WS	WS	WS	3	2
Wind storms				UHIE	WST	1	4
Urban Heat Island Effect					UHIE	2	3
Dust Storms						0	5

UF urban flooding, *WS* water stress, *WST* wind storm, *UHIE* urban heat island effect, *DS* dust storms

Table 5 Community-based appraisal of seasonal occurrence of climate change related disaster risks in Ambo town

Climate Hazards	**Jan.**	**Feb.**	**Mar.**	**Apr.**	**May**	**June**	**July**	**Aug.**	**Sep.**	**Oct.**	**Nov.**	**Dec.**
Urban Flooding												
Water Stress												
Urban Heat Island Effect												
Wind Storms												
Dust Storms												

3.6 *Urban Households Strategies to Adapt with Climate Change Related Disaster Risks in Ambo Town*

Key informants from Ambo town were asked to explain their personal experience with climate change related disaster risks and their adaptation strategies. Harvesting rainwater during rainy season and fetching water from Huluka River during dry season were reported as adaptation strategies of urban households to adapt to water stress/water shortage. Planting trees around home, growing grasses in their home, making traditional urban flooding passage around their home, maintaining and cleaning existing drainage channels with sense of ownership, cleaning their environment, taking appropriate care for their families during flooding season, and paying tax timely to the city administration, putting sand bags in front of their

Fig. 4 Some of the autonomous adaptation strategies to urban flooding disaster risk in Ambo town. *Source* Field Observation, July 2015

homes, covering ground with plastic sheets at market places, using rubber boots, rain coats, and umbrella during flooding season were reported as adaptation strategies for urban flooding by urban households in Ambo town (see Fig. 4). Using umbrella during hot season, wearing light cloths and shoes during hot season, undertaking livelihood activities under shading of trees, and consuming much cold drinking water were reported as adaptation strategies by urban households in Ambo town to adapt to urban heat island effect/increased urban heat.

4 Discussions

4.1 Climate Change Related Urban Disaster Risks

The climate change related disaster risks identified in Ambo town were: urban flooding, water stress/water shortage, urban heat island effect/increased urban heat, wind storm, and dust storms. Climate change related disaster risks, their impacts and adaptation strategies are discussed hereunder.

4.2 Urban Flooding Disaster Risk, Impacts, and Adaptation Strategies

The first priority climate change related disaster risk in Ambo town was urban flooding. Majority of the respondents from Huluka watershed where Ambo town is situated also reported that there is increasing trend in rainfall intensity during rainy period (July, August, and September) which is one of the major contributors of urban flooding (Parkinson 2003; Few 2003; Few et al. 2004; Tucci 2007; Douglas et al. 2008; Satterthwaite 2011; Jha et al. 2012; Kebede and Nicholls 2012; Wilby and Keenan 2012; Hambati and Gaston 2015; Messling et al. 2015; Yang et al. 2015).

Scholars of sustainable urban flooding risk management assert that urban land use planning plays vital role in urban flooding risk management (Zheng and Qi 2011; de Moel et al. 2014; Ezemonye and Emeribe 2014; Früh-Müller et al. 2014; Linnerooth-Bayer et al. 2014; Guerrin 2015; Idris and Dharmasiri 2015; Mashila 2014; EEA 2015). According to Mngutyo and Ogwuche (2013). Urban land use planning is concerned with the design and organization of urban space to guide and ensure the orderly development of settlements and communities. In other words, urban land use planning has become more vital as the society becomes more urbanized, ranging from producing blue print to more strategic approach of structure and local planning (Atedhor et al. 2011; UN-Habitat 2011; Wilby and Keenan 2012; Mngutyo and Ogwuche 2013; Murtaza et al. 2015; Umezuruike 2015).

4.3 Water Stress/Water Shortage, Impacts, and Adaptation Strategies

The second priority climate change related disaster risk reported by local communities in Ambo town was water stress/water shortage. Urban areas where there is a failure to address the impacts of climate change on water resources will leave their inhabitants vulnerable to a range of immediate acute and slow-onset disasters (Muller 2007; Fünfgeld 2010; Su et al. 2012; Wapwera et al. 2015). In other words, climate change has the potential to significantly alter river flow regimes in a river catchment which may seriously affect urban water supply (Hall and Murphy 2010).

Adaptation planning for water sector is a planning process of developing institutional and political capacities to ensure adequate water supply and water quality in the face of intensifying risks from climate and climate-related impacts (Nikitina et al. 2010; Sowers et al. 2011; Herrfahrdt-Pähle 2013). Urban adaptation strategies and discourses need to deal more strongly with processes and the knowledge base on how to improve adaptive capacities and adaptive planning, rather than focusing solely on a list of options to adjust physical structures and the built environment (Björklund et al. 2009; Hardoy and Pandiella 2009; Birkmann et al. 2010; van Buuren et al. 2014). Managing water demand, improving the efficiency of water use, and promoting conservation will be key ingredients in responding to climate-induced impacts on the water sector (Mata and Budhooram 2007; ECE 2009; Luthe et al. 2012; Porthin et al. 2013; Padgham et al. 2015; Sietz and Van Dijk 2015).

4.4 Urban Heat Island Effect/Increased Urban Heat, Impacts, and Adaptation Strategies

The third priority climate change related disaster risk identified in Ambo town was urban heat island effect (UHIE). January, February, March, April, and May were reported as the months when the urban heat island effect in Ambo town is grave.

While physical climate changes can impact upon both rural and urban areas, urban settlements generate unique local conditions that interact with heat events. Compared to rural areas, cities tend to have higher air and surface temperatures due to the urban heat-island effect: the tendency of cities to retain heat more than their surrounding rural areas (Yow 2007; UN-Habitat 2011). By increasing temperatures, urban heat-island effects can aggravate the heat-related negative implications of climate change and impose costly energy demands on urban systems as they attempt to adapt to higher temperatures. The degree of these effects is not uniform across cities. The physical layout of a city, its population size and density, and structural features of the built environment all influence the strength of the urban heat-island effect (Yow 2007; UN-Habitat 2011; Djibril et al. 2012; Umezuruike 2015). Scholars of climate change adaptation studies recommends incorporating vegetation in developed areas, reducing energy consumption, and installing green roofs or rooftop gardens as some of the best adaptation strategies for urban heat island effect (McKendry 2003; Yow and Carbone 2006; Yu and Hien 2006; Zomer et al. 2008; Chen et al. 2013; Demuzere et al. 2014; Laves et al. 2014; MacDonald et al. 2014; Murgida et al. 2014).

4.5 Wind Storms, Dust Storms, Impacts, and Adaptation Strategies

The fourth priority climate change related disaster risk identified in Ambo town was wind storm. The fifth priority climate change related disaster risk in Ambo town was dust storm. Majority of the respondents from Huluka watershed where Ambo town is situated also reported that there is increasing trend in wind intensity both in dry period and rainy period. January, February, March, April, and May were reported as the months when the wind storm and dust storm in Ambo town is severe.

Both wind storm and dust storm may be adapted to properly through the practice of ecosystem-based adaptation both in Ambo town and Huluka watershed. An Ecosystem is the dynamic complex of plant, animal and micro-organism communities and the nonliving environment interacting as a functional unit. It assumes that people are an integral part of ecosystems (Millennium Ecosystem Assessment

2005). Ecosystem-based Adaptation (EbA) integrates the use of biodiversity and ecosystem services into an overall strategy to help people adapt to the adverse impacts of climate change (Niemelä et al. 2010; Oteros-Rozas et al. 2014; EEA 2015). It includes the sustainable management, conservation and restoration of ecosystems to provide services that help people adapt to both current climate variability, and climate change (Millennium Ecosystem Assessment 2005; EEA 2015). Ecosystem-based Adaptation reduces vulnerability to both climate and non-climate risks and provides multiple economic, social, environmental and cultural benefits, including: disaster risk reduction, livelihood sustenance and food security, biodiversity conservation, carbon sequestration, and sustainable water management (Colls et al. 2009; Ngigi 2009; Ludi 2009; Vignola et al. 2009; Niemelä et al. 2010; Lantz et al. 2013; Wu et al. 2013; EEA 2015).

4.6 The Need to Mainstream Adaptation to Climate Change into Urban Land Use Planning and Management in Ambo Town

Mainstreaming adaptation to climate change related disaster risks into urban land use planning and management reduces the devastating consequences of unanticipated climate-related disaster risks, including costs that constitute significant drains on resources, thereby stifling the achievement of set goals (Klein et al. 2003; Wilson 2006; Kok et al. 2008; Wilson and Piper 2008; Yaro et al. 2010; Hurlimann and March 2012). It can also ensure that development programs and policies are not at odds with climate risks both now and in the future (Lindley et al. 2007; Huxtable and Yen 2009; Chinvanno 2011; Li 2012). There is a growing need for policy-makers, particularly in the ministries related to development such as in finance or planning, to better understand how climate change adaptation can be addressed in national and sub-national/regional planning processes, and through fiscal and investment decisions. For example, when making decisions on long-lived infrastructure, it may be more cost-effective to take adaptation needs into account earlier rather than later (Lebel et al. 2012; Faleiro et al. 2013; Ayers et al. 2014).

Mainstreaming adaptation into urban land use planning and management has been promoted as an effective way to respond to climate change in urban areas and the expected benefits for sustainable urban development in Ambo town include: avoided policy conflicts; reduced risks and vulnerability; greater efficiency compared with managing adaptation separately; leveraging the much larger financial flows in sectors affected by climate risks than the amounts available for financing adaptation separately, and easier to start with existing policies and practices, rather than creating new ones (Klein et al. 2003; Huxtable and Yen 2009; Chinvanno 2011; Oates et al. 2011; UNPEI 2011; Lebel et al. 2012).

5 Conclusion and Recommendations

Majority of the respondents in Huluka Watershed perceived decreasing trends of annual rainfall, wet period rainfall, and dry period rainfall. On the other hand, Majority of the respondents in Huluka watershed perceived increasing trends of rain intensity in wet period, hot period temperature, cold period temperature, heat intensity, and wind intensity in their area. Negative economic, social, and environmental effects of change in climatic parameters on livelihoods of households in Huluka Watershed and local adaptation strategies were also assessed. Climate change related disaster risks in Ambo town. Urban flooding disaster risk, water stress/water shortage, urban heat island effect/increased urban heat, wind storms, and dust storms were identified as the five critical climate change related disaster risks in Ambo town. Furthermore, autonomous adaptation strategies of urban communities to climate change related disaster risks were assessed.

In conclusion, there are convincing reasons to mainstream climate change adaptation into urban land use planning and management in Ambo town. Further investigation and characterization of each climate change related disaster risk in Ambo town with the help of geographic information system (GIS) and remote sensing techniques are of paramount importance in the mainstreaming process of climate change adaptation into urban land use planning and management. While it is commendable to appreciate the good start of urban greening and beautification by Ambo town administration and its municipality, the town administration has to take strategic actions to adapt Ambo town and its watershed to the negative impacts of climate change and climate variability.

The following strategic measures are forwarded to mainstream climate change adaptation into urban land use planning and management in Ambo town in the years to come:

- Awareness of urban local communities and other stakeholders on the benefits of mainstreaming climate change adaptation into urban land use planning and management should be created and increased for achieving sustainable development in Ambo town and Huluka Watershed;
- Community participation should be mainstreamed in the vulnerability assessment and adaptation planning to motivate them for active participation and benefit sharing from the urban development process;
- Vulnerable sectors, most vulnerable groups, and adaptive capacity of local communities should be assessed to mainstream climate change into different urban development activities in the town and in the watershed;
- Training and education opportunities on mainstreaming climate change adaptation into urban land use planning and management should be provided for urban land use planners to improve planning, implementation, monitoring and evaluation activities of mainstreaming;
- Stakeholders should be well informed about their roles and responsibilities in vulnerability assessment and adaptation process;

- Urban investment permits should include mainstreaming adaptation to climate change into urban development activities as key criteria;
- Mainstreaming climate change adaptation into urban land use planning should reduce the negative social, economic and environmental impacts of urban development and enable the most vulnerable groups like women, children, physically handicapped groups, and elderly adapt to climate change related disaster risks;
- Rural-Urban linkage in managing watersheds in rural villages should be strengthened to adapt Ambo town and neighboring rural villages to negative impacts of climate change and climate variability;
- Ambo town administration should properly monitor and evaluate climate adaptation projects in its development sector to check whether they meet their climate adaptation objectives, and what other benefits or adverse impacts they may have on the environment.
- The local urban government in Ambo town should be able to provide the available social services efficiently and effectively and mobilize urban local communities' resources to enable Ambo town and its watershed adapt to negative effects of climate change and climatic variability; and
- The adaptation measures should include participatory planting of indigenous and eco-friendly exotic trees and management both in urban and rural villages in Huluka Watershed.

Acknowledgements The authors wish to express their sincere gratitude to the International Science Foundation (IFS) for providing financial support (research grant) to the corresponding author to carry-out this research as part of his PhD study. Ambo University and the Ethiopian Institute of Architecture, Building Construction and City Development (EiABC) of Addis Ababa University deserve also our thanks for providing technical and additional financial support to the corresponding author in his PhD study. The authors are very much grateful also to all organizations and persons for their full cooperation in providing necessary data and information in the whole research process. Our special thanks also go to anonymous reviewers and editors for their efficient and effective management of the review and publication process of the manuscript.

Declaration of Conflict of Interest

The authors fully declare that they have no conflict of interest in publishing the manuscript.

References

Adem, A., & Bewket, W. (2011). *A climate change country assessment report for Ethiopia*. Addis Ababa: Epsilon International R&D.

Adem, A., & Guta, A. (2011). *Engendering climate change policy and practice in Ethiopia*. Addis Ababa: Forum for Environment.

Adger, W. N., Huq, S., Brown, K., Conway, D., Hulme, M., Adger, N., et al. (2002). Adaptation to climate change: Setting the agenda for development policy and research. *Progress in Development Studies, 3*(3), 179–195.

Ambo Town Administration Office. (2013). *Profile of Ambo Town* (Unpublished).

Atedhor, G. O., Odjugo, P. A., & Uriri, A. E. (2011). Changing rainfall and anthropogenic-induced flooding: Impacts and adaptation strategies in Benin City, Nigeria. *Journal of Geography and Regional Planning, 4*(1), 42–52.

Ayers, J. M., Huq, S., Faisal, A. M., & Hussain, S. T. (2014). Mainstreaming climate change adaptation into development: A case study of Bangladesh. *Wiley Interdisciplinary Reviews: Climate Change, 5*(1), 37–51.

Baloye, D. O., Adesina, F. A., & Kufoniyi, O. (2010). A spatial decision support system approach to sustainable physical development planning. *Journal of Geography and Regional Planning, 3* (9), 219–233.

Bartlett, S., Dodman, D., Hardoy, J., Satterthwaite, D., & Tacoli, C. (2012). Social aspects of climate change in urban areas in lowand middle-income nations. In D. Hoornweg, M. Freire, M. J. Lee, P. Bhada-Tata, & B. Yuen (Eds.), *Cities and climate change: Responding to an urgent agenda* (Vol. 2). Washington, DC: The World Bank.

Belinda Yuen, B., & Kong, L. (2012). Climate change and urban planning in Southeast Asia. In D. Hoornweg, M. Freire, M. J. Lee, P. Bhada-Tata, & B. Yuen (Eds.), *Cities and climate change: Responding to an urgent agenda* (Vol. 2). Washington, DC: The World Bank.

Biesbroek, G. R., Rob J. Swart, R. J., & van der Knaap, W. G. (2009). The mitigation–adaptation dichotomy and the role of spatial planning. *Habitat International, 33*, 230–237.

Björklund, G. Tropp, H., Harlin, J., Morrison, A., & Hudson, A. (2009). *Water adaptation in NAPAs: Freshwater in climate adaptation planning and climate adaptation in freshwater planning*. A UNDP Side publication to WWDR3. Stockholm: UNDP Water Governance Facility (WGF).

Birkmann, J., Garschagen, M., Kraas, F., & Quang, N. (2010). Adaptive urban governance: New challenges for the second generation of urban adaptation strategies to climate change. *Sustainability Science, 5*(2), 185–206.

Brooks, N., Adger, W. N., Barnett, J., Woodzard, & Lim, B. (2004). Assessing and enhancing adaptive capacity. In B. Lim, E. Spanger-Siegfried, I. Burton, E. Malone, & S. Huq (Eds.), *Adaptation policy frameworks for climate change: Developing strategies, policies and measures*. New York: United Nations Development Programme.

Central Statistical Agency of Ethiopia (CSA). (2007). *The 2007 Population and Housing Census of Ethiopia*. Addis Ababa: CSA.

Chappin, E. J., & van der Lei, T. (2014). Adaptation of interconnected infrastructures to climate change: A socio-technical systems perspective. *Utilities Policy, 31*, 10–17.

Chen, H., Zhu, T., Krott, M., & Maddox, D. (2013). Community forestry management and livelihood development in northwest China: integration of governance, project design, and community participation. *Regional Environmental Change, 13*(1), 67–75.

Chinvanno, S. (2011). *Mainstreaming Climate Change into Community Development Strategy: A critical opinion on climate change adaptation planning and case study in Thailand*. Bangkok, Thailand: Southeast Asia START Regional Center.

Colls, A., Ash, N., & Ikkala, N. (2009). *Ecosystem-based Adaptation: a natural response to climate change*. Gland, Switzerland: IUCN.

de Bruin, K., Goosen, H., van Ierland, E. C., & Groeneveld, R. A. (2014). Costs and benefits of adapting spatial planning to climate change: lessons learned from a large-scale urban development project in the Netherlands. *Regional Environmental Change, 14*(3), 1009–1020.

de Moel, H., van Vliet, M., & Aerts, J. C. (2014). Evaluating the effect of flood damage-reducing measures: a case study of the unembanked area of Rotterdam, the Netherlands. *Regional Environmental Change, 14*(3), 895–908.

Dede, O. M., Dikmen, Ç. B., & Ayten, A. M. (2012). A new approach for participative urban design: An urban design study of Cumhuriyet urban square in Yozgat Turkey. *Journal of Geography and Regional Planning, 5*(5), 122–131.

Demuzere, M., Orru, K., Heidrich, O., Olazabal, E., Geneletti, D., Orru, H., et al. (2014). Mitigating and adapting to climate change: Multi-functional and multi-scale assessment of green urban infrastructure. *Journal of Environmental Management, 146*, 107–115.

Denton, F. (2002). Climate change vulnerability, impacts, and adaptation: Why does gender matter? *Gender and Development, 10*(2), 10–20.

Djibril, K., Yun, Q., Ousmane, D., & Xiangrong, W. (2012). Processes and challenges of urban development in Cote d'Ivoire(Africa) with case study of Abidjan City. *Journal of geography and regional planning, 5*(13), 353–361.

Douglas, I., Alam, K., Maghenda, M., Mcdonnell, Y., McLean, L., & Campbell, J. (2008). Unjust waters: climate change, flooding and the urban poor in Africa. *Environment and Urbanization, 20*(1), 187–205.

Economic Commission for Europe (ECE). (2009). *Convention on the Protection and Use of Transboundary Watercourses and International Lakes: Guidance on Water and Adaptation to Climate Change*. New York and Geneva: United Nations.

European Environment Agency (EEA). (2015). *Exploring nature-based solutions: The role of green infrastructure in mitigating the impacts of weather- and climate change-related natural hazards*. Luxembourg: European Environment Agency.

Ezemonye, M. N., & Emeribe, C. N. (2014). Flooding and Household Preparedness in Benin City. *Nigeria. Mediterranean Journal of Social Sciences, 5*(1), 547–553.

Faleiro, F. V., Machado, R. B., & Loyola, R. D. (2013). Defining spatial conservation priorities in the face of land-use and climate change. *Biological Conservation, 158,* 248–257.

Few, R. (2003). Flooding, vulnerability and coping strategies: local responses to a global threat. *Progress in Development Studies, 3*(1), 43–58.

Few, R., Ahern, M., Matthies, F., & Kovats, S. (2004). *Floods, health and climate change: a strategic review*. Norwich: Tyndall Centre for Climate Change Research.

Früh-Müller, A., Wegmann, M., & Koellner, T. (2014). Flood exposure and settlement expansion since pre-industrial times in 1850 until 2011 in north Bavaria. *Germany. Regional Environmental Change, 15*(1), 183–193.

Fünfgeld, H. (2010). Institutional challenges to climate risk management in cities. *Current Opinion in Environmental Sustainability, 2*(3), 156–160.

Fünfgeld, H. (2012). *Local climate change adaptation planning A guide for government policy and decision makers in Victoria*. Melbourne: Victorian Centre for Climate Change Adaptation Research (VCCCAR).

Füssel, H. M. (2007). Adaptation planning for climate change: concepts, assessment approaches, and key lessons. *Sustainability Science, 2*(2), 265–275.

Gill, S. E., Handley, J. F., Ennos, A. R., & Pauleit, S. (2007). Adapting cities for climate change: The role of the green infrastructure. *Built Environment, 1978,* 115–133.

Guerrin, J. (2015). A floodplain restoration project on the River Rhône (France): analyzing challenges to its implementation. *Regional Environmental Change, 15*(3), 559–568.

Hall, J., & Murphy, C. (2010). Vulnerability analysis of future public water supply under changing climate conditions: a study of the Moy catchment, western Ireland. *Water Resources Management, 24*(13), 3527–3545.

Hallegatte, S., Henriet, F., & Corfee-Morlot, J. (2011). The economics of climate change impacts and policy benefits at city scale: a conceptual framework. *Climatic Change, 104*(1), 51–87.

Hambati, H., & Gaston, G. (2015). Revealing the Vulnerability of Urban Communities to Flood Hazard in Tanzania: A Case of the Dar es Salaam City Ecosystem. *International Journal of Geospatial and Environmental Research, 2*(1), 1–22.

Hardoy, J., & Pandiella, G. (2009). Urban poverty and vulnerability to climate change in Latin America. *Environment and Urbanization, 21*(1), 203–224.

Herrfahrdt-Pähle, E. (2013). Integrated and adaptive governance of water resources: the case of South Africa. *Regional Environmental Change, 13*(3), 551–561.

Hurlimann, A. C., & March, A. P. (2012). The role of spatial planning in adapting to climate change. *Wiley Interdisciplinary Reviews: Climate Change, 3*(5), 477–488.

Huxtable, J., & Yen, N. T. (2009). *Mainstreaming Climate Change Adaptation: A Practioner's Handbook*. Ha Noi, Vietnam: CARE International. in Vietnam.

Idris, S., & Dharmasiri, L. M. (2015). Flood risk inevitability and flood risk management in urban areas: A review. *Journal of Geography and Regional Planning, 8*(8), 205–209.

Jabareen, Y. (2013). Planning the resilient city: Concepts and strategies for coping with climate change and environmental risk. *Cities, 31,* 220–229.

Jabeen, H., Allen, A., & Johnson, C. (2012). Built-in resilience: Learning from grassroots coping strategies to climate variability. In D. Hoornweg, M. Freire, M. J. Lee, P. Bhada-Tata, & B. Yuen (Eds.), *Cities and climate change: Responding to an urgent agenda* (Vol. 2). Washington, DC: The World Bank.

Jha, A. K., Bloch, R., & Lamond, J. (2012). *Cities and flooding: A guide to integrated urban flood risk management for the 21st century.* Washington, DC: The World Bank.

Jones, L., Ludi, E., Carabine, E., Grist, N., Amsalu, A., Artur, L., et al. (2014). *Planning for an uncertain future: Promoting adaptation to climate change through flexible and forward-looking decision making.* London: ODI.

Kareem, B., & Lwasa, S. (2014). Gender responsiveness in infrastructure provision for African cities: The case of Kampala in Uganda. *Journal of Geography and Regional Planning, 7*(1), 1–9.

Kazmierczak, A., & Carter, J. (2010). *Adaptation to climate change using green and blue infrastructure: A database of case studies.* Manchester: University of Manchester.

Kebede, A. S., & Nicholls, R. J. (2012). Exposure and vulnerability to climate extremes: Population and asset exposure to coastal flooding in Dar es Salaam, Tanzania. *Regional Environmental Change, 12*(1), 81–94.

Kelman, I., Gaillard, J. C., & Mercer, J. (2015). Climate change's role in disaster risk reduction's future: Beyond vulnerability and resilience. *International Journal of Disaster Risk Science, 6,* 21–27.

Klein, R. J. T., Schipper, S. E. L., & Dessai, S. (2003). *Integrating mitigation and adaptation into climate and development policy: Three research questions* (Working Paper 40). Norwich: Tyndall Centre for Climate Change Research.

Kok, M., Metz, B., Verhagen, J., & Van Rooijen, S. (2008). Integrating development and climate policies: National and international benefits. *Climate Policy, 8*(2), 103–118.

Lantz, V., Boxall, P. C., Kennedy, M., & Wilson, J. (2013). The valuation of wetland conservation in an urban/peri urban watershed. *Regional Environmental Change, 13*(5), 939–953.

Laves, G., Kenway, S., Begbie, D., Roiko, A., Carter, R. W., & Waterman, P. (2014). The research-policy nexus in climate change adaptation: Experience from the urban water sector in South East Queensland. *Australia. Regional Environmental Change, 14*(2), 449–461.

Lebel, L., Li, L., Krittasudthacheewa, C., Juntopas, M., Vijitpan, T., Uchiyama, T., et al. (2012). *Mainstreaming climate change adaptation into development planning.* Bangkok: Adaptation Knowledge Platform and Stockholm Environment Institute.

Li, Y. (2012). Urban–rural interaction patterns and dynamic land use: Implications for urban–rural integration in China. *Regional Environmental Change, 12*(4), 803–812.

Lindley, S. J., Handley, J. F., McEvoy, D., Peet, E., & Theuray, N. (2007). The role of spatial risk assessment in the context of planning for adaptation in UK urban areas. *Built Environment, 1978,* 46–69.

Linnerooth-Bayer, J., Dubel, A., Sendzimir, J., & Hochrainer-Stigler, S. (2014). Challenges for mainstreaming climate change into EU flood and drought policy: Water retention measures in the Warta River Basin, Poland. *Regional Environmental Change,* 1–13.

Loret, E., & Ioannilli, M. (2012). Spatial analysis of the increasing urbanization trend in the Frascati wine designation of origin(DOC) area: A geostatistical approach. *Journal of Geography and Regional Planning, 5*(13), 346–352.

Ludi, E. (2009). *Climate change, water and food security. ODI Background Notes.* London: Overseas Development Institute.

Luthe, T., Wyss, R., & Schuckert, M. (2012). Network governance and regional resilience to climate change: Empirical evidence from mountain tourism communities in the Swiss Gotthard region. *Regional Environmental Change, 12*(4), 839–854.

Lwasa, S. (2010). Adapting urban areas in Africa to climate change: The case of Kampala. *Current Opinion in Environmental Sustainability, 2*(3), 166–171.

MacDonald, K., Sanyal, B., Silver, M., Ng, M. K., Head, P., Williams, K., et al. (2014). Challenging theory: Changing practice: Critical perspectives on the past and potential of professional planning. *Planning Theory & Practice, 15*(1), 95–122.

Mashila, T. (2014). *Spatial planning for climate change adaptation: Developing a climate change local area adaptation plan for Khayelitsha*. Cape Town: University of Cape Town.

Mata, L. J., & Budhooram, J. (2007). Complementarity between mitigation and adaptation: The water sector. *Mitigation and Adaptation Strategies for Global Change, 12*(5), 799–807.

McCarney, P. (2012). City indicators on climate change: Implications for policy leverage and governance. In D. Hoornweg, M. Freire, M. J. Lee, P. Bhada-Tata, & B. Yuen (Eds.), *Cities and climate change: Responding to an urgent agenda* (Vol. 2). Washington, DC: The World Bank.

McKendry, I. G. (2003). Applied climatology. *Progress in Physical Geography, 27,* 597–606.

Measham, T. G., Preston, B. L., Smith, T. F., Brooke, C., Gorddard, R., Withycombe, G., et al. (2011). Adapting to climate change through local municipal planning: Barriers and challenges. *Mitigation and Adaptation Strategies for Global Change, 16*(8), 889–909.

Messling, L., Corner, A., Clarke, J., Pidgeon, N. F., Demski, C., & Capstick, S. (2015). *Communicating flood risks in a changing climate*. Oxford: Climate Outreach.

Millennium Ecosystem Assessment. (2005). *Ecosystems and human well-being: Wetlands and water synthesis*. Washington, DC: World resources institute.

Mngutyo, I. D., & Ogwuche, J. A. (2013). Urban development, flood and disaster management: Challenges of contemporary urban planning perspectives. *International Journal of Innovative Environmental Studies Research, 1*(1), 1–6.

Muller, M. (2007). Adapting to climate change water management for urban resilience. *Environment and Urbanization, 19*(1), 99–113.

Murgida, A. M., González, M. H., & Tiessen, H. (2014). Rainfall trends, land use change and adaptation in the Chaco salteño region of Argentina. *Regional Environmental Change, 14*(4), 1387–1394.

Murtaza, K. O., Rather, M. A., Lateef, Z., Shah, V., & Kuchhay, N. (2015). Geospatial mapping of Srinagar City. *Journal of Geography and Regional Planning, 8*(6), 157–165.

Ngigi, S. N. (2009). *Climate change adaptation strategies: Water Resources management options for smallholder farming systems in Sub-Saharan Africa*. New York: The MDG Centre for East and Southern Africa, the Earth Institute at Columbia University.

Nielsen, J. Ø., & Reenberg, A. (2010). Cultural barriers to climate change adaptation: A case study from Northern Burkina Faso. *Global Environmental Change, 20*(1), 142–152.

Niemelä, J., Saarela, S. R., Söderman, T., Kopperoinen, L., Yli-Pelkonen, V., Väre, S., et al. (2010). Using the ecosystem services approach for better planning and conservation of urban green spaces: A Finland case study. *Biodiversity and Conservation, 19*(11), 3225–3243.

Nikitina, E., Ostrovskaya, E., & Fomenko, M. (2010). Towards better water governance in river basins: Some lessons learned from the Volga. *Regional Environmental Change, 10*(4), 285–297.

Oates, N., Conway, D., & Calow, R. (2011). *The 'mainstreaming' approach to climate change adaptation: Insights from Ethiopia's water sector. Overseas Development Institute Background Note*. London: ODI.

Ogato, G. S. (2013a). The quest for mainstreaming climate change adaptation into regional planning of least developed countries: Strategy implications for regions in Ethiopia. *Herald Journal of Geography and Regional Planning, 2*(2), 071–081.

Ogato, G. S. (2013b). The quest for mainstreaming climate change adaptation into urban development planning of Ambo Town, Ethiopia. *American Journal of Human Ecology, 2*(3), 103–119.

Oteros-Rozas, E., Martín-López, B., González, J. A., Plieninger, T., López, C. A., & Montes, C. (2014). Socio-cultural valuation of ecosystem services in a transhumance social-ecological network. *Regional Environmental Change, 14*(4), 1269–1289.

Padgham, J., Jabbour, J., & Dietrich, K. (2015). Managing change and building resilience: A multi-stressor analysis of urban and peri-urban agriculture in Africa and Asia. *Urban Climate, 12,* 183–204.

Parikh, J., Sandal, G., & Jindal, P. (2014). *Vulnerability profiling of cities: A framework for climate-resilient urban development in India* (Asian Cities Climate Resilience, Working Paper Series 8: 2014). London: International Institute of Environment and Development (IIED).

Parkinson, J. (2003). Drainage and stormwater management strategies for low-income urban communities. *Environment and Urbanization, 15*(2), 115–126.

Porthin, M., Rosqvist, T., Perrels, A., & Molarius, R. (2013). Multi-criteria decision analysis in adaptation decision-making: A flood case study in Finland. *Regional Environmental Change, 13*(6), 1171–1180.

Posey, J. (2009). The determinants of vulnerability and adaptive capacity at the municipal level: Evidence from floodplain management programs in the United States. *Global Environmental Change, 19*(4), 482–493.

Prabu, P. C., Wondimu, L., & Tesso, M. (2010). Assessment of water quality of Huluka and Alaltu rivers of Ambo, Ethiopia. *Journal of Agricultural Science and Technology, 13,* 131–138.

Reid, H., & Huq, S. (2014). Mainstreaming community-based adaptation into national and local planning. *Climate and Development, 6*(4), 291–292.

Reid, H., Haque, M., Kirkby, P., Roberts, E., Stubbings, A., Urquhart, P., et al. (Eds.). (2015). *Community based adaptation: Measuring and enhancing effective adaptation.* 9th International Conference 24–30 April 2015 Nairobi, Kenya.

Sanchez-Rodriguez, R. (2009). Learning to adapt to climate change in urban areas. A review of recent contributions. *Current Opinion in Environmental Sustainability, 1*(2), 201–206.

Sanderson, D. (2000). Cities, disasters and livelihoods. *Environment and Urbanization, 12*(2), 93–102.

Satterthwaite, D. (2011). *What role for low-income communities in urban areas in disaster risk reduction? Background Paper for UN-ISDR Global Assessment Report.* London: IIED.

Shanmugham, S., & Tekele, S. B. (2011). An assessment on the status of water supply and sanitation in Ethiopia: A case of Ambo town. *Journal of Sustainable Development in Africa, 13* (1), 23–43.

Sietz, D., & Van Dijk, H. (2015). Land-based adaptation to global change: What drives soil and water conservation in western Africa? *Global Environmental Change, 33,* 131–141.

Sowers, J., Vengosh, A., & Weinthal, E. (2011). Climate change, water resources, and the politics of adaptation in the Middle East and North Africa. *Climatic Change, 104*(3–4), 599–627.

Stein, A., & Moser, C. (2014). Asset planning for climate change adaptation: Lessons from Cartagena, Colombia. *Environment and Urbanization, 26*(1), 1–18.

Su, Y., Xu, J., Wilkes, A., Lu, J., Li, Q., Fu, Y., et al. (2012). Coping with climate-induced water stresses through time and space in the mountains of Southwest China. *Regional Environmental Change, 12*(4), 855–866.

Sussams, L. W., Sheate, W. R., & Eales, R. P. (2015). Green infrastructure as a climate change adaptation policy intervention: Muddying the waters or clearing a path to a more secure future? *Journal of Environmental Management, 147,* 184–193.

Taylor, A., & Peter, C. (2014). *Strengthening climate resilience in African Cities: A framework for working with informality.* Cape Town: African Centre for Cities, University of Cape Town.

Tucci, C. E. (2007). *Urban flood management.* Geneva: World Meteorological Organization.

Uittenbroek, C. J., Janssen-Jansen, L. B., & Runhaar, H. A. (2013). Mainstreaming climate adaptation into urban planning: Overcoming barriers, seizing opportunities and evaluating the results in two Dutch case studies. *Regional Environmental Change, 13*(2), 399–411.

Umezuruike, S. O. (2015). Physical planning administration in Nigeria: Abia State experience. *Journal of Geography and Regional Planning, 8*(3), 47–55.

United Nations Development Programme-United Nations Environment Programme Poverty-Environment Initiative (UNPEI). (2011). *Mainstreaming climate change adaptation into development planning: A Guide for practitioners.* Nairobi, Kenya: UNDP.

United Nations Human Settlements Programme (UN-HABITAT). (2006). *Addis Abba Urban Profile.* Nairobi: UN-HABITAT.

United Nations Human Settlements Programme (UN-HABITAT). (2008). *Ethiopia: Ambo Urban Profile*. Nairobi: UN-HABITAT.

United Nations Human Settlements Programme (UN-HABITAT). (2011). *Cities and climate change: Global report on human settlements*. Nairobi: UN-HABITAT.

Van Aalst, M. K., Cannon, T., & Burton, I. (2008). Community level adaptation to climate change: The potential role of participatory community risk assessment. *Global Environmental Change, 18*(1), 165–179.

van Buuren, A., Driessen, P., Teisman, G., & van Rijswick, M. (2014). Toward legitimate governance strategies for climate adaptation in the Netherlands: Combining insights from a legal, planning, and network perspective. *Regional Environmental Change, 14*(3), 1021–1033.

van den Brink, M., Meijerink, S., Termeer, C., & Gupta, J. (2014). Climate-proof planning for flood-prone areas: Assessing the adaptive capacity of planning institutions in the Netherlands. *Regional Environmental Change, 14,* 981–995.

Vasileiadou, E., Hisschemoller, M., Petersen, A. C., Hazeleger, W., Betgen, C., de Hoog, I., et al. (2014). Adaptation to extreme weather: Identifying different societal perspectives in the Netherlands. *Regional Environmental Change, 14,* 91–101.

Vignola, R., Locatelli, B., Martinez, C., & Imbach, P. (2009). Ecosystem-based adaptation to climate change: What role for policy-makers, society and scientists? *Mitigation and Adaptation Strategies for Global Change, 14*(8), 691–696.

Wapwera, S. D., Mallo, D. M., & Jiriko, G. J. (2015). Institutional framework and constraints in the urban and regional planning system in Jos Metropolis, Nigeria. *Journal of Geography and Regional Planning, 8*(10), 244–260.

Watson, V. (2009). 'The planned city sweeps the poor away …': Urban planning and 21st century urbanisation. *Progress in Planning, 72*(3), 151–193.

Wilby, R. L., & Keenan, R. (2012). Adapting to flood risk under climate change. *Progress in Physical Geography*, 1–31.

Wilson, E. (2006). Adapting to climate change at the local level: The spatial planning response. *Local Environment, 11*(6), 609–625.

Wilson, E., & Piper, J. (2008). Spatial planning for biodiversity in Europe's changing climate. *European Environment, 18*(3), 135–151.

Wu, J., Gong, Y., Zhou, J., Wang, X., Gao, J., & Yan, A. (2013). The governance of integrated ecosystem management in ecological function conservation areas in China. *Regional Environmental Change, 13*(6), 1301–1312.

Yamane, T. (1967). *Statistics: An introductory analysis* (2nd ed.). New York: Harper and Row.

Yang, L., Scheffran, J., Qin, H., & You, Q. (2015). Climate-related flood risks and urban responses in the Pearl River Delta, China. *Regional Environmental Change, 15,* 379–391.

Yaro, J. A., Dogbe, T. D., Bizikova, L., Bailey, P., Ahiable, G., Yahaya, T., et al. (2010). *Development and climate change: The social dimensions of adaptation to climate change in Ghana* (Discussion Paper Number 15, December 2010). Washington, DC; World Bank.

Yow, D. M. (2007). Urban heat islands: Observations, impacts, and adaptation. *Geography Compass, 1*(6), 1227–1251.

Yow, D. M., & Carbone, G. J. (2006). The urban heat island and local temperature variations in Orlando, Florida. *Southeastern Geographer, 46,* 117–141.

Yu, C., & Hien, W. N. (2006). Thermal benefits of city parks. *Energy and Buildings, 38,* 105–120.

Zheng, Z., & Qi, S. (2011). Potential flood hazard due to urban expansion in the karst mountainous region of North China. *Regional Environmental Change, 11*(3), 439–440.

Ziervogel, G., Taylor, A., Hachigonta, S., & Hoffmaister, J. (2008). *Climate adaptation in Southern Africa: Addressing the needs of vulnerable communities*. Stockholm: Stockholm Environment Institute.

Zomer, R. J., Trabucco, A., Bossio, D. A., & Verchot, L. V. (2008). Climate change mitigation: A spatial analysis of global land suitability for clean development mechanism afforestation and reforestation. *Agriculture, Ecosystems & Environment, 126*(1), 67–80.

A Geographic Information System as Support to the Healthcare Services of Nomadic Community, the Filtu Woreda Case Study

Elena Isotta Cristofori, Anna Facello, Alessandro Demarchi, Walther Camaro, Micol Fascendini and Alessia Villanucci

1 Introduction

Nomads and semi-nomads in the developing world are about 50–100 million, over 60% living in Africa (Sheik-Mohamed et al. 1999). About 8 million pastoralists are estimated to live in Ethiopia where pastoral land covers about 60% of the total land. Nomads have the least access to the healthcare delivery system, due to their traditional lifestyles entailing seasonal movements and spread settlement patterns. Therefore, pastoralists live in an uncertain environment resulting in a high vulnerability to the adverse effects of climate change (Kimaro et al. 2013). The increasing frequency of extreme weather conditions, such as flood and drought, will have in the future severe negative impacts on rural people and especially in key sectors like agriculture, water and human health.

Several initiatives have been implemented to identify gaps and best-practices that may help pastoralists to adapt to climate change effects in the Horn of Africa and the Sahel region (Frankenberger et al. 2012; Cervigni et al. 2015). Actions aiming to enhance adaptation and to increase resilience include the improvement of

E.I. Cristofori (✉) · A. Demarchi
Politecnico di Torino, DIST, Viale Pier Andrea Mattioli 39, 10125 Turin, Italy
e-mail: elena.cristofori@trimweb.it

E.I. Cristofori · A. Facello · A. Demarchi
TriM—Translate Into Meaning, Corso Sommeiller 24, 10128 Turin, Italy

A. Facello
National Research Council, Research Institute for Geo-Hydrological Protection, Strada delle Cacce 73, 10135 Turin, Italy

W. Camaro
ITHACA, via P.C. Boggio 61, 10138 Turin, Italy

M. Fascendini · A. Villanucci
CCM, via Ciriè 32/E, 10152 Turin, Italy

W. Leal Filho et al. (eds.), *Climate Change Adaptation in Africa*,
Climate Change Management, DOI 10.1007/978-3-319-49520-0_6

access to technologies, the use of technological resources for community working groups, the identification of climate variables, the creation of capacity to measure environmental variables and the creation of awareness about climate related consequences. Climate related issues such as access to health resources, diseases, food insecurity have to be analysed and prevented using a multidisciplinary approach promoting an effective exchange of know-how between scientific experts, humanitarian practitioners, governments and communities. Although the scientific community is currently able to provide innovative technologies essential for an accurate and quick analysis of wide amounts of data, scientific products are often not easily accessible for practitioners working close to local communities (Cullen 1990; Goggin et al. 2015). On the other hand analysis performed by operators who are not familiar with the science-technology but having a deep understanding of local communities, are essential to identify key issues and to enact strategies that meet nomadic populations requirements (Fazey et al. 2013). In addition, a crucial component needed to rise the community awareness is the ability to provide easy to understand information to both technical and non-technical people, including nomads, who are the final users of resilience and adaptation projects results.

Geospatial information technologies, such as Geographic Information Systems (GIS), Global Positioning Systems (GPS) and Satellite Remote Sensing (RS), offer a valuable help for the assessment of environmental issues and can significantly contribute to enhance the resilience of local communities for the following reasons. Firstly, GIS provide a flexible environment to store, analyse, manage and map spatial information integrating other typologies of georeferenced data provided as result of multidisciplinary analysis (Ashure et al. 1998; NRC 2002). Secondly, RS provides timely and reliable information of the dynamics of the Earth's surface and atmosphere which thus can be continuously monitored (Donoghue 2002; Melesse et al. 2007; Mather et al. 2009). Thirdly, GPS technologies enable technical and non-technical users to accurately track the position of relevant environmental and geographic features. In addition, the possibility to quickly integrate and visualize various information coming from different sources into one system, allows to quickly produce immediately understandable cartographic products that can address a variety of climate change related issues such as potential impacts on land use, health care, transportation or households (Ferrandino 2014; Duncan et al. 2014).

The main objective of this paper is therefore to present a GIS for the storage, analysis and visualization of environmental and social data collected in the framework of the Operational Research (OR) project "Enhance the health status of the Nomadic Pastoralists in Filtu Woreda, Liben Zone (Somali Regional State of Ethiopia)", using the potential of existing open-source geospatial technologies. The results are intended as a first step towards an operational integration of health information, geographic data and climatological data into a system able to make science more immediately accessible for a variety of users who perform multidisciplinary analysis aimed at identifying local communities adaptation priorities.

The following paragraphs present, firstly a description of the OR Project, secondly the area chosen as case study for the implementation of the project, thirdly the data used and the proposed methodology, fourthly the obtained results and finally some conclusions and future developments.

2 Background: The Operational Research Project

The OR project, funded by the Swiss Development Cooperation, was implemented from June 2015 to January 2016 by Comitato Collaborazione Medica (CCM) in partnership with the Somali Regional Health, Livestock and Pastoralists Development and Finance and Economic Development Bureaus of Ethiopia.

The healthcare delivery system among pastoral societies is extremely poor compared to rural places of non-pastoralists in Ethiopia. This is because of several factors generally associated with the nomadic lifestyle, including dispersed settlement patterns, seasonal mobility, and under-utilization of services even when and where they are available. Besides, health facilities in pastoralist communities are limited in number, understaffed and characterized by poorly organized service delivery (FMOH et al. 2011). These conditions cause higher infant, maternal and under-5 mortality rates than non-nomadic communities and high prevalence of waterborne and zoonotic diseases.

CCM has been working in Liben Zone since 2003, through actions aimed at strengthening the health system and improving the delivery of basic preventive and curative services, with a special focus on women and children. Capitalising the long-standing experience with nomadic populations, CCM is committed to identify effective strategies to address the still huge and not yet met health needs of pastoralist communities. Lessons learnt from the past experience have been revitalised into an Operational Research (OR) aimed at testing feasibility, efficacy and efficiency of the "One Health approach" in the district of Filtu (Ethiopia). According to the World Health Organization (WHO), one or more new infectious diseases have emerged every year since the 1970s (WHO 2007), with the majority being zoonosis that can be transmitted between animals and humans (Jones et al. 2008). These trends led to support an integrated and holistic approach to human, animal and environmental health, known as "One Health", which has received growing attention among policy makers, funders and practitioners seeking more effective prevention, control and treatment responses in an increasingly populous and globalized world.

The main purpose of the OR was to identify and analyse local pastoralists' needs, perceptions and behaviours towards human and animal health, in relation to the local socio-ecological context. Special attention was given to the strategies of adaptation to the environment—also in relation to climate change—and to the hindrances that prevent people to access the existing human and animal health facilities.

Since several years, climate change has been severely affecting the pastoralist livelihood, mainly due to frequent droughts and consequent decrease of grass, water and livestock (especially cattle, less resistant than shoats and camels). Pastoralists are not passive in front of climate changes: they actually enact different strategies to cope with new challenges, continuously adapting their lifestyle to environmental transformations and surviving in harsh contexts marked by the lack of basic resources. The OR aimed therefore at observing pastoralists life conditions and actions at grassroots level, in order to highlight best practices and strategies already in place and find gaps and hindrances to focus future development interventions on.

The OR was conceived and designed as the key preparatory activity of a One Health Project to be implemented in the area. Basing on its results, CCM will validate a new action that: (i) promotes an optimal common health for humans, animals and environment in the area of intervention; (ii) effectively responds to the needs of local nomadic pastoralists; and (iii) integrates its intervention with the current Ethiopian government strategies. In particular, the new project should increase the acceptability and accessibility of healthcare services to pastoralists and their animals, through the identification of effective and sustainable links between the pastoralist community and the existing national health and veterinary systems.

The OR has been structured in five steps:

1. Introduction of project proposal and operational research to concerned authorities and pastoralist communities, analysing objectives and methodologies through a participatory methodology.
2. Mapping of community and territory, retrieving basic and critical information on the social structure, territory and ecosystem of the area and drafting a strategic map for systematic data gathering.
3. Qualitative data collection on human health and animal health management and on the relations between pastoralists and their environment.
4. Data analysis, to guide the drafting of future One Health axis of intervention.
5. Presentation and discussion of the research results and future axis of intervention to concerned authorities in a Final Workshop, to ensure full alignment of the proposal in the local and regional plans; guarantee the acknowledgement of local communities involved in data collection; and collect suggestions and feedbacks to integrate within the future project proposal.

In line with the interdisciplinary dimension of the One Health approach, the OR has been conducted through a multidisciplinary, qualitative methodology combining medical anthropology, public health, veterinary medicine, human ecology and applied geography approaches with local knowledge and practices towards human and animal health. Qualitative data collection has been conducted through: (i) identification of a local guide-facilitator in each kebele, acknowledged by leaders and community members to ease the interaction between the OR team and the pastoralists families and individuals; (ii) collective meetings and focus group discussions with pastoral community representatives; (iii) semi-structured interviews with key-informants; (iv) participant observation of pastoralists and their livestock's everyday life; (v) collection of secondary data within concerned offices.

Due to the lack of existing maps of the district, the OR involved a cartographic component. The maps developed show the sites considered as most relevant from pastoralist community leaders and household members, combining information on the kind of human and animal health services, water sources and grazing areas, with the location of pastoralists temporary and permanent settlements visited. This allowed to enrich the qualitative analysis of pastoralists health seeking behaviours and transhumance paths with the spatial visualization of available resources. The combination of the two components enabled a better understanding of the pastoralists' perceived priorities and needs and the identification of existing gaps in the healthcare delivery services distribution and functioning.

3 Case Study Area

The project area is in Filtu Woreda, Liben Zone, Somali Region of Ethiopia. The woreda borders on the South with Dawa River—which separates it from Moyale and Udet districts—on the West with Dekasuftu district, which separates it from Oromia Region, on the North and East with the Ganale Dorya River—which separates it from the Afder Zone and on the Southeast with Dolo Odo district. Liben Zone has an estimated total population of 539,048 people, including 108,340 living in Filtu district, with about 49.9% women and 82.5% people leading a pastoral lifestyle.

The social system is based on clans kinship structure and communities are organized in patrilocal groups of households composed of close relatives linked by father-brother relations. During dry seasons, pastoralists household groups split, with members of the same family following different transhumance pathways in search of water and grass for their livestock. Movements and trajectories depend on the kind of animals owned, with camels going to farthest places (usually the two rivers, Genale and Dawa) and cattle and shoats rotating around main artificial and natural ponds and wells. During dry season, household groups reunite on the highlands, where rain and grass become available in larger quantity.

Sickness occurrence variates therefore depending on the season and is mainly related to the different environmental conditions faced by people and animals (i.e. climate, vegetation parasites and other pathogen agents). The kind and location of health and veterinary facilities and services used by pastoralists and their livestock also variate seasonally. Due to lack of transportation means and economic hindrances, animal and human health seeking behaviours appear strongly related to the distance of services and facilities from the environmental resources locally available and temporarily exploited (water points and grassing areas).

4 Data and Methods

In order to facilitate the storage and analysis of on-site collected data, a GIS environment that could support OR in identifying gaps and needs related to the access to health resources of nomadic communities has been generated. The possibility to integrate data collected in the field along with general reference data, RS information and weather satellite imageries allows to develop a dynamic visualization of the spatial distribution of the healthcare system (e.g. medical buildings, traditional healer, animal facilities, water resources, etc.). Moreover a GIS platform represents an effective tool to immediately overlay additional key features such as the roads network, essential to identify the best and fastest access route to health resources. In addition, the integration of environmental data such as extreme precipitation, permits to make nomadic communities informed about possible unreachable areas or at-risk facilities.

In the realization of the GIS, particular attention has been devoted to the analysis of user's requirements before designing the structure of the database. On the one hand data standards have been taken into account, in order to facilitate the interoperability of the information with others GIS, on the other hand some particular data have been adapted to the specific needs of the project and to the perception of users. Analysis and mapping have been performed using open-source data in an open-source environment. More specifically the QGIS software (released version 2.8.2) has been used.

4.1 Geospatial Data Collection

Three different typologies of geospatial data were needed to be integrated into the GIS:

1. A basemap cartography, to held a map depicting background reference information.
2. Several reference datasets, such as road networks, water resources, settlements, etc., to make spatial relationships between different types of data.
3. Meteorological data, to perform analysis of accumulated precipitation for different time frames.

Over the last decades, several open-source projects, based on the provision of web portals containing geo-referenced data, have been developed. As a consequence a fundamental source of geospatial information derives from open-source platforms such as OpenStreetMap (OSM) and Geonode. According to the needs of the OR, an open-source geospatial collection campaign have been performed to identify the type of information matching the specificities of the study area and the users. The following data have been downloaded, processed and integrated into the system (see Table 1):

Table 1 Open-source geospatial data integrated into the GIS

#	Dataset title	Dataset producer	Dataset type	Spatial resolution	Online resource
1	Landsat 8	NASA & USGS	Raster	30 m	http://earthexplorer.usgs.gov/
2	GADM	GADM	Vector (polygon)	–	http://www.gadm.org/
3	TRMM	NASA	Raster	0.25°	http://pmm.nasa.gov/data-access/downloads/trmm
4	GPM	NASA	Raster	0.1°	http://pmm.nasa.gov/data-access/downloads/gpm
5	GFS	NWS	Raster	0.25°	http://www.nco.ncep.noaa.gov/

1. Landsat 8 OLI/TIRS satellite imageries: four frames have been used to generate the basemap cartography for the entire overview case study area (path 166 row 56–57, path 167 row 56–57, percentage cloud cover <20%; time period September 2015).
2. Global Administrative Areas (GADM): to identify the administrative boundaries at national, regional and local level.
3. Tropical Rainfall Measuring Mission (TRMM) and Global Precipitation Measurement (GPM): used to produce the accumulated precipitation monitoring maps.
4. Global Forecast System (NOAA-GFS): used to produce the accumulated precipitation forecast maps.

However, the analysis of open-source data listed above did not provide a complete coverage of reference data needed and therefore an information extraction activity has been achieved in order to integrate into the GIS some additional data (see Table 2). Using Landsat 8 imageries, water resources and rivers network have been automatically extracted through a supervised classification while roads network and built-up areas have been manually extracted through photo-interpretation.

Table 2 Geospatial data produced and integrated into the GIS

#	Dataset title	Source	Dataset type	Dataset extraction technique
1	Water resources	Landsat 8	Vector (polygon and polyline)	Supervised classification
2	Road network	Landsat 8	Vector (polyline)	Photo-interpretation
3	Built-up areas	Landsat 8	Vector (polygon)	Photo-interpretation

4.2 On-Site Data Collection

In the first weeks of September 2015, several participatory meetings were conducted with representatives of local Governmental Offices (Health, Livestock Crop and Rural Development, Women Children and Youths Affairs Offices) and local NGO (Social Welfare Development Association), with the main scope of highlighting the hindrances related to the delivery of animal and human health services to pastoralist community and selecting the research areas. The selected kebele, located in different directions from Filtu town, were chosen as more relevant to the OR purposes and objectives based on the following criteria:

1. Presence of wider potential livestock grazing areas, indicating a high concentration of pastoralist settlements during the dry and the wet seasons;
2. Availability or lack of human health and animal health facilities and/or mobile services and related high or low level of performance;
3. Availability of natural/artificial water sources used by pastoralists and their livestock especially during dry season;
4. Best practices in gender empowerment activities, i.e. presence of local associations of women and promotion of mother-child and reproductive health.

The 8 selected sites are:

- *Melkal Libi* and *Melkahager* (respectively 21–23 and 35 km, north and north east of Filtu town);
- *Jayga-ad* and *Aynle* (42 and 70 km, southeast of Filtu town);
- *Mesajd* (17 km, east of Filtu town),
- *Haydimtu* (22 km, north west of Filtu town),
- *Bod Bod* and *Harabali* (respectively 130 and 40 km, south of Filtu town).

In the second half of September, CCM OR team conducted daily field missions in the selected kebele, to introduce the project and share the research aims and methodology during participatory meetings with leaders and elders of local pastoralist communities. During the missions, the selected areas were mapped through interviews with key-informants, direct observation activities and collection of geographic coordinates, focusing on:

1. Location, status and functionality of animal and human health facilities available on site,
2. Presence of mobile health workers delivering services to livestock and people,
3. Seasonality and geography of pastoralist migration paths in search of water and grass,
4. Availability and location of main grazing lands and water resources for livestock and people.

From the beginning of October 2015 to the end of January 2016, the OR team conducted a total of 31 field missions in the above mentioned kebele. Every mission lasted from 1 to 5 days, with overnights spent among the pastoralist community.

A total of 38 sites were visited, including rural kebele centres, permanent agro-pastoralists villages and temporary nomadic pastoralists household settlements.

During each mission, the OR team collected geographic coordinates of the main relevant sites visited, adopting pastoralist household members' perspective and assessing the availability and distribution of healthcare, water and grass resources.

More in detail, the geographic coordinates collected are related to:

- Healthcare and veterinary facilities distributed in rural areas, as Health Centers, Health Posts (in the Fig. 1, an example of Health Post is shown), Animal Health Posts and medical and veterinary private drug shops.
- Mobile professionals, both "traditional" (as Traditional Healers—providing care to animals and humans—and Traditional Birth Attendants) and "biomedical" or veterinary (as private health workers, Community Volunteers, Community Animal Health Workers and governmental Animal Health Technicians).
- Local administration buildings present in the kebele centers, included as relevant decision-making institutions.
- Open grazing lands and water resources as artificial ponds (collecting rain and used by pastoralists and livestock to fetch water during dry seasons), birka (cemented structures constructed by the government to collect rain for human consumption), wells (both hand dug wells, excavated by pastoralists family members, and structures provided by government and NGOs).

Precipitation and temperature data were also collected daily since the beginning of September until the end of December, either in Filtu Office or in fieldwork areas (see Fig. 2). The collection of meteorological variables has been considered important for the following three main reasons: firstly for the assessment of daily weather conditions, secondly to gather on-site data that can be compared with meteorological parameters obtained from earth observations and therefore enabling the validation of satellite estimates over the area and thirdly to raise the awareness of local people about the importance and the measurability of climatological conditions.

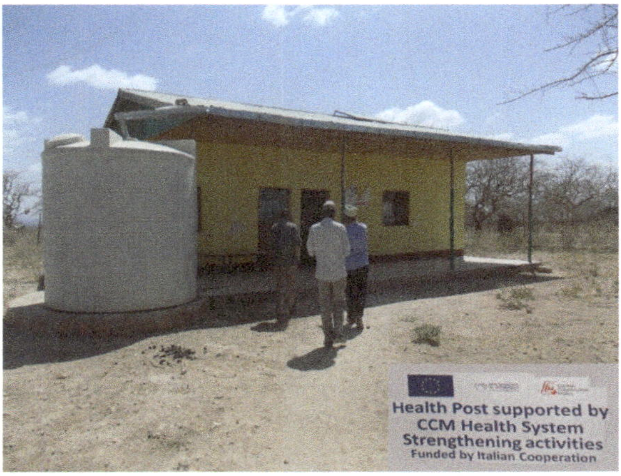

Fig. 1 Health post in Jayga-ad

Fig. 2 Participation of the community during precipitation measurements

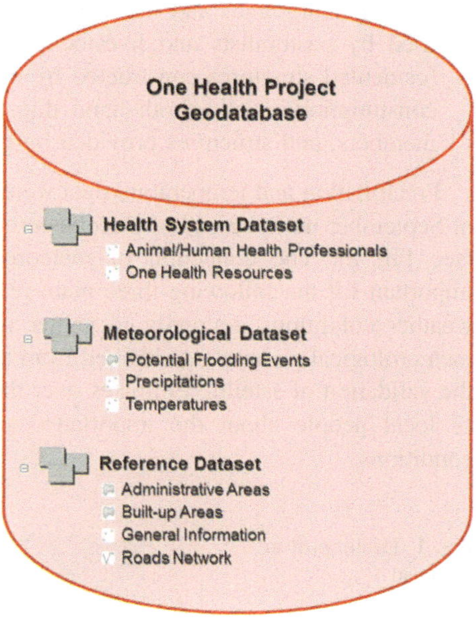

Fig. 3 One health project geodatabase structure

4.3 *Databases Organisation*

In order to integrate the collected information into a common environment, a geodatabase has been developed. The geodatabase provides flexibility in storing, edit and access the three following datasets: Health system, Meteorological and Reference datasets (see Fig. 3). In addition datasets are organized according to different feature classes representing the different data considered essential to perform the OR project analysis.

The structure of the geodatabase has been organized mainly following standard specifications and in particular the glossary produced by GeoNames (http://www.geonames.org/export/codes.html) has been considered as main reference, allowing to develop databases interoperable with other geospatial data platforms, such as OSM. On the other hand it has been necessary to adapt some of the feature classes to the specificity of data and analysis required by the project.

Therefore each dataset has been characterized by a set of common general attribute fields, and by a set of specific attributes able to describe the peculiarity of each one.

The set of common general attribute fields are:

- "Lat" and "long": the geographic coordinates of latitude and longitude that specify the precise location of the feature censed or extracted.
- "Elevation": height above the sea level of the feature censed or extracted.
- "Date": date when the feature has been censed or extracted.
- "Time": time when the feature has been censed or extracted.
- "Name": name of the feature censed or extracted.
- "Admin 1", "Admin 2", "Admin 3": respectively the primary administrative division of a country, the administrative subdivision of a first-order and second-order in which the feature is contained.

Some feature classes are then specified by some attribute fields:

1. "Animal/Human Health Professionals" (i.e. the feature class concerning the mobile professionals) is described by (see Table 3):
    - "Healthcare Branch": it distinguishes in three different branches of healthcare (Animal health, Human health and Intersectoral).
    - "Professionals Type": represents a classification concerning the typology of the professionals (Biomedical, Traditional or Veterinary).

Table 3 Database organisation for the "Animal/Human Health Professionals" feature classes

Healthcare branch	Professionals type	Professionals subtype	Description
Animal health	Traditional	THA	Traditional Healer for animal care
	Veterinary	AHT	Animal Health Technician
	Veterinary	CAHW	Community Animal Health Worker
	Veterinary	PVDS	Private Veterinary Drug Seller
Human health	Biomedical	CHV	Community Health Volunteer
	Biomedical	PHP	Private Health Practitioner
	Biomedical	PMDS	Private Medical Drug Seller
	Traditional	TBA	Traditional Birth Attendant
	Traditional	THH	Traditional Healer for human care
Intersectoral	Traditional	TH	Traditional Healer for animal or human care or both

Table 4 Database organisation for the "One Health Resources" feature classes

Feature class	Feature type	Description
Health facility	Animal Health Post	Facility that provides curative animal services (as drugs, treatments and vaccinations), run by a governmental AHT or a private CAHW
	Health Center	Primary healthcare facility that provides curative and preventive services and where health workers with medical training are available
	Health Post	Primary healthcare facility delivering health education and preventive actions, that supplies generic basic drugs and is run by Health Extension Workers with a nine months health education training on prevention and control methods
Environmental resource	Birka	Handmade structure for rainfall water collection constructed with cement and semi-covered
	Grazing land	Pastoral area
	Pond	Handmade small inland body of standing water collected from rainfalls
	Water Pipe Line	Systems of pipe lines collecting water from rivers or springs
	Well	Well with solar water pump
	Hand-dug well	Household well, hand-dug by the pastoralists

- "Professionals subtype": represents a further classification among the same typology of health professionals and practitioners.

2. "One Health Resources" (i.e. the environmental resources and the healthcare and veterinary facilities) is described by (see Table 4):

- "Feature Class": health facilities and environmental resources, as two key resources for the pastoralists community.
- "Feature Type": is a classification among the classes of resources and it could concern the type/capacity of the facility or the typology of the environmental resource.

3. "Precipitations" is described by:

- "Rain (yes/not)": occurrence of precipitations.
- "Rain (mm/24 h)": amount of precipitation recorded daily.

4. "Temperatures registered" is further described by the field "Temperature (°C)" which contains the values of temperature measured both as minimum (around 07:30 local time) and as maximum (between 12:00 and 13:00 local time).
5. "General information" (i.e. the services, facilities and other general features of the case study area) is described by (see Table 5):

- "Feature Code": it is an identification code for each category of item censed such as Place, Water, Transportation, Building, Landcover.

Table 5 Database organisation for the General information dataset

Feature code	Feature class	Description
Place	City	
	Town	
	Village	
	Mission	
	Settlements	
	Other	Any agglomeration of buildings where people live and work
Water	Aqueduct	A conduit used to carry water
	Canal	An artificial watercourse for draining or irrigating the land
	Lake	A large inland body of standing water (natural or artificial)
	Spring	A place where ground water flows naturally out of the ground
	River	A natural flowing watercourse, usually freshwater
	Stream	Body of water with a current, confined within a bed and stream banks (usually small river)
	Wells	Cylindrical holes, pits, or tunnels drilled or dug down to a depth from which water, oil, or gas can be pumped or brought to the surface
	Dam	A barrier constructed across a stream to impound water
Transportation	Road	An open way with improved surface for transportation of animals, people and vehicles
	Ancient road	The remains of a road used by ancient cultures
	Caravan route	The route taken by caravans
	Street	A paved urban thoroughfare
	Heliport	
	Trail	A path, track, or route used by pedestrians, animals, or off-road vehicles
Building	Government building	
	Diplomatic building	
	Residential building	
	Religious building	
	Medical building	
	Educational building	
	Hospital	
	Market	
	Farm	

(continued)

Table 5 (continued)

Feature code	Feature class	Description
	Camp	A site occupied by tents, huts, or other shelters for temporary use
	Refugee camp	A camp used by refugees
	Historical site	A place of historical importance
Landcover	Agricultural area	
	Pasture area	
	Savannah area	
	Forest	
	Urban area	
	Tribal area	

- "Feature Class": it is a characterization of the features observed adapted to the context of the case study area.

6. "Roads network" is further described by the field "HCT" which specifies the hierarchy of the road (i.e. Highway, Primary route, Secondary route, Local route).

5 Results

In their daily work, the health extension workers use to draw freehand maps to visualise important geographical features characterizing the kebele and ease the delivery of health education and promotion at household and village level. An example of freehand map is shown in Fig. 4, where Jayga-ad village is represented together with roads connecting different sub-kebeles.

Freehand mapping is considered an important tool to acquire knowledge about the localization of households, health facilities or road network and for the quantification of health and environmental resources. However freehand mapping has strong limitations in terms of time, operator subjectivity and spatial scale. In this framework, with the aim of mapping vast areas, including several different kebeles, and of performing a quick and standardized analysis of a big amount of data, the contribution of digital cartography becomes crucial.

Three different map templates, considered particularly relevant for the OR, have been therefore organized:

1. On-site data collection campaign
2. Healthcare resources and professionals distribution
3. Meteorological data

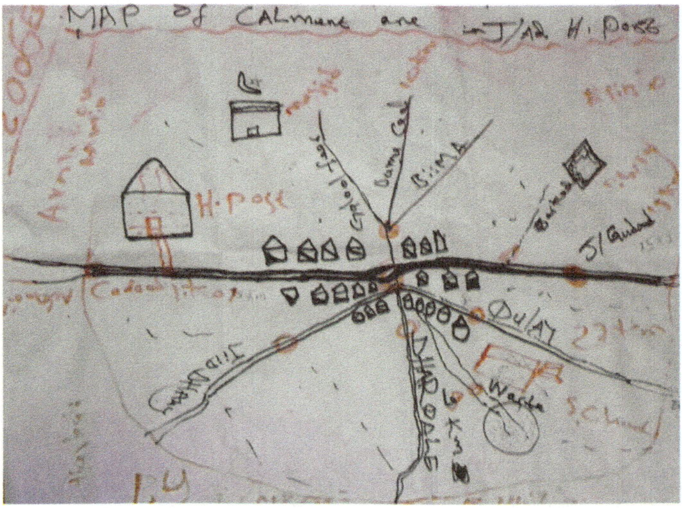

Fig. 4 Example of freehand map produced by HEW in governmental health facilities

End-users may thus easily update the database, integrate further geospatial information and update meteorological data producing updated maps characterized by the same structure.

In the following paragraphs some of the maps produced for the OR project are presented.

5.1 On-Site Data Collection Campaign Map

The on-site data collection campaign map offers an immediate overview of the whole data collection campaign divided by acquisition date and overlaid to the reference data within the overview case study area (see Fig. 5). This product has been particularly useful to summarize the step by step collection campaign work, to quantify and localize the activities performed and to efficiently plan the next steps.

5.2 Healthcare Resources and Professionals Distribution Map

This cartographic product is used to visualize the location of basic health resources and professionals for both human and animals along with the availability of further environmental resources such as water resources and grazing lands (see Fig. 6).

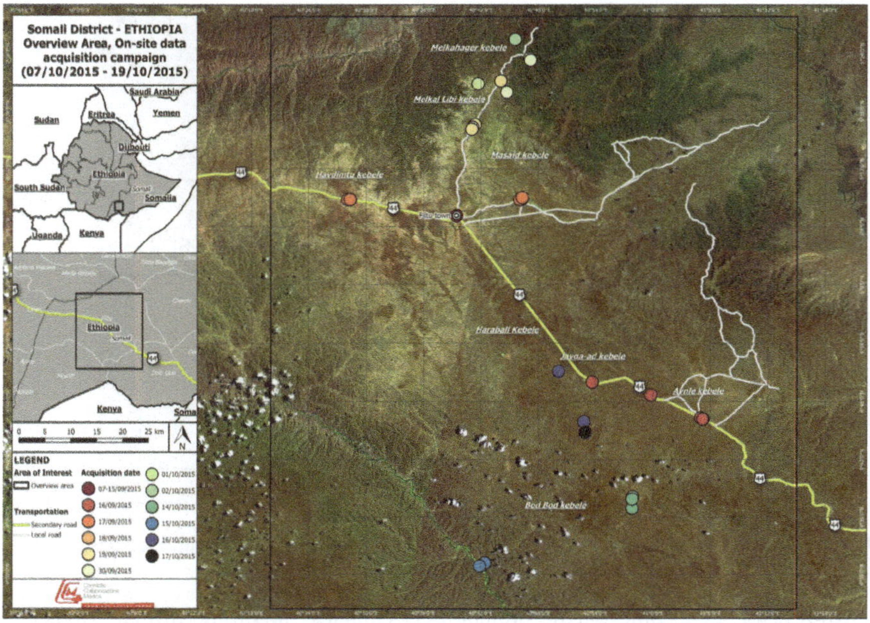

Fig. 5 On-site data collection campaign map for Filtu Woreda case study area. *Source of basemap* Landsat 8 data mosaic, frames provided courtesy of the U.S. Geological Survey

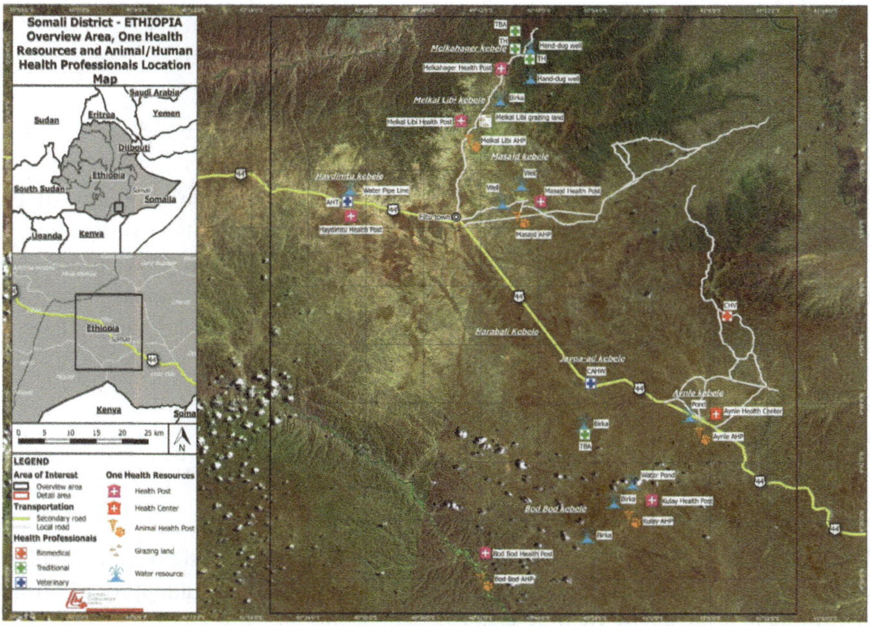

Fig. 6 One health resources and animal/human health professionals map for the Filtu Woreda area. *Source of basemap* Landsat 8 data mosaic, frames provided courtesy of the U.S. Geological Survey

It can be noted that different symbols have been used for depicting facilities and professionals in order to distinguish between mobile services and stable facilities. Symbols have been chosen with the aim to communicate meaningful and useful information to a wide range of end-users, both technical and non-technical, who have to easy and quickly interpret the map.

Including a labelling layer, a further description concerning the typology and the name of resources has been visualized.

5.3 Meteorological Maps

As far as the meteorological information is concerned, two different cartographic products have been produced:

- *Meteorological data collection maps*. These maps provide the visualization of daily temperature and precipitation records facilitating the quantification of the environmental variables. Precipitation records have been overlaid to the base cartography to geographically contextualize the data. In addition the comparison between precipitation measurements and GPM precipitation estimates has been performed both using mapping capabilities and representing the data into a table in order to perform the validation of satellite derived precipitation (see Table 6). The validation is essential in order to assess the accuracy and the reliability of satellite precipitation estimates over the area of interest.
- *Accumulated precipitation maps*. These maps have been provided to the OR team to ease the planning of the collection campaign activities during the unusual heavy rains period experienced between October and November 2015 and caused by the consequences of El Niño in the Region. Considering the precarious conditions of the local roads network, 7 or 14 days accumulated precipitation (elaborated using TRMM and GPM satellite imageries) and up to 1 week forecasted precipitation (elaborated using GFS models) have been used to provide the monitoring and the prediction of heavy precipitation over the area of interest (see Fig. 7). Accumulated precipitation maps have been particularly useful for the identification of areas with a potential high risk of flooding and therefore of areas where the collection campaign should have been postponed. This approach has resulted in a significant increase of the working activities planning efficiency.

6 Conclusions and Future Developments

The collection of geographic and meteorological data has been conducted both on field sites and in CCM Office with the active participation and involvement of all the OR team members (especially Medical and Veterinary Expert and Assistant

Table 6 Precipitation measurements correlation between in situ and satellite measures

Date	Location	Precipitation measured from in situ instrument (mm/24 h)	Precipitation revealed from GPM (mm/24 h)
18/09/2015	CCM Filtu office	6.2	0
29/09/2015	Melkahager HP	1.7	1.1
30/09/2015	Weshako Denan	Missing data	1.9
01/10/2015	Selah Meghen	missing data	0.1
02/10/2015	Melkahager HP	0.5	0
03/10/2015	CCM Filtu office	10.9	15.2
04/10/2015	CCM Filtu office	12.9	0
10/10/2015	CCM Filtu office	3.8	1.7
11/10/2015	CCM Filtu office	3.6	79.2
12/10/2015	CCM Filtu office	10.5	3
13/10/2015	CCM Filtu office	0.4	5.3
14/10/2015	Missing data	Missing data	0
14/10/2015	Kulay Health Post	Missing data	39.4
14/10/2015	Bod Bod Health Post	>10	41.2
16/10/2015	Private House	Missing data	0
17/10/2015	Darqale	Missing data	7
18/10/2015	Private House	>10.0	30.2
18/10/2015	CCM Filtu office	>10.0	47.1
19/10/2015	CCM Filtu office	1.8	12.2

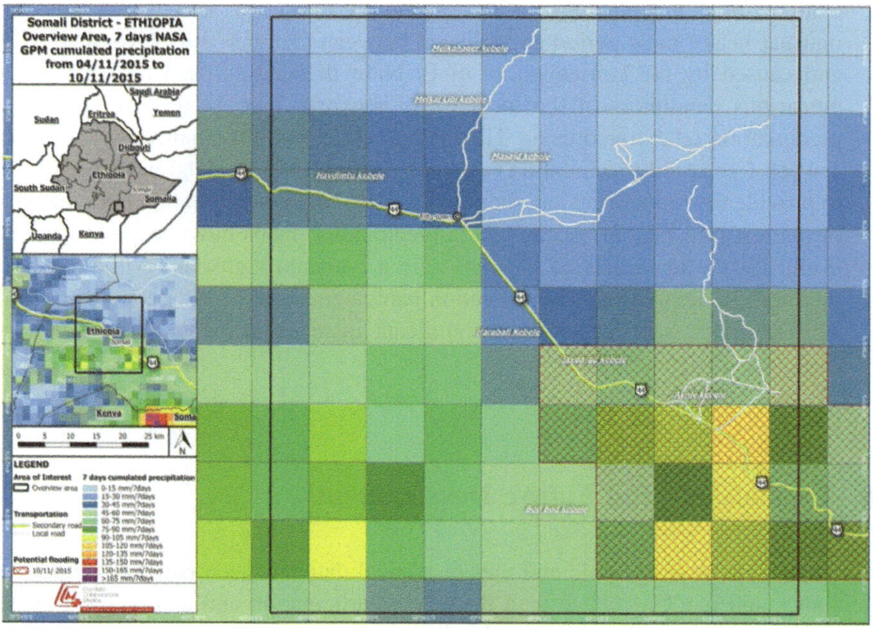

Fig. 7 7 days cumulated precipitation. Image credit NASA's GPM satellite

Project Coordinator) and support staff (District Health and Livestock Offices focal persons, local guides selected in each kebele and driver). Data collected have been recorded by CCM Project Coordinator and stored into the database presented in this paper. The particular organization of the database has allowed a quick upload of collected data and easily to link spatial data with information.

The data collection campaign received a positive welcome from pastoralists communities involved in the research that showed awareness of the incidence of climate change on their life conditions and demanded solutions and interventions to support them in facing current challenges. Concerning Human and Animal Health, the mapping of facilities distribution, structures and mobile services available can allow a better planning of future interventions, especially in relation to the identification of structural gaps between the medical and veterinary delivery systems and the pastoralists effective movements and distribution. Maps realized will also support the analysis of the pastoralists' movements and behaviours described during interviews and focus group discussions and directly observed on the field. Moreover, an operational use of measurement, monitoring and forecasting of precipitation and further climatological variables would help users and especially local communities to understand the potential of technologies and to enhance risk assessment capabilities.

However, a shortage of human and financial resources available and specifically allocated for the cartographic data collection campaigns has determined two main constraints.

On the one hand, regarding the on-site data collection, this shortage allowed only a partial mapping of the territory, limited to the specific selected sites visited by the OR team. People involved in the data collection worked as volunteers, conducting it parallel to their usual roles. This preliminary "test" showed the locals' proactive participation in this kind of activities: if conducted through proper funds and after proper trainings, similar mappings could be pursued covering larger portions of the territory and involving a wider number of features according to local stakeholders' needs and interests, widening their potential applications.

On the other hand, considering the geospatial data collection, this shortage of resources allowed to retrieve only open-source satellite imageries characterized by a medium-low spatial resolution. Instead, high quality satellite imageries would allow to hold a better basemap cartography and to perform more adequate information extraction activities providing so to retrieve complete and more precise reference datasets (such as local roads and pastoralists paths, streams and secondary water resources, landcover characterization, building footprints).

Finally, A process of sedentarization is increasing in the woreda due to several reasons, related to: the deterioration of nomadic living conditions caused by climate change; the consequent decrease of rainfalls, availability of grass, water and livestock; long-term consequences of local conflicts; and the input of governmental strategies aimed at creating resettled sites where basic services would be provided (education, water, health). A growing number of pastoralists is transforming its lifestyle, becoming agro-pastoralist; at the same time, the uncontrolled expansion of farming lands is causing the decrease of free grazing areas available to livestock, with the risk of future conflicts outbreak. As resulted from the OR, specific future actions should be planned and addressed to drought risks reduction and resource distribution for farming and livestock grazing. The possibility of using the proposed GIS as an operational platform for the integration of social, geographic and climate information would contribute to enhance the effectiveness and timeliness of these actions.

References

Ashur, S., Baaj, M. H., Crokett, B., & Abaza, K. (1998). GIS as a support tool for effective decision-making in engineering management: Two case studies from Arizona. In *First International Conference on New Information Technologies for Decision Making in Civil Engineering* (Vol. 1, pp. 601–612), 11–13 October, Ecole de Technologie Superieure, Montreal.

Cervigni, R., Liden, R., Neumann, J. E., & Strzepek, K. M. (2015). *Enhancing the Climate Resilience of Africa's Infrastructure: The power and water sectors. Africa Development Forum*. Washington, DC: World Bank.

Cullen, P. (1990). The turbulent boundary between water scientists and water managers. *Freshwater Biology, 24*(1), 201–209.

Donoghue, D. N. M. (2002). Remote sensing: Environmental change. *Physical Geography, 26*(1), 144–151.

Duncan, M., Crowley, K., Cornforth, R., Edwards, S., Ewbank, R., Karbassi, P., et al. (2014). Integrating science into humanitarian and development planning and practice to enhance community resilience. Full guidelines.

Fazey, I., Evely, A. C., Reed, M. S., Stringer, L. C., Kruijsen, J., White, P. C. L., et al. (2013). Knowledge exchange: a review and research agenda for environmental management. *Environmental Conservation, 40*(01), 19–36.

Federal Ministry of Health of Ethiopia (FMOH), World Health Organization (WHO), United Nations Children's Fund (UNICEF). (2011). *Lessons Learned from the Multi-Country Experience-Sharing Conference on Pastoralist Communities*, June 27–July 1, 2011, Addis Ababa, Ethiopia.

Ferrandino, J. (2014). Incorporating GIS as an interdisciplinary pedagogical tool throughout an MPA program. *Journal of Public Affairs Education, 20*(4), 529–544.

Frankenberger, T., Spangler, T., Nelson, S., & Langworthy, M. (2012). *Enhancing resilience to food security shocks in Africa* (Discussion Paper).

Goggin, C. L., Grieve, A. M., Summerell, G., Leys, J., Barrett, T., Oliver, I., et al. (2015). Working with practitioners: How scientists can develop effective relationships that produce an enduring legacy. In *The 21st International Congress on Modelling and Simulation (MODSIM2015)*, Gold Coast, Australia, 29 November–4 December 2015.

Jones, K. E., Patel, N. G., Levy, M. A., Storeygard, A., Balk, D., Gittleman, J. L., et al. (2008). Global trends in emerging infectious diseases. *Nature, 451*, 990–993.

Kimaro, E. G., & Chibinga, O. C. (2013). Potential impact of climate change on livestock production and health in East Africa: A review. *Livestock Research for Rural Development, 25* (7), Article #116.

Mather, P., & Tso, B. (2009). *Classification methods for remotely sensed data* (2nd ed.). Boca Raton(U.S.A.): CRC Press.

Melesse, A. M., Weng, Q., Thenkabail, P. S., & Senay, G. B. (2007). Remote sensing sensors and applications in environmental resources mapping and modelling. *Sensors, 7*(12), 3209–3241.

National Research Council (NRC). (2002). *Down to earth: Geographical information for sustainable development in Africa*. Washington, DC: The National Academies Press.
Sheik-Mohamed, A., & Velema, J. (1999). Where health care has no access: The nomadic populations of sub-Saharan Africa. *Tropical Medicine & International Health, 4*(10), 695–707.
World Health Organization (WHO). (2007). *World Health Report 2007, A Safer Future, Global Public Health Security in the 21st Century*. Geneva (Switzerland).

Advancing the Resilience of Rural People to Climate Change through Indigenous Best Practices: Experience from Northern Nigeria

Michael W. Musa and Sulaiman Umar

1 Introduction

As far back as the Rio Earth Summit in 1992 to the recently concluded Paris 21st Conference of the Parties (COP 21) in 2015, various development agencies have expressed urgent concerns and engaged in negotiations to help people in developed and developing countries deal with climate-related threats (IPCC 2007; IFAD 2008; FAO 2012; Dembowski 2015; Wikipedia 2016). At the United Nations Cancún agreements in 2010, it was estimated that at least US$ 100 billion per year will be needed by the year 2020 to assist poor rural people in developing countries reduce their vulnerability to climate change impacts (Mangani 2014; Wikipedia 2015). However, how much individual countries would contribute in mobilizing the needed fund or how the fund will be disbursed for developing countries to protect their poor rural people within the period anticipated is not quite clear.

In most parts of Nigeria, especially in the northern areas, climate change in not a new phenomenon to the rural producers. However, the devastating conditions of climate change are undoubtedly intensifying. Increasingly, the problems of poor soil fertility, high incidences of pest populations, inconsistent rainfall patterns leading to drought, crop failures and shortages of water for irrigation are becoming ever more common (Musa et al. 2013). Despite their vulnerability to climate change, these rural producers have over the years developed and implemented extensive indigenous best practices and adaptive strategies to cope with climate-related challenges.

M.W. Musa (✉) · S. Umar
Department of Agricultural Economics and Rural Sociology,
Faculty of Agriculture, Institute for Agricultural Research,
Ahmadu Bello University, PMB 1044, Samaru-Zaria, Kaduna State, Nigeria
e-mail: musamike@yahoo.com

S. Umar
e-mail: sulaimanumar@hotmail.com

W. Leal Filho et al. (eds.), *Climate Change Adaptation in Africa*,
Climate Change Management, DOI 10.1007/978-3-319-49520-0_7

Recently, a significant numbers of studies and reports have emerged documenting how rural producers in many parts of the world are responding to climate change and the implications for development (Meze-Hausken 2004; Nyong et al. 2007; Hunter 2007; FAO 2012; Musa et al. 2013; Spore 2014; Guillaume-Gentil 2014; Spore 2015; Umar and Musa 2015). Conversely, in Nigeria, there is insufficiency of information on how climate change is perceived and addressed within resource contexts by rural producers, and what role this can play in complementing development efforts. Also, the scientific, economic and social values that can be derived from integrating indigenous and modern best practices in climate change adaptation and mitigation have not been adequately unraveled for consideration in the design and implementation of Nigeria's climate change mitigation and adaptation policy. To a large extent, this accounts for rural producers not being sufficiently assisted by research, extension and advisory services in coping with the increasing threats of climate change. To advance the resilience of rural producers, viable options are needed. Redirecting attention towards the consideration of indigenous best practices adopted by rural producers over several years under changing circumstances can provide valuable insights which can offer long-term benefits. Therefore, well focused research is crucial in systematically identifying, documenting, selecting, improving upon and utilizing the potential aspects of indigenous best practices towards supporting rural producers in climate adaptation and mitigation.

Based on a recent study carried out, this paper describes the indigenous best practices employed against climate change for food security, by irrigation farmers in two agro-ecological zones of Katsina State in northern Nigeria. Specifically, the objectives of the study are to:

1. identify and describe the farmers' awareness and perception of climate change in the study area;
2. assess the indigenous best practices employed against climate change within resource contexts, and how they are complemented with modern coping strategies;
3. recommend further development, promotion and sustenance of indigenous best practices towards the resilience of the rural people in combating climate change impacts.

1.1 Justification and Limitations of the Study

At present, there is a high global demand for research and information on best practices which can boost development efforts in addressing climate change impacts in Africa. The study is intended to provide valuable information, which should be recorded, documented and made accessible for replication in other local communities, the wider African community, and the global information community. However, this study only examined the mitigation and adaptive strategies of irrigation farmers in northern Nigeria with regard to the management of soil, water and post-harvest under climate change. Women producers were not included due to inhibiting socio-cultural factors. The quality of information provided by this study, though useful is limited by the nature of data sources in terms of location-specificity, group size and representativeness of the rural producers interviewed. As such, cross-regional generalizations cannot be made.

2 Materials and Methods

2.1 Area of Study

The study was conducted in Katsina State which is considered in terms of agroecological peculiarities, population density of the larger part of northern Nigeria and one of the leading States in the production of cereal, legume crops and livestock husbandry. Katsina State which is located between latitudes 11°07′ and 13°22′ north and longitudes 6°52′ and 9°22′ east, has a total land area of 23,938 km^2, out of which 1.6 million hectares are arable. The projected population of the State was put at 7,452,629 in 2014 at a growth rate of 3.2% per annum (National Population Commission 2006).

Administratively, Katsina State is divided into 36 Local Government Areas (LGAs). The climate is semi-arid with two distinct wet and dry seasons. The rainy season is from May to October with the peak in August. Katsina state covers three agro-ecological zones comprising: the Sahel savannah zone in the extreme north, the Sudan savannah zone, and the northern guinea savannah zone in the south. Under the mandate of the Katsina State Agricultural and Rural Development Authority (KTARDA), the three agro-ecological zones are stratified into Zone I, Zone II and Zone III (see Fig. 1). Average annual rainfall is about 689 mm and mean annual temperature ranges between 26 and 28 °C. Soil types include upland and lowland soils. The upland soils comprise fine sand, clay and loam while lowland soils are characterized by a high year round water table. The major crops grown are maize, cotton, groundnut, millet, sorghum, cowpea and vegetables. The main ethnic groups are Hausa-Fulani with Islam as major religion (Umar and Musa 2015).

2.2 Sampling Procedure and Method of Data Collection

A multi-stage sampling procedure was applied in the conduct of the study. Zones I (Ajiwa rural community) and III (Dutsinma rural community) were purposively selected out of the three agricultural zones of KTARDA. From a sample frame of registered irrigation farmers in the study areas, 15% was selected via simple random sampling method which gave a total sample size of 200 respondents selected.

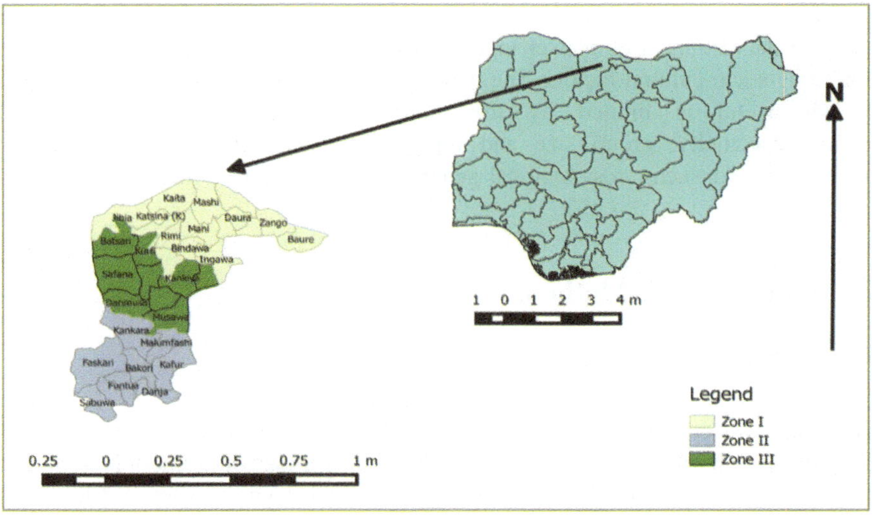

Fig. 1 Map showing the study zones in Katsina State, Nigeria

In contemporary research and development circles, high emphasis is now being placed on complementing conventional and participatory approaches to research and development (ILEIA 1988; Chambers 1999; IK and DM, 1998). Both are perceived to be mutually reinforcing. Thus, data was collected using a combination of conventional survey techniques (structured interview schedule, field observation, photographs and face-to-face interview) along with participatory methods and tools of data collection. These include: *Key informant method*: This was used to elicit information from selected irrigation farmers who were very knowledgeable and could provide the needed information); *Focus group discussion (FGD):* This involved discussion, observation and active listening to generate in-depth information in group situation; *Direct observation*; *Local quotes*: This was used to capture salient facts and people's realities that are expressed through local language either by metaphors or proverbs; *Semi-structured interviewing (SSI):* This entails using checklists and open-ended questions to thematically generate information that cannot be captured through standard routine of questionnaire. It allows for probing and triangulation so that information is generated in a participatory manner; *Case studies*: This was used to vividly describe the existing best practices of the irrigation farmers.

The application of this methodological approach in this study is significant in the sense that it gave respondents a greater freedom in describing their realities. Analysis of data was done using the *Statistical Programs for Social Scientists (SPSS).* Principally, this study is exploratory in nature. The information and data collected are largely qualitative. As such, the statistical tools used in data presentation are descriptive statistics by means of frequency distributions tables, percentages, figures and graphics. It is important to note this paper reports on the various views and perceptions of the rural people studied. The statements made and quotes reported in the paper are direct translations of the Hausa language as they were received.

2.3 Conceptual Clarifications

The terms 'indigenous' and 'best practices' whether applied to a wider or specific context, evokes different meanings to different people and therefore can be ambiguous. Thus, as used in this paper, the term 'indigenous best practices' refers to the sociological notion which describes the distinctive and peculiar skills, technologies, methods, knowledge or strategies pertaining to a people in a given territory, that has proven viable, effective and productive over many years under changing circumstances to a large extent, indigenous best practices are products of experiential learnings that are transferred from ancestral generations to successive generations through continuous adjustment to changing circumstances.

3 Results and Discussions

3.1 Irrigation Farmers' Awareness and Information Sources About Climate Change

The awareness and perceptions of climate change among the irrigation farmers' studied appears similar but their sources of information differ. Findings show that 98.1% of the irrigation farmers in Ajiwa (Zone 1) and 98.5% of the irrigation farmers in Dutsinma (Zone 111) indicated that they were aware of climate change and they know that it was affecting their agricultural activities. This finding corresponds with related studies that found that generally, farmers in northern Nigeria are aware and knowledgeable about changing climatic conditions (Musa 2006). Similarly, findings by Farauta et al. (2012) in three State of northern Nigeria reported that, 84% of farmers are aware of climate change and 79% of the farmers affirmed that they had knowledge of the changing climate, 81% of the farmers indicated that they had in various times encountered the incidences of climate change and 80.2% of the farmers indicated that they received information on climate change from various sources. In contrast, research by Ugwoke et al. (2012) claimed that although farmers are aware of climate change, they do not seem to know the causes. By implication, the study confirms that a large proportion of the irrigation farmers are well aware of the changes taking place in their environment and the consequences of such changes, but they still do not know the causes which is, of course of less importance to them. This study corroborates with the findings of Ugwoke et al. (2012).

Table 1 Distribution of irrigation farmers' according to sources of information on climate change

Source of information	Ajiwa zone (n = 105)	Dutsinma zone (n = 95)	Pooled data (n = 200)	Rank
Other farmers	99 (94.3)	93 (97.9)	192 (96.0)	1
Radio	95 (90.5)	82 (86.3)	177 (88.5)	2
Cooperatives	68 (64.8)	74 (77.9)	142 (71)	3
Market	69 (65.7)	58 (61.1)	127 (63.5)	4
Ext. agents	56 (53.3)	63 (66.3)	119 (59.5)	5
Print media	21 (20.0)	41 (43.2)	62 (31.0)	6
Television	32 (30.5)	24 (25.3)	56 (28.0)	7
Soothsayer	21 (20.0)	28 (29.5)	49 (24.5)	8
Phone	24 (22.9)	21 (22.1)	45 (22.5)	9
Internet	15 (14.3)	8 (8.4)	23 (11.5)	10

Note number greater than n = 200 due to multiple responses accepted. Figures in parentheses are percentages

Good information flow can contribute immensely towards adjusting to climate change crisis. With regards to sources of information, findings show that the irrigation farmers generate information on climate change differently. As depicted in Table 1, 96% of the irrigation farmers identified that other farmers (comprising of friends, relatives and neighbours) were the important sources of information on climate change, implying that information on climate change is largely disseminated informally in the study areas. For 88.5% of the irrigation farmers, the radio was indicated as their source of information on climate change. According to 71% of the irrigation farmers, information on climate change was derived from cooperative group members. For 63.5% of the irrigation farmers, the open market was the source of information while 59.5% of the irrigation farmers indicated getting information on climate change through government extension agents.

Though soothsayers (oracles and rainmakers that predict weather using supernatural powers) used to be a very important source of information on climate issues, however, the advent of Islam in the study area has diminished their relevance such that few (25%) of the farmers ever patronized the services of the soothsayers. With regards to the use of the internet and mobile phone as sources of information on climate change, as low as 12% of the irrigation farmers used the internet and 23% of the irrigation farmers relied on mobile phones as source of information. This indicates that the wide use of information and communication technologies (ICTs) for information on climate change is very little.

3.2 *Irrigation Farmers' Perception on Manifestations of Climate Change*

It was observed that the farmers' perception of climate change is value laden and based on self-descriptive criteria. The words, analogies used and images in the definitions, understanding and interpretations of climate change draws from individual and social group experiences of climate change attributes, characteristics, causes and effects. As depicted in Table 2, the farmers' perception of climate change is reflected by certain indicators assigned. For them, it is the intensity and the rate at which these indicators are occurring in recent times that show the climate is changing. According to 83% of the irrigation farmers, the most important manifestation of climate change in the area is the relatively longer dry seasons. Conversely, this means that shorter rainy season has negative consequences on agricultural production and thus, intensify the need for irrigation. There is a saying in the Hausa language which connotes when rains are normally expected to start. It goes: "*watan bakwai makarar rani, ko ba ruwa da alama*". Loosely translated it means: "*the seventh (lunar) month is the end of dry season, even if it does not rain, there would at least be signs that rainy season is about to start*". Going by this saying, the farmers believe that rains normally start within the seventh lunar month (each month having 28 to 31 days) after the last harvest, to the extent they used to practice what they termed "*bizne*", roughly translated as "implant". This is a practice whereby farmers sow their crops in the seventh lunar month though rains have not established, breaking the dry soil and putting their seeds, hoping for the arrival of rains. In recent years, to the dismay of the farmers, rains do not get established in their areas until late June. The irrigation farmers also identified erratic rain pattern as another indication of climate change. They cited examples with the recent rains that fell in some parts of the area in January of 2014. They also noted strange occurrences that signify that the climate is changing. As confirmed by the irrigation farmers, other indicators of climate change in the area include more intense and severe harmattan (72%), higher rates of drought (64.5%), increased pest incidence (52%), warmer temperature (39.5%), increased cases of flooding (37.5%) and increased disease incidence (36%).

This finding corroborates results obtained from related studies elsewhere. For example, from a survey of rural farming communities in three States of northern Nigeria (Kaduna, Kano and Jigawa States), Musa et al. (2013) reported that the rural farmers' perceptions of drought and observable changes were diverse, ranging from spiritual to physical indicators and causes. According to the study, 15% of the farmers gave value-laden description that the prevalence of drought incidence is a manifest feature of "*people's disobedient and immoral behaviours towards God for which they are receiving punishment*". But for 80% of the farmers drought and its early warning sign was perceived using indigenous indicators which include: the observation of the untimely fruiting of some tree species; the sudden migration of bird species called *Chilakowa* (Hausa word) in the middle of the rainy season; sudden rise in temperatures and intensive heat waves; reduction in soil fertility levels; rain falling between 4 to 7 pm (West Africa standard time), unusual appearance of disease and high insect pest populations. Similarly, studies conducted by Farauta et al. (2012) and Hir (2010) reported farmers' perceived manifestations of climate change to include: unusual early rains that are followed by weeks of dryness, higher temperature, loss of soil fertility, reduction in farm yields, high rate of disease incidence, delay in onset of rain, less rainfall, erratic rainfall pattern, long

Table 2 Distribution of irrigation farmer based on their perceptions of climate change

Perceptions of climate change	Ajiwa zone (n = 105)	Dutsinma zone (n = 95)	Pooled data (n = 200)	Rank
Longer dry season	89 (84.8)	77 (81.1)	166 (83.0)	1
Erratic rain pattern	90 (85.7)	69 (72.6)	159 (79.5)	2
Severe harmattan	71 (67.6)	73 (76.8)	144 (72.0)	3
Drought	70 (66.7)	59 (62.1)	129 (64.5)	4
Increased pest incidence	47 (44.8)	57 (60.0)	104 (52.0)	5
Warmer temperature	37 (35.2)	42 (44.2)	79 (39.5)	6
Flooding	38 (36.2)	37 (39.0)	75 (37.5)	7
Increased disease Incidence	39 (37.1)	33 (34.7)	72 (36.0)	8

Note number greater than n = 200 due to multiple responses accepted. Figures in parentheses are percentages

period of dry season, no or reduced harmattan, long period of harmattan and heavy and long period of rainfall. Another study in Southern Nigeria showed that farmers perceive deforestation, bush burning, gases released from industries, use of excessive chemicals in rice production, application of excess nitrogenous fertilizers, natural phenomena, violation of local custom and burning of firewood and farm residues (rice straws and husks) as causing adverse climate change to a great extent Nwalieji and Uzuegbunam (2012).

3.3 *Indigenous Best Practices in Adaptive Management of Soil Fertility*

Water availability and soil fertility are among one of the biggest drivers that provides the basis for the stability and continuity of rural food production systems. For the irrigation farmers, surviving under climate change, hard economic times and population pressures means the continuous dependence on innovative and viable adaptation practices. '*Fari*' is the Hausa language word for drought and is defined by the irrigation farmers as "*persistent and prolonged dry spell conditions during the rainy season for a period of four to six weeks*". As confirmed by all (100%) of the irrigation farmers, drought constitutes one of the most prevailing conditions of climate change risks affecting their agricultural activities. According to them, '*fari*' (drought) had been a long term "*evil in the past 20 to 40 years, with a reoccurring frequency of about 3 to 4 times in a period of 10 years to 12 years*". As obtains, the socioeconomic status of the irrigation farmers is largely determined by the ownership and access to land with good soil structure and fertility content. Their indigenous classification systems and management practices for land and soil types are based on many criteria. These include moral and spiritual worldview, characteristics of soil, value attributes, utility and seasonality as well as the perceived needs and problems. These criteria and classifications about good soils derives from evidence of 'best practices' resulting from many years of trial and error experimentation, re-adaptations, ancestral wisdom and shared experiential leanings. Within the context of on-going climatic changes, the age-old practice of using organic manure is still widespread.

Box 1. Best practice: Recycling garbage into organic manure for enhancing soil fertility under drought conditions

In recent times, the use of garbage organic manure has become increasingly popular among northern Nigeria farmers as an adaptive survival strategy. This is because it has been found to be very effective in restoring soil fertility as well as increasing agricultural production. Though it is expanding, the practice is yet to gain adequate institutional support from national and international development polices. Refuse dump sites comprise of putrefied contents, including rusty steel, metal, aluminum, plastic and polythene waste materials. They are sites to numerous contagions, including microbacterial activities and parasitic infections. While many people would avoid such places, for Mallam Abdul Saidu garbage dump sites hold prospects for increasing agricultural productivity and income admidst change. He is a farmer with over 20 years farming experience. For more than 8 years Mallam Abdul Saidu has intensively engaged in the tilling and recycling of garbage contents into organic manure for soil fertility maintenance, especially under increasing drought conditions linked to climate change (See Picture 1 and 2). Combustible materials such as polythene, plastics and other decomposed contents are burnt into ashes which becomes useful. Non-combustible objects such as rusty metal pieces, aluminium, broken glasses are sorted and sold to smelters for additional income. According to him and the other farmers who continuously use his product, organic manure from garbage sites have the potential capacity of retaining soil fertility for 8 years when used alone and up to 10 years in combination with mineral fertilizers, under drought or erratic rain fall conditions. Depending on the heap size of garbage dump site, Mallam Abdul Saidu produces 200–500 sackful of organic manure (See Picture 3 and 4). The price for a sack ranges between NGN300.00-NGN500.00 (2.00USD-3.00USD)* depending on the size of bag. As a low cost and effective internal input, organic manure from garbage has no side effects of soil acidity like inorganic fertilizers, does not alter the natural taste of farm products, is not cumbersome or complex to use, and offers opportunities for good market prices for crop.

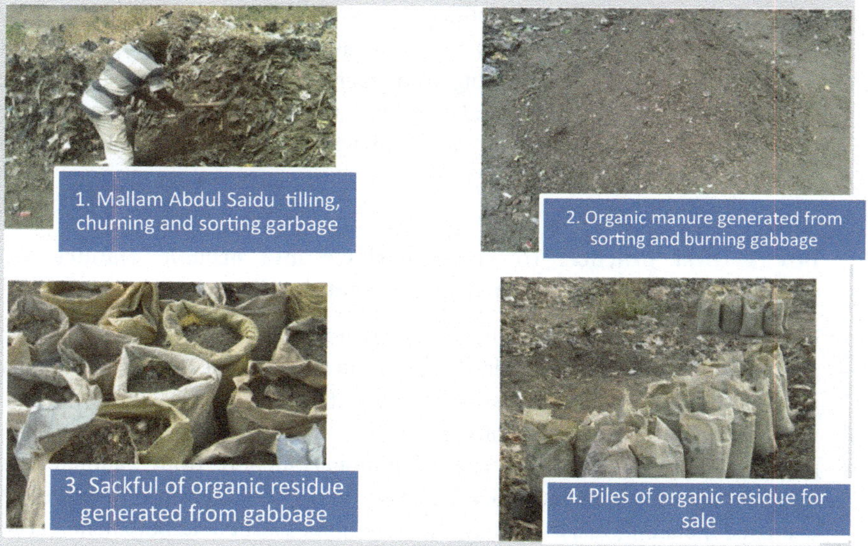

Photo: Michael W. Musa.
N1 = $0.00502 (current rate as in 2016) N = Nigerian Naira (NGN); $ = US Dollar (USD)

Due to conditions of erratic rainfall, high artificial scarcity, late or non-delivery of mineral fertilizers to designated warehouses in recent years, all (100%) of the irrigation farmers, confirmed that they are increasingly resorting to the use of organic manuring for rehabilitation of their farm lands, and enhancing rapid crop growth. In relation, 85% of the irrigation farmers indicated the high dependence and interactive use of organic and mineral fertilizers for enhancing soil fertility when they have access to it. Sources of organic manure include waste products of animals reared (such as poultry and cattle waste), composting and garbage dump sites. The case study in Box 1 show how garbage dump sites are potentials for deriving organic manure for enhancing soil fertility.

Even though there is hardly any farmer without one form of animal or the other, the quantities of organic manure produced are not sufficient to cover their lands. To overcome such constraints, some (55%) irrigation farmers resort to the establishment of complementary relationships with settled or roving pastoralists. Under this arrangement, farmer's crop residues provide sustenance for herder's livestock and livestock of pastoralists in turn provide organic manure for the fields. However, others seek for large quantities elsewhere. Before the onset of the rains, organic manure accessed by the farmers is spread on their farmland. When the rains set in, the organic residue fuses with soil and in the process release micro-nutrients into the soil. Farmers have observed that soil that has been well provided with organic manure retain moisture for about 20 days or more after short bout of rainfall and has the potential of sustaining soil fertility for 8 years without reapplication despite intensity of land usage. The use of organic manure offers a number of advantages to irrigation farmers and is an indigenous resource which is easily accessible, obtainable in large quantities and is management-friendly.

Other practices include, pre-season tillage and post-harvest mulching. Under post-harvest mulching practice, the irrigation farmers indicated that, they leave crop refuse after harvest and also cover their farms with grasses from unfarmed areas. Beside improving soil structure, the farmers indicated that this age-old practice reduce run-off water and allows more water to penetrate into the soil. With pre-season tillage, the farmers combine the use of modern technology by hiring tractor services to till their lands to allow for high absorption of water by soils during the rainy season. Resorting to this practice is in anticipation that rain for the year may not saturate the soil well enough for good crop growth. Also as indicated by the irrigation farmers, the techniques of tied ridges and bunds are also common practices of land management mainly to reduce run-off. In many instances, boulders and fertilizer bags filled with sands and pebbles are often used as bunds to reduce soil erosion due to water run-off.

Box 2. Best Practice: The 'rhumbu' storage system

The '*rhumbu*' is a simple indigenous storage and conservation best practice that is used to safe guard against postharvest losses due to pest incidences and drought conditions linked to climate change. It is an indigenous 'silo' or 'barn' made of easily accessible resources such as sticks, straws and vegetative twines. Crops such as millet, cowpea, sorghum, maize are stored by the irrigation farmers. In most instances before storage, the '*rhumbu*' is swept clean to get rid of any insect pest or microbial organism. In this regard, the '*rhumbu*' is fumigated using smoke that is emitted from hot charcoals placed inside it overnight. It is then swept clean the day following. Substances such as ashes, pepper powder and malodorous plant called by the Hausa word-'*Buzurun fadama or Buzurun fagge*' (*Hyptis spicigera*), are used in lining the bottom and sides of the barn. The crops are then placed inside by spreading them alternately layer after layer until all the crops are stored. The pungency of the substances keeps insect pests from attacking the stored crops. The '*rhumbu*' is then sealed by placing a strong lid which is then sealed with clay. Stored crops keep for about 10 months to over a year without insect infestation. The '*rhumbu*' can last about three years before depreciation. The '*rhumbu*' is an indigenous technology which has served the irrigation farmers in northern Nigeria for several decades and till today is still widely in use. The resiliency of this technology has earned it the reputation of being labelled a 'best practice'.

A collection of '*Rhumbu*' for post-harvest storage and crop preservation against drought.
Photo: Michael W. Musa

3.4 Adaptive Management of Post-harvest Under Drought

Efficient storage of food plays a vital role in the attainment of food security under drought conditions. Efficient storage of produce depends on a number of factors one of which is the availability of the structure to hold the produce (see Box 2). As regards the joint use of practices, 95% of the farmers indicated that they combined indigenous and modern practices in post-harvest pest management. In this case, the preservation of farm produce involves use of improved methods of agro-chemicals and the hermetic storage technique to complement indigenous practices of pest management Precautions are taken to ensure those products are well preserved, well dried and free from infestation of pests. The hermetic storage of cereals and legumes involve the use of plastic containers and hermetic (polythene) bags. This technique prevents further oxygen from gaining entrance into stored products, preventing insect growth and development thus, limiting infestation on the stored products.

Box 3. Best Practice: Vegetative management for storage and preservation

This vegetative best practice in the stacking of millet, sorghum and maize stalks in tree tops is a conservation measure for feeding livestock during periods of drought and for other household uses such as fencing and roof thatching. This age-old practice is also used by the farmers to safeguard stalks from destructive activities of pests such as rodents, termites and human pilferers.

Wood conservation technique against termite attack and for future use.
Photo: Michael W. Musa

Other forms of best practice used by the irrigation farmers include the innovative use of natural resources in storage and preservation vegetative resources. As depicted in Box 3, findings reveal that vegetative species such as trees were relied upon as effective ways of stacking and preservation stalks of sorghum, maize, millet and woody vegetation for future uses, especially for livestock feeding in periods of drought.

Box 4. Best practice: Shadoof irrigation system

Photo: Sulaiman Umar

3.5 *Adaptive Management of Water Under Drought*

In northern Nigeria, agricultural activities are carried out in two (rainy and dry) seasons and on major land types- the '*tudu*' and '*fadama*' which applies in the areas of study. The '*tudu*' are upland areas which is cultivable only during the rainy season while the '*fadama*' are lowland areas, seasonally flooded and natural depressions on adjacent low terraces where the water table is near the surface and can support crop production throughout the year. Though irrigated agriculture constitutes one of the most productive kinds of farming activities widely practised in northern Nigeria, investments for large-and-medium scale irrigation projects over the years by the national government have not adequately impacted. From a multiple response, 99.5% of the irrigation farmers indicated that, to harvest rain water and control water logging, the construction of trenches, wells, dams, water pits or basins are common practices. These practices are used to trap, direct flood water and to enhance good drainage on both *tudu* and *fadama* fields. Water retained comes in handy during the dry season. Water from wells, trenches and water pits are lifted manually and distributed to farmlands through an indigenous '*shadoof*' irrigation system (see Box 4). In addition, 75% the irrigation farmers indicated that they combine the use of generator pumping machines to propel and distribute water to their farms. Through these techniques, the irrigation farmers have been able to maintain continuous food production under stressful, risky and fluctuating conditions.

4 Conclusion and Recommendations

Rural producers in northern Nigeria have had years of intimate and site-specific knowledge of climate change and doing their best to adapt. As findings have shown, there is no doubt that the best practices used by the irrigation farmers for soil, water, postharvest and vegetative management have the potentials in coping with climate change disaster such as drought. These practices have proved to be entirely viable, easy to manage, very-cost effective and can easily be replicable elsewhere. Since there are elements of resiliency in the indigenous best practices of the rural producers, policy and scientific research would do well to identify, recognize and improve upon them. For instance, it would be good to expand the future of organic manuring, especially from garbage dump sites in a manner that allows for complementary use and mutual reinforcement with inorganic fertilizers. Likewise, investors and research scientist can begin to look into new directions of improving upon the '*rhumbu*' by identifying its limitations and providing solutions that do not jeopardize environmental, economic and socio-cultural fabrics of the rural communities in northern Nigeria.

To help rural people adjust well to climate change impacts, this paper recommends that climate change mitigation and adaptation support should begin with a deeper understanding of rural peoples' perspectives, their realities of climate change and their indigenous best practices. Above all,, national government, research institutes, private investors and donors interested in the design of cost-effective and site-specific project interventions are needed to scale-up community based best practices. In this context, the systematic documentation and utilization of the findings from this study will be of assistance.

Acknowledgements The authors wish to extend their sincere appreciation to the irrigation farmers studied in northern Nigeria and to the officials of the Katsina State Agricultural and Rural Development Authority (KTARDA) for their support.

References

Chambers, R. (1999). *Relaxed and participatory appraisal: Notes on practical approaches and methods*. Brighton, UK: IDS.

Dembowski, H. (2015). Climate talks. *International Journal of Development and Cooperation (D+C), 35*(12), 442.

Farauta, B. K., Egbule, C. L., Agwu, A. E., Idrisa, Y. L., & Onyekuru, N. A. (2012). Farmers' adaptation initiatives to the impact of climate change on agriculture in northern Nigeria. *Journal of Agricultural Extension, 16*(1), 132–144.

Food and Agriculture Organization [FAO]. (2012). Climate change: Climate risk management. http://www.fao.org/climatechange/49376/en/. Accessed on January 12, 2016.

Guillaume-Gentil, A. (2014). Climate-smart agriculture: A practical or idealistic approach? *Spore magazine, 172,* 4–5.

Hir, J. (2010). Sand dunes threaten Northern Nigeria. *Climate Change Group*. http://dailytrust.dailytrust.com/index.php?option=comcontent&view=article&id=4517sand-dunes-threaten-northern-nigeria, Accessed on December 29, 2015.

Hunter, L. M. (2007). Climate change, rural vulnerabilities and migration. http://www.prb.org/Articles/2007/ClimateChangeinRuralAreas.aspx, Accessed on December 15, 2015.

Indigenous Knowledge and Development Monitor [IKDM]. (1998). Editorial. *Indigenous Knowledge and Development Monitor, 6*(2), 1.

International Fund for Agricultural Development [IFAD]. (2008). Climate change: Building the resilience of poor rural communities. www.ifad.org/climate/factsheet/e.pdf. Accessed on 12 April, 2014.

Intergovernmental Panel on Climate Change [IPCC]. (2007). *Climate Change 2007: Impacts, adaptation and vulnerability. Contribution of Working Group II to the Fourth Assessment Report of IPCC*. Cambridge. UK: Cambridge University Press.

Information Centre for Low-External-Input and Sustainable Agriculture [ILEIA]. (1988) In *Proceedings of the ILEIA Workshop on Operational Approaches for Participatory Technology Development in Sustainable Agriculture*. Leusden: ILEIA.

Mangani, R. (2014). Unconditional support. *International Journal of Development and Cooperation (D+C), 41*(3), 102–104.

Meze-Hausken, E. (2004). Migration caused by climate change: How vulnerable are people in dryland areas? *Mitigation and Adaptation Strategies for Global Change, 5*(4), 379–406.

Musa, M. W. (2006). *Indigenous resource management systems (IRMS) among rural communities in north-west zone of Nigeria and their relevance for participatory poverty reduction* (Unpublished Ph.D dissertation). Ahmadu Bello University, Nigeria.

Musa, M. W., Hassan, M. B., Issa, F. O., & Aruhna, U. L. (2013). Sustaining food security under climate change risks: Options for institutional adjustments through agricultural extension. *Proceedings of the South African Society for Agricultural Extension (SASAE), 47,* 136–144.

National Population Commission. (2006). *Report of national population census*. Nigeria.

Nwalieji, H. U., & Uzuegbunam, C. O. (2012). Effect of climate change on rice production in Anambra State Nigeria. *Journal of Agricultural Extension, 16*(2), 81–91.

Nyong, A., Adesina, F., Elasha, B. O. (2007). The value of indigenous knowledge in climate change mitigation and adaptation strategies in the African Sahel. *Journal of Mitigation and Adaptation Strategies for global Change, 12*(5), 787–797. Accessed at www.Springerlink.com on 9 November, 2015.

Spore. (2014). Adapting to climate change. In *Spore magazine,* 172, 7.

Spore. (2015). Global warming: doing business in a time of climate change. In *Spore magazine special issue, 10*, 2–31.

Umar, S., & Musa, M. W. (2015). Determinants of the use of indigenous coping strategies against change among smallholder farmers in Katsina State, Nigeria. *Journal of Agriculture and Ecology Research International, 3*(1), 24–32.

Ugwoke, F. O., Nnadi, F. N., Anaeto, C. F., Aja, O. O., & Nwakwasi, R. N. (2012). Crop farmers' perception of and adaptation to climate change in Orlu agricultural zone of Imo State Nigeria. *Journal of Agricultural Extension, 16*(2), 212–223.

Wikipedia. (2015). 2010 United Nations climate change conference https://en.wikipedia.org/wiki/2010_United_Nations_Climate_Change_Conference. Accessed on 20 January, 2016.

Wikipedia. (2016). United Nations framework convention on climate change https://en.wikipedia.org/wiki/United_Nations_Framework_Convention_on_Climate_Change. Accessed on 29 January, 2016.

Deriving Useful Information from Bimonthly Global-Scale Climate Analysis for Climate Change Adaptation Over East Africa

Isaac K. Tetteh, Nana K.A. Appiah-Badu, Fredrick H.M. Semazzi and Olawale E. Olayide

1 Introduction

Effective planning and execution of climate change adaptations is expected to bring new lease of hope to the affected societies confronted with the negative tendencies of climate change. Success of this activity to some extent is contingent on a good understanding of climate variability and change. On the global scale, eclectic adaptation strategies have evolved, or continue to evolve, in response to the specific menace created by climate variability and change, irrespective of the background driving force—natural or anthropogenic.

Interestingly, East Africa, one of the most vulnerable regions to global environmental change, has benefited from implementations of adaptation strategies derived directly or indirectly from a plethora of scholarly research efforts (e.g., Ropelewski and Halpert 1987; Ogallo 1988; Barnston and Ropelewski 1992; Funk 2012; Smith and Semazzi 2014). Such efforts have been complemented by, for example, operationalization of seasonal forecasts of long (March-April-May: MAM) and short (October-November-December: OND) rains by the Greater Horn of Africa Climate Outlook Forum (GHACOF). GHACOF is organized by the

I.K. Tetteh (✉) · O.E. Olayide
Department of Theoretical and Applied Biology, Kwame Nkrumah University of Science and Technology (KNUST) Kumasi, Kumasi, Ghana
e-mail: iktetteh.cos@knust.edu.gh; isaac.tet@gmail.com

N.K.A. Appiah-Badu
Department of Physics, KNUST, Kumasi, Ghana

F.H.M. Semazzi
Marine, Earth, and Atmospheric Sciences, North Carolina State University, Raleigh, NC 27695-8208, USA

O.E. Olayide
Centre for Sustainable Development, University of Ibadan, Ibadan, Nigeria

W. Leal Filho et al. (eds.), *Climate Change Adaptation in Africa*,
Climate Change Management, DOI 10.1007/978-3-319-49520-0_8

IGAD (Inter-Governmental Authority on Development) Climate Prediction and Application Centre (ICPAC), Nairobi, Kenya. It typically uses consensus forecasts, which derive inputs from empirical climate research outcomes.

Empirical climate research on rainfall variability has documented the role of interannual variability of local and remote forcings of the East African seasonal climate, which are key precursors used in operational seasonal climate prediction scheme. Interannual variability has chiefly focused on climatic and dynamic features in the annual and seasonal cycles that govern the region's long and short rainfall variability. Three important ocean-atmosphere temporal scales, which may be identified in the interannual variability and have great influence on the region's seasonal rainfall patterns, are monthly, bimonthly, and seasonal timescales (e.g. Latif et al. 1999; Saji et al. 1999; Marchant et al. 2006; Smith and Semazzi 2014). As a general summary from these collections, the region's rainfall variability points to the Pacific ENSO and Indian Ocean, and to a lesser extent the Atlantic Ocean, and their associated atmospheric dynamics. These, however, do not rule out orographic/land surface forcing and vegetation feedbacks or dynamics of the region's climate variability.

Generally, the sub-Saharan African countries are lagging far behind the advanced counterparts in efficient utilization of adaptation technologies commonly applied to ameliorate the detrimental effects of climate variability and change (Washington et al. 2006; Akponikpè et al. 2010; Ford et al. 2015). These are due to many constraints such as the lack of adaptive capacities and capital resources. Addressing these constraints is beyond the scope of the current study. However, a critical issue that warrants consideration is the need for adaptation strategy enhancement that would rely on new climate information for sustainable resilience to climate shocks. The climate information should unequivocally describe or contain features—behavior, characteristics and evolution at different timescales, as they interact with other complex climate systems or subsystems. To the best of our knowledge, among the timescales in the interannual variability, bimonthly ocean-atmosphere features have been the least studied. It is therefore justifiable to explore the potential influence of these features on the region's rainfall variability, from a global stand point. This would improve predictive understanding of the region's rains, on the premise that current numerical and statistical model predictive skills are far from perfect. Also, it would contribute to effective adaptation management, especially of sustainable agriculture, climate-induced health problems, flood and drought disasters, and other climate-sensitive socioeconomic problems.

The trust of this study was to provide useful climate information needed by end users for incorporation into sustainable climate change developmental goals. Specifically, it was designed to carry out an analysis of bimonthly evolution of global ocean-atmosphere features and their relationship to the dominant modes of MAM seasonal rainfall variability during the climatologically prominent ENSO phase. It was expected that SST sectors and their associated upper level atmospheric circulation patterns would be delineated for improving climate change adaptive capacity.

2 Data and Methods

2.1 Data

The study utilized monthly gridded precipitation, global horizontal wind and SST data. The horizontal winds at 200 hPa level were sourced from the National Centers for Environmental Prediction-National Center for Atmospheric Research (NCEP-NCAR) reanalysis (Kalnay et al. 1996). The data, covering periods from January 1948 to March 2016 on a global grid of 2.5° × 2.5° spatial resolution, are continuously updated. The precipitation dataset, with spatial resolution of 0.5° × 0.5° resolution, was sourced from Climatic Research Unit (CRU). The data was recorded from about 4000 weather stations worldwide (land only) and covered the periods from January 1901 to December 2012 (Harris et al. 2014). This dataset is periodically updated. Extended Reconstructed SST (ERSST; Smith and Reynolds 2004) data was used. The dataset extended from January 1854 to March 2016, and has a spatial resolution of 2.0° × 2.0° on a global grid. This dataset is also continuously updated.

2.2 Analysis Techniques

2.2.1 Empirical Orthogonal Function (EOF) Analysis of MAM Rainfall

Standard EOF technique (Schreck and Semazzi 2004; Wilks 2006; Hannachi et al. 2007), based on the correlation matrix, was applied to CRU MAM precipitation over East Africa covering the domain of Bowden and Semazzi (2007). The data spanned 1951–2008, which consisted of nearly 60 years of continuous data. The analysis period was based on the framework of Tetteh (2012) over West Africa, which has been extended to investigate other hypotheses over the rest of the sub-Saharan climate systems. The EOF analysis was aimed at isolating and retaining dominant, spatially coherent patterns and their associated temporal structures of the seasonal long rains. The leading modes that satisfied the sampling error criterion in the eigenvalue spectrum were retained for further analysis (North et al. 1982). In this way, EOF was viewed as exploratory data analysis tool (Johnson and Weichern 2007).

2.2.2 Construction of Standardized Global SST and Upper Level Anomalies

To investigate the influence of bimonthly global ocean SSTs and their associated atmospheric structures during the climatologically active ENSO phase, the SSTs and horizontal winds (200 hPa) data were standardized to zero mean and unit variance, to put all the data on the same scale. This was first done by averaging the bimonthly fields before standardization for five consecutive two months period (December-January, January-February, February-March, March-April and April-May (DJ, JF, FM, MA and AM). This rolled over from 1950 to 2007 for December to 1951–2008 (from January through to May).

2.2.3 Lagged Heterogeneous Grid Point Correlations Between the Rainfall Modes and Standardized Global SSTs and Velocity Potential (Divergence) Computations

Lagged heterogeneous grid-point correlations were computed between each of the rainfall modes and the global SSTs for all the time lags until the rainfall season was over. The purpose was to evaluate, delineate and monitor the specific SST signals that were connected to the rains. A similar one was conducted for the horizontal winds (200 hPa)—zonal (u) and meridional (v), from which divergent circulation (Krishnamurti 1971) was computed, since this atmospheric flow played an important role in global monsoon dynamics (Trenberth et al. 2000).

3 Results

3.1 East African MAM Rainfall Variability

Figure 1 depicts the spatio-temporal patterns of the rainfall modes, in which four leading modes retained were based on the *delta-test* (North et al. 1982). These contributed to 34.2% of the total explained variance. The value signified low interannual variance of the long rains, which was consistent with previous studies (Ropelewski and Halpert 1987; Ogallo 1989). The respective contributions of precipitation EOFs 1, 2, 3, and 4 were 10.1, 9.1, 8.2, and 6.8%. Interestingly, all their EOF time series (Fig. 1e–h) demonstrated interannual variability but with differences in their amplitude of fluctuations. The precipitation patterns displayed distinct variations.

Figure 1a shows the eigenvector loadings of the precipitation EOF 1. Generally, a bipolar pattern was observed. The center of action depicted positive weights located over the greater portion of the domain, from south to northeast. The EOF 2 (Fig. 1b) also depicted a bipolar structure, but nearly antithetic to the EOF 1 pattern. Positive (negative) weights were centered over the northern (southern) sectors of the region. The EOF 3 loadings (Fig. 1c) were characterized by positive weights in the western and southern boundaries, whereas the remaining sectors contained the negative weights. The FOF 4 spatial patterns (Fig. 1d) showed small-scale positive weights confined to the horn area. Widespread negative weights were discernible in the rest of the region.

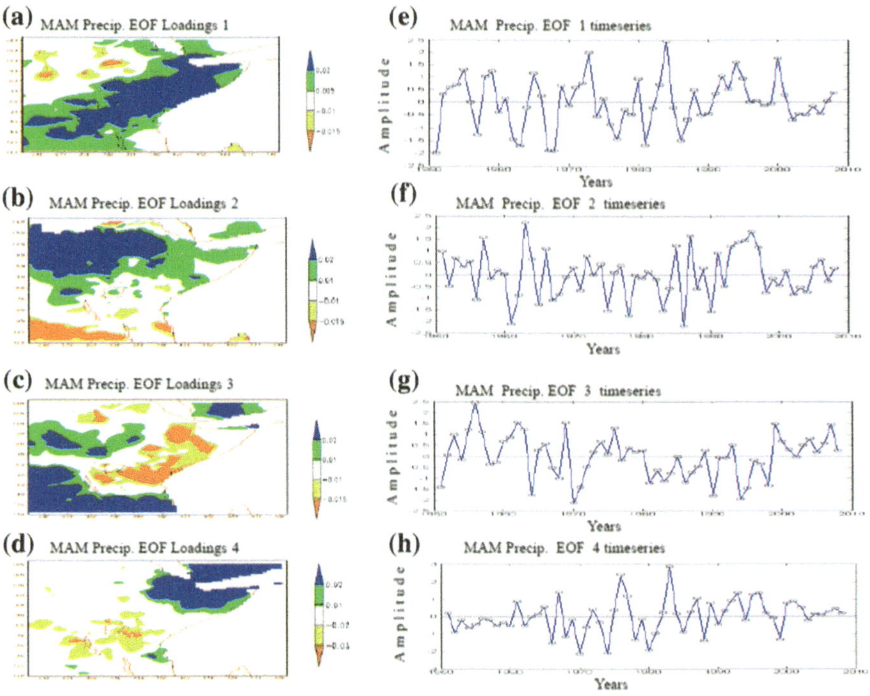

Fig. 1 Dominant modes of long rains (MAM) variability over East Africa. Panels **a–d** are the spatial patterns of the rainfall with corresponding legends. *Blue* and *green shades* depict areas of above-normal rainfall while all other shades correspond to below-normal rainfall. Panels **e–g** are the corresponding time series of the rainfall patterns

3.2 East African MAM Rainfall Modes and Global SST Relationships

In this section, grid-point correlations between the four rainfall modes and the standardized global SST anomalies are presented at various time lags (Figs. 2, 3, 4 and 5). Generally, the four modes responded differently to the Pacific ENSO, Atlantic and the Indian Oceans. Close observation showed that most of the Pacific Niños were represented. Specifically, these were Niño 3 (5° N–5° S, 150° W–90° W), Niño 3.4 (5° N–5° S, 170°–120° W) and Niño 4 (5° N–5° S, 160° E–150° W) regions. The analysis captured specific ones that had linkages to specific rainfall modes and these interactions were time-dependent.

It was noticed from Fig. 2 that the influence of global SST distributions in the tropical Pacific was strongest in DJ, but had the weakest association in AM. The DJ signal captured the climatologically active phase of ENSO, which is consistent with literature. Warm (cold) oceanic conditions in the Pacific enhanced (suppressed) convective activities associated with MAM precipitation EOF 1 mode. Other

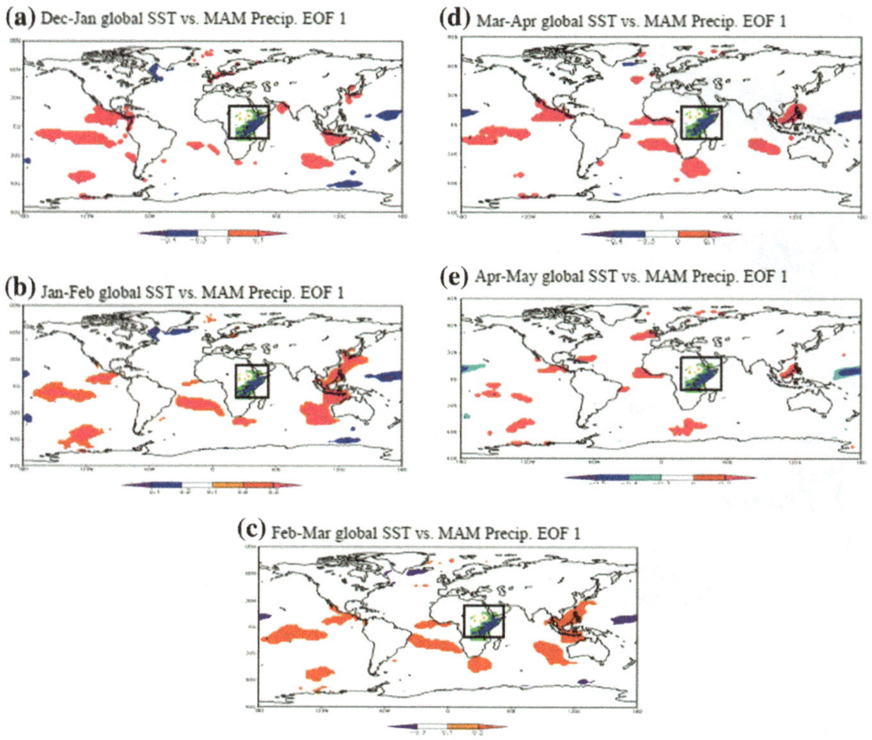

Fig. 2 Relationship between standardized bimonthly global SST anomalies and East African MAM precipitation EOF 1 time series at different time lags. Legend is for the range of correlation coefficient. Significant areas at 95% confidence level, using t-test, are *shaded*. The precipitation spatial patterns over the region are enclosed in *boxes*

features identified were the localized SST conditions in the Indian Ocean off the western seaboard of India and a more prominent one on the western-to-northern seaboard of Australia. The latter persisted up to MA, where it shifted to the vicinity of the Indian Ocean Dipole (IOD) domain (Saji et al. 1999; Marchant et al. 2006). The Indian Ocean features were in phase with the tropical Pacific conditions from DF to MA, but experienced a complete annihilation in AM. The tropical South Atlantic warming became discernible from JF up to MA and also had a good link to the precipitation mode. However, it started to weaken in AM, which indicated the onset of the cooling phase of the tropical South Atlantic Ocean—a vital driver of the West African climate.

Interestingly, precipitation EOF 2 mode in reality had no substantial relationship with the basinwide SST distributions over the study period (Fig. 3). A similar observation was made by Smith and Semazzi (2014), when they investigated the relationship between their precipitation EOF 2 mode and seasonal global SST conditions. The precipitation EOF 3 mode tended to display direct relationship with

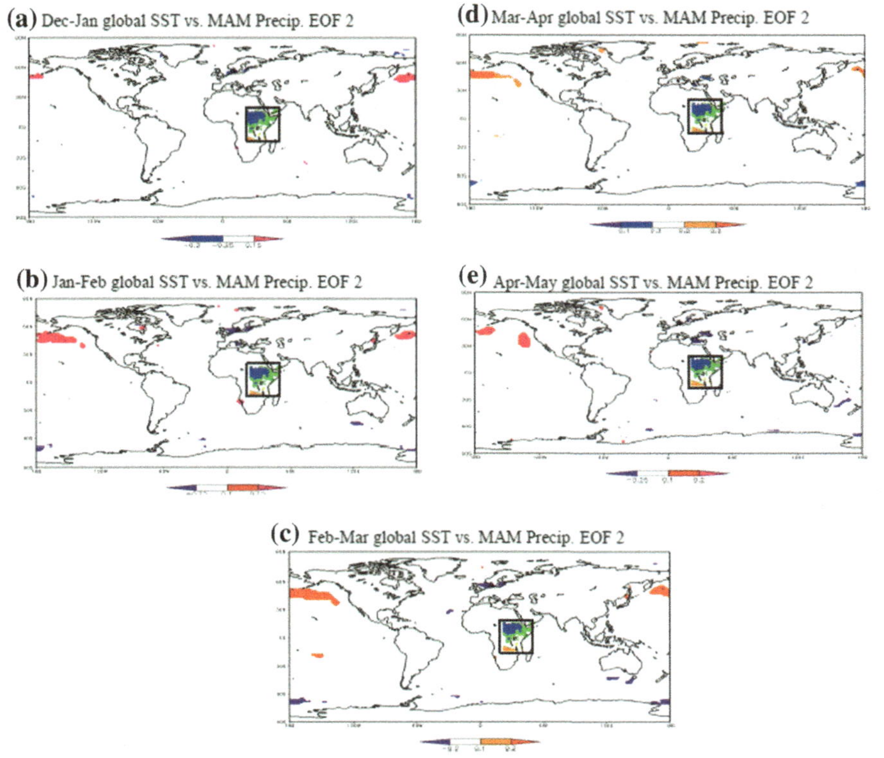

Fig. 3 Relationship between standardized bimonthly global SST anomalies and East African MAM precipitation EOF 2 time series at different time lags. Legend is for the range of correlation coefficient. Significant areas at 95% confidence level, using t-test, are *shaded*. The precipitation spatial patterns over the region are enclosed in *boxes*

persistent mid-latitude North Atlantic conditions (Fig. 4), which could be reminiscent of the Atlantic Multidecadal Oscillation (AMO). In contrast, the Pacific generally displayed an indirect relationship with the strongest (weakest) events captured in DJ (JF). No feature of climatological significance was observed over the Indian Ocean except a small-scale signal in AM, which resided in one of the arms of the IOD domain. Direct relationship between precipitation EOF 4 and the global SST was found to be impressively dominated by the Pacific (Fig. 5), peaking from MA to AM. The evolution of the Gulf of Guinea SST showed a progressive but abysmal development from DJ to MA, and thereafter decayed till the end of the rainy season. In tandem with this event was the detection of a North Atlantic SST pattern, also reminiscent of a small-scale AMO which persisted till the end of the study. The Indian Ocean's link to the precipitation mode was discernible from FM to AM.

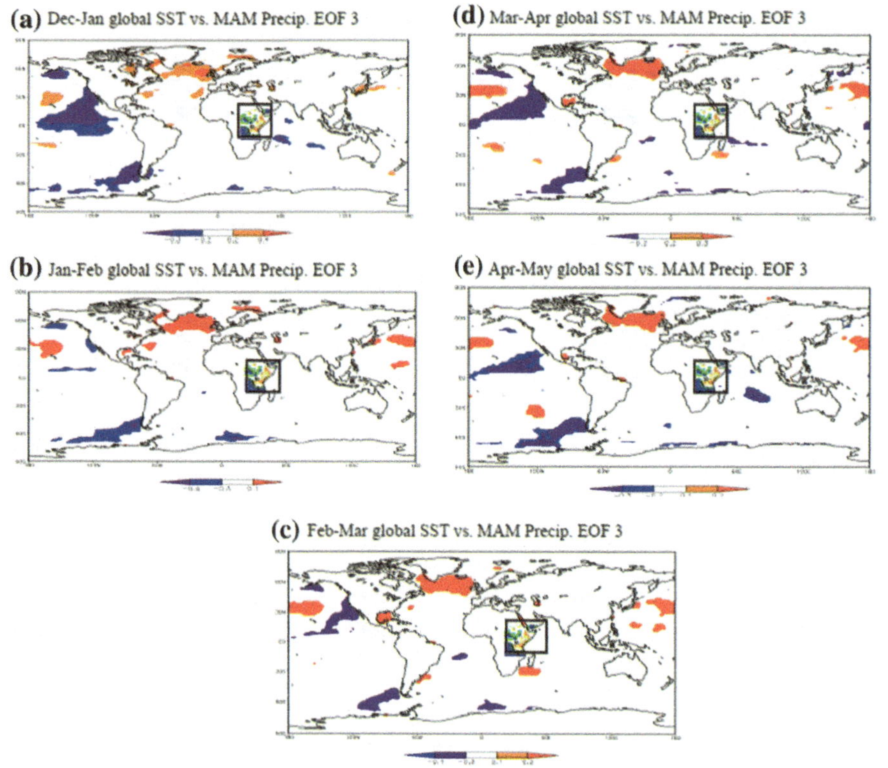

Fig. 4 Relationship between standardized bimonthly global SST anomalies and East African MAM precipitation EOF 3 time series at different time lags. Legend is for the range of correlation coefficient. Significant areas at 95% confidence level, using t-test, are *shaded*. The precipitation spatial patterns over the region are enclosed in *boxes*

3.3 Composites of Standardized Global-Scale Divergent Circulation Anomalies and East African MAM Precipitation Modes

On climate timescale, the boundary forcing emanating from, for example, the ocean surface drives atmospheric circulation. Since MAM precipitation EOF 2 seemed not to be well related to the global SST (Fig. 3), its atmospheric circulation would be excluded from this paper until this unusual observation is clarified through further studies possibly through numerical modeling. It is therefore instructive to focus on EOFs 1, 3, and 4, which is presented shortly.

Figures 6, 7 and 8 show the standardized global-scale velocity potential and divergent circulation anomalies at the upper atmosphere (200 hPa) in relationship to the MAM rainfall modes.

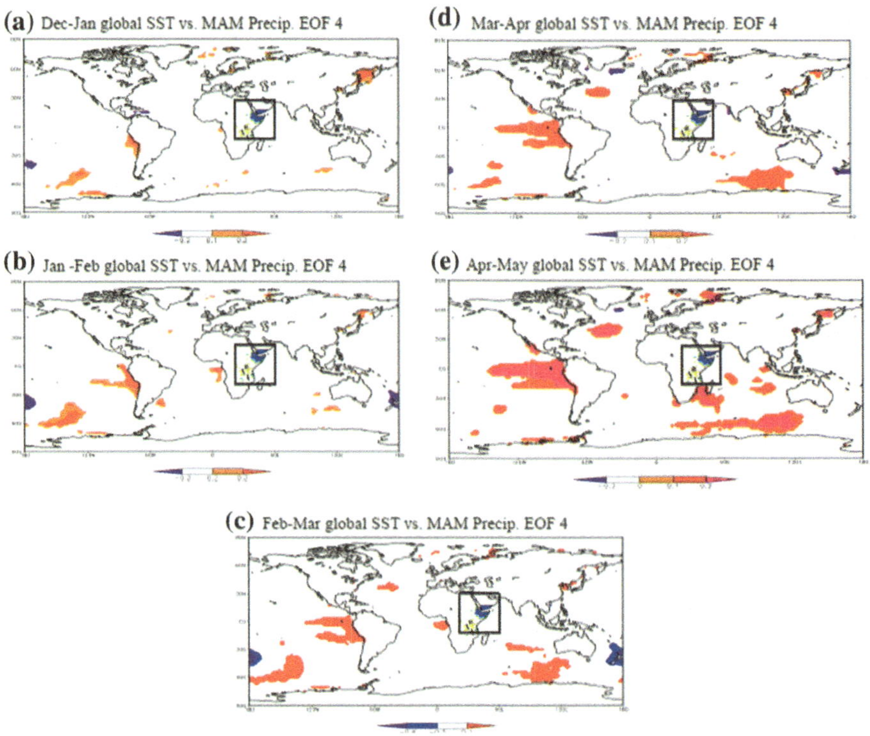

Fig. 5 Relationship between standardized bimonthly global SST anomalies and East African MAM precipitation EOF 4 time series at different time lags. Legend is for the range of correlation coefficient. Significant areas at 95% confidence level, using t-test, are *shaded*. The precipitation spatial patterns over the region are enclosed in *boxes*

With respect to MAM precipitation EOF 1, it was observed that upper level divergence was centered over the equatorial western Pacific from DJ to AM, but became weaker from MA to AM. Correspondingly, upper level divergence was located over Asia spreading to the Indian Ocean in DJ. In JF, there was a shift of the field more eastward over Asia, where it became more intensified. In tandem with this development was the formation of a smaller divergence center specifically located over the Indian Ocean. This signified convective development leading to rainfall over the ocean. However, a weak upper level convergence was found over GHA which suggested transport of weak subsidence anomalies of dry air mass from the upper atmosphere to the surface. This was indicative of a small rainfall deficit. In FM, the divergence center over Asia weakened, but the systems over the Indian Ocean and GHA persisted. In MA, an upper level divergence center, suggestive of Tropical East Jet (TEJ), was found over northern Africa. This co-occurred with a strong divergence center over the Indian Ocean. It appeared that the two centers were associated with a shift of the upper level convergence to the horn area, leading

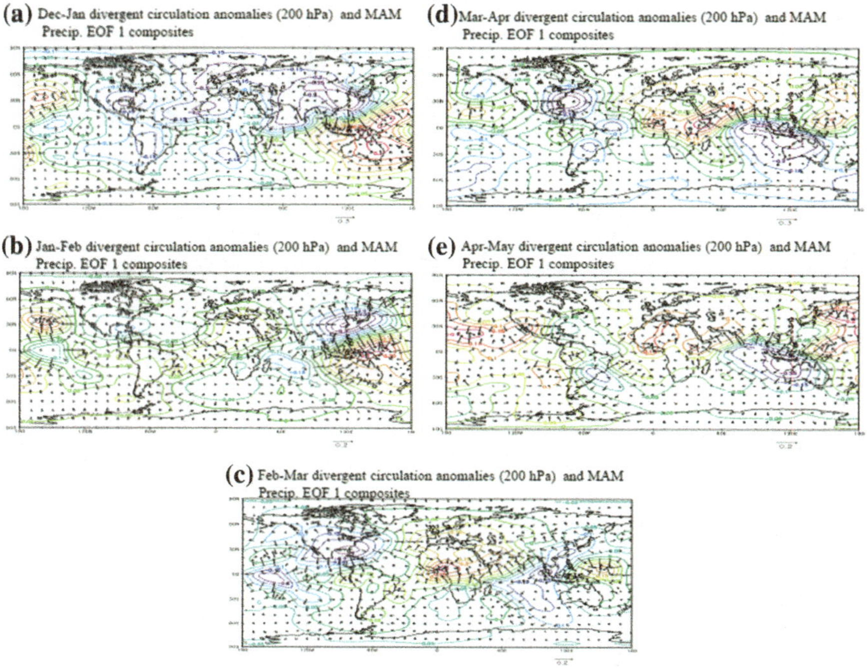

Fig. 6 Standardized bimonthly global divergent circulation anomalies and East African MAM precipitation EOF 1 mode composites at different time lags

to rainfall deficit. In AM, similar conditions persisted, but with disappearance of the TEJ-like feature.

In DJ and with respect to precipitation EOF 3, upper level divergence was located over equatorial western Pacific and over the GHA. There was a corresponding upper level convergence over the Indian Ocean (Fig. 7). Convective development leading to rainfall would occur over the region, in contrast to rainfall deficit over the Indian Ocean. Similar systems persisted from JF to MA, but the divergence center over the Pacific weakened. In AM, re-emergence of the divergent circulation was observed over the equatorial Pacific. While the upper level convergence over the Indian Ocean was maintained, the GHA divergence center disappeared. This was indicative of rainfall surplus (deficit) over the Pacific Ocean (GHA and Indian Ocean).

In DJ, the center of action of upper level convergence was located over equatorial western Pacific and Indian Oceans in relationship to precipitation EOF 4. No upper level divergence was detected over GHA, which was indicative of rainfall deficit. In JF, upper level divergence replaced convergence in the Pacific. This was accompanied by weak (strong) convergence over Indian Ocean (northern Africa), which persisted to FM. In MA, divergence developed over the equatorial central Pacific and the Indian Ocean, but a strong northern African convergence was

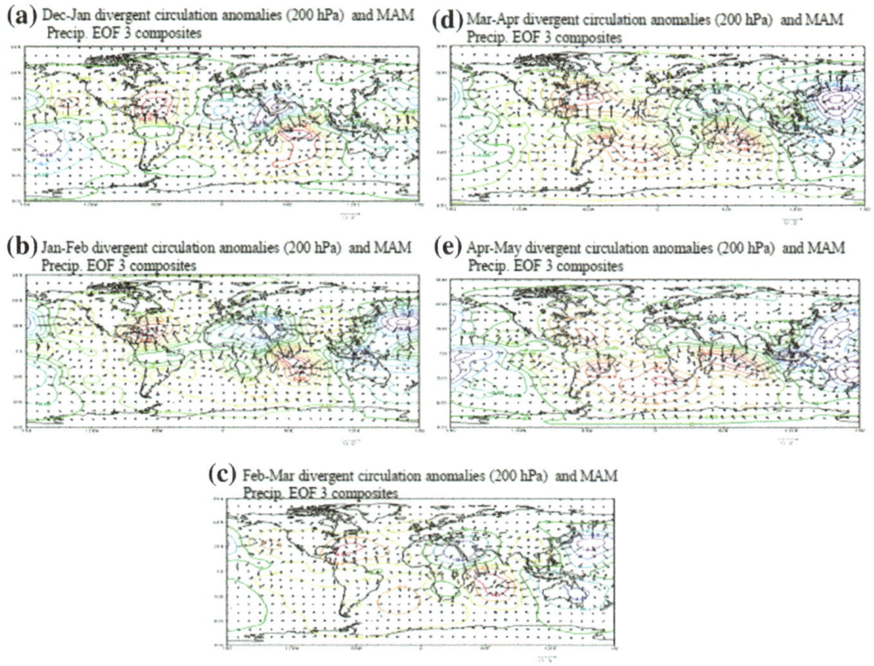

Fig. 7 Standardized bimonthly global divergent circulation anomalies and East African MAM precipitation EOF 3 mode composites at different time lags

observed. Finally, in AM the following observations were made: divergence over the central Pacific, convergent-divergent dipole system over the Indian Ocean, and disappearance of the northern African convergence.

4 Discussion

4.1 Standardized Global-Scale SST Anomalies and Their Associated Upper Level Divergent Circulation in Relationship to MAM Rainfall Modes

The boundary conditions of the ocean surface among other factors drive atmospheric circulations, including rainfall patterns, beyond weather timescale. In the study, four dominant MAM seasonal rainfall modes have identified. The spatio-temporal features were distinct but similar to earlier studies (e.g. Smith and Semazzi 2014). Several factors are known to contribute to MAM rainfall variability. Indeje et al. (2000) reported that local factors over the region played an important role. In a spatially remote sense, ENSO has been found to be the prime driver of the

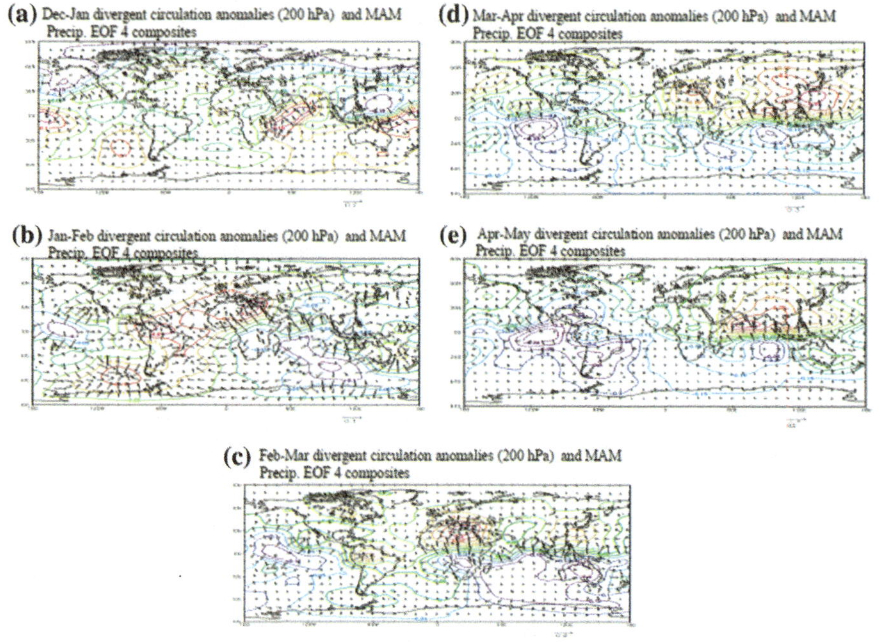

Fig. 8 Standardized bimonthly global divergent circulation anomalies and East African MAM precipitation EOF 4 mode composites at different time lags

interannual variability of the region's rainfall (Nicholson and Entekhapi 1986; Ogallo 1989).

In this study, even though the precipitation temporal patterns display low interannual variance (Ogallo 1989), their relationships with global SST distributions were distinct. In general, the modes responded differently to the Pacific, Atlantic, and Indian Oceans. They also differed with respect to specific oceanic sectors that corresponded to them. In particular, specific Niño regions were identified for each of the three precipitation modes as global ocean climate evolved with time. This underscored the importance of the Pacific, Atlantic, and Indian Ocean SST delineation for every timescale. In fact, the use of the bimonthly climate features has introduced a new picture in the oceanic conditions and their associated atmospheric structures modulating the rainfall variability. Comparison of this study with monthly and seasonal-scale features already studied over the same period (not shown) also supported the rationale of recognizing the important role of the bimonthly climate features in the East African climate system.

This study has revealed that over the Pacific key bimonthly warm (cold) oceanic features that drove or were in association with, the rainfall EOF 1 mode, representing El Niño (La Niña) tended to enhance (suppress) rainfall in GHA. The mechanism, however, was indirect and involved tropical Walker and Hadley circulations. The role of the tropical Atlantic, as with the Pacific, was also critical. It was observed that a warming (cooling) of the tropical Atlantic enhanced (suppressed) convective development over the region. This was pronounced when the Atlantic started warming up in the Northern Hemisphere (NH) spring. The Indian Ocean also played a role to some extent which had similar effects or associations as with the other two oceans.

The poor relatedness between rainfall EOF 2 and global SST distributions needed further attention. In fact, Smith and Semazzi (2014) made a similar observation in their study. This observation warranted verification using other high resolution datasets or numerical modeling studies. The results from Fig. 4 indicated a persistent and a direct AMO-like relationship with precipitation EOF 3. This oceanic mode is an essential part of the Atlantic Meridional Overturning Circulation, which has a great influence on the region's climate. Positive (negative) AMO phase was linked to anomalous wetness (dryness) of GHA, which was consistent with literature. However, the spatial extent of the effects of or associations with, the AMO were not expected to be homogeneous. In contrast, the Pacific as a whole indicated an inverse relationship with the rainfall mode, but the strongest (weakest) events detected in DJ (JF) emphasized the climatological significance of ENSO in NH winter. The competition between AMO-like feature and ENSO was important in further understanding of the region's climate variability. A localized climatic signal detected in AM over southwestern Indian Ocean was indicative of the association of the region's climate with IOD (Saji et al. 1999; Marchant et al. 2006). While ENSO is known to be prominent in NH winter, the direct relationship between precipitation EOF 4 and the Pacific SST (Fig. 5) rather peaked from MA to AM. This suggested that the Niño regions have their specific decay times. The AMO-like feature captured in the North Atlantic had similar role as the one detected for EOF 3.

The standardized divergent circulation anomalies associated with the time-evolving global SST distributions and with respect to the three rainfall modes have been investigated. The importance of global divergent circulations in monsoon dynamics has been reported (Trenberth et al. 2000). In this study, composite analysis revealed that divergence (convergence) and the centers of action at 200 hPa level provided useful information on how the bimonthly atmospheric flows could be applied to explain the precipitation patterns. The distinction of the circulation patterns (Figs. 6, 7 and 8) were based on their strengths, locations, and spatial extents. Diabatic heating is known to provide energy for driving such circulation patterns in the global tropics (Trenberth et al. 2000).

In the proximity of GHA, upper level divergence would lead to cloud formation and consequently precipitation, depending on other factors. Upper level convergence would be associated with transport of dry air mass from the upper atmosphere to the surface, which would indicate drought conditions.

4.2 Implications for Climate Change Adaptations

From this study, the region may be viewed as vulnerable to climate extremes based on the spatio-temporal patterns of the rainfall variability. The specific oceanic features identified could be used in improving the region's seasonal climate prediction. Where the patterns consistently showed extreme dry or wet conditions, it calls for policy-driven adaptation agenda to contain the excesses of climate change. These should be directed to climate-sensitive socio-economic problems such as food insecurity, flood/drought and associated disasters, loss of biodiversity, and climate-driven human diseases. The overall goal could be achievable if end users are prompted of the need to incorporate the outcomes of this study into the existing information required for adaptation strategies.

5 Conclusion and Recommendations

The study primarily focused on analysis of bimonthly global-scale climate features and their relationship to the dominant MAM rainfall modes over the GHA during the climatologically prominent phase of ENSO. EOF analysis isolated four dominant modes, which together contributed to 34.2% of the total explained variance. Showing distinct spatio-temporal patterns, their time series were all characterized by interannual variability. Each of the modes responded differently to the five consecutive two months standardized oceanic anomalies from DJ to AM. Specific oceanic features were noted as having an influence on or having an association with, the rainfall modes. The Pacific Niño regions were the prime factors, though the Atlantic and the Indian Oceans had their share. The associated divergent circulation displayed varied patterns in terms of their action centers, spatial extents and intensities, and these were specific to precipitation EOFs 1, 3, and 4 modes. These gave further guidance in understanding convective development or otherwise over the region.

Traditionally, statistical models rely on seasonal climate features for predicting seasonal rainfall patterns. On the basis of this study, it would be useful if the specific bimonthly predictors were incorporated into the prediction scheme, to enhance climate change adaptations specific to a climatic setting. Rainfall EOF 2 practically did not relate well with the global SST, a situation which required further research. This was substantiated with monthly and seasonal SST analyses over the same period of study. Furthermore, this observation was in consonance with an earlier study. It became clear from this study that the EOF 2 mode was perhaps governed by complex system involving land-ocean-atmosphere feedbacks, which warranted modeling studies. Alternatively, the authors have begun to explore the EOF 2 issue using different datasets. Finally, the current study is being extended by the authors to the short rains.

Acknowledgements The authors wish to acknowledge the following data sources: NCEP/NCAR reanalysis, CRU gridded precipitation data, and NOAA ERSST.

References

Akponikpè, P. B., Ge´rard, I. B., Michels, K., & Bielders. C. L. (2010) Use of the APSIM model in long term simulation to support decision making regarding nitrogen management for pearl millet in the Sahel. *In Journal of European Agronomy, 32,* 144–154.

Barnston, A. G., & Ropelewski, C. F. (1992). Prediction of ENSO episodes using canonical correlation analysis. *In Journal of Climate, 17,* 1316–1345.

Bowden, J. H., & Semazzi, F. H. M. (2007). Empirical analysis of intraseasonal climate variability over the greater Horn of Africa. *In Journal of Climate, 20,* 5715–5731.

Ford, J. D., Berrang-Ford, L., Bunce, A., McKay, C., Irwin, M., & Pearce, T. (2015). The status of climate change adaptation in Africa and Asia. *In Regional Environmental Change, 15,* 801–814. doi:10.1007/s10113-014-0648-2

Funk, C. (2012). Exceptional warming in the Western Pacific-Indian Ocean Warm Pool has contributed to more frequent droughts in Eastern Africa. *In Bulletin of American Meteorological Society, 7,* 1049–1051.

Hannachi, A., Jolliffe, I. T., & Stephenson, D. B. (2007). Empirical orthogonal functions and related techniques in atmospheric science: A review. *In International Journal Climatology, 27,* 1119–1152. doi:10.1002/joc.1499

Harris, I., Jones, P. D., Osborna, T. J., & Lister, D. H. (2014). Updated high-resolution grids of monthly climatic observations—The CRU TS3.10 Dataset. *In International Journal of Climatology, 34,* 623–642. doi:10.1002/joc.3711

Indeje, M., Semazzi, F. H. M., & Ogallo, L. J. (2000). ENSO signals in East African rainfall seasons. *In International Journal of Climatology, 20,* 19–46.

Johnson, R. A., & Wichern, D. W. (2007). *Applied multivariate statistical analysis* (6th ed.). New Jersey, USA: Pearson Education, INC.

Kalnay, E., Kanamitsu, M., Kistler, R., Collins, W., Deaven, D., Gandin, L., et al. (1996). The NCEP/NCAR 40-year reanalysis project. *In Bulletin of American Meteorological Society, 77,* 437–471.

Krishnamurti, T. N. (1971). Tropical east-west circulations during the northern summer. *In Journal of Atmospheric Sciences, 28,* 1342–1347.

Latif, M., Dommenget, D., Dima, M., & Grotzner, A. (1999). The role of Indian Ocean sea surface temperature in forcing East African rainfall anomalies during December–January 1997/98. *In Journal of Climate, 12,* 3497–3504.

Marchant, R., Mumbi, C., Behera, S., & Yamagata, T. (2006). The Indian Ocean dipole—the unsung driver of climatic variability in East Africa. *In African Journal of Ecology, 45,* 4–16.

Nicholson, S. E., & Entekhabi, D. (1986). The quasi-periodic behaviour of rainfall variability in Africa and its relationship to the Southern Oscillation. *In Archiv für Meteorologie, Geophysik und Bioklimatologie, 34,* 311–348.

North, G. R., Bell, T. L., Cahalan, R. F., & Moeng, F. J. (1982). Sampling errors in the estimation of empirical orthogonal functions. *In Monthly Weather Review, 110,* 699–706.

Ogallo, L. J. (1988). Relationships between seasonal rainfall in East Africa and the Southern Oscillation. *In International Journal of Climatology, 8,* 31–43.

Ogallo, L. J. (1989). The spatial and temporal patterns of the East African seasonal rainfall derived from principal component analysis. *In International Journal Climatology, 9,* 145–167.

Ropelewski, C. F., & Halpert, M. S. (1987). Global and regional scale precipitation patterns associated with the El Niño Southern Oscillation. *In Monthly Weather Review, 115,* 1606–1626.

Saji, N. H., Goswami, B. N., Vinayachandran, P. N., & Yamagata, T. (1999). A dipole mode in the tropical Indian Ocean. *In Nature, 401,* 360–363.

Schreck, C. J., & Semazzi, F. H. M. (2004). Variability of the recent climate of eastern Africa. *In International Journal of Climatology, 24,* 681–701.

Smith, T. M., & Reynolds, R. W. (2004). Improved extended reconstruction of SST (1854–1997). *In Journal of Climate, 17,* 2466–2477.

Smith, K. A., & Semazzi, F. H. M. (2014). The role of the dominant modes of precipitation variability over Eastern Africa in modulating the hydrology of lake victoria. *In Advances in Meteorology*, 11. Article ID 516762, http://dx.doi.org/10.1155/2014/516762

Tetteh, I. K. (2012). *West African seasonal climate variability and predictability*, PhD thesis, North Carolina State University.

Trenberth, K. E., Stepaniak, D. P., & Carbon, J. M. (2000). The global monsoon as seen through the divergent atmospheric circulation. *In Journal of Climate, 13,* 3969–3993.

Washington, R., Harrison, M., Conway, D., Black, E., Challinor, A., Grimes, D., et al. (2006). African climate change—Taking the shorter route. *In Bulletin of American Meteorological Society*, 1355–1366.

Wilks, D. S. (2006). *Statistical methods in the atmospheric sciences* (3rd ed.). New York, USA: Academic Press.

Adaptation to Climate Change in Egyptian Marginal Environments Through Sustainable Crop and Livestock Diversification: A Case Study

Hassan M. El Shaer and A.J. Al Dakheel

1 Introduction

One of the major challenges for Egypt Government is to balancing the food needs of fast growing population along with the decrease of severe climate change impacts on agricultural development. Northern Sinai region is of particular strategic, economic and social importance for the security and development of Egypt (El-Beltagy and Madkour 2012). The region is particularly vulnerable to environmental and climatic changes due mainly to local livelihoods dependent on marginal natural resources (SADS 2009; El Shaer 2015). These trends present enormous challenges to achieving food security and eradication of poverty of which are the top priorities on the national political agenda. Rural households engaged in subsistence agriculture and smallholder farmers are most vulnerable. To overcome such challenges, Egypt implemented El-Salam Canal, as a mega national agriculture project (240,000 ha) to create new communities along the canal particularly in northern Sinai region in order to re-charting Egypt's growing population map. From previous studies, farming community is largely composed of traditional small-scale farmers where soil and crop productivity, animal and human health are adversely affected by severe climate changes; it is facing many challenges such sand dune movement, increasing drought, a long hot summer with low and erratic winter rainfall and animal feed shortage (El Shaer 2010; El-Beltagy and Madkour 2012;

H.M. El Shaer (✉)
Animal Production and Rangelands Management, Egyptian Center of Excellence for Saline Agriculture (ECESA), Desert Research Center, 1 Matahaf El Mataria St., P.O. Box 11753, Mataria, Cairo, Egypt
e-mail: hshaer49@hotmail.com
URL: http://www.ecesa.gov.eg

A.J. Al Dakheel
International Center for Biosaline (ICBA), Dubai, UAE
e-mail: a.dakheel@biosaline.org.ae

W. Leal Filho et al. (eds.), *Climate Change Adaptation in Africa*,
Climate Change Management, DOI 10.1007/978-3-319-49520-0_9

Anon 2013). Experiences and skills of settled farmers in improving agriculture and animal productivity practices under these marginal conditions are humble. Such crucial challenges resulted from the impact of climate changes should find proper solutions otherwise the prevailed natural resources and farmers income will be drastically affected (El Shaer 2015). Enhancing plant production through transferring new integrated management approaches using marginal resources will ultimately contribute to the improvement of the livelihood and income of farmers in such regions (Anon 2013; Badran et al. 2013). Therefore, a five-year project was implemented in the region entitled: Adaptation to climate change in marginal environments through sustainable crop and livestock diversification (Anon 2015). The project was mainly targeting the improvement of livelihood and productivity of poor farmers in Sinai who only have access to marginal resources. The paper highlighted, in brief, the main project activities and outcomes related to introduce, evaluate and scale up environmentally forage—livestock systems; disseminate proper technologies for improving forage crops, animal production and capacity building for improving farmers livelihood resiliency to climate changes.

2 Project Objectives

The project aims globally at improving the livelihood of smallholder farmers in saline-affected environmentally areas along El Salam Canal, North Sinai region, Egypt through introducing and adapt new agricultural technologies for economic and sustainable crop-livestock production. Sahl El Tina area, as a severe marginal ecosystem in North Sinai region, was selected for the project implementation.

The main technical objectives of the project could be briefed as follows:

- Develop, evaluate and disseminate high-yielding, salinity-tolerant fodder crops and cereals seeds, cultivars and forage production and livestock production using the introduced integrated management packages (IMP).
- Conducting extensive on-farm evaluation and transfer the adopted technological IMP for the selected forage species (cereals for food and animal feed).
- Socio economic development at farm levels to evaluate the economics of all IPM adopted by the participated farmers.
- Enhancing information dissemination and capacity building for farmers.

2.1 The Main Key Project Activities and Outcomes

The project had been implemented from 2010 to 2015; mainly targeting the improvement of livelihood and productivity, resiliency to climate changes and income of poor farmers relying on marginal water and land resources.

The project outputs comprised of four key integrated components that covered several different achieved activities. They are, briefly, as follows:

1. Farmer based seed production technologies and delivery systems for large scale cultivar adoption developed and transferred.
2. Packages of efficient forage production and utilization suitable for marginal environment improved and disseminated to resources poor farmers.
3. Profitability and impact of the introduced forage-livestock production packages on the livelihood of poor farmers are assessed and quantified.
4. Enhanced women and men farmers skills in seed production, crop management and animal production.

The key project outcomes are briefly highlighted below, based on the achieved activities, as follows:

2.1.1 Description and Characterization of Sahl El Tina Resources

El Tina plain (so-called Sahl El Tina) area, North Sinai region, was selected, herein, to represent the most severe marginal environmental system in the region.

A Base Line Study of the targeted area was conducted to describe and characterize the prevailing natural resources through specific well designed questionnaires and several field visits (Anon 2013).

The general description of Sahl El Tina area indicated that it is located at Al Salaam Canal, on the eastern side of Suez Canal in North Sinai region. It lies in the north-western Mediterranean coast of Sinai, between 32° 350- and 32° 450-E and 31° 000- and 31° 250-N. It has a triangular shape, surrounded by the Suez Canal to the west, the Mediterranean Sea to the north and the northern Sinai sand sea to the south. The north-western corner of the El-Tina plain, south of Port-Fouad and directly east of the Suez Canal, is covered by the El-Malha Lake. It occupies a large triangle area of about 56 km^2 with a 14-km base and up to an 8 km maximum width. The southern part of the El-Tina plain was part of the Nile flood plain in the ancient time; composed of silty clay in-texture with salt crust in several locations. The southern delta plain is composed of muddy deltaic sediments; 1–2 m above sea level.

Irrigation water is obtained from mixed water (Nile water + drainage water) of El Salam Canal. The soil is characterized by severe salt affected, sandy to clay in texture, differs in profile depth and stratified profile layers (sand over clay or clay over sand). Salinity of soil and irrigation water varies between 12.5 and 15.6 dS/m and 1.6 and 2.3 dS/m, respectively (Abdel et al. 2011).

The total size of Sahl El Tina area is around 50,000 feddans; only 10,000 feddans are recently reclaimed and cultivated while the rest of lands are not being reclaimed due to high soil salinity. The project activities had been concentrated on the cultivated lands and owned by small farmers.

Table 1 summarizes the villages of Sahl El-Tina by area, type of investment, and number of settlers. Sahl El Tina comprises of 7 villages as follows: villages 1, 2, and 3 are averaged 5500 feddans and allocated for large investment; villages 4 and 7 are almost 4000 feddans for each and allocated for smallholders and finally villages 5 and 6 are allocated for medium and small holders.

The three Villages 4, 6, and 7 have the largest number of farmers (9000 farmers) which represents about 64% of the total population of Sahl El Tina area. However, only small proportion of such area is cultivated in the villages 1, 2, 3, and 5 with total area of 1700 feddans as shown in Table 1; while approximately half of the area is cultivated in the villages 4, 6, and 7 (Anon 2015).

With regard to gender profile, male properties range from 60 to 75%. High female property proportion is found in the four villages 4, 5, 6, and 7.

Concerning the cultivated lands in Sahl El-Tina area, Table 2 shows the total cultivated lands are amounted to be about 7870 fed. representing around 23% of the total area.

Concerning the crop pattern system in the region, the main winter crops in Sahl El Tina are: Sugar Beet, Wheat, Barley and Berseem which amounted of 3756, 2407, 790, and 748 Feddans, respectively. A small area is cultivated of vegetables (about 300 feddans) since saline soil and mixed water conditions are not convenient

Table 1 Villages of Sahl El Tina by area size, settlers and type of investment

Village	Area size (fed.)[a]	Type of investment	Settlers (No.)
1	5700	Large investments	1500
2	5300	Large investments	1200
3	5000	Large investments	1200
4	4000	Small farms	2500
5	3200	Medium enterprises	1150
6	6500	Small farms and medium enterprises	3000
7	4500	Small farms	3500
Total	34,200	–	14,050

Cited from Anon (2015), [a]a feddan = an acre

Table 2 Distribution of villages of Sahl El Tina by area size and cultivated lands

Village	Area size (fed.)	Cultivated lands (fed.)
1	5700	550
2	5300	200
3	5000	350
4	4000	2300
5	3200	600
6	6500	1870
7	4500	2000
Total	34,200	7870

Cited from Anon (2015)

Table 3 Number of livestock species in Sahl El Tina area

Type of livestock	Number, heads
Buffalos	1147
Cows	1710
Sheep	1077
Goats	351
Camels	210
Other	100

Cited from Anon (2015)

for vegetables cultivation. With respect to animal production, the larger numbers of livestock are cows, followed by buffalos, sheep and goats as shown in Table 3.

Therefore, it could be concluded that there are about 13,000 potential beneficiaries in the area; two-thirds of them are men and the other third are women could be involved in project activities. In addition, the poverty and inappropriate agriculture management practices beside the marginal soil and water resources are the constraints of agriculture development in this area.

Concerning the farmers' selection and participation, farmer's selection criteria was developed mainly to the target productive farmers. Pilot farms size is about 4 ha or less and should be distributed across the selected villages; pilot farmers should have interests to participate in Farmers Field Schools (FFS) activities and have both crop and livestock production activities. The accumulative number of beneficiaries of FFS activities reached about 2227 farmers by the last season of 2013/2014 (Diab 2015).

Concerning the project farmers beneficiaries, there are about 90% of them younger than 50 year old; 30.3% of them are illiterate. About half of beneficiaries (45%) have families with 8 members or more (Anon 2015). All of them are rural socialized; about 90% of them were received their lands before 2005. The majority of them (57%) using modern irrigation system for irrigation.

2.1.2 Enhance Crop Diversification and Identification of Better Adapted Crops to Climate Change Impact

Identifying climate resilient crop varieties and accessions that tolerate the marginal and saline conditions, climate change impact and restricting productivity is a key step in the development of integrated alternative or modified agricultural systems in such marginal environments of North Sinai region (Abdel et al. 2011; Badran et al. 2013). More than 20 forage species were introduced, screened and evaluated to identify genotypes with better stress tolerance and productivity under marginal conditions at both research and farm levels. The most important genotypes received from ICBA and selected through specific selection programs at Research Stations are shown in Table 4 as cited from Anon (2015).

Research station and farmers based field evaluation was carried out to select the best adapted genotypes and identified a list of winter and summer annual and

Table 4 Forage species genotypes received from ICBA and selected

Forage species	Received genotypes	Selected genotypes
Pearl millet	11	4
Sorghum	7	3
Barley	16	4
Safflower	54	15
Triticale	36	4
Fodder beet	16	4
Quinoa	3	3

Cited from Anon (2015)

perennial salt-tolerant forages constituting an integrated forage package that can secure forage resources year-round for multiplication and distribution to farmers (Anon 2013; Khalil 2013).

The selected plant species plants species were different varieties of: pearl millet, sorghum, triticale, barely, fodder beet, sunflower and safflower. Three verities from each of canola, quinoa and pearl millet were tested and produced at large scale at farmers' fields in order to produce great seed production. *Panicum turgedum*, *Kochia indica*, Rye grass, and Sudan grasses were also included among the evaluation and production.

The varieties selected certainly had higher yields that led to their selection by farmers (Badran and Moustafa 2014; Algosaibi et al. 2015).

Two nurseries were established at village No. 4 in Sahl El Tina for training farmers on planting and production of fodder shrubs seedlings then distributed to farmers. These fodder shrubs were: Sesbania, *Medicago arborium*, Kochia and *Acacia saligna*; they were produced and distributed to the participated farmers.

It is important to mention that the main fodder crop species that highly adopted by farmers in summer season were pearl millet followed by sorghum as annual forages and alfalfa as a perennial crop, in orders. The preferred winter annual crops were fodder beet, berseem, barley and triticale. Some fodder shrubs such as *Atriplex nummularia*, *Sesbania* and *Panicum turgedum* are also favorable by framers as feed materials (Anon 2013, 2015).

An integrated field management package (IMP) for seed production enhancement was developed and disseminated among the participated farmers. IMP guidelines were developed for each selected crop and extension performed for wider skill acquiring by farmers (Abdel et al. 2011; Algosaibi et al. 2015; Badran and Moustafa 2015). The IMP includes all aspects of practices for the purpose of improved seed production such as proper planting time, planting density and spacing, care for each crop type and field follow up for maximum seed setting. In field evaluation at farmers field level at the start of the project, the difference in yield between the project team managed fields with IMP practices, and the yield in farmers' fields were reached nearly 40%; while such yields were comparable by the end of the project (season 2014). It means that the skills of participated farmers to get high yields were improved.

Table 5 Farmers participated in seed production technology during the entire period of the project (2011–2014)

Year	No. participated farmers		Total participated farmers
	Direct	Indirect	
2011	16	43	59
2012	17	50	67
2013	23	115	138
2014	21	101	122
Total	77	309	386

Cited from Anon (2015)

The total number of farmers participated in seed production of the tested genotypes were increased from 59 farmers in 2011 up to 122 farmers in 2014. A total of 386 farmers participated in seed production technology during the entire period of the project (2011–2014) as summarized in Table 5 as cited from Anon (2015).

2.1.3 Develop and Disseminate Packages of Forage Production and Utilization

The production and dissemination of package (IMP) of efficient forage production and utilization suitable for marginal environment improved of climate change involved several activities carried out in Sahl El Tina area at research and Farms levels to refine and finalizes the full management practices packages (IMP) as well as improved integrated water management to improve water use efficiency. The objectives of these activities (IMP) were raising the skills and experiences of the participated farmers to enhance their capabilities and attitudes towards increasing the crops yield and to recognize the impact of improved management packages on crop yield, changes of soil salinity and water use efficiency.

Concerning developing and disseminating integrated management packages (IMP) of forage production and utilization technologies, sizable resources and technical efforts were allocated to develop IMP for crop production at the farm level and to develop similar packages for farm products enhancement and livestock products (Algosaibi et al. 2015; Badran and Moustafa 2015). The components of integrated management packages (IMP) included three main parts: (1) Soil management, (2) Irrigation and drainage management and (3) Forage crops field management. In general, the main management practices included: leaching requirements, rates, methods and types of fertilization, tillage, irrigation schedule, intercropping systems, cultivation method and rate and time of seeding. Irrigation using special time schedule for each crop was used to maintain a relatively high level of soil moisture to achieve periodic leaching of the soil. In this respect, identification of yield gaps and management changes/input to reduce the yield gaps was analyzed and the underlying causes were determined and addressed; some key factors in bridging the yield gaps were:

Table 6 The total number of participated farmers in growing the most adopted forage species during the project period

Year	Sorghum		Pearl millet		Egyptian clover		Fodder beet		Total
	Direct	Indirect	Direct	Indirect	Direct	Indirect	Direct	Indirect	
2011	4	16	6	24	2	8	0	0	70
2012	9	36	23	92	13	52	2	8	270
2013	0	0	40	160	30	120	8	32	450
2014	0	0	60	240	76	304	0	0	800
Total participation	13	52	129	516	121	484	10	40	1590

Cited from Anon (2015)

1. farm level management of irrigation and drainage, field management (particularly planting methods and crop care),
2. plant density and spacing between rows, and
3. fertilization programs to compensate for the weaker growth under stress.

The use of field management, proper fertilization regime and planting density led to increase in yield between 30 and 70% compared with the situation without the interventions (Algosaibi et al. 2015; Badran and Moustafa 2015).

The total number of participated farmers (both direct and indirect beneficiaries) in growing the most adopted forage species during the whole project period is summarized in Table 6. It is appeared that the direct and indirect participated farmers started with 70 farmers in winter and summer seasons of 2011 then progressively increased to 800 farmers in 2014 (Anon 2015).

Regardless the farmer's locations, the overall average yield of Berseem, fodder beet, pearl millet, and sorghum were approximately 88, 85, 129 and 90 ton/ha, respectively using the traditional practices while the corresponding values were 92, 92, 132 and 120 ton/ha using IMP in respective orders. Farmers preferred to grow pearl millet than Sorghum because it produced more yield than sorghum crop last years (Anon 2015).

2.1.4 Integrated Management Packages (IMP) for Enhanced Livestock Production in Marginal Environments

Feed availability plays an essential and effective role in livestock development, especially for small holder farmers who rely on crop-livestock systems for their livelihood (El Shaer 2010). In many situations securing fodder crops for animal accounts for up to 70% of the total cost of animal production investment. Therefore, efforts to reduce animal feed costs would help to increase the profits of the production system and farm income. The project focused on developing methods that can be applied at small scale farms to improve the farmer's potential in converting the produced forages into different feed forms that are more storable and with high nutritional value. The selected and adopted forage crops were fed to animals as an individual feed material or mixed with other feed ingredients as fresh materials or processed in different forms such as silage or feedblock. The potential salt tolerant fodder plants were: 1—Pearl millet (*Pennisetum americanum*), 2—Sorghum, 3—Fodder beet (*Beta Vulgaris*), 4—*Panicum turgedum*, 5—Alfalfa (*Medicago sativa*), 6—*Saltbush* (*Atriplex nummularia*), 7—*Kochia indica*, 8—*Leucaena leucocephala* and 9—*Sesbania sesban*. Series of experiments on nutritional and feeding values of the selected forage were conducted where effects on small ruminants growth, physiology, reproduction, intake, production and health were evaluated at research and farms levels (Afaf et al. 2010; Helal et al. 2013; Shaker et al. 2014). It is concluded that berseem, alfalfa, fodder beet, and all tested grasses species are good quality fodder crops. These grasses should be mixed with alfalfa or Leucaena to formulate balanced rations under saline conditions (El Shaer 2010). It is, also, concluded that Pearl millet, *A. nummularia, L. leucocephala and P. turgedum* are characterized by high yields and could have great potential as sources of livestock fodders; they contain enough protein and energy contents to cover their nutritional requirements (Ahlam et al. 2011; Helal et al. 2013).

The project emphasized the on-farm techniques for forage processing and utilization for better storability for enhanced feeding values. The key methods applied were: silage processing, feed blocks processing, covered stack feed manufacturing and biological treatments; these technologies were demonstrated for a total of more than 1500 farmers (Anon 2015).

2.2 *Dairy Processing Unit Establishment*

It is worthy to mention that milk yield was increased in the project areas as a result for increased forages productivity and availability. Value addition to farm products and income generation by rural family was addressed by the project through introducing techniques and methods to enhance the capacity of farmers in milk processing and dairy products (Anon 2015). Therefore, a Dairy Processing Unit (DPU) including Dairy Laboratory was established in village 7 at Sahl El Tina area, to process milk and to determine the hygienic and chemical quality of raw milk and dairy products all along with the value chain—solutions. The DPU is essential for increasing income of rural women as it is a better way to process buffalo, cow and goat milk and to improve of the milk value added. Six training courses for 45 women were held to train rural women on hygienic methods for production of yoghurt, cheese and other products for income generation.

2.2.1 Socio-economic Impact of Improved Production Systems on Farmers' Livelihoods in Marginal Environments

The impact of the integrated technologies and production packages developed and adopted by the farmers on the livelihood of poor women and men farmers in marginal environments were assessed. The economics of the integrated crop management packages introduced to marginal environment were evaluated by collecting socio-economic indicators, production, and soil and water information. The cost and benefits of the forage production packages based on the use of different types of marginal water were analyzed (ELdeep 2014; ELSayd and ELdeep 2014). Demonstration plots of successful production systems were established with full farmer participation, with particular attention to the participation of rural women. Enhancements of the productivity of forage-livestock systems in the project area were quantified and policy strategies recommended. On average, the new varieties and technology packages including management systems and processing harvests increased farmer income by at least 30% (in few cases up to 140%) since yields recorded an increase by the same rate and that added-value of the agricultural product can be enhanced especially through partial transformation to mixed forages, hay and silage or by livestock production as reported by ELdeep (2014). Socio-economic indicators, production, and soil and water information were collected following the scale up and farmers' adoption stage to verify that candidate technologies were profitable, sustainable and value adding. Results on assessment of economic value of cropping systems using salt-tolerant crops, achieved on several farmers' fields increased income by 70% compared to conventional practices (ELSayd and ELdeep 2014). The farmers participating to genotype evaluation through experiments carried out in their lands testified for the superiority of the selected genotypes of pearl millet and sorghum displaying forage yield at least 30% higher than that their conventional crops. Farms survey achieved on 500 farmers' fields showed that introducing of salt-tolerant forage in the farmer's cropping systems improved the farms income by 50% compared to conventional practices (ELdeep 2014). One of the most important project's successes was the rapid increasing of the participants' number who delivered and applied new technologies in growing salt tolerant fodders fed to animals (Anon 2015).

2.3 Adoption Impact Assessment

Development of farmers demonstration; awareness, adoption, and continuation of the innovations provided by IMP from 2010 to 2014 are summarized in Table 7. Regarding new varieties, pearl millet data refers to the cumulative progress of farmers' awareness to reach around 90% of farmers by 2014. Similarly, around half (56.8%) of farmers adopted and (48.8%) were eager to continue cultivating the crops (Anon 2015).

However, progress of farmers' awareness adoption and continuation of fodder beet were estimated at 67.7, 18.1 and 0.00%, respectively. Around two third of farmers became aware of both (77.2%) silage and (68.4%) covered stocks. Close to half of farmers (44.9%) adopted and welling to continue applying silage,

Table 7 The cumulative percentage of awareness, adopters, and continuers of the packages

Year	Pearl millet			Fodder beet			Silage production			Covered stocks		
	Awar	Adop	Cont	Awar	Adop	Cont	Awar	Adop	Cont	Awar	Adop	Cont
Before 2010	9.5	7.4	7.4	2.4	0.4	0.4	20.7	11.6	11.6	17.9	7.3	7.3
2010	36.2	17.7	17.7	20.3	5.7	5.7	41.4	21.4	16.4	33	14	9.8
2011	57.2	26	26	30.5	8.2	8.2	48.7	26	13.6	39	15.8	4.9
2012	84	45.3	37.2	63.5	17.7	11.9	66.7	37.9	25.5	57.9	26.3	15.4
2013	87.4	51.9	43.8	56.6	18.1	0.00	75	54.4	42	66.3	36.5	25.6
2014	87.4	56.8	48.8	67.7	18.1	0.00	77.2	57.2	44.9	68.4	39.6	28.8

Cited from Anon (2015)

Awar awareness, *Adop* adoption, *Cont* continuation

meanwhile, close to one third (28.8%) adopted and welling to continue applying covered stacks (Anon 2015).

2.3.1 Enhance Farmers' and Extension Staff Knowledge and Skills in Farm Management Under Marginal Conditions

Capacity building targeted large scale farmer-oriented seed production and farm management practices through demonstrations, participation, field days, and technology access. The project team implemented comprehensive capacity building program to develop technical knowledge of local staff in breeding/selection techniques of the all crops, and large scale seed production and other forage and livestock production management practices.

3 Conclusions

It is recommended that utilizing fragile ecosystem resources impacted by climate changes in North Sinai region in a more sustainable integrated model by adapting seeds, forage crops to be grown in saline environments integrated with livestock production. It would be an opportunity to contribute to the local socioeconomic development of small farmer's livelihood. It is important to replicate this model in similar marginal regions in Egypt or somewhere else to scale-up and disseminate high yielding forage/livestock production packages better adapted to the marginal environmental conditions.

Acknowledgements The authors, the National Project Principal investigator, and Project Regional Coordinator wishes to express their sincere gratitude and appreciation to the International Center for Bio-saline Agriculture (ICBA), Dubai, UAE and IFAD for their financial supports.

References

Abdel, R. S., Wassif, M. M., Ashour, I. A., & Mousa, A. A. (2011). Effect of using some soil conservation measures of productivity on salt affected soils at El Tina Plain, North Sinai. *Journal of Biology, Chemistry and Environmental Science, 6*(2), 73–94.

Abdou, A. R., Eid, E. Y., El-Essawy, A. M., Fayed, A. M., Helal, H. G., & El Shaer, H. M. (2011). Effect of feeding different sources of energy on performance of goats fed Saltbush in Sinai. *Journal of American Science, 7*(1), 1040–1050.

Algosaibi, A. M., El-Garawany, M. M, Badran, A. E., & Almadini, A. M. (2015). Effect of irrigation water salinity on the growth of Quinoa plant seedlings. Journal of Agricultural Science, *7*(8), 205–214.

Anon. (2013). *Annual report, project "adaptation to climate change in WANA marginal environments through sustainable crop and livestock diversification" IFAD, AFESD, IDB, OFID & ICBA, 2010–2013*. Cairo, Egypt.

Anon. (2015). *Final report: project "adaptation to climate change in WANA marginal environments through sustainable crop and livestock diversification" IFAD, AFESD, IDB, OFID & ICBA, 2010–2014*. Cairo, Egypt.

Badran, A. E., Abd El-Gawad, A. M., & Omar, S. A. (2013). Evaluation of some selected faba bean genotypes grown under drought conditions using some biofertilization treatments. *Egyptian Journal of Plant Breeding, 17*(5), 175–182.

Badran, A. E., & Moustafa, E. S. A. (2014). Drought resistance indices and path analysis in some wheat genotypes. *World Applied Sciences Journal, 30*(12), 1870–1876.

Badran, A. E., & Moustafa, E. S. A. (2015). Genetic parameters of some wheat (*Triticumaestivum* L.) genotypes using factorial mating design. *Journal of Agricultural Science, 7*(1), 101–105.

Diab, M. D. (2015). Learning impact of farmers field schools of integrated crop–livestock system in Sinai Peninsula, Egypt. *Annals of Agriculture Science* (in press).

El Shaer, H. M. (2010). Halophytes and salt-tolerant plants as potential forage for ruminants in the Near East Region. *Small Ruminant Research, 91*(1), 3–12.

El Shaer, H. M. (2015). Land desertification and restoration in Middle East and North Africa (MENA) Region. *Sciences in Cold and Arid Regions, 7*(1), 7–15.

El-Beltagy, A., & Madkour, M. (2012). Impact of climate change on arid lands agriculture. *Agriculture & Food Security, 1*, 3. doi:10.1186/2048-7010-1-3

ELdeep, S. M. (2014). The partial budget analysis for Sorghum farm in Sinai Peninsula, Egypt. *Annals of agricultural science, 59*, 77–82.

ELSayd, A. A., & ELdeep, S. M. (2014). The economic evaluation of extension activities in Forage crops production in Sinai Peninsula, Egypt. *Annals of Agricultural Science, 59*(1), 83–88.

Fayed, A. M., El-Essawy, A. M., Eid, E. Y., Helal, H. G., Abdou, A. R., & El Shaer, H. M. (2010). Utilization of Alfalfa and Atriplex for feeding sheep under saline conditions of South Sinai, Egypt. *Journal of American Science, 6*(1), 1447–1460.

Helal, H. G., Eid, E. Y., Nassar, M. S., & El Shaer, H. M. (2013). Some nutritional studies of four salt-tolerant fodder crops fed to goats under saline conditions in Egypt. *Egyptian Journal of Nutrition and Feeds, 16*(1), 65–77.

Khalil, R. M. A. (2013). Molecular and biochemical markers associated with salt tolerance in some Sorghum genotypes. *World Applied Sciences Journal, 22*(4), 459–469.

SADS. (2009). *Sustainable agricultural development strategy towards 2030*. Agricultural Research & Development Council. Arab Republic of Egypt, Ministry of Agriculture & Land Reclamation. October 2009.

Shaker, Y. M., Ibrahim, N. H., Younis, F. E., & El Shaer, H. M. (2014). Effect of feeding some salt tolerant fodder Shrubs mixture on physiological performance of Shami goats in Southern Sinai, Egypt. *Journal of American Science, 10*(2s), 66–77.

Forestry and Resilience to Climate Change: A Synthesis on Application of Forest-Based Adaptation Strategies to Reduce Vulnerability Among Communities in Sub-Saharan Africa

Vincent O. Oeba and Mahamane Larwanou

1 Introduction

The role of forests and tree-based cropping system on enhancing resilience and adaptation to climate change is increasingly gaining global attention. This is a major shift from what has been traditionally known on the importance of forests to ameliorate climate change through the process of carbon sequestration. Whereas mitigation to climate change is not occupying much of the discussions and attention, especially on developing world, forests and trees present both mitigation and adaptation opportunities whose integration maximizes the potential of reducing vulnerability of the rural poor to climate change. For instance, Reduction of Emissions from Deforestation and Forest Degradation (REDD+) has been recognized as a global mitigation mechanism that has a significant potential to contribute to adaptation by improving local livelihoods, enhancing biodiversity, conservation of ecosystem services and strengthening local institutions. Such adaptation benefits are argued to be realized if local forest people gain access to land and forest resources to enable them derive primary and secondary forest products and services. This requires a good enabling environment such as responsive and effective institutions and governance as well as integration of policies for adaptation and mitigation in forests at local, national, regional and international levels to build resilience through forestry sector (Thompson et al. 2009).

Forests and trees in the context of ecosystem based adaptation (EBA) approach have been also documented to build livelihood resilience in the face of climate

Manuscript submitted for oral presentation during the Symposium on Climate Change Adaptation in Africa Addis Ababa, Ethiopia, 21–23 February 2016.

V.O. Oeba (✉) · M. Larwanou
African Forest Forum, United Nations Avenue, P.O. Box 30677, Nairobi 00100, Kenya
e-mail: V.Oeba@cgiar.org

W. Leal Filho et al. (eds.), *Climate Change Adaptation in Africa*,
Climate Change Management, DOI 10.1007/978-3-319-49520-0_10

change through acting as: safety nets in times of emergency; source of products important for production and income diversification for farm household and rural families and source of employment among others. EBA is known to be cost effective, sustainable and potential in generating environmental, social, economic and cultural co-benefits. This is because EBA puts emphasis on conservation of ecosystems services through sustainable forest management (SFM) and forest adaptation. Therefore support of ecological infrastructure such as forests, mangroves and wetlands among others will significantly contribute to adaptation to climate change. Specifically, forests are often relied upon to: provide food for household or products to sell for survival during times of natural disasters such as floods, droughts or civil unrest; deliver clean water supply as well as protecting important household resources against landslides, erosion and land degradation; and enhance habitat of aquatic and terrestrial animals that provides range of products for household use that improves resilience of poor rural communities in the context of climate change among others (FAO 2013). Overall, building of resilience through forest based ecosystem approach emphasizes adjusting forest management to respond to the negative impacts of climate change and increase adaptation capacity of the vulnerable people to build and maintain resilient landscapes (Thompson et al. 2009).

In Africa, forests and woodland support livelihoods of the majority of the people directly and indirectly through provision of socio-economic benefits such as energy, food, timber, non-wood forest products, and ecological services among others (Chirwa et al. 2015a, b; Lundgren 2015). These benefits contribute to enhancement of resilience to climate change among the vulnerable communities of the African people. For example, Senegal has illustrated on how forestry provides both adaptation through environmental amelioration in farmed landscapes and mitigation through carbon stock enhancement (Kojwang and Larwanou 2015a, b). Also, the manifestation of resilience to climate change resides in the increase use of local knowledge and traditional adaptation strategies where specifically the local people become more aware on uses of a wide range of forest plant species. This enhances possibilities for meeting forest-depend people's needs by substitution of increasingly rare species with those that may become more abundant (Risto et al. 2009).

Notably, the introduction of participatory forest management (PFM) in the context of SFM in sub-Saharan Africa has resulted to multiple benefits to people living adjacent to forests as well as those who depend on rain fed agriculture. Specifically, through PFM activities, the communities adjacent to forests derive timber and non-timber forest products (NTFPs) that are part of income diversification strategies for many poor rural livelihoods facing increased climate variability and change. The use of NTFPs such as firewood, herbal medicines and thatch grass as part of their short term coping strategies and long term diversification of livelihoods under climate variability and change is on the increase (Pramova et al. 2012). The access of such benefits has created environment and culture of tree planting on farm in order to increase access of wood and non-forest/tree products to meets other demands from market industry. This has translated to progressive investment on farm forestry for provision of food, wood fuel and cash income that enhances coping mechanism to climate change among the rural poor communities in sub-Saharan Africa.

However, the role of the role of forests in reducing social vulnerability and building resilience of the ecosystems and people to climate change is rarely documented. Equally, the importance of forests and trees outside for the adaptation of people is not explicitly reflected in a number of policies in sub-Saharan African countries. This implies that forests are not well mainstreamed in the decision making processes on adaptation to climate change where the linkage between livelihoods and forests is strong. The objective of this study was therefore to identify and analyse forest-based adaptation strategies employed in sub-Saharan Africa that have the potential of enhancing resilience of vulnerable communities to the impacts of climate change, with a view of recommending them for up-scale and adoption. Specifically, this paper addresses the following in the context of enhancing resilience to climate change and improved livelihood: role of ecosystem based services in the context of SFM; forestry and food production; agro forestry systems and livelihoods; commercial forestry and economic development; and forestry based institutions and policies supporting adaptation to climate change in Africa.

2 Methodology

The study employed desk review approach in which relevant studies in Africa were mapped and synthesis was carried out to delineate role of forestry in improving resilience to climate change to various forest ecosystems and people. Specific information on the following areas were sought and analysed with an aim of amalgamating contribution of forest sector on building resilience and adaptive capacity to climate change in Africa: ecosystem services in the context of sustainable forest management; forestry and food production; role of agroforestry systems and livelihood; commercial forestry and economic development; and forestry based institutions and policies supporting adaptation to climate change in Africa. The review on ecosystem based approach in the context of SFM was advanced with analysis of Clean Development Mechanisms (CDM) and REDD+ project reports and associated data obtained from United Nations Framework Convention on Climate Change (UNFCCC) and forest carbon portal database as well as other open data sources.

Some of the key of areas of focus in this study, specifically on ecosystem services in the context of SFM and forest based institutions and legal frameworks were validated through a focused group discussion in selected 12 African countries (Table 1).

The respondents were organized into four to five groups and discussed a number of questions but not limited to the following: *What are the economic benefits associated with carbon project in your country for sustainable development? What lessons do you draw from carbon benefit sharing in your country in relation to low*

Table 1 Number of respondents in focused group discussion from different countries

No	Country	Number of respondents
1	Ethiopia	35
2	Zambia	21
3	Niger	34
4	Tanzania	29
5	Sudan	34
6	Zimbabwe	30
7	Kenya	54
8	Burkina Faso	35
9	Togo	33
10	Nigeria	52
11	Madagascar	30
12	Swaziland	30
		417

carbon for sustainable growth? What are the existing institutions, policy and legal frameworks that support forest carbon and sustainable forest management in your respective countries? The discussants presented the findings after group discussion to harmonize the findings. Content analysis approach was used to generate key themes from all responses across the country whose findings are presented in this study.

3 Results and Discussions

3.1 Role of Sustainable Forest Management and Ecosystem Services in Building Resilience to Climate Change

Sustainable forest management (SFM) is one of the novel approaches currently gaining momentum and receiving global attention as well as recognition because of its associated benefits to the society and biophysical systems. In this regard, SFM has become one of the overarching goals for the forestry sector applicable at international, regional, national and sub-regional levels. This is because SFM recognizes forests have important economic, environmental, social and cultural values that are needed in building resilience and adaptive capacity to climate change. As a result, a number of African countries are now developing national forest programmes (NFPs) and other strategies for SFM to address the following areas that will enhance forest resilience and that of the people: conservation and sustainable use of forest resources; conservation of forest biodiversity; improving forest health and vitality; enhancing protective functions of forest resources; strengthening socio-economic functions of forests and enhancing legal, policy and institutional framework.

It was evident from the synthesis of past studies that SFM is reported to have enhanced biodiversity which in return supports tourism industry; genetic resources for economic, social and environmental development; serving as an essential hydrologic regulator through its watershed functions that translates to enhancement of resilience to climate vulnerability (Fig. 1). Specifically, the implementation of sustainable forest management (SFM) in some countries has proved to be effective on management of water resources and reduction of soil degradation translating to increased crop yield and forest productivity. This has directly improved food security and resilience of the rural population to climate change and other external shocks. For example, the implementation of sustainable land management (SLM) practices, techniques and technologies in the context of SFM are currently promising solutions to Sahelian countries in Africa (Maisharou et al. 2015). Some of the notable benefits of SLM practices such as windbreaks include reduction of the velocity of the wind, improvement of microclimate by decreasing water evaporation from soil and plants, protect crops loss due to sand-shear of seedlings and increases the productivity of the crops. The other SLM practice that enhances resilience to climate change among the rural poor in the Sahelian countries is the use of farmer managed natural regeneration (FMNR). This practice does not only enable farmers to protect and manage regeneration of the native trees and shrubs among their crops but also provide alternative sources of livelihoods such as fuelwood, fodder, edible fruits and leaves (Maisharou et al. 2015).

SFM has entrenched the provision of forest ecosystem services such as payment of environmental services (PES) through different schemes in the African continent.

Fig. 1 Significance of SFM in enhancing hydrological and water shed functions in Congo Basin. *Source* Forest Department, Ministry of Lands, natural resources and Environmental Protection of the Government of Zambia, 2011

Some of the notable forest based schemes currently implemented sub-Saharan African countries include REDD+ and forest based CDM. The investments in these schemes are expected improve livelihoods and sustainable growth that is responsive to the impacts of climate change. The study revealed due to funding of REDD+ and CDM offset projects from climate finance, African countries have developed forest based national mitigation and adaptation strategies to secure funds for sustainable development (Oeba and Larwanou 2015). For example, through UN-REDD programme, countries such as Democratic Republic of Congo, Senegal, Tanzania, Kenya, Uganda, Madagascar, Mozambique, Ethiopia, Zimbabwe and Zambia secured funds, piloted and developed REDD+ projects with huge returns. Specifically, through REDD+ offset projects, DRC converted 4220 ha of degraded savannah land into forest plantations for sustainable supply of fuelwood and agricultural crops. This offered opportunity to reduce degradation and deforestation while alleviating poverty through local employment enhancement and community development activities (Fig. 2). The carbon credits generated from this project has also benefited the communities through construction of schools and health facilities providing education to local children and health services, respectively (Romero et al. 2013). Similar benefits have been reported in other countries where such projects have been established.

Fig. 2 Sustainable harvesting from Commercial eucalyptus forest plantation in South Africa. *Source* ICFR, 2010

Of the piloted REDD+ projects, majority of the activities focused on the following: promoting sustainable land use practices in forest adjacent communities; development of alternative livelihoods in forest adjacent communities; building local and national capacity and understanding of REDD+ mechanisms; contribution to national REDD+ strategies and policies and building village-level, local government and civil society organizational capacity towards understanding REDD+ in view of participating in the future, global forest carbon trading. These activities are expected to significantly enhance social and biophysical resilience to climate change.

The key economic and social benefits identified through implementation of REDD+ and forest based CDM project included the following: employment creation among experts and the community members on project formulation and implementation, respectively; improved income as a result of enhanced infrastructure on project sites; and increased investment as a result of income generated from carbon trading. For instance, the Kasigau phase II REDD+ project employed about 100 staff undertaking different duties such forest patrols, tree nursery attendants, cloth sewing in the eco-factory, greenhouse attendant for production of horticultural crops and eco-friendly charcoal producers. Kasigau Phase II REDD+ project also supports construction and equipping of health centres whereas activities promoted by Cameroon estuary mangroves promoted improved smoke stoves that have positive health implications. The social development from carbon offset projects correlates well with those from Asia, Latin America and Caribbean countries where some projects are providing medical insurance, built hospitals or health centres, provided access to ambulances and other related services. In some projects, like the case of Nhambita community project in Mozambique, farmers received direct payments that promoted their rural livelihood, an important characteristic of enhancing resilience to climate change. Similar economic returns from most of the carbon sequestration projects have been reported where local communities receives cash incomes as well as access to non-timber forest products (NTFPs) through forestry activities and adoption of cleaner energy (Jindal et al. 2008; Nyambura and Nhamo 2014; TNRF 2011; Romero et al. 2013). Energy based CDM offset projects are also viewed to enhance green economy and clean energy through improved investment on technology such as efficient cook stoves to reduce pressure on forests, hence enhanced resilience to forestry and people (Fig. 3).

3.2 *Forestry and Food Production in the Context of Enhancing Resilience to Climate Change*

Forests and trees outside forests play a major role in meeting food security and nutritional needs of the African people. Specifically, they contribute to dietary diversity and quality as well as addressing nutritional shortfalls. For instance, fuel,

Fig. 3 Rehabilitation of degraded woodlands in Zambia. *Source* Oeba, 2016

fodder and green fertilizer are essential to food production and nutrition for the poorest and most vulnerable groups such as children, women and marginalized communities. The non-timber forest products and agro forestry systems are important sources of revenue which contribute to food supply. In addition, forest and tree-based systems provide valuable ecosystem services that are essential for staple crop production and that of wider range of edible plants. Some of the notable services provided by forests and diverse tree-based cropping systems with landscape mosaic are the pollination of such important crops that supports livelihoods among the rural and urban dwellers as well as enhancing adaptability to broader range of environmental conditions important to climate change (Bhaskar-Vira et al. 2015).

There is evidence that the impacts of climate change are more pronounced to rural communities who have limited alternatives/adaptive capacity to climate variability as compared to those who live in urban areas. Such rural dwellers in most cases depend on forests products and services for income and nutrition. For instance, in Africa, many rural communities use non-timber forest products (NTFPs) for direct consumption or for trading when agriculture or livestock is affected by climate events. Some of the NTFPs include but not limited to food, fodder, medicines, gums and resins remain major source of income and nutrition to most rural people of Africa (Lundgren 2015).

The agricultural productions in most African countries are rain fed, which depend on the state of forests. This demand for improved forest restoration in order to provide required ecosystem services for enhanced resilience to climate change. The current trends of deforestation and forest degradation has increased vulnerability of forest adjacent communities and society beyond the forests, implying the role forests play adapting to climate change when in a health state. Its therefore vital to enhance SFM to enable forests and tree-based systems to provide a steady supply of wild and cultivated fruit, vegetables, seeds, nuts, oils, roots, fungi, herbs and animal protein, which complement more conventional staple diets derived from agricultural production systems, hence enhance resilience to climate change.

3.3 Role of Agroforestry Systems and Livelihood in the Face of Climate Change

The introduction of agro forestry systems in sub-Saharan African countries is not only enhancing food production but also building carbon stocks on farms resulting to climate change mitigation. In this perspective, agroforestry system acts as a good agent of adaptation and mitigation to climate change. This system enhances resilience to climate change through improved food production because of its ability to increase soil fertility through nitrogen fixation and build of soil organic matter. For instance: the promotion of integrated soil fertility management (ISFM) technologies such as use fertilizer tree species alone and combination with inorganic fertilizer such as *Calliandra calothyrsus*, *Sesbania sesban Tephrosia vogeli*, *Leucaena leucocephala*, *Crotalaria* spp., *Acacia* spp. in eastern and southern Africa; use of traditional woodlots in dryland western regional of Tanzania; traditional systems such as use of *Faidherbia albida* parkland systems in the Sahelian African countries among others improves soil fertility for increased crop yields in most parts sub-Saharan Africa (Chirwa et al. 2015a, b; Mugwe and Mugendi 1999). Overall, the use of such ISFM technologies have increased crop yield resulting to reduced months in a year households purchases food. This is a significant contribution to improved resilience of the rural poor farmers. In addition the benefits obtained from fodder obtained from prunings or harvested from *Calliandra calothyrsus* and other leguminous trees are fed to livestock yielding high milk production, thus diversifying sources of livelihood in household. This builds the capacity of farmers to respond to climate change and variability.

The surplus from the crop yield that is accessed through market channels increases income to the farmers and improves food security to most vulnerable people. It's evident that without such ISFM technologies in the African continent, the situation would be worse in current face of climate change where most of the African countries have demonstrated weak capacity to cope with negative impacts of climate change.

The returns from agroforestry systems practiced by different countries in Africa show significant contributions of tree-based system to abate impacts of climate change. For instance, in Botswana, drought-resistant fruit trees that are rich in vitamins are planted around villages and the trees are able to produce even during drought years, as well as providing an additional income when traditional crops fail due to poor weather. In Senegal, the cultivation of moringa trees that are very drought-resistant and tolerate a wide variety of soil types, they can be used to combat malnutrition by providing enriched food and by treating drinking water. In Shinyanga north of region Tanzania, traditional practices of conservation have been revived by a government initiative. This has encouraged vegetation regeneration and tree planting, 'ngitili' which has proven to help protect the environment, particularly against drought and aridity, and improve the livelihoods of communities in the region. In Zimbabwe, deep-rooted trees are used in agro-forestry operation in order to tap more moisture from a lower depth during the dry season, in order to increase the overall productivity of land. Different crop canopies use light efficiently, and the agro-forestry systems return large amounts of nutrients to the soil, as well as provide shelter against wind erosion. Also in Burkina Faso, afforested areas with Acacia trees protects against drought and aridity as well as provide fire-breaks hence enhancing resilience to climate change (Risto et al. 2009; Zomer et al. 2009; Torquebiau 2000; Torquebiau and Kwesiga 1996) (Fig. 4).

Fig. 4 Forests and diversification of livelihood sources to enhance resilience and adaptation to climate change, Coastal Kenya. *Source* Oeba, 2016

3.4 Building Resilience to Climate Change Through Commercial Forestry and Related Economic Development Initiatives

Forest and woodlands indirectly support economies to adapt to climate change by reducing the costs of climate related negative impacts (Maisharou et al. 2015). In this regard, forest ecosystems are therefore key assets for reducing vulnerability to the effects of climate change. Specifically, commercial forestry remains a source of employment and supports food production and livelihoods, creates employment, contributes to national incomes and supports trade and industry sector in most parts of African countries. For instance, in Niger, forest resources contribute to about 9.5% of the growth domestic product (GDP) in addition to forest resources serving as sources of raw materials for wood and non-wood forest resources. In the Sahelian sub-region of Africa, forestry and agricultural sector provide employment to more than 60% of the active population that has about 50 million people (Maisharou et al. 2015). Overall, the shift to commercial forestry is also largely attributed to increased demands for timber, biomass energy for both rural and urban dwellers, decline of prices for agricultural cash crops in most African countries (Lundgren 2015). For instance, in some of parts of eastern and southern Africa, farmers are earning better income from sale of trees than from crops. This has provided valuable support to households to respond to different challenges associated with climate variability and impacts of climate change.

Venture into commercial biomass energy in various parts of sub-Saharan Africa has not only created employment opportunities, but has increased revenue to the governments. For instance, in Kenya, charcoal industry represents an estimated annual market of USD 425 million and employs over 700,000 people who support over 2 million dependents. In Tanzania, charcoal industry generates an annual value of USD 650 million and creates at least 2 million full-time or part-time jobs to the locals. Similarly, tree-based products such as timber, pulp, poles, and construction posts among others are yielding good income to farmers and saw milling companies/industries (Lundgren 2015). Generally, various forms of commercial in the formal, primary and secondary forest operation activities are rapidly increasing resulting to public-private partnership in forestry based enterprises. Such enterprises have resulted to increased employment and source of income to man households, thus improving their adaptive capacity to climate change.

Overall, commercial forestry shows a significant potential in building resilience among the African people through diversification of sources of livelihood. This is expected to remain in the next century because Africa's population is rapidly increasing as well as urbanization. Whereas, this may pose a challenge on depletion of forest resources that will alter forest resilience, the promotion of SFM will remain the better approach on how to enhance resilience of the African people through forestry. Specifically, investment on forest certification, value addition and putting in place appropriate and effective policy and legal frameworks as well as strong institutions, the forest sector will continue to support African people to cope with impacts of climate change. The increased demand of forest products from Asian countries such as China, India as well as Middle East countries among others provide ready market for the forests produce that will directly increase income to different individuals and institutions in Africa. This will in return enhance diversification of sources of livelihoods strengthening African people with capacity to respond to myriad challenges of climate change now and the future.

3.5 *Forestry Based Institutions and Policies Supporting Adaptation to Climate Change in Africa*

The signing of UNFCCC and ratification of the Kyoto Protocol by most of African countries signaled need by African governments to align existing institutions and policies and develop new ones that will promote adaptation and mitigation programmes to climate change. However, majority of African countries have not developed appropriate institutions, policies and legal frameworks to support global initiatives that are beneficial to the African people. Currently, few countries such as Kenya, Ghana, South Africa, Democratic Republic of Congo, and Senegal are initial stages of developing policies related to climate change and carbon market. For instance, South Africa has developed working paper on climate change policy that outlines some important areas relevant to sustainable forest management, the same with Ghana and Senegal. Kenya has also advanced on developing various strategies, draft climate change policy and climate change bill to mainstream climate change into different sectors. It has also plans of developing REDD+ Strategy to guide its operations in the country to be in line with current draft on National Policy on carbon finance and emission trading. This draft policy aims to guide the setting up of legal, regulatory and institutional frameworks for developing and managing carbon trade for sustainable development. Specifically, the policy targets to: create a carbon sector which will tap into international climate change finances; support sustainable development programmes; provide employment and economic diversification; increase access to innovative research and technology; improve Kenya's balance of payments; and foster involvement of the private sector in carbon investment and trading. Overall, the policy envisions creating an independent carbon trade regulator similar to the Capital Markets Authority to oversee the market, ensuring financial transparency and probity while providing assurance and confidence to global funds, development partners, and private investors.

The establishment of such institutions, policies and legal frameworks will definitely benefit the forest sector through its internationally recognized schemes such as REDD+, CDM, Nationally Appropriate Mitigation Actions, National Adaptation Programmes of Actions (NAPAs) and Nationally Determined Contributions (NDCs) among others. In this regard, national policies should promote forest adaptation into the framework of sustainable forest management, and promote inter-sectoral coordination for linking forest and other sectors in adaptation policies.

This will define new ways of mainstreaming adaptation for forests and forests for adaptation to enhance resilience of ecosystems and people. African governments are therefore encouraged to develop national institutions that will build local governance mechanisms that engage local stakeholders to develop and sustain resilient forest systems that are beneficial to the people. Such institutions should include: social forest groups such as community forest associations; community based forest user groups; environmental youth groups; women based groups; and other groups for marginalized communities to access and use forest resources in sustainable manner. Strengthening formal and informal institutions through suitable ways such as training will encourage them to share information and experiences on forests and trees that end up building social resilience to climate change.

Finally, the afforestation and reforestation programmes have been recognized as NAPAs activities (Kojwang and Larwanou 2015a, b). This is expected to enhance resilience and mitigation to climate change. For instance, many NAPAs projects that include forestry focus on conservation, reforestation and restoration as a tool to sustain local livelihoods and preserve or restore ecosystem services. Some of the notable NAPAs activities in Africa include; Tanzanian NAPA that proposes tree plantations for improving the livelihoods of communities around Mount Kilimanjaro by providing alternative sources of income and food; expansion or establishment of forest protected areas like in Djibouti and Mali; Uganda NAPA includes a community tree-growing; Madagascar NAPA prioritizes transfer of the management of forests to the local level; Gambia and Sudan NAPAs focuses on agroforestry systems; Rwanda NAPA project recognizes the need of forests to maintain hydrological regimes, explaining why activities are centred on conservation and protection of lands against erosion and floods; and Mozambique NAPA projects focuses on coastal protection and islands to contain an activity on mangrove restoration (Thompson et al. 2009).

4 Conclusion and Recommendations

4.1 Conclusion

The identification of various forest-based adaptation approaches demonstrate capability of forestry contributing to enhancement of resilience to various ecosystems and people. In this regard, forest and tree-based ecosystems play a central role in improving Africa's adaptive capacity to climate change through sustainable forest management. Building national and local institutions that can support participatory and responsive decision making processes leading to equitable outcomes will be required in this process. This can only be achieved through sustainable forest management that provides a sound conceptual framework for building resilience. Also the implementation of SFM requires adequate technical knowledge and expertise as well as an enabling institutions, policy and legal framework to address resilience to climate change.

In conclusion, this study was limited forest based adaptation mechanisms and their contributions to improve resilience to the impacts of climate change. It mainly relied on qualitative data from focused group discussions and desk review where studies and publications were synthesized to identify the contributions of forests and trees to climate change adaptation. As a result, it was constrained with adequate collection of more quantitative primary data from appropriate sample to validate some statements from desk review. The number of countries sampled for primary data collection was not adequate due to financial constraints and time, even though many were covered through desk study approach.

4.2 Recommendations

The following recommendations will be valuable in supporting forest sector to enhance resilience to climate change.

4.2.1 Enhance Advocacy on Use of African Indigenous Traditional Knowledge (AITK)

There is a need to invest well on African indigenous traditional knowledge (AITK) to build synergies with scientific knowledge in order to enable rural communities strengthen their capacity to cope with climate change. This is important because with increase of urbanization in Africa, the young people have continued and are more likely to continue loss cultural values associated with land and forest management. This will translate to loss on livelihoods and biological diversity due to lack of codified AITK to the general public which has remained restricted to a few members of African indigenous families increasing vulnerability to inter-generational loss in the face of climate change. This will require creating platforms to enhance advocacy on AITK in order to enable African governments enact policies that will promote indigenous based approaches on improving capacity of forested landscape to provide ecosystem goods and services to the rural poor to cope with impacts of climate change.

4.2.2 Increase Level of Awareness on the Importance of Forests for the Adaptation

There is a need to invest on various platforms that will raise the profile of forestry on climate change adaptation and mitigation. Some of the key platforms will include but not limited to the following: different communication pathways; strengthening advocacy based organization; and learning and research based institutions to catalyze the role of forestry on enhancing resilience to climate change.

4.2.3 Integration of Forest Based Adaptation and Mitigation Strategies as Well as Other Agro-Ecosystem Interventions

The integration of adaptation and mitigation forest based intervention will enhance resilience and improve productivity of any ecosystem to supply the needed resources for better livelihoods. The integration of forest based and agro-ecosystems such as crops, grassland and livestock will increase resilience to climate change forests and wood lands, inland waters and coastal and marine ecosystems for better functionality and provision of essential goods and services to the people to cope with impacts of climate change and variability.

4.2.4 Development of Effective Institutional, Policy and Legal Frameworks

In order to strengthen the role of forest and tree-based systems to support livelihood and enhance resilience to climate change, there is a need to develop effective functional institutions, enact policies and put in place legal frameworks to support sustainable forest management in the face of climate change.

Acknowledgement The authors wish to express their sincere appreciation to the Government of Switzerland through Swiss Agency for Development and Cooperation (SDC) for providing financial support to implement African forests, people and climate change project in Africa that resulted to development of this paper. The authors also extend their sincere appreciation to Executive Secretary, of African Forest Forum (AFF), Prof. Godwin Kowero for the stewardship of the Secretariat and technical backstopping on the implementation of the project. The authors also acknowledge the support received from African governments where trainings were conducted and collection of primary data. Finally, feedback review provided that shaped this paper is highly appreciated.

References

Bhaskar, V., Christoph, W., & Stephanie, M. (Eds.). (2015). Forests, trees and landscapes for food security and nutrition. A Global Assessment Report. IUFRO World Series Volume 33. Vienna. 172 p.

Chirwa, P. W., Larwanou, M., Syampungani, S., & Babalola, F. D. (2015a). Management and restoration practices in degraded landscapes for Eastern Africa and requirements for upscaling. *International Forestry Review, 17*(s3), 20–30.

Chirwa, P. W., Larwanou, M., Syampungani, S., & Babalola, F. D. (2015b). Management and restoration practices in degraded landscapes of Southern Africa and requirements for upscaling. *International Forestry Review, 17*(s3), 31–42.

FAO. (2013). Forests, food security and gender: Linkages, disparities and priorities for action. *Background paper for the International Conference on Forests for Food Security and Nutrition*, FAO, Rome.

Kojwang, H. O., & Larwanou, M. (2015a). forestry-related input into relevant policies at regional and global levels: An African perspective on climate change. *International Forestry Review, 17* (s3), 92–102.

Kojwang, H. O., & Larwanou, M. (2015b). An overview of the nationally appropriate mitigation actions (NAMAs) and adaptation programmes of action (NAPAs) in Africa. *International Forestry Review, 17*(s3), 103–113.

Lundgren, B. (2015). *Forests and trees: Their roles and opportunities in Africa's economic development, food security and environmental health*. Nairobi, Kenya: African Forest Forum.

Maisharou, A., Chirwa, P. W., Larwanou, M., Babalola, F., & Ofoegbu, C. (2015). Sustainable land management practices in the Sahel: Review of practices, techniques and technologies for land restoration and strategy for up-scaling. *International Forestry Review, 17*(s3), 1–19.

Mugwe, J. N., & Mugendi, D. N. (1999). Use of leguminous shrubs for soil fertility improvement in sub-humid highlands of Kenya. In J. S. Tenywa, J. Y. K. Zake, P. Ebanyat, O. Semalulu, & S. T. Nkalubo, (Eds.), *Proceedings of the 17th Soil Science Society of East Africa conference* (pp. 194–198). Kampala, Uganda, 6–10 September 1999

Nyambura, B., & Nhamo, G. (2014). CDM projects and their impacts on sustainable development: A case study from Kenya. *Environmental Economics,* 5(1).

Oeba, V. O. & Larwanou, M. (2015). Opportunities and prospects of forest based CDM and REDD + projects for sustainable development in Sub-Saharan Africa. *Oral paper presented during fifth conference on climate change and development in Africa (CCDA-V); Africa sustainable development and climate change; prospects of Paris and beyond*, Victoria Fall, Zimbabwe, October 28–30, 2015.

Pramova, E., Locatelli, B., Djoudi, H., & Somorin, O. A. (2012). Forests and trees for social adaptation to climate variability and change. *WIREsClim Change, 3,* 581–596. doi:10.1002/wcc.195

Risto S., Alexander B., & Pia K. (Eds.). (2009). Adaptation of forests and people to climate change. A Global Assessment Report. P. IUFRO World Series Volume 22. Helsinki. 224 p.

Rohit, R., Swallow, B., & Kerr, J. (2008). Forestry-based carbon sequestration projects in Africa: potential benefits and challenges. *Natural Resources Forum, 32,* 116–130.

Romero, M. Z., Traerup, S., Wieben, E., Moller, L. R., & Koch A. (2013). Economics of forest and forest carbon projects-translating lessons learned into national REDD+ implementation. Technical report.

Tanzania Natural Resource Forum (TNRF). (2011). REDD realities: Learning from REDD pilot projects to make REDD work.

Thompson, I., Mackey, B., McNulty, S., & Mosseler, A. (2009). Forest resilience, biodiversity, and climate change. A synthesis of the biodiversity/resilience/stability relationship in forest ec retariat of the Convention on Biological Diversity, Montreal. Technical Series no. 43, 67 pages.

Torquebiau, E. (2000). A renewed perspective on agroforestry concepts and classification. *Comptes rendus de l'Académie des Sciences/Life Sciences, 323,* 1009–1017.

Torquebiau, E., & Kwesiga, F. (1996). Root development in a Sesbania sesban—maize system in Eastern Zambia. *Agroforestry Systems, 34,* 193–211.

Zomer, R. J., Trabucco, A., Coe, R., & Place, F. (2009). Trees on farm: Analysis of global extent and geographical patterns of agroforestry. Nairobi: World Agroforestry Centre, ICRAF: Working Paper No 89.

Health Vulnerability and Adaptation Strategies to Climate Change in Ethiopia

Belay Simane, Wakgari Deressa, Abera Kumie, Adugna Woyessa, Mirgissa Kaba, Girma Taye and Getachew Berhane

1 Introduction

The climate in Ethiopia is changing and projections suggest that the rate of change will increase in the future. Mean annual temperature has increased by 1.3 °C between 1960 and 2006, an average rate of 0.28 °C per decade (Conway and Schipper 2010; CRGE 2011). Daily temperature observations show increasing frequency of both hot days and hot nights. Climate models suggest that Ethiopia will see further warming in all seasons of between 0.7 and 2.3 °C by the 2020s and of between 1.4 and 2.9 °C by the 2050s (Conway and Schipper 2010).

Climate change is currently adversely impacting the health and lives of people around the world, particularly in low-income countries like Ethiopia (Patz et al. 2005; Portier et al. 2010). There are several mechanisms in which climate change impacts on health (WHO 2003). However, two main climatic impacts on health are evident from literatures: direct effect because of heat stress and weather related extreme events that results in increased morbidity and mortality, and an indirect effect that is climate-mediated change in the incidence of infectious diseases and deaths. Some impacts occur relatively immediately, while others depend on a succession of changes that may occur incrementally.

B. Simane (✉) · W. Deressa
College of Development Studies, Addis Ababa University, Addis Ababa, Ethiopia
e-mail: simaneb@yahoo.com; belay.simane@aau.edu.et

A. Kumie · M. Kaba · G. Taye
School of Public Health, Addis Ababa University, Addis Ababa, Ethiopia

A. Woyessa
Ministry of Health, Addis Ababa, Addis Ababa, Ethiopia

G. Berhane
College of Natural Sciences, Addis Ababa University, Addis Ababa, Ethiopia

W. Leal Filho et al. (eds.), *Climate Change Adaptation in Africa*,
Climate Change Management, DOI 10.1007/978-3-319-49520-0_11

The common direct health sector weather variability and climate change effects of human health in Ethiopia are also morbidity and mortality due to vector-borne infectious diseases like malaria, trypanosomiasis, onchocerciasis, schistosomiasis and leishmaniasis including the 2013/2014 phenomenon of yellow fever and dengue fever outbreak for the first time. Climate change is also directly affecting food and nutrition security, undermining current efforts to address under-nutrition. In Ethiopia, malnutrition particularly under nutrition is one of the public health problems. The major problems are protein-energy malnutrition and micronutrient deficiencies such as vitamin A, iron, and Iodine. These are responsible for morbidity and mortality among the general community although children and women are affected most (MOH 2014).

In Ethiopia, despite wide recognition of the impact of climate change on health, there is scanty information on the implication of climate variability on health making it difficult to develop evidence based adaptation strategy. The aim of this 'health vulnerability and adaptation to climate change assessment' is to assess climate variability and its impact on health in Ethiopia and develop national adaptation strategy aimed at reducing vulnerability in the near term by implementing programs that improve basic public health measures.

The objective of this study was to assess vulnerability of population and the health system in Ethiopia to the impacts of climate related risks and weather variability, and in response, develop appropriate national health adaptation strategies. The goal of this assessment is to contribute to the Ethiopia's Climate Resilient Green Economy (CRGE) strategy through developing the country's health sector NAP based on the long term data on health impacts and vulnerability to climate variables.

2 Methodology

Monthly mean weather data (maximum and minimum temperature and rainfall) for 19 years (1996–2014) were obtained from the National Meteorological Agency of Ethiopia (NMAE) to demonstrate the association of some of the epidemic diseases and climate in the country. In order to get health data and information comprehensive literature search or a systematic approach to literature retrieval from relevant sources was carried out using electronic, print or both sources. Data on some climate sensitive diseases with significant public health importance such as malaria, diarrheal diseases, meningitis, schistosomiasis and leishmaniasis were also collected. Data on human resources for health, health facilities and other infrastructures were also obtained from the health and health related indicators and ESPA+

Assessment of vulnerability to climate change mainly involves the exposure, sensitivity and adaptive capacity levels of a system in the presence of a specific impact e.g., rising frequency of climate sensitive diseases like malaria. The basic approach for the macro level study has been to compare the vulnerability index values of all the regions and identify the most vulnerable regions. The methodology places multiple indicators under the broad umbrella of the three factors which define vulnerability: exposure, sensitivity and adaptive capacity (Table 1).

The Health Vulnerability Index (HVI) assessment was adopted from the Livelihood Vulnerability Index (LVI) of the IPCC framework (HVI-IPCC) (Hahn et al. 2009; Simane et al. 2014). Regions are described in terms of their natural capital, social capital, financial capital, physical capital, and human capital. The HVI reorganizes determinants of health into new categories, which includes an explicit climate component, and is framed in a manner amenable to the use of secondary data from published documents. The HVI-IPCC maps the HVI components onto the three IPCC contributing factors to vulnerability—exposure, adaptive capacity, and sensitivity (Table 1). It used a balanced weighted average approach where each sub-component contributes equally to the overall index even though each major component of different livelihood assets includes a different number of sub-components. This weighting scheme could be adjusted by future users as needed.

2.1 Calculating the HVI

Because each of the sub-components is measured on a different scale, it was first necessary to standardize each as an index to a common scale. The equation used for this conversion was adapted from that used in the Human Development Index to calculate the life expectancy index, which is the ratio of the difference of the actual life expectancy and a pre-selected minimum, and the range of predetermined maximum and minimum life expectancy (UNDP 2007). Some sub-components such as the 'number of health professionals' are assumed to decrease vulnerability. In other words, we assumed that a region with more health professionals is less vulnerable than a region with less number. By taking the inverse of the crude indicator, we created a number that assigns higher values to regions with a lower number of health professionals. The maximum and minimum values were also transformed following this logic and Eq. (1) used to standardize these sub-components.

$$I_v = \frac{I_a - I_{min}}{I_{max} - I_{min}} \quad (1)$$

where

- I_v is the standardized value for the indicator,
- I_a is the value for the indicator I for a particular Region a,
- I_{min} is the minimum value for the indicator across all the Regions, and
- I_{max} is the maximum value for the indicator across all the Regions.

After each was standardized, the sub-components were averaged using Eq. 2 to calculate the value of each major component:

Table 1 Vulnerability factors, health determinants, profiles, and indicators used for HVI analysis using the IPCC framework

Vulnerability factors	Health status determinants	Profiles	Indicators	Units of measurements
Exposure	Climate	1. Climate	• Change in temperature • Change in precipitation	• Changes over time, °C • Changes over time, mm
		2. Hazard	• Occurrence of extreme events (drought +floods)	• No of population supported with PSPN • No of events and affected population over the last 20 years
Sensitivity	Natural capital	3. Ecosystem/geographic	• Suitability of the area for the CC sensitive diseases	• % of the area prevalent to CC sensitive health issues
		4. Demography	• Proportion of population who are vulnerable (young children, women and elderly)	• % of young children, women and elderly, exposed work force • % HHs in the exposed area
Adaptive capacity	Financial capital	5. Wealth (health care financing)	• Health care financing	• Wealth profile • Per capita government expenditure on health • Percentage budget of national budget allocated to health • Per capita government expenditure on health
	Physical capital	6. Technology and medicine	Critical systems, infrastructure and equipment safety	• Status of health facility systems such as electrical, telecommunication, water supply, waste management, fuel storage, medical gases, ventilation, equipment and supply, access
		7. Infrastructure	Health care	• Physical infrastructures status • No and type of health facilities • Health coverage • Safe water coverage and trend • Latrine coverage and trend

(continued)

Table 1 (continued)

Vulnerability factors	Health status determinants	Profiles	Indicators	Units of measurements
			Water and sanitation	• Physical infrastructures status • No and type of health facilities • Health coverage • Health care waste management
	Human capital	8. Community	Human resources for health	• Health professional (doctors, nurses, midwives) proportion per population by geographic area • Number of health extension workers per 5000 by admin unit • # of HDA (Health development army) per 5 HHs
	Social capital	9. Social	Social determinants of health and nutrition	• Male no education (%) • Female no education (%) • Safe water coverage (%) • Latrine coverage (%)

$$\boldsymbol{P_a} = \frac{\sum \boldsymbol{I_v}}{\boldsymbol{N}} \tag{2}$$

where P_a is the value for the profile in Region a and N is the number of variables in the profile.

Values for each of the nine components were then combined to obtain the Region level HVI using Eq. 3:

$$\boldsymbol{HVI_a} = \frac{\sum_{p=1}^{8} \boldsymbol{N_p P_{a,p}}}{\sum_{p=1}^{8} \boldsymbol{N_p}} \tag{3}$$

where HVI_a is the Health Vulnerability Index for Region a and N_p is the number of indicators in each profile.

The nine profiles are then combined according to the IPCC categorization scheme using Eq. 4:

$$CF_a = \frac{\sum_{p=1}^{f} N_p P_{a,p}}{\sum_{p=1}^{f} N_p} \tag{4}$$

where CF_a is an IPCC contributing factor [exposure (E), sensitivity (S), or adaptive capacity (A), f is the number of profiles associated with the contributing factor, and p is indexed to the profiles associated with the CF.

Finally, the HVI for Region a is calculated using Eq. 5:

$$HVI - IPCC_a = (E_a - A_a) * S_a. \tag{5}$$

The HVI-IPCC is scaled from −1 (least vulnerable) to 1 (most vulnerable) and is best understood as an estimate of the relative vulnerability of compared populations.

3 Results and Discussions

3.1 Trends in Weather/Climate Data

Ethiopia is a large complex country, with complex patterns of rainfall and livelihoods. In Ethiopia, higher elevations receive more rainfall than low arid areas and support agricultural livelihoods and higher population densities whereas the lowlands like receive minimal rainfall, and people generally support themselves by raising livestock (Fig. 1). The seasonality of rainfall varies in different areas of Ethiopia. In the eastern Somali region, rains come twice a year–during the March–June Belg season, and during the October–December Dryer season. In the south-central part of the country, most areas receive both Belg and summer

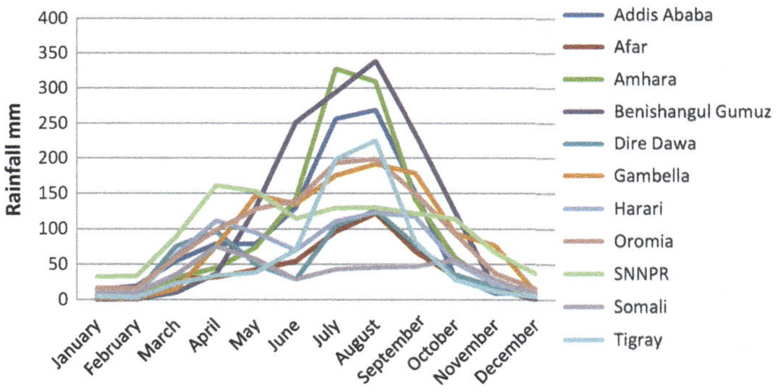

Fig. 1 Long-term monthly average rainfall distribution by region (1996–2014)

(June–September) Kiremt rains. Many farmers plant slowly maturing but high yielding 'long cycle' crops that grow during both the Belg and Kiremt seasons.

Ethiopia receives most of its rain between March and September. Rains begin in the south and central parts of the country during the Belg season, then progress northward, with central and northern Ethiopia receiving most of their precipitation during the Kiremt season. Rainfall totals of more than 500 mm during these rainy seasons typically provide enough water for viable farming and pastoral pursuits.

Annual rainfall trend analysis for the years 1996 to 2014 for all regions decreased by 4 (SNNP)–20 mm (Gambella) per season across country except Harari. Spatial patterns and changes in maximum and minimum temperature have been analyzed based on region-averaged maximum and minimum temperature data obtained from NMA for the period 1996–2014. Mean annual temperature has increased by 1.3 °C between 1960 and 2006, an average rate of 0.28 °C per decade. Warming has occurred across much of Ethiopia, particularly since the 1970s at a variable rate but broadly consistent with wider African and global trends (Conway and Schipper 2010). Daily temperature observations show significantly increasing trends in the frequency of hot days, and even higher increasing trends in the frequency of hot nights. Climate models suggest that Ethiopia will see further warming in all seasons of between 0.7 and 2.3 °C by the 2020s and of between 1.4 and 2.9 °C by the 2050s (CRGE 2011).

3.2 Vulnerability Assessment

Assessing the various components that contribute to climate change vulnerability is an important part of adaptation planning, and one of the first steps. Vulnerability assessments can assist in (1) determining the extent that climate change is likely to damage or harm a system and (2) adapting to the impacts of climate change. They provide hence a basis for identifying the most appropriate adaptation options. Vulnerability assessments are also important as they can provide evidence of the linkages between climate and development, improve understanding of specific risks and vulnerabilities in different localities, provide the opportunity for capacity building, and serve as a baseline analysis to monitor how risks may be influenced by a changing climate over time.

This national assessments of the potential effects of climate change on human health to better understand current vulnerability and to evaluate the country's capacity to adapt to climate change by modifying the health adaptive capacity or by adopting specific measures of public health. In addition, more national information and assessments are needed to feed into the international policy processes, such as that of the national communications to the UNFCCC and those from other climate change assessments.

The HVI analysis considers the three IPCC contributing factors to vulnerability —exposure, adaptive capacity, and sensitivity as major factors, the 9 profiles that determine the three major components in detail for the analysis as explained in the previous sections. The standardized values were used to calculate and explain the major factors' values only.

3.2.1 Exposure

The exposure component encompasses two profiles: climate and Hazard. The climate profile includes the annual average rainfall and temperature (both maximum and minimum). The hazard profile takes into account Average Emergency Beneficiaries (1997–2006 EC) and distribution of population in the climate sensitive disease exposed area by Region.

The combined standardized climate profile values ranges from 0.09 to 0.96 while the combined standardized Hazard profile values ranges from 0.04 to 0.58 (Fig. 2). The final values for the exposure levels ranged from 0.12 (AA) to 0.69 (Gambella). The analysis shows that Amara, Tigrai, Harari and SNNP regions are moderately exposed. A combination of moderate variation in climate and medium values in hazard profile are the probable reasons. The lowland emerging regions Afar, Gambella, Benishangul and Somali regions are highly exposed to changes to climate change.

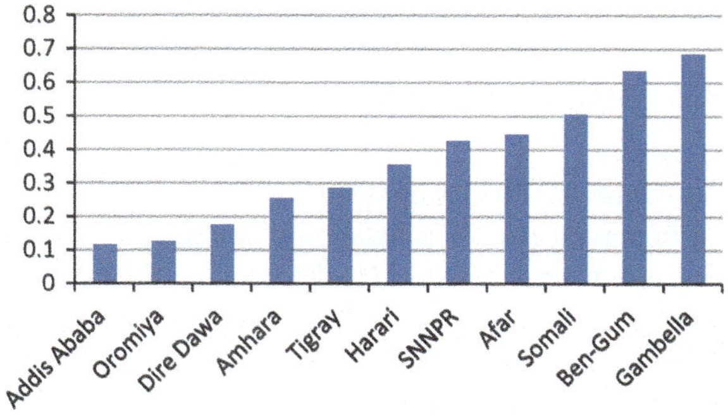

Fig. 2 Average standardized index values of exposure by region

3.2.2 Sensitivity

Sensitivity reflects the degree of response to a given shift in climate. As a result the area that are favorable to malaria, vulnerable population (Dependency Ratio) and the proportion of households living in a climate sensitive disease prone areas (malaria) are grouped under the sensitivity component (Table 15). As evident from the average sensitivity value, all the regions are highly sensitive to climate induced diseases with the exception of Addis Ababa (Fig. 3).

3.2.3 Adaptive Capacity

Adaptive capacity is the theoretical ability of a region or community to respond to the threats and opportunities presented by climate change. It is affected by a number of factors. It encompasses coping capacity and the strategies, policies and measures that have the potential to expand future coping capacity. Adaptive capacity denotes the capacity to cope up with the changes and adapt to changing conditions. It is dependent on several socio-economic factors such as finance, infrastructure, available health professionals and social determinants of health (Table 1).

The financial component values show that most regions have very low capacity as the values range from 0.63 (Dire Dawa) to 0.20 9 (Afar). Higher adaptive capacities are seen only in the main urban regions (Harari and Dire Dawa). Adequate funds are needed to maintain core health system functions, including in the case of a crisis. In addition to providing funds for core health and public health services (water/sanitation/environmental hygiene/disaster and health emergency preparedness), it is necessary to plan for insurance or replacement costs for health facilities and equipment lost or damaged due to extreme weather events.

Provision of infrastructure, facilities and services are highly interrelated and require increase at the same rate or higher to the level of population growth in order to provide the minimum health service package to the community. Health service delivery should combine inputs to provide effective, safe, good-quality health interventions in an efficient and equitable manner. In terms of physical facilities all the urban regions (AA, DD and Harari) have the highest indexes (more than 0.5). Addis Ababa has the highest physical capital (0.73). The three major regions have moderate values (Amhara (0.57), SNNPR (0.45) and Tigray (0.53)). Oromia and

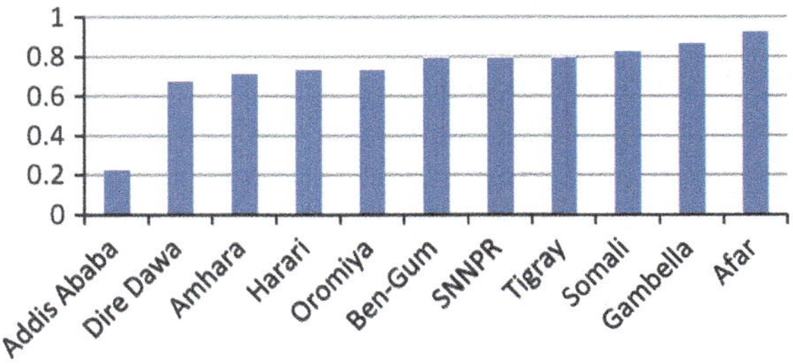

Fig. 3 Average standardized index values of sensitivity by region

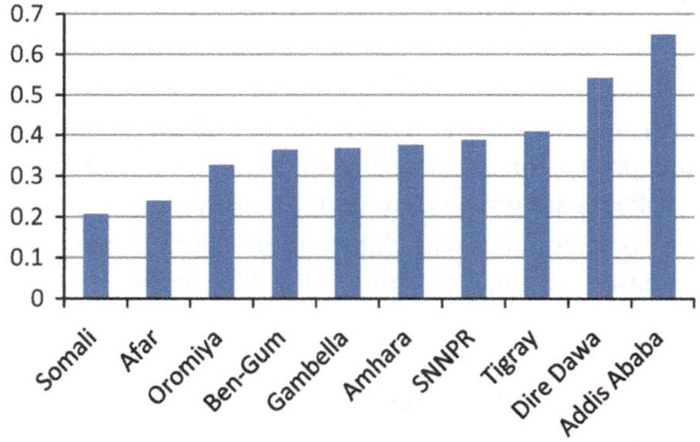

Fig. 4 Average standardized index values of adaptive capacity by region

the other emerging regions relatively low physical resource indexes. A range of medical products and technologies are needed to protect populations from climate-sensitive health conditions. These include medical equipment and supplies for emergency response, permanent and emergency health facility services, and technologies in health-supporting sectors such as water and sanitation and environmental hygiene.

A well-performing health workforce is needed to achieve the best health outcomes possible in view of climate change. This includes sufficient numbers and a mix of qualified, competent and productive staff to address all the climate sensitive health problems. In terms of human resources all regions have very low indexes with the exception of Harari. This is because of the low population number it has.

Social determinants included are demography, education and food security status of the respective regions. The standardized values ranged from 0.12 (Somali)—0.82 (AA). While the three urban regions have relatively better values, the four big regions have medium values. Afar and Somali have the least values. The average adaptive capacity values range from 0.21 (Somali) to 0.65 (AA) (Fig. 4). The finance, infrastructure, human and social profiles are found to be strong determinants of adaptive capacity. Addis Ababa and Dire Dawa regions have the highest adaptive capacity. While Somali and Afar regions have the lowest adaptive capacity, the rest do have intermediate capacities.

3.2.4 Health Vulnerability Index

The final vulnerability indexes for the regions have been calculated by combining all the three components of exposure, sensitivity and adaptive capacity. The results produce measures of exposure, sensitivity, and adaptive capacity, which all differ systematically across regions (Table 17). Based on the equations of LVI-IPCC (Eq. 5), it is evident that high values of exposure relative to adaptive capacity yield positive vulnerability scores (Simane et al. 2014). On the other hand, low values of exposure relative to adaptive capacity yield negative vulnerability scores.

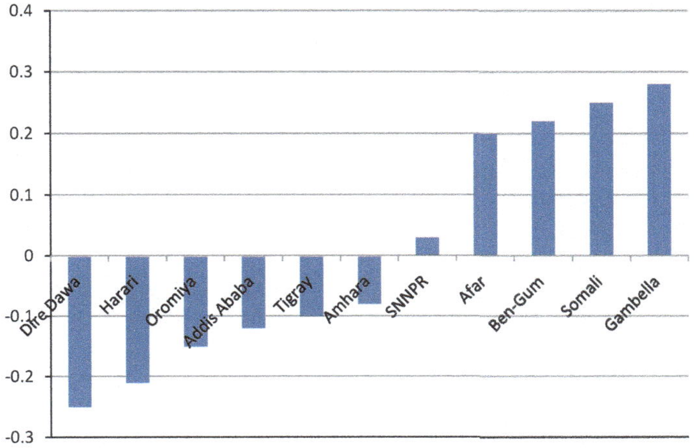

Fig. 5 Calculated indices for the health vulnerability index

Sensitivity acts as a multiplier, such that high sensitivity in a region for which exposure exceeds adaptive capacity will result in a large positive (i.e., high vulnerability) HVI-IPCC score. The values lie between −1 and +1. The lesser the value, the lesser is the vulnerability of the region.

The calculated HVI values ranged from −0.247 (Dire Dawa, less vulnerable) to 0.279 (Gambella, highly vulnerable) (Fig. 5). The final values are divided into four groups for ease of understanding the results for the policy makers in the health sector. The two urban regions (Dire Dawa and Harari) having less than −0.2 values are categorized as least vulnerable. Oromia, Addis Ababa, Tigray, and Amara regions with a value of −0.2 to 0.0 medium vulnerable. SNNP with a value of 0.033 is categorized as highly vulnerable and those with greater than 0.1 (Afar, Somali, Gambella and Benishangul Gumuz) are very highly vulnerable.

All the emerging regions are very highly vulnerable to the impacts of climate change sensitive health issues. This result indicates that an adaptive capacity is in deficit and high exposure relative to other regions. The opposite is true for the less vulnerable regions, in which adaptive capacity exceeds exposure and overall vulnerability is deemed to be low.

Putting in geographical context, the results of estimating HVI concludes that 0.6% of the total land mass (Harari and Dire Dawa) only is relatively least vulnerable to climate change sensitive diseases (Fig. 3). 49.95% of the land mass (Oromia, Addis Ababa, Amhara and Tigrai) are relatively moderately vulnerable. While 10.35% (SNNP) of the land mass is categorized as highly vulnerable, the rest 39.5% of the total land mass (Afar, Benishangul-Gumuz, Somali and Gambella) is categorized as having very high relative vulnerability (Table 2).

Consideration the population 0.74% of the total population (0.07 million) who live in Harari and Dire Dawa are relatively least vulnerable to climate change sensitive diseases while, 69.48% of the population (Oromia, Addis Ababa, Amhara

Table 2 Area coverage and population of Ethiopia by vulnerability classes

Vulnerability classes	Regions	Area coverage (km^2)/%		Population (mlns)/%	
Least vulnerable	Dire Dawa and Harari	**1901**	**0.6**	**0.635**	**0.74**
Moderately vulnerable	Oromia, Addis Ababa, Amhara and Tigrai	565,875	**49.95**	**59.562**	**69.48**
Highly vulnerable	SNNP	117,263	**10.35**	**17.403**	**20.3**
Very highly vulnerable	Afar, Benishangul-Gumuz, Somali and Gambella	447,855	**39.95**	**8.129**	**9.48**

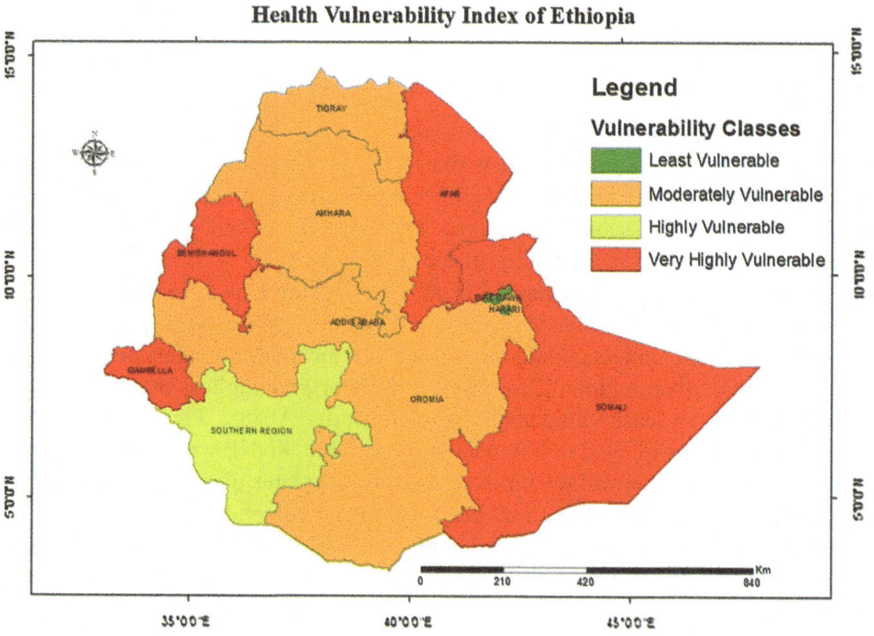

Fig. 6 Relative health vulnerability index ethiopia by region

and Tigray) are moderately vulnerable (Fig. 6). 20.3% of the total population (SNNP) is highly vulnerable. 9.48% of the populations are categorized as very highly vulnerable. These are people who live in the lowland regions of Afar, Benishangul-Gumuz, Somali and Gambella.

Climate change does not create new health problems but may worsen known clinical problems and alter geographic patterns of disease occurrence. Consequently, adaptation policy recommendations related to climate change and public health reflect the need to sustain and refine current measures to enhance their

Table 3 Suggested health adaptation options by vulnerability categories

No.	Adaptation options	Vulnerability Category			
		Low	Medium	High	Very high
1	Improve public health surveillance systems	***	***	***	***
2	Establish health and climate data management system	***	***	***	***
3	Strengthening early warning systems	***	***	***	***
4	Improved public health services	*	**	***	***
5	Improved water, sanitation, and hygiene system	*	**	***	***
6	Human resource development	*	**	***	***
7	Enhanced public awareness and attitudes	***	***	***	***
8	Targeted intervention to regional contexts by enhanced financial resources	*	**	***	***
9	Partnership, coordination and collaboration	***	***	***	***

*Less priority, **Priority, ***High priority

sensitivity to climate change. Health adaptation strategies recommended include improved public health surveillance systems; establishing health and climate data management system; strengthening early warning systems; improved public health services; improved water, sanitation, and hygiene system; human resource development; enhanced public awareness and attitudes; targeted intervention to regional contexts by enhanced financial resources (Table 3).

4 Conclusions and Recommendations

Climate change and variability is a cross-cutting issue where various institutions are expected to work together to set national priorities and implement in short and long-term. This assessment shows that, in Ethiopia, vulnerability to disease in connection to climate change is evident. Health vulnerability index (HVI) was developed following IPCC's variables: exposure, sensitivity and adaptive capacity as major factors to determine vulnerability. The values for the exposure levels ranged from 0.12 (Addis Ababa) to 0.69 (Gambella). While Amhara, Tigray, Harari and SNNP Regions are moderately exposed due to combination of moderate variation in climate and medium values in hazard profile are the probable reasons; Afar, Gambella, BenishangulGumuz and Somali regions are highly exposed to changes due to climate change. Sensitivity finding shows that all regions are highly sensitive to climate change induced diseases with the exception of Addis Ababa.

Adaptive capacity measured in terms of several socio-economic factors such as finance, infrastructure, available health professionals and social determinants of health reveals variation between regions characterized as urban and the rest. While urban centers were found to have relatively better adaptive capacity, the remaining region exhibited more or less similar state of low coping capacity.

Further analysis of Health Vulnerability Index shows rang from less vulnerable (−0.247)—Dire Dawa to highly vulnerable (0.279)—Gambella. While Oromia, Addis Ababa, Tigray, and Amahra regions were in the range of medium vulnerability, SNNP is found to be highly vulnerable and Afar, Somali, Gambella and BenishangulGumuz were very highly vulnerable.

Based on the preceding evidences, successful adaptation strategy for the health system needs to explore available opportunities and priority focuses of interventions in the face of looming challenges from climate change. In view of this, decision makers are expected to identify contexts where impacts in another sector such as water, agriculture and other sectors could adversely affect population health. Many of the possible health sector adaptation measures to climate change lie primarily outside the direct control of the health sector embedded such areas as sanitation and water supply, education, agriculture, trade, tourism, transport, development, urban planning, housing and so on. This calls for inter-sectoral and cross-sectoral adaptation strategies to reduce the potential health impacts of climate change and to optimize adaptive responses.

References

Conway, D., & Schipper, E. L. F. (2011). Adaptation to climate change in Africa: Challenges and opportunities identified from Ethiopia. *Global Environmental Change, 21*(1), 227–237.

CRGE Vision. (2011). *Ethiopia's vision for a climate resilient green economy*. pp. 28. Addis Ababa, Ethiopia.

Hahn, M. B., Riederer, A. M., & Foster, S. O. (2009). The livelihood vulnerability index: A pragmatic approach to assessing risks from climate variability and change—a case study in Mozambique. *Global Environmental Change, 19,* 74–88.

IPCC. (2007). Summary for policymakers. Climate change 2007: The physical science basis. In Solomon S, Qin D, Manning M, Chen Z, Marquis M, Averyt KB, Tignor M and Miller HL (Eds.), *Contribution of working group I to the fourth assessment report of the intergovernmental panel on climate change*. Cambridge, UK and New York, USA: Cambridge University Press.

Kelly, P. M., & Adger, W. N. (2000). Theory and practice in assessing vulnerability to climate change and facilitating adaptation. *Climatic Change, 47,* 325–352.

McSweeney, C., Lizcano, G., New, M., & Lu, X. (2010). The undp climate change country profiles. *Bulletin of the American Meteorological Society, 91,* 157–166.

MOH. (2014). Health sector development programme IV annual performance report of year 2006.

NAPA (Climate Change National Adaptation Program of Action of Ethiopia. (2007). *National meteorological agency* (p. 85). Ministry of Water Resources, Addis Ababa.

Patz, J. A., Campbell-Lendrum, D., Holloway, T., & Foley, J. A. (2005). Impact of regional climate change on human health. *Nature*, 438, 310–317 (17 November 2005) | doi:10.1038/nature04188.

Portier, C. J., Tart K. T., Carter S. R., Dilworth, C. H., Grambsch, A. E., Gohlke, J., et al. (2010). *A human health perspective on climate change: A report outlining the research needs on the human health effects of climate change*. Research Triangle Park, NC:Environmental Health Perspectives/ National Institute of Environmental Health Sciences. doi:10.1289/ehp.1002272, Available: www.niehs.nih.gov/climatereport.

Snover, A. K., Binder, W. L., Kay, J., Lopez, J., Willmott, E., Howell, D., et al. (2007). Preparing for climate change: a guidebook for local, regional, and state governments. *Environmental Health Perspectives, 117*(4), 617–623.

Simane, B., Zaitchik, B. F., & Foltz, J. D. (2014). Agroecosystem specific climate vulnerability analysis: Application of the livelihood vulnerability index to a tropical highland region. *Mitigation and Adaptation Strategies for Global Change, 21*(1), 39–65. doi:10.1007/s11027-014-9568-1.

UNDP. (2007). Human development report 2007/8. United Nations Development Programme 1 UN Plaza, New York, NY, 10017, USA. ISBN 978-0-230-54704-9.

USAID. (2012). A climate trend analysis of Ethiopia famine early warning systems network—Informing climate change adaptation series. Fact Sheet 2012–3053.

WHO (World Health Organization). (2003). Health impacts of climate extremes: in Climate change and human health: Risks and responses. Edited by Richard Wilkinson & Michael Marmot, Second edition 2003, 31 pages. ISBN 978 92 890 1401-4.

Potential for Scaling up Climate Smart Agricultural Practices: Examples from Sub-Saharan Africa

Kindie Tesfaye, Menale Kassie, Jill E. Cairns, Misiko Michael, Clare Stirling, Tsedeke Abate, B.M. Prasanna, Mulugetta Mekuria, Habtamu Hailu, Dil Bahadur Rahut, Olaf Erenstein and Bruno Gerard

1 Introduction

Agriculture in Sub-Saharan Africa (SSA) is predominantly rain-fed and smallholder farmers (>80%) are the primary producers of agricultural outputs in the region (AGRA 2014). Erratic weather patterns and extreme weather events exacerbated by the changing climate adds to the challenges faced by smallholder farmers in producing enough food to feed the ever growing population of the region (Shiferaw et al. 2014). While the farming communities are responding to these challenges, there is an intensive need for scaling-up adoption of climate smart agricultural practices (CSAs).

The term CSA represents a set of strategies that can increase resilience to weather extremes, adapting to climate change and decreasing greenhouse gas

K. Tesfaye (✉)
International Maize and Wheat Improvement Centre (CIMMYT),
Addis Ababa, Ethiopia
e-mail: k.tesfayefantaye@cgiar.org

M. Kassie
International Center of Insect Physiology and Ecology, ICIPE, Nairobi, Kenya

J.E. Cairns · M. Mekuria
CIMMYT, Harare, Zimbabwe

M. Michael · T. Abate · B.M. Prasanna
CIMMYT, Nairobi, Kenya

C. Stirling
CIMMYT, London, UK

H. Hailu
Ministry of Agriculture, Addis Ababa, Ethiopia

D.B. Rahut · O. Erenstein · B. Gerard
CIMMYT, El Batan, Mexico

W. Leal Filho et al. (eds.), *Climate Change Adaptation in Africa*,
Climate Change Management, DOI 10.1007/978-3-319-49520-0_12

(GHG) emissions from agriculture (Steenwerth et al. 2014). Producing food in a more 'climate smart' way has three advantages: providing food for an increasing population, maintaining food production under a changing climate, and reducing greenhouse gas emission from agriculture while absorbing carbon in vegetation and soil. By doing so, CSA contributes to economic development, poverty reduction and food security through maintaining and enhancing the productivity and resilience of natural and agricultural ecosystem functions, and reducing trade-offs involved in meeting these goals (Steenwerth et al. 2014).

The need for climate-smart agriculture has been emphasized by African leaders (since 2010) who recognized that food security, poverty and climate change are closely linked and should not be considered separately (Synnevag and Lambrou 2012). Among others, the need for early action to identify and scale up best practice in Africa has been highlighted by the leaders in the African Ministerial Conference on Climate-Smart Agriculture in 2011 (Synnevag and Lambrou 2012). It is believed that the threats of climate change to agriculture can be reduced by applying climate-smart technologies that can increase the adaptive capacity, resilience and resource use efficiency of farmers, and enhance the mitigation potential of agricultural landscapes (AGRA 2014).

Smallholder farmers across the continent have begun to embrace climate-smart farming approaches and technologies, but as the impacts of climate change become increasingly evident, they may need to adapt more quickly and more comprehensively (AGRA 2014). In many cases, reports on African smallholder farmers focus on constraints and limitations and rarely explain successes that can be scaled up and out. The main purpose of this paper is to highlight promising CSA practices in SSA, lessons learned and the potential for scaling up the practices to reduce vulnerabilities of smallholder farmers and their livelihoods to the impacts of climate change in the region.

2 Methodological Approach

The paper is organized based on secondary information from literature and primary data collected by the authors. There are several CSA practices in SSA, and the ones presented in this paper were selected based on their adoption potential and contribution to increased productivity/food security, adaptation and/or mitigation (AGRA 2014; Synnevag and Lambrou 2012). Most of the information on CSA practices in SSA was collected from literature (both published and unpublished), and after an extended review, the most promising CSA practices were selected, particularly in terms of initial uptake and showing both short- and long-term benefits to smallholder farmers.

An account on successful CSA practices in Ethiopia was also collected from semi-structured key informant interviews and focus group discussions with community leaders, and experts from the Bureau of Agriculture and Mekelle University in the Tigray Regional State and from agriculture and natural resource conservation experts from the Federal Ministry of Agriculture and Natural resources in 2014 and 2015. Moreover, primary data collected through the Drought Tolerance Maize for Africa (DTMA) and Sustainable Intensification of Maize–Legume Cropping Systems for Food Security in Eastern and Southern Africa (SIMLESA) projects were used.

3 Climate Smart Agricultural Practices in SSA

A review and analysis of potential climate-smart agricultural practices (CSAs) in SSA indicate that some CSAs are increasingly adopted by farmers and show potential for scaling-up. Some of these practices, which are reviewed below, include drought tolerant (DT) varieties, sustainable intensification, water harvesting and small-scale irrigation, crop associations, climate information and land, and water conservation and management.

3.1 Drought Tolerant (DT) Crop Varieties

Africa is a drought prone continent and climate change is expected to worsen the problem. It is likely that droughts/dry spells will be more frequent, more intense, and long lasting in the region under climate change (Masih et al. 2014; Nyasimi et al. 2014). Developing drought tolerant crop varieties has been one of the major strategies for managing water limitation in agriculture (e.g., Xoconostle-Cázares et al. 2010), and concerted efforts are underway in the region to develop climate resilient crop varieties. Several national and international research institutions have scored important gains in improving the drought tolerance of major grain crops such as maize, millet, cowpea, groundnut and sorghum (Shiferaw et al. 2014).

Maize is a major staple food for over 300 million people in SSA. However, the crop is threatened by changing rainfall patterns and increasing temperatures in many parts of the sub-regions, and will, on average, lose 12% of its productivity by 2050 (Tesfaye et al. 2015). Around 40% of maize-growing areas in the region face occasional drought stress with a yield loss of 10–25%, and about 25% of the maize crop suffers frequent drought with losses of up to 50% of the harvest (Fisher et al. 2015). This challenge requires developing, deploying and delivering drought tolerant maize varieties to smallholder farmers whose livelihood depended on the crop.

Through the Drought Tolerance Maize for Africa (DTMA) initiative, over 140 DT maize varieties have been released in 13 SSA countries since 2006 (Table 1). Currently more than 91 African seed companies are producing and marketing seed of these varieties to smallholders, and estimated 2 million hectares are planted with DT maize varieties every year, benefiting millions of smallholder households (Table 1). Under research fields, the new DT varieties have at least 30–40% yield advantage over commercial materials under severe stress, and similar performance under optimal conditions (Shiferaw et al. 2014). Farmers also reported a yield advantage of 20–30% from

Table 1 Number of drought tolerant maize (DT) varieties released, quantity of seed produced, estimated area covered, households and beneficiaries of drought tolerant maize in 13 African countries

Country	DT Varieties released[a]	DT seed (tons)	DT area ('000 ha)[b]	DT households ('000)[c]	DT beneficiaries ('000)[c]
Angola	7	511	204	51	358
Benin	6	132	53	13	92
Ethiopia	10	1544	618	154	1081
Ghana	12	79	32	8	55
Kenya	2	5050	2020	505	3535
Malawi	17	4416	1766	442	3091
Mali	7	98	39	10	69
Mozambique	9	855	342	86	599
Nigeria	21	3245	1298	325	2272
Tanzania	14	2376	950	238	1663
Uganda	7	1572	629	157	1100
Zambia	18	3422	1369	342	2395
Zimbabwe	13	7468	2987	747	5228
Total	**143**	**30,768**	**12,307**	**3077**	21,538

[a]Varieties were released between 2007 and September 2013; [b]1 ton enough for 40 ha; [c]Assumes that an average farmer plants 10 kg seed; average family size 7 people (*source* adapted from Abate 2013)

DT varieties under moderate drought conditions (Cooper et al. 2013). The released DT varieties have a combination of traits that allow them to thrive under drought stress conditions, which include shorter anthesis-silking interval (ASI), reduced bareness, reduced evapotranspiration, and functional stay green during grain filling (Bruce et al. 2002; Edmeades 2008). The new DT maize varieties also satisfies other requirements of the farmers such as cooking and milling and pest and disease resistance (Cooper et al. 2013). Because of their multiple attributes, some of the DT varieties are currently grown outside of the project countries.

The success of the initiative lies on the fact that it brought together a wide range of partners around a good CSA. It is led by the International Maize and Wheat Improvement Center (CIMMYT), and it is implemented in collaboration with the International Institute of Tropical Agriculture (IITA), national agricultural research systems (NARS), public and private seed companies, seed certification agencies, and farmers groups in Angola, Benin, Ethiopia, Ghana, Kenya, Malawi, Mali, Mozambique, Nigeria, Tanzania, Uganda, Ethiopia, Zambia, and Zimbabwe. The initiative is funded by the Bill & Melinda gates foundation, the Howard G. Buffett Foundation, the United States Agency for International Development (USAID) and the UK Department for International Development (DIFID). Availability of sustainable funding from donors, full participation of farmers in the process and the partnership between international, public and private actors help ensure that useful products are developed, multiplied and delivered to the farmers with the required inputs and market options.

The International Crops Research Institute for the Semi-Arid Tropics (ICRISAT) and national partner organizations have also made important gains in improving the drought tolerance of millet and sorghum in many regions. For example, about 34% of the millet area and 23% of the sorghum area in southern Africa have been planted with improved varieties in the last few years (CGIAR 2006). Moreover, several countries have released drought tolerant cowpea varieties with support from the International Institute of Tropical Agriculture (IITA) (CGIAR 2006).

3.2 Sustainable Intensification Practices

Sustainable Intensification Practices (SIPs) aim to increase output from the same cropped areas while reducing negative environmental impacts. SIPs are currently promoted across SSA and includes various agricultural practices like crop rotation, intercropping, conservation agriculture, integrated pest management, crop breeding, cropping system improvements, agroforestry and soil conservation, and livestock and fodder crops (AGRA 2014; Pretty et al. 2011).

Improving soil fertility and water management through a portfolio of SIPs greatly reduce crop failures, improve yield and minimize the impacts of climate risks (Shiferaw et al. 2014). Hence, sustainable intensification fosters more efficient resource use, and contributes to adaptation and mitigation through effects on farm productivity and incomes, and reduced GHG emissions per unit of product (AGRA 2014).

CIMMYT in collaboration with its partners is engaged in understanding SIPs adoption barriers and enabling conditions and in increasing yield and market access for African farmers through the Adoption Pathways and Sustainable Intensification of Maize–Legume Cropping Systems for Food Security in Eastern and Southern Africa (SIMLESA) projects. The projects aim to facilitate adoption and increase food security and incomes through improved productivity and resilience from more resilient and sustainable maize-based farming systems.

Four countries in eastern and southern Africa show promising levels of adoption of many SIPs with the level of adoption varying among countries (Fig. 1), particularly crop rotation, intercropping, minimum tillage, crop residue retention and soil and water conservation (SWC) with physical structures. Still adoption was diverse, and for instance, limited (<10%) for maize-legume rotation in Tanzania, minimum tillage in Malawi, SWC (ridges, stones and soil bunds) in Kenya and Tanzania. Adoption of conservation agriculture was limited in all the four countries (<7%, Fig. 1), often associated with inadequate awareness of the practice by smallholders, trade-offs in use of crop residues for other uses including feeding livestock, lack of suitable and low-cost zero-till equipment and chemicals for weed control, and delayed productivity benefits to smallholders (Corbeels et al. 2014; Shiferaw et al. 2014).

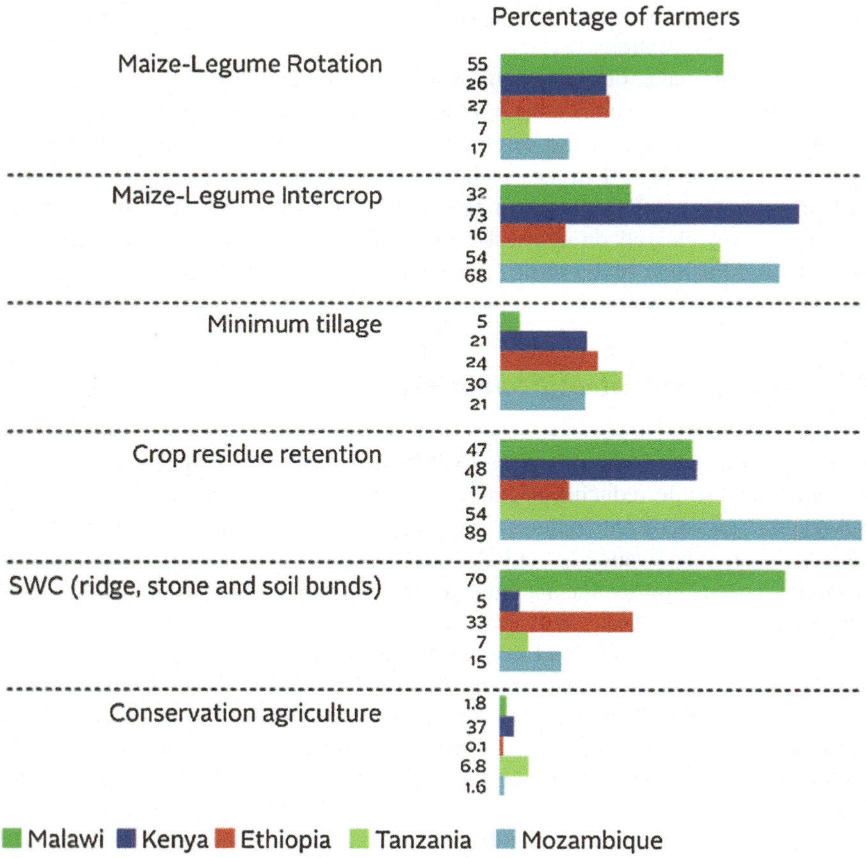

Fig. 1 Adoption of Sustainable Intensification Practices (SIPs) in Eastern and Southern Africa. *Source* Authors

Adoption of SIPs not only improves soil health and water retention but also increases and stabilizes the income of the adopters. For example, adoption of a given SIP in Ethiopia and Malawi increases income of farmers, and the additional income increases with a joint adoption of SIPs (Fig. 2). For instance, improved maize varieties increase incomes by 14–41% when they are jointly adopted with other SIPs. Similar results were found in Malawi where certain combinations of SIP options provided higher benefit than adopting them individually. These options also reduced the downside risk or risk of crop failures (Kassie et al. 2015; Teklewold et al. 2013).

Relatively higher adoption rates of SIPs in the project countries (Ethiopia, Kenya, Tanzania, Malawi and Mozambique) encouraged the SIMLESA initiative to expand its activities to spill over countries (including Rwanda, Uganda, and

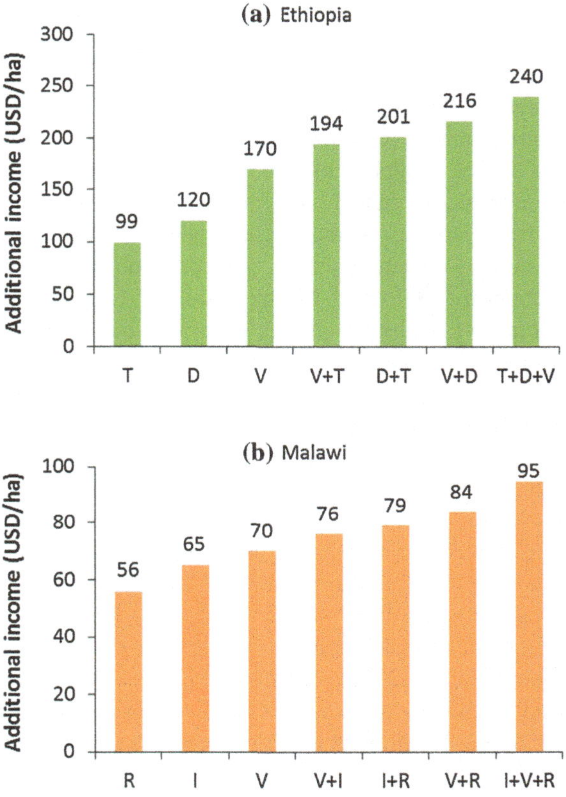

Fig. 2 Additional income from adoption of multiple sustainable intensification practices in (**a**) Ethiopia and (**b**) Malawi. T = minimum tillage (zero or one pass); V = improved maize varieties; D = crop diversification (legume-maize intercopping & rotation); I = legume-maize intercropping; R = legume-maize rotation. *Source* Authors

Botswana). The key elements that contributed to the success of the project are strong collaboration with NARS, a network of on-farm demonstrations, experience sharing among project countries and feedback, capacity building of national partners, and sustained funding commitment from donors.

3.3 Land and Water Conservation and Management

About 46% of Africa is affected by land degradation (a continuous loss of soil, and vegetation cover and nutrient depletion) and the highest rate of degradation occurs in SSA. It is estimated that losses in productivity of cropping land in SSA are in the order of 0.5–1% annually with a productivity loss of at least 20% over a period of 40 years (WMO 2005). In highly populated areas of Africa, short-term increase in agricultural production is often obtained through reduced fallowing, removal of vegetation from cropland, conversion of forest and woodlands on steep slopes into rangeland and marginal arable land which hasten land degradation (Nyssen et al. 2009b).

Climate and land degradation are closely interlinked affecting each other. The climate strongly influences environmental stocks (vegetation type, distribution, biomass and diversity) and flows (soil erosion, infiltration and the rate of decomposition of organic matter). The severity, frequency, and extent of erosion are set to be altered by changes in rainfall amount and intensity and changes in wind. Climatic stresses, which include high soil temperature, seasonal excess water, short duration low temperatures, seasonal moisture and extended moisture stresses, account for 63% of all the stresses on land degradation in Africa (WMO 2005), and failure to manage for climate variability has been a major source of degradation of natural resources.

Land use and land cover changes influence carbon fluxes and greenhouse gas (GHG) emissions which directly alter atmospheric composition and radiative forcing properties. Deforestation, biomass burning, conversion of natural to agricultural ecosystems, drainage of wetlands and soil cultivation emitted 136 gigatons of carbon globally since the industrial revolution. Land degradation aggravates CO_2-induced climate change through the release of CO_2 from cleared and dead vegetation and through the reduction of the carbon sequestration potential of degraded land (WMO 2005). Land use and land cover change is an important factor in determining the vulnerability of ecosystems and landscapes to environmental change (WMO 2005). Therefore, land management becomes a necessity to reverse land degradation, reduce GHG emissions, restoring ecosystem services and improving livelihoods in many areas (Haregeweyn et al. 2015; Nyssen et al. 2014, 2015a, b).

Land degradation is a major ecological and economical problem in Ethiopia (Haileslassie et al. 2005). It decreases soil fertility and leads to loss of biodiversity, lowering of water table, diminishing of grazing lands and an overall loss of agricultural productivity. All of these increase the vulnerability of degraded areas to cyclic droughts, famine and absolute neglect to care for natural resources (Nyssen et al. 2007). Land degradation causes not only environmental damage but also productivity and economic losses. For example, the annual economic loss is estimated at about 3% of GDP in Ethiopia and about 9.5–11% of GDP in Malawi; yield reduction due to soil erosion may range from 2–40% depending on the crop and location across Eastern Africa (Kirui and Mirzabaev 2014).

A good example that demonstrates the impacts of land degradation is the Tigray Regional State in North Ethiopia where land degradation and climate variability posed an existential threat to the livelihoods of smallholder farmers. The rugged landscapes of northern Ethiopia have been degraded by agriculture for at least three millennia (Nyssen et al. 2015a), and the degraded environment contributed to low agricultural production which exacerbated rural poverty (Edwards et al. 2007). In a few decades, however, the Tigray Region has become an example for concerted efforts to tackle land degradation using land rehabilitation measures such as construction of stone terraces and stone bunds, exclosures and forestation (Alemayehu et al. 2009; Haregeweyn et al. 2015; Mekuria et al. 2007; Nyssen et al. 2007, 2009b, 2015a).

3.3.1 Soil and Water Conservation in North Ethiopia

The agriculture sector in Ethiopia is highly vulnerable to climate variability and land degradation. The food shortages in Eastern Africa, including Ethiopia are largely due to the insufficient and erratic nature of rainfall and increasing soil (physical) degradation (Araya et al. 2015; Stroosnijder 2009). Cognizant of this, the Government of Ethiopian has initiated a massive community-based SWC program in the last two decades, and this has become effective in conserving soil and water and improving livelihoods particularly in the northern part of the country. A review of published reports in Ethiopia indicate that SWC interventions applied at various spatial scales are effective in erosion control, soil moisture conservation, vegetation regeneration and soil build up (Haregeweyn et al. 2015). SWC activities also enhance in situ infiltration during the short but intense storms and improve the situation with regard to flooding, soil erosion and groundwater (Nyssen et al. 2015a).

Integrated community based SWC work in the Tigray region over 15 years resulted in the construction of several conservation structures over vast areas of on-farm and degraded watersheds and landscapes (Table 2). Large efforts have been made to conserve soil and water through a range of techniques including stone bunds, soil bunds, runoff collection ponds (Fig. 3), grass buffer strips, in situ land surface management, check dams, and nutrient management (Alemayehu et al.

Table 2 Land and water management achievements in Tigray Regional State, North Ethiopia

SWC practices	Total quantity	Number of years
Farmland terraces		
Stone bund construction (km)	288,273	14
Soil bund construction (km)	54,715	14
Trench bund construction (km)	65,473	12
Stone faced trench bund (km)	51,048	8
Stone faced soil bund (km)	1320	4
Bund maintenance (km)	28,920	8
Degraded land rehabilitation		
Hillside terraces (km)	3755	10
Check dam construction (m^3)	3,103,561	14
Cut-off drain construction (m^3)	9,709,073	7
Cut-off drain maintenance (m^3)	144,000	1
Deep trench (km)	15,222	1
Exclosure (ha)	1,540,000	15

Adapted from Lemenih and Kassa (2014) and Adimassu et al. (2015)

Fig. 3 Physical soil and water conservation structures (stone bunds, soil bunds and runoff collection ponds) as part of the improved land management in Eastern Tigray. *Source* © Alemayehu et al. (2009)

2009; Nyssen et al. 2009b). These SWC interventions have not only a direct effect on the availability of green water that enhances in situ crop growth, but also on blue water such as improved springs and groundwater tables (Nyssen et al. 2015b). Infiltration of runoff in purposively constructed trenches on the surrounding mountain slopes contributes 30–50% of the total aquifer recharge in a plain irrigated with pumped groundwater in Ethiopia (Walraevens et al. 2009). Therefore, SWC has become an important environmental issue for enhancing food production and livelihood, both through on-site and off-site effects of conservation (Nyssen et al. 2015b). As indicated by Nyssen et al. (2009b), improved land husbandry is the only option in highly degraded environments with high pressure on the land; and such efforts are successful in places where decision making processes at different levels in society give the highest priority to the implementation of SWC and other land rehabilitation. Investments in land management and agricultural inputs in Ethiopia increased cereal food production per capita by 160% in 2005–2010 compared to in 1985–1990, and land management has become part of the agricultural system of the country, particularly in the north (Nyssen et al. 2015a).

3.3.2 Exclosure in North Ethiopia

Area exclosure is one of the most widespread forms of re-greening in Ethiopia today. It involves protecting degraded areas mainly through social fencing from any form of cultivation, cutting trees and shrubs, or grazing by livestock (Lemenih and Kassa 2014). In 1996, there were only about 143,000 ha of exclosure in Ethiopia.

Table 3 The major environmental benefits of exclosures

On-site	Off-site: adjacent land effects	Off-site: downstream effects
Improved soil fertility	Increase in soil nutrient content	Decrease in reservoir sedimentation
Increase in soil nutrient content	Increase in soil moisture content	Decrease in flood damage
Increase in soil moisture content	Increase in depth of top soil	Increase in the probability of new spring up-coming
Increase in depth of top soil	Increase in soil organic matter content	Decrease in salinization effect
Increase in soil organic matter content	Increase in rain water infiltration (decrease in run-off)	Decrease in provision of fertile sediments that may increase
Increase in rain water infiltration (decrease in run-off)	Increase in soil fauna population involved in nutrient recycling & soil aeration	Reduction in the cost of drainage maintenance
Increase in soil fauna population	Increase in wild life	
Increase in wild life	Enhanced biodiversity	
Enhanced biodiversity		

Source Balana et al. (2012)

However, the area under area exclosure has grown fast and by the end of 2013, exclosures covered 1.54 million ha in Tigray and 1.55 million ha in Amhara regional states (Lemenih and Kassa 2014). Although exclosures are effective in restoring vegetation, they also provide several environmental benefits; both on-site and off-site (Table 3). In north Ethiopia, such exclosures lead to restoration of soil fertility, soil physical and chemical properties, vegetation cover and composition, fauna, and water resources (Balana et al. 2012; Hadgu 2011; Mekuria et al. 2007; Nyssen et al. 2014, 2015a). The ecosystem services of exclosures (water availability, biodiversity, agricultural productivity, carbon sequestration, river regulation) in northern Ethiopia extend from local to global significance (Nyssen et al. 2009a; Schroter et al. 2005).

In addition to ecosystem and agricultural benefits, the rehabilitation of degraded areas through exclosures in north Ethiopia and elsewhere in the country has brought positive impact on the livelihoods of local people by providing access to fodder, income from cattle fattening, bee keeping, and vegetable and fruit production in the downstream areas (Deichert et al. 2014; Hadgu 2011; Tigabu et al. 2014).

Up to 30% of the Tigray Regional State has been rehabilitated from land degradation and about 1 million people have benefited from improved ecosystem services, productivity increase and better incomes (Hadgu 2011). Several factors contributed to the success of land rehabilitation and management in North Ethiopia. The major ones include community by-in and commitment, government support at all levels, strong partnerships among actors, recognition of community by-laws by the formal legal system, and experience sharing and feedback (Table 4). The

Table 4 Major factors that contribute to the success of land rehabilitation and management in the Tigray Regional State, North Ethiopia

No.	Major reasons for success	Source
1	Consensus on urgency of land degradation at all levels	Araya and Edwards (2006), Edwards et al. (2007), Nyssen (1998); key informant interview and focus group discussions
2	Demand driven	
3	Community commitment (40 h. free labour per household)	
4	Bottom-up planning (planning starts from community)	
5	Government committeemen at all levels	
6	Watershed and landscape approach	
7	Strong partnership between government and non-government actors (e.g., food for work safety net program; sustainable land management projects)	
8	Formulating and using community bylaws	
9	Experience sharing and feedback for scale up	

successes in North Ethiopia have led to expansion of land and water management to include many more communities in the regional state and in other parts of the country. The government has already adopted land rehabilitation and soil and water management as its main strategy for combating land degradation and poverty (Araya and Edwards 2006).

3.3.3 Farmer Managed Natural Regeneration in the Sahel

Farmer managed natural regeneration (FMNR) refers to a set of practices farmers use to foster the growth of indigenous trees on agricultural land (Haglund et al. 2011). FMNR is becoming a vital practice in increasing vegetation greenness in degraded environments such as the Sahel (Haglund et al. 2011; Rinaudo 2007; Tougiani et al. 2009). In Niger alone, more than 5 million hectares of degraded land has been restored with over 200 million trees established through FMNR (Cooper et al. 2013). A study indicates that about 26% of households in Niger practice a form of FMNR and adoption is strongly linked to factors like soil type, market access, and the education level of household head (Haglund et al. 2011). FMNR has several benefits to the adopting farmers including better access to firewood, improved soil fertility and soil structure, decreasing diseases, pests, soil erosion problems and rising water tables (Cooper et al. 2013). It also increases the density and diversity of trees on farmland which in turn increase income and crop diversity, and it is estimated that FMNR raises the annual gross income of the Sahel region by between 17 and 21 million USD (Haglund et al. 2011). These findings support the value of continued promotion of FMNR as an inexpensive means of enhancing rural livelihoods and adaptive capacity to climate extremes and an attractive alternative to reforestation efforts that are relying on tree planting. FMNR has spread across the Sahel in Burkina Faso, Mali and Senegal (Cooper et al. 2013), and in other countries such as Malawi, Zambia and Rwanda (AGRA 2014). Lessons from successful farmers and communities and increasing challenges from climate extremes and environmental degradation contributed for the wide adoption of FMNR in Niger and other regions (AGRA 2014; Cooper et al. 2013).

3.4 Water Harvesting

Climate change will affect all facets of society and the environment with strong implications for water and agriculture in SSA where poverty is intense (IWMI 2007). Achieving future food security in the region depends mainly on increasing production from rainfed agriculture, and hence water storage and control are important strategies to respond to climate change. In total, 874 million hectares of land in Africa could benefit from increased agricultural production by improving water management including rainwater harvesting and storage (Malesu et al. 2007). If only 15% of the rainwater in SSA were harvested, it would be more than enough to meet all the water needs of the continent (Malesu et al. 2007).

Water harvesting (WH) refers to a set of approaches which occupy an intermediate position along the water-management spectrum extending from in situ moisture conservation to water collection and storage for irrigation, which include small-scale systems that induce, collect, store and make use of local surface runoff for agriculture (Critchley and Gowing 2012). WH involves all methods of concentrating, storing and collecting rainwater/runoff for agricultural, domestic, environmental and industrial uses from roofs, ground surfaces as well as from ephemeral watercourses (Malesu et al. 2007). WH strategies may vary from direct runoff concentration in the soil for direct uptake by the crops, to collection and storage of water in structures (surface, sub-surface tanks, ponds and small dams) and aquifers for future productive uses. WH systems are commonly classified as rooftop catchment systems, in situ catchment systems and runoff catchment systems (Pachpute et al. 2009). Water supply in agricultural system can be effectively increased to meet crop water demands through a combination of different water harvesting systems. For example, in situ rain water harvesting systems increase crop yields by 30–50% in southern Africa; and different water harvesting systems (check dams, percolation ponds, farm ponds and small dams), with associated soil nutrient and crop management practices, have been found suitable to bridge the gap between potential and actual crop yields in SSA (Pachpute et al. 2009).

Water harvesting schemes that were developed during the droughts of the 1970s and 1980 provided lessons to promote sustainable development of dryland agriculture in the face of changing environmental conditions (Critchley and Gowing 2012). In fact, various technologies to harvest rainwater have been in use for over 4000 years and new ones are being developed all the time. In situ soil conservation and water harvesting as well as the construction of small reservoirs have led to decreased sediment yield and runoff coefficients at catchment scale, and increased water tables and have improved livelihood in Africa (Nyssen et al. 2015a).

Improving agricultural productivity in areas that depend on rainfall has the greatest potential to reduce poverty and hunger in SSA. Since yields are low in many rainfed systems, improving rainfed farming could double or quadruple yields and hence closing yield gaps promises huge social, economic, and environmental paybacks (IWMI 2007). Better management of rainwater, soil moisture, and supplemental irrigation is the key to helping the greatest number of poor people because it cuts the yield losses from dry spells, it gives farmers the security they need to risk investing in other inputs such as fertilizers and high-yielding varieties, and it allows farmers to grow higher value market crops, such as vegetables or fruits (IWMI 2007).

With a track record of being used for thousands of years for managing climate risks in rainfed agriculture, WH is a proven climate change adaptation measure of huge potential (Critchley and Gowing 2012). Successful water harvesting practices have been documented in Burkina Faso, Kenya, Rwanda, Uganda, Ethiopia, Tanzania and Zimbabwe (Abebe et al. 2012; Biazin et al. 2012; Critchley and Gowing 2012; Gowin and Bunclark 2013; Malesu et al. 2012). Implementation of rainwater harvesting may allow cereal-based smallholder farmers to shift to diversified crops, hence improving household food security, dietary status, and economic return (Biazin et al. 2012). In Ethiopia, for example, a 1% of additional harvested water increased the output of onion and tomato by 0.12 and 0.23%, respectively (Wakeyo 2012). This has prompted governments in SSA and even the World Bank to support WH implementations through budgetary commitments (Malesu et al. 2012).

3.5 Weather/Climate Agro-Advisory Services

Weather/climate agro-advisory refers to recommendations derived from weather/climate information that is transformed or translated using available agricultural knowledge that helps users along the agricultural information chain to make improved decisions for enhanced and sustained agricultural productivity (AGRA 2014). The adoption and success of CSA technologies and practices depend on the effective delivery of agro-advisory services. This requires the support of well-structured multidisciplinary and cross-sectoral collaborative approaches driven by an agreed framework to govern such collaboration.

In Mali, for example, farmers who received and applied climate agro-advisories increased their income due to a substantial production increase from maize, sorghum, pearl millet, groundnut and cotton. Farmers felt that they were exposed to lower levels of risk and hence confident about purchasing and using inputs like improved seed, fertilizer, and pesticides, all of which boost production (AGRA 2014; Hellmuth et al. 2007). This was possible because a multidisciplinary group comprising technical, development, and research experts drawn from the National Meteorological and Hydrological Services, the Ministry of Agriculture, agricultural research institutes, rural development agencies, farmers, and the media was established. Given the low literacy levels of farmers and the highly scientific level of climate products and information, the multidisciplinary group served as the interface between service providers and end users, and repackaged the climate data into useful information and advice for farmers, and made it available in multiple local languages (AGRA 2014).

Similarly, farmers who received 10 days agro-advisory services from before planting up to harvest in the Central Rift Valley of Ethiopia increased their productivity by up to 40% compared to those farmers who didn't receive the information (Personal communication with Girma Mamo, Agrometeorologist at Melkassa Agricultural Research Station).

Many African countries have developed meteorological and hydro-meteorological service agencies and the capacity to provide short-term and seasonal agricultural forecasts is growing over time (Hellmuth et al. 2007). However, a successful application of agro-advisory services at the farm level largely depends on the existence of communication channels which include conventional platforms such as radio, TV, and bulletins; farmer field schools; farmer-participatory climate workshops; and local climate information centres that together enhance the availability and accessibility of value-added climate information to smallholder farmers (Tall et al. 2014). Moreover, new innovative communication channels such as mobile phones, internet, and interactive voice-response systems could bridge the information gap related to conventional systems and spur a wider impact on the capacity of farmers to managing risks in agriculture. The number of people owning mobile phones in Africa is estimated at over 650 million (World Bank and African Development Bank 2012). This platform could support information dissemination on expected local weather and climatic conditions, local market prices, and other locally relevant information.

Besides agro-advisory services, the availability of long-term climate data at the national level and the capacity to access and install automatic weather stations helped piloting of weather index insurance as means of managing climate risks in some countries such as Ethiopia, Kenya, and Senegal (AGRA 2014; Tadesse et al. 2015).

4 Conclusion

Adoption of drought tolerant crop varieties and sustainable intensification practices are expanding in Africa's farming systems, particularly in eastern and southern Africa. The number of farmers that harvest water and grow short maturing crops and vegetables is also increasing in many countries in SSA. The demand for seasonal and intra-seasonal climate information has increased in some countries such as Ethiopia and Mali where farmers who received and applied integrated climate information increased their farm income by 10–80% compared to those who did not receive the information. Farmers in SSA are also widely involved in large-scale off-season soil and water conservation practices (e.g., Ethiopia), natural regeneration (e.g., Niger) and agroforestry practices (many countries). The experience from North Ethiopia shows that land rehabilitation has been accomplished over extensive areas despite a tenfold increase in population density. Community participation, commitment from local and national governments and strong partnerships among local, national and international development actors play an important role in scaling up CSA practices. Capacity building and the generation of new knowledge are essential for scaling up CSA in SSA. The presence of some successful smallholder CSA practices in SSA means that opportunities exist for cross-country learning and scaling-up by supporting farmers' efforts through exchange of knowledge, incentives and policies.

Acknowledgements This review paper was supported by the CGIAR Research Programs on Climate Change, Agriculture and Food Security (CCAFS), CRP MAIZE, Drought Tolerance Maize for Africa (DTMA) project funded by the Bill & Melinda Gates Foundation, Adoption Pathways and Maize–Legume Cropping Systems for Food Security in Eastern and Southern Africa (SIMLESA) projects funded by Australian Centre for International Agricultural Research (ACIAR). The views expressed in this paper are those of the authors and do not necessarily reflect the views of the donor or the authors' institutions.

References

Abate, T. (2013). *DTMA III highlights for 2012/13: An overview*. Presented at the Drought Tolerant Maize for Africa (DMA) Annual Meeting, 23–27 September 2013, Nairobi, Kenya.

Abebe, A., Lasage, R., Alemu, E., Gowing, J., & Woldaregay, K. (2012). Ethiopia: Opportunities for building on tradition—Time for action. In W. Critchley & J. Gowing (Eds.), *Water harvesting in Sub-Saharan Africa* (pp. 70–84). London.

Adimassu, Z., Langan, S., & Johnston, R. (2015). *Highlights of soil and water conservation investments in Ethiopia* (In Press).

AGRA. (2014). *Africa agriculture status report: Climate change and smallholder agriculture in sub-Saharan Africa (No. 2)*. Nairobi, Kenya: Alliance for a Green Revolution in Africa, AGRA.

Alemayehu, F., Taha, N., Nyssen, J., Girma, A., Zenebe, A., Behailu, M., et al. (2009). The impacts of watershed management on land use and land cover dynamics in Eastern Tigray (Ethiopia). *Resources, Conservation and Recycling, 53*(4), 192–198.

Araya, H., & Edwards, S. (2006). *The Tigray experience: A success story in sustainable agriculture* (Vol. 4). Penang, Malaysia: TWN.

Araya, T., Nyssen, J., Govaerts, B., Baudron, F., Carpentier, L., Bauer, H., et al. (2015). Restoring cropland productivity and profitability in northern Ethiopian drylands after nine years of resource-conserving agriculture. *Experimental Agriculture, 52*(2), 165–187.

Balana, B. B., Muys, B., Haregeweyn, N., Descheemaeker, K., Deckers, J., Poesen, J., et al. (2012). Cost-benefit analysis of soil and water conservation measure: The case of exclosures in northern Ethiopia. *Forest Policy and Economics, 15*(2012), 27–36.

Biazin, B., Sterk, G., Temesgen, M., Abdulkedir, A., & Stroosnijder, L. (2012). Rainwater harvesting and management in rainfed agricultural systems in sub-Saharan Africa—A review. *Physics and Chemistry of the Earth, 47–48*, 139–151.

Bruce, W. B., Edmeades, G. O., & Barker, T. C. (2002). Molecular and physiological approaches to maize improvement for drought tolerance. *Journal of Experimental Botany, 53*(366), 13–25.

CGIAR. (2006). *Drought tolerant crops for Drylands*. http://www.cgiar.org/web-archives/www-cgiar-org-impact-global-des_fact2-html/#top. Accessed on January 19, 2016.

Cooper, P. J. M., Cappiello, S., Vermeulen, S. J., Campbell, B. M., Zougmoré, R., & Kinyangi, J. (2013). *Large-scale implementation of adaptation and mitigation actions in agriculture*. CCAFS Working paper, No. 50. Copenhagen, Denmark.

Corbeels, M., de Graaff, J., Ndah, T. H., Penot, E., Baudron, F., Naudin, K., et al. (2014). Understanding the impact and adoption of conservation agriculture in Africa: A multi-scale analysis. *Agriculture, Ecosystems & Environment, 187,* 155–170.

Critchley, W., & Gowing, J. (Eds.). (2012). *Water harvesting in Sub-Saharan Africa*. London: Routledge.

Deichert, G., Krämer, F., & Schöning, A. (2014). *Turning degraded land into productive landscapes, Ethiopian highlands*. ETFRN NEWS 56, pp. 82–87.

Edmeades, G. (2008). Drought tolerance in maize: An emerging reality. In J. Clive (Ed.), *Global status of commercialized Biotech/GM Crops*. ISAAA Brief No. 39. Ithaca, NY: ISAAA.

Edwards, S., Berhan, T., Egziabher, G., & Araya, H. (2007). Successes and challenges in ecological agriculture: Experiences from Tigray, Ethiopia. In L. L. Ching, S. Edwards, N. E.-H. Scialabba (Eds.), *Climate change and food systems resilience in Sub-Saharan Africa*, pp. 231–294. Rome, Italy: FAO.

Fisher, M., Abate, T., Lunduka, R. W., Asnake, W., Alemayehu, Y., & Madulu, R. B. (2015). Drought tolerant maize for farmer adaptation to drought in sub-Saharan Africa: Determinants of adoption in eastern and southern Africa. *Climatic Change, 133*(2), 283–299.

Gowin, J., & Bunclark, L. (2013). Water harvesting experience in sub-Saharan Africa—Lessons for sustainable intensification of rainfed agriculture and the influence of available soils and rainfall data. *Geophysical Research Abstracts, 15,* 13640.

Hadgu, K. M. (2011). *Land rehabilitation and improved management: The case of Tigray, northern Ethiopia*. http://worldagroforestry.org/sites/default/files/drylands-seminar/DRYLANDSDAY1/PRESENTATIONS/TigrayEthiopia.pdf. Accessed on December 10, 2015.

Haglund, E., Ndjeunga, J., Snook, L., & Pasternak, D. (2011). Dry land tree management for improved household livelihoods: Farmer managed natural regeneration in Niger. *Journal of Environmental Management, 92*(7), 1696–1705.

Haileslassie, A., Priess, J., Veldkamp, E., Teketay, D., & Lesschen, J. P. (2005). Assessment of soil nutrient depletion and its spatial variability on smallholders' mixed farming systems in Ethiopia using partial versus full nutrient balances. *Agriculture, Ecosystems & Environment, 108*(1), 1–16.

Haregeweyn, N., Tsunekawa, A., Nyssen, J., Poesen, J., Tsubo, M., Meshesha, D. T., et al. (2015). Soil erosion and conservation in Ethiopia: A review. *Progress in Physical Geography, 39*(6), 750–774.

Hellmuth, M. E., Moorhead, A., Thomson, M. C., & Williams, J. (2007). Climate risk management in Africa: Learning from practice. In M. E. Hellmuth, A. Moorhead, M. C. Thomson, & J. Williams (Eds.), *Climate and society: Climate risk management in Africa: Learning from practice* (Vol. 1). Columbia University, New York, USA: IRI.

IWMI. (2007). *Water for food, water for life: A comprehensive assessment of water management in agriculture*. London: Earthscan.

Kassie, M., Teklewold, H., Marenya, P., Jaleta, M., & Erenstein, O. (2015). Production risk and food security under alternative technology choices in Malawi. Application of a multinomial endogenous switching regression. *Journal of Agricultural Economics, 66*(3), 640–659.

Kirui, O., & Mirzabaev, A. (2014). *Economics of land degradation in Eastern Africa*. ZEF Working Paper Series, No. 128. University of Bonn.

Lemenih, M., & Kassa, H. (2014). Re-greening Ethiopia: History, challenges and lessons. *Forests, 5*(8), 1896–1909.

Malesu, M., Khaka, E., Mati, B., Oduor, A., De Bock, T., Nyabenge, M., et al. (2007). *Mapping the potential of rainwater harvesting technologies in Africa: A GIS overview and atlas of development domains for the continent and nine selected countries*. Nairobi, Kenya.

Malesu, M. M., De Leeuw, J., & Oduor, A. (2012). *Water harvesting experiences from the SearNet (2003–2012)*. http://whater.eu/pluginfile.php/137/mod_page/content/37/Malesu_Waterharvesting ExperiencesfromtheSearnet2003-2012_IRC. Accessed on January 15, 2016.

Masih, I., Maskey, S., Mussá, F. E. F., & Trambauer, P. (2014). A review of droughts on the African continent: A geospatial and long-term perspective. *Hydrology and Earth System Sciences, 18*(9), 3635–3649.

Mekuria, W., Veldkamp, E., Haile, M., Nyssen, J., Muys, B., & Gebrehiwot, K. (2007). Effectiveness of exclosures to restore degraded soils as a result of overgrazing in Tigray, Ethiopia. *Journal of Arid Environments, 69*(2), 270–284.

Nyasimi, M., Amwata, D., Hove, L., Kinyangi, J., & Wamukoya, G. (2014). *Evidence of impact : Climate-smart agriculture in Africa*. CCAFS Working Paper No. 86, Copenhagen, Denmark.

Nyssen, J. (1998). Soil and water conservation under changing socio-economic conditions in the Tembien Highlands (Tigray, Ethiopia). *Bulletin de La Société Géographique de Liège, 35*, 5–17.

Nyssen, J., Descheemaeker, K., Haile, M., Deckers, J., & Poesen, J. (2007). *Lessons learnt from 10 years research on soil erosion and soil and water conservation in Tigray*. Tigray Livelihood Papers No. 7. Mekelle: Zala-Daget Project, Mekelle University, K.U. Leuven, Relief Society of Tigray, Africa museum and Tigray Bureau of Agriculture and Rural Development, 53 p.

Nyssen, J., Frankl, A., Haile, M., Hurni, H., Descheemaeker, K., Crummey, D., et al. (2014). Environmental conditions and human drivers for changes to north Ethiopian mountain landscapes over 145 years. *Science of the Total Environment, 485–486*, 164–179.

Nyssen, J., Frankl, A., Zenebe, A., Deckers, J., & Poesen, J. (2015a). Land management in the Northern Ethiopian Highlands: Local and global perspectives; past, present and future. *Land Degradation and Development, 26*(7), 759–764.

Nyssen, J., Frankl, A., Zenebe, A., Poesen, J., & Deckers, J. (2015b). Environmental conservation for food production and sustainable livelihood in Tropical Africa. *Land Degradation and Development, 26*(7), 629–631.

Nyssen, J., Haile, M., Naudts, J., Munro, N., Poesen, J., Moeyersons, J., et al. (2009a). Desertification? Northern Ethiopia re-photographed after 140 years. *Science of the Total Environment, 407*(8), 2749–2755.

Nyssen, J., Poesen, J., & Deckers, J. (2009b). Land degradation and soil and water conservation in tropical highlands. *Soil and Tillage Research, 103*, 197–202.

Pachpute, J. S., Tumbo, S. D., Sally, H., & Mul, M. L. (2009). Sustainability of rainwater harvesting systems in rural catchment of Sub-Saharan Africa. *Water Resources Management, 23*(13), 2815–2839.

Pretty, J., Toulmin, C., & Williams, S. (2011). Sustainable intensification in African agriculture. *International Journal of Agricultural Sustainability, 9*(1), 5–24.

Rinaudo, T. (2007). *The development of Farmer Managed Natural Regeneration*. http://fmnrhub.com.au/wp-content/uploads/2013/09/Rinaudo-2007-Development-of-FMNR.pdf. Accessed on January 29, 2016.

Schroter, D., Cramer, W., Leemans, R., Prentice, I. C. C., Araujo, M. B., Arnell, N. W., et al. (2005). Ecosystem service supply and vulnerability to global change in Europe. *Science, 310* (5752), 1333–1337.

Shiferaw, B., Tesfaye, K., Kassie, M., Abate, T., Prasanna, B. M., & Menkir, A. (2014). Managing vulnerability to drought and enhancing livelihood resilience in sub-Saharan Africa: Technological, institutional and policy options. *Weather and Climate Extremes, 3*, 67–79.

Steenwerth, K. L., Hodson, A. K., Bloom, A. J., Carter, M. R., Cattaneo, A., Chartres, C. J., et al. (2014). Climate-smart agriculture global research agenda: Scientific basis for action. *Agriculture & Food Security, 3*(1), 1–39.

Stroosnijder, L. (2009). Modifying land management in order to improve efficiency of rainwater use in the African highlands. *Soil and Tillage Research, 103*(2), 247–256.

Synnevag, G., & Lambrou, J. (2012). *Climate-smart agriculture: Possible roles of Agricultural Universities in a strengthened Norwegian climate change engagement in Africa*. Noragric Report No. 64, Department of International Environment and Development Studies, Noragric Norwegian University of Life Sciences, UMB.

Tadesse, M., Shiferaw, B., & Erenstein, O. (2015). Weather index insurance for managing drought risk in smallholder agriculture: Lessons and policy implications for Sub-Saharan Africa. *Agricultural and Food Economics, 3*, 26.

Tall, A., Hansen, J., Jay, A., Campbell, B., Kinyangi, J., Aggarwal, P. K., et al. (2014). *Scaling up climate services for farmers: Mission possible*. Learning from good practice in Africa and South Asia. CCAFS Report No. 13.

Teklewold, H., Kassie, M., Shiferaw, B., & Kohlin, G. (2013). Cropping system diversification, conservation tillage and modern seed adoption in Ethiopia: Impacts on household income, agricultural chemical use and demand for labor. *Ecological Economics, 93*, 85–93.

Tesfaye, K., Gbegbelegbe, S., Cairns, J. E., Shiferaw, B., Prasanna, B. M., Sonder, K., et al. (2015). Maize systems under climate change in sub-Saharan Africa: Potential impacts on production and food security. *International Journal of Climate Change Strategies and Management, 7*(3), 247–271.

Tigabu, M., Lemenih, M., Negash, M., & Teketay, D. (2014). Rehabilitation of degraded forest and woodland ecosystems in Ethiopia for sustenance of livelihoods and ecosystem services. In P. Katila, G. Galloway, W. de Jong, P. Pacheco, & G. Mery (Eds.), *Forests under pressure—Local responses to global issues* (pp. 299–313). Vantaa: International Union of Forest Research Organisations (IUFRO).

Tougiani, A., Guero, C., & Rinaudo, T. (2009). Community mobilisation for improved livelihoods through tree crop management in Niger. *GeoJournal, 74*(5), 377–389.

Wakeyo, M. B. (2012). *Economic analysis of water harvesting technologies in Ethiopia*. http://edepot.wur.nl/240909. Accessed on January 15, 2016.

Walraevens, K., Vandecasteele, I., Martens, K., Nyssen, J., Moeyersons, J., Gebreyohannes, T., et al. (2009). Groundwater recharge and flow in a small mountain catchment in northern Ethiopia. *Hydrological Sciences Journal, 54*(4), 739–753.

WMO. (2005). *Climate and land degradation*. http://www.wmo.int/pages/themes/wmoprod/documents/WMO989E.pdf. Accessed on January 24, 2016.

World Bank, African Development Bank. (2012). *The transformational use of information and communication technologies in Africa*. http://www.infodev.org/en/Publication.1162.html. Accessed on January 24, 2016.

Xoconostle-Cázares, B., Ramírez-Ortega, F. A., Flores-Elenes, L., & Ruiz-Medrano, R. (2010). Drought tolerance in crop plants. *American Journal of Plant Physiology, 5*(5), 241–256.

Ecosystem-Based Adaptation (EbA) as an Adaptation Strategy in Burkina Faso and Mali

Kennedy Muthee, Cheikh Mbow, Geoffrey Macharia and Walter Leal Filho

1 Introduction

One of the major threats to sustainable development is climate change (SDSN 2014). The world climate is still changing fast, posing serious challenges to sustainable livelihood and social-economic development, particularly in the developing countries. Climate change effects are evident in different sectors, such as environment, health, education, food security, energy, and inter alia (WWF 2009; Andrade et al. 2010), and they are a major risk to poor communities who lack the financial, institutional and technical capacity to adapt (Munang et al. 2014). Notably, a temperature rise beyond 2 °C can have devastating effects on crop production, water access, health and economic development (UNFCCC 2011a). This calls for different players such as governments, communities, institutions and individuals to recognize the urgency of addressing social, environmental and economic effects of climate change.

The United Nations Framework Convention on Climate Change (UNFCCC) has recognized that the Least Developed Countries (LDCs) are the most vulnerable to

K. Muthee (✉) · G. Macharia
Department of Environmental Science, School of Environmental Studies, Kenyatta University, 43844-00100 Nairobi, Kenya
e-mail: kenwahome@gmail.com

G. Macharia
e-mail: Machariag2001@gmail.com

C. Mbow
International START Secretariat, Washington, DC, USA
e-mail: cmbow@start.org

W. Leal Filho
Faculty of Life Sciences, Hamburg University of Applied Science, 21033 Hamburg, Germany
e-mail: walter.leal@haw-hamburg.de

W. Leal Filho et al. (eds.), *Climate Change Adaptation in Africa*, Climate Change Management, DOI 10.1007/978-3-319-49520-0_13

climate change effect, and has guided them in establishing the National Adaptation Programmes of Action (NAPA) (Pramova et al. 2012). NAPAs is a political instrument that helps the countries to identify and prioritize their most urgent adaptation needs, whose further delay can imply increased vulnerability and high costs later (UNFCCC 2010). Further, they provide an ideal starting point for country-specific adaptation initiatives through adaptation projects. Most recently, the UNFCCC parties have adopted the Intended Nationally Determined Contributions (INDCs) which are supposed to be addressing countries priority action for contributing voluntarily to mitigation efforts while supporting adaptation needs in developing countries (UNFCCC 2014).

Over the years, different adaptation approaches that can play a vital role in enhancing the NAPA's ability in promoting adaptation and sustainable development have emerged. One of them is through Ecosystem-based Adaptation (EbA), which seeks to promote societal resilience via ecosystems management and conservation (Colls et al. 2009). This approach recognizes the centrality of ecosystems in the adaptation process (Munang et al. 2014). Ecosystems, among others, maintain, strengthen and enrich different elements of life and livelihood on the planet (Capistrano et al. 2009). They support life on the earth through provision of ecosystem services (BirdLife International 2010), which are defined as the benefits that natural ecosystems provide to the society (Boyd and Banhzaf 2007; Prato 2008). These benefits are classified into four broad categories—supporting, provisioning, regulating and cultural (MEA 2005). The capacity of the ecosystem to deliver ecosystem services depends on its condition (healthy state) as well as the ability of the society to access it (Pramova et al. 2012).

The current study sought to (a) analyze the dimensions of the climate-change adaptation strategies, (b) Explore how ecosystem-based adaptation initiatives are incorporated in NAPA projects, and (c) draw lessons on EbA as a sustainable development strategy using Mali and Burkina Faso as case studies. It further recommended on how to improve the projects to harnesses ecosystem services, reduces negative impacts on the ecosystems, and promote social well-being.

2 The Concept of Ecosystem-Based Adaptation

EbA is defined as the approach of "*sustainably managing, conserving and restoring ecosystems to provide the services that allows people to adapt to climate change effects*" (Colls et al. 2009; IUCN 2015). These strategies utilize ecosystem services and biodiversity as a part of the community adaptation strategies to climate change effects (Gupta, Nair 2012; Munang et al. 2013a). The approach considers those adaptation projects that have both 'ecosystem face' and 'human face' on them. EbA recognizes the fundamental role played by the ecosystem services in the reduction of people's vulnerability to the effects of climate change (Vignola et al. 2009; UNFCCC 2011a).

2.1 Benefits of EbA

EbA strategies provide an array of institutional, social-cultural, ecological and economic benefits. The approach promotes restoration and protection of ecosystems; thus promoting healthy ecosystems (McGray et al. 2007) which in return acts as natural barriers to extreme weather conditions such as droughts, landslides, flooding, and extreme temperatures, among others (Andrade et al. 2010). A healthy ecosystem is resilient to the effects of climate change and ensures that communities continue to enjoy ecosystem services that they provide (Falkenburg et al. 2010). Livelihood support contributes towards poverty alleviation among communities. EbA approach leads to protection, restoration and management of ecosystems (Locatelli et al. 2008) which in return promotes conservation of biodiversity in addition to building the capacity of the people to adapt to climate change variability (Mercer et al. 2007), ultimately leading to sustainable development.

2.2 Principles of EbA

Several principles guide EbA strategies. Such includes promotion of multi-sectoral approach in the ecosystems management of different landscapes (Speranza et al. 2010). The EbA approach promotes collaboration and coordination of various sectors, communities and players that utilize ecosystem services (Richardson 2010; Delica-Willison and Gaillard 2011). The EbA functions at multiple spatial scales and landscapes such as local, sub-national, national and region (Cadag and Gaillard 2012). It is important to consider the complexity of ecosystems such as drivers to vulnerability, geopolitics involved in the ecosystems management, and trans-boundary nature of an ecosystem (McConney and Mahon 2005; Orlove et al. 2010). Lastly, it promotes participation, cultural appropriateness, accountability and embracing diversity in the project design and execution (Munang et al. 2013b).

2.3 Applications of EbA

The role and application of ecosystems in adaptation is recognized at the international level under the United Nations Framework Convention on Climate Change (UNFCCC), the Convention on Biological Diversity (CBD) and the United Nations Convention to Combat Desertification (UNCCD) (IUCN 2015). When one considers a global change context, it is widely known that climate change has a wide range of implications and impacts which are beyond those on ecosystems (Leal 2001). One special feature in EbA is the fact that, apart from the use of ecosystem services for people's adaptation, they can provide a concrete contribution towards increasing the resilience of these ecosystems to climate change (Locatelli and Pramova 2015). EbA can be used in areas as varied as:

(a) agriculture,
(b) water resource management,
(c) forest management interventions,
(d) biodiversity conservation and management.

In outlining the applications of EbA, one should not overlook the fact that we need new and enhanced adaptation approaches, so as to cope with the many problems and pressures posed by climate change, at different levels. One of them should be the engagement of decision-makers in the process, since the active participation of ecologists or environmentalists alone does not suffice (Vignola et al. 2009). Thus, there is a pressing need to engage various sectors in the planning and allocation of resources for adaptation action.

3 Materials and Methods

The study followed a survey approach. Two NAPAs (Gouvernement du Burkina Faso 2007; Gouvernement du Mali 2007) with a total of 31 projects were studied after which a summary was created in the form of a database for further analysis to reveal the adaptation patterns, extent of incorporation of EbA strategies, lessons learnt and prospects of EbA. The criteria for project inclusion and exclusion for analysis is summarized in Table 1.

4 Results and Discussions

4.1 Analysis of Adaptation Projects in the Study Areas

Thirty-one projects were examined from the two countries NAPAs (61% in Mali and 39% in Burkina Faso). They were categorized into different thematic (sectors)

Table 1 Criteria for projects selection

No.	Type of project	Description
1	Projects without ecosystem management	These projects do not mention any ecosystem management practice
2	Project with ecosystem management for environment	They are geared towards conserving ecosystem without mentioning the human benefits and well-being
3	Project with ecosystem management for both ecosystem resiliency and human adaptation (EbA)	These projects link ecosystem management, ecosystem services with human adaptation strategies and social well-being. They are defined as EbA projects in this study

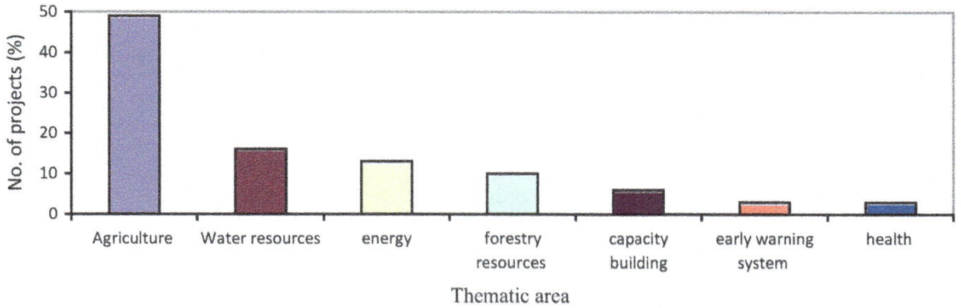

Fig. 1 Percentage number of projects per thematic area

Table 2 Percentage number of projects per geographical scope and scale

	Local	Sub-national	National
No. of projects (%)	13	61	26

areas that dictated the EbA approach they took. Most of the project studies (49%) fell into the agricultural sector, with Mali recording 53% and Burkina Faso 42% of the total projects. These results were consistent with studies by Richardson (2010), Pramova et al. (2012) and McGray et al. (2007) that singled out agriculture as the main area of focus in climate change adaptation due to the historical food insecurity in Africa. The water resources sector, energy sector, and forestry sector also recorded a significant percent of adaptation projects as summarized in Fig. 1.

The adaptation projects took different geographic scope and scale of execution. Three broad categories were established, local, sub-national and national scales. Most of the studies projects (61%) fell under the sub-national category, which referred to the projects that cut across two or more geographical locations (districts or regions) as well as projects that targeted ecosystems cutting across several geographical areas. Burkina Faso had the most of these projects at 75% as summarized in Table 2.

The adaptation projects also varied with the implementation duration (in years). The study established five categories based on the project's duration, ranging from less than one year to over five years. 68% of the projects studies were scheduled to be implemented within three years and 23% within two years. It is notable that 100% of the projects in Burkina Faso had three years implementation schedule.

Ten ecosystem services were established from the study and were categorized based on the MEA (2005) report—provisioning, support, regulatory and cultural services. The study established that 58% of the projects sought to provide provisioning services while regulatory services were provided by 21% of the total projects studied. Among the provisioning services, food and fodder topped the list at 50 and 29% respectively. The dominant support service was soil formation and fertility to support the provision of food and fodder, while the principal regulatory service was soil erosion/siltation control at 50%. These results were consistent with

Table 3 Summary of Ecosystem services identified

Provisioning services	Regulatory services	Support services
Food	Erosion/siltation/ sedimentation control	Soil formation/ fertility/productivity
Fodder	Disease control	
Habitats	Bushfire prevention	
Non- timber products	Pollution control	
Water services		

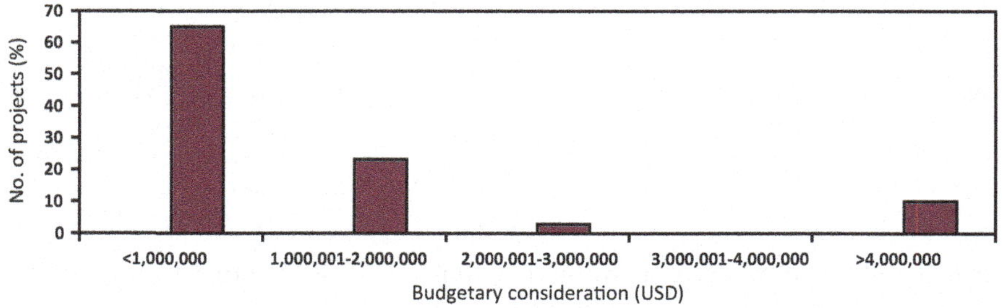

Fig. 2 Percentage number of projects per budgetary consideration

those of Pramova et al. (2012), which established that provisioning and regulatory services are dominant within the adaptation projects. Notably, there is also a clear pattern of the high priority of the number of projects within agricultural sector and the service they seek to provide, that is food and fodder provision (Table 3). The ecosystem services established are summarised below in Table 3 based on the Millennium Ecosystem Assessment (2005) classification.

The projects also varied with their budgetary considerations. The study defined five categories of budget between one USD and over four million USD. Most of the projects (64%) had a low budgetary consideration of one and one Million USD (92% and 47% in Burkina Faso and Mali respectively). Only 16% of the projects recorded highest budgetary consideration of over 4 million USD. The results concur with those of UNFCCC (2011b) and Wamunyima and Miga (2014) that point out budgetary and financial constraints as the major challenge facing the implementation of adaptation projects. The results are summarized in Fig. 2.

4.2 EbA Approaches Identified in the Projects

The EbA approach was largely dependent on the thematic area of the adaptation. Sixteen percent of the total projects studied mentioned the different way of employing ecosystem-based adaptation approach in their implementation activities and outputs as summarized in Table 4.

Table 4 EbA approaches established in different sectors

No.	Sector	Adaptation approach	Examples of activities established	Case studies
1	Agricultural sector	Integrated/sustainable agricultural practices to enhance food security and enhance ecosystem functioning	Rehabilitation of rangeland, agro-forestry practices, planting drought tolerant food crops	• The extension of improved varieties adapted to climatic conditions in major food crops (millet, sorghum, maize and rice) in Mali • Securing pastoral areas in the Sahel and East Burkina Faso
2	Water sector	Integrated watershed management to address food insecurity, reduce poverty, enhance watershed functionality, improve people's living conditions	Run-off control, restoration of watersheds and water points, rehabilitation of watersheds, reduce watershed degradation, reforesting catchment areas	• Catchment runoff, creation and restoration of water points in Mali • development and management of the pond Oursi in Burkina Faso
3	Cross-cutting sectors	Multi-sectored ecosystems management to reduce people's vulnerability to climate change effects	Improve soils productivity and fertility, control of soil and water erosion, rehabilitation of degraded lands, natural resources conservation, reforestation exercises	• Development action CES/DRS for agriculture, forestry and pastoralism projects in Mali
4	Forestry sector	Integrated forestry management	Promotion of non-timber products, promotion of reforestation, create new plantations and natural trees regeneration, biodiversity conservation	• Planning, management of natural formations and development of non-timber forest products (NTFPs) in Eastern Burkina Faso
5	Wildlife sector	Sustainable wildlife management	Species monitoring, ecological/eco-zones conservation, wildlife management, sustainable harvesting and co-management plans	• Towards promotion of management of wildlife and its habitat by local communities in the Mouhoun in Burkina Faso

5 Lessons, Prospects and Recommendations of EbA in Adaptation Projects

This study reveals several key lessons, prospects and recommendations on EbA and climate change adaptation strategies as reflected in the National Adaptation Programmes of Actions (NAPA) of the two countries.

The adaptation strategies are not uniform: they vary in terms of scope, duration, thematic area, ecosystem services provided and the EbA approach put in place. The key determinant is the thematic area (sector). A common pattern of the adaptation projects was drawn from the current study. The majority fell within the agricultural sector (49%) with low budgetary allocation (1–1 M USD), conducted within sub-national geographical and ecological coverage (61%). They had a medium implementation duration (68% within three years), and majority sought to enhance provisioning ecosystem services (58%). This variability informed the diverse EbA approach taken by these projects. The 'one suits all' adaptation strategy may not be appropriate for current and future EbA projects.

The EbA strategies should integrate indigenous and contemporary knowledge. Such includes sustainable agricultural practices using indigenous knowledge and crops (such as sorghum and millet) to ensure that food security and promote soil productivity. Basing the EbA approaches on the community's experience is likely to yield more participation and success. Studies such as Walmsley (2006), Mercer et al. (2010), and Speranza et al. (2010) all points out the need for community knowledge and participation integration in the adaptation processes.

There is a need for project contextualization based on the previous social, economic and environmental experiences within the project areas. Indeed, the whole NAPA process is based on the country-specific environmental changes that have affected the region's social and economic development. This approach aids in the prioritizing of the activities that are aimed at promoting ecosystem resiliency, promoting human wellbeing, and reduction of vulnerability, supporting ecosystem services and ultimately enhancing sustainable development.

The need for research and development of EbA cannot be understated. Research involves testing, refining and up scaling the EbA approach based on the local context. The project designers should employ tried and tested EbA approaches that seek to enhance both ecosystems and human wellbeing. There is also the need to document such initiatives for replication in other areas with similar geographical and ecosystems characteristics.

The incorporation of EbA and non-EbA activities in a project is also fundamental. Such includes the development of infrastructures to support the ecosystem services provided by these projects. To illustrate, the *planning, management of natural formation and development of the non-timber forest products* project would require non-EbA approaches such as development of roads to ensure that the communities make a living from their conservation and wise resource use exercises.

Adaptation activities should seek to integrate multi-sectoral and multi-stakeholder approach to meet the broad EbA objectives and principles. The designing and execution of these projects should incorporate social, economic, and ecological dynamics to yield multiple benefits. To realize this, the project designers and implementers should understand and balance between the benefits and trade-offs related to the execution of such projects.

6 Conclusions

The NAPAs provides an ideal entry point to the country-specific adaptation strategies. The focus is currently shifting from NAPA to INDCs creating the need to incorporate more EbA strategies in adaptation projects to meet both human and ecosystem needs; thus promoting sustainable development. The study shows a clear adaptation pattern, with more projects classified under agricultural sector, low budgetary consideration, medium implementation duration, and mainly seeking to provide provisioning and regulatory ecosystem services. The EbA strategies are considered in 16% of the studied projects and depend largely on the project's thematic area. Some of the strategies were based on agricultural practices, integrated watershed management, multi-sectored ecosystem management, intergraded forestry management and sustainable wildlife management. The analysis of the two NAPA projects also yielded several key lessons and recommendations on EbA. The community participation and local (indigenous) knowledge should be central in the designing and implementation of NAPA projects. The external knowledge, research, monitoring and evaluation are essential aspects of EbA, and should be incorporated in the future adaptation projects. Lastly, the adaptation projects should be contextualized according to the communities and ecosystems around them. The study concludes that incorporation of the diverse ecosystem based approaches in different thematic areas can promote sustainable development.

Acknowledgements The authors are very grateful to the BIODEV project funded by the Finland Government and the CGIAR Research Programs (CRP) 6.3 and 6.4.

References

Andrade, A., Herrera, B., & Cazzolla, R. (2010). *Building resilience to climate change: Ecosystem based adaptation and lessons from the field*. Gland, Switzerland: IUCN.

Boyd, J., & Banzhaf, S. (2007). What are ecosystem services? The need for standardized environmental accounting units. *Ecological Economics, 63*(2–3), 616–626.

BirdLife International. (2010). *Partners with nature: How healthy ecosystems are helping the world's most vulnerable adapt to climate change*. Zeist, The Netherlands: BirdLife International.

Cadag, J., & Gaillard, J. (2012). Integrating knowledge and actions in disaster risk reduction: The contribution of participatory mapping. *Area, 44,* 100–109.

Capistrano, D., et al. (2009). Science for managing ecosystem services: Beyond the Millennium Ecosystem Assessment. *Proceedings of the National Academy of Sciences of United States of America, 106*(5), 1305–1312.

Colls, A., Ash, N., & Ikkala, N. (2009). *Ecosystem-based adaptation: A natural response to climate change.* Gland, Switzerland: IUCN.

Delica-Willison, Z., & Gaillard, J. (2011). Community action and disaster. In B. Wisner, J. C. Gaillard, & I. Kelman (Eds.), *The routledge handbook of hazards and disaster risk reduction.* London: Routledge.

Falkenburg, J., Burnell, W., Connell, D., & Russell, B. (2010). Sustainability in near-shore marine systems: Promoting natural resilience. *Sustainability, 2,* 2593–2600.

Gouvernement du Burkina Faso. (2007). *Programme d'action national d'adaptation a la variabilite et aux changements climatiques (PANA Du Burkina Faso).* Available http://unfccc.int/resource/docs/napa/bfa01f.pdf

Gouvernement du Mali. (2007). *Programme d'action national d'adaptation aux changements climatiques.* Available from http://www.undp-alm.org/sites/default/files/downloads/mali_napa.pdf

Gupta, K., & Nair, S. (2012). Ecosystem approach to disaster risk reduction. New Delhi: National Institute of Disaster Management. Available http://unfccc.int/focus/indc_portal/items/8766.php

IUCN. (2015). *Ecosystem-based adaptation.* Available https://www.iucn.org/about/work/programmes/ecosystem_management/climate_change/eba/

Leal, F. (Ed.). (2001). *The economic, social and political aspects of climate change.* Berlin: Springer.

Locatelli, B., Kanninen, M., Brockhaus, C., Murdiyarso, D., & Santoso, H. (2008). Facing an uncertain future: How forests and people can adapt to climate change. CIFOR (Center for International Forestry Research). *Forest Perspectives, 5,* 1–86.

Locatelli, B., Pramova, E. (2015). *Ecosystem-based adaptation.* Indonesia: École thématique.

McConney, P., & Mahon, R. (2005). Size matters: Scaling management and capacity to achieve sustainability in SIDS. *FAO Fisheries Report, 782,* 293–306.

McGray, H., Hammill, A., Bradley, R. (2007). *Weathering the storm: Options for framing adaptation and development.* Washington, D.C: World Resources Institute (WRI).

Mercer, J., Dominey-Howes, D., Kelman, I., & Lloyd, K. (2007). The potential for combining indigenous and western knowledge in reducing vulnerability to environmental hazards in small island developing states. *Environmental Hazards, 7,* 245–256.

Mercer, J., Kelman, I., Taranis, L., & Suchet-Pearson, S. (2010). Framework for integrating indigenous and scientific knowledge for disaster risk reduction. *Disasters, 34,* 214–239.

Millennium Ecosystem Assessment. (2005). *Ecosystems and human well-being: Synthesis.* Washington, DC: Island Press.

Munang, R., Andrews, J., Alverson, K., & Mebratu, D. (2014). Harnessing ecosystem-based adaptation to address the social dimensions of climate change. *Environment (56), 1,* 18–24.

Munang, R., Thiaw, I., & Rivington, M. (2013a). Ecosystem management: Tomorrow's approach to enhancing food security under a changing climate. *Sustainability, 3,* 937–954.

Munang, R., Thiaw, I., Alverson, K., Mumba, M., Liu, J., & Rivington, M. (2013b). Climate change and ecosystem-based adaptation: A new pragmatic approach to buffering climate change impacts. *Current Opinion in Environmental Sustainability, 5,* 1–5.

Orlove, B., Roncoli, C., Kabugo, M., & Majugu, A. (2010). Indigenous climate knowledge in southern Uganda: The multiple components of a dynamic regional system. *Climatic Change, 100,* 243–265.

Pramova, E., Locatelli, B., Brockhaus, M., & Fohlmeister, S. (2012). Ecosystem services in the national adaptation programmes of action. *Climate Policy, 12*(4), 393–409.

Prato, T. (2008). Conceptual framework for assessment and management of ecosystem impacts of climate change. *Ecological Complexity, 4,* 329–338.

Richardson, R. (2010). Ecosystem services and food security: Economic perspectives on environmental sustainability. *Sustainability, 2,* 3520–3548.

Speranza, C., Kiteme, B., Ambenje, P., Wiesmann, U., & Makali, S. (2010). Indigenous knowledge related to climate variability and change: Insights from droughts in semi-arid areas of former Makueni District, Kenya. *Climatic Change, 100,* 295–315.

SDSN. (2014). *Solutions for sustainable agriculture and food systems.* NY: The Sustainable Development Solutions Network.

United Nations Framework Convention on Climate Change (UNFCCC). (2011a). *Ecosystem-based approaches to adaptation: Compilation of information.* Bonn, Germany: UNFCCC.

United Nations Framework Convention on Climate Change (UNFCCC). (2011b). *Assessing the costs and benefits of adaptation options: An overview of approaches.* Bonn, Germany: UNFCCC.

United Nations Framework Convention on Climate Change (UNFCCC). (2010). *National adaptation programmes of action (NAPA).* Available http://unfccc.int/national_reports/napa/items/2719.php

United Nations Framework Convention on Climate Change (UNFCCC). (2014). *Intended nationally determined contributions (INDCs).* Available http://unfccc.int/focus/indc_portal/items/8766.php

Vignola, R., Locatelli, B., Martinez, C., & Imbach, P. (2009). Ecosystems-based adaptation to climate change: What role for policy-makers, society and scientists? *Mitigation and Adaptation Strategies for Global Change, 14*(8), 691–696.

Walmsley, J. (2006). The nature of community: Putting community in place. *Dialogue, 25,* 5–12.

Wamunyima, M., Miga, W. (2014). *Assessment of adaptation financial flows in Zambia.* Consultancy report prepared for AFAI in Zambia, unpublished.

WWF. (2009). *WWF adaptation case studies for responding to climate change impacts.* Buenos Aires, Argentina: WWF Publication.

Building Adaptive Communities Through Integrated Programming: CARE Ethiopia's Experience with Climate Vulnerability and Capacity Analysis (CVCA)

Alebachew Adem, Karl Deering and Samuel Molla

1 Background

Climate change in Ethiopia: Rainfall variability and increased temperature over recent years have resulted in increased vulnerability in an already fragile Ethiopian environment. The pattern of rainfall in Ethiopia is experiencing both long term change and short term fluctuations. Mean annual temperature increased by 1.3 °C between 1960 and 2006, at an average rate of 0.28 °C per decade and 0.32 °C during JAS (McSweeney et al. 2007). Hot days (i.e. temperatures exceeded on 10% of days/nights of a specific region and season) have become increasingly frequent, as have hot nights. The number of cold days has decreased significantly throughout the year, except for December, January and February (DJF), and cold nights are much less common in all seasons (McSweeney et al. 2007).

Ethiopia's Vulnerability: Climate change is a significant threat to Ethiopia's development. Changing patterns and intensities of rainfall and increasing temperatures will have consequences for all Ethiopians, but especially for the more than 70 million poor people whose survival depends on rain-fed agriculture (crop and livestock). The reasons for Ethiopia's vulnerability are manifold. Its geographical location and topography entail high vulnerability to the impacts of climate change. Historically, Ethiopia has been prone to extreme weather variability. Since the early 1980s, the country has suffered seven major droughts—five of which have led to severe food insecurity—in addition to dozens of local droughts (World Bank 2010).

A. Adem (✉)
CARE USA, Addis Ababa, Ethiopia
e-mail: alebachew.adem@care.org

K. Deering
Poverty, Environment and Climate Change Network,
CARE International, Dar es Salaam, Tanzania

S. Molla
CARE Ethiopia, Addis Ababa, Ethiopia

W. Leal Filho et al. (eds.), *Climate Change Adaptation in Africa*,
Climate Change Management, DOI 10.1007/978-3-319-49520-0_14

Ethiopia's economy will remain highly vulnerable to exogenous shocks mainly because it is highly dependent on rain-fed small-scale agriculture which is subsistence-oriented and characterized by low inputs and outputs. Pastoral areas struggle with rangeland degradation, cyclical drought and historic under-investment. The population growth rate is 2.5% per annum, resulting in a doubling of the population in less than 30 years. Ethiopia's population is projected to reach 174 million by 2050.

Re evaluation of Approach: Climate change is prompting revision of existing strategies and the development of flexible models that allow programs to respond to the real and changing needs of rural populations. In order to plan effective adaptation actions, scientific climate change analysis is vital for broad context. However, at the local level, the most relevant information and knowledge often already exists or can be generated through local stakeholders' own analysis (ACCRA 2011). This entails facilitating analysis of vulnerability and adaptive capacity by members of communities themselves. In doing local stakeholder analysis CARE applies participatory values, processes and methods, to enable communities to articulate and enhance their own knowledge and understanding, and to plan action accordingly. CARE's approach to climate change adaptation is grounded in the knowledge that people must be empowered to transform and secure their rights and livelihoods. It also recognizes the critical role that local and national institutions, as well as public policies, play in shaping people's adaptive capacity.

CVCA: The CVCA helps communities and facilitating organizations to understand the implications of climate change for the lives and livelihoods of people. By combining local knowledge with scientific data, the process builds people's understanding about climate risks and adaptation strategies. It provides a framework for dialogue within communities, as well as between communities and other stakeholders. Recognizing that local actors must drive their own future, CVCA prioritizes local knowledge on climate risks and adaptation strategies in the data gathering and analysis process. The CVCA methodology is applied and adapted in Ethiopia to increase understanding among stakeholders, gather and analyze information to design climate-smart initiatives, as well as to integrate adaptation issues into local planning and decision-making.

Box 1: What's new and different about the CVCA?

There are a number of characteristics which make the CVCA process different from other forms of participatory learning and analysis. These include:

- **Focus on climate change**: The CVCA focuses on understanding how climate change will affect the lives and livelihoods of target populations. It examines hazards, vulnerability to climate change and adaptive capacity with a view to building resilience for the future. The types of tools suggested are tried-and-true Participatory Learning for Action (PLA) tools, but with a climate "lens". The tools are used to draw out issues, which are then examined in the context of climate change through guided discussion.

- **Analyzing conditions and hazards**: The CVCA attempts to combine good practices from analyses done for development initiatives, which tend to focus on conditions of poverty and vulnerability, and those done within the context of DRR, which tend to focus on hazards. The framework of the CVCA facilitates analysis of the information gained from both types of assessments from a climate change perspective.
- **Emphasis on multi-stakeholder analysis, collaborative learning and dialogue**: While the primary purpose of the CVCA is to analyze information, the methodology is designed to balance the research agenda with a process of learning and dialogue among local stakeholders. This can yield a greater understanding within communities of the resources available to them to support adaptation, and can promote dialogue among stakeholders on adaptation actions that make sense.
- **Focus on communities but also examines enabling environment**: Vulnerability to climate change can vary within countries, communities and even households. Therefore, adaptation requires context-specific activities, with strategies targeted to meet the needs of different vulnerable groups. At the same time, local and national policies and institutions play a critical role in shaping people's capacity to adapt to climate change. Thus, the CVCA process focuses on the community level but incorporates analysis of issues at regional and national level in an effort to foster an enabling environment for community-based adaptation.

Through the World Bank-supported Climate Smart Initiative (CSI), for example, CARE Ethiopia has conducted CVCA in 24 woreda, (comprising 212 watersheds) in six regions of Ethiopia. Through the—USAID Graduation with Resilience to Achieve Sustainable Development (GRAD) project, CARE has conducted CVCA in 16 woredas in PSNP highland woredas and in pastoral areas Pastoralist Areas Resilience Improvement through Market Expansion (PRIME) project conducted CVCA in 32 woredas 168 community group so our knowledge base and experience is significant.

2 CARE Ethiopia's Selected Projects

This report draws on analysis and evaluation undertaken by three projects implemented by CARE Ethiopia.

2.1 Climate Smart Initiative (CSI)

CSI is a multi-donor trust-funded component of two important food security programs in Ethiopia—the Productive Safety Net Program (PSNP) and the Household Asset Building Program (HABP). This second phase of CSI is designed to ensure that these programs are made climate smart through the systematic integration of the implications of climate change in program activities. It piloted climate smart activities and drew together lessons to ensure that the next generation of resilience building programs will enable the Government of Ethiopia to better manage risks related to climate change. CSI is implemented through the existing PSNP and HABP systems and structures.

2.1.1 The CSI CVCA Process and Achievements

CSI has been implemented on a pilot basis in 212 Kebeles in 24 Woredas, across six regions of Ethiopia. There were two rounds of analysis: in November 2013, the process was facilitated with 72 Kebeles as an initial pilot and capacity development exercise; in the remaining Kebeles it was conducted as part of the PSNP/HABP joint planning process in April 2014. The process was conducted by teams comprising Development Agents (DAs) and Woreda staff, including experts on natural resource management, early warning, crop extension and livestock production. Oversight of the process was provided by one PSNP and one HAPB staff in each region (in some cases, the regional experts also participated in the field work). Discussions were held with focus groups of women and men in each of the targeted Kebeles. The Focus Group Discussions used participatory tools to gather information on community perspectives and experiences. The tools used were: community mapping, historical timeline, seasonal calendar and vulnerability matrix. In addition, the process (in the second round) included a discussion on access to information and knowledge on climate change. The climate-smart planning guide provided detailed guidance on facilitation of FGDs and reporting of the results.

In this pilot phase, although the CVCA process was found to be highly resource- and labour-intensive, there is consensus among government decision-makers that the process was valuable. They identify a number of benefits that have accrued from the pilot processes:

2.1.2 Demand-Driven and Participatory Planning

- CSI's work on planning, including the CVCA methodology and building participatory planning capacity among officials at woreda and kebele level, and among communities and individuals themselves, constitutes what is probably the most important achievement of the programme.

- There is anecdotal evidence from the final CSI Regional Workshops that woredas and regions are committed to sustaining at least some of these methodological innovations (Dazé 2014a). Further, he second CVCA process resolved many of the time and cost issues experienced with the initial iteration. The process was revised with a view to integration with the Community based public work development guidelines, including tools to match interventions to CVCA findings and a prioritization tool to address the planning gap.

2.1.3 Climate-Smart Clusters

The cross-integration of PSNP and HABP activities has been hailed as CSI's second outstanding achievement, supporting a fundamental shift in thinking towards a 'livelihoods approach to climate adaptation' by which income-boosting activities are chosen in a risk-aware manner.

- Although PSNP 4 has been designed to integrate livelihood activities alongside public works, there is a risk that parallel processes will apply in practice, rather than the genuinely integrated approach developed by CSI. Sustainability will again depend on the political will to embed the approach in future PSNP 4 operations.

Figure 1 illustrates how CSI has worked to integrate climate-smart thinking into the PSNP and HABP project cycle. The practical meaning of 'climate-smart' is therefore of critical importance to assessing the achievements of CSI.

2.1.4 CSI CVCA Learning

- The Outcome Evaluation judges that CSI has been able to move some way towards making PSNP/HABP plans and implementation activities 'climate-smart', but a degree of ambiguity remains. Improved participatory Natural Resource Management planning, taking account of risk and vulnerability and the integration of measures to improve livelihoods and increase

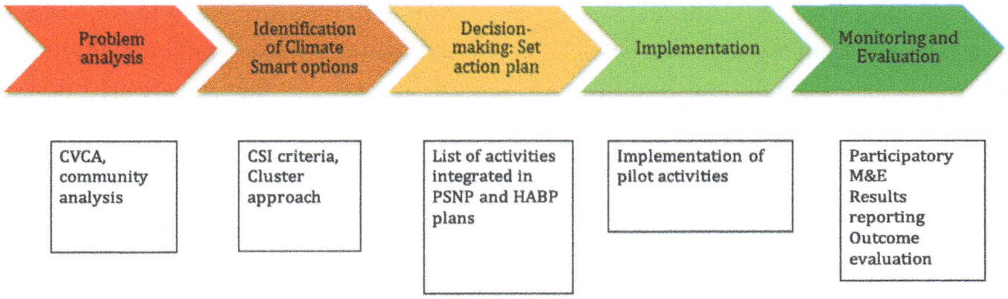

Fig. 1 CSI approach in PSNP/HABP

resilience, meant that the PSNP/HABP plans can be regarded as 'more climate-smart than before', but this is by no means absolute.
- The approach falls short in two critical ways:
  - The centrality of weather forecast and climate information, and contingency planning in the climate-smart approach cannot be overstated.
  - More fundamentally, the findings of the Participatory Monitoring and Evaluation (PME) process and the technical assessment of CSI's climate-smart approach suggest that officials and communities still struggle to articulate the meaning of 'climate-smart' as distinct from improved participatory and integrated NRM and livelihood planning. They are able to highlight the key differences in the CSI PSNP/HABP planning process in terms of: active participation of communities; climate change analysis in prioritizing activities, and integration of PSNP and HABP through cluster approaches.

2.2 *Graduation with Resilience to Achieve Sustainable Development (GRAD)*

The GRAD project aims to improve food security in rural communities in Ethiopia's highland areas by enhancing livelihood options, improving community and household resilience and strengthening the enabling environment to ensure scale and sustainability of the approach. As part of its overall approach, the project explicitly aims to support adaptation and build resilience to climate change. It is funded by the Feed the Future program of the United States Agency for International Development (USAID). GRAD works in 16 Woredas in four regions: Tigray, Amhara, Oromia and SNNPR. It is implemented by a consortium led by CARE and including Catholic Relief Service (CRS), ORDA, REST, Agri-Service, SNV and Tufts University.

2.2.1 GRAD CVCA Process and Achievements

For the GRAD project, the core of the CVCA process was FGDs with groups of women and men. These discussions used tools included hazard mapping, historical timelines, seasonal calendars and a vulnerability matrix to gather community knowledge on climate and livelihood linkages. The groups then developed impact chains to analyze the direct and indirect impacts of hazards on livelihoods, and discuss how people are currently responding to different impacts. Building on the impact chains, adaptation pathways were developed, which identified the most effective ways to manage the impacts. The tools were organized to build on one another and to provide openings to introduce to introduce the concept of climate change and to discuss observed and projected trends in temperature, rainfall and hazards. Consequently, the process not only provided critical information for project planning, but also served as an initial awareness raising exercise for the communities involved.

GRAD completed CVCA analyses with communities in 14 Woredas across its four operational regions (SNNPR, Amhara, Tigray, Oromia). The process involved 3 half-day sessions with each focus group, plus time for analysis and documentation of the information gathered by the analytical team. A validation session was also conducted to ensure that the interpretation of the information was correct from the community point of view.

According to GRAD intermediate report and Midterm Assessment Report for 2014,

1. 84% of households adopted at least two practices associated with climate change adaptation, up from 75% the year before, and 96% have adopted at least one practice. The three practices most often reported as being adopted were early maturing crop varieties (69%), moisture conserving practices (62%), and drought tolerant crop types and varieties (51%). The other practices reported were tree planting (48%), community based upland management (30%, also part of government activities), soil fertility enhancements (47%), forage crops ("hay making", 40%), no grazing areas ("area closure", 30%) and irrigation (23%).
2. All respondents interviewed in key informant interviews or in focus group discussions were well aware of climate change issues and adaptation measures, and they verified the levels of adoption found during GRAD MID term and assessment reports:

    - climate change adaptation activities are actually already part of current natural resource management and agricultural good practices being promoted widely in the country by the GFDRE. Few are actually "new" climate change adaptation activities for participating households. Those that are actually new include quality fuel saving stoves, crop insurance, rope and washer pumps, and new crops such as orange-fleshed sweet potatoes.
    - The GRAD Project used the CVCA to guide activities. While the focus in GRAD is primarily at the household level, some Village Economic and Social Associations also reported practicing tree planting and protection on public spaces such as around churches and around local water sources, part of an area enclosure approach to regenerate natural resources.
    - The mid term evaluation observed that forage crops and cut-and-carry practices are not yet widespread in project areas although they are increasing (30–40% in the IR Assessment for 2014).
    - In one REST area, some GRAD households were enrolled in a weather index-based crop insurance pilot managed by another organization. This program has run for two years without any pay-outs, however, so its effectiveness is still untested.
    - All households who desire and can afford new technologies such as fuel-savings stoves, rope and washer pumps, and short season crop inputs, expressed confidence that they will be able to obtain them by the end of

project. GRAD staff, however, are still serving as direct intermediaries between these households and the suppliers of the technologies, especially for the fuel-saving stoves and rope and washer pumps. This has implications on the sustainability and replicability of the input supply systems for these technologies.

2.2.2 GRAD CVCA Learning

The following were the key issues learned from the analysis process:

- GRAD communities are acutely aware of climate change. They describe changes in rainfall patterns, leading to increasing uncertainty in the timing and amount of rainfall, as well as more frequent droughts.
- Because crop and livestock production are the primary sources of food and income in the GRAD areas, livelihoods are highly sensitive to climate variability and change.
- Drought is the most significant concern from the point of view of communities, due to its devastating impacts on both crops and livestock. However, increasingly erratic rainfall is also cause for concern.
- Some people are proactively responding to the changes by planting early-maturing or drought-resistant seeds, diversifying their income sources and investing in savings. These strategies provide a buffer when crops fail and/or livestock are lost.
- Measures such as water harvesting, small-scale irrigation and soil and water conservation actions are helping to minimize impacts on critical resources such as water and agricultural land.

Other strategies, such as sale of firewood and charcoal and sale of livestock during crises when prices are low, are not sustainable and may actually undermine people's resilience for the future.

These insights have proven to be critically important for GRAD. Building on the analysis, the project is raising awareness of climate change, its implications for livelihoods and ways to adapt in its work with Village Economic and Saving Associations (VESAs). During group discussions, VESA members have shared their experiences and observations of climate change in their community, and discussed how they are responding to the challenges presented. In many cases, this has led the VESA members to develop Climate Change Adaptation Action Plans, which outline the actions they will take to manage climate risks and minimize the negative effects on their livelihoods. With support from GRAD, VESAs are monitoring the implementation of identified actions, sharing successes and lessons learned, and adjusting their plans accordingly (Dazé 2014b). This is a key element of building adaptive capacity—strengthening people's ability to plan, monitor results and changes, and identify the actions needed to manage risks and uncertainty in livelihoods (Dazé 2014c).

2.3 *Pastoralist Areas Resilience Improvement and Market Expansion (PRIME)*

The PRIME project aims to enhance pastoralists resilience to climate change through increased access to climate information, improving governance systems and piloting climate solutions. The project is working with more than 250,000 households in the Afar, Oromia and Somali Regions of Ethiopia.

2.3.1 PRIME CVCA Process and Achievements

In May and June 2013, the PRIME cluster teams conducted a total of 168 CVCA focus groups (86 female groups & 82 male groups). During this process 8 field guides or facilitation tools developed for supporting climate change adaptation. 25 action plans were developed, including 8 with government partners, 9 with male groups and 8 with female groups around disaster risk management, natural resource management, and livelihood adaptation strategies across 8 zones in three clusters. Approximately 258, 483 km^2, or nearly 26 million hectares of land mapped through participatory resource mapping and hazard mapping techniques using a systems-based approach. 25 community maps were produced, including 9 with women, 9 with men and 8 with government across three clusters.

2.3.2 PRIME CVCA Learning

Adopting an ecosystem or landscape based approach to NRM: Undertaking the CVCA analysis at the landscape or grazing system level, which incorporate multiple Kebeles, enabled the team to better understand the complex linkages between natural resource quality and availability, climate change impacts and pastoral people's responses to those impacts, including resource and conflict governance systems.

2.3.3 Making Climate Information Accessible and Usable

- Facilitating interaction among multiple stakeholders: Bringing together users of climate information (e.g., pastoralists, farmers), producers (e.g., climate scientists and traditional forecasters) and intermediaries (e.g., NGOs and agricultural extension services) enables different climate knowledge and interpretations to be understood, improves synergy and coordination amongst actors, and promotes consensus building and trust between diverse groups (Nurye 2016). Collective interpretation of climate information by concerned stakeholders enables locally-tailored climate information services and the joint production of new and innovative solutions to help manage an uncertain climate.

- Downscale and localize climate information so that it relates to local contexts and specific user needs—which is what people want, but is rarely available due to the poor coverage of local weather stations and historical information gaps.
- Combine different knowledge sources: Joint interpretation of local and scientific climate information unpacks the complexity of the information, and generates better understanding and trust in both sources of information and in the recommended actions based on the forecasts.
- Build flexibility into decision-making: Approaches involving two-way learning and feedback promote flexibility at all levels of decision-making, planning, resource allocation and action and enable systems that factor in uncertainty, respond to changing needs and identify new solutions, for example, creating advisories (locally relevant livelihood options based on scenarios of possible future climate impacts).

2.3.4 The Need to Address Socio-Cultural and Behavioral Barriers to Adaptation

- The adjustment of the Social Analysis and Action tool (SAA) is an important step in the evolution of CARE Ethiopia's thinking on climate change adaptation. Issues around decision-making, behavior and social and cultural norms—and how these influence reactions to shocks, stresses and change over time—are fundamental to adaptation, yet they are often neglected in adaptation initiatives and debates. This type of iterative tool is compatible with the process-oriented nature of adaptation, enabling participants to strengthen their analytical skills and incorporate new information and knowledge throughout the series of dialogues.

3 Implications for Programming

This section considers the implications of this learning for programming, with a focus on what needs to be done differently to ensure the integration of climate change into programming.

3.1 Climate-Related Hazards Can no Longer Be Treated as an Anomaly

Climate-related shocks and stresses are a defining feature of livelihoods in Ethiopia, across ecosystems and livelihood zones. While climate extremes such as droughts and floods are not a new phenomenon, community observations suggest that these shocks are occurring more frequently. The analysis also shows that recurrent shocks have undermined the household asset base and led to growing pressures on natural resources, leaving the poorest households increasingly vulnerable. With this trend expected to continue, it is imperative that programs that aim to enhance food security address climate-related hazards as characteristic of the context, rather than as an anomaly. This inevitably leads to an increased focus on measures that build resilience to shocks, in order to move people away from crisis-driven decision-making and coping strategies that undermine future potential. This has implications for both humanitarian action and longer-term development programming, demanding more effective and sustainable responses to crises and better integration of risk management in livelihoods and food security programming.

3.2 Climate Change Is not a 'Component' of Programming

Most of the CARE Ethiopia programs that have conducted CVCA analysis have done so because they have a specific climate change 'component' in their program, and in some cases, a specific budget line for climate change activities. While this makes sense in terms of project management and ensuring that this critical issue is explicitly addressed, a major learning outcome from CARE Ethiopia's analysis is that climate change cannot be tackled in isolation from livelihoods and food security. In addition to highlighting priorities for targeted action to address climate risks, climate change analysis must inform interventions by government and relevant actors in other areas, including humanitarian action, to ensure that efforts to promote sustainable livelihoods and improve food security are resilient and sustainable over the longer term and do not inadvertently increase vulnerability to climate change. This is particularly the case for women whose experience of climate shocks and stresses can be exacerbated by other existing vulnerabilities.

3.3 Climate Change Adaptation Is Inextricably Linked with Sustainable Natural Resource Management

It is clear from the CVCA analyses that the most direct manifestation of climate change from the perspective of communities is its impact on the availability and quality of water, pasture and fertile land for crop production. It is also apparent that people's responses to resource scarcity often result in further environmental degradation. Particularly in pastoral contexts, issues of access to and control over resources cut across communities and administrative borders, requiring a broader view of the linkages between sustainable NRM and climate change. In rural communities, the natural resource base will continue to be the foundation of livelihoods and ecosystem health will be a key determinant of resilience to climate extremes and capacity to adapt to change over time.

3.4 Adaptation to Climate Change Requires Options, not simply 'Solutions'

Development actors often endeavor to offer 'solutions' for reducing poverty, in the form of improved seeds, particular agricultural practices or access to markets, for example. The CVCA analyses have revealed that the biggest climate-related challenge people are facing is increasing uncertainty about rainfall and weather conditions. In the context of increasing uncertainty, there is a need to increase the range of options available to people to secure their livelihoods and manage risks. Programs must therefore move away from promoting particular strategies and focus efforts on opening up the options available to women and men to enable adaptive management of livelihoods, as well as contingency planning for times when primary strategies fail.

3.5 Informed Decision-Making Is Key to Adaptive Capacity

With more options comes the need to choose among them. Different livelihood options will make sense at different times, based on the weather conditions, seasonal forecasts, availability of resources to invest and market conditions, among other factors. For people to effectively manage climate-related risks to their livelihoods, they must first have access to the needed information, including early warnings, climate information and market information. Communication systems must be designed provide this information in ways that are relevant and equitably accessible for poor women, socially excluded groups and community members with mobility issues who may face barriers in access. However access is only a first step —in order to use the information effectively, they must also have the skills to analyze the information, weighing costs, benefits and risks, to make good decisions about how to invest their resources. Building these skills requires development actors to work with communities in new ways, facilitating learning processes that involve analysis, critical thinking and forward-looking decision-making.

3.6 Efforts to Increase Social Equity Must Underpin Adaptation Efforts

People's social positions, including those related to gender, have a strong influence over their vulnerability to climate change and the adaptation options available to them. Technologies such as drought-resistant seeds or irrigation systems are only useful to those who have land to cultivate and the power to decide how they will use it. Information is a critical resource for adaptation, which may or may not reach all members of the community depending on how it is communicated. Adaptation actions by one group may influence the availability of resources for another, making them more vulnerable to climate-related shocks and stresses. Consequently, equity in access to resources, opportunities and benefits must be an underlying principle in order to effectively build resilience and support adaptation. To support this, analysis of vulnerability and adaptive capacity must facilitate dialogue on the barriers facing particularly vulnerable social groups and identify specific actions to overcome these barriers and ensure equitable approaches to adaptation.

3.7 A Better Understanding of Cultural and Social Barriers to Adaptive Action Is Needed

Informed decision-making is a critical element of adaptive capacity, however it is important to also acknowledge the role that social and cultural norms play in determining who has the power to make decisions, how these decisions are made and what options are deemed appropriate. The CVCA analyses conducted by CARE Ethiopia have provided some insights into these dynamics, for example the cultural barriers to savings in pastoral communities and the role that prestige plays in decision-making. It also revealed inequalities in decision-making power within and between communities. These are helpful insights to inform how programs engage with stakeholders on decision-making processes related to adaptation, however a better understanding of these dynamics and how they relate to gender, ethnicity and socio-economic positions will enable more inclusive and effective approaches to build adaptive capacity.

4 Integrating the Analysis into Programs

CARE Ethiopia is actively making adjustments to its approaches to address issues emerging from the climate vulnerability and capacity analyses. The following program profiles provide snapshots of some of the changes in thinking and approaches that have resulted from this process.

4.1 Communicating the Implications of Climate Change for Income Generating Activities and Value Chains

GRAD used a simplified CVCA process to conduct its analysis, which was designed to inform the program's approaches to building community and household resilience. Using the CVCA results, GRAD is working with local stakeholders to address the challenges facing the most vulnerable households, for example through pilot implementation of irrigation systems and introduction of drought-resistant and early maturing crops.

The CVCA analysis has also proven very important in identifying and evaluating livelihood options. There are two main dimensions to GRAD's work on expanding livelihood options: engaging people in value chains and promoting alternative income generating activities (IGAs). Engaging poor women and men in agriculture-based value chains can make a significant contribution to improving household income and purchasing power. However, when production is affected by climate-related shocks and stresses, people may find themselves in a precarious situation, particularly if they have relied on credit to cover the costs of inputs. The CVCA analysis process yielded a better understanding of the potential risks associated with this type of activity, and led the GRAD team to conduct climate screenings on the value chains being promoted in order to deepen this awareness.

The process involved a review of each stage in the value chain, from inputs and production through post-harvest management, aggregation and processing to wholesale and retail marketing. The exercise also made clear the importance of combining work on value chains with other interventions in order to support people in developing livelihoods that are sustainable and resilient in a changing climate. Complementary activities identified by GRAD include facilitating access to seasonal forecasts for value chain participants to promote planning and ensure timely planting and ongoing access to water for red beans.

4.2 *Integrating Climate Change into Community Based Participatory Watershed Development*

CSI adapted the CVCA process which was designed as an entry point for integrating climate smart planning within the PSNP/HABP Household Asset Building Programme planning process, and conducted in line with the national guideline for Community Based Participatory Watershed Development (CBPWD).

Taking a watershed approach enables analysis of interactions between measures in different communities within a single watershed, ensuring that activities in neighboring communities are complementary. The PSNP and HABP programs also incorporate livelihoods interventions such as agro-forestry, forage development and improved farming practices. Implementation of actions occurs through community work parties, establishment of community groups and linkages with external sources of funding such as the PSNP and other development initiatives. The Development Agents and Woreda government representatives provide training and technical assistance for implementation of activities.

Taking a climate smart approach to watershed planning requires explicit integration of climate change considerations throughout the analysis and planning process. Pilot implementation of CVCA by CSI revealed a number of gaps in the CBPWD process with respect to climate-related risks, as well as the social dimensions of vulnerability, notably gender inequality. Integration of participatory climate change analysis into the process would begin to address these gaps, helping to ensure that activities are inclusive, sustainable over the longer term, that they support people in managing climate risks and that they don't inadvertently increase vulnerability to climate change.

4.3 Addressing Socio-Cultural and Behavioral Barriers to Adaptation

By using Social Analysis and action (SAA), PRIME enabled communities to identify connections between social and behavioral factors, vulnerability to climate change and food insecurity. The process explored issues such as herd management, the lack of savings culture and equitable decision making in pastoral communities, which represent an important obstacles to resilience building. Decision-making processes are unpacked to build understanding of attitudinal factors that govern how people respond to climate variability and change, and to highlight social dynamics that may lead to increased vulnerability, for example the pressure to maximize herd size and the resultant reluctance to sell livestock until a crisis occurs. Positive practices, for example rotational dry and wet season grazing and customary management of water points, will be highlighted and promoted, with emphasis on their value for risk management.

The adjustment of the SAA tool is an important step in the evolution of CARE Ethiopia's thinking on climate change adaptation. Issues around decision-making, behavior and social and cultural norms—and how these influence reactions to shocks, stresses and change over time—are fundamental to adaptation, yet they are generally neglected in adaptation initiatives and debates. This type of iterative tool is compatible with the process-oriented nature of adaptation, enabling participants to strengthen their analytical skills and incorporate new information and knowledge throughout the series of dialogues.

5 Integrating Climate Change Adaptation into Programming: CARE Ethiopia's Theory of Change

As shown in Fig. 2, each of the domains of change involves a number of different change pathways. These are described briefly below.

Fig. 2 CARE Ethiopia has placed increasing focus on integrating climate change adaptation into its programs over the last several years. This document introduces the organization's theory of change for climate change adaptation, which provides a complementary framework for its program strategies. This is a working framework, which will be refined and updated over time with further experience and learning on integrating adaptation into programs

5.1 *Social Norms and Cultural Practices Support Climate Change Adaptation*

CARE Ethiopia's analysis has determined that social norms and cultural practices may present barriers to adaptive actions. For example, social norms that limit women's access to information, resources and opportunities and undermine their decision-making power have a negative effect on their adaptive capacity, making them more vulnerable than their male counterparts. Cultural practices, such as the keeping of large herds of livestock to demonstrate social status in pastoral communities, may also exacerbate vulnerability to climate shocks and stresses.

Addressing these social and cultural barriers is a precondition for achieving change in the other domains.

5.2 Decision-Making Processes Are Informed, Forward-Looking and Adaptive

Climate change brings increasing risks and uncertainty for people in managing their livelihoods, income and food security. Consequently, there is no single technology or practice that will be effective under all conditions and through variability and change. As noted above, adaptation is an ongoing process, which involves day-to-day, seasonal and longer-term decisions about investments, activities and responses to shocks and stresses. For adaptive decision-making to occur, these decisions must be informed, taking into account the best information available on climate risks, uncertainties and changing conditions, as well as other relevant issues such as market prices.

5.3 Women and Men Have Flexible and Resilient Livelihood Options

The impacts of climate change manifest in effects on people's livelihoods, notably through crop losses and reduced productivity or loss of livestock. These effects have consequences for food and income security and may lead to actions that undermine longer-term adaptive capacity. Livelihoods are therefore an important entry point for building resilience of poor households. In a changing climate, efforts to improve livelihoods must emphasize flexibility and resilience, opening up the range of options available to people to anticipate and respond to shocks and stresses.

5.4 Institutions and Services Enable Adaptation

As noted above, the institutions and services that interact with communities play an integral role in enabling or constraining adaptation. These actors are themselves grappling with the new and evolving realities brought about by climate change, and are also in a process of adaptation. This domain is about supporting this process, to ensure that they have the necessary knowledge and capacities to support community adaptation efforts and that services provided are optimal in terms of enabling management of climate risks.

6 Conclusions

Integrating climate change into development programming adds a considerable layer of complexity to already complicated processes. CARE Ethiopia is in the midst of a longer-term learning process on how to do this and what difference it makes for the resilience and adaptive capacity of people and for the sustainability of its programs. The experiences and lessons summarized in this report are one output of this process, reflecting the organization's commitment to share its learning, even when many questions and challenges remain. The community profiles demonstrate that climate variability and change are critical issues affecting food and livelihood security in poor communities. They also reveal the underlying causes of vulnerability to climate-related shocks and stresses, notably those related to gender and social inequality, barriers in access to information and services and governance of natural resources and community affairs. The program case studies illustrate the difference that the analysis has made, providing a strong case for participatory analysis of vulnerability and adaptive capacity as a basis for integrating climate change considerations into programming in an equitable and empowering way.

References

Africa Climate Change Resilience Alliance (ACCRA). (2011). Understanding adaptive capacity at local level in Ethiopia.

Dazé, A. (2014a). A review of climate smart initiative climate vulnerability and capacity analysis process.

Dazé, A. (2014b). GRAD Community-Level Climate Change Analysis and Planning for adaptation, Synthesis report.

Dazé, A. (2014c). Using analysis of climate change vulnerability and adaptive capacity for program decision-making: Lessons from CARE Ethiopia's experience.

McSweeney, C., New, M., & Lixcano, G. (2007). Ethiopia—UNDP climate change country profiles.

Nurye, A. A. (2016). Blending science and community voices for multi-scale disaster risk reduction and climate resilience: A Participatory Scenario Planning Approach. *Ethiopian Journal of Environmental Studies and Management, 9* (Suppl. 1).

World Bank. (2010). *Ethiopia—economics of adaptation to climate change*. Addis Abeba: Wold Bank.

Impact Assessment of Climate Change on Crop Diseases Incidence and Severity in Nigeria

Timothy Ipoola Olabiyi, Emmanuel Ayomide Afolabi, Elizabeth Iyabo Adewole and Iyabo Olunike Omomowo

1 Introduction

Climate change refers to any change in climate over time, whether due to natural variability or as a result of human activity. The United Nations Framework Convention on Climate Change (UNFCC) defines climate change as a change of climate which is attributed directly or indirectly to human activity that alters the composition of the global atmosphere and which is in addition to natural climate variability observed over comparable time periods. Atmospheric carbon dioxide (CO_2), a major greenhouse gas, has increased by nearly 30% and temperature has risen by 0.3–0.6 °C. The Intergovernmental Panel on Climate Change (IPCC) predicts that with the current emission scenario, global mean temperature would rise between 0.9 and 3.5 °C by the year 2100. There are, however, many uncertainties that influence these predictions (IPCC 2007).

Climate change or global warming has become a new reality, with deleterious effects: seasonal cycles are disrupted, as are ecosystems; and agriculture, water needs and supply, and food production are all adversely affected. It also leads to sea-level rise with its attendant consequences, and includes fiercer weather, increased frequency and intensity of storms, floods, hurricanes, droughts, increased

T.I. Olabiyi (✉) · E.A. Afolabi
Department of Crop and Environmental Protection, Ladoke Akintola University of Technology, Ogbomoso, Nigeria
e-mail: tiolabiyi@lautech.edu.ng

E.I. Adewole
Nigeria Meteorological Agency, Ilorin International Airport, Ilorin, Kwara, Nigeria

I.O. Omomowo
Department of Pure and Applied Biology, Ladoke Akintola University of Technology, Ogbomoso, Nigeria

W. Leal Filho et al. (eds.), *Climate Change Adaptation in Africa*, Climate Change Management, DOI 10.1007/978-3-319-49520-0_15

frequency of fires, poverty, malnutrition and series of health and socio-economic consequences. Climate change is an environmental, social and economic challenge on a global scale (Scholze et al. 2006; Mendelsohn et al. 2006). The most devastating adverse impacts of climate change in Africa includes frequent drought, increased environmental damage, increased infestation of crop by pests and diseases, depletion of household assets, increased rural-urban migration, increased biodiversity loss, depletion of wildlife and other natural resource base, changes in the vegetation type, decline in forest resources, decline in soil conditions (soil moisture and nutrients), increased health risks and the spread of infectious diseases, changing livelihood systems, etc (Reilly 1999; Abaje and Giwa 2007).

The climate influences the incidence as well as temporal and spatial distribution of plant diseases. The main factors that control growth and development of diseases are temperature, light and water; similarly, these factors affect type and condition of host crop (Rosenzweig et al. 2001; Agrios 2005). The environment may affect plant pathogen, therefore, survival, vigor, rate of multiplication, sporulation, direction, distance of dispersal of inoculums, rate of spore germination and penetration can be affected (Su et al. 2004; De Wolf and Isard 2007; Kang et al. 2010). Plant diseases develop under a well-defined, optimal range of climatic variables (such as temperature, rain, relative humidity); however, the occurrence and severity of a disease in an individual plant is defined by the deviation of each climatic variable within the optimal range for disease development, thus climate affects all life stages of the pathogen and host (Agrios 2005). Despite the significance of weather on plant diseases, comprehensive analysis and information on how climate change influences plant disease is scarce. Evaluation of the limited literature in this area suggests that the most likely impact of climate change would be felt in three areas: in losses from plant diseases, in the efficacy of disease management strategies and in the geographical distribution of plant diseases. Climate change could have positive, negative or no impact on plant diseases. Most plant disease models use different climatic variables and operate at a different spatial and temporal scale than do the global climate models. This paper provides adequate information on the challenges that farmers experienced on crop production in relations to disease outbreak especially when climate variables changed. It therefore attempts to explain the correlation between climate variables and incidence/severity of diseases on plant. The paper answers the following research questions: How severe is crop diseases when climatic conditions are favorable or hash? Is there any correlation between plant disease incidence/ severity and climatic variables? The objective of the study was to assess the climatic dynamics on the incidence and severity of plant diseases for climate prediction purposes in Nigeria.

2 Study Area

This study covers the guinea savannah agro-ecological zone of Nigeria. The climate is hot, humid, and tropical with two distinct seasons. The dry season spans between November and February and raining season from March to November.

3 Data Collection

Data for the study were national aggregates on climate variables (cloud, temperature, relative humidity, sunlight intensity, and rainfall) from Nigerian Meteorological Agency, NIMET, Ilorin. Information on the incidence and severity of crop diseases (fungi, nematode, bacteria and virus), between 2010 and 2014, were obtained from both local and international journals and authors were adequately referenced. Average climatic factors between 2010 and 2014 were calculated while information on plant disease incidence and severity were collated and recorded.

4 Results

The information on the incidence and severity of plant diseases in Nigeria were cross-referenced with changes in climate variables over a period of 5 years (2010–2014). Average data on the cloud within the guinea savanna agro-ecological zone between 2010 and 2014 varied tremendously (Table 1). In 2010, August to October has the highest average cloud (7.1 okta) while the lowest average cloud (5.0 okta) was obtained in December. In 2011, the highest average cloud (7.0 okta) was obtained in July and August while the lowest for the year (4.0 okta) was obtained in January. In 2012, the highest average cloud (7.1 okta) was obtained in July and August while January had the lowest average cloud (6.8 okta). In 2013, highest average cloud (8.9 okta) was obtained in January while the lowest average cloud (6.8 okta) was obtained in December. In 2014, the highest average cloud (7.1 okta) was obtained in August while the lowest average cloud (6.7 okta) was obtained in December.

Average data on the temperature within the guinea savanna agro-ecological zone between 2010 and 2014 varied tremendously (Table 2). In 2010, highest

Table 1 Average cloud data between 2010 and 2014 (Okta)

	2010	2011	2012	2013	2014
January	6.9	4.0	6.8	8.9	7.0
February	7.0	6.6	7.0	7.0	7.0
March	7.0	7.0	6.9	6.9	7.0
April	7.0	6.9	6.9	7.0	7.0
May	7.0	7.0	7.0	7.0	7.0
June	7.0	7.0	6.9	7.0	7.1
July	7.0	7.1	7.1	7.0	7.0
August	7.1	7.1	7.1	7.3	7.1
September	7.1	7.0	7.0	7.1	7.0
October	7.1	6.9	7.0	6.9	6.9
November	6.9	6.8	6.7	6.9	6.9
December	5.0	7.0	6.9	6.8	6.7

Source Nigerian Meteorological Agency, Ilorin (2014)

Table 2 Average temperature data between 2010 and 2014 (°C)

	2010		2011		2012		2013		2014	
	Max	Mini	Max	Mini	Max	Mini	Max	Mini	Max	Mini
January	36.1	20.8	34.0	18.2	33.7	18.9	34.7	20.5	34.4	21.0
February	37.7	23.7	35.7	23.3	35.5	23.7	35.8	22.9	35.9	22.6
March	37.1	24.8	35.9	24.1	36.4	23.8	35.6	24.2	35.4	23.8
April	36.0	24.2	34.8	23.7	33.7	23.7	33.4	23.3	33.3	23.1
May	32.8	23.5	33.9	23.7	31.7	22.8	31.8	22.5	31.4	22.9
June	31.9	23.2	30.7	22.2	30.6	22.1	30.5	22.6	30.9	22.7
July	29.3	22.2	29.1	22.1	28.9	21.8	28.8	21.8	29.6	22.2
August	29.4	22.1	28.4	21.6	27.7	21.4	28.4	21.3	27.9	21.3
September	30.1	22.0	29.9	21.8	29.4	21.8	29.9	21.8	29.2	21.4
October	30.9	21.7	30.8	21.8	30.9	21.9	31.2	22.0	31.1	21.6
November	33.9	22.1	34.2	21.3	33.9	22.4	34.2	22.9	33.3	22.2
December	34.3	18.7	34.3	16.6	34.3	19.8	34.3	20.9	34.2	20.1

Source Nigerian Meteorological Agency, Ilorin (2014)

maximum temperature (37.7 °C) was obtained in February while the lowest minimum temperature (18.7 °C) was obtained in December (Table 2). In 2011, March had the highest maximum temperature (35.9 °C) while the lowest minimum temperature (16.6 °C) was obtained in December. In 2012, the highest maximum temperature (36.4 °C) was obtained in March while the lowest minimum temperature (18.9 °C) was obtained in January. In 2013, February had the highest maximum temperature (35.8 °C) while the lowest minimum temperature (20.5 °C) was obtained in January. Finally, in 2014, the highest maximum temperature (35.9 °C) was obtained in February while the lowest minimum temperature (20.1 °C) was obtained in December.

Average data on the relative humidity within the guinea savanna agro-ecological zone between 2010 and 2014 varied tremendously (Table 3). The highest relative humidity (88%) was obtained in August and September in 2010, while the lowest relative humidity (66%) was obtained in February and December. In 2011, August and September had the highest relative humidity (90%) while the lowest relative humidity (48%) was obtained in December. In 2012, the highest relative humidity (87%) extends from July to September while January had the lowest relative humidity (54%). In year 2013, the highest relative humidity (85%) was obtained in September while the lowest relative humidity (51%) was obtained in January. In 2014, August and September had the highest relative humidity (86%) while December had the lowest relative humidity (53%).

Average data on the sunlight intensity within the guinea savanna agro-ecological zone between 2010 and 2014 varied tremendously (Table 4). In 2010, the highest sunlight intensity (8.73 h) was obtained in December while the lowest (4.07 h) was obtained in July. In 2011, the highest sunlight intensity (8.70 h) was obtained in December while the lowest (3.75 h) was obtained in August. In 2012, the highest

Table 3 Average relative humidity data between 2010 and 2014 (%)

	2010	2011	2012	2013	2014
January	70	54	54	51	57
February	66	69	71	63	55
March	68	74	69	72	65
April	75	74	72	78	74
May	81	78	81	80	82
June	83	86	82	81	83
July	87	87	87	84	77
August	88	90	87	84	86
September	88	90	87	85	86
October	85	86	84	83	83
November	79	77	78	75	73
December	66	48	60	65	53

Source Nigerian Meteorological Agency, Ilorin (2014)

Table 4 Average sunlight intensity data between 2010 and 2014 (Hours)

	2010	2011	2012	2013	2014
January	8.05	7.21	6.78	8.30	8.03
February	8.41	7.48	6.53	7.03	7.50
March	5.34	7.50	6.27	7.51	7.30
April	7.51	6.95	6.97	7.10	7.63
May	6.93	7.38	5.43	7.40	6.33
June	6.10	5.90	6.94	5.84	6.10
July	4.07	4.77	4.35	4.70	5.10
August	4.55	3.75	2.99	3.68	3.85
September	4.74	4.25	4.19	5.10	4.28
October	5.43	6.30	5.97	7.13	5.60
November	8.23	8.42	7.55	8.70	7.20
December	8.73	8.70	8.41	8.73	7.40

Source Nigerian Meteorological Agency, Ilorin (2014)

sunlight intensity (8.41 h) was obtained in December while the lowest (2.99 h) was obtained in August. In 2013, December had the highest sunlight intensity (8.73 h) while August had the lowest (3.68 h). In 2014, the highest sunlight intensity (8.03 h) was obtained in January while the lowest (3.85 h) was obtained in August.

Average data on the rainfall within the guinea savanna agro-ecological zone between 2010 and 2014 varied tremendously (Table 5). In 2010, highest rainfall (267.3 ml) was obtained in September while the lowest rainfall (0.0 ml) was obtained in January and December. In 2011, September had the highest rainfall (247.5 ml) while the lowest (0.0 ml) was obtained in January, February, November, and December. In 2012, the highest rainfall (230.0 ml) was obtained in September while the lowest rainfall (0.0 ml) was obtained in January. In 2013, March had the highest rainfall (384.0 ml) while the lowest rainfall (0.0 ml) was obtained in

Table 5 Average rainfall data between 2010 and 2014 (ml)

	2010	2011	2012	2013	2014
January	0.0	0.0	0.0	4.8	9.7
February	0.0	23.6	22.1	5.5	28.6
March	29.4	23.1	4.0	384.0	52.2
April	73.5	21.1	138.2	223.4	408.8
May	95.1	122.8	138.2	121.7	232.3
June	89.3	212.9	152.8	148.4	305.2
July	95.1	93.1	120.7	110.6	332.3
August	143.8	203.4	123.4	51.4	225.3
September	267.3	247.5	230.0	266.5	511.7
October	148.3	202.4	153.7	308.8	379.4
November	11.4	0.0	8.7	0.0	154.0
December	0.0	0.0	25.2	6.8	0.0

Source Nigerian Meteorological Agency, Ilorin (2014)

November. Finally, in 2014, the highest rainfall (511.7 ml) was obtained in September while the lowest rainfall (0.0 ml) was obtained in December.

Variation in the incidence and severity of plant diseases come as a result of variations of climate variability from time to time. These were illustrated in Tables 6 and 7. In 2010, cloud (6.8 okta), temperature (27.7 °C), relative humidity (78%), sunlight intensity (6.5 h) and rainfall (105.9 ml) resulted to 15.7–30.0% incidence of powdery mildew, 1.02–6.07% yam tuber galls prevalence, 10–45% incidence of potato bacterial infection. Also, it leads to 10–44.5% viral infection on potato, and 9.3% incidence of virus disease on sole cropped cowpea.

In 2011, cloud (5.7 okta), temperature (26.1 °C), relative humidity (76%), sunlight intensity (6.6 h) and rainfall (127.8 ml) resulted to 7.35% average incidence of leaf rot disease on aloe. It also resulted to higher virus incidence in rainforest agro-ecology (79–90%) while it resulted to low virus incidence in guinea savannah agro-ecology (43–73%). In 2012, cloud (6.9 okta), temperature (26.1 °C), relative humidity (76%), sunlight intensity (6.0 h) and rainfall (101.5 ml) resulted to highest incidence of anthracnose disease (48%) and lowest incidence of 37.33%. Also, it leads to maximum occurrence of root gall (82.53%) and minimum occurrence (21.73%). In 2013, cloud (7.2 okta), temperature (26.7 °C), relative humidity (75%), sunlight intensity (6.8 h) and rainfall (148.4 ml) resulted to 1.7–5% incidence of anthracnose on okra, 1.7–26.7% anthracnose incidence on eggplant, 16.7–31.3% leaf spot incidence on okra, 10–33.3% incidence of leaf spot on eggplant. It resulted also to 1.3–21.3% and 1.3–25% incidence of bacteria wilt on okra and eggplant respectively. Also, this results to high incidence of maize streak virus (40–100%), high incidence of sugarcane mosaic disease (83%) and also low incidence of maize dwarf mosaic virus (5%). In 2014, cloud (7.0 okta), temperature (26.2 °C), relative humidity (73%), sunlight intensity (6.4 h) and rainfall (240 ml) resulted to 25.68% highest incidence of dry rot disease of yam, 8.40% incidence of *Meloidogyne* tuber gall disease, it also resulted to 16–97% virus incidence on pepper.

Table 6 Incidence and severity of crop diseases between 2010 and 2014

Year	Fungi	Nematode	Bacteria	Virus
2010	Incidence of powdery mildew ranged from 15.7–30.0% (Bem et al. 2010)	Highest prevalence of yam tuber galls is 6.07% while the lowest is prevalence is 1.02% (Ogara and Bina 2010)	Occurrence of bacterial infection ranges from 10–45.58% on potato (Ndor 2010)	Occurrence of viral infection on potato ranges from 10–44.5%. Incidence of virus disease on sole cropped cowpea is 9.3% (Ndor 2010; Shatu et al. 2010)
2011	There was high incidence of fungi infection on rice (Makun et al. 2011)	No record	No record	Virus incidence were higher in the rain forest agroecology (79–90%) while it is low in the guinea savanna agroecology (43–73%) Bello et al. (2011)
2012	Incidence of anthracnose disease is 48% while the least disease incidence is 37.33% (Awa et al. 2012; Kawuri et al. 2012)	Maximum occurrence of root gall is 82.53% and the minimum is 21.73% (Niranjan et al. 2012)	No record	Cucumber mosaic virus is severe on cucurbit crops and weeds ranging between 10 and 30% (Ayo-John et al. 2012)
2013	Incidence of anthracnose on okra is 1.7–5% while that of eggplant 1.7–26.7%. Incidence of powdery mildew ranged from 15.7–30.0% Bem et al. (2013)	No record	Incidences of leaf spot on okra and eggplant are 31.3 and 33.3% while the lowest are 16.7 and 10% respectively. Incidences of bacteria wilt on okra and eggplant are 21.3 and 25% while the lowest are 1.3 and 1.3% respectively (Anjorin et al. 2013)	Incidence of maize streak is between 40–100%. Sugarcane mosaic disease has the highest incidence (83%) while Maize dwarf mosaic virus has the lowest incidence (5%) (Akinbode et al. 2013)
2014	Dry rot disease of yam has highest incidence of 25.68% (Anjorin et al. 2014)	*Meloidogyne* tuber gall disease has incidence of 8.40% (Anjorin et al. 2014)	No record	The highest virus incidence is 97% while the lowest is 16% on pepper (Aliyu 2014)

Table 7 Mean of the climatic data per year

Year	Cloud (okta)	Temperature (°C)	Relative humidity (%)	Sunlight intensity (hours)	Rainfall (ml)
2010	6.8	27.7	78	6.5	105.9
2011	5.7	26.1	76	6.6	127.8
2012	6.9	26.1	76	6.0	101.5
2013	7.2	26.7	75	6.8	148.4
2014	7.0	26.7	73	6.4	240.0

5 Discussion

Climate change poses a threats to the disease severity, incidence and control. As climate variables continue to change in Nigeria new pathogens may emerge causing new symptoms and infections. Climatic variables that favors disease severity and incidence are high rainfall, cloud cover, and relative humidity coupled with low temperature and sunlight intensity which are most experienced during rainy season in Nigeria. High (increasing) temperatures had been attributed to contribute to a poleward migration of ranges of many organisms (Parmesan 2006). An increase in temperature by 2.4 °C was reported to result (Cayan et al. 2008) in an increased amount of new pests and diseases species, as the range of many pests and diseases have historically been limited by cold winter minimums in California. Changes in climate variables can aid in the severity of plant and crop diseases. A local increase in rainfall due to changes in climate factors might be responsible for the increase in plant diseases. Sturrock et al. (2011) reported that increased plant diseases on the forests plant have been attributed to climate change. It was further stressed that as climate changes, the effects of forest diseases on forest ecosystems also change and that there were relationships between climate variables and several forest diseases, as well as current evidences of how climate, host and pathogen interactions are responding or might respond to climate change.

6 Conclusions

No doubt climate change is an important phenomenon that affects the incidence and severity of plant diseases with resultant effects on agricultural production. Temperature in conjunction with rainfall (precipitation) and other climatic factors determine the incidence and severity of disease on plants. However, the effect could be positive or negative. Notably, plant disease incidence and severity are on the high side during rainy season (May to October) when the rainfall, cloud cover and relative humidity are high and temperature and sunlight intensity are low. These condition is favorable for the diseases causing micro-organisms (pathogens). It is a period when the pathogens do multiply with resultant high plant disease incidence and severity. Thus, change in climate variables affect the actual, spatial and temporal distribution of plant diseases. However, the magnitude of these effects remains controversial. Coincidentally, these condition that is favorable to pathogenic organism are also among the preferred conditions for plant growth and developments. The food security could therefore be at risk due to a change in the incidence of plant diseases as a result of climate change. Given the importance of food in human wellbeing and the importance of agriculture in the functioning of national and regional economies, the menace of plant diseases occasioned by climate change should be tackled with all seriousness it deserves.

References

Abaje, B., & Giwa, P. (2007). *Urban flooding and environmental safety: A case study of Kafanchan Town in Kaduna State*. A Paper Presented at the Golden Jubilee (50th Anniversary) and 49th Annual Conference of the Association of Nigerian Geographers (ANG) Scheduled for 15th–19th October, 2007 at the Department of Geography, University of Abuja, Gwagwalada, Abuja.

Agrios, G. N. (2005). *Plant pathology* (5th ed., p. 922). Dana Dreibelbis Publisher. USA: Elsevier Academic Press.

Akinbode, O. A., Ogunniyan, D. J., Olasoji, J. O., Ajijiola, S., Anjorin, F. B., Olakojo, S. A., et al. (2013). Survey of resurgence of maize diseases in South Western and Kwara States of Nigeria. *International Journal of Agriculture and Forestry, 4*(6), 451–458.

Aliyu, T. H. (2014). The incidence, severity and occurrence of four viruses infecting pepper (*Capsicum* spp.) in the Southern Guinea Savannah Agro-ecological Zone of Nigeria. *Agriculturae Conspectus Scientificus, 79*(4), 233–237.

Anjorin, T. S., Jolaoso, M. A., & Golu, M. T. (2013). A survey of incidence and severity of pests and diseases of okra (*Abelmoschus esculentus* L. Moench) and eggplant (*Solanum melongena* L.) in Abuja, Nigeria. *American Journal of Research Communication, 1*(11), 333–349.

Anjorin, T. S., Nwokocha, O. V., & Sanni, A. D. (2014). Morphological characteristics and incidence of diseases on white yam (*Dioscorea rotundata* L. Poir) tubers in Abuja, Nigeria. Nature and Science, *12*(7), 58–64.

Awa, O. C., Osunlaja, S. O., Owowru, O., & Sosanya, O. (2012). First report of fruit anthracnose in mango caused by *Colletotrichum gloeosporioides* in Southwestern Nigeria. *International Journal of Scientific & Technology Research, 1*(4), 30–34.

Ayo-John, E. I., Olorunmaiye, P. M., Odedara, O. O., Dada, O. B., Abiola, K. O., & Oladokun, J. O. (2012). Notulae Scientia Biologicae, *6*(3), 321–325.

Bello, O. B., Ganiyu, O. T., Wahah, M. K. A., Azeez, M. A., Abdulmaliq, S. Y., Ige, S. A., et al. (2011). Yield and disease reactions of quality protein maize varieties in the Southern Guinea Savanna Agro-Ecology of Nigeria. *International Journal of Agriculture and Forestry, 2*(5), 203–209.

Bem, A. A., Igbawundu, J. T., Terna, P. T., Bem, S. L., Akesa, M., & Fadimu, O. Y. (2013). preliminary evaluation of incidence and severity of powdery mildews of some crop plants in Makurdi, Benue State, Nigeria. *International Journal of Advanced Biological Research, 3*(4), 519–523.

Bem, A. A., Oluma, H. O. A., Nwantiki, A. O., & Agede, A. J. (2010). Some fungi diseases associated with Tomato (*Lycopersicon esculentum* L.) in Benue State, Nigeria. *Biotropic Research International Journal, 2*(1), 51–58.

Cayan, D. R., Luers, A. L., Franco, G., Hanemann, M., Croes, B., & Vine, E. (2008). Overview of the California climate change scenarios project. *Climatic Change, 87*(Suppl 1), 1–6.

De Wolf, E. D., & Isard, S. A. (2007). Disease cycle approach to plant disease prediction. *Annual Review Phytopathology, 45,* 203–220.

IPCC. (2007). Summary for policy makers in: Climate Change 2007: Impacts, adaptation and vulnerability. In N. L. Parry, O. F. Canziani, J. P. Palutikof, van Der Lindenand, & C. E. Hansen (Eds.), *Contribution of a working group 11 to the 4th assessment report of the intergovernmental panel on climate change* (pp. 7–22). Cambridge, UK: Cambridge University Press.

Kang, W. S., Yun, S. C., & Park, E. W. (2010). Nonlinear regression analysis to determine infection models of *colletotrichum acutatum* causing anthracnose of chili pepper using logistic equation. *Plant Pathology Journal, 26*(1), 17–24.

Kawuri, R., Suprapta, D. N., Nitta, Y., & Homma, T. (2012). Destructive leaf rot disease caused by *Fusarium oxysporum* on *Aloe barbadensis* Miller in Bali. *Agricultural Science Research Journal, 2*(6), 295.

Makun, H. A., Dutton, M. F., Njobeh, P. B., Phoku, J. Z., & Yah, C. S. (2011). Incidence, phylogeny and mycotoxigenic potentials of fungi isolated from rice in Niger State, Nigeria. *Journal of Food Safety, 31*(3), 334–349. doi:10.1111/j.1745-4565.2011.00305.x

Mendelsohn, R., Dinar, A., & Williams, L. (2006). The distributional impact of climate change on rich and poor countries. *Environment and Development Economics, 11,* 159–178.

Ndor, D. C. (2010). Survey of Potato (*Solanum tuberosum l*) bacterial and viral diseases in Bokkos Local Government Area of Plateau State Nigeria. *International Journal of Emerging Knowledge, 1*(4), 64–68.

Niranjan, K., Adamu, M. A., Isah, K. M., & Lawal, A. F. (2012). A survey of vegetable fields for root-gall disease In Niger State, Nigeria. *PAT, 10*(1), 17–27.

Ogara, I. M., & Bina, R. N. D. (2010). Prevalence and persistence of galls and the galling nematodes *Meloidogyne* spp. on yams in Nasarawa state, Nigeria. *Acta SATECH, 3*(2), 10–13.

Parmesan, C. (2006). Ecological and evolutionary responses to recent climate change. *Annual Review of Ecology Evolution, and Systematics, 37,* 637–669. doi:10.1146/annurev.ecolsys.37.091305.110100

Reilly, J. (1999). What does climate change mean for agriculture in developing countries? A comment on Mendelsohn and Dinar. *World Bank Research Observer, 14,* 295–305.

Rosenzweig, C., Iglesias, A., Yang, X. B., Chivian, E., & Epstein, P. (2001). Climate change and extreme weather events: implications for food production, plant diseases and pests. *Global Change and Human Health, 2,* 90–104.

Scholze, M., Knorr, W., Arnel, N. W., & Prentice I. C. (2006). A climate-change risk analysis for world ecosystems. *Proceedings of the National Academy of Sciences, 103*(35), 116–120.

Shatu, A., Alegbejo, M. D., Boniface, K., Olalekan, O. B., Robert. A., & Lava-Kumar, P. (2010). Distribution and incidence of viruses infecting yam (*Dioscorea* Spp.) in Nigeria. *Global Journal of Bio-Science and Biotechnology, 1*(2), 163–167.

Sturrock, R. N., Frankel, S. J., Brown, A. V., Hennon, P. E., Kliejunas, J. T., Lewis, K. J., et al. (2011). Climate change and forest diseases. *Plant Pathology, 60,* 133–149. doi:10.1111/j.1365-3059.2010.02406.x

Su, H., van Bruggen, A. H., Subbarao, K. V., & Scherm, H. (2004). Sporulation of Bremia lactucae affected by temperature, relative humidity, and wind in controlled conditions. *Phytopathology, 94,* 396–401.

Part II
Institutional Experiences on Adaptation, as well as Case Studies, Examples of Projects and of Good Practice

A Synthesis of Smallholder Farmers' Adaptation to Climate Change in Southern Africa: Averting Adaptation Vacuum

Obert Jiri and Paramu Mafongoya

1 Introduction

Resource poor smallholder farmers in southern Africa, who are natural resource-dependent, bear the disproportionate burden of climate change impacts (Mendelsohn 2009). In southern Africa, the effects of climate change are already being experienced, with noticeable disastrous consequences for rural households (Adger et al. 2013). It is important to note, however, that the rural poor have been successfully coping with climate change and variability in the past (Kelly and Adger 2000). The success of these historical coping and adaptation strategies in the future depends on exogenous formal and informal rural institutions. It is important to realize that coping and adaptation to climate change should be highly localized in order to be effective (Stringer et al. 2009).

2 Impacts of Climate Change in Smallholder Farming Systems

The impacts of climate change and variability on smallholder agriculture have been abundantly reported in literature, with the early literature and research focusing primarily on the vulnerability of the sector (Speranza 2010). The conclusion from most studies is that the degree of vulnerability of the smallholder agricultural sector to climate change and variability is dependent on local environmental and management factors. These include local soil fertility, type of crop grown, extent of

O. Jiri (✉) · P. Mafongoya
School of Agricultural, Earth and Environmental Sciences,
University of KwaZulu-Natal, Scottsville, Pietermaritzburg, South Africa
e-mail: obertjm@gmail.com

W. Leal Filho et al. (eds.), *Climate Change Adaptation in Africa*,
Climate Change Management, DOI 10.1007/978-3-319-49520-0_16

knowledge and awareness of climate change, the extent of support from government and other external agencies, and the ability of key stakeholders (at the national, local, and household level) to address climate concerns (Osman-Elasha et al. 2006). However, the increased uncertainty surrounding climate change and its impacts presents an additional problem that smallholder farmers have to address (Jiri et al. 2015a). For example, in Zimbabwe, poor soil quality, financial constraints, and lack of access to markets can constrain agricultural productivity to begin with, regardless of climate effects (Rurinda et al. 2014). Climate change, thus, represents an additional burden that, for smallholder farmers, translates into production risks associated with crop and livestock failure, and hunger.

Climate change and variability, particularly changes in temperature and rainfall, will alter the distribution of agro-ecological zones (Martin et al. 2000). The timing and length of growing seasons will be affected by changes in soil moisture and content in diverse ways. In southern Africa, it is expected that higher temperatures will adversely affect growing conditions (Lobell et al. 2008). Irrigation availability and demand will also be affected by both changes in temperature and precipitation. Reduction in rainfall is likely to intensify further water exploitation for agriculture, competing with non-agricultural uses. An increase of potential evapotranspiration is likely to intensify drought stress and associated extreme events such as heatwaves (Solh and Van Ginkel 2014).

Increased climate variability and change would also cause carbon dioxide effects with a positive impact due to, for example, greater water use efficiency and higher rate of photosynthesis. Literature dealing with the agronomic effects of climate change indicate that carbon dioxide concentrations may increase by up to 57% by 2050 (Jat et al. 2012). Increasing carbon dioxide concentrations in the atmosphere are important as they increase the rate of photosynthesis and water use efficiency. These effects are strongest for C3 plants such as wheat, rice, and soybean. However, carbon dioxide enrichment is also positive for C4 plants such as the common cereals like maize and sorghum. IPCC (2013) estimate that a doubling of carbon dioxide concentrations would lead to yield improvements ranging from 10 to 30%. However, while higher atmospheric concentrations of CO_2 will, by reducing evapotranspiration, improve water use efficiency of crops and increase the rate of photosynthesis, the net result may be moderated by costly pest and weed infestations (Le Quesne et al. 2010).

Table 1 shows the impacts of climate change on smallholder farming systems at field and natural environmental levels as well as the impacts on human health and non-agricultural activities. The impacts of climate change at the local level largely depends on vulnerability and adaptive capacity of the smallholder farmers before the climate event. Smallholder farmers in southern Africa have a repertoire of adaptive strategies which they have used to cope with the vagaries of climate change and variability in the past (Speranza 2010). These strategies include agronomic and livestock options to deal with low rainfall and drought. Other coping strategies could also include non-agricultural options such as selling of assets to cope with drought.

Table 1 Direct impacts of climate change on smallholder farming systems

Impact	Field/organism	Landscape/environment	Human health	Infrastructure and non-agricultural employment
Extreme events	• Increased droughts • Increased chances of failed growing seasons • Increased outbreaks of rainfall-related animal diseases	• Increased soil erosion, resulting from floods	• Increase in waterborne diseases during floods	• Destruction of infrastructure in drylands and adjacent urban areas through floods
Increased variability in rainfall and temperature	• Specific effects on crop development e.g. shorter growing seasons	• Increased soil erosion, resulting from heavier erosiveness of rainfall	• Greater risks of food insecurity • Shifts in climate-related disease prevalence	• Damage to roads, resulting from heavier erosiveness of rainfall
Shift in temperature and rainfall	• Higher crop water demands, resulting from warming • Decreased rainfall in some areas • Shorter growing periods • Direct carbon dioxide fertilization • Declining yields, especially maize • Increased heat stress to cattle	• Remobilization of sand dunes • Increased encroachment of woody species on rangeland • Decreased river flows, less water available for irrigation • Shifts in boundary of tsetse fly/trypanosomiasis	• Shifts in malaria and meningitis	• Decreased tourism revenue and employment in tropical countries, resulting from warming and water shortages

Source Modified from Morton (2007)

3 Smallholder Farmer Perceptions to Climate Change

At the local smallholder farmer level, the specific nature of climate change impacts continues to be uncertain, even though the general impacts of climate change will be striking and long-lasting if the current trends continue. Smallholder farmers, however, perceive changes in long term rainfall and temperature (Jiri et al. 2015b). It is important to note that the basis of most coping and adaptation strategies employed by smallholder farmers are based on their perceptions of climate variability particularly in the short term. These farmers need information on when it will rain in their local area. Such information is obtained from their indigenous knowledge of their local area. Indigenous knowledge information in southern Africa is derived from long term observations of animal behaviour, tree penology and cloud and wind observations (Ajani et al. 2013).

While differences remain in the criteria used to define seasonal phenomena by both farmers and scientists, there is a significant overlap between indigenous and scientific knowledge regarding weather and climate forecasts (Kalanda-Joshua et al. 2011). This makes indigenous knowledge potentially useful for scientific forecasting, particularly in tracking change. Moreover, both local and scientific knowledge in weather forecasting are produced through observation, experimentation and validation, suggesting that there is a convergent point between the two kinds of knowledge. However, indigenous knowledge is generally criticized as lacking any regulations and includes a measure of spirituality that is not considered in scientific forecast (Folke 2004). Therefore, there is potential for integration of indigenous knowledge used by farmers and scientific knowledge to enable devising of robust adaptation strategies at the local level.

Smallholder farmers are often prepared to integrate new information into their traditional forecasting methods as demonstrated by these farmers readiness to engage, discuss and use modern scientific forecasts (Kalanda-Joshua et al. 2011). This could allow designing of forecasts that would be in sync with farmers' priorities and more acceptable to smallholder farmers. For instance, climate scientists use coarse spatial analysis that does not address the risks in drier sub-regions within relatively moist regions, providing an opportunity for integration of indigenous knowledge of spatial variability in climate patterns for the identification of areas exposed to drought risk.

4 Agricultural Vulnerability to Climate Change in Smallholder Farming Systems

Increased climate change and variability in southern Africa calls for agriculture and development efforts that consider vulnerability, adaptive capacity and resilience in order not to worsen the plight of the smallholder farmers (Masih et al. 2014). It is critical to attend to the frequency, intensity and timing of impacts to understand how they affect adaptive capacity. The dynamics of risk exposure are crucial to understand how they determine sensitivity and adaptive capacity of rural households.

Vulnerability to climate change is a function not only of biophysical outcomes related to variations and changes in temperature, precipitation, topography, and soils, but also of socio-political and institutional factors (Kelly and Adger 2000). Other social attributes such a gender, ethnic affiliation and age are also closely related with vulnerability. The degree to which they are associated with vulnerability tends to depend on location factors. This depicts that although climate change is a global phenomenon, adaptation to climate impacts inevitably is local (Corobov et al. 2013).

Even though individual households and communities have, in the past, used different adaptation strategies in the face of climate variability, their adaptive capacity depends on climate information shared between the community and their interaction with external institutions. Successful adaptation may depend on specific institutional arrangements because adaptation does not occur in an institutional vacuum.

Climate change adaptation usually involve practices that provide diverse income earning activities. Diversification can be achieved at individual, household, and community levels. Different activities are pursued according to season, in response to bad seasons, and sequentially so that farmers shift among cropping, livestock, and use of wild resources. In dryland farming systems, common in the smallholder farming areas, adaptation strategies include use of multiple crops varieties and livestock species; intensified use of labour at critical times of the year (Speranza 2010); on-farm food storage and prudent management of soil and water resources.

5 Why Are We Concerned with Adaptation?

Climate change has and is occurring. Anticipation and consideration of the current and anticipated climate change impacts motivates adaptation. The vulnerabilities in the smallholder farming sector require rapid and adequate support for alleviation against impacts of climate change. This support includes increased investment in adaptive capacity and increasing resilience to climate change. Adaptation, therefore, aims to build climate resilient livelihoods and communities by creating a buffer against climate-induced shocks and disturbances (Shiferaw et al. 2014). Sustainable adaptation happens when the nature of prevalent shocks at the local level is considered and a variety of strategies formulated to cope and adapt to the shocks. An integrated approach that combines local indigenous knowledge systems and meteorological data would lead to development of sustainable adaptation and increase resilience at the local level (Jiri et al. 2015a, b). Future vulnerabilities depend not only on climate change, but also on current adaptation pathways (Eriksen et al. 2011).

5.1 Conceptualizing Adaptation

The Intergovernmental Panel on Climate Change (IPCC) defines adaptation as a basic adjustment in natural or human systems to climate change. Adaptation is often described as an adjustment in ecological, social or economic systems in response to actual or expected climatic shock and impacts. It refers to changes in processes, practices or structures to moderate or negate potential damages or to take advantages of opportunities associated with changes in climate (Smit et al. 2001). However, adaptation cannot be treated as an isolated event divorced from other policy and institutional imperatives (Wheeler 2011). In relation to agriculture in southern Africa, it is critical to understand terms related to adaptation at local smallholder farming levels. These include coping, resilience and failure to adapt (Fig. 1). The terms are used to describe the nature and extend of adaptation options employed by smallholder farmers in the face of climate change and variability. Coping are those actions taken to counter those disturbances that individuals or a community can tolerate without adverse impacts on livelihoods or the environment. Coping strategies are usually deployed when there is a decline in food sources, and these are regarded as involuntary responses to disaster or unanticipated failure in major sources of survival (Thornton et al. 2007). The term 'coping' is usually applied to short-term measures used by farmers who experience a loss or reduction

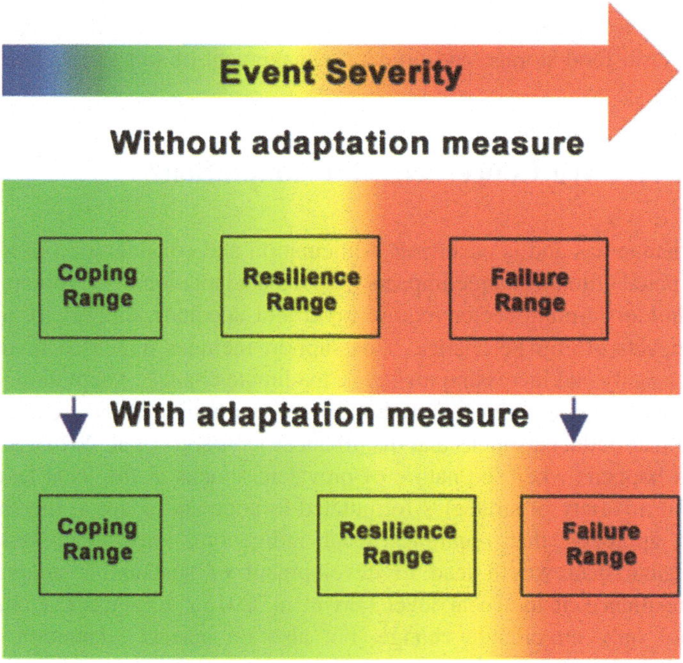

Fig. 1 Variation in adaptation with event severity (adapted from Know Climate Change 2016)

in their direct access to food when harvests fail, or to workers who, for whatever reasons, lose their employment and income and therefore face the threat of hunger or even starvation. In a study in Chiredzi, Zimbabwe, on climate change adaptation, Mutambara et al. (2013), observed farmers engaging in short-term employment at a local large sugarcane estate as a coping strategy to avert hunger. Thus, coping also refer to the ways in which people deal with seasonal food stress (Vincent and Cull 2013). It should, however, be noted that there are overlaps between coping and adaptation in the sense that coping may lead to adaptation.

The difference between coping and adaptation is usually in temporal comparisons. Coping strategies are associated with short term interventions whereas adaptive strategies are deemed medium term and more permanent interventions (Wilhite et al. 2014). Coping strategies may become adaptive strategies when people are forced to use them over a run of bad years and across seasons rather than just at the worst time of the year. Although not all coping strategies are erosive, they also can erode household assets, with the risk of households sinking into destitution and degrading the environment. Coping and adaptation are interlinked so that the way households cope with crises either may enhance or may constrain their future coping strategies, as well as their possibilities to adapt in the longer term (Wilhite et al. 2014). For example, if drought losses force households to sell off productive assets (such as cattle) or send their productive labour to work for others, the households will find it increasingly difficult to restore their former levels of activity and income.

Out-migration by individuals or entire households is also being used increasingly to cope with climate-related stresses, generating significantly different impacts on home and receiving areas (Black et al. 2013). This migration has led to substantial remittance flows back to the family home and those flows can then be used for buying food and farm equipment as well as investing in shops and buying land.

Resilience defines the magnitude of damage individuals or a system can tolerate, and still autonomously return to its original state (Gitz and Meybeck 2012). Failure to adapt occurs when the magnitude of damage is such that a system can no longer tolerate it without adverse impacts. For instance, when smallholder farmers employ climate smart cropping options but still fail to get a yield as a result of lack of adequate rainfall in that particular season or location (FAO 2014). In an experiment that evaluated a basket of drought tolerant legume and cereal crops in the smallholder communities of Chiredzi, Zimbabwe, there was no rain and the suggested climate smart crops suffered maladaptation (Jiri et al. 2016—in print). The severity of a shock at the local level depends on the ability of the system to cope, and the inbuilt resilience of the system.

In the absence of adaptation strategies, resilience and coping decrease and failure range increase (Fig. 1). Thus, people become more vulnerable to climate change shocks and impacts. With adaptation, however, with increasing severity, coping and resilience increase and failure to adapt decreases as people become less vulnerable (Adger and Barnett 2009). A system or individuals with adaptation measures would be more resilient to shocks in case a severe event takes place, compared to a system without adaptation measures.

5.2 Types of Adaptation Measures

Adaptation measures employed by smallholder farmers vary depending on prevailing factors. Generally adaptation measures can be classified depending on the timing and aim of their employment. Accordingly, adaptation can be reactive or planned actions (Brooks and Fisher 2011; Smit et al. 2000; Fig. 2). Reactive adaptation actions are those actions that are taken by smallholder farmers after the initial impacts of climate change have become evident. Anticipatory adaptation occurs before the impacts have occurred.

Planned adaptation is the result of deliberate policy decision, based on observed or expected climate change impacts. This usually follows a top-down approach as those in authority try to influence decisions and actions at local level. Autonomous adaptation actions, on the other hand, are usually taken by individuals or communities as they independently adjust to their own perceptions about climate risk and impact. These actions, which include various coping strategies, are usually short term and are normally reactive.

Adaptation can also be of spatial scope (local, regional, national), form (e.g. technological, financial, institutional), and degree of necessary change (incremental, transformational) (Smit et al. 2000).

In a study which analyzed micro-level analysis of adaptation that focused on tactical decisions farmers make in response to seasonal variations in climatic, economic and other factors, Nhemachena et al. (2014) indicated that adaptation decisions are influenced by a number of socio-economic factors that include

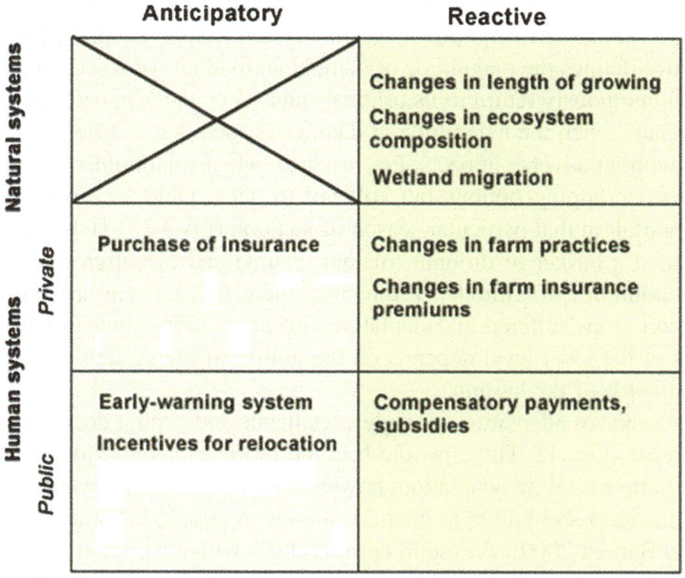

Fig. 2 Types of adaptation actions employed by farmers (adapted from Smit and Pilifosova 2003)

Table 2 Main farm-level adaptation strategies in Southern Africa (% of respondents)

Adaptation	Southern Africa	South Africa	Zambia	Zimbabwe
Different varieties	11	5	13	15
Different crops	4	4	6	3
Crop diversification	9	6	9	12
Different planting dates	17	7	5	38
Diversifying from farming to non-farming activity	8	5	11	7
Increased use of irrigation/groundwater/watering	9	18	5	6
Increased use of water conservation techniques	5	6	3	7
Number of observations	1719	236	829	654

Source Nhemachena et al. (2014)

household demographic characteristics, household resource endowments, access to information and availability of formal institutions. Farm-level decision making occurs over a very short time period usually influenced by seasonal climatic variations, local agricultural cycle, and other socio-economic factors. Adaptation is important for farmers to achieve their farming objectives such as food and livelihood security, and high incomes. It would also reduce potential negative impacts that are associated with changes in climatic and other socio-economic conditions. Table 2 shows some of the farm level adaptation decision employed by smallholder farmers in southern Africa.

5.3 Approaches for Adaptation Strategies

Adaptation strategies employed by farmers are different depending on climatic stimuli and intervening conditions or non-climatic stimuli. The different stimuli influence the sensitivity of a particular system and the nature of adjustments or adaptation required. As a result adaptation measures need to consider socio-economic and institutional arrangements at a particular locality. Impacts of climate change are quite different depending on the socio-economic disposition of the farmers, and may require different adaptive responsive, both in the short and in the long term. The appropriateness of a particular adaptation strategy is highly dependent on time and place as they are influenced by key cultural and indigenous observations and indicators at the local level. These indigenous observations, while sometimes robust, are usually peculiar to a local area or region.

There are generally two approaches to adaptation (Stringer et al. 2009). The first is an approach that advocates for actions that reduce existing vulnerability. The use of early warning systems, for instance, means individuals and communities are able to employ anticipatory adaptation. The second approach is to mainstream climate change into existing activities. Mainstreaming ensures that future vulnerability to climate change is countered by considering climate change in decision making. This is the trend in most developing countries where development is a priority. This approach is particularly useful where climate change may increase the risk of failure of assets.

It is vital, therefore, to increase resilience, coping and adaptive capacity of natural and human systems, so as to prepare them for future variability and extremes due to climate change.

5.4 *The Role of Institutions to Smallholder Adaptation*

Local institutions are fundamental to facilitate adaptation to climate change and variability in the future. Institutional arrangements structure risks and sensitivity to climate hazards, facilitate or impede individual and collective responses and shape the outcome of coping and adaptation strategies (Ziervogel et al. 2006). Understanding how these institutions operate in relation to climate change impacts is important in designing interventions that can influence adaptive capacity and adaptation options. Local institutions shape the effects of climate impacts in three ways: they influence how households are affected by climate impacts; they shape the ability of households to respond to climate impacts and pursue different adaptation practices; and they mediate the flow of external interventions in the context of adaptation (Majule et al. 2013).

5.4.1 Local Institutions Shape the Impact of Climate Change and Variability on Communities

A single climate event will have different effects on the livelihoods of communities, depending on the nature of local governance and local institutional arrangements. For instance, reduced precipitation by 10% in a given year in southern Africa will have a less negative impact on farmers who have access to irrigation compared to those who rely on rainfed agriculture (Bryan et al. 2009). The negative effect of crop failure would be minimised if farmers have more access to climate and environment-related institutions as opposed to where institutional access is stratified and information is monopolized by fewer people.

5.4.2 Local Institutions Shape the Way Communities Respond to Climate Change and Variability

Institutions provide the framework within which households choose adaptation strategies by linking individuals with other communities. In many southern African communities, for instance, strong institutional norms around labour sharing will reduce the ability of households to adapt by out-migrating (Zuma-Netshiukhwi et al. 2013). Closely knit social networks make it easier to undertake communal pooling of resources as an adaptation strategy. Communities that lack access to capital and infrastructure may be unable to use exchange strategies to cope with climate variability. Without access to markets, communities may be forced to adopt storage of harvests as an adaptation response and invest resources into storage infrastructure (Paavola 2008).

5.4.3 Local Institutions Are the Intermediaries for External Support to Climate Change Adaptation

Institutions are the conduit through which external interventions help or undermine existing adaptation practices. All external interventions need local institutional support to leverage the impact of interventions (Majule et al. 2013). Local institutional partners' involvement strengthens the effectiveness of external interventions. However, local institutions, which are usually informal, are rarely supported by government and external interventions (Ziervogel et al. 2014). This is because when external support is provided, it is channeled through formal institutions. When external public institutions get involved in adaptation practices, their relationships are more often with formal local civic institutions.

5.5 Modes of Smallholder Farmers' Adaptation to Climate Change

The impacts of climate change and variability can be classified in a variety of ways: short-term or long-term, those due to sudden disasters or those due to slow changes in trends, predictable or unpredictable (Biagini et al. 2014). However, at household level, it is important to examine how climate-related risks affect livelihood capabilities over time, across space, across asset classes, and across households. These four types of risks to livelihoods define the major conceptual categories of the ways climate variability and change threaten the ability of households to adapt and secure livelihoods.

Households and communities have developed strategies to adapt to climate variability. Smallholder farming communities in southern Africa have already experienced many forms of extreme climate events. Over time, they have developed a range of adaptive responses to cope with climatic threats to their livelihoods (Biagini et al. 2014). When climate hazards are repeated, the distinction between short-term and long-term adaptation (that is, coping versus adaptation) breaks down. Thus, there is need to address adaptation strategies as they assist coping. Local adaptation responses to climate change and variability can be classified into five categories.

5.5.1 Mobility

This adaptation response addresses movements of various kinds in response to climatic risks and climate-induced scarcities. It is a common adaptation strategy used by smallholder households and communities, particularly in dry areas, such as in southern Africa (Mertz et al. 2009). It pools risks across space, and is successful in combination with clear information about the spatial and temporal distribution of rainfall. In the context of climate change, mobility sometimes has been viewed as a maladaptation, in which climatic stresses lead to involuntary migrations. However, mobility also is a way of life for large groups of people in semiarid regions, and a long-standing mechanism to deal with spatiotemporal variations in rainfall and range productivity (Marin 2010). For agricultural populations, mobility often can be the last resort in the face of climate risks and disruption of livelihoods. To address the needs of mobile populations, the role of information in tracking human and livestock movements, and the mobile provision of basic services such as health, education, credit, and marketing of animal products, are especially important to reinforce adaptive capacity.

5.5.2 Storage

This adaptation strategy entail the storage of surpluses is an effective measure against future livelihood failures. Smallholder farming households have indigenous storage infrastructures for seeds and harvested crops, and have developed procedures for drying fruits for storage. When combined with well-constructed infrastructure, low levels of perishability, and a high level of coordination across households and social groups, it is an effective measure against drought (Meinke et al. 2006). As an adaptation strategy to avert climate risk, storage is relevant to individual farmers and communities to address food and water scarcities.

5.5.3 Diversification

This strategy is usually used in relation to on- and off-farm employment opportunities, assets usage and consumption strategies. Diversification into different farm management practices and crop varieties, and using a combination crop and livestock farming are common diversification responses to climate risk (Paavola 2008). Diversification can also occur in relation to productive and non-productive assets, consumption strategies, and employment opportunities. It is reliable to the extent that benefit flows from assets are subject to uncorrelated risks. Diversifying households typically gives up some returns in exchange for the greater security provided by diversification. Individuals using this strategy would be willing to live with some level of poverty in exchange for reduction in vulnerability.

5.5.4 Communal Pooling

This refers to adaptation responses involving joint ownership of assets and resources. It entails sharing of wealth, labour, and/or incomes from particular activities across households. It also involves mobilization and use of resources that are held collectively during times of scarcity. It pools risks across households (Berkes 2007). Although communal pooling can occur in combination with mobility, storage, and diversification, its key indicator is joint action by community members with the aim of pooling their risks and resources. Communities in dryland areas, for example, increase water rationing and/or often prohibit the consumption of certain foods and forest products except during times of famine or long-term rainfall failure.

5.5.5 Market Exchange

Market exchange adaptation strategy is perhaps the most common and versatile mechanism for adaptation. Such a mechanism can substitute for the other four practices when smallholder farmers have access to markets (Paavola 2008). However, this requires well-developed institutions to facilitate market access. In the absence of institutional mechanisms that can ensure equity, the smallholder farmers are less likely to benefit from purely market exchange–based adaptation.

All adaptation practices in smallholder farming areas may succeed if given the necessary depend institutional support. Adaptation never occurs in an institutional vacuum. Institutional and social factors also play a major role in shaping the extent to which smallholder households and communities are vulnerable to different climatic risks. This highlights the importance of mainstreaming adaptation at and across institutional levels. Where successful, these responses pool uncorrelated risks associated with flows of benefits from different classes of assets owned by households and economic agents. They also can enable a shift away from more risky economic strategies to other, less risky ones.

5.6 Limitations and Constraints to Adaptation

Adaptation studies looking at the agricultural sector considered autonomous farm level adaptation (Nelson et al. 2009). The literature mainly focus on adaptation benefits, such as yield or welfare increases, or decreases in the number of affected people. Adaptation costs are generally not considered in most studies (Nelson et al. 2009). Studies also reveal that adaptation plans and actions have limits and barriers to adaptation as they try to address climate change impacts. High adaptive capacity may not automatically translate into successful adaptations to climate change (Adger et al. 2011). This section looks at some of the barriers to climate change adaptation.

5.6.1 Physical and Ecological Limits

Ecological studies indicate that the resilience of socio-ecological systems to climate change and variability will depend on the rate and magnitude of climate change. However, there may be critical thresholds beyond which some systems may not be able to adapt without radically altering their functional state (Klein et al. 2014). Increased climate change and variability may lead to changes in the physical environment of an area that limit the possibilities for adaptation (Smith and Tol 1998). For example, further reduction in rainfall without the option for irrigation would limit smallholder agriculture and adaptation possibilities, with potential options being limited to migration.

5.6.2 Technological Limits

Technological innovations can be a critical adaptation strategy to climate variability and change. New technologies can be developed to adapt to climate change, and the transfer of appropriate technologies to developing countries forms an important component of the UNFCCC (Kurukulasuriya et al. 2006). However, there are also potential limits to technology as an adaptation response to climate change. For instance, some adaptations may be technologically possible, but may not be economically feasible or culturally desirable.

5.6.3 Financial Limits

At the local level, individuals and communities can be constrained by the lack of adequate financial resources in implementing certain adaptation strategies. Financial poverty is a factor that limits use of seemingly inexpensive climate smart agricultural options, such conservation farming, while limited public finances contribute to choices by government extension workers to give low priority to measures that would reduce vulnerability to climate-related agricultural risks (Gall 2013). In field surveys and focus groups, farmers often cite the lack of adequate financial resources as an important factor that constrains their use of adaptation measures which entail significant investment, such as irrigation systems, improved or new crop varieties, and diversification of farm operations.

5.6.4 Informational and Cognitive Limits

Uncertainty about future climate change and variability combines with individual and social perceptions of risk, opinions and values to influence judgment and decision-making concerning climate change (Pulwarty and Sivakumar 2014). Interpretations of climate change risk are context specific and human cognition can limit adaptation responses (Kalanda-Joshua et al. 2011). Four main view points on informational and cognitive constraints on individual adaptation responses to climate change are noticeable from literature. These include knowledge of climate change causes, impacts and possible solutions may not lead to adaptation; differing perceptions of climate change risks; perceptions of vulnerability and adaptive capacity; and fear of adaptation (Kalanda-Joshua et al. 2011).

5.6.5 Biological Limits

Biological characteristics affect the capacity of organisms to cope with increasing climate stress through acclimation, adaptation, or behaviour as well as influencing migration decisions and patterns (Kilroy 2015). The biological capacity for migration among non-human species is linked to characteristics such as fecundity, phenotypic and genotypic variation, dispersal rates, and interspecific interactions. Additional research is needed to clarify the capacity of species and communities to migrate in response to a changing climate.

5.6.6 Economic Limits

Long-term trends in economic development as well as current economic situations can have important influence on the capacity for adaptation. The effects of economic constraints vary depending on the vulnerability of the sector to climate change. Smallholder economies that are composed of climate-sensitive sectors such as rainfed agriculture are particularly vulnerable to the effects of climate change and would encounter greater constraints on their capacity to adapt (Nhemachena and Hassan 2010).

5.6.7 Human Resource Limits

The success of household and community efforts to adapt to climate change is dependent on humans who are the primary agents of change. Human resources provide the foundation for climate information gathering, technology use, and leadership for implementation of adaptation policies and measures (Bauer and Scholz 2010). Meinke et al. (2006) concludes that there is need to build institutions for human resources development in the climate change discipline.

5.6.8 Social and Cultural Limits

Social and cultural factors can also limit adaptation as they can influence perceptions to risk (Gall 2013; Jiri et al. 2015b). These social and cultural factors can influence the distribution of vulnerability and adaptive capacity among different societal elements (United Nations 2011).

5.6.9 Institutional and Governance Limits

There would be need for resource mobilization, decision making structures, and policy implementation for adaptation to be successful (Pye-Smith 2011). Institutional capacity is key in promoting and influencing adaptation in the small-holder farming sector. Leary et al. (2006) indicated that adaptation efforts are associated with the extent to which institutions prioritize climate change and variability, for instance.

6 Conclusion

The vulnerability of environmental and human systems to climate change and variability, and their ability to adapt to climate hazards, needs combined efforts from experts from a wide range of fields, including climate science, development studies, disaster management, health, social science, policy development and economics. Developed conceptual models of vulnerability and adaptation need to be adapted to local smallholder farmer situations. Researchers from different backgrounds must develop a common language so that vulnerability and adaptation research can move forward in a way that integrates these different traditions in a coherent yet flexible fashion, allowing researchers to assess vulnerability and the potential for adaptation in a wide variety of different contexts. This would lead to climate change adaptation and avoid an adaptation vacuum in southern Africa.

References

Adger, W. N., & Barnett, J. (2009). Four reasons for concern about adaptation to climate change. *Environment and Planning A, 41*(12), 2800–2805. doi:10.1068/a42244

Adger, W. N., Brown, K., Nelson, D. R., Berkes, F., Eakin, H., Folke, C., et al. (2011). Resilience implications of policy responses to climate change. *Climatic Change, 2*(September), 757–766. doi:10.1002/wcc.133

Adger, W. N., Quinn, T., Lorenzoni, I., Murphy, C., & Sweeney, J. (2013). Changing social contracts in climate-change adaptation. doi:10.1038/nclimate1751

Ajani, E. N., Mgbenka, R. N., & Okeke, M. N. (2013). Use of indigenous knowledge as a strategy for climate change adaptation among farmers in sub-Saharan Africa: Implications for policy. *Asian Journal of Agricultural Extension Economics & Sociology, 2*(1). Retrieved from www.sciencedomain.org

Bauer, S., & Scholz, I. (2010). Adaptation to climate change in Southern Africa: New boundaries for sustainable development? *Climate and Development, 2*(February 2013), 83–93. doi:10.3763/cdev.2010.0040

Berkes, F. (2007). Understanding uncertainty and reducing vulnerability: Lessons from resilience thinking. *Natural Hazards, 41*(2), 283–295. doi:10.1007/s11069-006-9036-7

Biagini, B., Bierbaum, R., Stults, M., Dobardzic, S., & Mcneeley, S. M. (2014). A typology of adaptation actions: A global look at climate adaptation actions financed through the global environment facility. *Global Environmental Change, 25*, 97–108. doi:10.1016/j.gloenvcha.2014.01.003

Black, R., Adger, W. N., & Arnell, N. W. (2013). Migration and extreme environmental events: New agendas for global change research. *Environmental Science & Policy, 27*(January 2016), S1–S3. doi:10.1016/j.envsci.2012.09.010

Brooks, N., & Fisher, S. (2011). Tracking adaptation and measuring development, TAMD. *Change*. Retrieved from www.iied.org

Bryan, E., Deressa, T. T., Gbetibouo, G. A., & Ringler, C. (2009). Adaptation to climate change in Ethiopia and South Africa: Options and constraints. *Environmental Science and Policy, 12*(4), 413–426. doi:10.1016/j.envsci.2008.11.002

Corobov, R., Sîrodoev, I., Koeppel, S., Denisov, N., & Sîrodoev, G. (2013). Assessment of climate change vulnerability at the local level: A case study on the Dniester river basin (Moldova). *The Scientific World Journal, 2013*, 173794. doi:10.1155/2013/173794

Eriksen, S., Aldunce, P., Bahinipati, C. S., Martins, R. D., Molefe, J. I., Nhemachena, C., et al. (2011). When not every response to climate change is a good one: Identifying principles for sustainable adaptation. *Climate and Development, 3*(1), 7–20. doi:10.3763/cdev.2010.0060

FAO. (2014). *FAO success stories on climate-smart agriculture,* 28. Retrieved from http://www.fao.org/3/a-i3817e.pdf

Folke, C. (2004). Ecology and society: Traditional knowledge in social–ecological systems. *Ecology and Society, 9*(3), 7. Retrieved from http://www.ecologyandsociety.org/vol9/iss3/art7/

Gall, M. (2013). From social vulnerability to resilience: Social vulnerability and adaptation measuring progress toward disaster in Fragile States. *InterSecTions, "Interdisciplinary Security ConnecTions"*. Retrieved from http://www.ehs.unu.edu/file/get/11125.pdf

Gitz, V., & Meybeck, A. (2012). Risks, vulnerabilities and resilience in a context of climate change. *Building Resilience for Adaptation to Climate Change in the Agriculture Sector,* (2006), 19–36. Retrieved from http://www.fao.org/docrep/017/i3084e/i3084e00.htm

IPCC. (2013). Climate change 2013: The physical science basis. Summary for policymakers. *IPCC*, 1–29. doi:10.1017/CBO9781107415324

Jat, R. A., Craufurd, P., Sahrawat, K. L., & Wani, S. P. (2012). Climate change and resilient dryland systems: Experiences of ICRISAT in Asia and Africa. *Current Science, 102*(12).

Jiri, O., Mafongoya, P. L., & Chivenge, P. (2015a). Indigenous knowledge systems, seasonal "quality" and climate change adaptation in Zimbabwe. *Climate Research, 66*, 103–111. doi:10.3354/cr01334

Jiri, O., Mafongoya, P. L., & Chivenge, P. (2015b). Smallholder farmer perceptions on climate change and variability: A predisposition for their subsequent adaptation strategies. *Journal of Earth Science & Climatic Change, 6*(5). doi:10.4172/2157-7617.1000277

Kalanda-Joshua, M., Ngongondo, C., Chipeta, L., & Mpembeka, F. (2011). Integrating indigenous knowledge with conventional science: Enhancing localised climate and weather forecasts in Nessa, Mulanje, Malawi. *Physics and Chemistry of the Earth, 36*(14–15), 996–1003. doi:10.1016/j.pce.2011.08.001

Kelly, P. M., & Adger, W. N. (2000). Theory and practice in assessing vulnerability to climate change and facilitating adaptation. *Climatic Change, 47*(4), 325–352. doi:10.1023/A:1005627828199

Kilroy, G. (2015). A review of the biophysical impacts of climate change in three hotspot regions in Africa and Asia. *Regional Environmental Change*, 1–12. doi:10.1007/s10113-014-0709-6

Klein R. J. T., Midgley, G. F., Preston, B. L., Alam, M., Berkhout, F. G. H., Dow, K., et al. (2014). *16. Adaptation opportunities, constraints, and limits. Assessment report 5—Climate change 2014: Impacts, adaptation, and vulnerability. Part A: Global and sectoral aspects*. Paul Watkiss Johanna Wolf Cambridge University Press.

Know Climate Change. (2016). *Adaptation and mitigation—Adaptation*. Retrieved February 23, 2016, from http://know.climateofconcern.org/index.php?option=com_content&task=article&id=147

Kurukulasuriya, P., Mendelsohn, R., Hassan, R., Benhin, J., Deressa, T., Diop, M., et al. (2006). Will African agriculture survive climate change? *Source: The World Bank Economic Review, 20*(3), 367–388. Retrieved from http://www.jstor.org/stable/40282223

Le Quesne, T., Matthews, J. H., Von der Heyden, C., Wickel, A. J., Wilby, R., Hartman, J., et al. (2010). *Flowing forward: Freshwater ecosystem adaptation to climate change in water resources management and biodiversity conservation*. Water Working Notes. Retrieved from http://www-wds.worldbank.org/external/default/WDSContentServer/WDSP/IB/2010/12/02/000334955_20101202041330/Rendered/PDF/582130NWP0P11710WN281FlowingForward.pdf

Leary, N., Baethgen, W., Barros, V., Burton, I., Canziani, O., Downing, T. E., et al. (2006). *A plan of action to support climate change adaptation through scientific capacity, knowledge and research*. AIACC Working Papers. Retrieved from www.aiaccproject.org

Lobell, D. B., Burke, M. B., Tebaldi, C., Mastrandrea, M. D., Falcon, W. P., & Naylor, R. L. (2008). Prioritizing climate change adaptation need for food security in 2030. *Science, 319* (February), 607–610. doi:10.1126/science.1152339

Majule, A. E., Stathers, T., Lamboll, R., Liwenga, E. T., Swai, E., & Chipungu, F. (2013). Enhancing capacities of individuals, institutions and organizations to adapt to climate change in agricultural sector using innovative approaches in Tanzania and Malawi. *World Journal of Agricultural Sciences, 1*(July), 220–231.

Marin, A. (2010). Riders under storms: Contributions of nomadic herders' observations to analysing climate change in Mongolia. *Global Environmental Change, 20*, 162–176. doi:10.1016/j.gloenvcha.2009.10.004

Martin, R. V, Washington, R., & Downing, T. E. (2000). *Seasonal maize forecasting for South Africa and Zimbabwe derived from an agroclimatological model*.

Masih, I., Maskey, S., Mussá, F. E. F., & Trambauer, P. (2014). A review of droughts on the African continent: A geospatial and long-term perspective. *Hydrology and Earth System Sciences, 18*, 3635–3649. doi:10.5194/hess-18-3635-2014

Meinke, H., Nelson, R., Kokic, P., Stone, R., Selvaraju, R., & Baethgen, W. (2006). Actionable climate knowledge: From analysis to synthesis. *Climate Research, 33*(1), 101–110. doi:10.3354/cr033101

Mendelsohn, R. (2009). The impact of climate change on agriculture in developing countries. Retrieved January 28, 2016, from http://www.tandfonline.com/doi/pdf/10.1080/19390450802495882

Mertz, O., Mbow, C., Reenberg, A., & Diouf, A. (2009). Farmers' perceptions of climate change and agricultural adaptation strategies in Rural Sahel. *Environmental Management, 43*(5), 804–816. doi:10.1007/s00267-008-9197-0

Morton, J. F. (2007). The impact of climate change on smallholder and subsistence agriculture. doi:10.1073/pnas.0701855104

Mutambara, J., Jiri, O., Jiri, Z., & Makiwa, E. (2013). Agricultural training post land reform in Zimbabwe: Implications and issues. *Online Journal of African Affairs, 2*(2), 38–45.

Nations, U. (2011). *The social dimensions of climate change*.

Nelson, G. C., Rosegrant, M. W., Koo, J., Robertson, R., Sulser, T., Zhu, T., et al. (2009). *Climate change: Impact on agriculture and costs of adaptation—Appendix 1*.

Nhemachena, C., & Hassan, R. (2010). Measuring the economic impact of climate change on African agricultural production systems. *Climate Change Economics, 1*(1), 33–55. doi:10.1142/S2010007810000066

Nhemachena, C., Hassan, R., & Chakwizira, J. (2014). Analysis of determinants of farm-level adaptation measures to climate change in Southern Africa. *Journal of Development and Agricultural Economics, 6*(5), 232–241. doi:10.5897/JDAE12.0441

Osman-Elasha, B., Goutbi, N., Spanger-Siegfried, E., Dougherty, B., Hanafi, A., Zakieldeen, S., et al. (2006). *Adaptation strategies to increase human resilience against climate variability and change: Lessons from the arid regions of Sudan*. AIACC Working Papers. Retrieved from www.aiaccproject.org

Paavola, J. (2008). Livelihoods, vulnerability and adaptation to climate change in Morogoro, Tanzania. *Environmental Science and Policy, 11*(7), 642–654. doi:10.1016/j.envsci.2008.06.002

Pulwarty, R. S., & Sivakumar, M. V. K. (2014). Information systems in a changing climate: Early warnings and drought risk management. *Weather and Climate Extremes, 3*, 14–21. doi:10.1016/j.wace.2014.03.005

Pye-Smith, C. (2011). Farming's climate smart future: Placing agriculture at the heart of climate-change policy. *Farming's Climate-Smart Future*, 36. Retrieved from http://agris.fao.org/agris-search/search.do?recordID=GB2013201598

Rurinda, J., Mapfumo, P., Van Wijk, M. T., Mtambanengwe, F., Rufino, M. C., Chikowo, R., et al. (2014). Sources of vulnerability to a variable and changing climate among smallholder households in Zimbabwe: A participatory analysis. *Climate Risk Management, 3*, 65–78. doi:10.1016/j.crm.2014.05.004

Shiferaw, B., Tesfaye, K., Kassie, M., Abate, T., Prasanna, B. M., & Menkir, A. (2014). Managing vulnerability to drought and enhancing livelihood resilience in sub-Saharan Africa: Technological, institutional and policy options. *Weather and Climate Extremes, 3*, 67–79. doi:10.1016/j.wace.2014.04.004

Smit, B., Burton, I., Klein, R. J. T., & Wandel, J. (2000). An anatomy of adaptation to climate change and variability. *Climate Change*. doi:10.1023/A:1005661622966

Smit, B., & Pilifosova, O. (2003). *Adaptation to climate change in the context of sustainable development and equity*. Retrieved February 23, 2016, from http://alapaap.observatory.ph/resources/IPCC/TAR/wg2/pdf/wg2TARchap18.pdf

Smit, B., Pilifosova, O., Skinner, M., Wandel, J., Canada, B. S., Canada, M. S., et al. (2001). Adaptation to climate change in the context of sustainable development and equity. *Working Group II: Impacts, Adaptation and Vulnerability*, 879–906.

Smith, J. B., & Tol, R. S. J. (1998). *Handbook on methods for climate change impact assessment and adaptation strategies*.

Solh, M., & Van Ginkel, M. (2014). Drought preparedness and drought mitigation in the developing world's drylands. *Weather and Climate Extremes, 3*, 62–66. doi:10.1016/j.wace.2014.03.003

Speranza, C. I. (2010). Resilient adaptation to climate change in african agriculture. *Studies*.

Stringer, L. C., Dyer, J. C., Reed, M. S., Dougill, A. J., Twyman, C., & Mkwambisi, D. (2009). Adaptations to climate change, drought and desertification: Local insights to enhance policy in Southern Africa. *Environmental Science and Policy, 12*(7), 748–765. doi:10.1016/j.envsci.2009.04.002

Thornton, P. K., Boone, R. B., Galvin, K. A., BurnSilver, S. B., Waithaka, M. W., Kuyiah, J., et al. (2007). Coping strategies in livestock-dependent households in East and Southern Africa: A synthesis of four case studies. *Human Ecology, 35*, 461–476. doi:10.1007/s10745-007-9118-5

Vincent, K., & Cull, T. (2013). *Climate change and food security in Southern Africa: Implications of theoretical development for the promotion of sustainable equitable development*, 1.

Wheeler, D. (2011). *Quantifying vulnerability to climate change : Implications for adaptation assistance*. Working Paper 240 January 2011, New York 4164000416 (January 2011), 53. Retrieved from http://www.cgdev.org/content/publications/detail/1424759

Wilhite, D. A., Sivakumar, M. V. K., & Pulwarty, R. (2014). Managing drought risk in a changing climate: The role of national drought policy. *Weather and Climate Extremes, 3*, 4–13. doi:10.1016/j.wace.2014.01.002

Ziervogel, G., New, M., Archer Van Garderen, E., Midgley, G., Taylor, A., Hamann, R., et al. (2014). Climate change impacts and adaptation in South Africa. *WIREs Clim Change, 5*, 605–620. doi:10.1002/wcc.295

Ziervogel, G., Nyong, A., Osman, B., Conde, C., Cortés, S., & Downing, T. (2006). *Climate variability and change: Implications for household food security*. AIACC Working Papers Climate Variability and Change: Implications for Household Food Security 1. Retrieved from www.aiaccproject.org

Zuma-Netshiukhwi, G., Stigter, K., & Walker, S. (2013). Use of traditional weather/climate knowledge by farmers in the South-Western free State of South Africa: Agrometeorological learning by scientists. *Atmosphere, 4*(4), 383–410. doi:10.3390/atmos4040383

Farmers' Livelihoods Vulnerability to Climate Variability and Change in Didesa Basin Southern Part of Abay Basin, Ethiopia

Chala Dechassa, Belay Simane and Bamlaku Alamirew

1 Introduction

Climate related evidences shows that there is an increase in frequency of extreme events such as drought, floods and different crop and animal pests in relation to climate changes where, historical records show an increase in global temperature since the late 19th century. Intergovernmental Panel on Climate Change (IPCC) predicts an increase in temperatures between 0.3 and 0.7 °C over the next two decades and an increase of 0.3–4.8 °C by the end of the 21st century, depending upon emission scenarios (Collins et al. 2013; Kirtman et al. 2013). The changes were with different manifestation like an increase in temperatures, frequency of cold days, cold nights, and decreased in coolness, while frequency of hot days, increased hot nights, and heat waves (Kirtman et al. 2013).

Africa is one of the most vulnerable continents to climate variability and change due to high reliance on climate sensitive sector (e.g. agriculture) and low adaptive capacity (Boko et al. 2007; Adger et al. 2007) result from poverty. By the 2050s, 350–600 million Africans will be at risk of water stress, extensive floods, drought,

C. Dechassa (✉)
Rural Livelihoods and Development, College of Development Studies,
Addis Ababa University, Addis Ababa, Ethiopia
e-mail: cdechassa@yahoo.com

B. Simane
College of Development Studies, Center for Environment and Development,
Addis Ababa University, Addis Ababa, Ethiopia
e-mail: simaneb@yahoo.com

B. Alamirew
College of Development Studies, Center for Rural Development,
Addis Ababa University, P.O. Box 1176, Addis Ababa, Ethiopia
e-mail: bamlakalamirew@gmail.com

W. Leal Filho et al. (eds.), *Climate Change Adaptation in Africa*,
Climate Change Management, DOI 10.1007/978-3-319-49520-0_17

famine, and loss of life specifically, sub-Saharan Africa (Arnell 2004; IPCC 2007; Lobell et al. 2011).

In relation to the changes, IPCC (2007) reported that, rain fed crop yields will decline by 10–20% by 2050 and crop revenues could fall by 90% by 2100. Ethiopia, with an estimated 96 million people live on less than $1.25 a day (Belay et al. 2013) where, agriculture is a major economic sector employing 85% of the labor force and contributing 40% to the countries' national GDP considered the most vulnerable due to climate related hazard and risks (Belay et al. 2013; Boko et al. 2007; Conway and Schipper 2011; Cooper et al. 2008; Thomas et al. 2007).

Von Braun (2007), for instance, identified that a 10% decrease in seasonal rainfall from the long-term average generally translates into a 4.4% decrease in the Ethiopian food production. Study by World Bank (2010), projects that unless steps to build resilience are effective, climate change will reduce Ethiopia's GDP growth by 2–6% by 2015, and up to 10% by 2045 between 0.5 and 2.5% and between 0.5 and 2.5% per year (EPA 2011). Such impacts would be greatest in SSA and Ethiopia for many reasons (Patt et al. 2009; Deressa et al. 2009). First, high reliance on climate sensitive sectors (agriculture). Second, due to unsustainable patterns of land use practices this could exacerbate climate change. Third, lack the necessary capital to invest in adaptation as well as maintain stability in society when physical harm occurs (Patt et al. 2009; IPCC 2007).

According to Simane et al. (2014) due to high projection of significant future climate change in Ethiopia in the coming decades; aggregate national vulnerability result does not capture the complexity of vulnerability at agro ecological level leads agricultural productivity remains challenged. This is particularly true in the Didessa basin where the soil and terrain, diverse topography, socio-economic, prevalence of climate variability, environmental conditions and sensitivity of agriculture and ecosystem exhibited in land, water resources degradation and biodiversity loss.

Different studies have developed an index to determine the level vulnerability on farmers to climate extremes using the "vulnerability as expected poverty" approach (Deressa et al. 2009; Swain and 2011). Using Livelihoods Vulnerability Index (LVI) and LVI-IPCC (Hahn et al. 2009; Entwire et al. 2013; Can et al. 2013; Simane et al. 2014). They used several variables into major components to capture the level of exposure, adaptive capacity and sensitivity to climate change impacts.

The present study used Agro Ecological Zones (AEZ) as unit of analysis to examine farmer's livelihoods vulnerability to climate variability and change considering the effect of agriculture relevant climate factors, sensitivity and adaptive capacity in the Didessa basin of the Blue Nile River. Livelihood Vulnerability Index (LVI) framed within the Livelihood Vulnerability Index—Intergovernmental Panel for Climate Change(LVI-IPCC) categorization (Hahn et al. 2009; Simane et al. 2014) and Sustainable Livelihood Approach (SLA) (Birkmann 2006) used in the context of the study area.

2 Materials and Methods

2.1 *Descriptions of Study Area*

This study was conducted in the Didessa basin of the Blue Nile basin located in the southern most part in 2015 on 450 farm households in Ethiopia. Astronomically it is located 36° 02′ and 36° 46′ east longitude and between 7°43′ and 8° 13′ north latitude. Mean annual rain fall 1586 mm having only one summer rainy season. The mean max-min temperature is 30 and 11.45 °C. respectively (NMSA 2015) Altitude ranges between 1720 to 2088 m a.s.l excluding mountains of greater than 3500 m a.s.l (Buzuneh 2011) (Fig. 1).

2.2 *Study Design, Data Source, Sampling Techniques, Procedure and Size*

The study was applied a cross-sectional quantitative study design, supplemented by qualitative study, to assess the level of farmers vulnerability to climate variability across agro ecological zones of Didessa basin. The study was agro ecological zone based conducted in the western Ethiopia. Consequently four districts were selected in such a way that each class in the sample proportion for each AEZ in the entire basin. Eighteen (18) villages representing AEZ were selected and systematic random sampling was then used in selecting twenty five (25) households from each AEZ totaled 450 respondents.

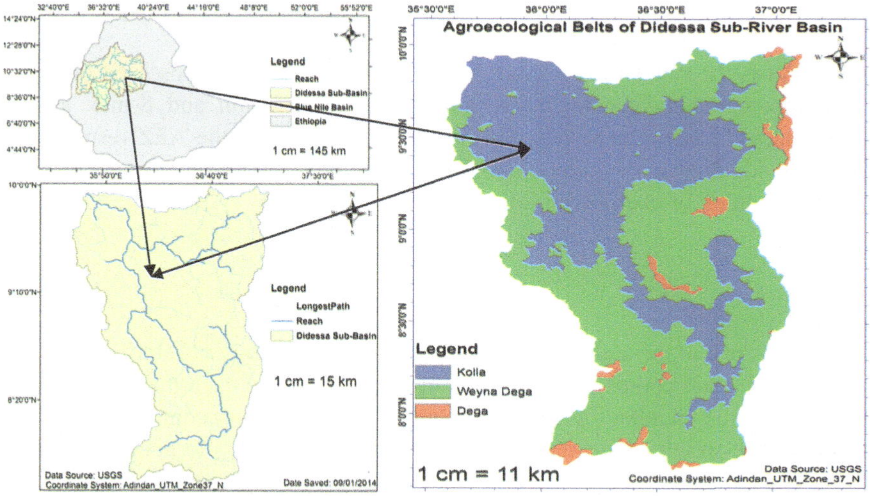

Fig. 1 Map of Didessa basin, boundary, agro-ecological setting. *Source* United State Geological Survey, 2014

The data for this research was collected using standard questionnaires prepared for the survey in May 2015. The collected data includes the ten major components in the livelihoods vulnerability index (LVI) analysis namely: socio demographic profiles of the households, ecosystem, infrastructure, health, wealth, social network and institution, livelihoods and other relevant information. Besides, exposure data were obtained from national meteorological service agency proxy station. To verify the quantitative data, qualitative data were collected through focused group discussion (FGD). Finally, two sets of analysis were undertaken: calculation of a balanced weighted average LVI and calculation of a LVI based on the IPCC framework (IPCC 2007).

2.3 Calculating the LVI: Composite Index Approach (Model)

Drawing up on the SLA and the LVI-IPCC approach (Hahn et al. 2009; Simane et al. 2014) the study apply a balanced weighted average approach (Sullivan et al. 2002). Each subcomponent contributes equally to the overall index even though each major component includes a different number of sub-components. In LVI analysis, first transform the raw data into appropriate measurement unit such as percentage, ratio and indices. Second, standardizing the sub components measured on different scale using Eq. (1). To standardize a main component, the quotient of the difference between the actual score and the minimum value and the difference between the maximum and minimum values obtained from the total sample was calculated.

$$\textbf{Index Ivi} = \frac{Sr - Smin}{Smax - Smin} \tag{1}$$

where IVi is the standardized value for the indicator, Sr is the observed (average) sub-component indicator for agro-ecology of r, and S-min and S-max are the minimum and maximum values, for the indicator across all the AEZ respectively. Third, standardizing, the sub-component indicators are averaged using Eq. (2) to obtain the index of each major component:

$$\textbf{Mr} = \frac{\sum Ivi}{N} \tag{2}$$

where M_r is one of the ten major components for agro-ecology r (Table 1); index Iv_i represents the sub components, indexed by i, that make up each major component and "N" is the number of sub-components in each major component. Fourth, combines the weighted averages of all the major components to generate the LVI score. The number of indicators of which it is compressed to determines the weights of each major component (Wmi). Values for each of the ten major components for an agro-ecology are calculated and averaged using Eq. (3) to obtain the AEZ-level LVI:

Table 1 Vulnerability Index Components: key components used to generate the vulnerability index across the agro ecological zone

LVI-IPCC	Capitals	Major components	Sub-component (Indicator)	Hypothesis
Exposure	*Climate*	Change in temp, and RF	Change in temperature Change in precipitation Occurrence of extreme events (drought, floods)	Increase in temperature and decrease in precipitation increase vulnerability
Sensitivity	*Natural Capital*	Ecosystem	Land use system, Land cover change, Land suitability, Land management (SWC) Irrigation potential	The more forest cover and suitable land and access to irrigation the less sensitive
		Agriculture	Annual total production (inverse) changes in productivity, diversity of crop species, agricultural diversity index	The more productivity, diversification the less sensitive
Adaptive capacity	Financial capital	Wealth	Farm size of the HHs (inverse), Number of livestock(inverse), saving at HH level, loans, non-agricultural income	The more the wealthy status the more the adaptive capacity
	Physical capital	Technology	Insecticide and pesticide supply, fertilizer supply, improved seeds supply, irrigation potential	The more access to technology the more the adaptive capacity
		Infrastructure	Access to all-weather roads, schools, veterinary services, market, saving and credit, electricity and telephone	The more access to infrastructure the more the adaptive capacity
	Human capital	Community	Sex of HH heads, education level, health services, radio owner ship availability of extension skill/training	The more the information the more the adaptive capacity

(continued)

Table 1 (continued)

LVI-IPCC	Capitals	Major components	Sub-component (Indicator)	Hypothesis
		Livelihood strategy	Average agric. livelihood diversity [1/(no. of agric. activities + 1)], Percent of HHs dependent on agric as major source of income, Percent of HHs dependent on non-farm activities as source of income	The higher the agricultural livelihoods diversity the lesser the vulnerability
	Social capital	Social network	Availability of bylaws, number of no-working days/month, tradition of working together	The more farmers are members with less no-working days and the more the tradition of working together the better
		Socio demographic	Dependency ratio, female headed household, average family members in the households	The higher the dependence ratio the higher, the vulnerability

Source Hahn et al. (2009), Simane et al. (2014) and profiles indicators tailored to the study area

$$LVIr = \frac{\sum_{P=1}^{10} WMiMri.}{\sum_{P=1}^{10} WMri} \tag{3}$$

where, LVIr is the Livelihood Vulnerability Index for AEZ_r. Following Eqs. (1)–(3), Hahn et al. (2009) calculated the major components of LVI based on the LVI-IPCC vulnerability categorization (Exposure (E), Sensitivity(S) and Adaptation (A)) using the following equation:

$$\mathbf{CF_r} = \frac{\sum_{P=1}^{f} WMiMri.}{\sum_{p=1}^{f} WMri} \tag{4}$$

where, CFr, is the LVI-IPCC defined contributing factors for agro-ecology of r, Mri, are the major components for the agro-ecology of r, indexed by i, Wi, the weights of each major components, f is the number of the profiles associated contributing factors and p is indexed to the profiles associated with the CF. Finally, the researchers developed the LVI-IPCC vulnerability categorization (see Table 1) using Eq. 5

$$\text{LVI-IPCCr} = (e_r - a_r) * S_r \quad (5)$$

where, LVI-IPCC_r is the LVI for agro-ecology *r* expressed using the IPCC vulnerability framework, *er* is the calculated exposure score for agro-ecology *r*, a_r is the calculated adaptation capacity score for agro-ecology *r* and *sr* is the calculated sensitivity score for agro-ecology *r*. We scaled the LVI-IPCC from −1 (least vulnerable) to 1 (most vulnerable) and is best understood as an estimate of the relative vulnerability of compared populations in agro-ecology.

3 Results

3.1 Livelihoods Vulnerability Index Components in the Didessa River Basin

In this section we provide an analysis of the key profiles used to generate the vulnerability index, highlighting the key profiles and where they differ between agro-ecological zones. The study participants' backgrounds are shown in Table 2. A total of 450 farm households participated in the study and of those studied 86.88% were male and 13.12% were female. In terms of age category, the majority of respondents 81% are within the active working age group (31–65), while 32 respondents 4% are above 65 years of age. About, 396 (88%) were married and the remainder were single, divorced, or widowed. Education levels are low with many of the respondents 30.22% being illiterate, approximately 16%, were attended adult education, and about 42.66% attended primary cycle of schooling (1–4 level).

Table 2 Demographic characteristics of the study population (*n* = 450)

Characteristics	Category	Highland	Midland	Lowland
Gender	Male headed household	90	80	94
	Female headed household	10	20	6
Marital status	Married	95	82	91.33
	Not married	1	3.5	3.33
	Divorced	1	3	4.66
	Widowed	3	11.5	0.666
Education	No education/illiterate	26	26.5	38
	Adult education (reading and writing)	34	12	9.33
	Primary school	30	51.5	39.33
	Secondary school and above	10	10	13.33

Source Survey result, 2015
*All values in the table presented in percent (%) in a given AEZ

3.1.1 Exposure Profile: Natural Disaster and Climate Variability

The exposure analysis result shows high score of climate variability and change in the lowland with 0.501 followed by 0.423 and 0.414 in highland and midland respectively. When the main components were reviewed by sub-components (i.e. indicators), lowland agro ecology was most vulnerable in terms of exposure to climate variability (0.501). Farmer perceptions toward climate variables in the basin indicated that temperature is increasing and rainfall decreasing though slight variation across agro ecological zone. People living in different agro-ecological system socio economic, cultural, farming system and soil are believed to perceive climate change differently (Simane et al. 2012; Gutu et al. 2012; Sewagegn 2011). The exposure profiles includes undulating and steeply sloping farmlands, low soil fertility level due to frequent degradation, soil erosion, below average rain and mounting temperature (Table 3). As a result, agricultural productions decreased and threaten the lives and livelihoods of farmers in the basin. Similar studies identified such indicators as the measurement of exposure (Deressa et al. 2009; Fussel 2010; Gutu et al. 2012).

3.1.2 Sensitivity Profiles: Ecosystems and Agriculture

Ecosystem

The dominant farming systems in the Didessa basin are shaded coffee/livestock farming system and cereal/livestock mixed farming system. The sensitivity of the agro ecologies in the basin ranges from 0.302 to 0.28. The highland agro-ecology shows, high sensitivity score (0.31) for ecosystem than the midland (0.256) and lowland (0.2) agro-ecology. This is due to failure to take care of natural resources (e.g. soil and water), shortage of land, inadequate livelihoods option, population pressure, fuel demand (socio economic) and low level of awareness on natural re-sources management. Natural capital and vulnerability to climate change are tightly linked (Bankoff et al. 2004). The greater the level of dependence of a household on the ecosystems and the greater sensitivity of natural resources, such as farming, forestry, the higher their vulnerability to climate change and vulnerability level varies depending on the contribution of natural resources to their

Table 3 Annual average monthly temperature and precipitation in the surveyed AEZ

AEZ	Altitude (a. s.l)	Area coverage (%)	Temperature		Av. Monthly precipitation	Average annual precipitation (mm)
			Maximum	Minimum		
Kola	1000–1450	47.2	31.27	13.92	133.00	900–1500
WD	1500–2450	45.8	30.6	10	136.77	1500–1850
Dega	2229–2870	7	23.74	8.66	167.25	1850–2750

Source Authors calculation from NMA (2015)

livelihoods. However, the key informant from the Didessa basin stated that the rate of ongoing forest distraction is higher than the previous periods due to increased population pressure people to occupying the forest lands without restriction.

Water is usually sourced by women and young girls hence distant water sources increases the time burden of household chores and affects time for care in the case of women, and school attendance in the case of the girl child. In this regards, water source, the average time taken to reach is found to be highest in the lowland. The highland and Lowland agro-ecology reported the highest percentage (32.3 and 31) of households that do not have a consistent water supply. These households become even more vulnerable during the dry season when most natural water sources tend to dry up. Households in the highland reported storing water as compared to those in the midland and lowland agro-ecology. Inverse household water letter per day indicated (0.016) in highland (0.024) in the midland and (0.047) in the lowland. Besides, in terms of irrigation use the highland agro-ecology better off with 17% and the soil and water conservation practice is high in the highland (73%) than midland and the lowland.

Agriculture: Farm Land size

The average land holdings were 1.86 ha although the farm land size differs across agro-ecology of the Didessa basin. In the highland 1.39 ha, in lowland 2.39 and in the midland 1.79 ha which is above the national average household farm size 1.22 ha (CSAa 2012). However, about 41.21, 32.37, 31.38% of farm households in the highland, midland and lowland households respectively owning less than one hectare. The vulnerability level of household in the study area shows that households who owned a better size of land were less vulnerable to climate change (inverse average farm size LVI score HL (0.089), ML (0.066) and LL (0.047). This is because the bigger the size of land, the higher the diversity of crops that are grown, the better their access to improved technology, and the better their access to financial services and extension services. Accordingly, the score value those households with more farm sizes will have high adaptive capacity and less vulnerable to climate impacts as compared to those with less sized farms. This is similar with the finding of Gutu et al. (2012) on a time series analysis of climate variability and its impacts on food production in north Shewa zone in Ethiopia.

Moreover, according to the information from the agronomist of the study area, the number of plots a farmer operates at different locations play vital role in reducing vulnerability because of differences in climate variability across agro ecological zones, within the same agro ecological zones and even differences exhibited for those households with more number of farm plots are less vulnerable as compared to those with single or fewer plots.

3.1.3 Adaptive Capacity

Infrastructure

Access to major indicators of infrastructure are comparable to national average figures and do not vary significantly across agro-ecological zone (Table 4). It is worth noting that 70.83% of the households have access to agricultural extension services and about 72.07% of household's access to health services within 5 km of their homes a figure that does not vary significantly across agro-ecological zone.

Technology

The percentage of farmers who have access to different farm technology (irrigation, insect side and pest sides, fertilizers use and improved seeds) were 65 and 11% in highland, midland and lowland agro-ecology. The differences in technology profiles between AEZ were attributed primarily to differences in the agro-ecology, soil and terrain, farming system and use of chemical fertilizer, improved seed as well as irrigation potential. Farmers who have better tropical livestock unit (TLU) will have better access to farm technologies. But the application of these inputs was limited even among farms in a better wealth status due to sky rockets of the farm technologies prices. In addition, the total amount of chemical fertilizer applied is low, even among farmers that report using some fertilizer. The use irrigation on their land in the agro-ecology has not statistically significant due to low experience and low access to irrigation technology which still goes with purchasing power. The improved seed use is different among the agro ecology where the highland and midland farmers do use more improved seeds on average compared to lowland. The improved varieties use rate is high in the midland agro-ecological zone than in the

Table 4 Access to basic services and indicators of infrastructure within 5 km of the home, as reported in household surveys in the Didessa basin

Indicators	Proportion of farmers with access (%)		
	Highland	Midland	Lowland
Road (access to all weather roads)	100	99.5	94
Primary school	96	94	82.66
Veterinary services	88	98	93
Access to market (km)	100	99.5	98
Access to health services	74	81.3	60.9
Access to credit	97	98.5	92.6
Access to institutions	95	91	91
Access to electricity	10.10	12.06	33.55
Access to telephone	54.73	32.5	41.37

Source Survey result, 2015

highlands due to the favorable topography and climatic variables for the intensive crop production.

Wealth profile: Access to credit

The wealth profile score showed that access to credit was a significant determinant of vulnerability level. In terms of access to credit, there is no significant difference in the three agro ecological zone. Farmers in the category of high access to credit have high probability to reduce vulnerability as compared to those who do not have access within that category. However farmers claimed the bureaucracy for access to cash for purchase of agricultural inputs, which resort farmers to borrowing from local lenders at exorbitantly high interest rates. This is similar to Shewmake's (2008) finding that households with poor access to credit are more vulnerable.

Land Holding and Livestock Ownership

The average land holding is 1.86 ha, with 93.06% of the farmers owning their land. Of this, 80.88% got land ownership certification (Table 5). Among respondents, 83, 49.74 and 47.65% reported that their land holding size has decreased over the past 20 years in highland, midland and lowland respectively, due to land degradation and loss of soil fertility, while 6, 9.23 and 16.78% of the households in the three agro-ecology highland, midland and lowland have reported an increase in agriculturally useable land holdings as a result of clearing forest lands and grazing areas and redistribution of land. Lowland agro-ecology statistically significant larger average land holding size (2.39 ha) compared to the highland and midland due to possibility to expand to communal land and absence of land ownership certificate especially in the northern part of the basin.

In terms of livestock ownership, the highland agro ecological zone has a higher level of livestock ownership than the mid highland and lowland zones. The inverse average No of live stock unit (TLU) of LVI score shows highland was less vulnerable (LVI score = 0.0097) compared to lowland and midland 0.0182 and 0.013)

Table 5 Land and livestock holdings size in the Didessa basin across agro ecological zone

Description	Highland	Midland	Lowland
Land ownership	100	93	52
Average land holding size (ha)	1.39	1.79	2.39
Changes in farmland size over the last 20 years	6	9. 23	16.78
• Increase	83	49. 74	47.65
• Decrease	11	41. 03	35.57
• No change			
Possession of livestock in tropical livestock unit (TLU)	6.7	4.77	3.57

Source Survey result, 2015

*All figures are in % within the given AEZ, except for average land holding size and TLU

due to better number of livestock with verities. This is due to the facts that a farm land in the midland and lowland are better for agricultural production and the highland suitable for livestock grazing land due to its mountainous and not conducive for cultivation and favorable in terms of access to livestock health service.

3.1.4 Community Profile

Agricultural Experiences

The result revealed that in the lowland farmer's agricultural experience were less than mid and highland farmers with an average year of experience around 21.77 years, 23 years and 24.88 years, respectively. Many years of experience have shown a lower level of vulnerability to climate impact as compared to those households with very few years of experience. Hence, experienced farming households have an increase likelihood of choosing change planting date, planting different crops and varieties, apply various farm management practices and techniques that can be used in the face of anticipated climate change which confirm the findings of (Nhemachena and Hassan 2007; Gutu et al. 2012).

Education and Access to Farm Extension Training

Farmers in the Didessa agro-ecological basin had access to extension training with 89.7, 68.3 and 54.5% in highland, midland and lowland agro-ecology respectively. Farm households with an access to formal education greatly contribute to climate change adaptation and reduce vulnerability in the basin. Extension services have the potential to influence farmers' decision to change their farming practices in response to climate change (Nhemachena and Hassan 2007; Maddison 2007).

Health

In highland only 12.1%, the least households reported ill health as a result of better livelihood options and government and private health facilities. The agro-ecological zone with most households reported ill health was lowland zone (26%), because of lack of health facilities and distance from the health centers. As a result, family member had to miss school or work in the past 6 month due to illness was high in lowland (18.5%) which result low participation of family members in agricultural activity, decline of the agricultural production and affect adaptive capacity of farmers in the agro-ecology. Average malaria exposure prevention index was high in the lowland (0.485) than the highland (0.159) and midlands (0.314). Lowland households reported as being more vulnerable to malaria exposure due to less access health services (60.9%) to get bed net and high mounting temperature relative to the highland and mid lands (74 and 81.3%) respectively.

Livelihoods Strategy

The livelihood strategies of the households in the Didessa basin of Blue Nile basin were diverse which were different across agro-ecology due to difference in their knowledge and experience to disaster exposure. The common livelihoods strategies employed in the basin includes growing crops, raising animals, collecting natural resources and family member or members go to other areas for work. Almost all the agro ecological zone households relying solely on agriculture for income the result indicated that about 95, 93 and 87% of households in HH, ML and lowland respectively. Although, the Lowland had higher vulnerability scores for two of the Livelihood Strategies indicators, Lowland agro-ecological zone households are better in traveling outside the community to work 0.15 as livelihoods strategy compared to 0.067 and 0.09 in highland and midland agro ecology.

Average agricultural livelihood Div. Index which was calculated as the inverse of (the number of agricultural livelihood activities +1) reported by a household the highland and mid land basically relay on crop and animal raring where as the lowland agro ecology relay on hunting and forests products in addition to the crop animal raring. Consequently the lowland agro ecology was less vulnerable (with LVI score of 0.25) compared to the highland and midland (score, 0.33). However, But when the three sub-components were averaged, the overall livelihood strategies index was lowest in lowland (0.42) and highest in midland (0.45) and highlands (0.449). This is because of its relative favorable environmental situation such as health, infrastructure, schools and access to market to work in the highland and mid highland and relatively better options of livelihood activities and agricultural diversification option. Consequently, the lowland agro-ecology was relatively more vulnerable than the rest agro-ecology due to limited livelihoods strategy.

Socio Demographic and Social Network Profile

The issue and the role of population growth and family size in development, in general, and climate change adaptation and vulnerability in particular is largely unresolved. The family size of the households in the study area differs across different agro-ecology due to differences in access to infrastructure such as health center, education and extension services. As a result, about 54, 26.5 and 30.66% of farm households have family size of greater than 6 members in highland, midland and lowland agro-ecology respectively. Large family size was assumed to be the source of labor, skills and strong social capital to adapt to changing climate situation (Deressa et al. 2011) and enable a household to accomplish various agricultural tasks especially at the peak seasons.

However, large family sizes have negative impacts on the households in the Didessa basin. This is because the available livelihoods opportunities to family members are very much limited and only one or two of a household member usually engage in productive livelihood activities that can support the family plus members of the households whose ages are less than 14 and greater than 65 age categories are also not active participants. Similar study confirmed that in west Arsi zone of Ethiopia children, women and large sized families are affected mostly by the climate change events (Abate 2009). Therefore, households with large family size are highly vulnerable to climate change induced risks which is true in the highland area of the basin where access to land and livelihoods option is limited compared to the two agro ecological zones.

Social capital measures varied little between agro-ecology. Households reported borrowing money more frequently and receiving more in-kind assistance from family and friends relative to the number of times they lent money or provided assistance in the past month is higher in highland than the midland and lowland households (borrow: lend ratio: 0.02, 0.06, 0.16; receive: give ratio: 0.2, 0.247, 0.27) respectively. Generally, social bonds in the basin is great due to the cultural and tradition showing decreased compared to the past two three decades. Borrowing and lending money indicate the financial assistance households receive in cash and kind from their social network and households that borrow money more than they lend are more vulnerable (Hahn et al. 2009). Overall, the lowland (0.44) households were more vulnerable than the highland and midland in social networks profile score (0.44) and (0.40) respectively.

4 Discussion

4.1 Livelihood Vulnerability Index: Practical Implications in the Didessa Basin

The standardized values of all the ten major contributing profiles are presented in spider diagram (Fig. 2). The spider diagram ranges between 0 (least) and 0.9 (highest). Index value should be interpreted as a relative value to be compared with the study sample only. Lowland agro-ecology is most vulnerable in terms of socio-demographic profile (0.268), social network (0.44) and technology (0.11). The highland agro-ecology is most sensitive in terms ecosystem (Land use system, Land cover change, Land suitability land management (SWC), irrigation,) (0.313) while the lowland agro-ecology is most sensitive in terms of agriculture (0.517) followed by highland (0.43). In terms of percent of technology access the midland and highland were superior to the lowland (65, 58 and 11%). Overall, the LVI score for highland, midland and lowland are (0.43, 0.422 and 0.392) respectively

4.2 IPCC-LVI Contributing Factors in the Didessa Basin

The LVI-IPCC is computed by grouping the ten major profiles into three factors, namely exposure (made up of one major component), sensitivity (two major

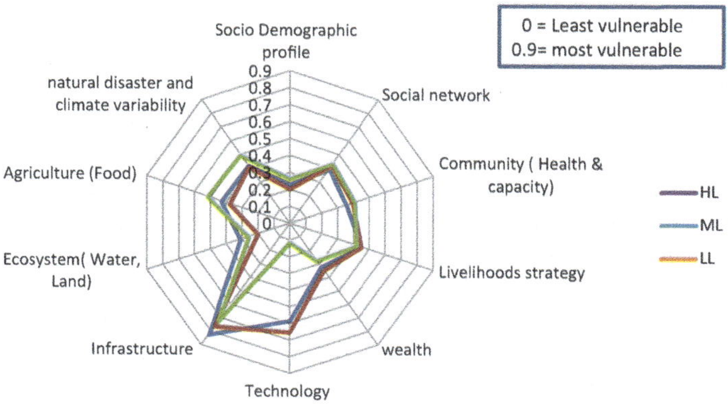

Fig. 2 Spider Diagram of the Major livelihoods vulnerability contributing profiles for Didessa basin across agro-ecology. *Source* computed from Survey, 2015

Table 6 Calculated indices for contributing factors and the Livelihood Vulnerability Index under the LVI-IPCC framework

AEZ	Exposure	Sensitivity	Adaptive capacity	LVI-IPCC
Highland	0.423	0.36	0.461	−0.012
Midland	0.414	0.28	0.462	−0.013
Lowland	0.501	0.374	0.434	0.067
Average	0.446	0.318	0.453	0.014
Max	0.501	0.374	0.463	0.067
Min	0.414	0.28	0.434	−0.013
SD	0.4784	0.04917	0.01646	

Source Survey result, 2015

Fig. 3 LVI of the different agro ecologies in the Didessa basin, *Blue* Nile basin Ethiopia. *Source* Survey result, 2015

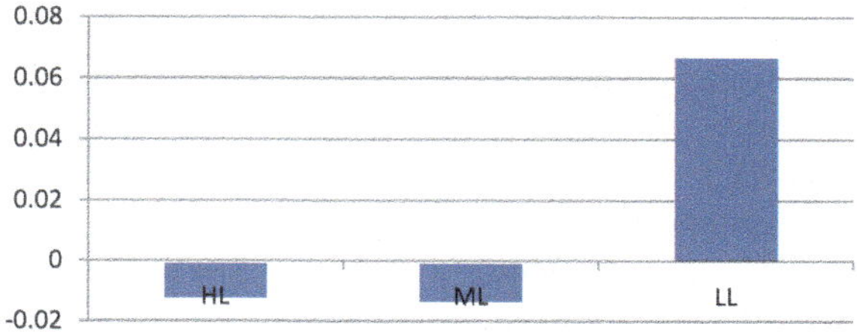

components) and adaptation capacity (seven major components). Calculating indices for the three factors are represented in Table 6. Index values are interpreted as relative values to be compared within the study sample only. The LVI-IPCC is on a scale from −1 (least vulnerable) to 1 (most vulnerable) (Fig. 3).

The calculated LVI-IPCC indices ranged from 0.067 to −0.013 (Table 6). Lowland is more exposed (0.501) to climate change impacts than HL (0.423) and ML (0.414). Based on socio demographic, livelihoods, and social networks, wealth, infrastructure, technology, and community profile, midland and highland ecosystem showed a higher adaptive capacity (0.462 and 0.461) relative to the lowland (0.374). However, considering for the ecosystem (water,) and agriculture (food) lowland (0.374) and highland (0.302) may be more sensitive to climate change impacts than midland (0.28). The overall LVI-IPCC scores indicate that lowland households are more vulnerable than highland and midland households (0.067 versus −0.012, −0.134, respectively). This study result, is similar with Simane et al. (2014) where 62% of the total land mass (lowland, valley fragmented AES 1 and the mountainous highland AES 5) is categorized as having high relative vulnerability in Choke Mountain ecosystem.

5 Conclusions and Recommendation

The LVI and LVI-IPCC offers a framework to evaluate and understand relative climate variability and changes vulnerability at household to community level in Didessa river basin. This chapter presented the LVI-IPCC as alternative methods for assessing vulnerability of farmers in different AEZs to climate variability and change. The LVI method provides a detailed depiction of factors driving household livelihood vulnerability in a particular study area. The LVI-IPCC result indicated that the lowland agro-ecology is most vulnerable to climate change due to high level of exposure and relatively low level of adaptive capacity compared to midland. The finding of the study will have important policy relevance that could enable smallholder farmers in lowland agro-ecology to better adapt to the effects of climate change and variability and to develop programs to strengthen the most vulnerable sectors.

The study recommends that increasing adaptive capacity to climate variability and change to the range of climate extremes that they experience (drought, floods wheatear related shocks). The study also recommends the flexible application of LVI-IPCC as tools for the climate related analysis and impact assessment by substituting the value of the indicator that is expected to change and recalculating the overall vulnerability index. The study LVI might be used to project future vulnerability, for example under simple climate change scenarios.

Acknowledgements This research was supported in part by Addis Ababa university climate change thematic research project. We would like to thanks the anonymous reviewers of the climate change adaptation in Africa for their constructive comments and suggestion to improve the manuscript.

Conflict of interest The authors declare that no conflict of interest

.

References

Abate, F. (2009). Climate change impact on livelihood, vulnerability and coping, mechanisms: A case study of West-Arsi Zone, Ethiopia.

Adger, W. N. (2007). Vulnerability of global environmental change 16 (3): 268–281. Agriculture: Ricardian approach. World Bank Policy Research Paper No. 4342.

Arnell, N. W. (2004). Climate change and global water resources: SRES emissions and socio-economic scenarios. *Global Environmental Change, 14*(1), 31–52.

Bankoff, G., Frerks, G., & Hilhorst, D. (2004). *Mapping vulnerability: Disasters, development and people*. London: Earth Scan.

Simane B., Zaitchik B. F., & Mesfin D. (2012). Building climate resilience in the blue Nile/Abay Highlands: A framework for action international journal of environmental research and public health. ISSN 1660-4601 www.mdpi.com/journal/ijerph.

Simane B., Zaitchik B. F., & Foltz, J. D. (2014). Agro ecosystem specific climate vulnerability analysis: Application of the livelihood vulnerability index to a tropical highland region. *Mitigation and Adaptation Strategies for Global Change* doi:10.1007/s11027-014-9568-1.

Belay T., Hengsdijk, H., Rötter, R., Kahiluoto, H., Asseng S., & Van Ittersum, M. (2013). Adapting to climate variability and change: Experiences from cereal-based farming in the Central Rift and Kobo Valleys. *Ethiopia Env'tal Management, 52*, 1115–1131. doi: 10.1007/s00267-013-0145-2.

Birkmann J. (2006). Measuring vulnerability to remote disaster-resilient societies: Conceptual frameworks and definitions. In J. Birkmann (Ed). *Measuring Vulnerability to natural hazards: Towards disaster resilient societies.* (pp. 9–54). United Nations University Press.

Buzuneh A. (2011). Flow regime and land cover changes in the Didessa sub-basin of the Blue Nile River, South-Western Ethiopia: Combining empirical analysis and community perception. M. Sc. Thesis Swedish University of Agricultural Sciences. Uppsala, Sweden.

Boko, M., Niang, I., Nyong, A., Vogel, C., Githeko, A., Medany, M., et al. (2007). Africa. In M.L. Parry, O.F. Canziani, J. P. Palutikof, D. Van Linden (Eds.).

Can, N. D., Tu1, V. H., & Hoanh, C. T. (2013). Application of livelihood vulnerability index to assess risks from flood vulnerability and climate variability—a case study in the Mekong Delta of Vietnam.

Collins, M., Knutti, R., Arblaster, J., Dufresne, J.-L., Fichefet, T., Friedlingstein, P., et al. (2013). Long-term climate change: Projections, commitments and irreversibility In Midgley, (Ed). *Contribution of Working Group I to the Fifth Assessment Report of the IPCC*. (pp. 1029–1136). Cambridge University Press, doi:10.1017/CBO9781107415324.024.

Conway, D., & Schipper, E. L. F. (2011). Adaptation to climate change in Africa: Challenges and opportunities identified from Ethiopia. *Global Environmental Change, 21*(1), 227–237.

Cooper, P. J. M., Dimes, J., Rao, K. P. C., Shapiro, B., Shiferaw, B., & Twomlow, S. (2008). Coping better with current climatic variability in the rain-fed farming systems of sub-Saharan Africa: An essential first step in adapting to future climate change? *Agriculture, Ecosystems & Environment, 126*(1), 24–35. doi:10.1016/j.agee.2008.01.007.

CSA. (2012a). Agricultural sample survey 2011/2012, vol IV: Report on land utilization. Addis Ababa, Ethiopia: Central Statistical Agency.

Deressa, T. T., Hassan, R. M., & Ringler, C. (2011). Perception of and adaptation to climate change by farmers in the Nile basin of Ethiopia. *The Journal of Agricultural Science, 149*(1), 23.

Deressa, T. T., Hassan, R. M., Ringler, C., Alemu, T., & Yesuf, M. (2009). Determinants of farmers' choice of adaptation methods to climate change in the Nile Basin of Ethiopia. *Global Environmental Change, 19*(2), 248–255.

Etwire, P. M., Al-Hassan, R. M., Kuwornu, J. K., & Osei-Owusu, Y. (2013). Application of livelihood vulnerability index in assessing vulnerability to climate change and variability in Northern Ghana. *Journal of Environment and Earth Science, 3*(2), 157–170.

EPA. (2011). Ethiopia's climate-resilient green economy (CRGE). Addis Ababa, Ethiopia: Federal Democratic Republic of Ethiopia.

Fussel, H. (2010). How inequitable is the global distribution of responsibility, capability, and vulnerability to climate change: A comprehensive indicator-based assessment. *Global Environmental Change*, *20*, 597–611. IFPRI Discussion Paper 00804. Washington, DC USA.
Gutu, T., Bezabih, E., & Mengistu, K. (2012). Analysis of vulnerability and resilience to climate change induced shocks in North Shewa. *Ethiopia. Journal of Agricultural Science, 3*(6), 871–888.
Hahn, Micah B., Riederer, Anne M., & Foster, Stanley O. (2009). The Livelihood vulnerability index: A pragmatic approach to assessing risks from climate variability and change a case study in Mozambique. *Global Environmental Change, 19*, 74–88.
IPCC. (2007). Climate change: Impacts, adaptation, and vulnerability. In Contribution of Working Group II to the Fourth Assessment Report (Ch. 9). Cambridge, UK: Cambridge University Press.
Kirtman, B., Power, S. B., Adedoyin, J. A., Boer, G. J., Bojariu, R., Camilloni, I., et al. (Eds.). (2013) Near-term climate change: projections and predictability. In: Climate Change 2013: working group i to the fifth assessment report of the IPCC. United Kingdom and New York, USA: Cambridge University Press, Cambridge.
Lobell, D. B., Bänziger, M., Magorokosho, C., & Vivek, B. (2011). Nonlinear heat effects on African maize as evidenced by historical yield trials. *Nature Climate Change, 1*(1), 42–45. doi:10.1038/nclimate1043.
Maddison, D. (2007). The perception of and adaptation to climate change in Africa. In *CEEPA*. Discussion paper no. 10. centre for environmental economics and policy in Africa. Pretoria, South Africa: University of Pretoria.
Nhemachena, C. and Hassan, R. (2007). Micro-level analysis of farmers' adaptation to climate change in Southern Africa. In *IFPRI Discussion Paper No. 00714*. Washington DC: International Food Policy Research Institute.
NMA. (2015). *National metrological agency-didessa basin climate variables secondary data*. Ethiopia: Addis Ababa.
Patt, A., Klein, R. J., & de la Vega-Leinert, A. (2005). Taking the uncertainty in climate-change vulnerability assessment seriously. *Comptes Rendus Geoscience, 337*(4), 411–424.
Sewagegn A. (2011). The economic impacts of climate change on crop agriculture and the role of agricultural technologies towards better adaptation: A case in Choke Mountain watersheds of Ethiopia (Thesis). Addis Ababa: Addis Ababa University.
Shewmake S. (2008). Vulnerability and the impact of climate change in South Africa's Limpopo River Basin.
Sullivan, C., Meigh, J. R., & Fediw, T. S. (2002). *Derivation and testing of the water poverty index phase 1*. UK: Final Report. Department for International Development.
Swain, M., & Swain, M. (2011). Vulnerability to agricultural drought in Western Orissa: A case study of representative blocks. *Agricultural Economics Research Review*, *24*, 47–56.
Thomas, D. S. G., Twyman, C., Osbahr, H., & Hewitson, B. (2007). Adaptation to climate change and variability: Farmer responses to intra-seasonal precipitation trends in South Africa. *Climatic Change, 83*(3), 301–322. doi:10.1007/s10584-006-9205-4.
Von Braun, J. (2007). *The world food situation: new driving forces and required actions*. (p. 18). Washington, D.C.: International Food Policy Research Institute.
World Bank. (2010). In: Economics of adaptation to climate change, Ethiopia. https://openknowledge.worldbank.org/handle/10986/12504 License: CC BY 3.0 Unported.,www.iwmi.org/Publications/IWMI_Research_Reports/index.aspx. Accessed March 24, 2014.

Sustainable Urban Transport: Improving Mobility Conditions for Vulnerable Road Users in Sub-Saharan Africa

Elke Sumper and Marketa Barker

1 Introduction

Urbanization is a spreading phenomenon worldwide. In 2007 already 50% of the earth's population lived in urban areas (Lexas 2015), a share that is expected to rise to 66% by 2050 (United Nations 2015). In the last few decades urbanization in developing countries caught up unprecedented dimensions. With the growing population transport demand increases. At the same time energy consumption and greenhouse gas emissions rise rapidly. Africa is by far the smallest producer of CO_2 emissions worldwide (Fig. 1) at the moment. The Intergovernmental Panel on Climate Change (Sims et al. 2014, p. 603) expects that transport demand per capita in developing countries increases at a faster growing rate in the next decades though. Aggressive mitigation measures are requested to prevent transport emissions from rising above the rates of other energy sectors in 2050.

Energy consumption and CO_2 emissions are growing fastest in the transport sector (Meyer et al. 2012). From 1971 to 2006 global transport energy use rose steadily between 2 and 2.5% per year, closely paralleling growth in economic activity around the world (Ajanovic et al. 2012, p. 1). Transport accounts for about 28% of global energy demand (GEA 2012, p. 49) and one quarter of energy-related greenhouse gas emissions in the European Union (European Commission 2016). Over 70% of these emissions are caused by road transport—almost solely by motorized individual traffic (Fig. 2).

Cities suffer the most from overpopulation, heavy traffic, poor air quality, noise pollution and accidents. Urban traffic produces up to a quarter of traffic caused greenhouse emissions and is responsible for nearly 70% of all traffic accidents (European Commission 2011, p. 8).

E. Sumper (✉) · M. Barker
Factum Chaloupka & Risser OG, Institute for Traffic and Social Analysis Vienna, Vienna, Austria
e-mail: elke.sumper@gmx.at

W. Leal Filho et al. (eds.), *Climate Change Adaptation in Africa*,
Climate Change Management, DOI 10.1007/978-3-319-49520-0_18

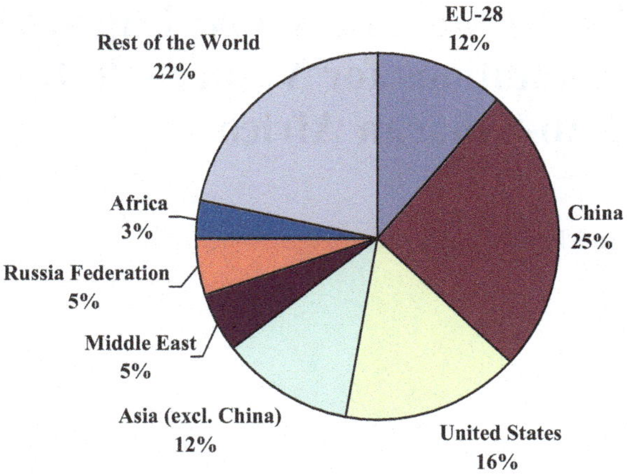

Fig. 1 CO_2 emissions worldwide 2012 (European Commission 2015 p. 18)

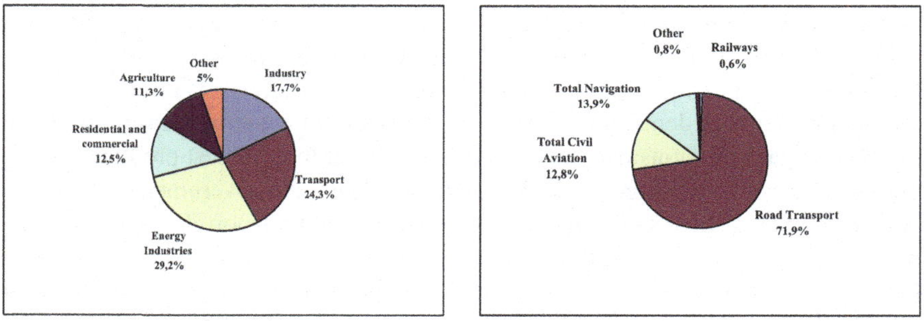

Fig. 2 EU greenhouse gas emissions by sector and by transport mode 2012 (European Commission 2016)

The goal of this paper is to discuss measures for a significant improvement of conditions for vulnerable road users in Sub-Saharan Africa, in order to build a future oriented, economically and ecologically feasible and socially balanced traffic system.

With regard to the CO_2 reduction target non-motorized transport modes are preferable to motorized ones. At the same time walking or cycling are often connected with very low prestige and considered as transport modes of the poor. This view has to be changed as well as the popular focus on the expansion of the motorized traffic.

In Sub-Saharan Africa it is not so much the change of choice rather than the change of attitudes toward sustainable transport modes. Addis Ababa for instance has an exemplary high share of non-motorized traffic (NMT). Moreover the design of the road traffic environment has to consider the needs of vulnerable road users specifically because the vast majority in the population are pedestrians.

The following steps are taken to reach the goal of the paper:

- First of all, mobility behaviour and its motivation will be examined for a better understanding of the choices people make.
- Second, a few indicators which are relevant for the urban mobility behaviour will be described (situation of Vienna and Addis Ababa).
- Third, the authors will develop measures from a traffic psychological point of view which aim at increased protection of vulnerable road users. By focusing on the weakest group, a more equal traffic system shall be promoted and, at the same time, awareness for transport modes with low greenhouse gas (GHG) emissions and low energy consumption shall be raised.
- Finally, the good practice examples of how to improve NMT modes will show that (even little) efforts can make a difference.

This paper is not based on one empirical study about influencing traffic behaviours. The authors use a heuristic approach to derive measures and solutions. They refer to empirical knowledge out of their working environment of traffic sociology and psychology of the last decades. The paper discusses why and how changes should be implemented from a social scientific point of view considering psychological aspects as well.

2 Levels of Influence on Mobility Behaviour

"Mobility" for the purpose of this paper is used as physical and circular mobility which refers to daily travels of households and their members (Herry et al. 2011, 87).

The road traffic system is a system of relations and behaviours. Road users' behaviours do not arise from simple, distinct and deterministic cause-effect relations. There are interdependencies between individual actors, between individuals and their environment and also between individuals and vehicles. In order to understand those interdependencies and to deduce specific measures, factors on the five levels of the Diamond in Fig. 3 are considered:

- The individual itself (and its characteristics; for instance its internal processes like motives, attitudes, experiences)
- The vehicle (and its characteristics such as physical and technical aspects-size, height, speed, power; technical features like assistance systems etc.)
- Infrastructure (including road and space characteristics like road width and-surface, etc.)
- Society including existing laws and traffic rules and also public discussions like for instance discussed conflicts between cyclists and car drivers etc.
- Interaction between road users including non-verbal gestures like for instance waving to say 'thank you' but also verbal articulations, for example an angry car driver screaming at the crossing pedestrian etc.

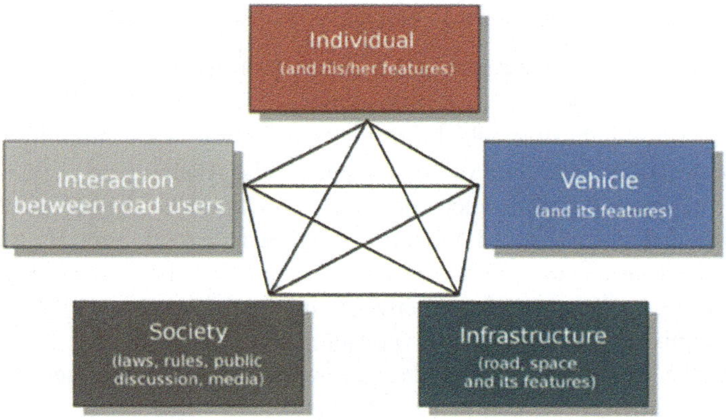

Fig. 3 Diamond—levels of influence on traffic behaviour (Risser 2000)

The factors of all five areas in the Diamond interact with each other. Changes in one area usually cause changes in one or more of the other areas.

3 Mobility in Addis Ababa and Vienna

When interpreting mobility differences between cities, countries and continents various factors have to be considered (see also Diamond Fig. 3). The modal split is crucial for analysing mobility behaviours and participation in road traffic. The numbers of driver license holders, out-of-house mobility, numbers of trips per person, trip distances, durations of trips, purpose of trips (work, spare time, shopping, pick-up services etc.) are examples for important details that mobility surveys commonly collect.

Mobility data in general are rather difficult to obtain. Mobility surveys are not conducted frequently and the methods of data collection vary from country to country, even from survey to survey. Therefore comparisons between mobility data have to be treated with caution. The historical, social and economic situation (gender, household incomes, land use patterns, traditions) need to be taken into consideration. Comparisons between cities, countries and continents are difficult but necessary, if target-group-specific measures shall be developed.

In the following a few key figures of mobility such as the modal split, the household income and accidents will be described for Addis Ababa and Vienna and the differences in their development plans will be pointed out.

The modal split of passenger traffic or passenger transport mode share "refers to the percentage of passenger journeys or trips by the main mode of transport" (LTA 2011, p. 60).

3.1 *Mobility in Vienna: A Few Indicators*

Vienna, with about 1.8 million inhabitants (City of Vienna 2015a), is the largest city and the capital of Austria.

3.1.1 Modal Split in Vienna

Latest modal split data of Vienna (City of Vienna 2015b) show a 39% share for the public transport, 28% of motorized individual traffic (27% cars + 1% motorbike), 26% of walking and 7% cycling (Fig. 4).

3.1.2 Household Income Distribution

64% of Austrian households in 2014 are one or two-person households. The average household income is 3404 Euros/month. 10% of all Austrian households do have an income less than 1139 Euros/month. (Statistik Austria 2015a, own calculations).

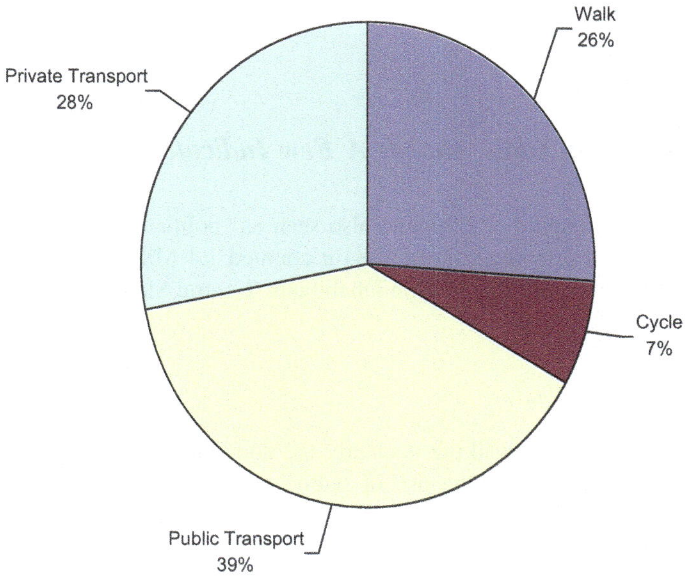

Fig. 4 Modal split of Vienna 2014 (City of Vienna 2015b)

3.1.3 Motor Vehicle Accidents

Fatal traffic accidents in Austria in the past two decades are on the decline. In the year 2000 4925 accidents happened in Vienna, 34 of them ended deadly. 2014 more accidents happened in total numbers (5802) but less people (21) died on the roads. About 20% of the accident victims were pedestrians (Statistik Austria 2015b).

3.1.4 Past and Future Ambitions

Since 1993 the public transport share in Vienna rose constantly every year from 29% (Wiener Stadtwerke 2015) to the current share of 39%. Motorized individual traffic on the other side was reduced significantly from 36% in 2001 (Herry et al. 2007) to 27% in 2014 (City of Vienna 2015b). Cycling experienced an upward trend for the last two decades, whereas walking stayed on almost the same level with a slight decreasing tendency in the last few years (Wiener Stadtwerke 2015; City of Vienna 2015b). Ongoing efforts, however, point in one direction: Vienna actively promotes public transport use, cycling and walking. STEP 2025 (City of Vienna 2014), the urban development plan for Vienna, actually plans a long term reduction of motorized individual traffic (MIT) up to the aspired goal of 15% in 2030. The modal split target for 2025 is "80:20"—80% of all trips shall be covered by public transport, walking or cycling, whilst the current MIT share shall be reduced to 20%.

3.2 Mobility in Addis Ababa: A Few Indicators

Addis Ababa, the capital of Ethiopia, also seen as "political capital" of the continent, is a very fast growing city. In 1984 it counted 1.4 Million inhabitants, while the latest census reported 3.3 Million inhabitants (Central Statistical Agency 2015).

3.2.1 Modal Split

In Addis Ababa (Fig. 5) walking is clearly the dominant mode of transport with a share of 60%, followed by the use of public transport ([mini-]buses and taxis), chosen for 27% of all trips. Cycling (7%) as well as private transport by car or motorbike (6%) are rather rarely used modes.

The figure indicates that 94% of all trips are covered by sustainable transport modes. Compare to Vienna's targeted 80% in 2025!

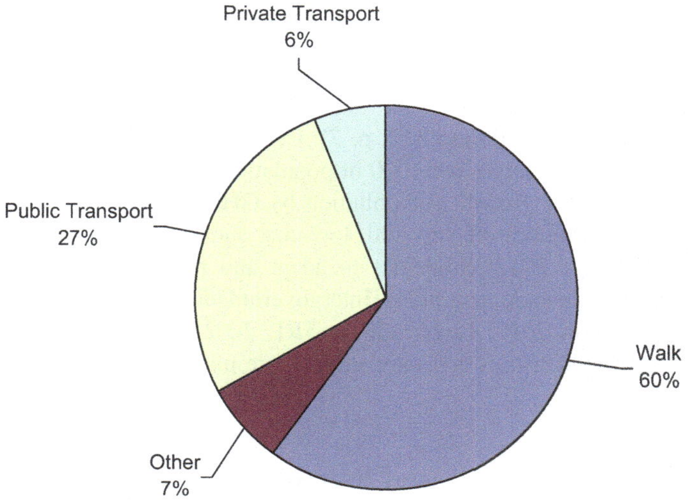

Fig. 5 Modal split Addis Ababa (Nyarirangwe 2008)

3.2.2 Household Income

The census in 2007 counted an average monthly household income of about 31 Euros ($\sim$725 Ethiopian Birr) with an average household size of 4, 1 persons. Over 60% of the population of Addis Ababa live below the poverty line with an income less than 20 Euros per household per month (Bogale 2012).

3.2.3 Accidents

Against the background of the high modal split share of non-motorized road users and the marginal number of car drivers, Ethiopian accident figures assume alarming proportions. Vulnerable road users such as pedestrians, children, seniors or disabled people are the most frequent victims. (Materu et al. 1999) Data from the Addis Ababa Police Commission indicate 391 deaths out of 3003 car accidents in 2013/14 (Abdu 2015). Half of road traffic accident deaths in Ethiopia are pedestrians (IRIN 2011) other studies indicate even higher fatality rates under pedestrians (e.g. Akloweg et al. 2011).

3.2.4 Future Prospects

Due to the expansion of Addis Ababa the demand for public transport services is also growing rapidly. Car ownerships do not increase at the same pace as the growth of the population (TRL 2002). Private transport modes like cars or motorcycles, are too expensive for the vast majority of the people. Therefore the role of public transport will gain importance and, of course, walking will remain one of the most commonly used modes of transport.

Forecasts though predict a modal shift of 15% moving from walking to the motorized transport (Wondimu 2012, p. 7). Car ownership at the moment is still very low (below 100 vehicles per 1000 of population) but it is expected to increase rapidly with economic growth. Air pollution by GHG emissions is a serious concern given the distribution of very old, low cost and poorly maintained vehicles. (Voukas and Palmer 2012) Strategies to adapt and mitigate climate change are discussed in urban development plans (Intergovernmental Panel on Climate Change 2014; CRGE Vision 2012; Bogale 2012; TRL 2002) but exact numbers which relate to the future measures and interventions are missing.

4 Measures to Promote and Support Sustainable Urban Transport

Taking into account the continuing urbanization, cities have to handle serious challenges to achieve substantial reductions of greenhouse gas emissions. Crucial targets regarding the transport sector are (AAR 2014, p. 922):

- Car traffic avoidance (reduction of driven distances).
- Modal shift to sustainable transport modes such as public transport and NMT.
- Implementation of "Zero-Emission-Vehicles" and renewable energy sources.

Talking about Sub-Saharan Africa, the authors add one target:

- Maintaining the generally very high share of non-motorized road users, under better conditions though (i.e. make it a bonus instead of a problem).

To achieve those targets different kinds of measures or their combination have to be implemented. Political measures like fiscal political instruments, spatial planning and legal measures, technological innovations for conventional and new alternative resources have controlling functions for instance. Infrastructure is another effective tool. Efforts are made worldwide to reduce GHG emissions. Psychological factors are underestimated respectively ignored systematically though. Mobility behaviour is a very complex interdependency between individuals, their environment and the vehicle.

The following elaborations concentrate on areas where psychology has to be implemented to influence and/or change and to promote sustainable mobility behaviours.

4.1 Design of Public Space and Traffic Environment

Infrastructure affects traffic behaviour strongly. Here a few examples (from highly regulated street spaces in Vienna):

- If people can choose between two ways, they will take the shorter one.
- Cars would not be used in a city where no parking space is offered.
- If red phases of pedestrian lights are too long people will not wait for the green light.
- Broad and straight streets induce speeding-car drivers would not accept tempo limits of 30 km/h.

Measures on the infrastructural level consist of elements improving communication between traffic users, of constructional speed reducing road design, of arrangements reminding of rules (section control for instance), self explaining roads and alike (Risser 2011).

Regarding the adaptation and mitigation of environmental damage actions on the infrastructural level need to focus specifically on traffic calming. Besides the fact that traffic calming measures have great impact on traffic safety, GHG emissions of the motorized individual traffic, too, can be reduced significantly by lowering the speeds.

Speed limits are especially effective if they optimize the flow of traffic. Less acceleration and braking manoeuvres reduce consumption and emissions as well as abrasion and fine dust resuspension (Umweltbundesamt 2016). In Austria for instance, displays on urban motorways remind car drivers to reduce speed, in cases of increased risk of exceeding the permitted emission and fine dust limits.

Speed limits only make sense if violations are sanctioned. A fine is one possible way. Immediate negative effect e.g. by speed bumps is another. Latter makes the driver feel physically the exceeded speed (Fig. 6). Drivers have to slow down and other traffic users like pedestrians who want to cross are noticed better. Visual stimuli (Fig. 7) and aesthetically appealing design are further possibilities to induce desired mobility behaviour.

Fig. 6 Speed bump in Vienna

© Factum OG

Fig. 7 Lane divider in Vienna

© Factum OG

Fig. 8 Shared space on Sonnenfelsplatz in Graz

© City of Graz_Heike Falk

For car drivers making swift headway might be the primary interest. For pedestrians, aesthetic aspects, safety issues or an easy access to infrastructure are additional decisive factors. Important in this context is to keep walking distances as short and direct as possible, otherwise illegal behaviours are prompted or people start avoid walking altogether (Risser 2015). Green zones with plants, sitting benches or fixtures for play or rest upgrade public space and invite people to stay and socialize.

Shared space is a concept where all road users equally move in open space. Traffic signs, traffic lights, pedestrian islands etc. are reduced to a minimum; the only regulation is "to give way to the right". Providing space and opportunity for communication and interaction is the core principle of shared space and it secures a smooth traffic and movement flow in this zone. By removing the channelling effect of streets, the motorized traffic loses its dominance and the manifold meaning of traffic mixed with social life, culture and history of the place restores its balance (Fig. 8).

The shared space philosophy is based on respect for each other and mutual consideration on the road. By introducing intentional insecurity and ambiguity all players are forced to make eye contact and interact.

4.2 Social Marketing and Communication

Regulations, laws, financial incentives and urban infrastructure-in the past and nowadays-produce and promote pre-eminently car dependant mobility. Mobility management is one promising method to counteract this unfortunate trend and to make passenger transport environmentally friendlier instead.

Mobility Management is built on the cooperation of all parties involved-transport companies, enterprises, communities, tourist organisations and affected individuals (VCÖ 2003). It includes economic instruments and measures of awareness raising activities. Marketing or campaigns are appropriate tools to promote specific attitudes and behaviours. They have to take into account the inhomogeneity of road user groups. Both "car drivers" and "pedestrians" are not homogeneous units. Individuals in all these groups find themselves in different stages of problem awareness or willingness to change behaviour. Marketing and campaigning tools have to consider those varying levels and have to start at the particular stage where the individuals stand (Chaloupka-Risser et al. 2011).

There are numerous psychological models which provide hints how attitudes and behavioural patterns can be influenced. Most promising are holistic approaches addressing people on different levels. One of those, the Social Marketing Model by Kotler and Armstrong (1996), is a communication model, which considers the promoting of an idea as a process in which the needs and wishes of individuals are targeted. This has been proven to be successful in terms of influencing decision making processes of individuals. The following policies need to be addressed:

- **Information policy**: learn as much as you can about needs and interests of your target group(s) and key players
- **Product and distribution policy**: develop adequate (technical) solutions that promote desired behaviour, easy-to-adopt strategies which enforce the desired behavioural patterns; display the products appropriately
- **Incentive and pricing policy**: link the desired behavioural patterns to positive (extrinsic) stimuli in order to motivate individuals to change their usual behavioural patterns
- **Communication policy**: inform the target group about the benefits and that their needs and interests will be satisfied if they adopt this particular behavioural change.

5 Good Practice Examples

5.1 Traffic Calming Zone in Vienna

One example for a very successful traffic calming project is the "Shared-space Mariahilferstraße" in Vienna. This is one of the main shopping streets. It suffered heavily from congestion, air pollution and traffic noise. In 2014/15 it was successfully transformed into a traffic calmed encounter zone. The actual infrastructure design is characterized by an easy accessible and safe path network, free space to interact, shop, sit, rest, eat and drink or play. It includes consume free areas and comfortable street furniture. It accommodates all road users-pedestrians, cyclists, car drivers, old and young. The re-modelling of the street was done in a participatory process. The residents were involved in the planning (information policy) and the project "Mariahilferstraße" has been frequently discussed in public media (product and distribution policy-communication policy) (Figs. 9 and 10).

5.2 *Improving Pedestrian Mobility in Addis Ababa*

During a summer workshop (ETH-EiABC) in 2014 a group of Swiss and Ethiopian architecture and urban planning students developed a number of low-cost

© Herbert Ortner

Fig. 9 Mariahilfer Straße BEFORE traffic calming

© Stadt Wien/b+b orso.pitro

Fig. 10 Mariahilfer Straße AFTER traffic calming

interventions that, without great changes or constructional measures, could be applied to the existing traffic organization and infrastructure while bringing about a progressive and environmentally friendly and lasting improvement of the situation of pedestrians.

The project targeted the wide area of Churchill Road, one of the main traffic arteries of the city, its transportation hub-the minibus station and its busy surroundings with numerous open stalls and sellers. In the planning and designing process the uniformly bad infrastructure for pedestrians along the road (no maintenance, malfunctioning pavements, misappropriation of space like street vending, parking, material storage during construction work) and the poorly equipped and crowded waiting area of the minibus station were remodelled. The old muddy island was replaced with a new, hard ground surface, it was extended by 3 m and equipped with new seats for waiting passengers (Figs. 11 and 12). The plan envisioned: planting of new trees providing shade from sun and rain for the passengers; artificial trees serve as weather protection until the trees are fully grown; seats encircle the trees; new adjustable weather protection (pull-out/fold-in roof) for

Fig. 11 Minibus stop in Addis Ababa

© Ernst Bosina

Fig. 12 Model for rain protected bus stop with sitting facilities

© Pablo Acebillo

the sellers and their customers along the road to the minibus station; clean and improved sanitary facilities free of charge next to the minibus station.

The task at hand was to find a bottom-up solution and to design a safe and serviceable zone for pedestrians and passengers; the issues that needed to be considered were: number of people and minibusses, waiting passengers, number of pedestrians on the street and sidewalk, number of people crossing at red, usage of the crosswalk, missing weather protection for street sellers and buyers, lack of seating and sanitary facilities at the minibus stop etc.

The advantage of the bottom-up planning approach is that the existing situation is taken, necessary interventions-small and cost-effective-are made, and the impact of the interventions is great, while the existing structure and organization remains intact. The proposed interventions lead to temporary and later permanent solutions.

6 Conclusion

Industrial countries pursue relentlessly and with remarkable (financial) efforts the objective of distinctive reduction of motorized individual traffic; at the same time non-motorized traffic, public transport modes and multimodality are strongly promoted.

Developing countries start from a different position. They can benefit from the great advantage of having very high modal split shares for NMT. Therefore the aim is to show how this generally sustainable mobility behaviour can be made more attractive and individuals voluntarily choose it.

For decades the car was the criterion of "good life", of success, a highly coveted status symbol. This is still the case for many people. Especially in developing countries walking or cycling are deemed to be "transport modes of the poor". This image has to be upgraded. With the policies of social marketing and awareness raising campaigns, non-motorized transport modes can be promoted and communicated as means of the future. They are sustainable and cost-effective, they are physically and psychologically healthy, consume very little space and enhance quality of life. This message has to be communicated through different channels-governments, public authorities, politicians, media, educational institutions and others. Sustainable mobility behaviour has to become an attractive choice people actively make.

In order to support this change of attitude and to make it safe, infrastructure needs to reflect respect for vulnerable road users, the vast majority in Sub-Saharan countries.

The following table gives an overview of measures on different levels of influence on traffic behaviour based on the "Diamond" (in Chap. 2) (Table 1).

Improving preconditions for vulnerable road users and the image of NMT in countries like Ethiopia and cities like Addis Ababa can prevent them from sliding

Table 1 Measures on the different levels of influence

	Skills	Knowledge	Willingness
Individual	(Target group specific) training	Information, education	Social marketing
Communication with other traffic users	"Learning by doing"; concepts like "shared space" or "self explaining roads" promote communication	Knowledge transfer e.g. informal norms like "imitating"	Mobility management; role models; social marketing; e.g. awareness raising campaigns
Society, Structures		Profound information in media about NMT- e.g. people killed by cars, preconditions for vulnerable road users etc.; no one-sided presentations in favour of motorized traffic	Upgrade image of NMT; subsidies for NMT and public transport; taxes for motorized traffic; spatial planning
Vehicle	User friendliness, telematics; techno-logical innovations including practical low-tech solutions	User-friendly operation instructions, telematics	Vehicle characteristics: type, design, power and speed, equipment; technological innovations
Infrastructure	Easy use of infrastructure; reducing traffic signage to a minimum to avoid confusion (e.g. shared space)	Traffic signals-look at the needs of vulnerable road users; rules; sanctions for violations;	Traffic calming; speed reducing road design; attractive design inviting to stay in public spaces; green zones; parking space management etc.

into mobility and traffic conditions extremely difficult to make undone. A sustainable traffic system able to provide a safe, available and accessible mobility for all citizens can be achieved with comparatively little efforts before motorization takes over and captures the majority of the population.

References

AAR14 Austrian Assessment Report. (2014). *Österreichischer Sachstandsbericht Klimawandel*. Vienna: Austrian Academy of Sciences.

Abdu, B. (2015). Ethiopia among worst in Traffic Safety. In *Addis fortune*. http://addisfortune.net/articles/ethiopia-among-worst-in-traffic-safety/, accessed on February 16, 2016.

Ajanovic, A., Dahl, C., Schipper, L. (2012). Modelling transport (energy) demand and policies—An introduction. In *Energy policy*, vol. 41, February 2012. Elsevier, pp. 3–14.

Akloweg, Y., Hayashi, Y., Kato, H. (2011). The effect of used cars on African road traffic accidents: A case study of Addis Ababa, Ethiopia. *In International Journal of Urban Sciences, 15*(1), 61–69.

Bogale, Y. A. (2012). *Evaluating transport network structure: Case study in Addis Ababa, Ethiopia*. Master Thesis. University of Twente, Enschede.

Central Statistical Agency of Ethiopia. (2015). *Census reports*. http://www.csa.gov.et/, accessed on January 5, 2016.

Chaloupka-Risser, C., Risser, R., & Zuzan, W. (2011). *Verkehrspsychologie*. Vienna: Grundlagen und Anwendungen, Facultas.

City of Vienna. (2014). *STEP 2025 Urban development plan. Thematic concept*. https://www.wien.gv.at/stadtentwicklung/strategien/step/step2025/, accessed on January 4, 2016.

City of Vienna. (2015a). *Wien in Zahlen*. MA 23—Wirtschaft, Arbeit und Statistik (Ed.). www.statistik.wien.at, accessed on January 4, 2016.

City of Vienna. (2015b). *Modal Split 2014: Radfahren in Wien immer beliebter*. https://www.wien.gv.at/rk/msg/2015/02/10006.html, accessed on January 4, 2016.

CRG Climate Resilient Green Economy Vision. (2012). *Ethiopia's vision for a climate resilient green economy*. The Federal Democratic Republic of Ethiopia Environmental Protection Authority.

European Commission COM. (2011). *White paper. Roadmap to a single European transport area—Towards a competitive and resource efficient transport system*. Brussels. http://ec.europa.eu/transport/themes/strategies/doc/2011_white_paper/white_paper_com%282011%29_144_en.pdf, accessed on February 10, 2016.

European Commission. (2015). *EU energy in figures*. Statistical Pocketbook. Luxembourg. https://ec.europa.eu/energy/sites/ener/files/documents/PocketBook_ENERGY_2015%20PDF%20final.pdf, accessed on February 10, 2016.

European Commission. (2016). *Reducing emissions from transport*. http://ec.europa.eu/clima/policies/transport/index_en.htm, accessed on February 12, 2016.

ETH-EiABC Workshop 2014. (2015). *Addis Ababa, Ethiopia. Improving pedestrian mobility through bottom-up strategies*. Final Report, Zurich/Addis Ababa.

GEA. (2012). *Global energy assessment—Toward a sustainable future*. In Technical Summery. Cambridge University Press, Cambridge, UK and New York, USA and the International Institute for Applied System Analysis, Laxenburg, Austria.

Herry, M., Sedlacek, N., & Steinacher, I. (2007). *Verkehr in Zahlen*. Bundesministerium für Verkehr, Innovation und Technologie Abteilung V/Infra 5 (ed.). Vienna.

Herry, M., Sedlacek, N., & Steinacher, I. (2011). *Verkehr in Zahlen*. Österreich. Bundesministerium für Verkehr Innovation und Technologie Abteilung II/Infra 5 (Hrsg.): Vienna.

Intergovernmental Panel on Climate Change (IPCC). (2014). Climate change 2014: Synthesis report. In Core Writing Team, R. K. Pachauri & L. A. Meyer (Eds.), *Contribution of Working Groups I, II and III to the Fifth Assessment Report of the Intergovernmental Panel on Climate Change*. IPCC, Geneva, Switzerland, pp. 151.

IRIN humanitarian news and analysis. (2011). *Ethiopia: Government moves to address road-accident toll*. http://www.irinnews.org/report/94165/ethiopia-government-moves-to-address-road-accident-toll, accessed on February 15, 2016.

Kotler, P., & Armstrong, G. M. (1996). *Principles of Marketing, Prentice Hall Publishers*.

Lexas Information Network. (2015). *Länderdaten*. Urbanisierung. http://www.laenderdaten.de/bevoelkerung/urbanisierung.aspx, accessed on January 9, 2016.

LTA Land Transport Authority. (2011). Passenger transport mode shares in world Cities. In *Journeys Issue 7*, November 2011. Singapore: LTA Academy, pp. 60–70.

Materu, J., Paradza, G., & Mushamba, S. (1999). Mobility in Africa. In *Bremen Initiative on business and the municipality*. Report of the African Consultative Workshop held in Mombasa, Harare.

Mercer. (2015). *In Wien ist die Lebensqualität weltweit am höchsten*. http://www.mercer.at/newsroom/In-Wien-ist-die-Lebensqualitaet-weltweit-am-hoechsten.html, accessed on January 4, 2016.

Meyer, I., Kaniovski S., & Scheffran, J. (2012). Scenarios for regional passenger car fleets and their CO_2 emissions. In *Energy policy. Modeling transport (energy) demand and policies*, Vol. 41, February 2012. Elsevier, pp. 66–74.

Nyarirangwe, M. (2008). Integrating multi-modal transport service provision in tourism cluster development in Addis Ababa. In *Proceedings of the 27th Southern African Transport Conference (SATC 2008)*. Pretoria.

Risser, R. (2000). *Measuring influences of speed reduction on subjective safety*. ICTCT Workshop on Traffic Calming; New Delhi.

Risser, R. (2011). Gestaltung der Verkehrsumwelt, pp. 217–224. In C. Chaloupka-Risser, R. Risser, & W. Zuzan (Eds.), *Verkehrspsychologie*. Grundlagen und Anwendungen. Vienna: Facultas.

Risser, R. (2015). A Philosophy of traffic Calming. In *Psychologie in Österreich 2*. Themenschwerpunkt Gestaltung der Umwelt, Vienna, pp. 148–156.

Sims, R., Schaeffer, R., Creutzig, F., Cruz-Núñez, X., D'Agosto, M., Dimitriu, D. … Tiwari, G. (2014). Transport. In O. Edenhofer, R. Pichs-Madruga, Y. Sokona, E. Farahani, S. Kadner, K. Seyboth, … J. C. Minx (Eds.), *Climate change 2014: Mitigation of climate change. Contribution of working group III to the fifth assessment report of the intergovernmental panel on climate change*. Cambridge, UK and New York, USA: Cambridge University Press.

Statistik Austria. (2015a). *Verfügbares Haushaltseinkommen in Österreich nach Haushaltstyp*. http://www.statistik.at/web_de/statistiken/menschen_und_gesellschaft/soziales/haushalts-einkommen/index.html, accessed on January 6, 2016.

Statistik Austria. (2015b). *Unfallgeschehen nach Bundesländern*. Zeitreihen 1992 bis 2014. http://www.statistik.at/web_de/statistiken/energie_umwelt_innovation_mobilitaet/verkehr/strasse/unfaelle_mit_personenschaden/index.html, accessed on January 6, 2016.

TRL Transport Research Laboratory. (2002). *Scoping study*. Urban Mobility in Three African Cities. Addis Ababa, Dar es Salaam, Nairobi. Sub-Saharan Africa Transport Policy Program Working Paper No. 70.

Umweltbundesam. (2016). *Höhere Geschwindigkeit führt zu höherer Umweltbelastung*. http://www.umweltbundesamt.at/umweltsituation/verkehr/fahrzeugtechnik/pkw/tempo/, accessed on January 24, 2016.

United Nations, Department of Economic and Social Affairs, Population Division. (2015). *World Urbanization prospects: The 2014 revision*, (ST/ESA/SER.A/366).

VCÖ. (2003). *Mobilitätsmanagement bringt mehr Verkehrssicherheit und Effizienz im Verkehr*. Medieninformation. http://wko.at/up/enet/mobilitaetpresse.pdf, accessed on January 6, 2016.

Voukas, Y., & Palmer, D. (2012). Sustainable transportation in East Africa. The Bus rapid Transit Evolution in Addis Ababa, Ethiopia. In *Conference CODATU XV. The role of urban mobility in (re)shaping cities*, October 2012 in Addis Ababa, pp. 22–25.

Wiener Stadtwerke. (2015). *Modal split—Verkehrsmittelwahl der Wienerinnen und Wiener*. http://www.nachhaltigkeit.wienerstadtwerke.at/daseinsvorsorge/oepnv/modal-split.html, accessed on January 5, 2016.

Wondimu, A. (2012). Sustainable Urban mobility planning for coping challenges of Urban transportation in the fast urbanizing Ethiopia. In *Conference CODATU XV. The role of urban mobility in (re)shaping cities—Addis Ababa*. http://www.codatu.org/wp-content/uploads/Wondimu-Abeje-ARTICLE-Codatu-XV-2012-EN.pdf, accessed on January 6, 2016.

Influence of Climate Variables on Vector and Prevalence of Bovine Trypanosomosis in Tselemti District, North West Tigray, Ethiopia

Amdework Zekarias, Habtamu Taddele and Amanuel Zenebe

1 Introduction

Ethiopia with its great variation in climate and topographic features possesses the largest livestock populations in Africa. Ethiopia's estimated ruminant livestock population consists of 56.71 million cattle, 29.33 million sheep, 29.11 million goats and 1.16 million camels. The cattle population in Tigray is about 3.5 million (CSA 2012).

Among other livestock, cattle provide direct benefits like meat and milk, and hide. The indirect benefits from cattle include manure for replenishing soil fertility and restoring nutrients, and animal traction. However, their production potential and productivity is influenced by several factors. Of these factors animal diseases like trypanosomosis make the greatest influence. Trypanosomosis is a vector-borne disease that can be transmitted both biologically and mechanically. Most diseases associated with tabanids are transmitted mechanically (Desquesnes and Dia 2003; Desquesnes et al. 2009).

In Africa, over 60 million heads of cattle and equivalent numbers of other livestock species are at risk of contracting the disease trypanosomosis and over three million cattle and other livestock are lost every year. Thirty five million doses of

A. Zekarias · H. Taddele (✉) · A. Zenebe
Institute of Climate and Society (ICS), Mekelle University, Mekelle, Ethiopia
e-mail: hbtmtaddele@gmail.com

A. Zekarias
Livestock Unit, Tigray Bureau of Agriculture and Rural Development, Mekelle, Ethiopia

H. Taddele
College of Veterinary Medicine, Mekelle University, Mekelle, Ethiopia

A. Zenebe
Department of Land Resource Management and Environmental Protection,
College of Dryland Agriculture and Natural Resources, Mekelle, Ethiopia

W. Leal Filho et al. (eds.), *Climate Change Adaptation in Africa*,
Climate Change Management, DOI 10.1007/978-3-319-49520-0_19

trypanocides are administered each year to protect livestock in tsetse infected areas (Holmes et al. 2004; Van den Bossche and Delespaux 2011). Direct losses due to Trypanosomosis are estimated between US$ 1–1.2 billion each year whereas the indirect impact of the disease on agriculture in sub-Saharan Africa exceeds this amount. Reports of the MoA (Ministry of Agriculture 2005) of Ethiopia show that the annual losses to the national economy due to direct and indirect impact of trypanosomosis on agricultural and livestock production are estimated to exceed US $200 million. Uncontrolled movement of animals from place to place worsens the incidence of African Animal Trypanosomosis (Selby et al. 2013). Apart from the cyclical transmission of trypanosomosis by the *Glossina species*, it is also highly considered that mechanical transmission is a potential threat to livestock productivity in some parts of Ethiopia. Studies conducted in Tigray region of Ethiopia indicated a significant prevalence of mechanically transmitted bovine trypanosomosis (Tigray Bureau of Agriculture and Rural Development 2008; Abebayehu et al. 2011).

Vector-borne diseases are highly influenced by spatial and temporal changes in the environment. Climate has been established as an important determinant in the distribution of vectors and pathogens. Climate variability and the breeding activity of the vectors are considered one of the important environmental contributors to the transmission of vector-borne diseases like trypanosomosis (Sean et al. 2012). Every change in climate has different impacts on the fly's natural activity. The 2001 Intergovernmental Panel on Climate Change (IPCC) report highlights the impact of a rising temperature on the reproduction and maturity of infective agents as well as the survival rate of the vectors, thereby influencing the disease transmission (IPCC 2001). Several studies and reports emphasized the effects of climate both on pathogen and vector habitat suitability through changes in temperature, precipitation, humidity and wind patterns. In line to this, it has been projected that climate warming over the next century is expected to have a large impact on the interactions between pathogens, vectors, and their animal and human hosts. A change in climatic variables could alter vector development rates, shift their geographical distribution and alter transmission dynamics (Gilles et al. 2005; IPCC 2007a; Tabachnick 2010; Sean et al. 2012).

Several studies on climate change and its negative impact to agricultural productivity in Ethiopia are highly discussed but most of them are from crop productivity perspective; that is the contribution of livestock in the agricultural sector is neglected. However, the livestock contributes significantly to the agricultural productivity especially in countries like Ethiopia where agriculture is highly dependent on livestock. The livestock sector is one of the most vulnerable sectors to climate change. Livestock diseases, particularly vector-borne diseases are highly influenced by the climate variables such as temperature and rainfall. However, there is paucity of information on the linkage between climate variables, vector distribution and disease prevalence under Ethiopian condition. Hence, research works on impacts of climate change on livestock productivity and associated diseases are important to design appropriate adaptation strategies at country level. Therefore, the present study was designed to identify disease causative agent at species and vectors at genus level; Assess impact of climatic variables on vector population dynamics and distribution; Determine present spatial distribution of trypanosomosis in relation to agro-ecological classes; and Predict climate variables for the three terms (2013–2099) and disease distribution. Moreover, the indigenous knowledge on climate change impacts and their adaptation strategies were assessed.

2 Materials and Methods

2.1 Description of the Study Area

2.1.1 Location

The present study was conducted from December 2013 to September 2014 in five peasant associations (PAs) in Tselemti district, located in North West Tigray about 385 km far from the regional capital, Mekelle. It is located between 37.77 N and 38.71 N latitude, and 13.39 E and 13.89 E longitude. Based on altitude it is divided in three agro climatic classes, with 2.6% highland, 19% midland and 78.4% lowland (Fig. 1).

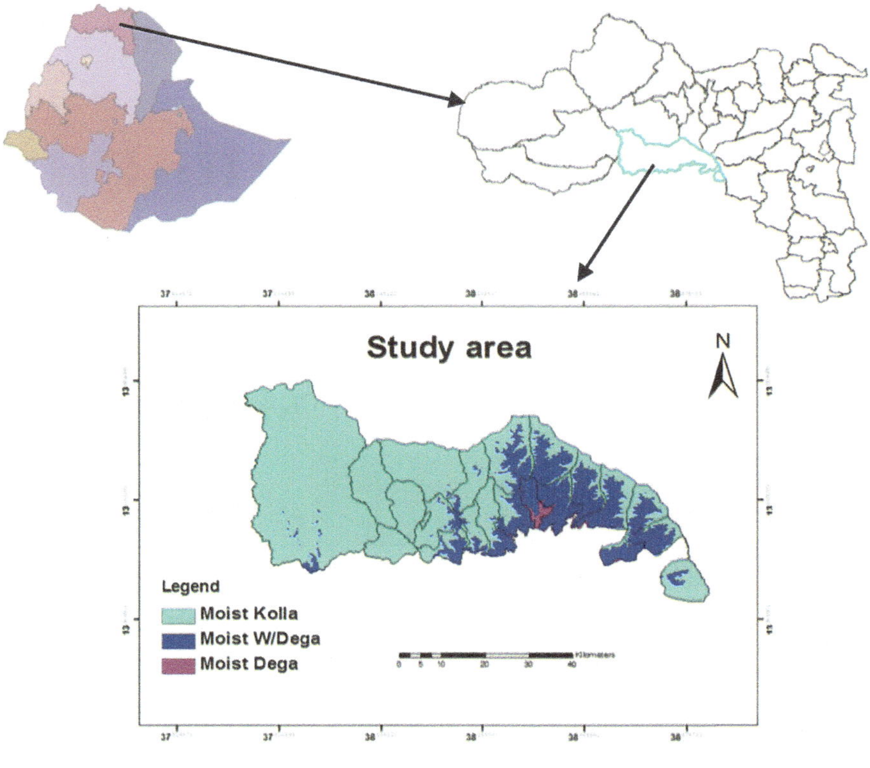

Fig. 1 Map of the study area

The ruminant livestock population of the district is estimated about 163,925 cattle, 1999 sheep and 181,027 goats. The district is divided into 22 PAs ('Tabias') with an estimated human population of 165,024. Agriculture is the mainstay of the livelihood of people with a mixed farming system. Livestock plays an integral role for agricultural activity in the district (Tselemti District Office of Agriculture and Rural Development 2013).

2.1.2 Physical Characteristics

According the records of National meteorology service agency, the study area has summer rainfall (June–September) and dry season (December–April). The highest raining months are June, July and August and the range of rainfall is between 750 and 1250 mm. The minimum and maximum temperature registered in the district is 15 and 30 °C, respectively. The average maximum temperature and rainfall registered during the years 1983 to 2001 is presented in Figs. 2 and 3, respectively.

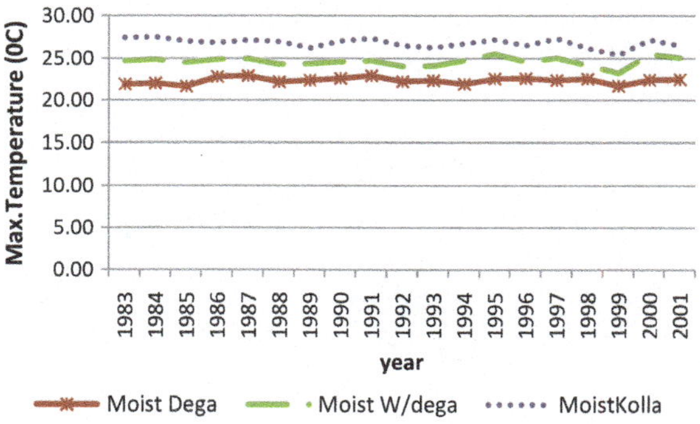

Fig. 2 Average maximum temperature of Tselemti (1983–2001)

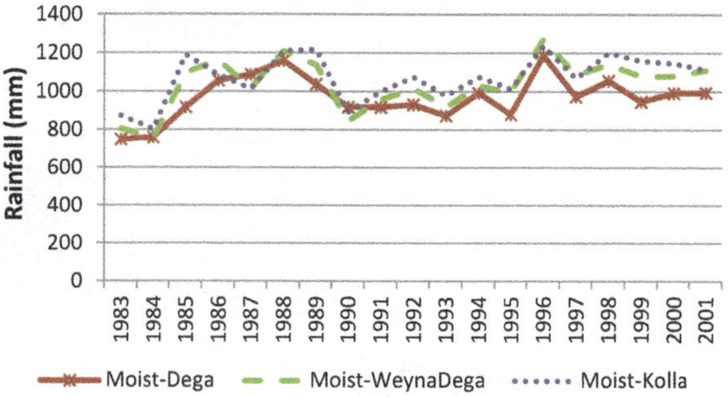

Fig. 3 Annual rainfall of Tselemti (1983–2001 EC)

2.2 *Research Methods*

2.2.1 Research Design

Stratified sampling technique was used by dividing the area into three agro-climatic groups: lowland (500–1500 m.a.sl), mid-land (1500–2300 m.a.sl) and highland (2300–3200 m.a.sl). Simple random sampling approach was used to select PAs from the three acro-climatic zones. Accordingly, one PA from highland and mid-land and three from lowland agro-climatic zones were selected based on proportion and prior information of the area with regard to livestock population, disease situation and vector distribution.

2.2.2 Sample Size and Sampling Techniques

Sample size was determined according to (Thrusfield 2007) where 95% confidence interval and 5% precision level was maintained. To calculate the sample size, prevalence of 2.66% previously reported in the area was considered (Abebayehu et al. 2011). Accordingly, a total number of 40 cattle per PA were required for the study. However, to increase the accuracy and reliability of the result, 100 cattle per PA (a total of 500 cattle) were considered to take blood samples for assessing the prevalence of the disease. Standard traps of tsetse flies were deployed to collect vectors. Acetone and two weeks old cow urine was used as bait to attract biting flies. Traps were maintained for 72 h and flies were collected, stored in dry fly box and transported to the regional laboratory for further identification.

2.2.3 Parasitological and Vector Study

Determination of parasitaemia was conducted at field level. Identification of trypanosome parasites at species level and vector identification at genus level was conducted at Tigray regional veterinary laboratory according to standard procedures (CLSI 2000; Garcia 2007; Mullens and Rogers 2009).

2.2.4 Climate Data Collection

Data on climatic variables (30 year temperature and rainfall data) was obtained from the National Meteorology Agency to assess the influence of these climate variables on the prevalence and distribution of the disease, and vectors across different agro-climatic classes.

2.3 Analysis of Disease Prevalence, Vector Distribution and Climate Data

Arc GIS software (version 9.3) was used for spatial disease and vector distribution analysis. Relationships of agro climatic classes with diseased and non-diseased animals were analyzed using X^2. For descriptive analysis SPSS software (version 16.0) was used. To downscale large scale predictor variables SDSM software was used (Philip and Ralph 2009). The future time series was constructed by the statistical downscaling technique using predictor variables of HadCM3A2a and HadCM3B2a. The predictor variables used for SDSM model input were downloaded from Canadian Institute for Climate Studies (CICS 2013) for model output of HadCM3. The predictor variables are supplied on a grid basis by selecting African window and site location on the grids.

The prediction of future climate scenarios was done by (i) downscaling regional climate data using SDSM version 4.2 which involves selection of climate predictor variables best correlated with daily observed precipitation and temperature series at Tselemti for the baseline period 1983–2001, and (ii) scenario generation, i.e. predictions using HadCM3 data predictors and two scenarios A2a and B2a to obtain synthetic precipitation (rainfall) and temperature for three future time periods (2013–2043, 2044–2074 and 2075–2099).

2.4 Indigenous Knowledge Assessment on Climate Change Adaptation

The indigenous knowledge of farmers on climate change impacts on livestock and their adaptation strategies were assessed using a focus group discussion tool. This assessment was done as a supplemental survey to the major objectives of the present study.

2.5 Limitation of the Study

The present study was limited to assess how climatic variables affect vector distribution and disease prevalence in the study area using predictor variables of HadCM3A2a and HadCM3B2a, but it does not use other climate models for comparison. Moreover, laboratory based investigation on how these climate variables affect the vector growth and multiplication is not part of this study.

3 Results

3.1 *Prevalence and Identification of Trypanosome Species and Vectors*

The present findings indicated an overall prevalence of 8% (40/500). In the present study *T. vivax* was isolated from all smears.

A total of 6211 flies were caught from the five PAs of the three agro-climatic classes. Of the total caught flies *Stomoxis* and *Tabanus* flies accounted for 45.8% (2842), 2.6% (163), respectively while the rest 3206 (51.6%) belong to other fly genera (Table 1). Nevertheless, both at field level data collection and laboratory analysis, no tsetse fly recorded in the study areas.

The highest fly population was caught in lowland (70.7%) followed by midland (19.4%). In the present study, trypanosomosis disease causing vectors together account 48.4% of the total caught flies. Statistically, the above findings clearly show that there is significant variation in the number of trypanosome vectors (somoxis and tabanus) between lowland and highland ($p = 0.000$) and highland and midland ($p = 0.007$). However, no significant variation was observed between the fly population caught in midland and lowland agro-climatic classes ($p = 0.418$) (Table 2).

Table 1 The distribution of biting flies in different agro-climatic zones

Agro-climate	Tabia	Trap no	Caught biting fly by genus			Total fly population	% by Tabia
			Stomoxis	Tabanus	Others		
Highland	Mayami	3	253	8	354	615	9.9
Midland	Adi-Wesen	3	579	33	595	1207	19.4
Lowland	Mai-Ayni	3	655	35	752	1442	23.2
	Tsaeda - kerni	3	685	46	745	1476	23.8
	Mezekir	3	670	41	760	1471	23.7
Total		15	2842	163	3206	6211	
% by genus			45.8	2.6	51.6		

Table 2 Statistical analysis vis-à-vis stomoxis and tabanus along agro-climatic zones

Agro-ecology	Trapped fly population		
	Highland	Midland	Lowland
	261	612	2132
Highland	–	$p = 0.007$	$p = 0.000$
Midland	$p = 0.007$	–	$p = 0.418$
Lowland	$p = 0.000$	$p = 0.418$	–

3.2 *Spatial Distribution of Trypanosomosis in Relation to Agro-Climatic Classes*

The spatial distribution of bovine trypanosomosis with respect to the different agro-climatic classes is summarized in Table 3 and Fig. 4.

Table 3 Prevalence of trypanosomosis within a given agro-climatic class

Agro-climatic zones		Altitude	Sampled cattle	Trapped flies	Diseased cattle	Prevalence (%)	X^2	*p*-value
Highland		2450	100	261	0	0.0	3.046	0.00
Midland		1620	100	612	3	3.0		
Lowland	Mai-Ayni	1392	100	690	9	9.0	2.286	0.00
	Mezekir	1133	100	711	12	12.0		
	Tsaeda kerni	1103	100	731	16	16.0		

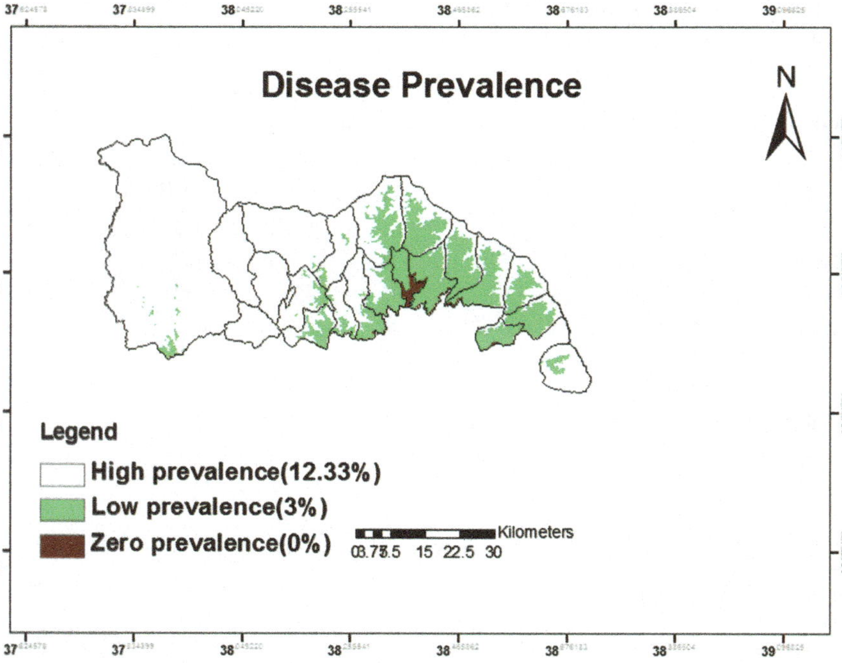

Fig. 4 Spatial distribution of trypanosomosis in the study area

3.3 *Predicting Climate Variables for the Three Terms (2013–2099) and the Disease*

The prediction of trypanosomosis with respect to climatic variables was done indirectly through the distribution of the biting flies which are the mechanical transmitters of the disease. This is because the vectors (biting flies) are directly affected by climatic variables (temperature and rainfall) that vary with altitude and other climatic conditions. The data agreement test of the SDSM showed strong agreement between observed 1983–2001 output versus modeled data value generated using scenarios of HadCM3A2a and HadCM3B2a for temperature (Fig. 5) and rainfall (Fig. 6).

After observing the agreement, prediction was done by looking the change. The change was expressed as the difference between baseline period (1983–2001) and future monthly results. Positive change indicates increase from the baseline period value while a negative change indicates decrease from the baseline.

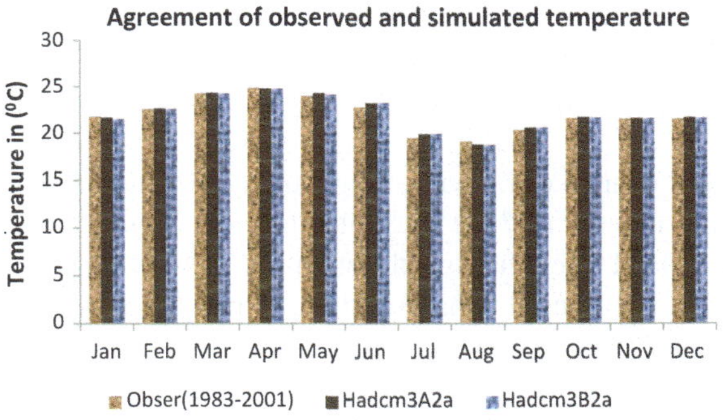

Fig. 5 Comparison against observed and model simulated temperature

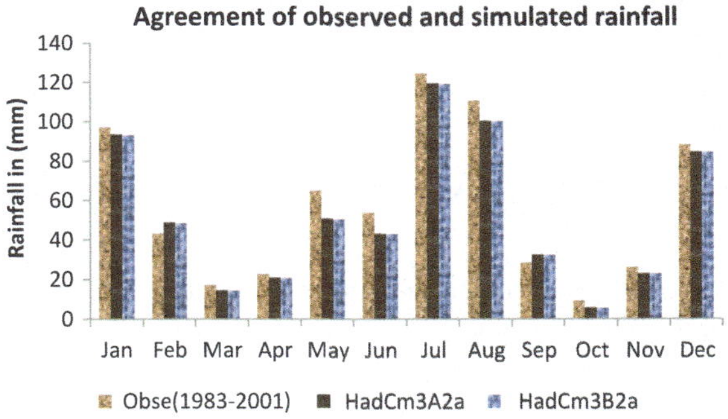

Fig. 6 Comparison against observed and model simulated rainfall

3.3.1 Projected Change in Monthly Minimum and Maximum Temperature

According to this projection, an increase of maximum temperature throughout the year in all agro-climatic classes is expected in both scenarios. The temperature increment in A2a scenario falls in the range of +0.39–0.90, +0.54–1.26, and +0.55–1.49 °C for near term (2020), midterm (2050) and end of 21st century (2080), respectively. While temperature increase for B2a ranges from +0.38–1.07 °C, +0.50–1.23 °C, and +0.54–1.46 °C for 2020, 2050 and 2080 in the same order (Fig. 7). For both scenarios (A2a and B2a) a uniform increment of temperature was observed. In general the temperature increment from March to May, and September was high.

Similarly, the change of minimum temperature showed an increasing trend in all agro-climatic classes. Generally increasing temperature is expected in the range of +0.10 to +0.5 °C, +0.12 to +0.6 °C, +0.16 to +0.7 °C (A2a senario) and 0.07–0.48 °C, 0.10–0.58 °C and 0.16–0.68 °C (B2a senario) for the three terms (Fig. 8). The expected temperature increment in May is high whereas the increment in October is low.

3.3.2 Projected Change of Rainfall

The change of rainfall in highland agro-climatic class shows an increasing trend in the months of June to September and February while increment in midland and lowland goes through the months of April to September. However, a decreasing trend of rainfall is predicted in the months of January, March, October, November and December in all agro-climatic classes in both scenarios. For example, in midland the predicted increase in rainfall for both scenario (A2a and B2a) in the months of April to September, and in February is expected in the range of +3.1 to 25.66, +1.3 to 28.13 and +2.56 to 31.93% for 2020, 2050 and 2080, respectively and in January, October to December is expected to decrease in the range of −0.2 to −6.6, −1.7 to −11.9 and −2 to −12.54 in the same order (Fig. 9).

3.4 Indigenous Adaptation Strategies to Climate Change

A focus group discussion was organized with selected farmer groups in order to understand their perception about climate change and its impacts on the livestock sector. Farmers responded that they observed rainfall reduction and seasonal shift, and increment in temperature in their district and peasant associations ('Tabias'). Moreover, they indicated that the change in temperature and rainfall is causing

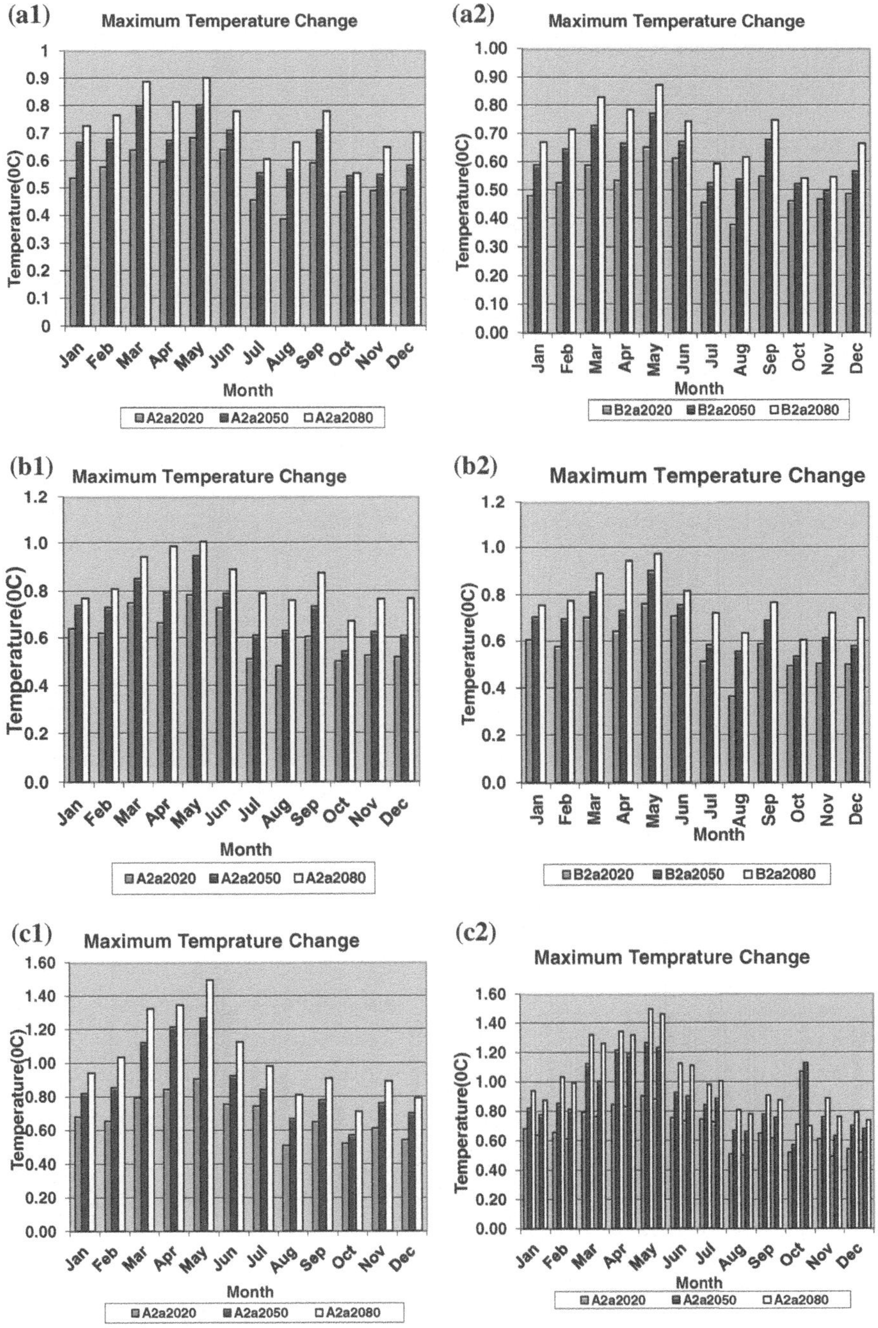

Fig. 7 Change in maximum temperature in the three agro-climatic zones (In the figures **a**, **b** and **c** represent for highland, midland and lowland)

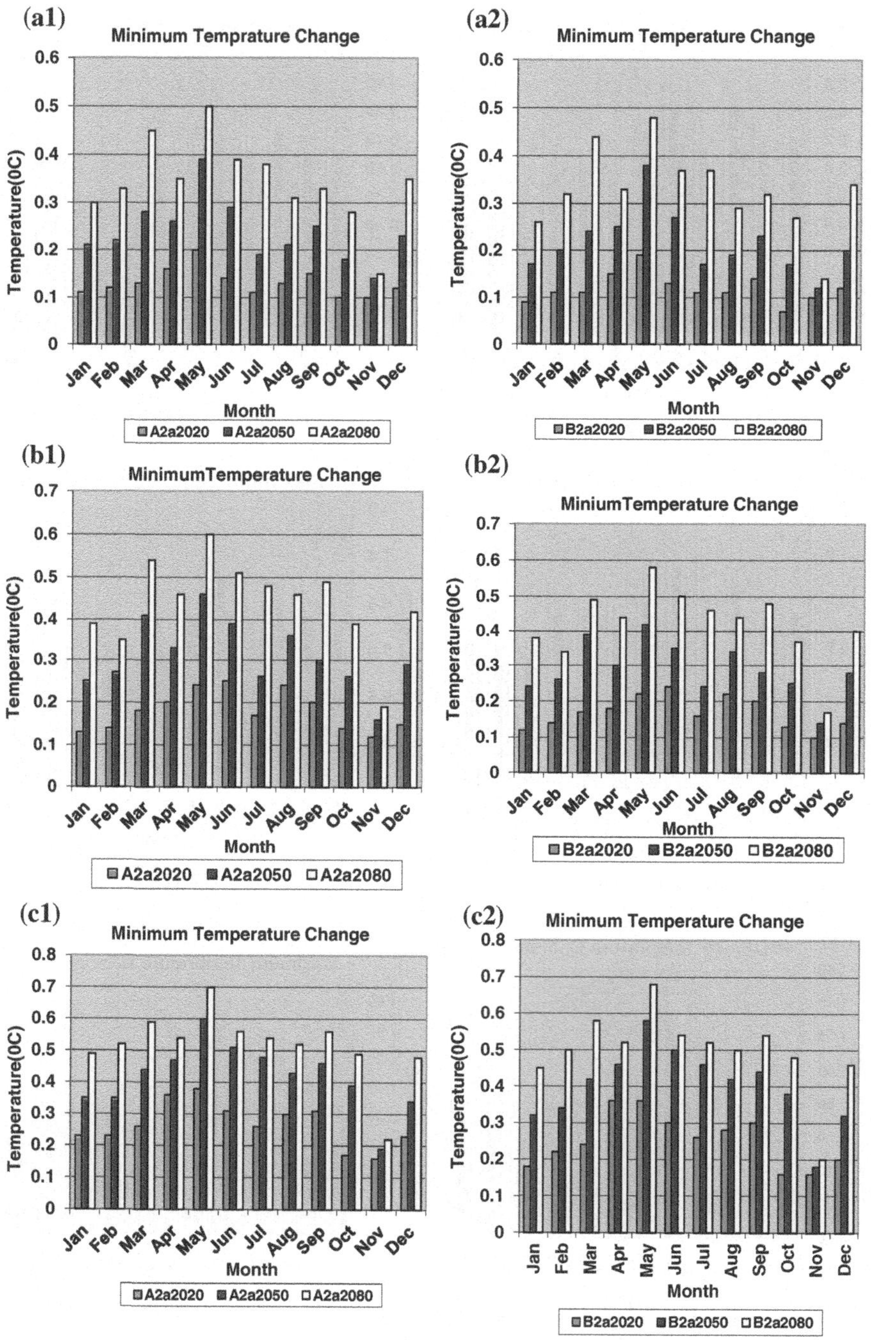

Fig. 8 Change in minimum temperature in three of agro-climatic zones (in the figures **a**, **b** and **c** represent highland, midland and lowland)

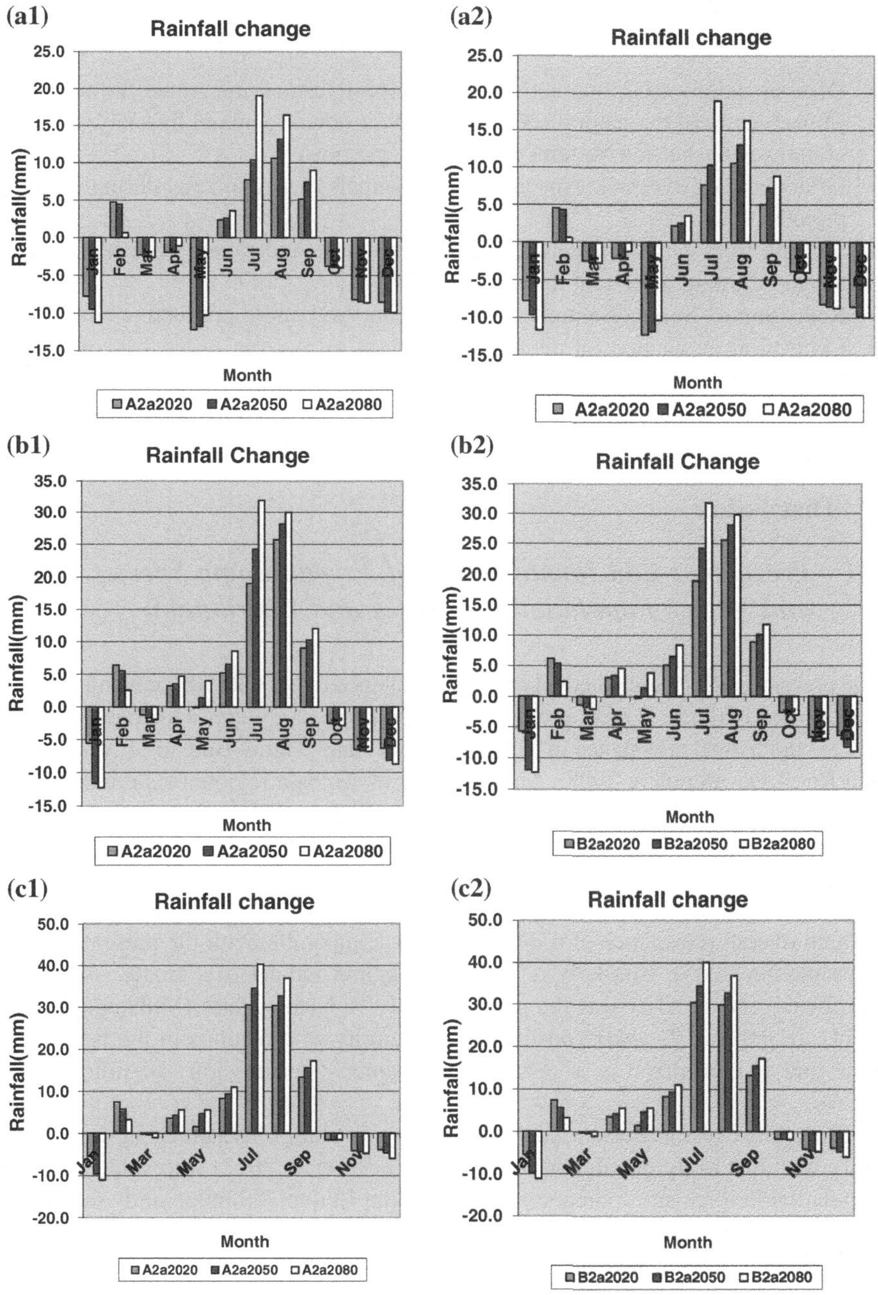

Fig. 9 Change in rainfall in three agro-climatic zones (in the figure **a**, **b** and **c** represent highland, midland and lowland)

impact in their livestock production, productivity and health. The major indigenous climate change adaption strategies by the local communities include:

- Diversification of farming activities (i.e. crop-livestock integrated farming)
- Conservation of the natural ecosystem and introducing mixed livestock farming systems (i.e. stall-fed systems and pasture grazing)
- Destocking of livestock especially large animals like ruminants during drought periods
- Shifting from cattle production to small ruminants and camel production which are more drought tolerant animals
- Reduction of livestock numbers (i.e. keeping only more productive animals)
- Controlling movement of animals from place to place in order to avoid introduction of the disease in the free areas.

4 Discussion

4.1 Prevalence and Identification of Trypanosome Species, and Vector Population Dynamics and Distribution

The present findings indicated an overall prevalence of 8% (40/500) which is higher than earlier findings; 2.66% prevalence in the same study area (Abebayehu et al. 2011), and 5.78% prevalence in Tigray regional state (Tigray Bureau of Agriculture and Rural Development 2008). Similar study in Amhara regional state indicated a prevalence of 6.6% (Cherenet et al. 2004). The higher prevalence of the disease in the present study compared to previously conducted studies could be due to the variation in agro-ecology and the changing climate in recent years that favors the spread of the disease. A study conducted in Sudan identified *T. vivax* in all smears with an overall prevalence of 6–15% which is comparable with the present finding (Husameldin 2005). Similarly a study in Uganda indicated a 15.3% of overall prevalence and *T. vivax* was the predominantly isolated species (Muhanguzi et al. 2014). In general, *T. vivax* constitutes the majority of infections in the tsetse free zone and is considered as a result of mechanical transmission (D'Amico et al. 1996).

The results of the present study on vector population dynamics showed similarity with that of fly population records (Mekonen et al. 2012) who studied the fly population in three agro-climatic zones in central Ethiopia. Similar study in Turkey showed the influence of altitude on the seasonal activity of horse fly (Altunsoy and Kilic 2012). Among the two mechanical transmitter flies of trypanosomosis, the population of stomoxis biting flies was higher than tabanus flies within agro-climatic classes. These findings are in agreement with earlier studies conducted in Ethiopia (Sinshaw et al. 2006; Addisalem and Afere 2012; Mekonen et al. 2012).

A study in Nigeria also showed for a higher number of stomoxis flies compared to tabanus flies (Ahmed et al. 2005).

4.2 Spatial Distribution of Trypanosomosis in Relation to Agro-Climatic Classes

The incidence of the disease in lowland is directly related with altitude of the location where the highest incidence was observed in the lower altitude. Thus, the present finding indicates as subcategory's altitude gets higher, its fly population decreases thereby decreasing the number of diseased cattle. The present finding is in agreement with previous findings where the incidence of trypanosome vector is affected by the environment (Desta et al. 2013). Due to the changing climate in recent years, the incidence of these biting flies is increasing in the mid-highland and highland areas. This in turn facilitates the transmission of vector borne diseases like trypanosomosis (van den Bossche and Coetzer 2008; Tilahun et al. 2014). Variation in the prevalence of trypanosomosis in different agro-ecologic zones has also been reported by other researchers (Abdalla et al. 2008; Yacob et al. 2010). Research finding in Nigeria showed the negative influence of altitude on African Animal Trypanosomosis (AAT) where the risk of an animal acquiring AAT at 1350 m altitude is 0.35 times less than 800 m (Majekodunmi et al. 2013). Given to Ethiopian condition lowland areas are suitable for livestock production due to the availability of pasture and water. However, due to the higher prevalence of Trypanosomosis the livestock production in these areas is seriously challenged (Abebe 2005).

4.3 Relationship of Projected Climate Variables Vis-à-Vis Biting Flies and Diseases

In general, the results of the present finding indicate an increasing trend of temperature and rainfall in all agro-climatic classes in wet season (June–September). Further, there is an increasing trend in midland and lowland in the months of April to May. The findings of the present study are in agreement with the reports of other researchers (Kebede et al. 2013; Abraha 2014). Furthermore, the IPCC report indicates an increasing trend of rainfall in the Eastern African countries (IPCC 2007b) and an increase in global average surface temperature between 1.1 and 6.4 °C (IPCC, 2007a) for the period between 2001 and 2100. Temperature has a relation with fly activity; high population was caught in the lowland (27 °C), midland (25 °C) and highland (21 °C) in the month of December. A previous study also reported that high fly population density Peaks up in wet season (July–September) and rainfall is one factor responsible for breeding of these flies. Flies lay their eggs during rainy season, hatch their egg and finally increase their number (Mekonen et al. 2012). In addition, a study conducted in Nigeria revealed that optimum temperature for the rapid reproduction and distribution of flies lay between 24.1 and 28.6 °C. In addition, this report indicated as initiation of mating occurs at 19–20 °C, hunting and biting by females began at 24 to 25 °C (Ahmed et al. 2005). Moreover, it has been indicated that male pupae storage temperature has effect on the emergence, mating and survival rate of the *Glossina palpalis gambiensis* vectors (Mutika et al. 2014).

Moreover, it is explained that temperature had an effect on production of eggs (optimum temperature range 15.5–18.3 °C) and embryonic development (21–29.4 °C) (Roberts 1980). Likewise the interval between blood-meal and oviposition decrease with increasing temperature. A study in Croatia showed the influence of temperature, rainfall and flood duration on the numbers of adult horse fly population (Mikuška et al. 2012) and a study in Norway indicated high numbers of tabanus flies during the summer season (June–August), which indicates the effect of temperature on flying activity (Bergersen et al. 2004). Similar Studies conducted in Sudan (Abdalla et al. 2005; Rahman 2005) indicated high incidence of trypanosomosis following seasons of high rainfall and when biting flies were abundant. A study in Ethiopia (Sinshaw et al. 2006) also reported higher density of biting flies during the late rainy season which resulted in higher prevalence of *T. vivax.* All the above studies justify the strong link between the population of biting flies and incidence of *T. vivax* infection.

4.4 Indigenous Adaptation Strategies to Climate Change

IPCC (2001) described adaptation to climate change as adjustment in natural or human systems in response to actual or expected climatic stimuli and their effects which moderates harm or exploits beneficial opportunities. The climate change adaptation practices exercised by the local community in the study area are also within the recommended adaptation strategies by several experts and organizations (FAO 2007; Sidahmed 2008; Thornton et al. 2008).

5 Conclusion and Recommendations

T. vivax was the only trypanosome species identified in the study area. This is also evidenced by the existence of biting flies such as *Stomoxis* and *Tabanus* in all agro-climatic classes which are the vectors for the mechanically transmitted trypanosomosis. There was a significant variation in the distribution of the biting flies in the three agro-climatic zones which is linked with temperature and rainfall difference. The climate model prediction also showed for the influence of the climate variables (temperature and rainfall) on the reproduction and biting fly activity thereby influencing the disease distribution. The predicted increase in temperature in all the months is important for shortening the lifecycle, increasing fly population and flying activity whereas the increase in rainfall gives suitable environment for larva development. The present study concludes that there will be high incidence of trypanosomias infection in the lowland due to the increasing temperature and rainfall. Moreover, there could be a higher risk of disease incidence in the midland where the current disease prevalence is lower and highland where the disease prevalence is zero unless climate change adaptation and mitigation measures are considered. This finding also gives an indication for the possible increase in the incidence of vector-borne diseases into the highland areas of the region due to migration of biting flies. There is a need to assess and implement appropriate climate change adaptation and mitigation measures to reduce the incidence and control diseases like trypanosomosis. Moreover, community engagement and consideration of the indigenous knowledge on climate change adaptation is important in designing integrated disease control strategy.

Acknowledgements This research project received financial support from Open Society Foundation Project (OSF) at Mekelle University. Therefore, the researchers would like to extend their appreciation and thank to the OSF project coordinators for the financial support. Institute of Climate and Society of Mekelle University and Tigray Bureau of Agriculture and Rural Development are also acknowledged for their material and technical support. We are also thankful to Ethiopian Meteorological Agency for provision of climate data, and Tselemti district veterinary officers for their assistance.

References

Abdalla, M., Siham, E. S., & Amet, O. B. (2005). Trypanosoma vivax infection in Sudanese cattle in central Sudan. *Journal of Animal and Veterinary Advances, 4,* 945–948.

Abdalla, M. A., Amel, O. B., & Seham, E. S. (2008). Seasonal prevalence of bovine trypanosomosis in Abubugar district, central Sudan, Sudan. *Journal of Veterinary Science and Animal Husbandry, 47,* 1–5.

Abebayehu, T., Eset, H., & Berhanu, M. (2011). Mechanical transmitted Bovine trypanosomosis in Tselemti Wereda, North West Tigray, Ethiopia. *Agricultural Journal, 6,* 10–13.

Abebe, G. (2005). Trypanosomosis in Ethiopia: Review article. *Ethiopian Journal of Biological Sciences, 4,* 75–121.

Abraha, A. A. (2014). *Modeling hydrological responses to changes in land cover and climate in Geba river Basin, Northern Ethiopia.* Belgium: Institute of Geographic Sciences and Physical Geography.

Addisalem, H. B., & Afere, C. A. (2012). Prevalence of Bovine trypanosomosis in Addisamba and Amarit District of West Gojjam Zone, Amhara Regional State. *American Journal Scientific Research, 7,* 12–117.

Ahmed, A. B., Okiwelu, S. N., & Samdi, S. M. (2005). Species diversity, abundance and seasonal occurrence of some biting flies in Southern Kaduna, Nigeria. *African Journal of Biological Research, 8,* 113–118.

Altunsoy, F., & Kilic, A. Y. (2012). Seasonal abundance of horse fly (Diptera: Tabanidae) in Western Anatolia. *Journal of Entomological Research Society, 14,* 95–105.

Bergersen, R., Straumfors, P., & Nilssen, A. C. (2004). The distribution of horse flies (Diptera: Tabanidae) in North Norway. *Norwegian Journal of Entomology, 51,* 3–26.

Cherenet, T., Sani, R. A., Panandam, J. M., et al. (2004). Seasonal prevalence of bovine trypanosomosis in a tsetse-infested zone and a tsetse-free zone of the Amhara Region, north-west Ethiopia. *Onderstepoort Journal of Veterinary Research, 71,* 307–312. doi:10.4102/ojvr.v71i4.250

CICS, Canadian Institute for Climate Studies. (2013). *Predictor variables used for SDSM model input.* http://www.cics.uvic.ca/scenarios/sdsm/select.cgi. Accessed May 17, 2013.

CLSI. (2000). *Laboratory diagnosis of blood-borne parasitic diseases: Approved guideline.* CLSI document M15-A. PA: Wayne.

CSA. (2012). *Federal democratic Republic of Ethiopia agricultural survey data: Report on Livestock and Livestock characteristics.* Ethiopia: Addis Ababa.

D'Amico, F., Gouteux, J. P., Le Gall, F., & Cuisance, D. (1996). Are stable flies (Diptera: Stomoxyinae) vectors of Trypanosoma vivax in the Central African Republic? *Veterinary Research, 27,* 161–170.

Desquesnes, M., Biteau-Coroller, F., Bouyer, J., et al. (2009). Development of a mathematical model for mechanical transmission of trypanosomes and other pathogens of cattle transmitted by tabanids. *International Journal of Parasitology, 39,* 333–346.

Desquesnes, M., & Dia, M. L. (2003). Trypanosoma vivax: Mechanical transmission in cattle by one of the most common African tabanids, Atylotus agrestis. *Experimental Parasitology, 103,* 35–43.

Desta, M., Menkir, S., & Kebede, A. (2013). The study on tsetse fly (Glossina species) and their role in the trypanosome infection rate in Birbir valley, Baro Akobo River system, western Ethiopia. *Journal of Veterinary Medicine and Animal Health, 5,* 186–194.

FAO. (2007). *Adaptation to climate change in agriculture, forestry, and fisheries: Perspective, framework and priorities.* Rome.

Garcia, L. S. (2007). *Diagnostic medical parasitology* (5th ed.). Washington, DC: ASM Press.

Gilles, J., David, J. F., & Duvallet, G. (2005). Temperature effects on development and survival of two stable flies, Stomoxys calcitrans and Stomoxys niger (Diptera: Muscidae), in La Réunion Island. *Journal of Medical Entomology, 42,* 260–265.

Holmes, P. H., Eisler, M. C., & Geerts, S. (2004). Current chemotherapy of animal trypanosomiasis. In I. Maulidn & P. H. M. M. Holmes (Eds.), *The trypanosomiasis* (pp. 431–444). UK: CABI International Wallingford.

Husameldin, O. E. (2005). *Prevalence and ranking of Bovine trypanosomiasis in unity state, Sudan by participatory epidemiological, clinical and laboratory testing.* University of Khartoum.

IPCC. (2001). *Climate change 2001: The scientific basis.* Cambridge: Cambridge University Press.

IPCC. (2007a). Climate change 2007: Impacts, adaptation and vulnerability: Working group II contribution to the Fourth Assessment Report of the IPCC Intergovernmental Panel on Climate Change. *Assessment, 1,* 976. doi:10.2134/jeq2008.0015br

IPCC. (2007b). *Climate change 2007—The physical science basis: Working group I contribution to the fourth assessment report of the IPCC (Climate Change 2007).* Cambridge University Press.

Kebede, A., Diekkrüger, B., & Semu, A. M. (2013). An assessment of temperature and rainfall change projections using a regional and a global climate model for the Baro-Akobo Basin, Nile Basin, Ethiopia. *Journal of earth Science and Climate Change, 4,* 1–12.

Majekodunmi, A. O., Fajinmi, A., Dongkum, C., et al. (2013). A longitudinal survey of African animal trypanosomiasis in domestic cattle on the Jos Plateau, Nigeria: Prevalence, distribution and risk factors. *Parasites and Vectors, 6,* 239. doi:10.1186/1756-3305-6-239

Mekonen, A., Leake, D., & Getachew, G. (2012). Distribution, seasonality and relative abundance of stomoxys flies in selected districts of central Ethiopia. *World Applied Science Journal, 19,* 998–1002.

Mikuška, A., Krčmar, S., Radović, A., & Mikuška, T. (2012). The influence of temperature, precipitation and floods on the development of horse fly populations (TABANIDAE) in the alluvial habitats of the Danube River In Croatia. *Polish Journal of Ecology, 60,* 395–406.

Ministry of Agriculture. (2005). *Ministry of agriculture monthly health status report: Ministry of agriculture and rural development, veterinary series, epidemiology unit*. Addis Ababa, Ethiopia: MEGA Printing Press.

Muhanguzi, D., Picozzi, K., Hattendorf, J., et al. (2014). The burden and spatial distribution of bovine African trypanosomes in small holder crop- livestock production systems in Tororo District. *Parasites and Vectors, 7,* 1–10. doi:10.1186/s13071-014-0603-6

Mullens, B. A., & Rogers, D. M. (2009) Horse flies and deer flies (Tabanidae) and Muscide flies (Muscidae). In G. Mullen & L. Durden L (Eds.), *Medical and veterinary entomology* (2nd ed., pp. 263–301). Elsevier.

Mutika, G. N., Kabore, I., Parker, A. G., & Vreysen, M. J. B. (2014). Storage of male Glossina palpalis gambiensis pupae at low temperature: Effect on emergence, mating and survival. *Parasites and Vectors, 7,* 465. doi:10.1186/s13071-014-0465-y

Philip, W., & Ralph, L. (2009). *Downscaled climate change data from the HADCM3 and ECHAM5 models on Rainfall and temperature for Ethiopia and Kenya report.*

Rahman, A. H. A. (2005). Observations on the trypanosomosis problem outside the tsetse belts of Sudan. *Revue Scientifique et Technique, 24,* 965–972.

Roberts, R. H. (1980). The effect of temperature on the duration of oogenesis and embryonic development in Tabanidae (Diptera). *Journal of Medical Entomology, 17,* 8–14.

Sean, M., Sourva, S., Kyle, W. T., & Holly, V. (2012). Predicting the effect of climate change on African trypanosomiasis: Integrating epidemiology with parasite and vector biology. *Journal of Research Society, 9,* 817–830.

Selby, R., Bardosh, K., Picozzi, K., et al. (2013). Cattle movements and trypanosomes: Restocking efforts and the spread of Trypanosoma brucei rhodesiense sleeping sickness in post-conflict Uganda. *Parasites and Vectors, 6,* 1. doi:10.1186/1756-3305-6-281

Sidahmed, A. (2008). Livestock and climate change: Coping and risk management strategies for a sustainable future. In *Livestock and global climate change conference proceeding*. Tunisia.

Sinshaw, A., Abebe, G., Desquesnes, M., & Yoni, W. (2006). Biting flies and Trypanosoma vivax infection in three highland districts bordering lake Tana, Ethiopia. *Veterinary Parasitology Journal, 142,* 35–46.

Tabachnick, W. J. (2010). Challenges in predicting climate and environmental effects on vector-borne disease episystems in a changing world, Florida. *Journal of Experimental Biology, 213,* 946–954.

Thornton, P., Herrero, M., Freeman, A., et al. (2008). *Vulnerability, climate change and Livestock —Research opportunities and challenges for poverty alleviation*. Kenya: Nairobi.

Thrusfield, M. (2007). *Veterinary epidemiology* (3rd ed.). Singapore: Black Well Publishing.

Tigray Bureau of Agriculture and Rural Development. (2008). *Prevalence study of trypanosomosis in Western and Northwestern part of Tigray*. Tigray, Ethiopia: Mekelle.

Tilahun, Z., Jiregna, D., Solomon, K., et al. (2014). Prevalence of Bovine Trypanosomosis, its vector density and distribution in Dale Sadi District, Kellem Wollega Zone, Ethiopia. *Acta Parasitologica Globalis, 5,* 107–114.

Tselemti District Office of Agriculture and Rural Development. (2013). *Tselemti district agriculture and rural development office annual report*. Tigray, Ethiopia: Tselemti.

van den Bossche, P., & Coetzer, J. A. W. (2008). Climate change and animal health in Africa. *Revue Scientifique et Technique, 27,* 551–562.

Van den Bossche, P., & Delespaux, V. (2011). Options for the control of tsetse-transmitted livestock trypanosomosis. An epidemiological perspective. *Veterinary Parasitology, 181,* 37–42.

Yacob, H., Efrem, D., Basu, A., & Hagos, A. (2010). Bovine trypanosomosis in Gimbi district of Western Oromia, Ethiopia. *Animal Biology, 60,* 123–131.

Promoting Climate Change Adaptation in Developing Countries: The Urban Planning Opportunities in Resilience Building

Davidson S.A. Alaci

1 Introduction

Climate change and urbanization aptly manifest the nexus of environment and development. The duo is among the most remarkable phenomena in the history of mankind. The relationship between the two represents significant socio-ecological issues. On one hand is the rise in urban form, as cities have become the defining ecological phenomenon of the 21st century, and on the other hand, climate change has assumed a real threat to the socio-ecological sustainability of urban areas (Satterthwaite et al. 2009).

Climate change according to the Intergovernmental Panel on Climate Change (2007) is change in the state of the climate (*climate itself being the average weather condition of a place over a considerably long period usually 30–35, years or more*) identifiable by changes in the mean and/or the variability of its properties and persisting for an extended period—a decade or longer. Some of the most visible of such properties are climatic elements of rainfall and temperature, which changes often manifest in the form of thermal heat, urban heat island, drought, excessive rainfall, flooding among many others. The United Nations Framework Convention on Climate Change (2013) corroborated further that climate change is a change of climate due directly or indirectly to anthropogenic activities that alter the composition of the global atmosphere and which is in addition to natural climate variability observed over comparable time periods. Apart from agriculture, the environment in general is an area of key impact and the urban environment is highly vulnerable, because it epitomizes intensification of anthropogenic activities.

The consensus in most literature is that climate change impact on developing countries will be severe. Several studies and reports have also exonerated Africa as

D.S.A. Alaci (✉)
Department of Urban and Regional Planning, Faculty of Environmental Science, Benue State University, P.M.B. 102119 Makurdi, Benue State, Nigeria
e-mail: dralacidsa@gmail.com

W. Leal Filho et al. (eds.), *Climate Change Adaptation in Africa*,
Climate Change Management, DOI 10.1007/978-3-319-49520-0_20

a culprit in Green House Gas (GHG) emission; one plausible determinant of climate change, but nowhere is as much vulnerable to impact as Africa. This is because in most instances on the continent, concern for climatic issues are not made key components of policies issues (UNDP 2006; Ogboi 2012). In addition there is the overwhelming dependence of the economies on natural resources, low incomes and an inherent low adaptive capacity (Zakieldeen 2009; NISER 2010). Satterthwaite (2008) reported that, most cities in Africa, Asia and Latin America and the Caribbean will experience more heat waves in the coming years. The impact of climate change is epitomized by changes in environmental attributes such as rising temperatures and heat waves, flooding, drought, among others. Urban vulnerability to climate change is expected to increase in the years to come as both urbanisation and climate change are not abating. According to the Intergovernmental Panel on Climate Change (2012), climate change is expected to increase the incidence of extreme weather events through the altering of climatic conditions, inducing greater climatic variability and increasing extreme weather events like floods, drought, rising temperature among others. Urban vulnerability to climate change is interpreted to mean the degree to which an urban system is susceptible to the adverse effects of climate change or is unable to cope with it or adapt to it.

There are different dimensions to the effects of climate change and adaptation approach. According to Iliya et al. (2012), some of the ravaging effects of climate change are in land degradation, noting that increasing dryness do not only affect livelihood but would require interventions that are locally based and sustainable. Earlier Osman-Elasha (2009) had explored the link between climate change and sustainable development in terms of vulnerability, adaptation and adaptive capacity. It submitted that, since climate change is a constraint to development, and sustainable development is a key to capacities for mitigation and adaptation, sustainable development and climate change should be addressed together.

The need for proactive measures from government and civil society from developing countries and have been stressed by Satterthwaite et al. (2009). They argued that the lives and livelihood of hundreds of millions of people will be affected by what is done and or what is not done in low income urban areas with respect to climate change over the next five to ten years. Writing on the theme 'Adapting to Climate Change in Urban Areas: The Possibilities and Constraints in Low- and Middle-income Nations'. It posits that while the need for municipal government to act is well taken in the developed world, the same is not happening in low income countries. In fact, many African governments are ill-prepared to cope with the negative consequences of climate variability despite firm commitments at various forums (Kula et al. 2013). Therefore there is the need to rapidly build adaptive capacity and to develop a stronger evidence base for local adaptation strategies.

Consequently, Efe and Eyefia (2014) recommended the need for a more proactive urban planning with emphasis on green city and green roof options to curb the rising incidence of urban heat island. The study recognized the occurrence of urban heat island in Benin-city, Nigeria and seeks urban planning interventions. In terms of coping strategies, Gyampoh et al. (2009) had earlier identified strategies by rural communities in the basin of the River Ofin in Ghana to cope with climate change. One such strategy is creating awareness of the effects of deforestation around water bodies. Caprì et al. (2015) situates Green walking networks for climate change adaptation as a tool that can support planning and design of walking networks. The study adopted an integrated approach, where transport and land-use planning concepts were combined with outdoor thermal comfort and network accessibility. Both Kula et al. (2013) and Gyampoh et al. (2009) focused on land use and green infrastructure, as key in climate change intervention. In fact, Kula et al. (2013) established a clear link between the network of urban green and climate change mitigation and recommended an imaginative Green infrastructure programme. In a study of the role of green infrastructure in the mitigation of climate change in the urban realm, Mell and Davies (2009) suggested an integrative approach to landscape management. Furthermore Gill et al. (2007) discussed how in a changing climate, the functionality provided by urban green space becomes increasingly important. It examined the potential of green infrastructure in adapting cities to climate change, concluding that green space needs to be strategically planned to ensure that their functionality is properly understood and conserved. This is fundamental to ecologically friendly and sustainable development in urban areas.

The connection between urban agglomerations and changing climate resulting from anthropogenic activities notwithstanding, not many studies on climate change do focus on urban planning leverages in resilience building. Previous studies in this direction have either been largely silent about how town planning operations can contribute to climate change adaptation or have not considered it to be of significant research focus. This study is a departure from the traditional issues of conceptualization as was done by Olatunde and Alaci (2010, 2012); Ogboi (2012), even when mitigations and adaptation are focused, in the case of Edorado et al. (2011), Eke (2011) and Anselm et al. (2011) and impact studies Ademola (2011) and Adefila (2011) it does not cascade to town planning operations, especially within the ambit of physical planning. In instances where urban planning had been espoused as Caprì et al. (2015), it was not in the perspective of Africa and Kula et al. (2013) was largely a clarion call to duty. Thus a clear research gap exists, needing studies that clearly explain urban planning opportunities in the context of climate change resilience building. This is the background setting of this research that adopts Lokoja, Kogi State, Nigeria as a case study.

Lokoja is medium-sized urban centre in Nigeria. It is both the administrative and commercial capital of Kogi state. Lokoja is located between Latitude. 7° 45′N–7° 51′N and Longitude 6° 41′E–6° 46′E., at the confluence of Rivers Niger and Benue in Nigeria (Fig. 1).

The original settlers in Lokoja were the Bassa-Nges and Oworos. The Bassa-Nges arrived around 1760 and the Oworos later in 1831 (Akamisoko 2002). It was about this time that the missionaries started arriving Lokoja. The other indeginous tribes found in Lokoja are the several Nupe language groups; Kakanda, Kupa, Ganagana and Egan. Since assuming metropolitan status from pre-independence days, Lokoja is now home to many Nigerian ethnic groups Igala,

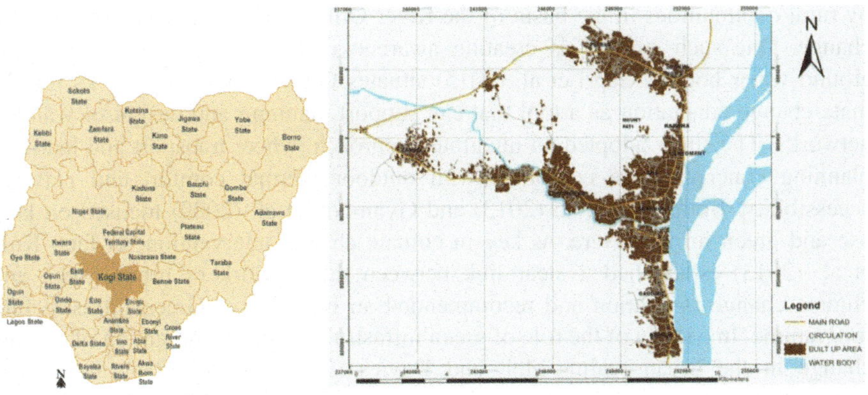

Fig. 1 Location map of the study area: (Kogi state in the context of Nigeria and Lokoja showing built up area and circulation. *Source* MoHUD Lokoja

Hausa, Egbura, Yoruba, Igbo, Tiv, Idoma, just to mention but few. The Yorubas have been a dominant group since the days of Kwara state while the Igalas have also become a dominant group following the creation of Kogi state in 1991 and the attendant migration. Lokoja is therefore cosmopolitan in nature with an estimated population of over 127,139 (Lokoja Master Plan 2009).

The climate of the study area falls into the koppen-Aw climatic group that is, it is the warm continental type. Rainfall starts in the month of March/April and it extends to October/November, often with a short break in August. The mean annual temperature scarcely fall below 30.7 °C, February and March are the hottest months, Ifatimehin et al. (2010). Several environmental attributes, events and changes in Lokoja are suggestive of the impact of climate change. The level of discomfort arising from heat wave as is openly discussed by residents of Lokoja is a serious issue of concern; in fact, incidence of rising urban temperature in Lokoja is clearly observable. These environmental challenges are further reinforce by the increasing anthropogenic activities. Therefore the key research question here is what instrument exists in the urban planning practice that can leverage climate change resilience building.

Studies like this are important because Urban Planning is saddle with task of providing an aesthetically pleasing and functional environment. According to Agbola (1985), urban and regional planning is aimed at providing a more satisfying environment, where residents can reside, work, and pursues other ambitions enhancing human dignity and leading to the attainment of a richer and more satisfied life. It is in line with this aim, that development control applies various instrument of control in urban development. Existing instruments in Nigeria including zoning, building line regulation, density control, building height regulation and type of material for construction, do not address the task of building urban resilience to climate change. Yet, Dodman et al. (2013) noted, the urban area with its high population density are areas of vulnerability with high impact.

Additional and improved knowledge is thus required to reduce the consequences of risk associated with climate change especially in low and middle income countries. In Africa studies on urban resilience to climate change are important because, while the continent faces the greatest environmental, health and poverty challenges, it has the least capacity to deal with them (UNEP 2002). Climate Change Adaptation planning promotes resilience in the face of hazards. It is concerned with reducing communities' vulnerabilities and increasing resilience. Of particular interest is the fact that planning in Nigeria appears to have very little to offer in terms of cutting down the GHGs emissions, yet the effect occurs.

Therefore, reducing vulnerabilities and exposure to natural hazards including climate change is critical to achieving sustainable development in the context of Sustainable Development Goals (SDGs). Climate conditions sustain human life; they provide elements which are necessary for the reproduction of labor power bought as a commodity by capital. Consequently, the intensification of climate change is expected to have adverse effects on capitalist firms and national economies which will (at least in part) register as increases in costs, values, and prices, resulting in changes in profits, rents, and wages (Vlachou and Konstantinidis 2010). This will ultimately affect quality of life because urban areas are the nexus of capitalism and national economies.

In the light of these issues, the goal of this study is to examine the operational opportunities in urban planning that can build resilience to climate change impact in Lokoja, Nigeria. This study limits itself analyzing how urban planning instrument can and should address climate change adaptation and resilience building. Scientific analysis of data to establish the facts of climate change, including analysis of meteorological data will not form part of the study, because this study is premised on the existing facts/knowledge on climate change impact as contained literature. The study hypothesize that temperature values do not show significant differences across locations differentiated on the basis of ecosystem elements as climate moderating ingredient.

2 Methodology

This study adopted the survey design. Primary and secondary data sources were used. The primary data were derived through a structured questionnaire and interviews, while the secondary data were elicited from reports, manuals, journals and excerpts from the Nigerian Meteorological Agency (Nimet). Rainfall and temperature data was gathered from published works for a 30-year period (1981–2010). In order to explain the pattern of climate change, the data was disaggregated into decades and compared over the period of the record. Other secondary data sources include government documents such as reports and manual including the interim development order of the state. The reports and manuals were obtained from the Ministry of Housing, Land and Urban Development (MoHLUD) and the Kogi State Town Planning and Development Board (KOTPDB), Information derived from these sources helped in assessing instruments within the realm of urban planning that can leverage climate change resilience building.

The questionnaire was structured into 3 sections of 20 questions of which eight were open-ended and the rest were closed ended questions. Multi-stage sampling technique enabled 430 respondents spread equally across 10 selected neighbourhoods to be sampled out of a total of 21,190 households in the study area. The sampling methods for selecting the neighbourhoods and respondents (household heads) were stratified and simple random sampling techniques respectively. The interviews used purposive sampling to target two Directors of relevant public agencies in Kogi State namely: Ministry of Lands, Housing and Urban Development and the Town Planning Development Board (TPDB). Ambient air temperatures were measured using thermometers at ten (10) designated locations. The ten neighbourhoods were further divided into two groups of five each. In one group temperature records were taken in areas with outstanding green/tree cover (defined as community of trees), there were five (5) of such. In the second set of neighbourhoods, temperature values were taken from locations that clearly contrast with the first set (i.e. characteristically devoid of trees). The temperature of the ambient air was taken continuously for three months beginning in the last week of April and ending in the last week of July. Convenient structures like wooden stands were identified and on such structures the thermometer was hung at height of between 1.5 and 2.0 m above the ground level. This is to reduce the impact of direct ground radiation on the reading. Temperature readings were done at 7 a.m. and 2 p.m., and the average was recorded for each day. The duration of 12 weeks was considered long enough for a reliable trend analysis between locations with two different sets of environmental conditions and the choice of the time was to be able to account for the extremely high and low temperature periods.

Data obtained from respondents is summarized and presented using frequency tables, percentages and averages because they present actual facts devoid of ambiguity. Secondary temperature and rainfall data were grouped into a 3 decades period and on that basis deduction were made. Relevant quantitative analysis or statistical test is also used to further elucidate on the data set. Thus, data analysis is a combination of qualitative and quantitative techniques upon which inferences were drawn. The temperature figures obtained in particular is presented on a table, while Student t-test and analysis of variance (ANOVA) statistics is used to determine the level of spatial variation. Therefore, the Student t-test and analysis of variance ANOVA statistics are relied upon to test and analyses the hypothesis earlier formulated. The student t-test is a parametric test statistics for testing hypothesis when data are in ratio or interval. The utility of student t-test lies in its ability for testing hypothesis with small samples ($n < 30$) and large samples ($n > 30$). The student t-test is an appropriate test for establishing the relationship between the differences between the mean of samples, especially in determining if two sets of data are significantly different from each other. Student t-test is suitable for one mean value or for comparing two. The formula for t-test is given as

$$t = \frac{(x\bar{a} - x\bar{b})}{\sqrt{\frac{o' a^2 + o' b^2}{na + nb}}}$$

Student t-test is well known for studies involving changes and impact. For example Franklin et al. (2014) applied the student t-test on a study on the empirical evidence of climate change effects on rice production in the northern region of Ghana. Similarly Arun et al. (2014) relied on the student t-test to examine the Impact of climate change on rainfall over Mumbai using Distribution-based Scaling of Global Climate Model projections. To further strengthen the result, the data was further subjected to ANOVA. This is so because ANOVA in particular is a form of statistic heavily used in the analysis of experimental data. Analysis of variance ANOVA statistics is used in the analysis of comparative experiments, those in which only the difference in outcomes is of interest. The formula for one-way ANOVA test is:

$$SSW = SST - SSB$$

$$SST = \sum X^2 - \frac{(\sum X)^2}{N}$$

$$SSB = \frac{(\sum X_1)^2}{N} \times \frac{(\sum X_2)^2}{N} \times \frac{(\sum X_3)^2}{N} \times \frac{(\sum X_n)^2}{N}$$

The entire statistical manipulation was conducted within the environment of Statistical Product for Service Solution (SPSS). The use of the SPSS is because it helps to make a research work more scientific and result replicable and reliable. Both the student t-test statistics and analysis of variance were used to analyse the data at 0.5% significant level.

3 Result and Analysis

3.1 Climate Change Effects in Lokoja

The Evidence of Climate Change based on the experience of some environmental attributes in Lokojais examined using surrogates of temperature, rainfall and flood events. The general observation of respondents is that the pattern of climate change as expressed in such elements as flood events, ambient air temperature as well as rainfall provides evidence of climate change (Table 1).

While about seventy six (76%) percent of the respondents feel strongly that flood events have been on the increase, some four hundred and six (406) representing 94.4% agreed that ambient air temperature (expressed locally as heat) is on the increase. However, a slight departure from this general trend was observed in the

Table 1 Climate change based on some environmental attributes in the last 5–10 years

(a) Increasing flood event				
Variable	Yes	No	Don't know	Total
Freq	330	57	43	430
%	76.7	13.2	10	100
(b) Rising ambient air temperature (heat)				
Variable	Yes	No	Don't know	Total
Freq	406	13	11	430
%	94.4	03	2.6	100
(c) Increasing rainfall				
Variable	Yes	No	Don't know	Total
Freq	201	167	62	430
%	46.7	38.8	14.4	100

Source Field survey, 2015

Table 2 Mean decadal temperature values for Lokoja

S/n	Decade	Mean value
1	1981–1990	28.03
2	1991–2000	28.19
3	2001–2010	28.08

Source Adapted from Audu 2012

case of rainfall element. In its case, although majority of the respondents' i.e. 46.7% agreed that rainfall in the last five to ten years has been on the increase, an equally high percentage of 38.8% do not think so, while some 14.4% claimed not to know whether rainfall is on the increase or not. This position notwithstanding, the overall implication of the information on the table is that climate change evidence registers strongly. This is very clear from the assertion on temperature and flood event and to a lesser extent rainfall. To further buttress this fact, data on temperature variation for a 30 years period is presented on Tables 2 and 3.

Table 2 is indicative of a changing trend in the temperature values of the study area. Although, the increasing values fluctuate, the general pattern is that of increasing trend, for example the difference between the first and second decade is +0.16 °C, between the second and the third is −0.11 °C and between the third and first decade is +0.5 °C, so that in the last three decades, overall it's been some form of increase. This trend is indicative of climate change, the pattern of increase is expected to progress in the years to come and this will firmly establish the case of climate change. The situation of rainfall is similar. This is derived from the next two tables; Table 4, Pattern of Rainfall (mm), 1981–2010 (30 years) and Table 5.6 on Annual Rainfall, Mean and Rain days, 1981–2010 (30 years) as adapted from Audu (2012).

In Lokoja, rainfall amount tend towards increasing pattern (Table 4), the first pointer is the clear disappearance of the August break. Rainfall amount in the month of August have not been particularly low, in fact, going by the average amount of rainfall over the last 30 years, the month of August has remained the

Table 3 Mean decadal temperature values differences for Lokoja

S/n	Decade	Value difference °(C)
1	2nd decade–1st decade	0.16
2	2nd decade–3rd decade	0.11
3	3rd decade–1th decade	0.05

Source Adapted from Audu 2012

Table 4 Pattern of rainfall (mm), 1981–2010 (30 years)

Month	Rainfall	Mean
January	33.1	1.1
February	241.2	8.0
March	661.3	22.0
April	3128.7	104.3
May	4698.3	156.6
June	5081.9	169.4
July	5988.5	199.6
August	6120.1	204.0
September	6443.7	214.8
October	4003.7	133.5
November	87.1	2.9
December	17.1	0.6
Total	36,504.7	101.4

third rainy month in a year. Table 5, further demonstrates the fact that climate change is progressively real in terms of rainfall amount. Although the number of rain days varies across the years and decades, the rainfall amount is rather on the increase. Of the top five (5) rainy days, 4 of them are in the last 2 decades i.e. from 1991 to 2000 and 2001 to 2010. Similarly, a review of the top ten (10) rainy days also shows that apart from 1999 with the highest rainfall amount of 1767.1 and average of 147.3, the other leading rainfall years are mainly in the last decade. For example 2006 = 1684.1 (140.3); 2009 = 1653.3 (137.8), 2007 = 1501.4 (125.1) are among the top 5 rainy years in the period under review. Another emerging and interesting pattern in the rainfall values in the last 30 years is the fact that years with the highest rainfall total do not necessarily reflect the years with the highest rainy days. The years 2006 and 2009 have the highest rainfall total in the period under review, but the two years are not the years with the highest number of rainy days. The years 1989, 1991, 1994, 1995 and 2007 are the years with the highest number of rainy days but not the highest rainfall total. The years with fewer rainy days but higher rainfall total are likely to produce more intense rainfall impact including inducing flood incidence. The state of affairs is enough to induce flood experience.

The synopsis arising from the temperature and rainfall data shows that,the ten-year rainfall readings for the city indicate significant increases in both mean annual rainfall (MAR) and number of rainy days (NRD) records. The MAR and NRD for the 1981–1990, 1991–2000, and 2001–2010 decades are 1135.49 mm (79.4 days), 1251.74 mm (74.3 days) and 1263.26 mm (81.9 days) respectively.

Table 5 Annual rainfall, mean and rain days, 1981–2010 (30 years)

Year	Total rainfall (mm)	Mean (mm)	Rain days
1981	1144.6	95.4	88
1982	804.5	67	53
1983	853.7	71.1	60
1984	1147.7	95.6	80
1985	965.7	80.5	84
1986	1281	106.8	87
1987	1170.8	97	71
1988	1330.9	110.9	88
1989	1519.7	126.6	91
1990	1136.3	94.7	92
1991	1492.7	124.4	98
1992	1083.1	90.3	82
1993	995.6	83	73
1994	1194.9	99.6	93
1995	1291.1	107.6	91
1996	1240.7	103.4	80
1997	1410.5	117.5	83
1998	1031	85.9	80
1999	1767.1	147.3	90
2000	1010.7	84.2	71
2001	1003.8	83.7	77
2002	1276	106.3	90
2003	923.6	77	73
2004	1335.4	111.3	84
2005	939.4	78.3	72
2006	1684.1	140.3	87
2007	1501.4	125.1	98
2008	1239.5	103.3	73
2009	1653.3	137.8	83
2010	1076.1	89.7	82

Source Adapted from Audu (2012)

The ten-year mean annual temperature for 1981–1990, 1991–2000 and 2001–2010 decades are 28.03, 28.19 and 28.08 °C, correspondingly.

Rainfall and temperature are among conventional attributes for drawing conclusion on climate change and climatic variability. Therefore this finding is in harmony with earlier studies that also relied on rainfall and temperature in establishing the case of climate change. See for example Olatunde and Alaci (2012) and Abaje et al. (2012), Umar (2012) all examined the situation of the incidence of climate change and relied on the parameters of monthly rainfall, rainy days and minimum and maximum temperatures and concluded that there are drier conditions in Nigeria as evident from these variables of weather and climate.

3.2 Local Climate Change Resilience Strategies

The strategies used to build local resilience to climate change are a blend of coping, adaptation and mitigation measures. The options proposed were made up of tree planting, wider house windows (ventilation), not paving the surface of homes/offices, ensuring green/plant cover among others. The opinion of respondents is summarized on Table 6.

Observed local coping strategies or approaches to counter climate change effect are all build around temperature and to a lesser extent rainfall. These include life style modification (23.7%), urban greening such as flower/tree planting, green cover and environmental beautification (59%), and building modifications (16.9%) particularly providing extra larger window sizes. See Plate 1.

The pictures of buildings with extra-large window sizes and deliberate tree planting by individuals, here windows are inserted from the second or third block after foundation/base level. This against the convention of inserting/cutting window from the fourth or fifth block as shown on Plate 2, where conventional window on buildings are high above the ground and so wide.

All the existing strategies can leverage town planning operations to the extent of building resilience. The principal strategy is urban greening which revolves around planting trees, ensuring green cover and environmental beautification all emphasizing the productive role of vegetal resources in replenishing earth oxygen. This finding is in line with the conclusion of an earlier study by Gill et al. (2007), who contended that the overall ability for building resilience is strengthened with a strong base in a network of green areas because it's an asset with strong and efficient ability to mitigate and adapt to the rising temperatures and other extreme weather events associated with climate change.

The specific areas of intervention are in; natural cooling effect to mitigate the urban 'heat island' a community of plant cover also provides room for sustainable urban drainage to absorb excess rainfall and reduces the likelihood of flood occurrence. Green spaces are efficient and cost-effective 'soak away' for rain water

Table 6 Approaches to building resilience to climate change

Approach	Frequency	Percentage
Life style modification Increasing the rate of bathing Avoid paving surfaces in homes/houses Use of air conditioner and ceiling fan	163	23.7
Urban greening Planting trees Ensuring green cover Environmental beautification	409	59
Building modification	117	16.9
Extra-large window sizes Multiple room windows	689	100

Source Field survey, 2015

Plate 1 Samples of buildings with extra-large window sizes and deliberate tree planting

Plate 2 Pictures of buildings with conventional window height

and a reservoir for grey water storage. They also provide spaces to grow food using sustainable methods, such as organic cultivation. This contributes to enhancing biodiversity, and provides job and educational opportunities. It will also act as carbon sink; as vegetation is able to reduce the effects of air pollution and to store carbon. It is an attractive, cooler and shaded outdoor area in hotter summers, readily accessible from homes. It also helps to reduce carbon emissions and as a cooler and shaded outdoor area in hotter seasons, green covered areas are attractive and readily accessible from homes. In view of the assertions in the preceding section, a closer look at the situation of environment with trees (ecosystem elements) and environment without trees is contained in the next section.

3.3 Spatial Variation in Temperature Values

In last section, urban greening in general and tree planting in particular was captured as observed climate change resilience building mechanism in Lokoja. The extent to which tree planting and general environmental greening can have effect on the micro climate of sampled locations in the neighborhood is the subject of further analysis. Accordingly data on ambient air temperature collected is presented in Table 7 and compared with Nimet data in Table 8.

Table 7 Mean temperature of sampled neigbourhoods in Lokoja (April–July)

Neigbourhood	Temperature (°C)
1. Ganaja	32.02
2. Gaduma	31.9
3. Kabawa/Galile	32.53
4. Cantonment	32.5
5. Sarikin Noma	32.78
6. Lokongoma	27.25
7. GRA/New Layout	29.74
8. Adankolo	28.98
9. Felele	28.63
10. Otokiti/ZangoDaji	29.32

Source Field survey, 2015

Table 8 Lokoja mean monthly temperature

Month	Temperature (°C)
January	27.9
February	28.85
March	29.3
April	28.75
May	27.55
June	26.55
July	25.9
August	26.95
September	26.15
October	26.75
November	27.1
December	26.45

Source Nimet Station Lokoja 2014

The locations with ecosystem elements like trees and other green cover will normally experience lower temperature values and on the basis of such parameter human comfort would be higher in the ecosystem based neighbourhoods. The average monthly temperature for the whole city compares favourably with the disaggregated values for sampled neighbourhood. Thus while the lowest value for the sampled neighbourhood is 27.25 °C that of the monthly mean obtained from Nmet is 25.9 °C. It is important to note that whatever slight variation observed between the overall monthly temperature values and those obtained from the sampled neighbourhood could have been due to variation in the length of time taking for the recorded observation. The experimented temperature record did not factor in night temperatures which may have lowered the values obtained from Nimet. Notwithstanding, the overall implication of this finding is that a close and positive relationship tend to exist between community of tree elements and ambient air temperature. So that areas in the city characterized with tree clusters would constitute city resilience elements. A good example is the image of some neighbourhoods as seen on Plate 3.

Generally, an environment characterized with trees and general green cover as seen on Plate 3 will have its solar input moderated by the green cover element. This is the situation in the experimented sites of GRA/Newlayout, Lokongma, Adankolo, Felele and Otokiti/Zangodaji where temperature values were all found to be lower than 30 °C. The human comfort or discomfort arising from heat wave would obviously be moderated compared with situation found in Cantonment, Kabawa/Galile neighbourhood as is seen on Plate 4. An environment with little or no vegetation is opens to so many environmental attacks. Unfortunately, due to many reasons as enumerated earlier, the urban landscape of Lokoja is increasingly tending towards a green less environment, as every available space is built-up.

The state of affairs as seen in cantonment and kabawa/Galilie area is the larger image of the city of Lokoja. The environmental implication of this is monumental, especially in the context of climate change impact. Here both rainfall impact and sunshine would be direct and severe, thus situation like this are potent generators of high ambient air temperature and its attendant heat wave. The issue of high temperature is a very common area of human discomfort and a major concern in the study the area. Tree planting is a major instrument used in moderating this impact and as seen earlier, significant temperature variation exist between areas characterized with community of trees and the areas without. The extent to which this can continue to be held as truth is subjected to rigorous scientific manipulation through the testing of hypothesis. This is the subject of the next section.

Data on ambient air temperature shows that value margin vary from 5 °C to as low as 0.02°C with lower values in sites with ecosystem elements. From the t-statistics, a significant difference ($t = 34.515$, $df = 118$, $p < 0.05$) was observed between sites where trees are planted and where trees are not. From the ANOVA statistics a significant difference ($F = 10.249$, $df = 4$, $p < 0.05$) was observed between the average temperature reading across the different sites in the location where trees were not planted. With a multiple range test (i.e. the least significant difference LSD) it was discovered that except for the sites 1 (Ganaja) & 2 (Gaduma), site 3 (Kabawa/Galile) & 4 (Cantonment) and sites 5(SarkinNoma/ Nataco), every other combination of the sites are significantly different. Similarly

Plate 3 Images of parts of GRA/New layout with green cover in Lokoja. *Source* Field survey, 2015

Plate 4 Images of parts of cantonment and Kabawas Galile in Lokoja. *Source* Field survey, 2015

with respect to sites with trees, the ANOVA statistics showed significant difference ($F = 13.546$, $df = 4$, $p < 0.05$) in the average temperature reading, this is not unexpected as tree population was found not to be the same. Therefore, the null hypothesis that "There is no significant difference in temperature variation on the basis of locations with and without ecosystem elements", is hereby rejected. This means that that a close and inverse relationship exists between community of trees (ecosystem elements) and ambient air temperature. This strengthens existing local resilience strategies and planning institutions responses, which are a compendium of urban greening forms.

This position is corroboration with earlier findings and recommendations of similar studies. In fact, tree planting in particular have been advocated by Olatunde and Alaci (2012) and Abaje et al. (2012), Umar (2012) where they called for proactive measures to be put in place to enhance mitigation and adaptation efforts in climate change, particularly measures that ensure the replenishment of the lost vegetal cover, as this could help in averting the effects of increased albedo. Since tree cluster constitute resilience elements, as established by the research hypothesis and confirmed from earlier studies, the next section examines resilience possibilities in the planning instruments available to the MDA.

3.4 Climate Change Adaptation Measures in the Practice of Urban Planning

An analysis of climate change adaptation measures in the practice of urban planning in Lokoja is derived from the review of functions and activities of the MDAs in urban planning and management. Accordingly, the MDAs were represented by the Ministry of Lands, Housing and Urban Development Board (MoLHUD), with core departments as Urban and Regional Planning; while the relevant parastatal is The Kogi Town Planning and Development Board (KOSTPDB). The extent to which these organizations can and are involve in climate change issues is the subject of the next section.

The Department of Urban and Regional Planning is basically responsible for the even and orderly physical development of the State. This is achieved largely through the formulation and implementation of Land Use Planning Policies in the state; Monitoring, preparation and implementation of urban and subject plans; Beautification of towns; Guiding Town Planning Development Board in their operations; Recommending town planning views on plots for the granting of Right of Occupancy as well as assessing and processing requests on plots requiring change of use and Provision of technical assistance and town planning information for adoption by user organisations, students and other individuals. The attributes of climate change adaptation in the operating planning instruments can thus be summarized as. Formulation and implementation of land use planning policies; monitoring, preparation and implementation of urban plans; and city beautification among others.

Pursuant to the goal and objectives of the Department of Urban and Regional Planning, the department has produced a total of one hundred and twenty layout/schemes since the creation of the state in 1991. Majority of these schemes are residential involving seventeen thousand, two hundred and twenty three (17,223) plots. Of the residential layouts, eleven (11) are fully developed government housing estates. In the eleven government housing estates, there are limited provisions of green areas on the layout plan; these were largely limited to their occurrences in land uses designated as setbacks. The status of almost all of such green areas is either been encroached upon or converted to other uses. Most of what exist as green areas is setbacks to river courses and by their location; it is probably the only use to which they can be put. Others are incidental greens or undeveloped plots, and occasionally may serve as avenue for urban agriculture. The implication of this development is that urban greening as a deliberate climate change resilience building opportunity is never factored into the urban plans.

The KOSTPDB was established by the Kogi State Government Edict No. 5 of 1991. The objectives were to plan, promote and secure the orderly control over development and use of land throughout the State. This is carried out with guidelines as contained in the Annual volume of laws of Kogi state and the Interim Development order. These are the documents that provides guide to developers in the state on the board requirement for any physical development approval. The Interim Development Order stipulates clear percentage area developable in the case of both residential plots and other land uses. The summary is contained on Table 9.

Table 9 is a summation of the content of the interim development order as per what ratio of a given plot should be developed. The percentage or ratio developable

Table 9 Interim development order building plot percentage

Ratio of % developable (%)	Density	Remark (%)
33.5	Low density Residential Areas	Silent on the use of 66.5
50.5	Medium Density Residential Areas	Silent on the use of 49.5
60	High Density	Silent on the use of 40
75	Office, Commercial and industrial	Silent on the use of 25

Source Authors Compilation, 2015

ranges from 33.5% in the case of low residential density to 60% in high residential density areas and 75% in other uses like industrial, commercial and institutional uses. Observation shows that developers take the silence on the remainder portion to mean tacit approval for further development. Thus rather than have functional open spaces and green areas within the plot areas, we are rather having more buildings to the extent that some residential plots are developed up to 80–90% (Alaci 2012).

The situation is further compounded as a close scrutiny of the interim development order requirements for building approval is very silent on the issues of urban greening or landscaping. The document contains clear guidelines and requirements for the approval of new development; accordingly, a residential development typically would require a developer to submit to the KOSTPDB the following documents: 4 copies of the building plan (A3 size); a survey plan; either a Site Analysis Report (SAR) or Environmental Impact Assessment Report (EIAR); land title document; drainage plan, in addition for a high rise building/story building/2 floors and above, a developer would also require to submit: an attestation letter from the supervising engineer; soil test report; calculation sheet and structural design sheet. All documents must be endorsed by relevant professionals. The requirements as enumerated have no place for landscaping or any activity that can stimulate the promotion of urban greening whatsoever.

Ordinarily, in a climate change resilience conscious environment, where tree planting have been incorporated in urban planning, the expectation would be for building plan approval and all development for that matter to make provision for urban greening. Such an arrangement would cause physical development to positively affect urban greening, such that number of buildings and tree numbers will positively correlate. And in that way urban planning will be responding to climate change resilience building. In fact, the foregoing present opportunities that can leverage climate change resilience building with clear opportunities to expand the frontiers of urban greening.

3.5 Planning Urban Green Spaces Expansion in Lokoja

In view of the statutory role of the KOSTPDB and the urban and regional planning department of the ministry, this section examines the changes taking place in the green component of the urban space. It is examined in the context of the growth of two anthropogenic urban components (population and Built-up area) in Lokoja versus green area. The two urban anthropogenic elements are central to the growth and expansion of urban centres. This is because urban growth and spatial expansion find expression in the changing dimensions of the population and the built-up area. Table 10, present the changing dimension of the different urban land use from 1987 to 2005.

As shown in Table 10, the year 1987 and 2001, experience relative growth in the town. The human induced land use sectors were increasing, while the naturally

Table 10 Lokoja land use changes and growth rate 1987–2005 (km)

Land use	1987	2001	Amt of change	2005	Amt of change
Bare surface	0.25	2.43	2.18	2.52	0.09
Built-up area	0.34	9.17	8.83	10.31	1.14
Cultivated area	18.22	37.25	19.03	38.18	0.93
Natural vegetation	42.21	10.13	−32.08	7.90	−2.23
Water body	2.8	4.6	1.8	4.40	−0.2

Source Extracted from Alaci et al. (2011)

occurring sectors were declining except for water body. Natural vegetation declined with about 32.08 km in Lokoja between 1987 and 2001. The rather large amount of natural vegetation lost between 1987 and 2001 can easily be accounted for as weak planning instrument and enforcement. This is more so as the period was characterized by the influx of people to the town as a result of the transformation of Lokoja in 1991 to a state capital. The pattern of land-use change continued into the years after 2001. On the average the urban green elements (vegetation) declined at about −4.0 km per annum, but the built-up area increased at about +1.0 km per annum. This development contradicts what the operational objectives of the KOSTPDB and the Urban and Regional Planning department of the ministry ought to have achieved in the area of urban greening if the operational objectives are religiously followed and it is in this context that urban planning can leverage climate change resilience building.

4 Conclusion and Policy Implications

The many fronts of climate change effects—temperature, rainfall and flooding calls for an all-inclusive urban planning response to resilience building. This study has revealed that urban planning can and is responding to climate change adaptation through resilience building measures; unfortunately, urban planners do not see climate change resilience opportunities in urban planning. But opportunities for resilience building appropriately abound. Some of the measures are pseudo coping strategies to the changing circumstance of climate by the people, while others are actual documented procedurals. With the reality of climate change and its impact, the need to make these procedurals sacrosanct and institutionalized other coping strategies is now most imperative; and will form a key urban planning response to climate change resilience building.

The first and most crucial step to take is the review of the interim development order of the state. As it stands the interim development order has stipulated what fraction of allotted plot should be develop, but it is silent on what should be done with the rest. This is an opportunity area for building climate change resilience as developers can be made to turn such areas into green spaces including planting trees. The review should also make residential area landscaping mandatory.

Plate 5 A home with adequate tree presence

One way to do it is to include tree planting and residential area greening as part of the requirement for building plan approval. At the moment tree planting or landscaping it is not part of the requirements for building plan approval.

The example presented on Plate 5 demonstrate clear incorporation of trees in residential apartment and in this way, it will be possible to positively correlate rise in number of buildings with number of trees. This in addition to strengthening other urban greening elements including, making urban greening a mandatory component of urban master plans, making green infrastructure as cardinal part of urban planning and preparing and implementing urban green master plan among others. The urban green master plan, apart from making provision for the conventional urban parks and gardens, the peculiarity of fragile ecosystem would be accounted for, just as ecologically-optimum use of such special areas like wetlands, river courses would be harnessed. In addition, it will make provision for, greening induced and environmentally friendly land uses including urban agriculture. The totality of all these will be the enhancement of the provision for greening induced and environmentally friendly land uses including urban agriculture with climate change resilience abilities and this can be replicate across Africa.

This study addressed itself to the call by Efe and Eyefia (2014) for proactive urban planning studies with emphasis on green city and green roof options to curb the rising incidence of urban heat island in Nigeria. The findings of the study are supported by similar result as can be seen in Caprì et al. (2015) where Green walking networks is situated as a tool for climate change adaptation. The place of urban planning in climate change resilience building is largely under studied especially in this part of the world, but urban planning can and should be explored in the quest for resilience building. Therefore, urban and regional planning research results from universities and research institutes in future should further deepen studies in place making emphasizing on urban greening as an integral part of the right mix in place and non-place urban realm across Africa. This is with a view to eventually making urban green a way of life and thus a climate change resilient Nigerian and African cities.

References

Abaje, I. B, Ati, O. F., & Iguisi, E. O. (2012). Changing climatic scenarios and strategies for drought adaptation and mitigation in the Sudan-Sahelian ecological zone of Nigeria. In M. A. Illiya, M. A. Abdulrahim, & I. M. Dankani, (Eds.), Climate change and sustainable development in Sub-Saharan Africa. A publication of the Department of Geography, Usumanu Dan Fodio University Sokoto-Nigeria.

Adefila J. O. (2011). An assessment of global warming effects on socio-economic development in Sub-Saharan Africa. In *Proceedings of International conference of the Nigerian Meteorological Society* (*NmetS*) 13–17th November, Ahmadu Bello University Zaria.

Ademola, J. S. (2011). Climate change: An assessment of ground water quality of selected borehole in Zaria. In *Proceedings of International Conference of the Nigerian Meteorological Society* (*Nmets*) 13–17 November, Ahmadu Bello University Zaria.

Agbola, T. (1985). The concept, evaluation and role of development control and planning administration in Nigeria. Paper presented at the National Workshop on Planning Administration in Nigeria. NISER/CURP, 24–26 June.

Akamisoko. (2002). Samuel Ajayi Crowther in the Lokoja Area. Ibadan: Safer.

Alaci, D. S. A., Amujabi, F. A., Baba, A. N., & Daniel, O. (2011). Spatial growth assessment with remote sensing data for central Nigeria. In *Journal of Agriculture and Social Sciences.* ISSN Print: 1813-2235; ISSN Online: 1814-960X. http://www.Fspublishers.Org. J. Agric. Soc. Sci. 7:1–6.

Arun, R., Kean, F., Thomas, B., Jonas, O., & Lars, B. (2014). Impact of climate change on rainfall over Mumbai using Distribution-based Scaling of Global Climate Model projections. *Journal of Hydrology: Regional Studies*, 107–128.

Audu, E. B (2012). An Analytical View of Temperature in Lokoja, Kogi State, Nigeria. *International Journal of Science and Technology. IJST Volume* 2 No. 12, December 856–859. http://www.ejournalofsciences.org.

Caprì, S., Ignaccolo, M., Inturri, G., & Le Pira, M. (2015). Green walking networks for climate change adaptation. *Transportation Research Part D.* http://dx.doi.org/10.1016/j.trd.2015.08.005.

Dodman, D., Brown, D., Francis, K., Hardoy, J., Johnson, C., & Satterthwaite, D. (2013). Understanding the nature and scale of urban risk in low-and middle-income countries and its implications for humanitarian preparedness, planning and response. Human Settlements Group International Institute for Environment and Development (IIED) http://pubs.iied.org/10624IIED.html; http://www.dfid.gov.uk/r4d/Output/191454/Default.aspx.

Edorado, C., Sabrina, M., & Tania, M. (2011). Comparing mitigation policies in five large cities: London, New York City, Milan, Mexico City and Bangkok in CITIES AND CLIMATE CHANGE: Responding to an urgent Agenda. In D. Hoornweg, M. Freire, M. J. Lee, P. Tata & B. Yuen (Eds.), The World Bank, Washington, DC.

Efe, S. I., & Eyefia, O. A. (2014). Urban warming in Benin City, Nigeria. *Atmospheric and Climate Sciences*, *4*, 241–252. http://dx.doi.org/10.4236/acs.2014.42027; http://www.scirp.org/journal/acs.

Eke, O. P. (2011). Climate change: Perception adaptation and mitigation. In *Proceedings of International Conference of The Nigerian Meteorological Society* (*Nmets*) 13–17 November, held at The Ahmadu Bello University Zaria.

Franklin, N. M., Daniel, B. S., & Yaw O.-A. (2014). Empirical evidence of climate change: Effects on rice production in the Northern Region of Ghana. British Journal of Economics, Management & Trade *4*(4), 551–562. SCIENCEDOMAIN *international* www.sciencedomain.org.

Gill, S., Handley, J. F., Ennos, R., & Pauleit, S. (2007). Adapting cities for climate change: The role of Green Infrastructure. *Built Environment, 33*(1), 97–115.

Gyampoh, B. A., Amisah, S., Idinoba, M., & Nkem, J. (2009). Using traditional knowledge to cope with climate change in rural Ghana in Unagylva, Vol. 60:1–2 in international/rural of forest and forest industries published by Food and Agriculture Organization of the united nation. http://dx.doi.org/10.1080/09613218.2014.8582083/.

Ifatimehin, O. O., Ishaya, S., & Ujo, F. (2010). An analysis of temperature variations using remote sensing approach in Lokoja Area, Nigeria. *Journal of Production Agriculture and Technology (PAT), Nassarawa State University Keffi PAT December 6* (2), 35–44.

Iliya, Baba and Opponkumi (2012), Iliya,, M.A, Mohammed-Baba, T.A and OpponKumi (2012) Climate change, land degradation, forced migration and farmers pastoralist conflict in North Western Nigeria: Observation and extrapolation. In Climate Change and Sustainable Development in Sub-Saharan Africa. In M. A. Illiya, M. A. Abdulrahim, & I. M Dankani, (Eds.), Geography Department, Usumanu Dan Fodio University Sokoto-Nigeria.

IPCC. (2012). Managing the risks of extreme events and disasters to advance climate change adaptation: A special report of working groups I and II of the Intergovernmental Panel on Climate Change. In C. B. Field, V. Barros, T. F. Stocker, D. Qin, D. J. Dokken, K. L. Ebi, M. D. Mastrandrea, K. J. Mach, G. K. Platter, S. K. Allen, M. Tignor, & P. M. Midgley (Eds.), Cambridge/New York: Cambridge University Press. http://ipcc-wg2gov/SREX/IMAGES/UPLOADS/srex-All_FINAL.pdf.

IPCC. (2007). Climate change 2007: Impacts, adaptation and vulnerability contribution of working group II to the fourth assessment report of the intergovernmental panel on climate change. In M. L. O. F. Parry, Canziani, J. P. Palutikof., P. J. Vander Linden, C. E. Hanson, (Eds.), Cambridge, UK, New York, US: Cambridge University Press.

Kula, N., Haines, A., & Fryatt, R. (2013). Reducing vulnerability to climate change in Sub-Sahara Africa: The need for Better evidence in PLOS/Med 10 (1) January doi:10.1371/Journal.pmed. 100374.

Lokoja Master Plan. (2009). The Lokoja Master Plan 2010–2035, prepared by Aims Consultants Abuja for the Government of Kogi State.

Mell, R. M., & C. Davies. (2009). Exploring the role of GI in the mitigation of climate change in the urban realm: Climate change: Global risks, challenges and decisions IOP Conference series: Earth and Environmental science 6.

NISER. (2010). *Socio-economic impact of climate change in Nigeria: NISER Review of Nigerian Development, 2009/2010.* Ibadan Nigeria: The Nigerian Institute of Social and Economic Research (NISER).

Ogboi, K. C. (2012). An overview of climate change: Causes, processes and manifestation. In *Lead Paper presented at 14th Mandatory Professional Development Programme (MCPDP) of the Nigerian Institute of Town Planners (NITP) Theme Climate*: Challenges for Physical Planning in Nigeria.

Olatunde, A. F., & D. S. A. Alaci, (2010). Temporal characteristics of drought in the Sudano-Sahelian region of Nigeria and its implications for inland waterway transportation. In A. A. Ogidiolu, S. D. Musa, & O.O. Ifatimehin (Eds.), Contemporary issues in infrastructural development and management in Nigeria: A conference proceedings. Published by Department of Geography and Planning, Kogi State University, Anyigba.

Olatunde A. F., & Alaci, D. S. A. (2012). Is the trend and persistence of drought in the Sudan Sahel Region of Nigeria an indication of climatic change? In M. A. Illiya, M. A. Abdulrahim, & I. M. Dankani (Eds.), *Climate change and sustainable development in Sub—Saharan Africa.* Usumanu Dan Fodio University Sokoto—Nigeria: A publication of the Department of Geography.

Osman-Elesha, B. (2009). Climate Change Impacts, Adaptation and Links to Sustainable Development in Africa. In Unasylva; International Journal of Forestry and forest industries. Vol. 60 No. 1–2 Food and Agriculture Organisation (FAO) of the United Nations.

Satterthwaite, D. (2008). Cities' Contribution to Global warming; notes on allocation of Green House Gas Emissions. *Environment and Urbanization, 20*(2), 539–550.

Satterthwaite, D., Huq, S., Reid, H., Pelling, M., & Lankao. (2009). Adapting to climate change in Urban areas: The possibilities and constraints in low and middle-income nations. In J. Bicknell, D. Dodman, & D. Satterwaite (Eds.), Adapting cities to climate change: Understanding and addressing the development challenges. Earthscan, UK.

Umar, A. T. (2012). Climate change in Nigeria: Evidence from meterological parameter. In M. A. Illiya, M. A. Abdulrahim, & I. M. Dankani, (Eds.), Climate change and sustainable development in sub-Saharan Africa. A publication of the Department Geography, Usumanu Dan Fodio University Sokoto-Nigeria.

UNDP. (2006). Human Development Report: Beyond Scarcity; power. *Poverty and the global water crisis*. New York: The United Nations Development Program.

UNEP. (2002). Environment for Development: Annual Report. United Nations Environment Programme.

UNFCCC. (2013). United Nations Framework Convention on Climate Change, Retrieved from Unfcc.int/essential background/convection on 5/12/2013.

Vlachou, A. & Charalampos K. (2010). Climate change: The political economy of Kyoto flexible mechanisms 42: 32 Review of Radical Political Economics. doi:10.1177/0486613409357179 http://www.sagepublications.com.

Zakieldeen, S. A. (2009). Natural resource management adaptation to climate change: A vulnerability assessment for Sudan. International Institute for Environment and Development (IIED).

Promoting Farmers' Resilience to Climate Change: An Option of the N'Dama Cattle in West Africa

Olawale Festus Olaniyan and Modupe Orunmuyi

1 Introduction

Animal husbandry plays vital roles in the lives of smallholder farmers who depend on it for their daily livelihood. Up to 600 million smallholder farmers in the tropics rely on livestock to reduce their vulnerability to adverse environmental impacts (Steinfeld et al. 2006). Besides, livestock production is particularly an option for the resource-poor people in the marginal areas where cultivation of crop may not be feasible (Godber and Wall 2014). There is therefore a close linkage among many aspects of livestock production, farmers' resilience, and climate change.

However, studies on the potentials of livestock in the context of climate change management are still limited. Particularly, the roles of locally adapted livestock breeds in dealing with the impacts of climate change have not been adequately addressed in the scientific literatures. Previous scholarly work rather focused on the crop sector, and there is a partial neglect of livestock farming potential in enhancing the resilience of entire agricultural system (Weindl et al. 2015). Using the largest portion of land and consuming up to 60% of the world's biomass (Weindl et al. 2015), it is expected that livestock sector especially in the developing countries will change in terms of system, species and breeds required for different future scenarios of climate change. Livestock production therefore needs to be at the centre of discussions on agriculture and climate change management.

O.F. Olaniyan (✉)
University of the Gambia and International Trypanotolerance Centre,
P.O. Box 3530, Serrekunda, The Gambia
e-mail: ofolaniyan@hotmail.com

M. Orunmuyi
Department of Animal Production and Health, Federal University
Oye-Ekiti, PMB 373, Oye-Ekiti, Nigeria
e-mail: modupe.orunmuyi@fuoye.edu.ng

W. Leal Filho et al. (eds.), *Climate Change Adaptation in Africa*,
Climate Change Management, DOI 10.1007/978-3-319-49520-0_21

There are still some research gaps concerning appropriate strategies for climate change adaptation and mitigation in Africa. However, climate change requires urgent, consistent and rigorous researches (Herrero and Thornton 2013) and a range of future climate adaptations options could be explored (Weindl et al. 2015). One of those options in building smallholder farmers' resilience may include the use of locally adapted livestock breeds. This paper, based on systematic review of literature and field experience, aims to review and highlight some of the climate-related challenges which confront smallholder livestock farmers in the developing countries. The paper put forward some evidences that the N'Dama cattle, a local breed found in more than 16 West African countries can enhance farmers' resilience to climate change. Future research and policy needs for optimizing the use of this breed in the context of climate change were also suggested.

2 Materials and Methods

2.1 Study Area

This paper focuses on West Africa which is the origin of N'Dama cattle. West Africa is made up of 17 independent countries. It covers an area of 6140,000 km^2 which is up to 20% of the entire Africa. Literacy level of farmers is low and many countries are classified as low-income. The region is characterized by rugged terrain (Nunn and Puga 2012). Climatic zones include the tropics as well as sahel while its hot northernmost part is sparsely populated. Its agroecology includes humid rain forest, tropical grass savannah and semi-arid. Nomadism, transhumance, crop-livestock integration are common livestock farming systems in this region. A significant population of the taurine cattle (*Bos taurus)* and Zebu (*Bos indicus)* are found in this region. Livestock farming system is mainly extensive and animals are exposed to many adverse external factors such as disease and low-quality feed especially in the dry season. The small-scale producers in this region are already facing the challenges of climate variability such as drought and heat stress (Challinor et al. 2007; Thornton et al. 2009). Concerning the N'Dama cattle, the breed is found in different ecological zones and geographical areas (Fig. 1).

2.2 Review of Literatures

Peer-reviewed articles, scientific reports, and grey literature from organizations including Intergovernmental Panel on Climate Change and Food and Agriculture Organization were reviewed. The research objectives were first defined and this step was followed by a draft of the full paper's outline. Keywords used in searching for relevant literatures were derived from research objectives of this study. In addition to published articles, online databases such as Domestic Animal Genetic Resources

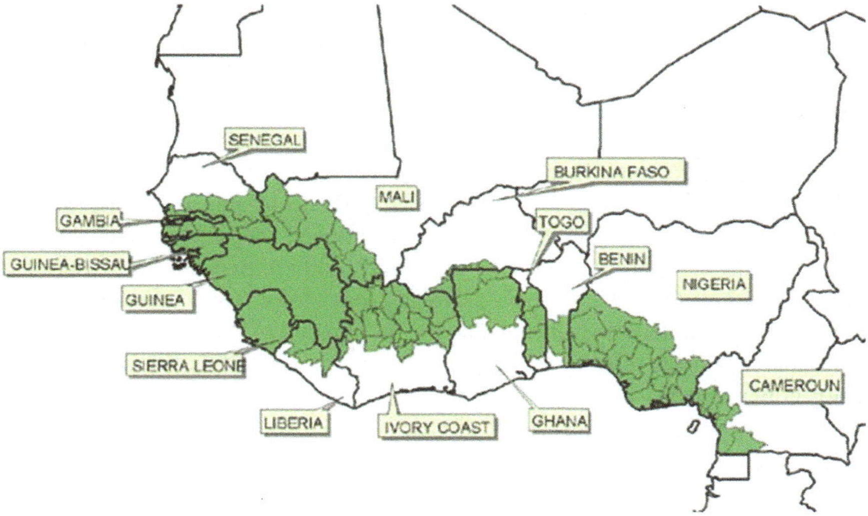

Fig. 1 N'Dama cattle distribution in West and Central African countries (Adapted from ILRI 2011)

Information System (DAGRIS) and Animal Genetics Training Resource (ILRI) were consulted to acquire more information on the N'Dama cattle. New and existing information were presented with texts, illustrations and tables.

3 Results

3.1 Climate-Related Challenges of Livestock Production in West Africa

3.1.1 Spread of Vector-Borne Diseases

Perry et al. (2013) who reviewed current drivers of disease dynamics indicated that endemic diseases in the poor countries would expand. Even though livestock diseases cannot be easily narrowed to one causative factor, there are evidences that climate change is one of the drivers which will affect disease incidence and trajectories in the developing countries.

Concerning West Africa, the tsetse fly transmitted disease commonly called trypanosomosis is already a threat to ruminant production and farmers' livelihood in many countries. Grace et al. (2009) in a survey carried out in Mali, Guinea, and Burkina Faso reported that trypanosomosis was the most important disease mentioned by the interviewed farmers. Also, Kima et al. (2015), reported that 73% of respondents in the sub-humid zone of Burkina Faso experienced increased animal morbidity and mortality as a result of vector-borne diseases. Trypanosomosis is both sensitive to climate (McDermott et al. 2002) and habitat change (Baylis and Githeko 2006). For instance, Kima et al. (2015) reported a southward shift of 25–150 km in the tsetse flies zone of Burkina Faso due to a combination of changes in rainfall and population density. Being more sensitive to ecological change, it is anticipated that trypanosomosis will react to demographic factors such as bush clearing and deforestation than drivers of climate change.

Tick-borne diseases including anaplasmosis, babesiosis, and cowdrioisis are also of economic burden to livestock production although they are only endemic to certain regions of West Africa (Baylis and Githeko 2006). Because some aspects of the tick's life cycles are sensitive to climate, this vector could be driven to new areas under the influence of climate change. As reported by Yatoo et al. (2012) in a study that examined effects of climate change on animal health and disease, vector-borne diseases closely linked with the environment are more sensitive to climate change than other groups of disease.

3.1.2 Prolonged Drought

The drought-related challenges associated with livestock farming under climate change scenarios are summarized in Table 1.

3.1.3 Rising Temperature

In the context of developing countries, there are various effects (Table 2) of high temperature on livestock and their farming systems.

3.1.4 Contribution to Emission of Greenhouse Gases (GHG)

Discussion on how farm animals contribute to climate change through emission of methane, carbon dioxide and nitrous oxide is gaining momentum in the literatures (Baylis and Githeko 2006; Steinfeld et al. 2006; Thornton et al. 2009; FAO 2015). Potential sources of GHG emission in extensive and nomadic pastoral farming

Table 1 Direct and indirect effects of drought on livestock farming

Effects of drought	References
Decreased plant biomass, crop, and dry matter yield	Hopkins and Del Prado (2007)
Increased mortality rate in the national herds	Thornton et al. (2009)
Outbreak of diseases such as anthrax, Rift Valley Fever and bluetongue	Baylis and Githeko (2006)
Decreased C_4 grass production	IPCC (2007)
Threat to agropastoralists' livelihood and food security	IPCC (2007)

Table 2 Effects of high temperature on livestock and production system

Effects of high temperature	Other predisposing factor	References
Heat-related stress	High relative humidity	FAO (2015)
Reduced production and reproductive activities	Extreme weather events, reduced feed intake	Thornton et al. (2009), FAO (2015)
Increased mortality of animals	Heat wave	Sirohi and Michaelowa (2007)
Poor quality forage	Drought	FAO (2015)
Decreased conception rates	Relative humidity	Amundson et al. (2006)
Changes in host-pathogen relationship	Other environmental stress factors	Baylis and Githeko (2006), Yatoo et al. (2012)
Threats to the existing adaptive systems of animals	Extreme weather events	IPCC (2014)

systems which are mainly practised in West Africa include livestock husbandry, feed production, meat and milk processing, marketing, distribution, cold storage of products, and manure handling. Emission of GHGs from those sources has both environmental and socioeconomic consequences. If emission is not abated, it may undermine livestock production and livelihood of smallholder farmers.

However, there are potentials for mitigating GHGs originating from livestock production sector. Recommended options include improved livestock production system and management (Valin et al. 2013; Havlík et al. 2014; Weindl et al. 2015); lower consumption of animal products (Nelson et al. 2014); use of manure for biogas and energy production and as organic fertilizer on the farms.

3.2 *Promoting Livestock Farmers' Resilience*

3.2.1 The N'Dama Cattle Breed

N'Dama cattle is a transboundary breed presently found in more than 16 countries of West and Central Africa. Common names for the breed include Boyenca, Malinke, N'Dama petite, Gambian longhorn (DAGRIS 2007). The breed is found in different ecological zones and geographical areas (Fig. 1). In the following sections, some of the desirable traits of the breed which makes it one of the choices that could enhance resilience of livestock farmers are highlighted.

Coat Colour

The typical coat colour of an N'Dama cattle is light to dark brown (Fig. 2) but there could be grey, black, light red, dun or chestnut patches around the head, belly and lower part of the tail (ILRI 2011).

Fig. 2 An N'Dama cattle herd in the Gambia

Influence of coat colour genes on body measurements, heat tolerance and haematological parameters was reported by Decampos et al. (2013). Although focusing on West African Dwarf (WAD) sheep, the authors established that black coat colour had the highest significant effect on rump height (57.80 ± 1.29 cm) and tail length (22.10 ± 0.89 cm). On the other hand, WAD sheep with brown coat colour were able to tolerate heat stress than the sheep with different coat colours. There is presently no study that examines the relationship between coat colour tolerance in the N'Dama or other tropical breeds of cattle. However, the brown colour as found in the N'Dama cattle is expected to improve animal's tolerance to increased temperature associated with climate change.

Body Size and Shape

According to Porter and Kearney (2009), size and shape are among the relevant factors which determine warm-blooded animals' thermoneutral zone in terms of minimizing energy and water. In another study by Huey et al. (2012), the role of body size in determining endothermal mammal's vulnerability to climate warming was also emphasized. The study indicated that given behavioural, physiological and genetic data, we can predict the organisms that would be most at risk as a result of change in environmental variables. N'Dama cattle is a medium-sized cattle breed with thick neck and relatively small head (Fig. 2). The horn on average, is about 60 cm long while shoulder height is 100 and 120 cm for cow and bulls respectively (ILRI 2011). Such a small body conformation of the N'Dama cattle confers certain advantage in terms of vulnerability to climate change. While thermal comfort zone will be affected by many factors, the tropical breeds such as the N'Dama cattle will be better adapted to heat and reduced feed availability (FAO 2015). There is no precise estimate of the N'Dama population size but it is large in Guinea and Gambia where more than 90% of the livestock farmers still keep the breed. However, Williams et al. (2008) cautioned that even though large population sizes can reduce extinction rate of certain biological species, the same rationale is doubtful in the context of climate change.

Disease Tolerance

The State of the World's Animal Genetic Resources for Food and Agriculture published by the Food and Agriculture Organization clearly indicated that certain cattle breeds are tolerant or resistant to diseases and parasites (FAO 2007). N'Dama cattle's typical tolerance to African trypanosomosis and tick-borne diseases is evident (FAO 2007; Grace et al. 2009; Perry and Grace 2009). The N'Dama cattle will be an important asset in tackling the burden of those tropical diseases on animals and farmers. Utilizing the breed especially in the smallholder settings of West Africa will eliminate the use of disease-control chemicals which are not environmental friendly. The use of a tolerant and hardy breed such as the N'Dama cattle will increase farmers' income through reduction in purchase of drugs. The breed will continue to play important roles in supporting resource-poor farmers' livelihood (Olaniyan 2015) and in enhancing their resilience to the likely impacts of Climate Change in West and Central Africa where they are mainly raised in an extensive system.

Utilization of Low-Quality Feed

Depending on animal species and production system, resource-use and intensity of greenhouse gas emission will be modified by feed quality and efficiency (Herrero et al. 2013). While low-quality forage is better utilized by the Zebu cattle, the taurine including the N'Dama cattle can better convert feed to animal products (FAO 2015). Given that C_4 grasses in the tropics will increase under an elevated temperature (IPCC 2007), there is a need for livestock farmers to rely on animals that can survive on low-quality breed. In a situation where pasture or rangeland is infested with pathogens or vectors, raising a disease-tolerant breed such as the N'Dama cattle will be important as a coping strategy.

4 Discussion

4.1 Alternatives to Extensive Livestock Farming System

In the context of climate change, there would be a threat to animal husbandry and the growing human population which depends on animal products for their subsistence. The fate of extensive livestock production system which predominates in the West African countries is therefore subject to many challenges. To deal with

Table 3 Optional livestock production systems to enhance farmers' resilience

Recommended systems	Opportunities	References
Integrated crop and grassland management	Reduced harmful environmental effects, increased food production, biodiversity, and ecosystem services	Soussana and Lemaire (2014)
Mixed crop-livestock systems	Better resource-use and adaptation cost efficiency	Herrero et al. (2009), Weindl et al. (2015)
Industrial livestock production	Balanced livestock production with livelihood and environment	Herrero et al. (2009), Havlík et al. (2014)
Rangeland-based livestock production	Increased agricultural production and rural income; reduced impacts of agriculture on climate	Jones and Thornton (2009), Weindl et al. (2015)

such challenges, alternative adaptation strategies in technology and farming systems (Table 3) may be required.

4.2 Farmers' Decision Paradigm

Farmers' choice of a livestock species, farming system, and the way they will respond to various climatic conditions are closely linked to their socioeconomic situations. The choice to keep or sell livestock in a smallholder farming system could be based on market factors rather than coping strategy (Morton 2007). An N'Dama cattle farmer may decide to keep different livestock species within a herd as a way of adapting to various ecological niches, climate variability and labour availability. The rationale and decision making processes of smallholder livestock farmers in the context of climate change still needs to be further researched. However, adaptation and also, mitigation options in the context of climate change in developing countries should be based on a detailed framework that assesses impacts and trade-offs (Thornton and Gerber 2010).

Managing the effects of climate change on livestock production and farmers also requires participatory and quality decision making. Community-based adaptation plans and programs involving different groups of stakeholders may be important in this regard. Multi-stakeholder public policy design and implementation has a potential to leverage sustainable livestock management as well as adaptation and mitigation costs. Concerning choice among alternative farming systems, Steinfeld et al. (2006), recommended policy options that limit opening up of rangeland and reduce stocking of animals in densely populated livestock areas. However, the abilities of African countries to facilitate appropriate adaptation strategies was questioned by Challinor et al. (2007).

4.3 Meeting Research Needs

The reasons for low research and development of climate technological innovation in some developing countries include inadequate finances and poor capacity to implement research activities (Adenle et al. 2015). In addition to strengthening of existing regional organizations, research mandates on climate change adaptation and mitigation should focus on agriculture, meteorology, anthropology, and economics (Millner and Washington 2011). Sociology, food and feed science, animal and human nutrition are also highly important and should not be neglected.

For the N'Dama cattle, research on how to improve the breed's adaptive traits in the context of future climate change is required. Funding and coordination of such research activities would be the responsibilities of organizations such as Economic Community of West African States (ECOWAS) and African Union Interafrican Bureau on Animal Genetic Resources (AU-IBAR). Unilateral and multilateral collaborations of research institutions and universities within and outside this region are crucial. Generating new information through climate research is not sufficient in itself, getting such research findings to the end users which are the vulnerable farmers is also important (Perry et al. 2013). Research outcomes can be disseminated to farmers through radio and television programmes, local authorities, annotated leaflets, and informal group discussions. Uncertainty should be carefully communicated because it determines how climate information will be eventually utilized by the stakeholders (Patt and Dessai 2005).

5 Conclusion and Recommendation

This study identified some challenges of climate change in the context of developing countries. For livestock farming in the West African region, there would be a change in system and breed in response to climate change related events. N'Dama cattle which originated from that region is one of the animal breeds which farmers will rely upon in order to enhance their resilience and adaptation potentials. This paper further showcased that N'Dama cattle breed will be important for farmers because of its certain services and products, disease tolerance, small body size and shape, coat colour, and ability to utilize low-quality feeds. However, alternatives to extensive livestock farming system being practised in West Africa need to be explored. In order to maximize the potentials of N'Dama cattle in enhancing the resilience of smallholder livestock farmers, the paper recommends that certain favourable policies and research programs need to be pursued by the West African governments and other relevant stakeholders.

References

Adenle, A. A., Azadi, H., & Arbiol, J. (2015). Global assessment of technological innovation for climate change adaptation and mitigation in developing world. *Journal of Environmental Management, 161,* 261–275.

Amundson, J. L., Mader, T. L., Rasby, R. J., & Hu, Q. S. (2006). Environmental effects on pregnancy rate in beef cattle. *Journal of Animal Science, 84*(12), 3415–3420.

Baylis, M., & Githeko, K. A. (2006). The effects of climate change on infectious diseases of animals. http://tna.europarchive.org/20100630051843/http:/www.foresight.gov.uk/Infectious Diseases/t7_3.pdf. Accessed on 15 Dec 2015.

Challinor, A., Wheeler, T., Garforth, C., Craufurd, P., & Kassam, A. (2007). In assessing the vulnerability of food crop systems in Africa to climate change. *Climatic Change, 83*(3), 381–399.

DAGRIS. (2007). Domestic animal genetic resources information system. In: S. Kemp, Y. Mamo, B. Asrat, & T. Dessie (Eds.). International livestock research institute, Addis Ababa, Ethiopia. http://dagris.ilri.cgiar.org. Accessed on 10 Dec 2015.

Decampos, J. S., Ikeobi, C. O. N., Olowofeso, O., Smith, O. F., Adeleke, M. A., Wheto, M., et al. (2013). Effects of coat colour genes on body measurements, heat tolerance traits and haematological parameters in West African Dwarf sheep. *Open Journal of Genetics, 3*(4), 280–284.

FAO. (2007). The state of the world's animal genetic resources for food and agriculture. Food and agriculture organization. http://www.fao.org/docrep/010/a1250e/a1250e00.htm. Accessed on 8 Dec 2015.

FAO. (2015). Coping with climate change: The roles of genetic resources. Food and Agriculture Organization. http://www.fao.org/3/a-i3866e.pdf. Accessed on 10 Dec 2015.

Godber, O. F., & Wall, R. (2014). Livestock and food security: Vulnerability to population growth and climate change. *Global Change Biology, 20*(10), 3092–3102.

Grace, D., Randolph, T., Affognon, H., Dramane, D., Diall, O., & Clausen, P. H. (2009). Characterisation and validation of farmers' knowledge and practice of cattle trypanosomosis management in the cotton zone of West Africa. *Acta Tropica, 111*(2), 137–143.

Havlík, P., Valin, H., Herrero, M., Obersteiner, M., Schmid, E., Rufino, M. C., et al. (2014). Climate change mitigation through livestock system transitions. *Proceedings of the National Academy of Sciences of the United States of America, 111*(10), 3709–3714.

Herrero, M., Havlík, P., Valin, H., Notenbaert, A., Rufino, M. C., Thornton, P. K., et al. (2013). Biomass use, production, feed efficiencies, and greenhouse gas emissions from global livestock systems. *Proceedings of the National Academy of Sciences of the United States of America, 110*(52), 20888–20893.

Herrero, M., & Thornton, P. K. (2013). Livestock and global change: emerging issues for sustainable food systems. *Proceedings of the National Academy of Sciences of the United States of America, 110*(52), 20878–20881.

Herrero, M., Thornton, P. K., Gerber, P., & Reid, R. S. (2009). Livestock, livelihoods and the environment: Understanding the trade-offs. *Current Opinion in Environmental Sustainability, 1* (2), 111–120.

Hopkins, A., & Del Prado, A. (2007). Implications of climate change for grassland in Europe: Impacts, adaptations and mitigation options: a review. *Grass and Forage Science, 62*(2), 118–126.

Huey, R. B., Kearney, M. R., Krockenberger, A., Holtum, J. A. M., Jess, M., & Williams, S. E. (2012). Predicting organismal vulnerability to climate warming: roles of behaviour, physiology and adaptation. *Philosophical Transactions of the Royal Society of London. Series B, Biological Sciences, 367*(1596), 1665–1679.

ILRI. (2011). Animal genetics training resource. International Livestock Research Institute. http://agtr.ilri.cgiar.org/agtrweb/. Accessed on 10 Dec 2015.

IPCC. (2007). Summary for policymakers. Intergovernmental panel on climate change. In P. J. van der Linden, C. E. H. M. L. Parry, O. F. Canziani, & J. P. Palutikof (Eds.). *Climate change*

2007: Impacts, adaptation and vulnerability (pp. 7–22). Cambridge: Cambridge University Press.

IPCC. (2014). Climate change 2014: Synthesis report. Contribution of working groups I, II and III to the fifth assessment report of the intergovernmental panel on climate change. In R. K. Pachauri, & L. A. Meyer (Eds.). Intergovernmental panel on Climate Change, Geneva, Switzerland.

Jones, P. G., & Thornton, P. K. (2009). Croppers to livestock keepers: Livelihood transitions to 2050 in Africa due to climate change. *Environmental Science and Policy, 12*(4), 427–437.

Kima, S. A., Okhimamhe, A. A., Kiema, A., Zampaligre, N., & Sule, I. (2015). Adapting to the impacts of climate change in the sub-humid zone of Burkina Faso, West Africa: Perceptions of agro-pastoralists. *Pastoralism, 5*(1), 16. doi:10.1186/s13570-015-0034-9

McDermott, J. J., Kristjanson, P. M., Kruska, R. L., Reid, R. S., Robinson, T. P., Coleman, P. G., et al. (2002). *The African trypanosomes* (Vol. 1). Boston: Kluwer Academic Publishers.

Millner, A., & Washington, R. (2011). What determines perceived value of seasonal climate forecasts? A theoretical analysis. *Global Environmental Change, 21*(1), 209–218.

Morton, J. F. (2007). The impact of climate change on smallholder and subsistence agriculture. *Proceedings of the National Academy of Sciences of the United States of America, 104*(50), 19680–19685.

Nelson, G. C., Valin, H., Sands, R. D., Havlík, P., Ahammad, H., Deryng, D., et al. (2014). Climate change effects on agriculture: economic responses to biophysical shocks. *Proceedings of the National Academy of Sciences of the United States of America, 111*(9), 3274–3279.

Nunn, N., & Puga, D. (2012). Ruggedness: The blessing of bad geography in Africa. *Review of Economics and Statistics, 94*(1), 20–36.

Olaniyan, O. F. (2015). Sustaining N'Dama cattle for the resource-poor farmers in The Gambia. *Bulletin of Animal Production and Health in Africa, 63*(1), 83–92.

Patt, A., & Dessai, S. (2005). Communicating uncertainty: lessons learned and suggestions for climate change assessment. *Comptes Rendus Geoscience, 337*(4), 425–441.

Perry, B., & Grace, D. (2009). The impacts of livestock diseases and their control on growth and development processes that are pro-poor. *Philosophical Transactions of the Royal Society of London. Series B, Biological sciences, 364*(1530), 2643–2655.

Perry, B. D., Grace, D., & Sones, K. (2013). Current drivers and future directions of global livestock disease dynamics. *Proceedings of the National Academy of Sciences of the United States of America, 110*(52), 20871–20877.

Porter, W. P., & Kearney, M. (2009). Size, shape, and the thermal niche of endotherms. *Proceedings of the National Academy of Sciences of the United States of America, 106*(2), 19666–19672.

Sirohi, S., & Michaelowa, A. (2007). Sufferer and cause: Indian livestock and climate change. *Climatic Change, 85*(3–4), 285–298.

Soussana, J. F., & Lemaire, G. (2014). Coupling carbon and nitrogen cycles for environmentally sustainable intensification of grasslands and crop-livestock systems. *In Agriculture, Ecosystems and Environment, 190,* 9–17.

Steinfeld, H., Gerber, P., Wassenaar, T., Castel, V., Rosales, M., & de Haan, C. (2006). Livestock's long shadow. Food and agriculture organization. http://www.fao.org/docrep/. Accessed on 7 Dec 2015.

Thornton, P. K., & Gerber, P. J. (2010). Climate change and the growth of the livestock sector in developing countries. *Mitigation and Adaptation Strategies for Global Change, 15*(2), 169–184.

Thornton, P. K., van de Steeg, J., Notenbaert, A., & Herrero, M. (2009). The impacts of climate change on livestock and livestock systems in developing countries: A review of what we know and what we need to know. *Agricultural Systems, 101*(3), 113–127.

Valin, H., Havlík, P., Mosnier, A., Herrero, M., Schmid, E., & Obersteiner, M. (2013). Agricultural productivity and greenhouse gas emissions: trade-offs or synergies between mitigation and food security? *Environmental Research Letters, 8*(3), 035019. doi:10.1088/1748-9326/8/3/035019

Weindl, I., Lotze-Campen, H., Popp, A., Mueller, C., Havlik, P., Herrero, M., et al. (2015). Livestock in a changing climate: Production system transitions as an adaptation strategy for agriculture. *Environmental Research Letters, 10*(9), 0940.

Williams, S. E., Shoo, L. P., Isaac, J. L., Hoffmann, A. A., & Langham, G. (2008). Towards an integrated framework for assessing the vulnerability of species to climate change. *PLoS Biology, 6*(12), 2621–2626.

Yatoo, M. I., Kumar, P., Dimri, U., & Sharma, M. C. (2012). Effects of climate change on animal health and diseases. *International Journal of Livestock Research, 2*(3), 15–24.

An Approach to Use Earth Observation Data as Support to Water Management Issues in the Ethiopian Rift

Sabrina Bonetto, Anna Facello, Elena Isotta Cristofori, Walther Camaro and Alessando Demarchi

1 Introduction

Water represents a natural and invaluable resource for people and for the equilibrium of the planet.

Recently, climate change assessment studies have indicated an alteration in the global hydrological cycle (IPCC 2007, 2014). Likely this variation can have great impacts on regional water resources, affecting both ground and surface water supplies for domestic, industrial and agricultural uses (Mahdy and Al-Najar 2015). Particularly in water-stressed areas, people and ecosystems are vulnerable to decreasing precipitation due to climate change (IPCC 2007). In this framework, changes in rainfall (amount, frequency, intensity and distribution) directly affect the magnitude and timing of runoff and the intensity of floods and droughts, causing stresses on water, agricultural and human health (IPCC 2007, 2014).

This study focuses on the area of the central Ethiopian Rift, where the water resources of lakes are intrinsically linked with human health, food production and economic livelihoods (Ayenew 2007). Three main basins are present: Awash basin (lake Koka, Beseka, Gemari, Abe, etc.), the Main Ethiopian Rift (MER) lakes

S. Bonetto (✉)
Department of Earth Sciences, University of Turin,
Valperga Caluso 35, 10125 Turin, Italy
e-mail: sabrina.bonetto@unito.it

A. Facello
National Research Council, Research Institute for Geo-Hydrological Protection,
Strada delle Cacce 73, 10135 Turin, Italy

E.I. Cristofori · A. Demarchi
Politecnico di Torino, DIST, Viale Pier Andrea Mattioli 39, 10125 Turin, Italy

E.I. Cristofori · W. Camaro · A. Demarchi
ITHACA, Via P.C. Boggio 61, 10138 Turin, Italy

W. Leal Filho et al. (eds.), *Climate Change Adaptation in Africa*,
Climate Change Management, DOI 10.1007/978-3-319-49520-0_22

region (lakes Zway, Langano, Abiyata and Shala) and southern basin (lake Awassa, Abaya, Chamo and Chew-Bahir) (Alemayehu et al. 2006) (Fig. 1).

The proposed approach allows the implementation of a systematic database regarding lakes extent in relation to precipitation and monitoring methodologies that local and national Authorities can use for a proper water management.

Volcano-tectonic activity has conditioned the formation and evolution of these lakes and, particularly in the Late Quaternary, their evolution was mainly controlled by climate change during the intense climatic fluctuations of the last 100,000 years (Benvenuti et al. 2002). As described in literature, the water surface extent of most lakes was modified according to climatic conditions of the region, with particular regard to precipitation trends of the adjacent highlands (Alemayehu et al. 2006; Ayenew 2007).

Due to the environmental, social and economic significance of these inland water bodies, scientific community and governmental offices pay attention to their

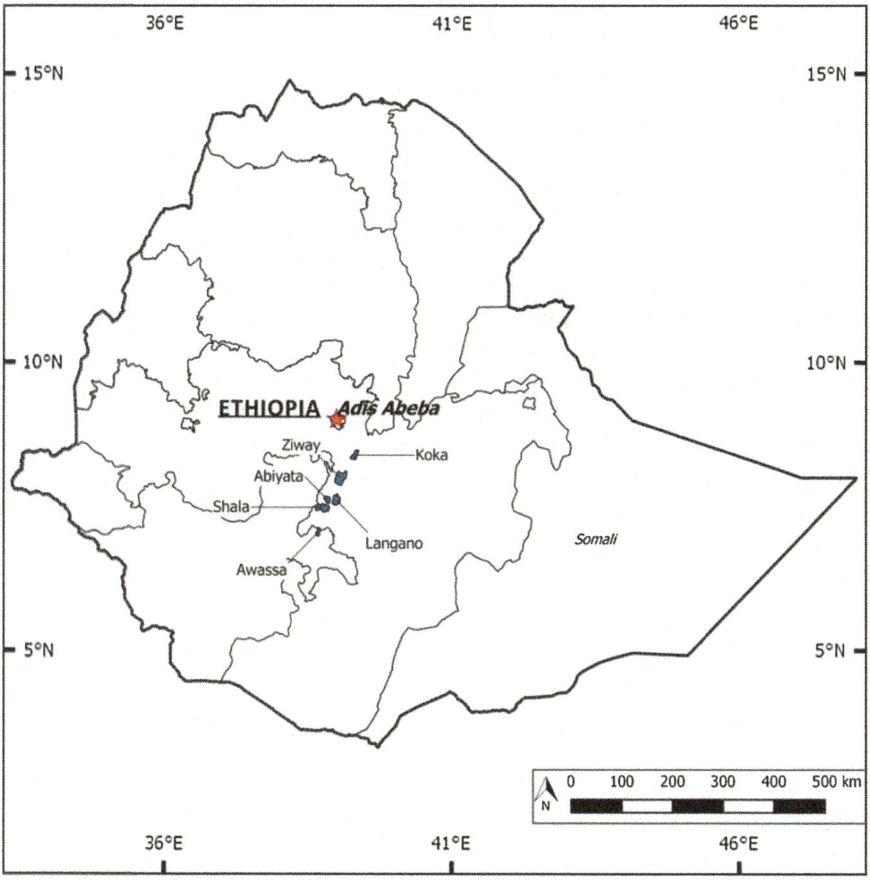

Fig. 1 Location map

evolution in order to adopt proper management practises, strategies and regulations for the preservation and use of the lake water resources (Ayenew 2007).

Usually, data related to surface water features and dynamics of surface water are obtained through ground based measurements. However, due to economic and political reasons, in many cases data are limited and often not able to provide adequate or continuous observations. Recent studies have shown that remote sensing instruments provide a huge amount of data that have become extensively used for detecting and extracting surface water and depict its changes. Several image processing techniques have been introduced for the extraction of water features data from satellite observations (McFeeters 1996; Ouma and Tateishi 2006; Xu 2006; Munyaneza et al. 2009; Nath and Deb 2010; Cretaux et al. 2011; McFeeters 2013; Tang et al. 2013).

The satellite observation includes also precipitation-based products. These data are possible means for quantifying the rainfall in developing countries or remote locations, where conventional rain gauge data are sparse (Huges 2006; Shaban et al. 2009; Stisen and Sandholt 2010).

Remote sensing techniques can contribute to assure a complete and systematic database of cartographic, hydro-meteorological, climatological records, in addition to in situ, historical and current data, to produce meaningful and useful information for environmental protection and management. The availability of these datasets might increase the quantity and quality of information on water management, regional network enhancement, climate change impact on the lake system and surrounding human activities. Often, the available ground based observations are lacking and data can be neither harmonised nor expressed in the same reference frame. Remote sensing hence could offer an opportunity to improve their (Cretaux et al. 2011).

In particular, this work shows one of the potential applications of remote sensing data. A comparative analysis of rainfall climatology from January 2000 to December 2015 was performed using Tropical Rainfall Mapping Mission (TRMM) data (Scheel et al. 2011; Huffman et al. 2010). Secondly multi-temporal Landsat ETM+ and OLI data were processed in order to detect spatial-temporal changes of some lakes of the Ethiopian Rift (Ouma and Tateishi 2006; Rockni et al. 2014). A qualitative comparison between rainfall and water lakes extent was finally pointed out.

2 Geological, Hydrogeological and Physiographic Setting

The Ethiopian Rift Valley is the northern sector of the East African Rift System and it may be divided into two parts: the Afar depression to the north, which includes the Danakil Depression in the northeast, and the Main Ethiopian Rift (MER) in the south (Gizaw 1996).

The study area lies in the MER, which is a NNE-SSW trending segment bordered by the Ethiopian Plateau to the west and the Somali Plateau to the east by means of evident fault slopes belonging to the main Rift Valley System Fault (Benvenuti et al. 2002).

The area is composed by three physiographic zones: the lowlands, corresponding to the graben area of the rift floor, the transitional escarpments and the highlands of the horst areas associated to the plateau.

The MER is mainly made of volcanites and pyroclastic rocks (Abebe et al. 1998; Benvenuti et al. 2002; Boccaletti et al. 1999). Large areas of the lowlands are covered by volcano-lacustrine and fluvio-lacustrine deposits mainly consisting in clay, silt, sand and gravel interbedded layers, with various thickness up to few hundred of meters (Boccaletti et al. 1999) (Fig. 2).

Several tectonic events have influenced its geological and geomorphological evolution and abundant parameters indicate the presence of ongoing volcanic and tectonic activities in the area (Boccaletti et al. 1998; Furi et al. 2011).

The faults in the MER are parallel and sub-parallel to the NE–SW trending rift axis (Woldegabriel et al. 1990). Active faulting within the rift valley causes a geothermal circulation of hot fluids, probably due to the persistence of magma at shallow depth. The magma heat would induce geothermal and fumarolic activities (e.g., at Aluto-Langano) and the occurrence of high temperature thermal springs bordering the lakes (Di Paola 1970; Chernet 1982; Gizaw 1993; Rango et al. 2010).

The floor of the rift is characterized by the presence of many lakes; most of them took origin when volcanoes stopped their activity and their evolution were marked by transgressive and regressive phases following each other. Because of the remnants of an ancient lacustrine basin all around the Zway, Langano, Abiyata and Shala lakes (MER), Benvenuti et al. (2002) hypothesize the presence of a unique wide lake in the area actually covered by the Late Quaternary fluvio-lacustrine sediments. Some lakes are connected by a drainage system consisting both in few large perennial rivers originating from the highlands and many seasonal streams (Ayenew 2007) and are intimately linked with the groundwater system (Ayenew 1998).

Lake Ziway receives surface inflow from Meki and Ketar rivers that drain the western and eastern highlands respectively. The Abijata Lake receives water from the Ziway Lake via the Bulbula River and seasonally by the Langano Lake via the Horakelo River, whereas Lake Shala forms a separate basin (Rango et al. 2010).

The lakes of the MER are also in connection with the lakes of the Awash basin (Koka, Beseka, Gemari, Abe) to the north and the southern basin (lake Awassa, Abaya, Chamo and Chew- Bahir). Hydrologically, the basins form separate units, but hydrogeologically they form a unique system within the Rift due to the underground interconnection by NE–SW aligned regional faults (Alemayehu et al. 2006).

The main source of water to the rift lakes and rivers is the rainfall in the eastern and western highland, which strongly controls the discharge of feeder rivers.

The climate is humid to sub-humid in the highlands and semi-arid in the Rift. The area is characterized by a short rainy season (called the "Belg") from March to April and a long rainy season (called the "Kiremt") from June to September. The remaining of the year is generally dry. The annual rainfalls range from 650 to

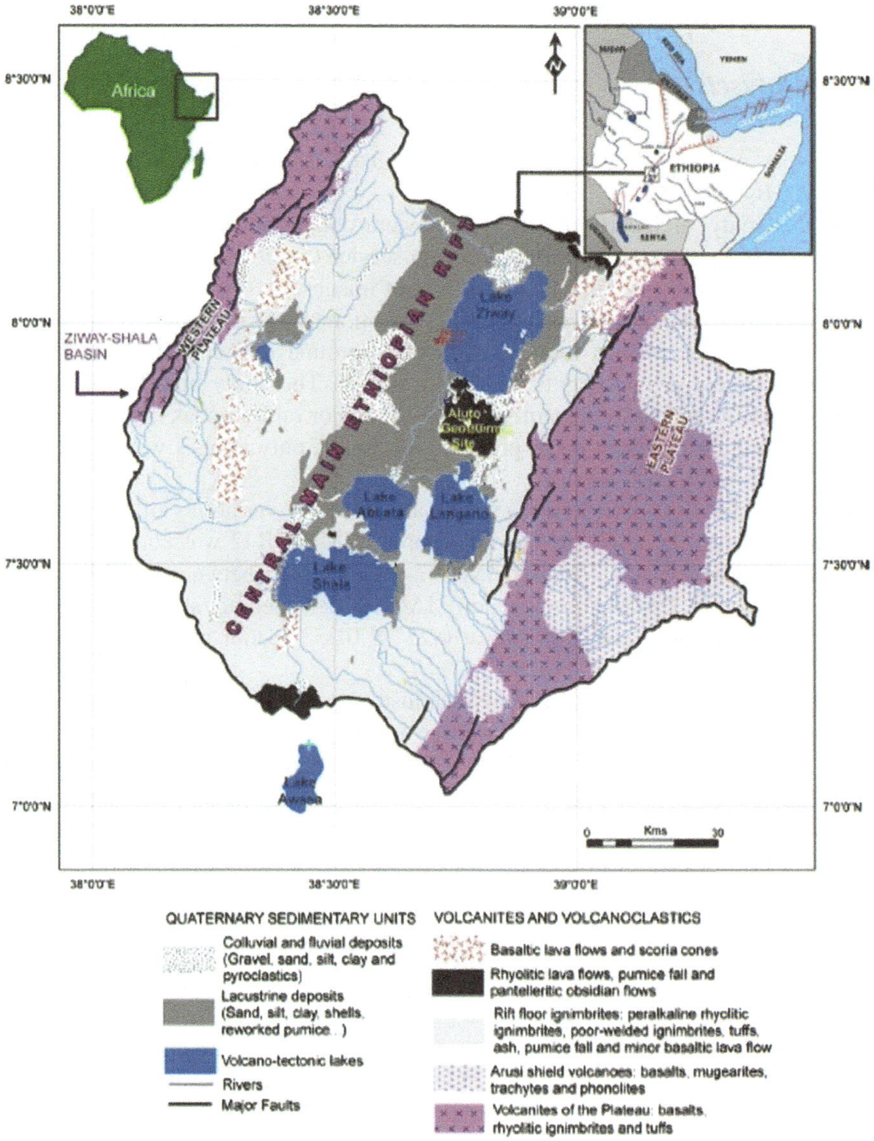

Fig. 2 Simplified geologic map of the MER study area (modified from Rango et al. 2010)

700 mm on the lowlands to 1150–1200 mm on the highlands (Le Turdu et al. 1999; Ayenew 1998; Vallet- Coulomb et al. 2001; Alemayehu et al. 2006). The rift floor is usually rainfall deficit with annual rainfall less than the evaporation. (Alemayehu et al. 2006).

3 Dataset and Methodology

3.1 Datasets

Two types of remotely sensed data were utilized: Landsat ETM+ and OLI and TRMM 3B42.

In this study, archived Landsat images were obtained from the US Geological Survey (USGS), through EarthExplorer web-service (http://earthexplorer.usgs.gov). Table 1 presents the specifications of Landsat images.

A limitation of optical remote sensing for monitoring is cloud-cover. In such situations, satellite images cannot measure the ground surface radiance and consequently cannot provide continuous information. The study area is affected by cloud-cover from June to September, when the major rainfall peak occurs. In order to avoid problems related to the presence of clouds, only images with cloud coverage <20% were considered.

With the aim to characterize the spatial and temporal variability of precipitation, the most recent version of the TRMM dataset TRMM 3B42 was chosen (Romilly and Gebremichael 2011). The TRMM is a joint mission between NASA and the Japan Aerospace Exploration Agency (JAXA), designed to study and monitor tropical rainfall. Data are available at 0.25° × 0.25° spatial resolution, with 3 h revisit time (Huffman et al. 2010).

The mission includes 5 instruments, but just 2 are normally used for precipitation measurement purpose: TRMM Microwave Imager (TMI) and Precipitation Radar (PR). Both datasets cover the latitude band 50° N-S for the period from 1998 to present (Huffman et al. 2010). In this study the TRMM 3B42 daily average product has been defined as the most relevant tool to identify the temporal variability.

Table 1 Specifications of Landsat ETM+ and OLI data, period 2000–2015

Satellite	Sensor	Path/Row	Resolution (m)	Wavelength (μm)
Landsat 7	ETM+	168/54 168/55	30	Band 1: 0.45–0.515 Band 2: 0.525–0.605 Band 3: 0.63–0.69 Band 4: 0.75–0.90 Band 5: 1.55–1.75 Band 7: 2.09–2.35
Landsat 8	OLI	168/54 168/55	30	Band 1: 0.435–0.451 Band 2: 0.452–0.512 Band 3: 0.533–0.590 Band 4: 0.636–0.673 Band 5: 0.851–0.879 Band 6: 1.566–1.651 Band 7: 2.107–2.294 Band 9: 1.363–1.384

3.2 Methodology

The first step involved the analysis of rainfall climatology from January 2000 to December 2015 both at pixel and at basin scale for different chosen basins. Watersheds boundaries were obtained from Hydrological data and maps based on SHuttle Elevation Derivatives at multiple Scales (HydroSHEDS) (Lehner et al. 2006). The watersheds considered are: id 67343, id 76858, id 77424, id 78057 (Fig. 3), related to the main lakes of the MER.

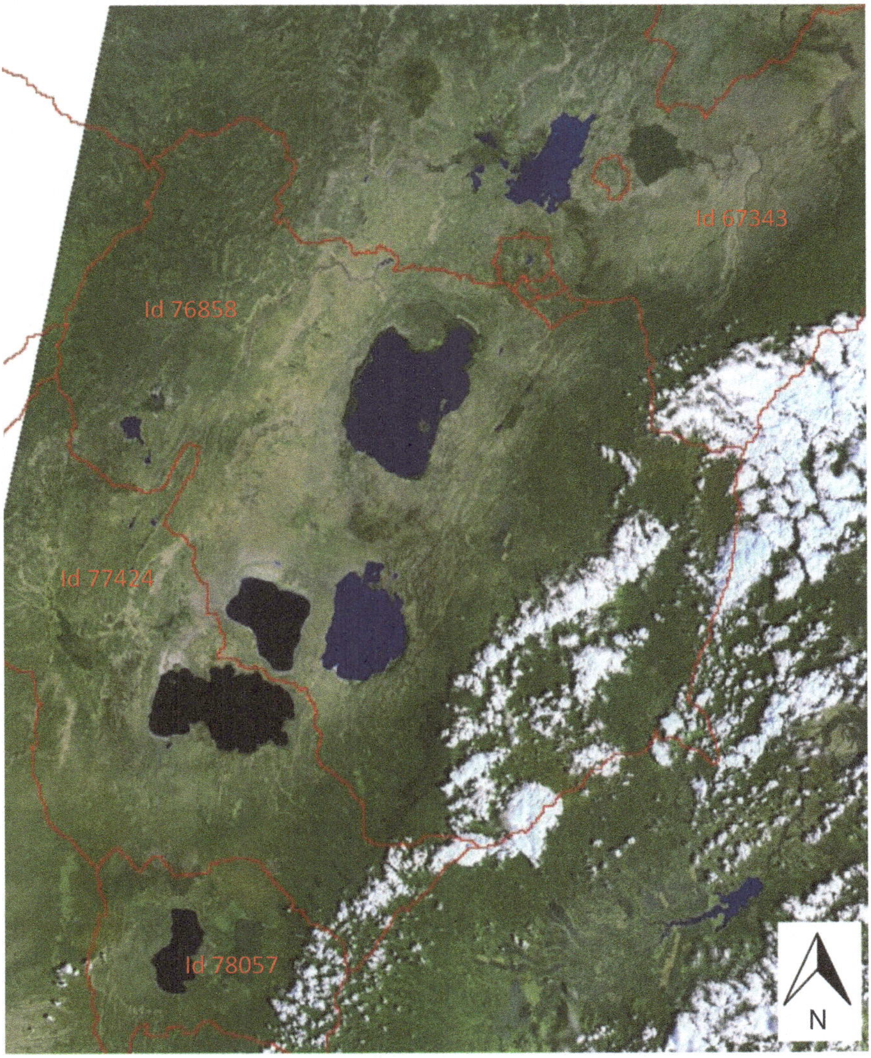

Fig. 3 Watersheds (*Source* HydroSHEDS—Hydrological data and maps based on SHuttle Elevation Derivatives at multiple Scales)

Secondly the variations of the lakes shorelines during different months of the year was analysed. The chosen years are related to the satellite images availability.

The analysis of precipitation allows derivation of the rainfall seasonality over the area of interest. This information is essential in order (i) to assess the inter-annual variability of precipitation during the analysed period and (ii) to choose the best periods for the water extent analysis. All these information allow the assessment of existing correlations between the gradient of precipitation and the water extent variations.

The TRMM 3B42 have been extensively used, at local and regional scales, for monitoring intensity and position of different rainfall events. Usually, the information is presented as adequate spatial resolution, data are available in near real time and/or in historical datasets observation that allows the identification of the climatology trends in the study areas (Huffman et al. 2010).

In order to have an accurate estimation of the temporal distribution of the rainfall from TRMM gridded dataset, an analysis based on temporal window of 15 days-time has been implemented. Firstly the precipitation trend was derived over the different watersheds, dividing the total precipitation values over each basin by the number of pixels contained within the basin area. Secondly the cumulative rainfall gridded datasets was generated considering the 15 days-time frames from 2000 to 2015. Based precipitation products allow derivation of the rainfall average datasets and also the precipitation trends for each analysed year. Thirdly, the precipitation anomalies were derived by the difference of accumulated precipitation and the average accumulated precipitation for each year.

Concerning the Landsat data, in order to prepare the input satellite images for further processing, a set of pre-processing steps were performed, including the conversion to surface reflectance values, mosaicking and corregistration. In regard to the surface reflectance step, Landsat 7 and Landsat 8 Surface Reflectance data were considered; these data are available on-demand through EarthExplorer (Masek et al. 2006; U.S. Geological Survey 2015a, b). Subsequently, the images have been mosaicked to generate new images covering the entire study area and corregistrated. Finally, the water surface in each image was extracted in order to define the water extended in relationship with the interannual and seasonal variations of rainfall.

The Normalized Difference Water Index (NDWI) was first proposed by McFeeters (1996) to detect surface waters in wetland environments and to allow the measurement of surface water extent. This method was applied to many different contexts and published in several works (Chowdary et al. 2008; Ouma and Tateishi 2006; Rokni et al. 2014; McFeeters 2013; U.S. Geological Survey 2013).

The NDWI is calculated using the following equation:

$$\mathrm{NDWI} = (\mathrm{Band}\,2) - (\mathrm{Band}\,4)/(\mathrm{Band}\,2) + (\mathrm{Band}\,4)$$

where Band 2 is the green band in surface reflectance and Band 4 is the near-infrared (NIR) band in the surface reflectance.

In order to detect the variations of the water extent of analysed lakes during the period 2000–2015, the NDWI was applied. The NDWI index was calculated from the Landsat ETM+ and OLI images. The values of NDWI greater than zero are assumed to represent water surfaces, while values less than, or equal, to zero are assumed to be non-water surfaces (McFeeters 1996).

Generally, NDWI thresholds vary depending on the proportions of subpixel water/non-water components, such as lake water quality (Ji et al. 2009). In this study, in order to establish appropriate thresholds for clearly identifying water features, the threshold value for pure water proposed by Ji et al. (2009) is chosen (NDWI ≥ 0.015).

4 Results and Discussion

4.1 Precipitation Climatology

The analysis of precipitation climatology was performed for all the watersheds selected.

Figure 4 shows the results of the 15 days-time average precipitation in the four watersheds analysed.

The four curves show the typical double maxima rainfall trend, according to the Ethiopian climatology. The first maximum is recorded between April and May while the second one is observed between July and September.

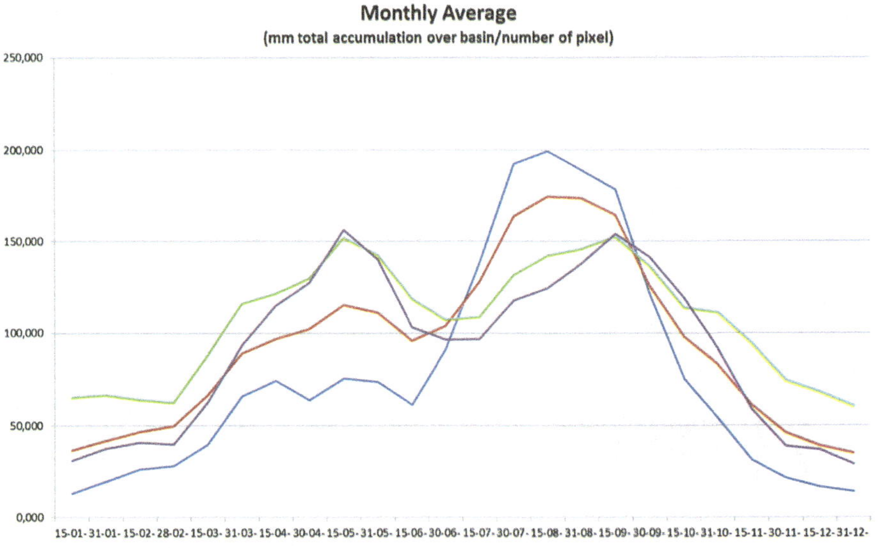

Fig. 4 Monthly average precipitation calculated for each watershed during the period 2000–2015

According to the observed precipitation evolution, it was decided to perform the water extent analysis considering the identified dry and wet periods: the months of January and October were chosen as the most relevant for the analysis.

The cumulative rainfall and the anomalies of precipitation was analysed for each watersheds. Due to the extension and the number of lakes included, only the results of basin id 76858 are reported below.

Figure 5 shows precipitation anomalies. It can be noted that 2000, 2001, 2002, 2004 and 2009 show several negative anomaly periods. Also the end of 2007, 2010 and 2013 and beginning of 2008, 2011 and 2013 show significant negative anomalies. On the other hand positive anomalies are observed especially in the beginning of 2010, end of 2011 and during the almost all 2012. Moreover, 2005 and the 2007 show small positive anomalies.

A more in depth analysis of precipitation anomalies was performed, considering the time period analysed for the water extent. The analysed years (2002, 2005, 2007, 2010, 2012, 2014) were defined on the basis of cloud cover and spatial and temporal continuity in the acquisition. According to the months considered relevant for the analysis of water extent (January and October), each anomalies analysis started in November (i.e. from November 2001 to November 2002, etc.) in order to consider meaningful rainfall data for the water extent evaluation on the January of each studied years. The goal is to obtain values and trends that can be used to verify if some relationships between the gradient of precipitation and the variation of the water extent can be found. An eventual lack of correlation between these values

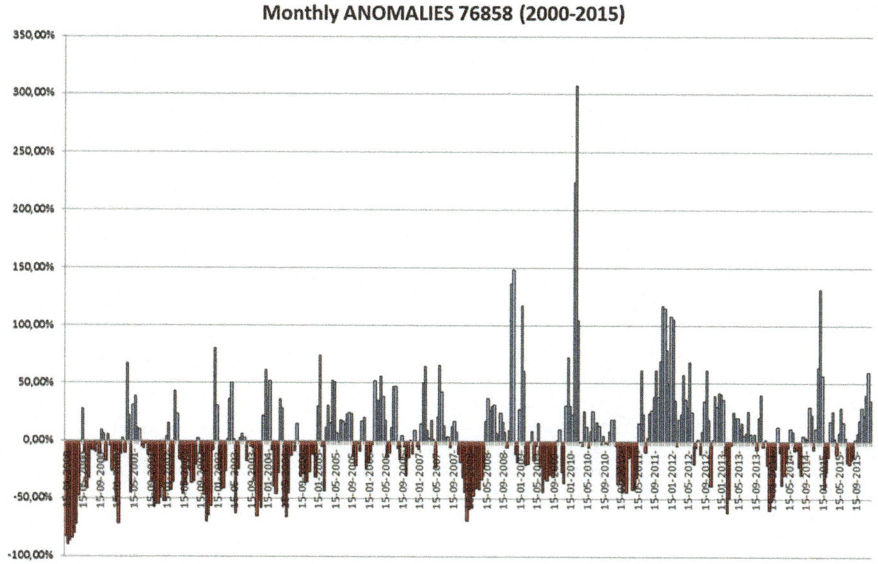

Fig. 5 Monthly anomalies calculated using TRMM data for the watershed id 76858 considering the TRMM historical archive from 2000 to 2015

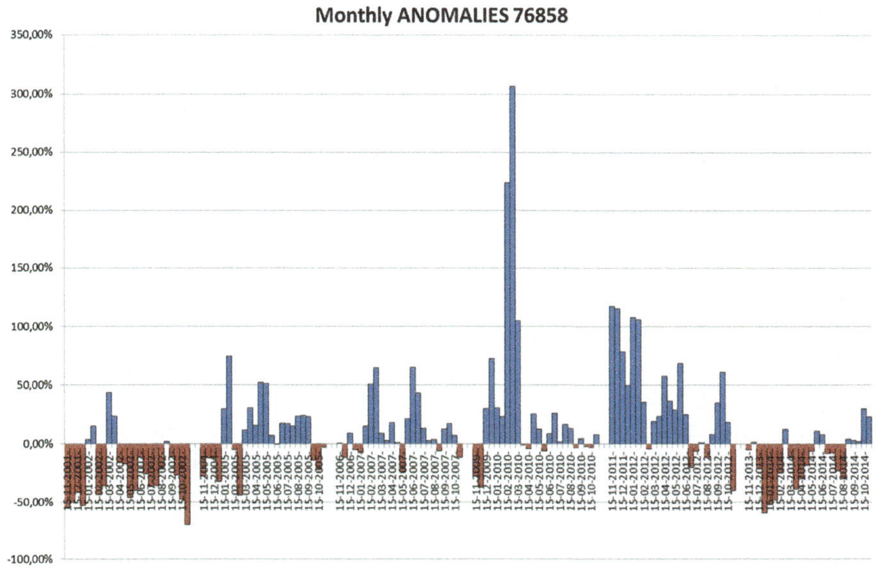

Fig. 6 Monthly anomalies data for the watershed id 76858, considering only the water extent in the analysis years (2002, 2005, 2007, 2010, 2012, 2014)

might also help managers to verify if additional causes for the variation of the water extent exist.

Figure 6 shows the precipitation anomalies for the basin id 76858.

Considering 2002, a negative anomaly is observed just before the beginning of the year, then a slight positive anomaly is recorded during January and finally the following months are characterized by below average precipitations. In 2005 above average rainfalls are recorded in the beginning of the year and after February till October, negative anomalies characterize February and October. The 2007, 2010 and 2012 were mostly affected by above average rainfall while 2014 shows prevalence of negative rainfall anomalies.

4.2 Water Surface Observation from Satellite

The water surface extent was obtained by NDWI using a threshold (NDWI $\geq$ 0.015) suggested in literature in the selected regular periods (January and October of 2002, 2005, 2007, 2010, 2012, 2014).

In Fig. 7, an example of image dataset (Landsat ETM+ of 08 October 2002) obtained after the pre-processing is shown together with the results of the NDWI. The only surface water, which was resulted difficult to extract, is the Lake Shala. This is probably related to the hydro-geochemical characteristics of the lake. All analysed lakes are alkaline and characterized by sodium bicarbonate type waters;

the lake Shala is the most alkaline and rich in geothermal spring in the surrounding (Alemayehu et al. 2006), affecting the reflectance values in the green and NIR bands.

Once identified the proper parameters for the surface water extraction, multi-temporal analysis in the shoreline definition was carried out. An example of multi-temporal analysis results is reported in Fig. 8, where a detail of the Lake Abiyata shoreline monitoring is represented.

Table 2 reports the surface extension changes of each monitored lakes.

Table 3 provides the water surface variation for the watersheds id 76858 by analogy with the precipitation analysis. The total surface extension does not include the Lake Shala due to the problems mentioned above.

A qualitative analysis was conducted by comparing the precipitation trends, the rainfall anomalies values and the variation of water extent.

In 2002, the records of two lakes are missing, negative anomaly are observed from January till October. This anomaly reflect the water extent variation of the Abyata Lake, whereas the Langano Lakes shows almost no variations and the Awassa Lake have an anomalous behaviour in respect to the rainfall.

Up to 2005 abundant negative monthly anomalies are present, whereas since 2005 positive anomalies increase. In 2005, lakes do not record important variations, except for the Koka Lake and the Abyata Lake, which shows a decrease in water extent in October.

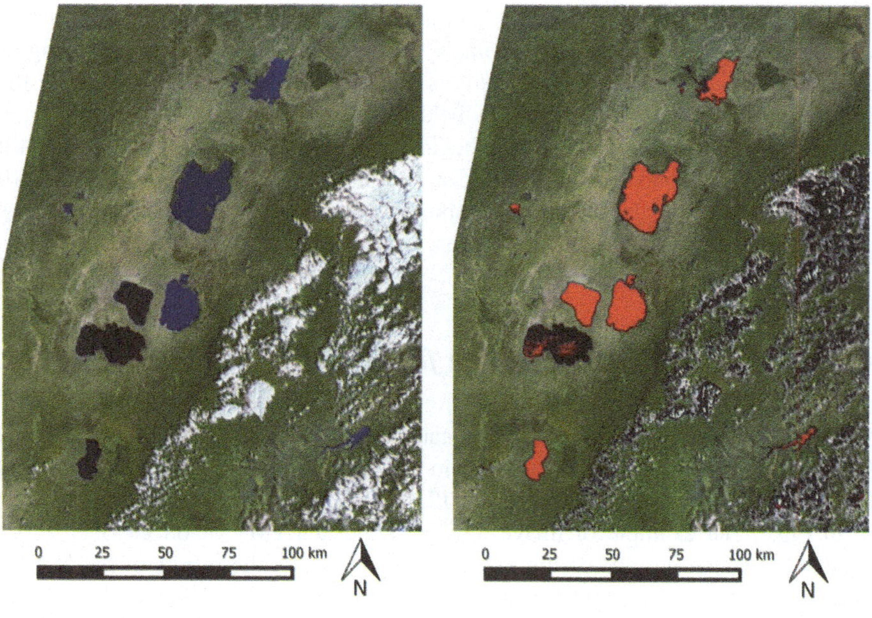

Fig. 7 The Landsat ETM+ 08 October 2002, RGB-753 false color composite (*left*) and the normalized difference water index (NDWI ≥ 0.015) (*right*)

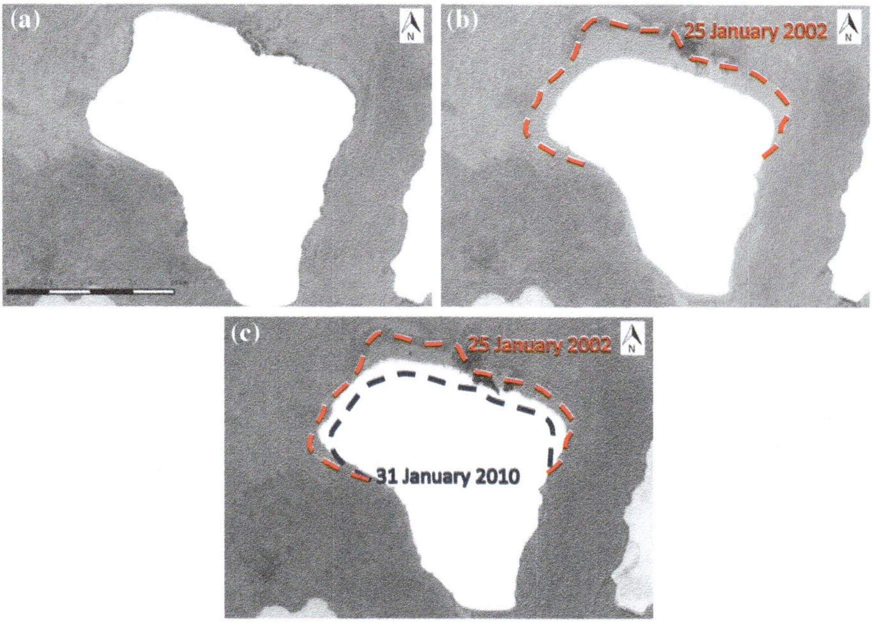

Fig. 8 A detail of the results of NDWI, Landsat ETM+ 25 January 2002 (**a**), 31 January 2010 (**b**) and 26 January 2014 (**c**) (Lake Abiyata)

Table 2 Extension of each lakes monitored in established periods (January and October of 2002, 2005, 2007, 2010, 2012, 2014)

		Surface extension (km^2)				
		Lake Abiyata	Lake Langano	Lake Ziway	Lake Awassa	Lake Koka
2002	Jan.	162	227	–	87	–
	Oct.	156	228	410	90	136
2005	Jan.	104	226	409	89	144
	Oct.	95	226	410	89	156
2007	Jan.	107	227	410	89	150
	Oct.	119	229	414	90	164
2010	Jan.	113	224	–	87	103
	Oct.	122	227	413	88	157
2012	Jan.	125	226	413	88	144
	Oct.	122	221	412	87	148
2014	Jan.	129	225	374	87	–
	Oct.	128	227	408	89	147

During 2007, 2010 and 2014 the water extent variation matches with the precipitation seasonal trends, giving higher values in October than in January. While 2007 and 2010 show mostly positive anomalies, in the 2014 mostly negative

Table 3 Surface water variation in watershed id 76858 calculated for established periods (January and October of 2002, 2005, 2007, 2010, 2012, 2014)

		Watershed id 76858 (MER) (surface extension km^2)
2002	Jan.	–
	Oct.	794
2005	Jan.	739
	Oct.	731
2007	Jan.	744
	Oct.	762
2010	Jan.	–
	Oct.	763
2012	Jan.	764
	Oct.	755
2014	Jan.	728
	Oct.	763

anomalies are present. In 2012 a good relationship exists among the precipitation records and the water extent showing a higher water extent in January than in October due to the significant above average rainfall occurred at the end of 2011. Correlations on the Lake Koka are less meaningful due to the anthropogenic control on this artificial basin.

5 Conclusion

The evaluation of the water storage variations is critical for studies about the impact of climate change and human activities on the terrestrial water resources. Changes in lakes water levels can reflect water mass balance of a basin and are related with climate parameters, such as precipitation and temperature (Cretaux et al. 2011; Zhang et al. 2011).

This paper illustrates an example of methodology for the monitoring of surface water bodies extension and precipitation trends using open-source Earth observation data (Landsat and TRMM). The remotely sensed data are particularly valuable for performing analysis in areas hardly reachable or where little amount of information and resources are available. Advantages consist in the easier and faster collection of data, avoiding long and sometime dangerous field campaigns that often require expensive field instrumentations. Moreover, geospatial technologies allow an immediate global vision of extended areas of interest (i.e. watersheds scale or more).

In this work, remotely sensed data were used to provide information about the spatiotemporal changes of rainfall and water extent over the central Ethiopian Rift in the period 2000–2015. The cumulative rainfall and the anomalies of precipitation were analysed together with the variation of water extent of the main lakes of the MER using NDWI. Data were subsequently qualitative compared.

The results provide quite a good correlation between the water surface extent and the precipitation trends. The variation in lakes extent is particularly evident since 2009 (particularly on 2010, 2011, 2012), when positive anomalies are predominant and rainfall abundant with well-defined dry/wet season. Abiyata is the lake that shows the better link between precipitations and water extent. The Awassa Lake does not show important seasonal variation in water extent, whereas the Lake Koka is characterised by some anomalies in the water extent variation in relation to rainfall, probably due to the anthropogenic control on this artificial basin.

The approach proved the reliability of remote sensing data and its potentiality in the assessment of water surface monitoring. Both rainfall and water extent analysis are an essential sources of information needed by decision makers to implement water conservation plans at national, regional or local scale.

Nevertheless, water extent reduction can also be influenced by other causes, as for example anthropogenic actions (deforestation, irrigation, water pumping, etc.), natural factors (surface-groundwater relationship and water balance) and further climatological parameters (i.e. evapotranspiration). Satellite data can similarly contribute to the definition of the role that such factors play on the water surface extent. For example, with regard to the land-uses, the satellite data can easily and rapidly quantify the variation in the extension of agricultural and urban areas, the deforestation increase or point out the presence of mineral exploitation and anthropic water channels.

Due to the influence of the water lakes on the lifestyle of the people living in the area (industry, agriculture, fishing and human consumption) (Ayenew 2007), the proposed approach allows the implementation of a systematic database regarding lakes extent in relation to precipitation and allow the local and national authorities to use monitoring methodologies for a proper water management. The availability of open source data, the versatility and the possible purposes of the method allow the application of this method in other areas characterized by similar problems.

The present paper shows possible applications of modern methodologies in the climate change research and represents an example of the first step analysis in a complex framework were other factors have to be considered. A possible further improvement of the proposed research might include additional information, such as land uses and monitoring of further climatological variables.

Acknowledgements Landsat Surface Reflectance products courtesy of the U.S. Geological Survey Earth Resources Observation and Science Center.

References

Abebe, B., Boccaletti, M., Mazzuoli, R., Bonini, M., Tortorici, L., & Trua, T. (1998). *Geological map of the Lake Ziway—Asela region (Ethiopian Rift). 1:50000 scale*. C.N.R., ARCA, Firenze, Italy.

Alemayehu, T., Ayenew, T., & Kebede, S. (2006). Hydrochemical and lake level changes in the Ethiopian rift. *Journal of Hydrology, 316*(1–4), 290–300.

Ayenew, T. (1998). *The hydrogeological system of the lake district basin, Central Main Ethiopian rift*. Published PhD thesis, Free University of Amsterdam, The Netherlands, p. 259.

Ayenew, T. (2007). Water management problems in the Ethiopian rift: Challenges for development. *Journal of African Earth Sciences, 48*, 222–236.

Benvenuti, M., Carnicelli, S., Belluomini, G., Dainelli, N., Di Grazia, S., Ferrari, G. A., et al. (2002). The Ziway-Shala lake basin (main Ethiopian rift, Ethiopia): A revision of basin evolution with special reference to the Late Quaternary. *Journal of African Earth Sciences, 35*, 247–269.

Boccaletti, M., Bonini, M., Mazzuoli, R., Bekele, A., Piccardi, L., & Tortoricci, L. (1998). Quaternary oblique extensional tectonics in the Ethiopian Rift (Horn of Africa). *Tectonophysics, 287*, 97–116.

Boccaletti, M., Bonini, M., Mazzuoli, R., & Trua, T. (1999). Pliocene-Quaternary volcanism and faulting in the northern Main Ethiopian Rift (with two geological maps at scale 1:50,000). *Acta Vulcanologia, 11*(1), 83–97.

Chernet, T. (1982). *Hydrogeology of the lakes region, Ethiopia (Lakes Ziway, Langano, Abiyata, Shala and Awasa)*. Report Ethiopian Institute Geological Survey, Addis Abeba, Ethiopia, p. 97.

Chowdary, V. M., Vinu Chandran, R., Neeti, N., Bothale, R. V., Srivastava, Y. K., Ingle, P., et al. (2008). Assessment of surface and sub-surface waterlogged areas in irrigation command areas of Bihar state using remote sensing and GIS. *Agricultural Water Management, 95*, 754–766.

Crétaux, J.-F., Jelinski, W., Calmant, S., Kouraev, A., Vuglinski, V., Bergé-Nguyen, M., et al. (2011). SOLS: A lake database to monitor in the Near Real Time water level and storage variations from remote sensing data. *Advances in Space Research, 47*, 1497–1507.

Di Paola, G. M. (1970). *Geological-geothermal report on the central part of the Ethiopian Rift Valley*. Report Ethiopian Institute Geological Survey, Addis Abeba, Ethiopia, p. 46.

Furi, W., Razack, M., Abiye, T. A., Belachew, D. L., & Ayenew, T. (2011). Fluoride enrichment mechanism and geospatial distribution in the volcanic aquifers of the Middle Awash basin, Northern Main Ethiopian Rift. *Journal of African Earth Sciences, 60*(5), 315–327.

Gizaw, B. (1993). Aluto-Langano geothermal field, Ethiopian Rift Valley: Physical characteristics and the effects of gas on well performance. *Geothermics, 22*(2), 101–116.

Gizaw, B. (1996). The origin of high bicarbonate and fluoride concentration in waters of the Main Ethiopian Rift Valley, East African Rift System. *Journal of African Earth Sciences, 22*(4), 391–402.

Huffman, G. J., Adler, R. F., Bolvin, D. T., & Nelkin, E. J. (2010). The TRMM Multi-satellite Precipitation Analysis (TMPA). In F. Hossain & M. Gebremichael (Eds.), *Satellite rainfall applications for surface hydrology* (Chapter 1, pp. 3–22). Springer.

Hughes, D. A. (2006) An evaluation of the potential use of satellite rainfall data for input to water resource estimation models in southern Africa. *Climate Variability and Change—Hydrological Impacts, 308*, 75–80.

IPCC. (2007). Climate change 2007: The physical science basis: Contribution of Working Group I to the Fourth Assessment Report of the Intergovernmental Panel on Climate Change. Cambridge. In S. Solomon, D.Qin, M. Manning, Z. Chen, M. Marquis, K. B. Averyt, M. Tignor, & H. L. Miller (Eds.), (p. 996) Cambridge, United Kingdom and New York, NY, USA: University Press.

IPCC. (2014). Climate Change 2014: Synthesis Report. Contribution of Working Groups I, II and III to the Fifth Assessment Report of the Intergovernmental Panel on Climate Change. In R. K Pachauri & L. A. Meyer (Eds.), (p. 151) Geneva, Switzerland: IPCC.

Ji, L., Zhang, L., & Wylie, B. (2009). Analysis of dynamic thresholds for the normalized difference water index. *Photogrammetric Engineering and Remote Sensing, 75*, 1307–1317.

Le Turdu, C., Tiercelin, J. J., Gibert, E., Travi, Y., Lezzar, K., Richert, J., et al. (1999). The Ziway-Shala lake basin system, Main Ethiopian Rift: Influence of volcanism, tectonics, and climatic forcing on basin formation and sedimentation. *Palaeogeography, Palaeoclimatology and Palaeoecology, 150,* 135–177.

Lehner, B., Verdin, K., & Jarvis, A. (2006) *HydroSHEDS technical documentation*. Washington, DC: World Wildlife Fund US. http://hydrosheds.cr.usgs.gov

Mahdy, J., & Al-Najar, H. (2015). Climate change and its impact on domestic water consumption in Sub-urban regions in the Gaza Strip. *International Journal of Climate Change Strategies and Management, 7*(1), 3–16.

Masek, J. G., Vermote, E. F., Saleous, N. E., Wolfe, R., Hall, F. G., Huemmrich, K. F., et al. (2006). A Landsat surface reflectance dataset for North America, 1990–2000. *IEEE Geoscience and Remote Sensing Letters, 3*(1), 68–72. doi:10.1109/LGRS.2005.857030

McFeeters, S. K. (1996). The use of the Normalized difference water index (NDWI) in the delineation of open water features. *International Journal of Remote Sensing, 17,* 1425–1432.

McFeeters, S. K. (2013). Using the Normalized Difference Water Index (NDWI) within a geographic information system to detect swimming pools for mosquito abatement: A practical approach. *Remote Sens, 5,* 3544–3561.

Munyaneza, O., Wali, U. G., Uhlenbrook, S., Maskey, S., & Mlotha, M. J. (2009). Water level monitoring using radar remote sensing data: application to Lake Kivu, central Africa. *Physics and Chemistry of the Earth, 34,* 722–728.

Nath, R. K., & Deb, S. K. (2010). Water-body area extraction from high resolution satellite images-an introduction, review, and comparison. *International Journal of Image Processing (IJIP), 3*(6), 265–384.

Ouma, Y. O., & Tateishi, R. (2006). A water index for rapid mapping of shoreline changes of five East African Rift Valley lakes: An empirical analysis using Landsat TM and ETM+ data. *International Journal of Remote Sensing, 27*(15), 3153–3181.

Rango, T., Bianchini, G., Beccaluva, L., & Tassinari, R. (2010). Geochemistry and water quality assessment of central Main Ethiopian Rift natural waters with emphasis on source and occurrence of fluoride and arsenic. *Journal of African Earth Sciences, 57,* 479–491.

Rokni, K., Ahmad, A., Selamat, A., & Hazini, S. (2014). Water feature extraction and change detection using multitemporal Landsat imagery. *Remote Sensing, 6,* 4173–4189.

Romilly, T. G., & Gebremichael, M. (2011). Evaluation of satellite rainfall estimates over Ethiopian river basins. *Hydrology and Earth System Sciences, 15,* 1505–1514.

Scheel, M. L. M., Rohrer, M., Huggel, C., Santos Villar, D., Silvestre, E., & Huffman, G. J. (2011). Evaluation of TRMM Multi-satellite Precipitation Analysis (TMPA) performance in the Central Andes region and its dependency on spatial and temporal resolution. *Hydrology and Earth System Sciences, 15,* 2649–2663.

Shaban, A., Robinson, C., & El-Baz, F. (2009). Using MODIS images and TRMM data to correlate rainfall peaks and water discharges from the Lebanese Coastal Rivers. *Journal of Water Resource and Protection, 1*(4), 227–236.

Stisen, S., & Sandholt, I. (2010). Evaluation of remote-sensing-based rainfall products through predictive capability in hydrological runoff modeling. *Hydrological Processes, 24*(7), 879–889.

Tang, Z., Ou, W., Dai, Y., & Xin, Y. (2013). Extraction of water body based on Landsat TM5 imagery—A case study in the Yangtze river. *Advances in Information and Communication Technology, 393,* 416–420.

U.S. Geological Survey. (2015a). Product guide—Provisional landsat 8 surface reflectance product. Version 1.8, December 2015.

U.S. Geological Survey. (2015b). Landsat 4–7 climate data record (CDR) surface reflectance. Version 6.1, December 2015.

US Geological Survey (USGS) and US Department of the Interior. Normalized Difference Water Index (NDWI). http://deltas.usgs.gov/fm/data/data_ndwi.aspx. Accessed on May 3, 2013. Page Last Modified: November 2016

Vallet-Coulomb, C., Legesse, D., Gasse, F., Travi, Y., & Chernet, T. (2001). Lake evaporation estimates in tropical Africa (Lake Zeway, Ethiopia). *Journal of Hydrology, 245,* 1–18.
Woldegabriel, G., Aronson, J., Walter, R. (1990). Tectonic development of the Main Ethiopian rift. *Geological Society of America Bulletin, 102,* 439–458.
Xu, H. (2006). Modification of normalised difference water index (NDWI) to enhance open water features in remotely sensed imagery. *International Journal of Remote Sensing, 27,* 3025–3033.

Climate Change Threatens Major Tourist Attractions and Tourism in Serengeti National Park, Tanzania

Halima Kilungu, Rik Leemans, Pantaleo K.T. Munishi and Bas Amelung

1 Introduction

Nature-based tourism in Tanzania is largely conducted in national parks. Serengeti is the keystone national park and one of the most well-known world heritage sites. The land cover diversity and the vast endless Savannah plains in the park host the last remaining wildlife migration of about 1.3 wildebeests (Sinclair et al. 2015). The park contributes substantially to the national economy compared to other fifteen national parks (Eagles and Wade 2006) and provides a significant employment opportunities for Tanzanians (Melamari 1996). Nature-based tourism in Serengeti is thus very important in Tanzania.

The rainfall patterns in Serengeti are very variable and its wildlife species are generally well adapted to these variations. A quarter of its main tourist attractions (i.e. wildebeest migration) is triggered by rainfall (Boone et al. 2006; Musiega et al. 2006), and two-thirds is driven by rain-fed food availability (Mduma et al. 1999; Boone et al. 2006) and drinking water (Walonski and Gereta 2001; Strauch 2013). Thus, the significance of rainfall variations in Serengeti's tourism is powerful and should not be undermined.

The timing, amount, and duration of both short and long rains are fundamental determinants of the diversity of major tourist attractions and day-to-day tourism activities in Serengeti. The normal rainfall seasons (i.e. March to May and October to December) in southern Serengeti provide the conducive growth environment for short grass which contain high nutritional content required for both calves and

H. Kilungu (✉) · R. Leemans · B. Amelung
Environmental Systems Analysis Group, Wageningen University,
P.O. Box 47, 6700 AA Wageningen, The Netherlands
e-mail: halima.hassan@wur.nl

P.K.T. Munishi
Department of Forest Biology, Sokoine University of Agriculture,
P.O. Box 3010, Morogoro, Tanzania

W. Leal Filho et al. (eds.), *Climate Change Adaptation in Africa*,
Climate Change Management, DOI 10.1007/978-3-319-49520-0_23

lactating wildebeests (Walonski et al. 1999; Walonski and Gereta 2001). As a result, each year wildebeests give birth synchronously with almost half a million calves being born in between February and March in southern short grasslands (Estes 1976). The synchronised breeding results to large aggregations of wildlife that form a significant attraction to tourists. In case the synchronised breeding coincides with drought or shifted rainfall seasons, the survival for both calves and lactating mothers is reduced (Estes 1976), consequently the likeliness of tourists observing large wildlife aggregations is less guaranteed.

Between 1900s and 1970s, a series of droughts of different magnitudes have occurred in the Serengeti ecosystem with a minimum return period of ten years (FAO 2010). Due to these extreme events and associated diseases (e.g. rinderpest), the wildebeest population dropped to 200,000 in 1950s and the migration stopped due to smaller and fragmented wildebeest population. Migration resumed in post-1960s (Mduma et al. 1999). Recently, a severe drought caused the wildebeest population in the Kenyan Amboseli ecosystem to collapse by more than 85% (FAO 2010; MEMR 2012). In 2010, its population numbered only 3000 animals out of over 15,000 animals a year before (Ogutu et al. 2011; MEMR 2012; Ogutu et al. 2013). These climatic (and or weather) events affected the perceived attractiveness and in turn tourism. Recent studies (Hastenrath 1984; Hemp 2005, 2009; IPCC 2007a; Munishi et al. 2009, 2010) show that since 1980s, the rate of rainfall and temperature variations has exceeded those of the past 100 years. The increased climate variability has likely changed the return period of extreme events in Serengeti National Park (SENAPA), and likely interfered tourism seasonality and natural land-cover that supports tourism.

Climate indirectly affects nature-based tourism by impacting the physical resources (i.e. land-cover) that define the nature and quality of natural environments on which tourism depends (Scott 2005). Land cover heterogeneity and the diversity of land patches harbour a variety of tourist attractions and provide a range of habitats and food for large aggregations of wildebeests in Serengeti. Any changes in the properties of land-cover could negatively influence tourism by reducing wildlife diversity and thus its perceived attractiveness. According to Turpie and Siegfried (1996) and Sandra et al. (2005), changes in land-cover is the major factor that reduces the diversity of tourist attractions and diminishes the attractiveness of most protected areas in South Africa. Savannah grasslands, which are the major vegetation cover in the Serengeti and a pillar of tourist attractions, are also highly sensitive to short-term rainfall variations and highly vulnerable to long-term changes (Vanacker et al. 2005). Habitat modification and loss are the major consequences of land-cover change (Balmford et al. 2001; Brooks et al. 2002; Duerksen and Snyder 2005). The ecological consequences of land-cover changes are well studied in SENAPA (Trager and Mistry 2003; Sinclair et al. 2007; Sharam et al. 2009) in contrast to its relevance for tourism management.

In summary, climate is a powerful driver of wildlife migration and likely a driver of tourism in SENAPA. Thus, understanding the relationship between climate and tourism and consequences of climate change on tourism becomes a pressing issue to tourism planning and informing policy. In recent decades, there is a growing concern that climate change threatens wildlife migrations in East Africa (NAPA

2007; Ogutu et al. 2011, 2012, 2013). The increased frequency and severity of droughts and floods that are expected to occur will modify vegetation growth and hence wildlife's food availability (IPCC 2012). These changes are expected to influence the Serengeti's attractiveness and tourism, though the consequences of climate-change on Serengeti's tourism are poorly studied. This knowledge gap is becoming more important as the economic significance of tourism is growing and the impact of climate change on tourism is becoming more apparent.

This study analyses the consequences of the long-term climate variability and change, and associated land-cover change on the major tourist attractions and tourism in SENAPA. Specifically the study addresses two major questions. (1) What is the relationship between rainfall, tourist visits, and tourist attractions? (2) How have contemporary climate variability and change affected tourist attractions? To answer these two questions, the study (1) assessed the relationship between rainfall variations, tourist visits, and wildebeest migration, (2) analysed the impact of rainfall and temperature variability and change in this relationship (3) assessed the impact of climate-driven land-cover change on tourist attractions and its consequences on tourism for the past 40-years (i.e. from 1970 to 2010). The results of these analyses were combined to suggest possible adaptation measures for tourism in SENAPA. This study strongly builds on the relationship between wildebeests and rainfall that was developed by Boone et al. (2006).

The next sections of this paper provide a brief description on the methods; the case study, data used and the methodology for analyses. Then, the results are presented briefly followed by a detailed discussion on the consequences of recent climatic trends and land-cover change on major tourist attractions and tourism. The conclusion highlights on necessary adaptation measures to reduce adverse impact of climate change on tourism in SENAPA.

2 Materials and Method

2.1 Description of the Study Site

The focus of this study is the Serengeti National Park. The park covers almost fifty percent of the major Serengeti-Mara Ecosystem. The park is largely operating under a '*closed managed system model*' which means almost ninety percent of its ecosystems are legally protected and surrounded by other protected ecosystems (Fig. 1). The park boundaries lie between 34° to 36° E, 1° 15′ to 3° 30′S and stretches over Northern Tanzania and Southern Kenya. As a result, the park's attractions are unique as it falls in the Somali-Masai centre of plants and animal endemism (White 1983). Its location between Lake Victoria in the West and Lake Eyasi in the South strongly influences the wildlife's food availability and makes the wildlife-based tourism reliable in dry season. The Great Rift Valley to the East and the Ngorongoro Crater to the South enhance the plant and animal diversity, in turn this makes the park the most attractive in Africa. Altitude ranges from 920 to

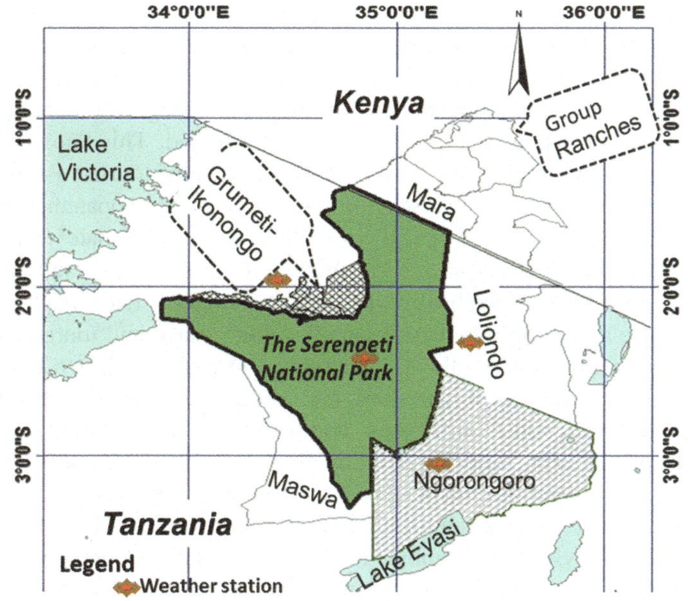

Fig. 1 A map of the Serengeti-Mara Ecosystem showing the Serengeti National Park as enclosed by other protected areas and weather stations used

1850 m ASL, and mean annual temperature varies from 15 to 25 °C. This situation provides a conducive temperature for the wildlife safari drive. Annual rainfall varies from 450 mm in the lee-wards of the Ngorongoro highlands to 1050 mm closer to the shores of the Lake Victoria (Schaller 1972; McNaughton 1979, 1983; Sinclair 1979; Sinclair et al. 2000). Rainfall is strongly seasonal and occurs in two distinct rainy seasons: a long rainy season from March to May and a shorter rainy season in October to December. The slight seasonal rainfall variation between North and South enhances the cyclical annual wildlife migration that attracts tourist worldwide. The weather stations close to the park provided this study with quality and historical rainfall and temperature data (1970–2010) for climate change analyses. The weather stations (Fig. 1) are owned by Tanzania Meteorological Agency and include; Mugumu, representing the northwest zone, Seronera, representing the central zone, Ngorongoro Crater, representing the southern zone, and Loliondo, representing the eastern zone of SENAPA.

2.2 Data Collection and Analyses

2.2.1 Climate and Tourist Arrivals Data

To assess the relationship between seasonal variations in rainfall, tourism and wildebeest migration calendar, the study used monthly rainfall and tourist visitation (i.e. number of person visits) data for the years 2000–2010, and wildebeest migration calendar as influenced by rainfall based on Boone et al. (2006). The study used EXCEL to calculate the correlation coefficient (r) to assess the strength of the relationship between rainfall and tourist visitations whereby −1 indicates a perfect negative and +1 indicates a perfect positive relationship.

Furthermore, this study aimed to assess how the long-term rainfall variability and change threatens the major tourist attractions and tourism. To achieve this objective, the study used forty years rainfall data (i.e. 1970–2010) to analyse the rainfall trend in the different zones within SENAPA. The statistical test (p-value) was used as an indicator of a significant change. In addition, the study assessed the trends for different climatic seasons that correlate with tourist visitation patterns in order understand the influences of rainfall on wildebeest migration and in turn tourism. These seasons were; the long rain season of March to May (i.e. MAM), the dry season (June, July, August and September, JJAS), the short rain season of October to December (i.e. OND) and the transition dry season (January and February, JF).

The analysis of seasonal rainfall aimed to assess how climate variability and change affect tourist attractions and day-to-day tourism activities. Increasing rainfall was used as an indication of food availability, consequently concentration of large aggregations of wildlife migration that in turn influencing the focus of tourism in these areas. Decreasing rainfall means that wildebeests will be scattered searching for food and water, in this case, tourism is expected to be chaotic because locating large aggregations of wildebeest will be difficult. In addition, we compared the amount of rainfall received in peak months to see if seasonal shifts exist and integrated it into the migration calendar stability and tourism seasonality. Temperature data were subjected to trend analyses to assess the warming trends and its influence on tourism. The mean monthly temperature variations were compared with the traditional comfort temperature (i.e. normal temperature in 1970s) used for the wildlife safari drives.

The results of annual rainfall and temperature trends were finally superimposed on wildebeest population data. The aim was to assess how climate variability and change have influenced the major tourist attractions (i.e. wildebeests) in SENAPA.

2.2.2 Environmental and Other Qualitative Data

Landsat TM7 and Enhanced Thematic Mapper plus (ETM+) images were used to assess the land-cover change as an indicator of food change for wildlife and changes in the perceived attractiveness. Change detection techniques in ArcGIS were performed to assess land-cover change. The study applied the Normalised Difference Vegetation Index (NDVI) to assess the extent of land-cover change (Pettorelli et al. 2005; Zurlini et al. 2006). The resulting land-cover maps of 1984, 1995, and 2009 were used to indicate changes in suitable areas for wildebeests. Image used for this analysis were for July 1984, July 1995, and October 2009 which are periods that experienced severe droughts (FAO 2010) and can represent possible climate-change impacts. The images were freely acquired from GLOVIS, the USGS Global

Visualization Viewer (www.glovis.usgs.gov/). Wildebeest population raw data for 1971–2010 were acquired from Grant Hopecraft (grant.hopcraft@glasgow.ac.uk).

The information collected during field observation and key informant interview from Serengeti park authority (i.e. Tourism and ecology wardens), the hotel managers, and hot air balloon safari manager in March–April 2013 hinted on the consequences of already felt climate-change impacts on tourism and coping strategies in place.

3 Results

3.1 Relationship Between Climate, Tourism and Tourist Attractions

Our findings indicate that tourist visits in SENAPA are marked in two distinct seasons with low rainfall. During the dry season months of June to September and December to February (i.e. high tourism season), tourist visits were found to be high (~20,000 per month). The rainy seasons in March to May and October to November (i.e. low tourism season) experienced the lowest visits of about 5000 per month (Fig. 2). These findings demonstrate that natural climate in particular rainfall is an important factor shaping tourism seasonality in SENAPA though institutional factors like the summer or the end of the year holidays in the tourist's country of origin likely to play a role also. The climate-tourism dependency was also proven

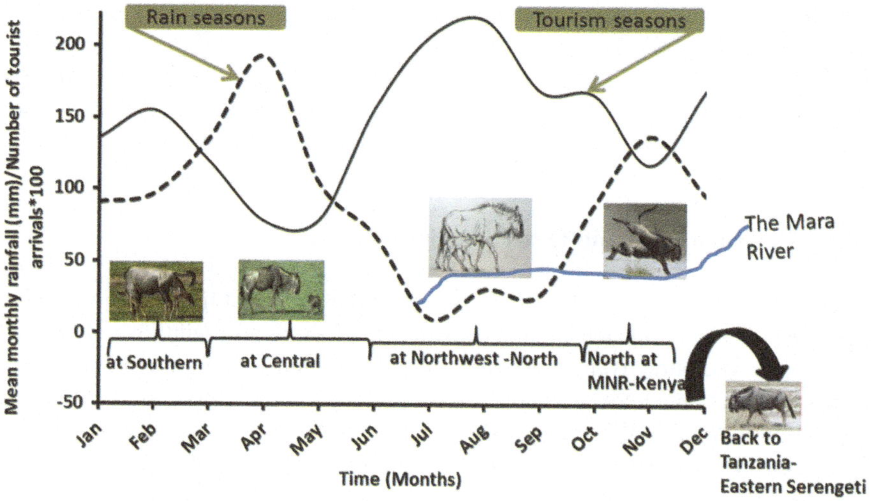

Fig. 2 The relationship between rainfalls, tourism seasons, and known wildlife migration viewing areas within SENAPA from January to December (the *photos* represent the location of migrating wildebeest in a year)

by a high negative correlation coefficient (r = −0.8) between rainfall and tourism visitations (Fig. 2). Integrating rainfall-tourism relationship with the known wildebeest migration calendar give the insight that the December to February tourism is mainly conducted in the Southern zone. This coincides with the synchronous wildebeests breeding. June to September tourism occurs in the Northwest Serengeti along the Mara River where wildebeests feed during the dry season. Tourism in South-central Serengeti (i.e. Seronera, Lake Ndutu, and Magadi) is mainly conducted in March to May where wildebeests feed during this period (Fig. 2).

3.2 Consequences of Climate Change on Tourist Attractions and Tourism Seasons

Rainfall in SENAPA has become highly variable despite the fact that long-term climate trends show no significant change ($p > 0.05$). Lack of such trend does not mean no impacts rather highlight the high climate variability in a short-term (i.e. monthly). Some months received unusual heavy rains (i.e. climate anomalies) that masked the trend. In addition, rainfall received in different seasons either in the Northern or Southern zones has changed significantly (Fig. 3). In the Northwest

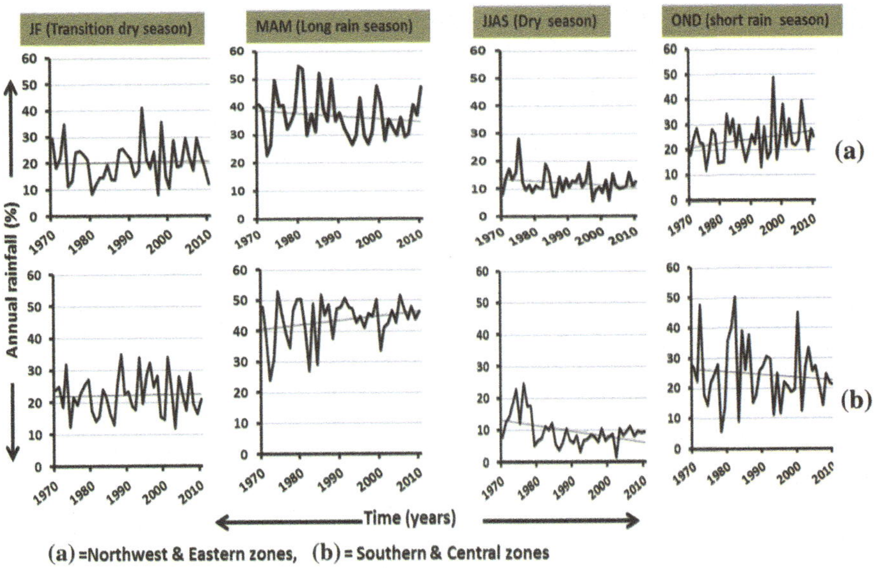

Fig. 3 Changes in rainfall received in various rain seasons from 1970 to 2010 in the Serengeti National Park

Table 1 The rainfall seasonal variability and peak shifts in SENAPA from 1970 to 2010

Amount of rainfall (%) falling in long, short and transition dry season										
		Long rain season (MAM)			Dry season JJAS	Short season (OND)		NDJ	Transition dry season JF	
Zone	Time	Mar	Apr	May		Oct	Nov	Dec	Jan	Feb
Northwest	1970s	8	27	13		6	11	10	9	13
	2000s	13	18	22		8	10	11	10	6
Eastern	1970s	18	24	12		3	7	9	12	10
	2000s	15	17	10		4	12	14	13	13
Southern	1970s	14	24	8		5	10	12	11	10
	2000s	16	18	15		3	5	15	14	11
central	1970s	8	16	9		6	8	9	10	12
	2000s	18	15	10		4	14	10	12	9

Blue circles =peak months, **red arrows** =decrease in rainfall in peak months in 1970s, **blue arrows** =increase in rainfall in non peak months in 2000s

Rainfall in non-peak months in 2000s are increasing at the expense of decreasing rainfall in peak months in 1970s. No specific peaks in 2000s as in 1970s . This study refer the situation as less predictability of rainfall seasons, prolonged dry season and droughtness in Serengeti

and Eastern Serengeti, the amount of rainfall received in the short rainy season (i.e. OND) has increased by 8%, while in South-central Serengeti decreased by 5%. In transition dry season (i.e. JF) no substantial change is shown either in the North or South. The long rain season (i.e. MAM) decreased by 5% in the Northwest-Eastern, and increased by 6% in the Southern-Central zone. In the dry season (i.e. JJAS), rainfall in both zones has decreased by 3% (Fig. 3).These changes have resulted in a substantial decrease in rainfall during peak months of April or November in 1970s making them non-peak months in the 2000s. Contrary to this, March or May for long rains and December/January for short rains received more rains relative to 1970s (Table 1).

If the results in Figs. 2, 3, and Table 1 are suggestive and rainfall is the cue for migration and driver of the Serengeti's attractiveness, its implications for tourism management will be substantive. The results likely to imply a high possibility of migration staying less or not proceeding further to the North (i.e. Masai Mara National Reserve in Kenya) or arrive earlier in the eastern Serengeti because of the increased OND rainfall in this zone. In turn, this will likely improve tourism in the eastern zone. In addition, migration likely to arrive earlier in central Serengeti and delay further going to the northwest that what is actually happening today. The delayed northwest movement is supported by increased MAM rains in central and deceased MAM rains in the northwest. This situation in turn, affects December to February tourism in the southern zone, delays the synchronized breeding in the South, and improves tourism in the central zone. The high tourism season (i.e.

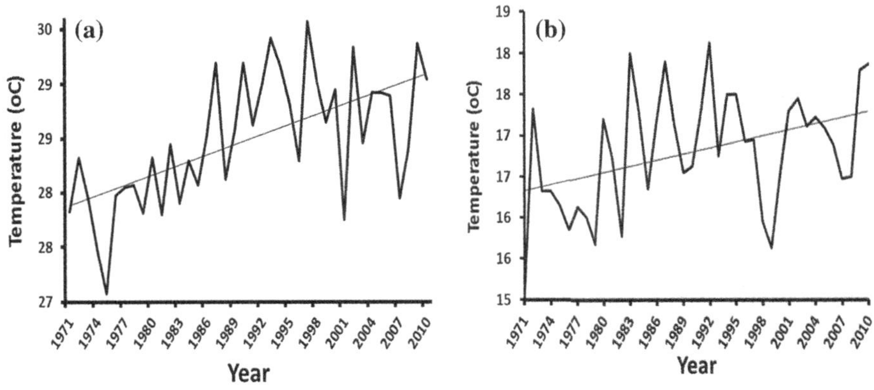

Fig. 4 Temperature trends in high tourism season (JJAS) **a** maximum and **b** minimum temperature

JJAS) largely conducted in Northwest along the Mara River is likely to be chaotic and unreliable. The decreased rainfall likely reduces the Mara River flow and in turn, decreases the foraging areas on riparian ecosystems. This situation results into large aggregations of migrating wildebeest roaming randomly searching for food and water. The SENAPA tourism officials also acknowledged that for quite sometime now migration arrives in the central early than expected and delay further movement to the West and North. For example, in 2012, the migration did not proceed north to Mara National Reserve in Kenya (Ihucha 2012).

Our results show that the park has warmed by 0.8 °C (i.e. approximately 0.2 °C per decade) over the past 40 years (Fig. 4). The warming has accelerated since 1980s, where the temperature has been well above the average (i.e. 23.1 °C). The year 1999 has been the coldest in Serengeti since 1978. The overall mean monthly minimum and maximum temperature during the high tourism season that ranged between 17.5 and 28.1 °C in 1970s have shifted to 18.3 and 28.7 °C in 2000s. The monthly temperatures in this season sometimes exceed 30 °C (Fig. 4).

Since 1990s, drought frequencies have increased and become more intermittent (Fig. 5). The wildebeest population dropped substantially soon after drought and recovered afterwards when rainfall increased. In a severe drought such as the 1993–1999, the population took longer to recover. The El Niño flooding seems destructive as well, but the wildebeest population recovered sooner than in the severe drought event. The population of wildebeests has stabilised at 1.3 million individuals which seems to be its ecological carrying capacity. However, if the observed drought frequencies continue a drop in wildebeest population could be expected. Drought has been the major driver that decreases the size of wildebeest aggregations thus reducing the attractiveness of the park. This situation threatens the future of migration driven form of tourism in SENAPA.

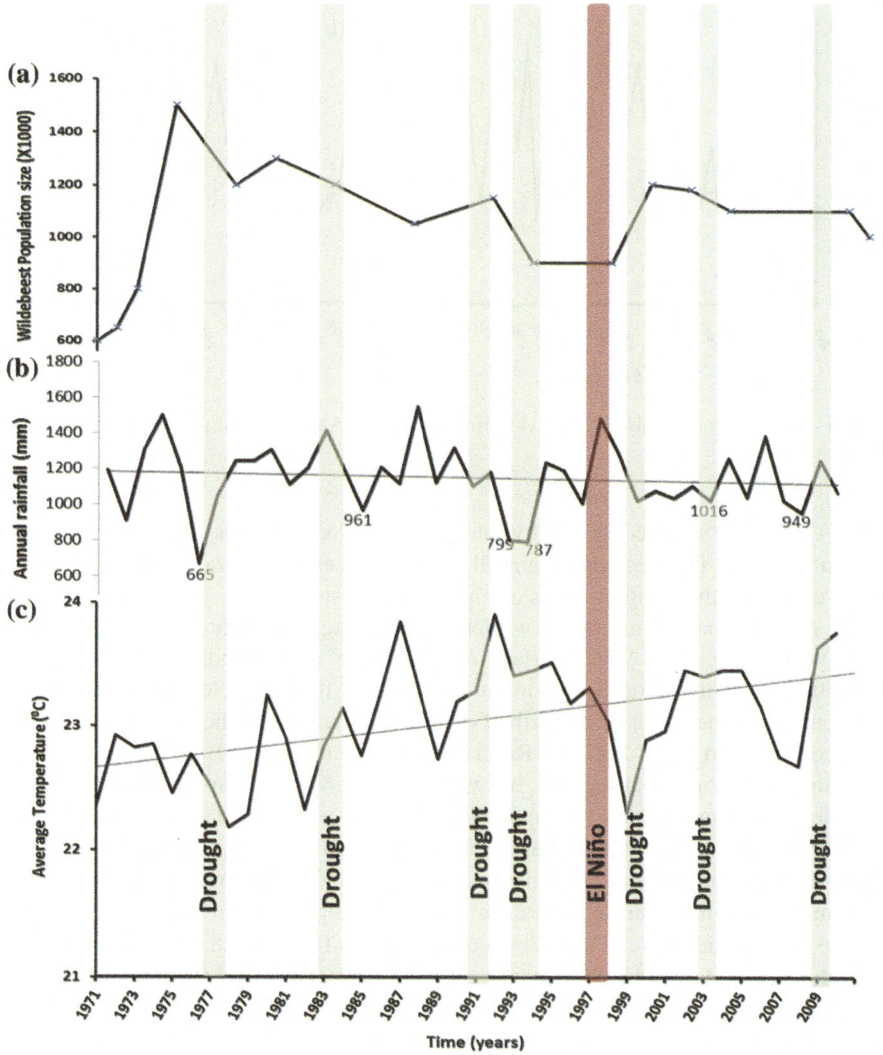

Fig. 5 Wildebeest population dynamics as influenced by climate variability and change from 1970 to 2010 [wildebeest population (**a**), rainfall variability and change (**b**), and temperature variability and change (**c**)]

3.3 *Consequences of Land-Cover Changes on Tourist Attractions in SENAPA*

Savannah grasslands cover more than fifty percent of the park, though our results show that grassland cover has increased by 21%. Furthermore, woodlands decreased by 87%, while riparian forests decreased by 30% and the inland lakes

Table 2 Changes in land cover within SENAPA from 1970 to 2010

Cover types	Area (km^2)			% change
	1984	1995	2009	(1984–2009)
Grassland	8194.6	8062.5	9915.4	21
Open woodland	2099.2	2843.4	272.9	−87
Forest	53.9	55.0	37.8	−30
Lakes	3.7	3.2	3.1	−14
Thickets	2424.2	2439.8	2181.8	−10
Bush land/shrubs	1987.3	1359.2	2351.9	1

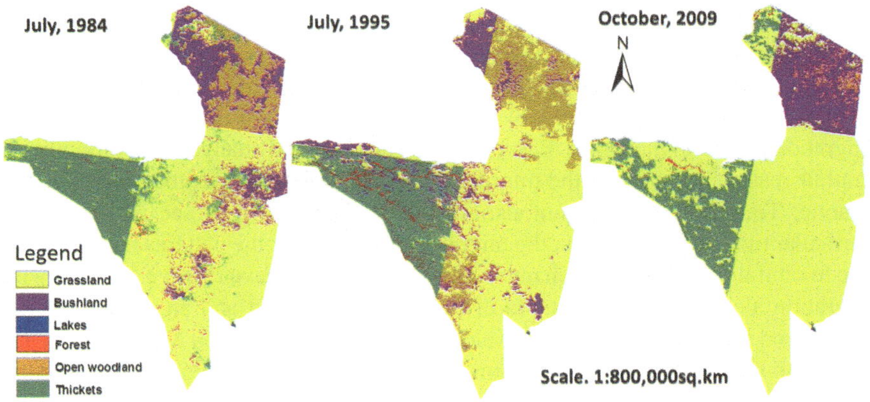

Fig. 6 Changes in land-cover in the Serengeti National Park from 1970 to 2010

(i.e. Ndutu and Magadi) shrank by 14%. These results signify that the park is becoming more homogenous, a situation threatening the diversity of tourist attractions for which the park is famed. In addition, drinking water for wildlife within the park is becoming a pressing issue (Table 2; Fig. 6).

4 Discussions

In this paper, the climate-tourism relationship and consequences of climate change on tourism in the Serengeti National Park have been explored. The findings indicated that climate, tourist visits, and wildebeest migration are closely connected. As such the major tourism activity is largely migration viewing termed '*migration chasing tourism*' in this paper. This form of tourism has already been affected by seasonal rainfall variability and change, reduced rainfall in some seasons, increased temperature by 0.8 °C and changes in land-cover. Tourism in Serengeti is about time (i.e. perfect time to see the migration) and place (where is the migration). Time is of central importance, especially as tourism is generally defined in terms of the use of time (i.e. tourism seasonality) and in turn tourism activities are severely constrained by time availability (Dietvorst and Ashworth 1995). Patterns of tourist visits to many protected areas are typically seasonal, with peak periods occurring in relatively narrow windows of time (Baum and Lundtorp 2001; Hadwen et al. 2008). Despite the fact that traditional institutional arrangements (i.e. holidays) especially for overseas visitors have substantial influences in determining tourism seasonality, various studies have commented on the role of climate (and/or weather) and critical ecological processes, such as animal migration and plant flowering in influencing tourism and recreation in natural areas (de Freitas 2003; Keller et al. 2005; Hadwen et al. 2011).

Accommodation bookings in Serengeti, are based on time (i.e. traditional migration calendar) and location of the accommodation facility (e.g. along the migration routes or close to a hippo pool). In other words, it may seem like '*view the great migration at your hotel window*'. In fact, when migration is in the South or central zones, accommodations in this zone are fully booked and vice when the migration moves to the North, East or West. The observed changes and variability in rainfall seasons have made the migration chasing tourism sometime unreliable and chaotic. Tourists book accommodations where they expect to see migration at a particular time only to find that the migration is delayed or the patterns have changed due to rainfall variability and change. As a result, tourists have to drive far in viewing migration a situation that seems to disturb tourists comfort and an added cost (i.e. money and time) (Kimaro and Kihwele, Personal communication March 2013). The delay and changing migration patterns are the major climate-tourism challenge that the park has to adapt and deal with. At the moment, this challenge may be seen as of less concern. However, in future decades, this may result into either abandonment of some of the accommodation facilities as they cannot offer what they promised to tourists or increases the costs of accommodations to cater for unnecessary disturbances that may arise. If this disturbances increase, either tourism in Serengeti will likely be expensive or tourist may look for alternative destinations. The economic impact of this situation is overwhelming as the park contributes substantially to TANAPA's income and the country at large (Melamari 1996; Eagles et al. 2006).

Serengeti National Park has warmed about 0.2 °C above the observed 0.6 °C global mean temperature rise over the last three decades (Hulme et al. 2001; Hansen et al. 2006). Mean monthly temperature in high tourism season (June, July, August, and September) has increased while the daily temperature sometimes reaches above 30 °C, which would be expected in the lower coastal regions. This rise likely interferes with traditional comfort temperature ($\sim$15 to 25 °C) adapted for the Serengeti safari drive in the 1970s and 1980s. According to Maddison (2001) the maximum daytime temperature for tourism activities should ideally be close to 30 ° C, while Lise and Tol (2002) argue that the optimal minimum daily temperature would be around 21 °C. Understanding the effect of temperature variability and change is important for tourism in Serengeti though Abegg et al. (1998) argue that in global tourism the effects of temperature variability have long been taken for granted because of their supposedly long-term stability. Temperature comfortability (i.e. thermal comfort) is one of the key tourism climatic index that may drive tourists either to choose or not to choose a destination (Amelung and Viner 2006), or guarantee a certain tourism activity to be performed.

Large aggregations of wildlife, which are the major attractions in Serengeti have been decreasing due to increased drought frequencies and persistence, and abnormal rains. These results support various studies (e.g. Ogutu et al. 2008, 2012) that climate change is happening and affects the Serengeti ecosystem defining tourism. Our findings differ from (TMA 2007; Munishi et al. 2009, 2010) that areas with bimodal rainfall patterns like SENAPA will experience increased rainfall of 5–45%. However, our findings concur with Hulme et al. (2001), Michael (2006), IPCC (2007b) that under intermediate warming, December to February rainfall in Eastern Africa for which Serengeti is part will likely increase by 5–20% before the end of the 21st century. All these changes have aggravated the situation interfering with large aggregation of wildebeest that tourist goes for. Wildebeests and other iconic wildlife have been starving to death due to drought and thirst. In 1993, almost half of the wildebeest population and 70% of buffaloes starved to death due to drought (Mduma et al. 1999; Gereta et al. 2003; Sinclair et al. 2007), while in 2010, almost 10,000 wildebeest drowned to death in Mara River (Farouky 2007).

Wildebeest population has never resumed 1.5 million individuals reached in 1977 due to droughts and floods (Mduma et al. 1999). Although droughts in Serengeti are common, they have been more destructive in post-1980s as their frequencies increased and likely there has not been adequate time for the ecosystem to adjust, adapt and build resilience. The massive losses in wildlife have major consequences in conservation and tourism. If wildebeest population decreases to a point of no migration and end up revoking the world heritage site status, this situation will negatively affect tourism as the wildlife migration is used as a competitive advantage and self-tourism marketing strategy. Not only in Serengeti, droughts have been a major driver that changes the attractiveness of most national parks and other protected areas by removing tourism flagship species in East Africa. For instance, in just two weeks, severe drought killed 60–80 hippos in Masai-Mara National Reserve in 2006 (Bogonko and Lee 2006) and left the 'attractive pools' as a 'graveyards' filled up with rotting hippo carcases.

Land cover heterogeneity is a basis for diverse tourist attractions in SENAPA. The presence of a large proportion of grasslands has contributed to the long-term existence of large aggregations of migrating wildlife, which in other parts of the world has become extinct. Large and diverse grasslands nourished by rainfall variations provide adequate food all-year round for migrating wildebeest, an important attraction for tourists within Serengeti. The current land-cover change creates high uncertainty for migration and tourism in future decades. In the 1990s, Sinclair (1995) noticed that more than thirty percent of vegetation cover in Serengeti has changed. Our current results suggest that more than seventy percent of the Serengeti's land cover have changed. This means there is likely less virgin land cover remaining in the park. The consequences of such drastic land-cover change on the attractiveness of the park are numerous. Wildebeest's population will either continue to increase due to increase in grasslands as food won't be a limiting factor or decrease due to habitat loss and/or fragmentation (Mduma et al. 1999; Ogutu et al. 2012).

Moreover, increased grasslands suggest drier conditions, which mean less forage in turn, decrease the population. These changes might have positive effects now for tourism as wildebeest population increases, though in the long-run the number of wildebeests might drop rapidly due to food shortage. Short-lived grasses are replacing trees and other woody plants, which are used as a substitute feed in critical dry seasons. Wildebeests rely heavily on grass for their forages. However, in times of scant grasses, they usually subsist on foliage from trees and shrubs. When they cannot find suitable forage, they wildebeests migrate long distances in search of forages. Changes in land cover not only affects wildebeests but also other tourist flagship species. Between 1966 and 2006 the Serengeti ecosystem has suffered drastic losses of 6–16 species of fruit bird eaters due to 70–80% loss of the Grumeti riparian forest cover (Sharam et al. 2009). If the loss of bird species is cumulatively added to 30% loss in riparian forest in this study, the impact on tourism, especially bird watching will be extensive. The Grumeti riparian forests not only serve as a bird watchers paradise, but also the only area where few Black and White Colobus Monkey population exist and a refugee for migrating wildebeests in critical dry season.

Climate change has already affected tourism in Serengeti. As adaptation measures hot air balloon safari was introduced to maximise migration chasing tourism experiences. However, for the last ten years, flying schedules for the baloons have been cancelled more often due to sudden changes and unfavourable weather conditions (Gereta, Personal communication March 2013). In addition, mobile camping was introduced to match with changing migration patterns and/or delays (Kimaro et al., Personal communication March 2013). These adaptation measures seem the best though this study sees it as temporary measures for transition in tourism resulting from changes in tourist attractions in response to the observed climatic conditions. Not only in the Serengeti but also in the rest of the world, tourism seasons are expected to shift as a response to tourism resource change due to climatic changes (Amelung et al. 2007). For Serengeti, all weather roads given the black cotton soil type likely be the best adaptation option to take the park though this transition period. Other adaptation measures would include reducing human induced impacts on the park and avoid activities that block migration routes, securing sufficient drinking water during the dry season by improving water storage facilities such as dams and underground sources and maintenance sufficient forage areas while avoiding wild fires.

4.1 Limitations of the Study

This study used short-term tourist arrivals data that were available. In addition, the tourist arrivals data did not indicate the types of activities a tourist participated. As a result, all tourists were assumed interested in viewing wildebeest migrations. Availability of long-term tourist arrival data linked with activities that a tourist undertaken in a particular place would improve the future studies of this nature. Despite lack of long-term tourist arrivals data, this study provided insights that tourism in the Serengeti is highly seasonal and climate change is the major threat.

5 Conclusions

Climate, tourist attractions (i.e. wildebeest migration), and tourist visits are closely connected in Serengeti. Changes in climate have directly influenced tourism seasonality and indirectly influenced natural attractiveness of the park by changing the tourism flagship species and changing the natural landscape. There has been a substantial land cover change from heterogeneous towards more homogeneous conditions (i.e. increasing grasslands at the expense of other land-covers). This change threatens the diversity of tourist attractions that the park is famed for. Decreasing rainfall and increasing temperature have led into more drier conditions and increased aridity resulting into shortage of drinking water for wildlife. These changes have a potential to influence the distribution of wildlife and in turn, affects tourism. Natural factors such as climate, wildlife and land cover diversity can, in principle, establish and maintain tourism in Serengeti National Park. However, if strategies for managing these factors are not adequate, the park will gradually lose its touristic appeal, despite the increasing number of tourist visits annually. Adapting tourism to climate change impact requires active and integrated management approaches that improve the park's attractiveness. Thus, the climate-tourism insights provided by this study are of significant importance for tourism planning to maintain the Serengeti's natural attractiveness and day-to-day tourism activities of which climate change has of recent become a threat.

References

Abegg, B., König, U., Bürki, R., & Elsasser, H. (1998). Climate impact assessment in tourism. In J. H. Hohnholz (Ed.), *Applied geography and development* (pp. 81–93). Tübingen: Institute for Scientific Co-operation.

Amelung, B., Nicholls, S., & Viner, D. (2007). Implications of global climate change for tourism flows and seasonality. *Travel Research, 45*(3), 285–296. doi:10.1177/0047287506295937

Amelung, B., & Viner, D. (2006). Mediteranian tourism: Exploring the future with the tourism climatic index. *Sustainable Tourism, 14*(4), 349–366.

Balmford, A., Moore, J. L., Brooks, T., Burges, N., Hansen, L. A., Williams, P., et al. (2001). Conservation conflicts across Africa. *Science, 291,* 2616–2619.

Baum, T., & Lundtorp, S. (2001). Seasonality in tourism: An introduction. In T. Baum & S. Lundtorp (Eds.), *Seasonality in tourism* (pp. 1–5). Oxford, UK: Elsevier Science Ltd.

Bogonko, B., & Lee, M. (2006, January 16). News about planet Earth: The Drought Kills Hippos in Kenyan Wildlife Reserve. *Terra Daily.*

Boone, R. B., Thigood, R. B., & Hopcraft, J. G. (2006). Serengeti wildebeest migratory patterns modeled from rainfall and new vegetation growth. *Ecology, 87*(8), 1987–1994.

Brooks, T. M., Mitttermeier, R. A., Fonseca, C. G., Rylands, A. B., Konstant, R., Flick, P., et al. (2002). Habitat loss and extinction in the hotspots of biodiversity. *Conservation Biology, 16,* 909–923.

de Freitas, C. R. (2003). Tourism climatology: Evaluating environmental information for decision making and business planning in the recreation and tourism sector. *International Journal of Biometeorology, 48,* 45–54.

Dietvorst, A. G. J., & Ashworth, G. J. (1995). Tourism and spatial transformations. In G. J. Ashworth & A. G. J. Dietvorst (Eds.), *Tourism behaviour and the importance of time-space analysis* (pp. 163–181). Wallingford, UK: CAB International.

Duerksen, C., & Snyder, C. (2005). *Nature friendly communities.* Washington, D.C., USA: Island Press.

Eagles, P. F. J., & Wade, D. J. (2006). Tourism in Tanzania: Serengeti National Park. *Bois et Forêts Des Tropiques, 290*(4).

Estes, R. D. (1976). The significance of breeding synchrony in the wildebeest. *East African Wildlife Journal, 14,* 135–152.

FAO. (2010). Analysis of climate change and variability risks in the smallholder sector: Case studies of the Laikipia and Narok Districts representing major agro-ecological zones in Kenya. In O. O. Gordon, A. Jaspat, & S. Charles (Eds.), *Environmental and natural resources working paper* (Vol. 41). Department of Resource Surveys and Remote Sensing (DRSRS) in collaboration with the Food and Agriculture Organization of the United Nations, Rome, Italy.

Farouky, J. (2007, October 9). Global warming: Bad news for Gnus. *Times.*

Gereta, E. J., Walonski, E., & Chiombola, E. A. T. (2003). Assessment of the environmental, social and economic impacts on the Serengeti Ecosystem of the Developments in the Mara River Catchments in Kenya (pp. TANAPA/Frankfurt Zoological society). Serengeti/Arusha, Tanzania.

Gereta, J. (Personal communication March, 2013-Hot air balloon safari manager). *Consequences of climate change on the Hot air Balloon safaris in the Serengeti National Park.*

Hadwen, W. L., Arthington, A. H., Boon, P. I., Brett, T., & Christine, S. F. (2011). Do climatic or institutional factors drive seasonal patterns of tourism visitation to protected areas across diverse climate zones in Eastern Australia? *Tourism Geographies: An International Journal of Tourism Space, Place and Environment, 13*(2), 187–208. doi:10.1080/14616688.2011.569568

Hadwen, W. L., Hill, W., & Pickering, C. M. (2008). Linking visitor impact research to visitor impact monitoring in protected areas. *Journal of Ecotourism, 7,* 1–7.

Hansen, J., Ruedy, R., Sato, M., & Lo, K. (Eds.). (2006). *NASA Goddard Institute for Space Studies and Columbia University of Earth Institute*, New York, NY, USA.

Hastenrath, S. (1984). *The Glacier of Equatorial East Africa: Solid Earth Sciences Library.* Dordrecht, Holand: D. Reidel Publishing Company.

Hemp, A. (2005). Climate change-driven forest fires marginalize the impact of ice cap wasting on Kilimanjaro. *Global Change Biology, 11,* 1013–1023. doi:10.1111/j.1365-2486.2005.00968.x

Hemp, A. (2009). Climate change and its impact on the forests of Kilimanjaro. *African Journal of Ecology, 47,* 3–10.

Hulme, M., Doherty, R., Ngara, T., New, M., & Lister, D. (2001). African climate change: 1900–2100. *Climate Research, 17,* 145–168.

Ihucha, A. (2012, July 25). Tanzania denies blocking wildebeest migration to Maasai Mara. *The East African Newspaper.*

IPCC. (2007a). *Climate Change 2007: The physical science basis. Contribution of working group 1 on the AR4. Report of the IPCC.* Cambridge, UK, and New York, USA: Cambridge University Press.

IPCC. (2007b). Summary for policymakers: In climate change 2007. In M. L. Parry, et al. (Eds.), *Impacts, adaptation and vulnerability. Contribution of working group II to the fourth assessment report of the intergovernmental panel on climate change* (pp. 7–22. Retrieved 2014-2010-2020.). Cambridge, UK, and New York, NY, USA: Cambridge University Press.

IPCC. (2012). *Managing risks of extreme events and disasters to advance climate adaptation. A special report of working group I and II of the IPCC*. UK, New York, USA: Cambridge University Press.

Keller, M. C., Fredrickson, B. L., Ybarra, O., Cote, S., Johnson, K., Mikels, J, et al. (2005). A warm heart and a clear head: The contingent effects of weather on mood and cognition. *Psychological Science, 16*, 724–731.

Kimaro, G., & Kihwele, E. (Personal communication March, 2013). *Impacts and implications of climate change on wildlife ecology and tourism in Serengeti National Park.*

Lise, W., & Tol, R. S. J. (2002). Impact of climate on tourist demand. *Climatic Change, 55*(4), 429–449.

Maddison, D. (2001). In search of warmer climates? The impact of climate change on flows of British tourists. *Climatic Change, 49*(1/2), 193–208.

McNaughton, S. J. (1979). Serengeti Migratory Wildebeest: Facilitation of energy flow by grazing. *Science, 191*, 92–94.

McNaughton, S. J. (1983). Compensatory plant growth as a response to herbivores. *Oikos, 40*, 329–336.

Mduma, S. A. R., Sinclair, A. R. E., & Hilborn, R. (1999). Food regulates the Serengeti wildebeest: A 40-year records. *Journal of Animal Ecology, 68*, 1101–1122.

Melamari, L. (1996). *Financing and management of national park and protected areas for tourism*. Paper presented at the seminar on finance and management of wildlife parks for tourism (July 20–August 2), Arusha, Tanzania.

MEMR. (2012). Mapping wildlife dispersal areas and migratory routes/corridors (part1): Southern Kenya Renglands (draft version March, 2012). Ministry of Environment and Mineral Resources (MEMR), Nairobi, Kenya.

Michael, C. (2006). *Climate change impacts on East Africa. A review of scientific literature* (pp. 12). Switzerland; WWF-Gland.

Munishi, P. K. T., Kihupi, N. I., Nindi, S. J., Mpeta, E., Tilya, F. F., Chang'a, L., et al. (2009). *Profiling the climate of Tanzania for climate change mitigation and adaptation* (pp. 1–53). Unpublished Report, Sokoine University of Agriculture (SUA), Morogoro, Tanzania.

Munishi, P. K. T., Shirima, D., Jackson, H., & Kilungu, H. (2010). *Analysis of climate change and its impacts on productive sectors, particularly agriculture in Tanzania*. United Republic of Tanzania: Ministry of Finance and Economic Affairs.

Musiega, D. E, Kazadi, S.-N., & Fukuyama, K. (2006). A framework for predicting and visualizing the East African wildebeest migration-route patterns in variable climatic conditions using geographic information system and remote sensing. *Ecological Research, 21*, 530–543. doi:10.1007/s11284-006-0175-9

NAPA. (2007). *National Adaptation Programme*. United Republic of Tanzania: Vice President Office, Division of Environment.

Ogutu, J., Owen-Smith, N., Piepho, H.-P., Kuloba, B., & Edebe, J. (2012). Dynamics of ungulates in relation to climatic and land use changes in an insularized African savanna ecosystem. *Biodiversity Conservation, 21*, 1033–1053.

Ogutu, J., Owen-Smith, N., Piepho, H.-P., & Said, M. (2011). Continuing wildlife population declines and range contraction in the Mara region of Kenya during 1977–2009. *Journal of Zoology, 285*(2), 99–109.

Ogutu, J., Owen-Smith, N., Piepho, H.-P., Said, M., Kifugo, S., Reid, R., et al. (2013). Changing wildlife populations in Nairobi National Park and adjoining Athi-Kaputiei Plains: Collapse of the migratory wildebeest. *Open Conservation Biology Journal, 7*, 11–26.

Ogutu, J. O., Piepho, H. P., Dublin, H. T., Bhola, N., & Reid, R. S. (2008). Rainfall influences on ungulate population abundance in the Mara-Serengeti ecosystem. *Journal of Animal Ecology, 77*, 814–829.

Pettorelli, N., Vik, J. O., Mysterud, A., Gaillard, J. M., Tucker, C. J., & Stenseth, N. C. (2005). Using the satellite-derived NDVI to assess ecological responses to environmental change. *Trends in Ecology & Evolution, 20*, 503–510.

Sandra, J. V., Yadvinder, M., Dominic, M., Jim, W., & Salman, H. (2005). Valuing the impacts of climate change on protected areas in Africa. *Ecological Economics, 53*(1), 21–33. doi:10.1016/j.ecolecon.2004.7.024

Schaller, G. B. (1972). *The Serengeti Lion. A study of predator-prey relations*. Chicago, Illinois, USA: University of Chicago Press.

Scott, D. (2005). Global environmental change and mountain tourism. In S. Gössling & C. M. Hall (Eds.), *Tourism and global environmental change: Ecological, economic, social and political interrelationships* (pp. xiii–227). New York: Routledge.

Sharam, G. J., Sinclair, A. R. E., Turkington, R., & Jacob, A. L. (2009). The savanna tree Acacia polyacantha facilitates the establishment of riparian forests in Serengeti National Park, Tanzania. *Tropical Ecology, 25*, 31–40. doi:10.1017/S0266467408005683

Sinclair, A. R. E. (1979). The Serengeti environment (Chap. 2). In A. R. E. Sinclair & M. Norton-Griffiths (Eds.), *Serengeti: Dynamics of an ecosystem* (pp. 31–45). Chicago, Illinois, USA: University of Chicago Press.

Sinclair, A. R. E. (1995). Serengeti past and present. In A. R. E. Sinclair & P. Arcese (Eds.), *Serengeti II: Dynamics, management, and conservation of an ecosystem* (pp. 3–30). Chicago, Illinois, USA: University of Chicago Press.

Sinclair, A. R. E., Mduma, S. A. R., & Arcese, P. (2000). What determines phenology and synchrony of ungulate breeding in the Serengeti? *Ecology, 81*(8), 2100–2111.

Sinclair, A. R. E., Mduma, S. A. R., Hopcraft, J. G. C., Fryxell, J. M., Hilborn, R., & Thitgood, J. S. (2007). Long-term ecosystem dynamics in the Serengeti: Lesson for conservation. *Conservation Biology, 21*(3), 580–590. doi:10.1111/j.1523-1739.2007.00699.x

Sinclair, A. R. E., Metzger, K. L., Mduma, S. A. R., & Fryxell, J. M. (Eds.). (2015). *Serengeti IV* (Vol. IV). Chicago: The University of Chicago Press Book.

Strauch, A. M. (2013). The role of water quality in large mammal migratory behaviour in the Serengeti. [Review]. *Ecohydrology, 6*(3), 343–354. doi:10.1002/eco.1279

TMA. (2007). Climate change variations in Tanzania: Tanzania Meteorological Agency (TMA).

Trager, M., & Mistry, S. (2003). Avian community composition of kopjes in a heterogenous landscape. *Oecologia, 135*, 458–468. doi:10.1007/s00442-003-1204-9

Turpie, J. K., & Siegfried, W. R. (1996). The conservation-economic imperative: Securing the future of protected areas in South Africa. *Africa Environment and Wildlife, 4*(3), 35–37.

Vanacker, V., Linderman, M., Lupo, F., Flasse, S., & Lambin, E. (2005). Impact of short-term rainfall fluctuation on interannual land cover change in sub-Saharan Africa. *Global Ecology and Biogeography, 14*, 123–135.

Walonski, E., & Gereta, E. J. (2001). The water quality and quantity as the factors driving the Serengeti ecosystems Tanzania. *Hydrologia, 458*, 169–180.

Walonski, E., Gereta, E. J., Borner, M., & Mduma, S. A. R. (1999). Water, migration and the Serengeti ecosystem: Understanding the mechanism that control the timing of the wildebeest migration may prove vital to successful management. *Animal Science, 87*, 526–533.

White, F. (1983). *The vegetation of Africa, a descriptive memoir to accompany the UNESCO/AETFAT/UNSO vegetation map of Africa* (Vol. 20, pp. 1–356). Natural Resource: UNESCO.

Zurlini, G., Riitters, K. H., Zaccarelli, N., Petrosillo, I., Jones, B. K., & Rossi, L. (2006). Disturbance patterns in a socio-ecological system at multiple scales. *Ecological Complexity, 10*, 119–128.

Tools and Procedures for Seasonal Downscaling of Climate Forecasts for Use by Farmers Over the Greater Horn of Africa: A Case Study for Western Kenya

Jasper Batureine Mwesigwa, Gilbert Ongisa Ouma, Jully Odhiambo Ouma, Ismael Mulama Lutta, Samuel Gathara Mwangi and Laban Ayieko Ogallo

1 Introduction

Climate information is one of the fundamental inputs to the Agricultural Planning and Decision Making process (ICPAC 2015; Mwesigwa et al. 2012). It is needed at every stage of Agricultural development, and is central to the entire agricultural transformation process. Specifically, information on seasonal climate variability is very important for agricultural planners and decision makers at all levels of the agricultural value chain (Hansen et al. 2006). However, in Sub-Saharan Africa, limited community-specific (downscaled) climate information is a major hindrance to appropriate agricultural planning and decision making. In the Greater Horn of Africa region, climate change has greatly contributed to the extreme vulnerability of the agriculture sector to seasonal climate variability (IPCC 2013; Morton 2007). More extreme precipitation changes over Eastern Africa such as droughts and heavy rainfall events have been experienced more frequently during the last 30–60 years (IPCC 2013). The agriculture sector continues to rely heavily on

J.B. Mwesigwa (✉) · J.O. Ouma · I.M. Lutta · L.A. Ogallo
Inter-Governmental Authority on Development's Climate Prediction and Applications Centre (ICPAC), P.O. Box 10304, Nairobi 00100, Kenya
e-mail: mjbatureine@gmail.com

J.B. Mwesigwa · G.O. Ouma · I.M. Lutta
Institute for Climate Change and Adaptation, University of Nairobi, P.O. Box 30197, Nairobi 00100, Kenya

G.O. Ouma · J.O. Ouma
Department of Meteorology, University of Nairobi, P.O. Box 30197, Nairobi 00100, Kenya

S.G. Mwangi
Kenya Meteorological Department (KMD), P.O. Box 30259, Nairobi 00100, Kenya

W. Leal Filho et al. (eds.), *Climate Change Adaptation in Africa*, Climate Change Management, DOI 10.1007/978-3-319-49520-0_24

seasonal rainfall but with minimal irrigation taking place in a few other areas (Siebert et al. 2006). The risk of loss of livelihoods and income due to extreme climatic events and associated impacts is particularly impacting on rural farmers and pastoralists in the arid and semi-arid regions. Farmers are always uncertain about the upcoming season particularly regarding the risks and opportunities that come with the season (Hassan and Nhemachena 2008) due to limited availability and access to reliable, downscaled, seasonal climate information tailored for specific communities and users' needs. Continuous generation of timely climate forecast followed by proper downscaling and dissemination to farming communities is a sure way of contributing to climate risk management and climate change adaptation by such vulnerable farmers.

Several tools are available for use to downscale seasonal climate forecasts into easily understandable climate information (Hansen et al. 2006). However, only robust and easy to use tools and procedures are desirable, particularly in the greater horn of Africa region where most intermediaries (information disseminators) do not have adequate skills in interpreting probabilistic forecasts.

This study used simple but robust tools and procedures such as Fact-Fit, GeoCLIM, Analogue Year approach and GeoWRSI to comprehensively interpret seasonal, monthly and 10-day forecasts generated from Kenya Meteorological Department (KMD) between 2012 and 2014. Downscaled forecasts were further disseminated to three pre-selected communities to test their applicability in their farming activities. All the tools used were complementary and produced the desired adequate and easily understandable sets of information and products on which participating farmers relied for planning and decision making. Precise interpretation promoted clear understanding of the forecasts and provided a common ground between climate information providers and users.

This case study serves an important step towards enhancing community based adaptation for improved food security and livelihoods through proper application of quality climate information.

2 Materials and Methods

2.1 Study Area

Three pilot locations in Western Kenya were chosen for evaluating the applicability of downscaled climate forecast products. These included Reru (a usually dry area yet close to Lake Victoria with subsistence agriculture as their key livelihood activity); Nganyi (which has been running indigenous-knowledge based climate forecasting for many generations) and Nyahera (a peri-urban community with high agro-business potential) (Fig. 1). These communities lie within Kisumu and Vihiga counties which are among Kenya's 'bread baskets' yet they usually experience false start of seasonal rains, making information on true onset (and therefore on when to

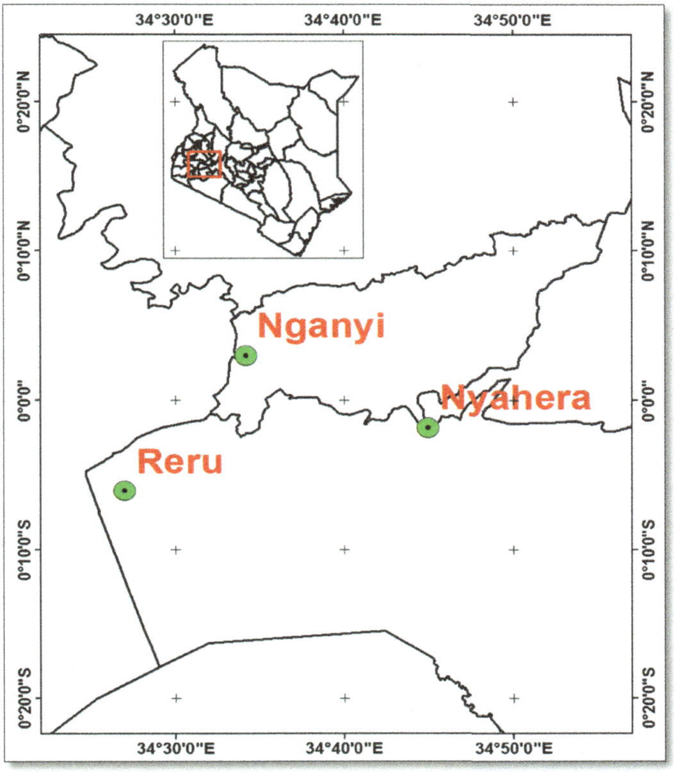

Fig. 1 Case study area in Western Kenya

plant) very critical for successful crop production in the county. This area also experiences bi-model form of climate with two distinct season, i.e. March–May and September–November, which provides an opportunity of producing food throughout the year. The area receives between 800 and 1200 mm of rainfall annually and lie at altitudes of between 800 and 1400 m above sea level. Annual average temperatures range between 20 °C (minimum) and 35 °C (maximum).

This work was done between 2012 and 2014. In this paper, results from only one season (i.e. SOND 2012) are presented. In the case of Kenya and particularly in the study areas; this season was very important in many aspects. It was one of the good seasons succeeding the 2009–2011 drought in the Horn of Africa region during which over 13 million people across the region were affected and left food insecure, according to many reports such as those by USAID (https://www.usaid.gov/crisis/horn-africa). Therefore highly favourable rainfall conditions were desirable as these could support improved agricultural productivity and replenishing of livestock resources such as water and pastures. On the other hand, negative impacts such as increased risk of flooding, hail storms, landslides, soil erosion and other extreme climatic events in the risk prone areas were expected to suffice. Similar work was done and comparable outcomes achieved from other seasons respectively.

2.2 *Seasonal Climate Forecast Downscaling*

ICPAC produces three regional consensus Greater Horn of Africa (GHA) Seasonal Climate Outlooks for March–April–May, June–July–August, and September–October–November–December seasons for all member countries (www.icpac.net). The regional outlooks are majorly probabilistic in nature and are so generalized with low spatial resolution of approximately 50 km. These regional forecasts spell out areas within the region that are likely to receive near normal to below normal rainfall as well as those areas with increased probability for near normal to above normal rains, and any areas that are expected to remain dry throughout the forecast period. However, due to limiting spatial resolution, further downscaling by ICPAC member countries' National Meteorological and Hydrological Services (NMHSs) is done to produce more detailed national and sub national level forecasts. In Kenya, national and sub national downscaling is done by KMD (www.meteo.go.ke/). This study utilised downscaled forecasts from ICPAC and KMD (Figs. 2 and 3) to test the capability of three tools and procedures to produce relevant, tailored, location-specific downscaled climate information that could be relied upon by grassroots farming communities to plan and manage their farming activities.

This involved testing of Fact-Fit and GeoCLIM tools and the analogue year methodology for their capability to further interpret low resolution, probabilistic forecasts into more understandable variables like expected rainfall amounts, onsets and cessation dates (duration of the season), rainfall distribution (in space and time) including an analysis of potential occurrence and intensity of destructive rainfall events, such as harmful dry spells, storms, etc. (see research design in Fig. 4). These seasonal climatic variables were necessary at the beginning of the season as they served as a basis for strategic decision making by farmers.

As the seasons progressed, WRSI tool was used to monitor crop health and surplus soil moisture upon which relevant advisories were issued to trigger tactical decision making by farmers. Appropriate dissemination channels were employed which enabled communities under study to receive, interpret and appropriately utilize downscaled information on time.

2.2.1 Fact-Fit Tool

The method used by this tool to convert rainfall probability forecast to amounts is the Gamma distribution parameters (Husak et al. 2011; Min et al. 2009). It first generates historical gamma distribution for the season of interest after which the forecast is digitized and the probabilities used to convert it to rainfall amounts. Most of the probabilistic forecasts are issued in the form of tercile-based categorical

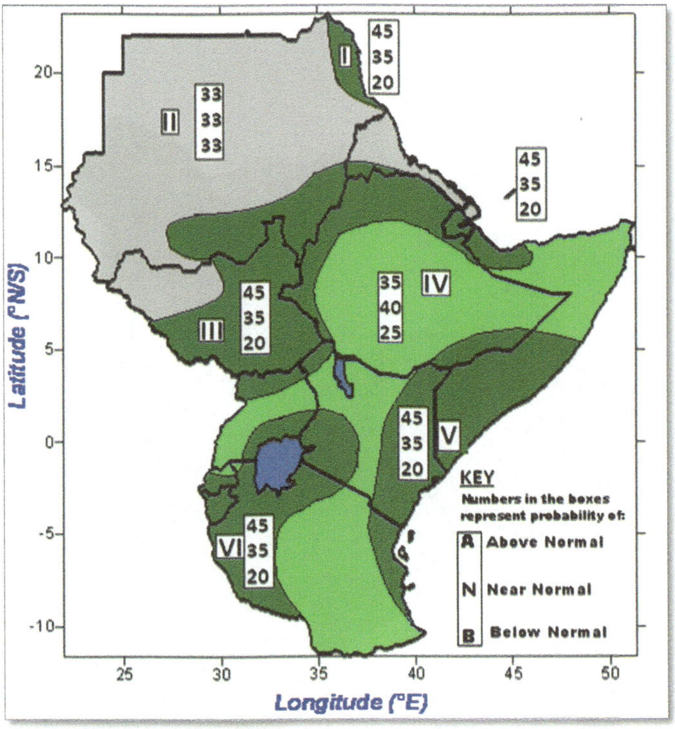

Fig. 2 Sept–Dec 2012 consensus seasonal forecast for GHA (*Source* ICPAC)

probabilities (Husak et al. 2011; Min et al. 2009), that is, below-normal (BN), near-normal (NN), and above-normal (AN). The tool is then used to estimate the tercile probabilities BN, NN, and AN. With each category, the forecast probability is estimated as a portion of the cumulative probability of the forecast sample associated with it. A user can generate probability maps converted to rainfall amounts and indicating locations that are likely or unlikely to attain a give threshold conducive to crop production (Husak et al. 2011). In this study, Fact-Fit was used as a downscaling tool since the resolution could easily be increased to get the forecast at community level (Figs. 5 and 6a, b).

2.2.2 GeoCLIM Tool

GeoCLIM software (http://chg.geog.ucsb.edu/tools/geoclim/) facilitates climatological analysis of historical rainfall and temperature. It was developed by USGS FEWNET in support of the USAID PREPARED and Global Climate Change activities. In this study, GeoCLIM was used to blend and grid historical observation and satellite data and for seasonal forecast interpretation through

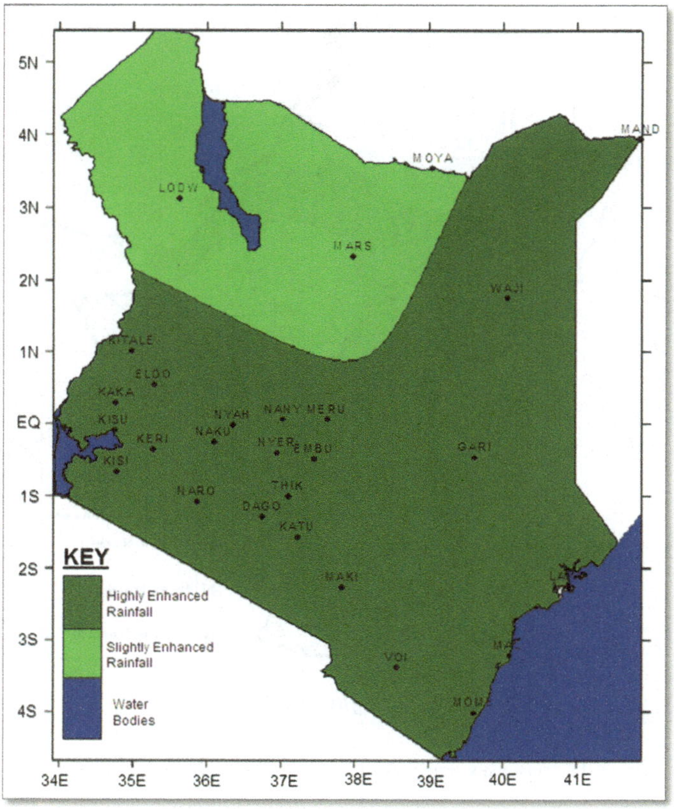

Fig. 3 Sept–Dec 2012 seasonal forecast for Kenya (*Source* KMD)

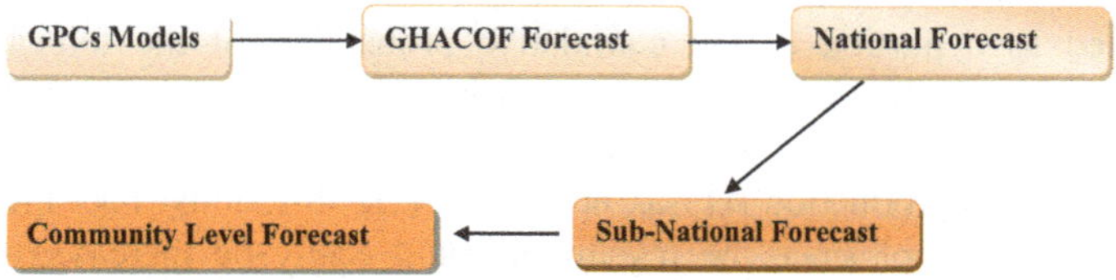

Fig. 4 Research design

intra-seasonal time-series analysis, i.e. mapping of seasonal baselines (Long Term Averages, LTAs), mapping of observed seasonal amounts and crop suitability mapping based on analogue years (Figs. 7, 8a–d, 9 and 10).

Long term seasonal averages were computed based on the 1981–2010 baseline as recommended by WMO. Historical daily datasets (from 1959) for Kisumu and Kakamega weather stations were provided by KMD. The gridded datasets were 10-daily with a 5 km resolution.

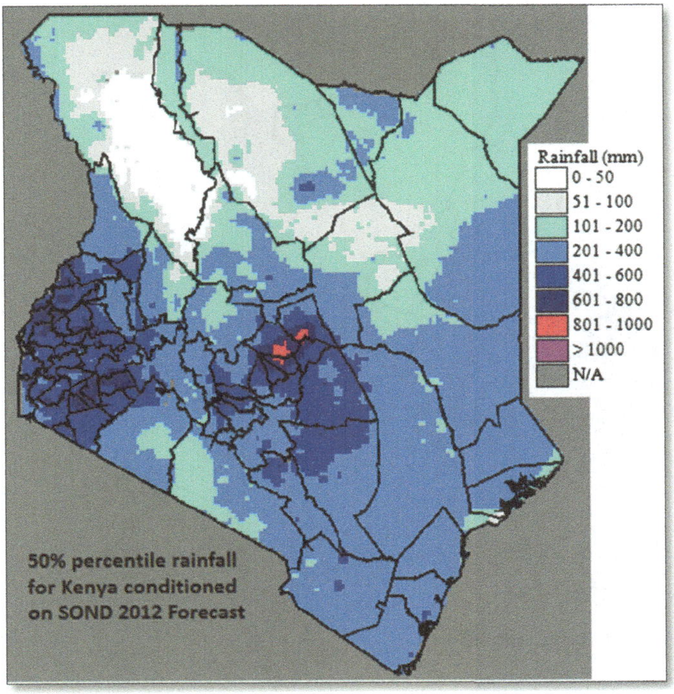

Fig. 5 Kenya SOND 2012 rainfall forecast downscaled into amounts using Fact-Fit tool

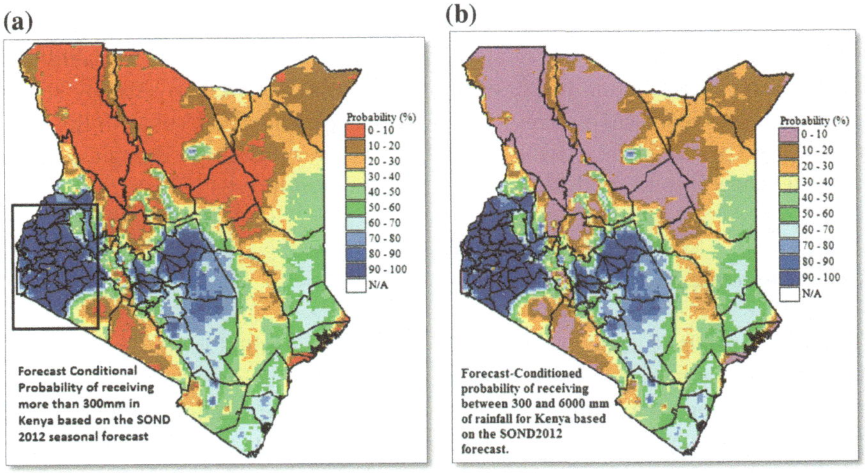

Fig. 6 Kenya SOND 2012 downscaled forecast showing areas with: **a** probability of receiving more than 300 mm and **b** probability of receiving between 300 and 600 mm of rainfall during the Sept–Dec 2012 season

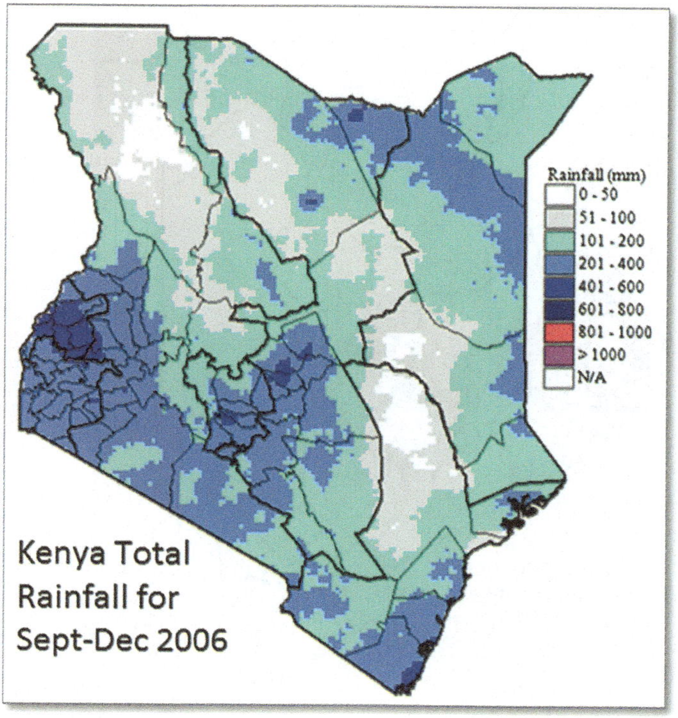

Fig. 7 Kenya total rainfall for SOND 2006 showing expected total SOND 2012 rainfall distribution

GeoCLIM provided an advantage over Analogue years for its capability to produce spatial distribution maps, including for areas not well covered by station data over time. GeoCLIM was also used for climate change and vulnerability trend analysis for specific areas. Finally, GeoCLIM was very useful in drought analysis due to its capability to produce Standard Precipitation Indices (SPIs).

2.2.3 Analogue Year Analysis

Currently, analysis of analogue years is by far the best method for predicting onsets and cessation dates and for identifying potentially harmful extremes like dry spells and heavy precipitation events. This study made use of the analogue years provided by ICPAC and KMD for each season, to estimate onsets, seasonal progression, distribution in time and cessation dates using daily seasonal point data. Analyses were done using Excel (Microsoft Office Inc). Graphs of specific analogue year alongside historical averages were produced (Fig. 12). The downside of this methodology was that quite old analogue years were more difficult to comprehend by farmers as many of them would not memorise how the season started or

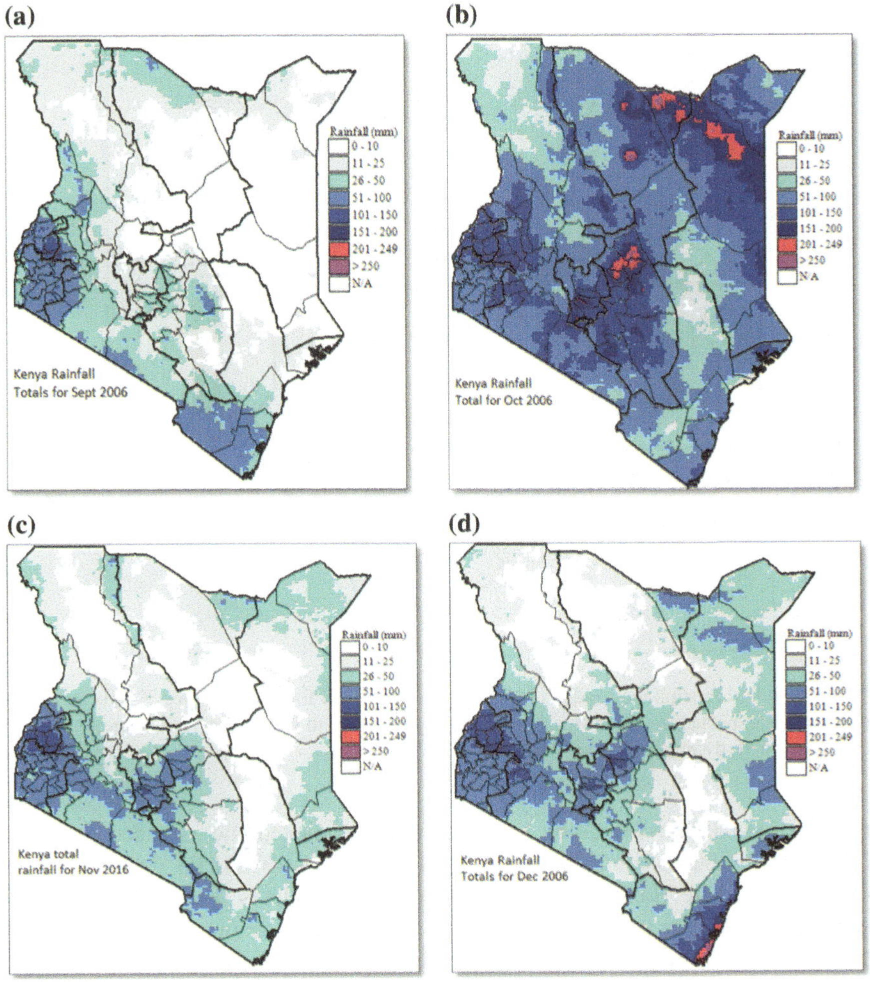

Fig. 8 a–d Total monthly rainfall distribution for September, October, November and December 2006 used to downscale SOND 2012 forecast

progressed; however, newer analogues years were quite easy to comprehend by farmers as they still could remember how seasons started and progressed.

2.2.4 WRSI Analysis

As the seasons progressed, water requirements, water deficit and surplus water at different crop growth stages, crop phenology, and other parameters were monitored using Geo Spatial Water Requirement Satisfaction Index (GeoWRSI), a stand-alone visual basic implementation of the spatially explicit water requirement satisfaction

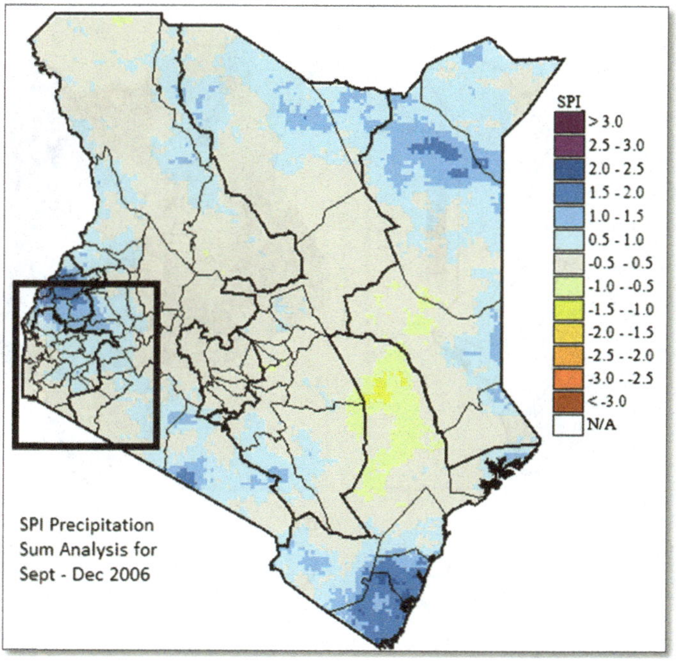

Fig. 9 Standard Precipitation Index (SPI) Analysis for Kenya for the Sept–Dec 2006 season. This figure shows expectation of a normal (SPI range: −0.5 to 0.5) to wet season (SPI range: 0.5 to 1.5) for the entire study area (see *box*) during SOND 2012

index—originally developed by FAO but recently adapted and extended by USGS (http://chg.geog.ucsb.edu/tools/geowrsi/). This tool uses CHIRPS data downloadable from Climate Hazards Group (CHG). WRSI products (Figs. 13a, b and 14) were very useful for farmers in making quick tactical decisions like protecting the growing crop from potentially harmful hazards (like dry spells), improving on management aspects to maximise use of good weather (fertilizer application, crop protection, weeding) or avoiding extra investment costs if losses were perceived to happen.

2.2.5 Verification of Tools Used and Farmers' Benefits from Use of Downscaled Climate Forecast Information

Observed rainfall distribution (amounts both in space and in time) for SOND 2012 were plotted and/or mapped using Excel and GeoCLIM to produce graphs and maps respectively. Farmers who used the forecasts were also monitored and their maize and sorghum yields per unit area computed, and compared to baselines (Fig. 15). In addition, participating farmers were consulted on their view about the

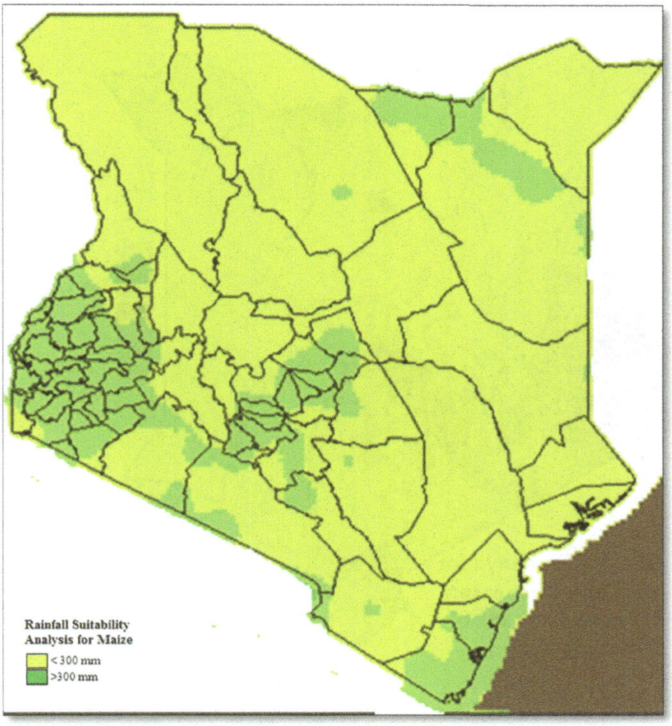

Fig. 10 Crop suitability analysis of SOND 2012 forecast for Kenya showing areas where the expected rainfall is enough to support maize production (*green*) and those that would most likely not support maize production (*yellow*), provided that other environmental conditions are suitable. Maize water requirement was set at 300 mm

applicability and benefits of downscaled, location specific climate information as compared to national and sub national level climate outlooks.

2.3 Tools and Procedures for Downscaling

Currently within the GHA region, ICPAC and partners downscale forecasts from Global Producing Centres (GPCs) to produce regional consensus seasonal climate outlooks during the GHA Climate Outlook Forums (usually termed as GHACOFs). The regional Climate outlooks are quite generalised with very low (approx. 50 km) resolution (www.icpac.net).

Upon receipt of regional climate outlooks, the NMHSs of ICPAC member states further downscale them to national and (in some cases) sub-national levels. These are quite more detailed but still lack community specificity among other qualities. It is therefore imperative that the national level forecasts undergo further downscaling

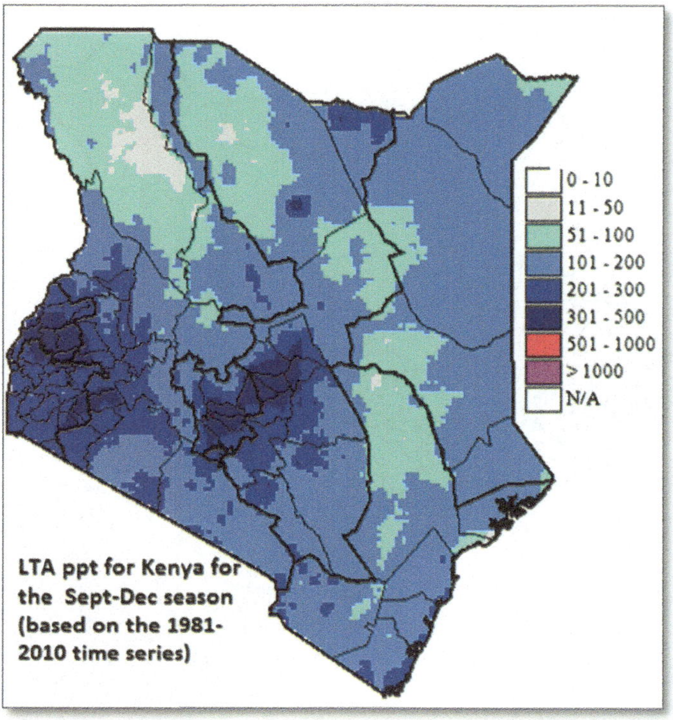

Fig. 11 Long term average (LTA) precipitation for Kenya for the Sept–Dec season based on the 1981–2010 time series

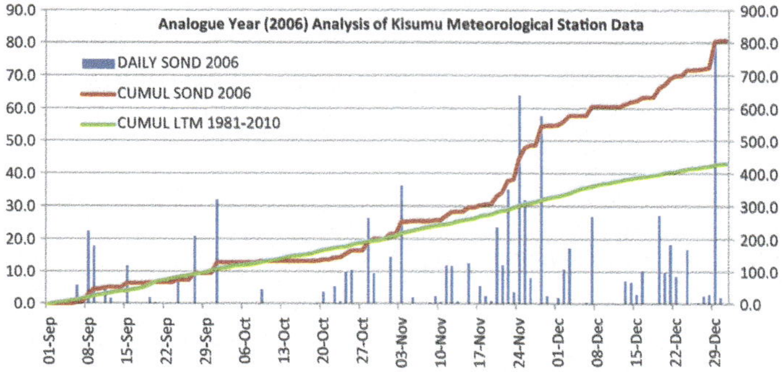

Fig. 12 Analogue year analysis of Kisumu meteorological station (Kenya) data (OND 2006) showing expected onsets (Mid October) and cessation (end of December) dates, seasonal distribution, number and intensity of rainfall events, distribution and length of dry spells as downscaled forecast for SOND 2012 season. This analysis did not show potentially harmful dry spells (defined as a period stretching for ≥10 days receiving ≤25%LTA)

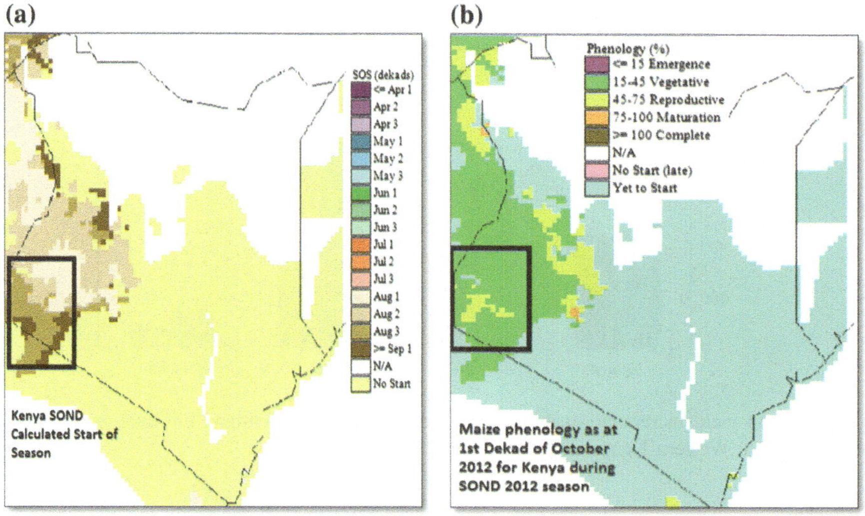

Fig. 13 Start of season map (**a**) and crop phenology for maize (**b**) for Kenya for the SOND 2012 season, generated from GeoWRSI tool

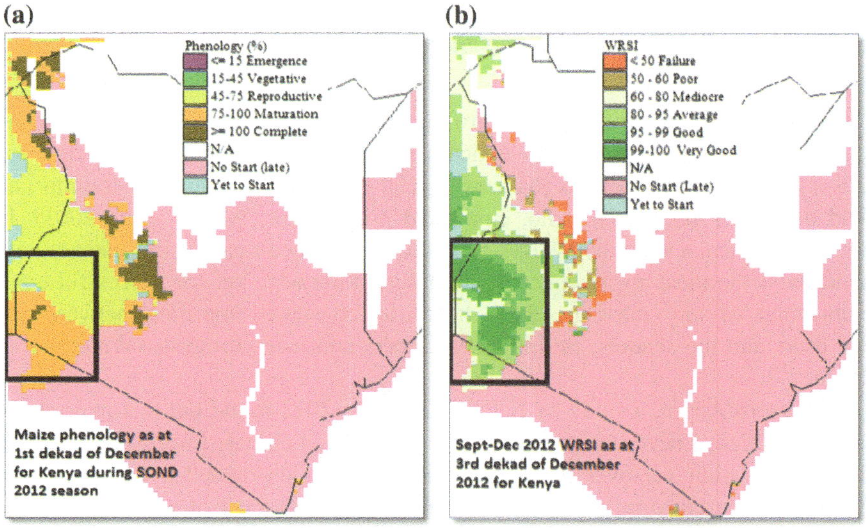

Fig. 14 Maize crop phenology as of 1st dekad of December 2012 (**a**) and Sept–Dec 2012 WRSI for Kenya (**b**) for the SOND 2012 season, generated from GeoWRSI tool

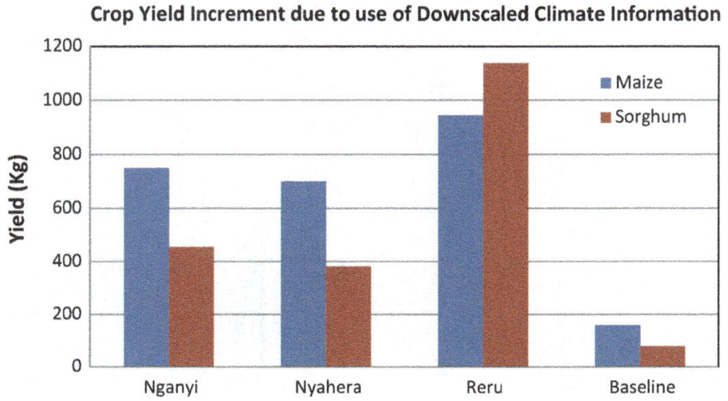

Fig. 15 Crop yield increment attributable to use of downscaled climate information in SOND 2012 season in Western Kenya

so as to make them comprehensible both to intermediaries (communicators) and end users. The tools and procedures presented in this paper are therefore very useful for translating low resolution, probabilistic forecasts into simpler, easy to understand charts, maps, tables and narrative for direct application across the entire GHA region.

2.4 Dissemination of Downscaled Information

Dissemination of location-specific agro-meteorological information and advisories to target farmers in Sub-Saharan Africa has mainly been done through five channels, i.e. community meetings (barazas), social groupings, field demonstrations and field days (ICPAC 2015), radio, email and print-media (Walker et al. 2001). In Southern Africa, a study by Walker et al. (2001) revealed that most farmers did not understand the meaning in concepts of climate forecasts. Yet for successful communication of any message, there needs to be shared meaning between the end-users and the senders; in this case, farmers and meteorologists (Walker et al. 2001).

In Eastern Africa, a study by ICPAC (2015) in 2012 found out that majority of farmers in Kenya never used to receive any simplified climate forecast information for their location yet over 90% of farming households owned a mobile phone. Farmers also used to misinterpret all forms of forecasts most of which used to be broadcast frequently on radio. By 2012, the most common form of disseminating climate forecasts in East Africa was through internet (website), followed by press releases. Through these channels, in addition to the complexity of the un-downscaled climate forecasts, there were enormous limitations on information comprehension and therefore assimilation by grassroots farmers.

3 Results

3.1 SOND 2012 Seasonal Characteristics

The September to December 2012 constituted an important rainfall season over the equatorial and southern sectors as well as the southern parts of the northern sector of the GHA region. Both the regional consensus climate outlook and its downscaled product for Kenya for the September to December 2012 rainfall season indicated increased likelihood of above to near normal rainfall over much of the GHA during this period.

In the study area, comprehensive interpretation of the seasonal forecast and proper dissemination of this information was done so as to motivate farmers to take advantage of the benefits that come with a good season and to mitigate the negative impacts that could come along with such seasons.

3.1.1 Downscaling of Probabilistic Forecasts into Rainfall Amounts Using Fact-Fit Tool

The expected rainfall totals and the associated probabilities of the study areas receiving adequate rainfall within suitable water requirement thresholds for maize and other annual crops were clearly evaluated and mapped based on the probabilistic Kenya seasonal climate forecast for SOND 2012 using Fact-Fit (Figs. 5 and 6a, b). These were very useful for suitability mapping.

3.1.2 Interpretation of Seasonal Characteristics Using Analogue Year Methodology

The year 2006 was provided by ICPAC and KMD as the best match for the season under study (SOND 2012). Analysis of SOND 2006 revealed the following characterised:

- Early onset (mid October) thus favouring early/timely planting (Fig. 12);
- Generally good rainfall distribution (both in time and space);
- Rains were expected to continue into December 2012 (approximately 17 weeks);
- Seven potentially heavy rainfall events (≥30 mm/day) which increased chances of flooding, possibility of hail storms, landslides and severe soil erosion in the risk prone areas;

These seasonal characteristics obtained from analysis of analogue year in Excel spreadsheet gave us an indication of what was expected during SOND 2012 around or near Kisumu Meteorological Station. However, this information was not representative of the entire study area especially those areas that are at ≥5 km far from

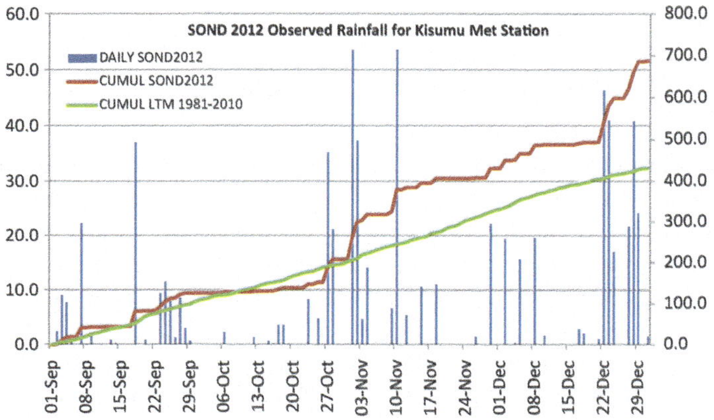

Fig. 16 Verification of SOND 2012 forecast: observed rainfall distribution for Kisumu Meteorological station

Kisumu station. It was thus necessary to use another tool to further analyse the analogue year for purposes of generating spatially disaggregated forecast products.

Verification of SOND 2012 forecast revealed clear congruence with analogue-year based seasonal characteristics including eight heavy rainfall events, slightly fewer number (but heavier) rainfall events, slightly late (last dekad of October), fairly good distribution, and same cessation periods at the end of September (Fig. 16). Two potentially harmful dry spells were evident (between 19–28 November and 9–16 December); however, given they came in between heavy rainfall events, they were not quite detrimental to crop performance.

3.1.3 Interpretation of Seasonal Characteristics Using GeoCLIM Tool

GeoCLIM tool produced spatial maps of expected seasonal (SOND 2012) amounts (Fig. 7), disaggregated monthly forecast totals (Fig. 8a–d), seasonal SPI (Fig. 9), crop suitability (Fig. 10) and SOND LTA (Fig. 11) through the analysis of 2006 analogue year. This information formed the basis for planning of planting by the farmers.

3.1.4 Crop Growth and Development Monitoring Using the GeoWRSI Tool

Start of season, progressive crop phenology, water requirement satisfaction index and other WRSI products were continuously monitored both at decadal and monthly basis. End of season WRSI was also calculated. WRSI input parameters (precipitation and potential evapotranspiration) were concurrently assessed every dekad. The outputs of this tool were essential for identifying critical periods of water deficits. Such information was used to trigger tactical management options including irrigation.

3.2 Farm-Level Benefits from Use of Downscaled Climate Forecast Information

3.2.1 Crop Yield Increment

The yield of both Maize and sorghum significantly increased to 4-fold when farmers used climate information (Fig. 15). Sorghum production significantly increased during SOND season (also known as Kenya short rains season) than in MAM (Long rains) season. Production of cassava and sweet potatoes also improved but not as significant as the cereals. One major output though from the root tubers was the increased availability of planting material (i.e. sweet potato vines and cassava cuttings respectively). The performance of pulses such as beans (except for climbing varieties), green grams and groundnuts) unfortunately did not significantly improve. Farmers attributed to this static performance to use of improved (new) but un-adapted varieties. On the other hand, a multiplier factor of 1:120 (seed to produce or grain) was realized for sorghum. This represented a 200% increment from their baselines.

3.2.2 Social-Economic Benefits

Some of the social benefits of community-based climate services were increased food production and improved food storage and sale of surplus food which resulted into improved food security and income generation. In addition, there were less household (husband-wife) conflicts and separation because of the availability of adequate food at home. Farmers also reported higher returns to production when climate information was incorporated into their agricultural planning which encouraged them to till larger pieces of land. It also encouraged neighbouring farmers to start practicing climate smart agriculture.

3.2.3 Environmental Benefits

Whenever the forecast indicated likelihood of enhanced rainfall with potential to cause floods, landslides and soil erosion, farmers were alive at quickly constructing soil and water conservation structures, opening up water channels to lead away flood waters, and most importantly planting of plantation crops (like bananas, coffee, sugarcane) as well as agroforestry trees and woodlots. In cases when little water was expected in the season, farmers were able to harvest water and use it for irrigation and domestic use.

4 Discussions

4.1 Seasonal Climate Downscaling

Severe weather impacts to agriculture and other climate-sensitive sectors can significantly be avoided if clear, simple, and easily understandable climate information is availed to farmers in time to enable them make appropriate farming decisions. The current regional and national forecasts are not simplified enough to enable direct use by the grassroots farmer. In addition, there is currently minimal or no formal community-level climate services in sub-Saharan Africa. The tools and procedures presented in this paper open up opportunities for climate scientists, agrometeorologists and other intermediary climate information providers and communicators an opportunity to clearly unpack the common, highly probabilistic forecasts into simpler, comprehensible information for direct application by end users. The study demonstrated that appropriate use of downscaled climate information by farmers leads to improved food production, food security and community livelihoods and significantly contributes to climate change adaptation.

The work presented in this paper, therefore provides an important, reliable procedure for simplifying climate information using simple but robust tools and procedures. It significantly contributes to climate change adaptation in a sense that it presents an important step for climate information producers and communicators the opportunity to produce and avail quality climate information for immediate adoption by vulnerable communities. Though the case study was done in Kenya, the tools and procedures presented in this paper are equally important and are highly applicable in other similar agro-ecologies across sub Saharan Africa and beyond.

4.2 Benefits of Use of Downscaled Information

Climate information has become a fundamental input to the decision making process across all development sectors (ICPAC 2015; Mwesigwa et al. 2012). In the Agriculture sector, for example, it is necessary at all stages of Agricultural development and transformation process including climate change adaptation. It is necessary before, during and even after commencement of any development effort.

Downscaling of climate forecasts presents the users with an opportunity to appropriately interpret the forecasts so they can plan and make informed decisions. It provides answers to such salient questions as: when will it start raining? When Will it stop? Where will it rain? For how long will it be raining? How much water is expected during the entire season? Will there be any potentially harmful extreme events (like storms, dry spells, etc.) during the season—when and for how long? Once the user has such information, they are well prepared to make clear strategic decisions about what technologies to deploy, when, and how.

Acknowledgements This work was partly supported by The Rockefeller Foundation through a grant to IGAD Climate Prediction and Application Center (ICPAC) in 2011. The grant was aimed at strengthening the capacity of ICPAC in climate prediction and information dissemination for improved agricultural production and food security to enhance adaptation to climate variability and change. We are very grateful to The Foundation. We are also indebted to the friendly collaboration and support from Kenya Meteorological Department (KMD) and University of Nairobi (UoN). Similarly, we are grateful to the case-study host communities and farmers especially in Reru, Nganyi and Nyahera in western Kenya for their tireless contributions to this study. Finally, we acknowledge and are grateful to the support from Prof. Maria Onyango from Jaramogi Oginga Odinga University of Science and Technology (JOUST) and Dr. Gordon Wayumba from The Technical University of Kenya (TUK).

References

James, H. W., Challinor, A., Ines, A., Wheeler, T., & Moron, V. (2006). Translating climate forecasts into agricultural terms: advances and challenges. *Climate Research, 33*, 27–41.

Hassan, R., & Nhemachena, C. (2008, March). Determinants of African farmers' strategies for adapting to climate change: Multinomial choice analysis. *AfJARE, 2*(1).

Husak, G. J., Michaelsen, J., Kyriakidis, P., Verdin, J. P., Funk, C., & Galu, G. (2011). The forecast interpretation tool—A Monte Carlo technique for blending climatic distributions with probabilistic forecasts. *International Journal of Climatology, 31*(3), 461–467.

ICPAC. (2015). *Community based climate services in support of improved food security, livelihoods and resilient community: Experiences from four pilot communities in Kenya.* A Publication by IGAD Climate Prediction and Applications Centre (ICPAC), July, 2015. Accessed on December 15, 2015 at: http://www.icpac.net:8082/modules/board/files/Climate_Information_Package_ICPAC_Case_Study.pdf

IPCC. (2013). Summary for policymakers. In T. F. Stocker, D. Qin, G.-K. Plattner, M. Tignor, S. K. Allen, J. Boschung, et al. (Eds.), *Climate change 2013: The physical science basis. Contribution of Working Group I to the Fifth Assessment Report of the Intergovernmental Panel on Climate Change*. Cambridge, United Kingdom and New York, NY, USA: Cambridge University Press.

Min, Y. M., Kryjov, V. N., & Park, C. K. (2009). A probabilistic multimodel ensemble approach to seasonal prediction. *Weather and Forecasting, 24*(3), 812–828.

Morton, F. J. (2007). The impact of climate change on smallholder and subsistence agriculture.

Mwesigwa, J. B., Onyango, M., Oduor, J., Mwangi, S., Ayieko, J., Onyango, M., et al. (2012). New approaches for improving farmers' livelihoods through timely and proper use of weather forecast information: A pilot case in Kenya.

Siebert, S., Hoogeveen, J., & Frenken, K. (2006). Irrigation in Africa, Europe and Latin America —Update of the Digital Global Map of Irrigation Areas to Version 4. *Frankfurt Hydrology Paper 05*. Institute of Physical Geography, University of Frankfurt, Frankfurt am Main, Germany and Food and Agriculture Organization of the United Nations, Rome, Italy.

Walker S., Mukhala, E., van den Berg, W. J., & Manley, C. R. (2001). *Assessment of communication and use of climate outlooks and development of scenarios to promote food security in the Free State Province of South Africa.* Available online at: www.wmo.int

Adaptation Pathways for African Indigenous Vegetables' Value Chains

Silke Stöber, Winifred Chepkoech, Susanne Neubert, Barnabas Kurgat, Hillary Bett and Hermann Lotze-Campen

1 Introduction

Climate change generally poses unequivocal risks for food systems, and in particular for tropical agro-climatic zones (ACZ) in sub-Saharan Africa, due to their high exposure and low adaptive capacities (Niang et al. 2014; FAO 2015a). Food production in sub-Saharan Africa largely depends on smallholder rain-fed agriculture, highly vulnerable to seasonal shifts in precipitation patterns and extreme weather events (Lotze-Campen 2011; Anyah and Qiu 2012). Increasing variations in precipitation patterns, temperature rises and increased frequency and severity of weather-related extremes consequently cause heat and water stress and shortened cropping seasons, leading to yield reductions (Burke and Lobell 2010). Model-based estimates using the Agricultural Model Intercomparison and Improvement

Symposium on Climate Change Adaptation in Africa 2016 "Fostering African Resilience and Capacity to Adapt", Addis Ababa, Ethiopia, 21st–23rd February 2016.

S. Stöber (✉) · W. Chepkoech · S. Neubert · B. Kurgat
Faculty of Life Sciences, Centre for Rural Development (SLE),
Humboldt-Universität zu Berlin, Robert-Koch-Platz 4, 10115 Berlin, Germany
e-mail: silke.stoeber@agrar.hu-berlin.de

W. Chepkoech
e-mail: sangwinfred@gmail.com

H. Bett
Department of Agricultural Economics and Agribusiness Management,
Egerton University, Njoro, Kenya

H. Lotze-Campen
Potsdam Institute of Climate Impact Research (PIK), Potsdam, Germany

H. Lotze-Campen
Department of Sustainable Land Use and Climate Change, Faculty of Life Sciences,
Humboldt-Universität zu Berlin, Berlin, Germany

W. Leal Filho et al. (eds.), *Climate Change Adaptation in Africa*,
Climate Change Management, DOI 10.1007/978-3-319-49520-0_25

Project (AgMIP) indicate production declines for the main staple foods, maize and beans, of up to 20% for East Africa (Ramirez-Villegas and Thornton 2015). A crop simulation meta-analysis indicates that even with incremental adaptation measures maize crop losses cannot be avoided, with severe consequences for sub-Saharan African food systems (Challinor et al. 2014). There are no comparable simulation models to predict the risks for tropical horticulture in the face of climate change (Ayyogari et al. 2014; Midmore 2015). Compared to cereal crops, fruits and vegetables have very fixed climatic requirements for their physiological processes, resulting in high sensitivity to high temperatures or low soil moisture (Masinde and Stuetzel 2005; Ngugi et al. 2007; Muthomi and Musyimi 2009; Adebisi-Adelani and Oyesola 2013; Ayyogari et al. 2014). A changing climate also influences the nutritional value of vegetables: less ascorbic acid due to water deficiency and a sharp decrease in iron content due to the CO_2 fertilisation effect have been reported (Jain et al. 2007; Luoh et al. 2014).

Parts of sub-Saharan Africa are exposed to multiple stressors, and have therefore been termed hotspots of climate change, sharing a triple burden of (1) high exposure to the effects of climate change, (2) high poverty rates, and (3) high population densities (Müller et al. 2014). This study focuses on one of the hotspots, the densely-populated regions of Kenya, located in or adjacent to the Lake Victoria region. Kenya, despite being a middle income country, shows high rates of chronic food insecurity, with an estimated 30% of children under the age of 5 being stunted. According to Grace et al. (2012) increased stunting correlates with increased temperature and decreased rainfall in Kenya. High urbanisation and population growth together with crop yield reductions will exacerbate food insecurity in Kenya, where a doubling of the population to 97 million, with the rate of urban dwellers increasing from 25 to 46%, is projected for 2050 (FAO 2015b). A global simulation on regional undernourishment in the face of climate change estimates increased stunting rates in East and South sub-Saharan Africa by 55% (Lloyd et al. 2011).

Vegetables and fruits are protective foods, being rich in micronutrients and therefore a viable solution to fight undernourishment and hidden hunger. Studies show that 80% of vitamin A consumption in African diets derives from vegetables and fruits (Ruel 2001). Kenya's vegetable intake, at 88 kg/capita/year is high compared to other sub-Saharan countries. However, vegetable consumption among poorer rural households is lower compared to the better-off urban population (Okado 2001). African indigenous vegetables (AIVs) have the potential to improve food security in the face of climate change, for several reasons. AIVs are mainly produced by resource-poor smallholders, are nutrient-denser than exotic vegetables, have various health benefits, contribute to identity and authenticity, and offer a range of agronomic advantages (Ngugi et al. 2007; Abukutsa-Onyango 2010). There is a growing demand for AIVs, and planting areas have increased from 17,000 to 40,000 ha over 3 years (HCDA 2015). African nightshade (*Solanum scabrum*), spiderplant (*Cleome gynandra*), amaranth (*Amaranthus* spp.) and recently cowpea leaves (*Vigna ungulculata*) are the most common (HCDA 2013, 2015). Overall, there are more than 210 species with nutritional value (Maundu et al. 1999). Compared to spinach,

AIVs contain twice the amount of protein and 1.5–2 times as much vitamin A and more than 4 times as much vitamin C (Oniang'o et al. 2008; Yang and Keding 2009; Abukutsa-Onyango et al. 2010; Luoh et al. 2014). 100 g of fresh AIV contain 100% of the daily requirements of iron, vitamins, and calcium, and 40% of protein (Lenné et al. 2005; Abukutsa-Onyango 2010; Keatinge et al. 2010; Afari-Sefa et al. 2012). The potential of AIVs is largely underutilised as they are often overlooked in food policies and programs due to their antiquated image as a "poor man's crop" or "backward" food (Abukutsa-Onyango 2010; Gevorgyan et al. 2015). This attribute largely derives from the colonial past, when exotic vegetables were highly promoted and indigenous species entirely neglected.

Climate change is a global phenomenon, whereas vulnerabilities are highly contextual. Adaptations are expected to be carried out by local people in their specific settings (Sada et al. 2014). The need for adaptation action is gaining more importance, as highlighted in the IPCC AR5, which differentiates between incremental and transformative adaptation (Noble et al. 2014). Incremental adaptation practices are adjustments addressing proximate causes by building resilience into specific systems. Transformative adaptation pursues broader and systematic change by addressing the underlying roots of vulnerability. In agricultural and food value chain systems incremental adaptation includes a range of climate-smart, no-regret activities from the crop, land and water management spectrum, whereas transformative adaptations in the AIV value chain system include increased social inclusiveness, bargaining power, access to markets, information, land, and water resources. The process of adjustment is based on local decision making and therefore often referred to as adaptation pathways (Wise et al. 2014). Adaptation pathways are trajectories of no-regret actions, whether incremental or transformative, in a given adaptive space. A precondition for suitable adaptation action is the awareness of local decision makers, such as smallholder farmers, of local climate variation risks and sensitivities. Farmers' perceptions of weather and system sensitivity are therefore an entry point for planning farm-level adaptation practices (Teka et al. 2013).

The triple-win framework of climate-smart agriculture (CSA) aims at (1) sustainably increasing agricultural productivity to boost incomes and food security; (2) building resilience to climate change; and (3) reducing greenhouse gas (GHG) emissions from agriculture (FAO 2011, 2013; World Bank et al. 2015). Climate-smartness has been recently operationalised as a group of criteria (weather, water, nitrogen, carbon, energy, and knowledge) for assessing various farm-level practices in a number of countries, including Kenya (World Bank and CIAT 2015). Climate-smartness would serve as a reference model for proposing adaptation pathways for AIV value chains.

Overall, climate variability and change pose currently unknown risks to AIV value chains. Neither their sensitivity, nor the adaptation strategies of AIV farmers are adequately known. The aim of this study is to document local perception of climate risks and its impact on AIV systems, particularly agronomic sensitivities in the wet and dry seasons. For three distinct ACZs, semi-arid, semi-humid, and humid, site-specific features are broken down. The study answers four specific questions: (1) How do farmers perceive climate variability and change in their local situation and how closely do these perceptions match historical weather data? (2) How sensitive are the various AIVs to changing climate in dry and wet seasons? (3) Which farm-level adaptation strategies are pursued in different agro-climatic zones? (4) Which factors are hindering the implementation of adaptation strategies, i.e. what are the adaptation gaps?

2 Materials and Methods

2.1 *Study Area*

Kenya is divided into seven ACZs based on vegetation characteristics, amount of rainfall, and soil ecological potential. The study was conducted in Kakamega, Nakuru and Kajiado counties, representing the humid, semi-humid, and semi-arid zones, respectively (Table 1 and Fig. 1). The high to medium potential areas are the humid, sub- and semi-humid ACZs. They allow arable agriculture because they have an annual rainfall of more than 800 mm (MAFAP 2013). The low potential areas are the arid and semi-arid lands. AIV production is concentrated in the high and medium potential areas, as vegetables need well-watered soils. Horticultural land is prevalent in Western Kenya, as in Kakamega (Table 1).

2.2 *Data Sources*

The study used a mixed method approach, combining quantitative and qualitative data sets collected in the three ACZs. The different data sets come from a representative household panel survey conducted by the HORTINLEA project in 2014 (Kebede et al. 2015). In addition, in-depth information on climate perception and

Table 1 Study area

County	Kakamega	Nakuru	Kajiado
Agro-climatic zone	Humid	Semi-humid	Semi-arid
Rural-urban character	Rural	Peri-urban	Peri-urban
Horticultural land in ha/% of crop land	8,627/3.4%	33,734/0.1%	3,494/0.03%
Population	1,660,651	1,603,325	687,312
Poverty rate (%)	53	40.1	11.6
Mean temperature max (°C)	29	20	34
Mean temperature min (°C)	18	15	22
Mean precipitation p.a. (mm)	2,000	800	500

Source HCDA (2013), ASDSP (2014), CIA (2013)

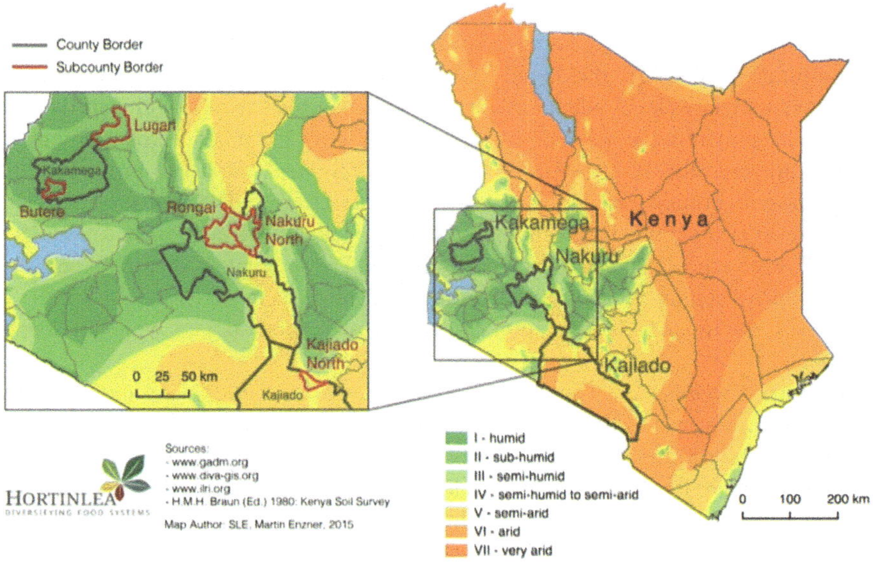

Fig. 1 Agro-climatic zones in Kenya and main study areas (sub-counties)

the sensitivities of AIVs to climate variability and change was gathered in focus group discussions (FGDs) with farmers. Farmers' perceptions of climate change were compared with empirical historical weather data obtained from the Kenya Meteorological Office (KMO).

The household panel survey data were analysed for three counties (Table 1), consisting of 610 growers, among whom the humid and semi-humid ACZs are well represented with 590 growers. The results for the semi-arid ACZ Kajiado must be interpreted carefully, as the sample contains only 20 respondents. The household survey section about "weather perceptions and effects of climate change" contains mostly categorical variables, for which cross tabulation procedures have been run to display contingency tables and their associations with Chi-Square test and Cramer's V (Figs. 3 and 6).

The household survey results on perceptions of climate change and adaptation strategies were compared to the information gained during 18 FGDs in which 189 AIV farmers participated. Local extension officers, trained in the research design and non-biased interview techniques, facilitated the discussions. Most of the participating farmers (60%) cultivate AIVs in a semi-commercial way, i.e. they produce for their own consumption and sell the surplus on local markets. 32% produce only for home consumption and 8% primarily for sale. 68% of the farmers were female, and 32% were male. Each discussion started with shifts and variability in temperature and precipitation. All assertions were recorded. Then, a sensitivity ranking of all AIV species grown by the farmers was conducted. The ranking was done for each season separately. Agreement was reached among farmers on the species which are least sensitive, which come second, third, and which are most sensitive to rainy and dry seasons respectively. Agreement was only reached after all arguments for their ranking decisions were provided. The adoption rate of climate-smart farm-level adaptation strategies in land and water, soil fertility, and crop management were assessed by simply adding up the practices of each farmer from a prepared list. At the end of the FGD, all farmers were asked to identify the three most important gaps in adapting to climate change. These statements were categorised into knowledge, technology, institutional and funding gaps, following a generic structure as proposed by UNEP (2014). The qualitative data obtained was processed by coding procedures according to content analysis (Mayring 2015). In this study, the impacts of adaptation strategies and adaptation gaps were not deeply explored.

Historical weather data on monthly mean temperature in °C and monthly precipitation in mm were obtained from the KMO. For the period 1980–2014, trends were analysed for three reference weather stations: Kakamega town for the humid zone Kakamega, the Jomo Kenyatta International Airport (JKIA) for the semi-arid zone Kajiado, and Nakuru town for the semi-humid zone Nakuru. As data were only available as monthly averages, it was not possible to exactly determine onsets and cessations as well as the intensity of rainfall. For exact rainfall distribution analysis daily data would have been needed.

3 Results

3.1 Farmers' Perceptions of Weather and Climate Variability and Change

Farmers in the **humid zone**, **Kakamega**, report a regular rainfall pattern until the year 2000, with two pronounced rainy seasons from March to May and September to December. This common pattern no longer holds, since a majority of the 90 participants have observed three major weather changes. Farmers agree on more overall rainfall, more unpredictable and more intense rainfalls. The increased frequency of hailstorms has been reported in six out of eight locations. "*If it rains, it pours down in a short time, and then for several days, we suffer from serious dry spells.*" (FGD 11) "*In the past, rains started in February, but now it rains throughout the entire year.*" (FGD 8) These observations are supported by the results from the household survey (Fig. 3), where 70% of the 373 AIV farmers stated more overall rainfall, and 18% indicated unpredictable and extreme rainfall events.[1] The perception of temperature change is less clear. Most farmers assert an increase in day and night temperatures, particularly during the rainy season, and

[1]The Cramer's V coefficient of 0.305 indicates a strong association between rainfall perceptions and ACZ.

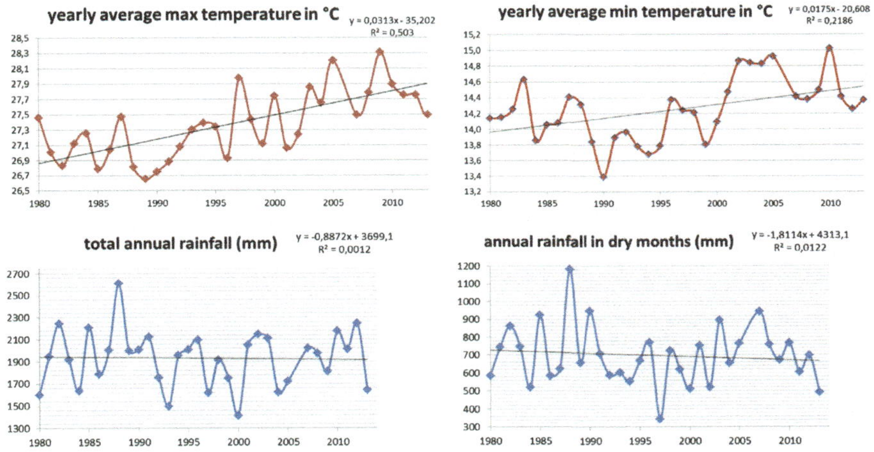

Fig. 2 Temperature and rainfall trends 1980–2013 Kakamega (humid zone)

more hot days. "*The days can be very hot now, that is different from the past.*" (FGD 7) "*In the past we were able, but now we cannot work in the field anymore at 2 PM, it is just too hot.*" (FGD 5) Farmers state that extremes in both directions are more pronounced. The results of the household survey also indicate varying perceptions,[2] such as hotter dry seasons and more hot days ranked first by 28% of respondents, followed by the opposite observation of longer cool seasons by 24.6%, and cooler dry seasons by 22.5%.

Historical data show a rather insignificant rainfall trend in terms of total annual rainfall (Fig. 2), and therefore do not support farmers' assertion of increased annual rainfall. In some monthly timelines, which are not illustrated here, monthly mean rainfall data confirm an upward trend, particularly during the short rains (September, December). For temperature, farmers' assertions of a temperature increase are supported by historical increases in maximum and minimum temperature of 0.4 and 1 °C, respectively (Fig. 2).

In the **semi-arid zone, Kajiado,** all 31 farmers consistently assert increased temperatures, increased frequency of very hot days, and a sharp decline in, and more unreliable rainfall. "*In the past, when I walk up the hill to the fields, it was warm, but now it is sometimes so hot, that I need to rest.*" (FGD 3) Likewise, the household survey reveals that the majority of farmers perceive higher and more extreme temperatures and less rainfall (Fig. 3). Overall, farmers' assertions are in line with the trends. Historical weather trends strongly support the perceived increase in night and day temperatures (Fig. 4). Rainfall shows a sharp decline with a slight rising trend in the dry months, indicating a less distinct rainy season.

[2]The Cramer's V coefficient of 0.209 indicates a medium association between temperature perceptions and ACZ.

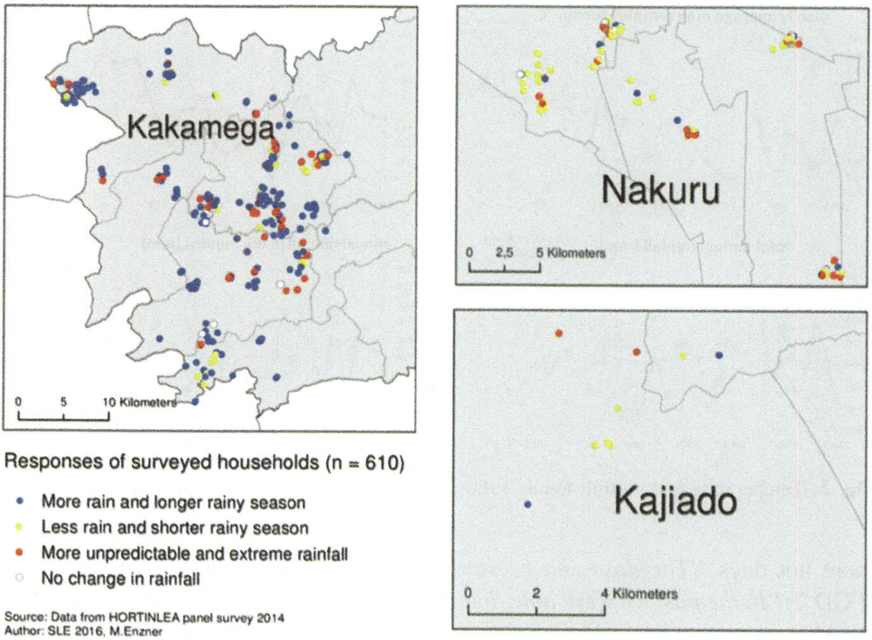

Fig. 3 Farmers' perceptions of changes in rainfall

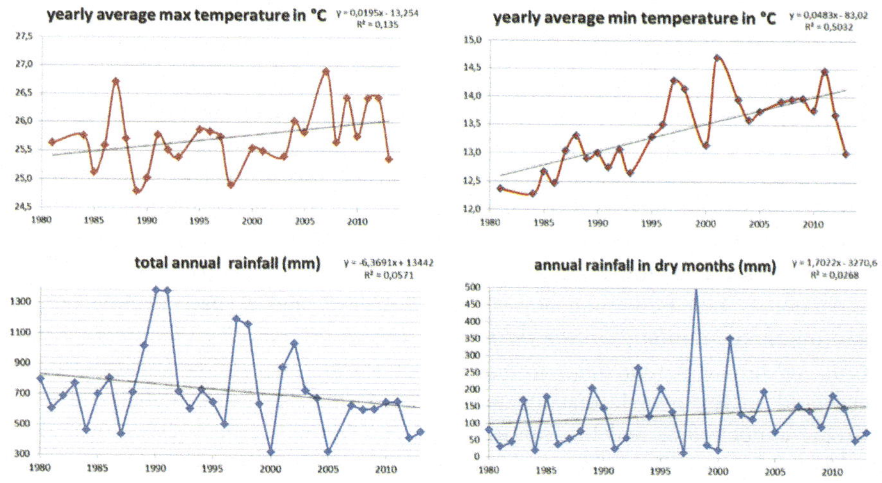

Fig. 4 Temperature and rainfall trends, JKIA (Kajiado, semi-arid zone)

In the **sub-humid zone, Nakuru**, the majority of the 68 participating farmers describe climate change as increased unreliable and unpredictable rainfalls and more frequent and severe dry spells with hotter temperatures. "*During the dry*

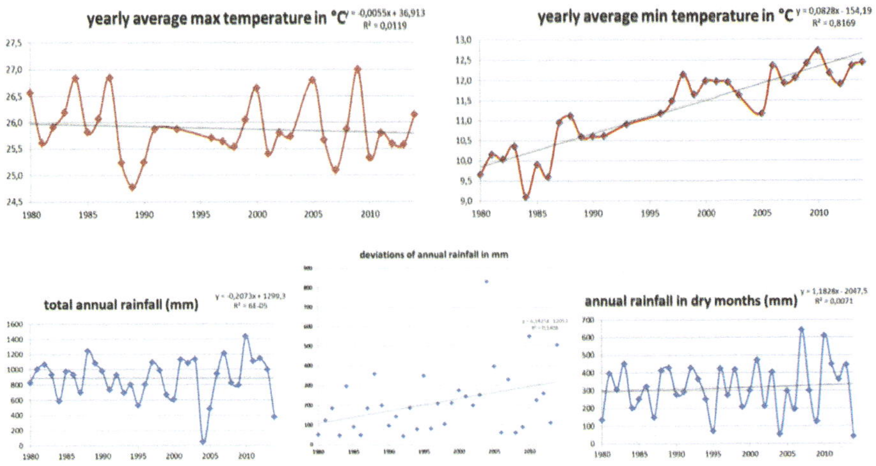

Fig. 5 Temperature and rainfall trends Nakuru (sub-humid zone)

season we abandon farming, as for the last 15 years, there have been increased droughts and unpredictable dry spells." (FGD 13) "*The temperature has increased most during the night*". (FGD 17) The representative household survey results match farmers' assertions, as many of the 183 farmers in Nakuru perceive hotter dry seasons (41.8%) and less rainfall (57.1%). Nakuru's historical weather data support a decreasing trend in total annual rainfall (Fig. 5). Figure 5 also reveals increased variability in rainfall from year to year, corresponding closely to farmers' perceptions of more unreliable and unpredictable rains.

3.2 AIV Sensitivities

As climate variability and change is real, it is important to know how AIVs respond to these changes and how sensitive they are to them. The sensitivity ranking (Table 2) provides a comprehensive overview. The table lists AIV species according to their importance on farms, as measured in size of plots per household survey. Overall, plot sizes are very small, with medians of 0.1 acre (~400 m^2). Nine different AIV species are grown in the study area, of which nightshade, cowpea, spiderplant and amaranth are the most important. The greatest diversity is found in Kakamega, where farmers grow up to eight species per farm, followed by Nakuru with six, and Kajiado with four species. Sensitivities to too much rain, water logging, dry spells, water stress and more pests and diseases are most common. The most tolerant "*survivor plants*" grown across all ACZs are pumpkin leaves and wild amaranth. Slenderleaves, jute mallow and Indian spinach are regionally important. Generally, AIVs thrive better in rainy seasons. The legumes cow pea and slenderleaves tolerate dry conditions. 15 out of 18 focus groups ranked cow peas and slenderleaves as most resistant to dry spells. Spiderplant is rather sensitive, as it is susceptible to too much and too little rain, and needs a lot of experience to grow, "*it depends on farmer's hands.*" (FGD 7) Nightshade tolerates extreme rainfall and performs well under wet conditions. However, in the dry season or during dry spells, nightshade is often affected by various pests and diseases. The results of Table 2 reveal that AIVs are not particularly sensitive, especially in the rainy season. Spiderplant, however, is more sensitive in both seasons, and requires more attention and knowledge.

3.3 Adaptation Strategies

According to the household survey, 88% of all farmers feel that climate variability and change affect their livelihoods, with the highest impacts in rural, humid Kakamega.[3] Major impacts are lower yields and more crop failure (67%), while impacts on villages, transport infrastructure and health are not prominent (2.2%). 18.8% of farmers in humid ACZs also point to positive effects of climate change, such as higher yields and less crop failure. Given the high impacts, farmers use different adaptation strategies.

Figure 6 shows that in Kajiado and Nakuru less than 50% of farmers claim that they adapt to climate change, while in rural Kakamega almost 90% of farmers pursue adaptation. Crop diversification activities are most popular (67%), whereas on-farm investments (irrigation, dams, terraces, tree planting, and ponds) and off-farm activities like non-farm employment or migration are rare strategies, due to the fact that investments and off-farm activities require capacities and resources that most farmers lack. On-farm investments are found more often in the better-off peri-urban Kajiado,[4] as farmers are more integrated into markets and cannot produce vegetables at all without irrigation due to the dry weather conditions.

The results from FGDs reveal in more detail which climate-smart farm-level adaptation strategies are applied by the farmers. Table 3 lists the adoption rates of water, land and crop management practices, which considerably differ across the ACZs. An adoption rate of more than 60% is considered high, 30–60% medium and less than 30% low. Only 10% of farmers use water management technologies; simple methods like the watering can or buckets prevail. Rainwater harvesting technologies are widely adopted only in the semi-arid county Kajiado. Few sustainable land management practices are applied; due to sloped fields, terracing in Kakamega is widespread. Integrated soil fertility management has medium adoption rates, with manure and compost being more widespread. Crop management

[3]The Cramer's V coefficient of 0.248 indicates a medium association between impact of climate change on livelihoods and ACZ.

[4]The Cramer's V coefficient of 0.280 indicates a medium to high association between the different adaptation strategies to climate change and the counties.

Table 2 AIV relevance and sensitivities to dry spells and heavy rainfall

English/Latin/local name	No. of plots[a]	Area[b]	Sensitivities in rainy season[c]	Sensitivities in dry season[c]
African nightshade *Solanum scabrum* Managu	915 (28.1%)	0.187/0.1	*Low* Rarely affected by pests and diseases, except blight and powdery mildew; blight is increasing problem, esp. in cold season (July); tolerates waterlogging; responds well to manure	*High* Affected by aphids, sugar ants, spider mites, nematodes, esp. in high temperatures; shallow roots very sensitive to drought, wilts quickly; gets bitter under water stress; doesn't germinate well in dry spell
Cowpea *Vigna unguiculata* Kunde	607 (18.6%)	0.196/0.1	*Medium* (standing variety) Cannot withstand heavy rains; doesn't thrive when it is too cold; susceptible to a wide range of diseases and pests (leaf rust, aphids, white flies, black spot)	*Very low* (creeping variety) Drought resistant; no pests and diseases
Spiderplant *Cleome gynandra* Sage, sageti	563 (17.3%)	0.195/0.1	*Very high* Highly affected by heavy rains, flooding not tolerated, difficult to grow	*High* Affected by spider mites, white flies; affected by water stress, gets bitter and wilts; needs irrigation; difficult to grow
Amaranth *Amaranthus* spp. (*mainly blitum, lividus, graecizans*) Terrere	534 (16.4%)	0.168/0.1	*Very low* Grows everywhere; likes wet conditions; resistant to all pests and diseases; thrives also on poor soils; responds well to organic fertilizer	*Low* Resistant to drought; no pests and diseases
Slenderleaf *Crotalaria brevidens* Miro, mitoo	51 (1.6%)	0.06	*Low* Does not suffer from diseases and even less from pests	*Very low* Drought resistant due to structure and size of leaves and robust deep tap root; no diseases and pests
Jute mallow *Corchorus olitorius*	5 (0.2%)	0.06	*Very low* No pests and diseases; mucous content is bitter and keeps off pests	*Low* Deep root system is drought tolerant

(continued)

Table 2 (continued)

English/Latin/local name	No. of plots[a]	Area[b]	Sensitivities in rainy season[c]	Sensitivities in dry season[c]
Murere, mwenda, murenda				
Pumpkin leaves *Cucurbita moschata duchesne* Massaiveve, liro, seveve	12 (0.4%)	0.06	*Very low* No pests and diseases; weeds do not spread, as stems meander; tolerates a wide range of soils, grows under shade and in sunshine, tolerates high amounts of water	*Low* Self-mulching, as the leaves cover the soil; little evaporation; fairly drought tolerant, but needs water for germination; host for, but not affected by white flies
Indian spinach *Basella alba* Nderema	0	0	*Very low* No pests and diseases; tolerates high amounts of water	*Very low* Survives in any conditions; resistant to dry spell

[a]As per household survey total/in percent of all plots
[b]Mean/median area (in acre) per household as per household survey
[c]Sensitivity ranking in 18 focus group discussions (189 farmers)
Source Own compilation

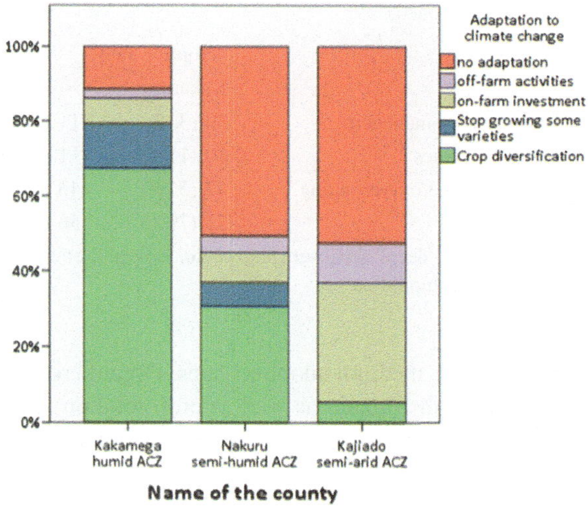

Fig. 6 Adaptation to climate change (n = 588). *Source* Data from HORTINLEA household panel survey 2014

Table 3 Adoption rate of farm-level adaptation strategies in AIV production

	Adoption rate			
	Kakamega	Nakuru	Kajiado	3 counties
Water management	*8%*	*3%*	*31%*	*10%*
Rainwater harvesting	6 (7%)	1 (1%)	27 (87%)	34 (18%)
Watering cans, buckets	25 (28%)	4 (6%)	14 (45%)	43 (23%)
Drip irrigation with bottles	5 (6%)	0	7 (23%)	12 (6%)
Drip irrigation with pipes	0	1 (1%)	5 (16%)	6 (3%)
Sprinkler	0	6 (9%)	2 (6%)	8 (4%)
Water pans	9 (10%)	1 (1%)	3 (10%)	13 (7%)
Land management	*21%*	*14%*	*9%*	*17%*
Agroforestry	11 (12%)	31 (46%)	1 (3%)	43 (23%)
Terraces	51 (57%)	10 (15%)	6 (19%)	67 (35%)
Raised seedbeds/double digging	7 (8%)	2 (3%)	7 (23%)	16 (8%)
Trash line	19 (21%)	0	0	19 (10%)
Retention ditch	6 (7%)	6 (9%)	0	12 (6%)
Soil fertility management	*40%*	*37%*	*30%*	*37%*
Crop residue mulching	45 (50%)	34 (50%)	15 (48%)	94 (50%)
Composting	62 (69%)	29 (43%)	9 (29%)	100 (53%)
Cover cropping	38 (42%)	3 (4%)	3 (10%)	44 (23%)
Advanced fertilising[a]	8 (9%)	1 (1%)	1 (3%)	10 (5%)
Ash	52 (58%)	29 (43%)	4 (13%)	85 (45%)
Manure	10 (11%)	56 (82%)	24 (77%)	90 (48%)
Crop management	*39%*	*42%*	*19%*	*37%*
Special seed varieties	19 (21%)	21 (31%)	0	40 (21%)

(continued)

Table 3 (continued)

	Adoption rate			
	Kakamega	Nakuru	Kajiado	3 counties
Integrated pest management	30 (33%)	11 (16%)	0	41 (22%)
Organic remedies	10 (11%)	11 (16%)	23 (74%)	44 (23%)
Mixed cropping/intercropping	47 (52%)	43 (63%)	7 (23%)	97 (51%)
Crop rotation	71 (79%)	56 (82%)	0	127 (67%)

[a]Micro-dosing, "deep" fertiliser, phosphate, Bokashi, effective microorganisms (EM)
Source Own compilation

practices have medium adoption rates. Organic remedies to build resistance to pests and diseases in Kajiado as well as crop rotation and mixed cropping/intercropping of AIVs with other species in Nakuru and Kakamega are widely practiced.

3.4 Adaptation Gap

Although various climate-smart AIV production technologies are practised in all three ACZs, only a few are widespread. Adaptation is linked to challenges in institutions and infrastructure, capacities, technologies, and finance. During FGDs, farmers provided more than 270 suggestions on how to close this adaptation gap. 25% of the solutions are improved funding for water systems and inputs along the value chain. Another 24% of solutions are providing extension services. Improvements to weak and non-systematic support in market integration (20%) and access to local water systems (15%) are also suggested. Only 15% of the solutions concern technology gaps, most frequently the lack of certified seeds, appropriate small-scale dryer and freezer systems, and effective pest management practices.

4 Discussion

4.1 Perceptions of Climate Variability and Change

This research supports the climatological evidence of continuously increasing temperatures, with most farmers (85%) confirming this trend. In fact, between 1960 and 2003 Kenya has experienced an average annual temperature increase of 1 °C (Met Office 2011; GoK 2012). In the semi-arid zone, temperature increases are most pronounced and during discussions farmers unequivocally stressed a sharp temperature increase.

Insignificant historical rainfall trends in terms of total annual amount do not match the change perceived by a majority of farmers (67%), with the exception of the declining trend in the semi-arid ACZ. Given the differences in rainfall perceptions even within a small region, it is suggested that rainfall patterns and changes are very site-specific. In one village of Kakamega, farmers highlighted the increased occurrence of hailstorms, while the neighbouring village did not suffer from hailstorms at all. County averages and monthly averages do not have sufficient resolution to reflect farmers' assertions about rainfall changes. Recent studies in Kenya and Ghana also showed that farmers perceive temperature increases and rainfall decreases even though climatological evidence do not show declining rainfall trends (Bryan et al. 2013; Amadou et al. 2015). The authors argue that rain-fed dependent farmers associate climate change with the variability of rainfall, its irregular onset and cessation, changing intensities and dry spells, which do not necessarily influence the annual or monthly total amount. Similarly, Arku (2013) and Nzeadibe et al. (2012) argue that farmers' perceptions of climate change reflect the fact that rainfall is the most important constraining factor in rain-fed farmers' decisions about cropping patterns. The findings of this research suggest that farmers' perceptions of changes in rainfall reflect a change in rainfall patterns, particularly shifts in onset and distribution, rather than changes in annual amounts.

4.2 AIV Sensitivities in Dry and Wet Seasons

A key determinant of AIV vigour is a reliable rainfall pattern, which has become a constraint in all ACZs. A lack of rainfall and extreme heat cause wilting, attract pests, and reduce yields. However, more frequent and extreme precipitation events result in high moisture conditions, causing favourable environments for diseases, destroying fields, and leaching nutrients. Overall, sensitivities are least pronounced in rainy compared to dry conditions. The farmers call jute mallow, slenderleaves, pumpkin leaves, cowpea and amaranth "*survivor plants*", showing a wide tolerance to temperature and precipitation extremes, and being least affected by pests and diseases. The majority of farmers state that the most popular African nightshade (32% of market share within the AIV market) (Irungu et al. 2011) and spiderplant are more sensitive to dry spells than other AIV species. Due to a range of agronomic advantages, AIVs are very popular among smallholders. AIVs perform well under harsh climatic conditions, are not susceptible to pests and diseases, and have a very short growing period of 3–4 weeks (Abukutsa-Onyango 2010; Biovision 2015; Prota4u 2015). Studies emphasise AIV drought tolerance and their resistance to pests and diseases compared to exotic vegetables (Muhanji et al. 2011; Luoh et al. 2014).

4.3 Adaptation Strategies

Soil fertility (using manure, ash, composting, and mulching) and crop management practices (rotation and mixed cropping) are the most widespread. They can be carried out fairly autonomously, as they depend only on the individual smallholder's decisions. Complex interventions in water and land management, however, require joint planning efforts and more financial support. According to Chesterman and Neely (2015), rainwater harvesting methods are widely promoted, though unlike the semi-arid Kajiado, farmers in Nakuru and Kakamega have never been exposed to this technology. To speed up adaptation in water management, better coordination and additional support would be needed. The findings indicate that farmers consider most sustainable agricultural practices as effective adaptation strategies, even though extension services have not promoted them explicitly as adaptation strategies. Within the scope of this study it was not possible to further assess the impact of all adaptation strategies used.

Low sensitivities during the rainy season make AIV production very easy and lead to oversupply and limited market potential. Amaranths, for example, are hardly purchased in the rainy season in rural areas, as they grow abundantly in home gardens or are collected freely outside the homestead. In the dry season, the opposite is the case, giving amaranths high market potential. A study in Tanzania therefore concludes that AIVs are an attractive commercial crop exclusively to be promoted in the dry season (Weinberger and Msuya 2004). The high market potential with scarce supply and higher prices in the dry season is coupled with higher sensitivities, and therefore requires improved adaptation packages for dry season AIV value chains. Commercialisation needs to consider the danger of biodiversity losses, as commercial production concentrates on only a few marketable species (nightshade varieties and spiderplant). The trade-offs of commercial production for on-farm AIV diversity have been already reported for farms around Nairobi (Irungu et al. 2011).

4.4 Adaptation Gap

Apart from the availability of improved certified vegetable seed and post-harvest technologies, adaptation gaps are linked to knowledge, funding, and institutional support. A number of CSA practices, recognized in the Kenya Climate Change Action Plan (2013–2017), are relevant for AIV production: drought-tolerant crops, water harvesting, drip irrigation, integrated soil fertility and pest management, and agroforestry (GoK 2012). These practices are not explicitly promoted in Kenya's agricultural strategy and not yet reflected in the agricultural sector budget. Funding for climate-smart adaptation is provided by a small number of stakeholders, including three relevant ministries—agriculture, environment, water and irrigation—research institutes, and NGOs. The technical document of the Kenya Climate-Smart Agriculture Programme 2015–2030 (Chesterman and Neely 2015) provides a good basis for developing funding opportunities for adaptation, but specific measures for AIVs are not included.

5 Conclusion and Future Research

Climate change is expected to have a significant impact on food security in Kenya. As nutrient-dense food, AIVs play an important role in fighting hidden hunger. While tropical vegetables require very specific water and temperature ranges, there is still little evidence of climate change having an impact on vegetable production. Farmers' understanding of climate change is highly associated with changes in rainfall. Farmers are mainly concerned with changes in rainfall distribution, which includes intensity, onset and cessation of rainfall, with significant differences between agro-climatic zones ranging from more and longer rains, to less or more unpredictable rainfall. Historical rainfall trends show little significant change in total amount, but differ considerably across the zones. It is recommended to analyse regional-scale historical trends more profoundly by analysing daily data and testing various parameters, such as increased variability, delayed onset and increased intensity of rainfall in order to confirm farmers' perceptions from the climatological perspective.

Many AIV species tolerate a wide spectrum of climate variability and are therefore considered insensitive to climatic variations. This is particularly true for the rainy season, while in the dry season, some AIVs are more affected by the consequences of climate variability, as some AIVs suffer from pests and diseases and water stress, particularly the marketable nightshade and spiderplant species. It is suggested to conduct market research on the potential of the so-called survivor plants, which are common on farms, but have a low market share (jute mallow, Indian spinach, local amaranth, pumpkin leaves, and slenderleaves). They play an important role in the mixed cropping system as they contribute to protection against pests and diseases.

Farmers' adaptation practices, such as crop diversification, crop rotation, simple irrigation with watering cans, tend to be of an incremental character. The same applies to a number of sustainable soil fertility management practices, as they are promoted by local extension services. These results underline the need for climate-smart strategies beyond autonomous adaptation, as the latter is not sufficient to increase resilience and productivity and to reduce trade-offs in smallholder AIV production. Only few farmers are able to invest in water and land management and are integrated well enough into social networks to be able to participate in commercial AIV value chains. The adaptation gaps include lack of funding, extension services, market integration, water resources, certified seed and post-harvest technologies. One potential adaptation pathway to promote AIV value chains is suggested by promoting a mixture of commercial AIV species in the dry season. This pathway would comprise a package of adaptation strategies starting from quality seed, efficient water use technologies, integrated pest management, and market empowerment. In this study, the climate-smartness of farm-level adaptation strategies wasn't assessed. In order to measure the impacts and trade-offs of adaptation, it is suggested to evaluate adaptation pathways against climate-smart criteria as proposed by the World Bank and CIAT (2015). Since AIV value chains are not yet considered in climate change adaptation policies, it is suggested to continue with a mix of qualitative and quantitative research, and to include farmers, extension services and policy makers in a participatory research process to jointly assess the costs and benefits of climate-smart adaptation pathways. Policy makers and practitioners being aware of the costs and benefits of AIV value chain adaptation will also contribute to developing socially inclusive policies and practices for fair and ecologically sustainable AIV value chains.

Acknowledgements The authors would like to express their special gratitude to the German Federal Ministry for Education and Research and the German Federal Ministry for Economic Cooperation and Development for providing financial support to the HORTINLEA (Horticultural Innovation and Learning for Improved Livelihoods in East Africa) project and enabling the researchers to conduct this research. Special thanks to the farmers for patiently sharing their knowledge and experiences concerning AIV. Thanks to all colleagues and two anonymous reviewers for providing data, valuable support and helpful advice for the paper.

References

Abukutsa-Onyango, M. (2010). African indigenous vegetables in Kenya: Strategic repositioning in the horticultural sector, Nairobi, Kenya.

Abukutsa-Onyango, M., Kavagi, P., Amoke, P., & Habwe, F. O. (2010). Iron and protein content of priority African indigenous vegetables in the lake victoria basin. *Journal of Agricultural Science and Technology, 4*(4), 67–69.

Adebisi-Adelani, O., & Oyesola, O. B. (2013). Fruit vegetable farmers' perception of climate change and adaptation strategies in Gombe State, Nigeria. International Society for Horticultural Science (ISHS), Leuven, Belgium.

Afari-Sefa, V., Tenkouano, A., Ojiewo, C. O., Keatinge, J. D. H., & Hughes, J. d'A. (2012). Vegetable breeding in Africa: Constraints, complexity and contributions toward achieving food and nutritional security. *Food Security, 4*(1), 115–127.

Amadou, M. L., Villamor, G. B., Attua, E. M., & Traoré, S. B. (2015). Comparing farmers' perception of climate change and variability with historical climate data in the Upper East Region of Ghana. *Ghana Journal of Geography, 7*(1), 47–74.

Anyah, R. O., & Qiu, W. (2012). Characteristic 20th and 21st century precipitation and temperature patterns and changes over the Greater Horn of Africa. *International Journal of Climatology, 32*(3), 347–363.

Arku, F. S. (2013). Local creativity for adapting to climate change among rural farmers in the semi-arid region of Ghana. *International Journal of Climate Change Strategies and Management, 5*(4), 418–430.

ASDSP. (2014). Agricultural Sector Development Support Programme, County Fact Sheets, Kenya. http://www.asdsp.co.ke/

Ayyogari, K., Sidhya, P., & Pandi, M. K. (2014). Impact of climate change on vegetable cultivation—A review. *International Journal of Agriculture, Environment and Biotechnology, 7*(1), 145–155.

Biovision. (2015). Infonet Biovision. http://www.infonet-biovision.org/node/28271

Bryan, E., Ringler, C., Okoba, B. O., Roncoli, C., Silvestri, S., & Herrero, M. (2013). Adapting agriculture to climate change in Kenya: Household strategies and determinants. *Journal of Environmental Management, 02*(114), 26–35.

Burke, M., & Lobell, D. (2010). Climate effects on food security: An overview. In *Climate change and food security: Adapting agriculture to a warmer world* (Vol. 37, pp. 13–30). Dordrecht, Heidelberg, London, New York: Springer.

Challinor, A. J., Watson, J., Lobell, D. B., Howden, S. M., Smith, D. R., & Chhetri, N. (2014). A meta-analysis of crop yield under climate change and adaptation. *Nature Climate Change, 4* (4), 287–291.

Chesterman, S., & Neely, C. E. (2015). Evidence and policy implications of climate-smart agriculture in Kenya. CCAFS working paper no. 90. CGIAR Research Program on Climate Change, Copenhagen, Denmark.

CIA. (2013). *The world factbook 2013–14*. Washington, DC: Central Intelligence Ageny.

FAO. (2011). *Climate-smart agriculture: A synthesis of empirical evidence of food security and mitigation benefits from improved cropland management*. Rome, Italy: Food and Agriculture Organization of the United Nations.

FAO. (2013). *Climate-smart agriculture sourcebook*. Rome, Italy: Food and Agriculture Organization of the United Nations.

FAO. (2015a). *Climate change and food systems: Global assessments and implications for food security and trade*. Rome: Food and Agriculture Organization of the United Nations.

FAO. (2015b). FAOSTAT online database, Rome, Italy.

Gevorgyan, E., Cronjaeger, P., Elsen, M., & Gefäller, L. (2015). *Connecting innovators, making pro-poor solutions work*. Berlin, Germany: SLE Publication Series.

GoK. (2012). *National climate change action plan 2013–2017*. Nairobi, Kenya: Government of Kenya.

Grace, K., Davenport, F., Funk, C., & Lerner, A. (2012). Child malnutrition and climate in Sub-Saharan Africa: An analysis of recent trends in Kenya. *Applied Geography, 35*(1–2), 405–413.

HCDA. (2013). *USAID-KHCP horticulture performance report 2010–2012*. Kenya: Ministry of Agriculture, Department of Horticultural Crops Development Authority.

HCDA. (2015). *National horticulture validated report*. Kenya: Ministry of Agriculture, Department of Horticultural Crops Development Authority.

Irungu, C., Mburu, J., Maunde, P., Grum, M., & Hoeschle-Zeledon, I. (2011). The effect of market development on-farm conservation of diversity of African leafy vegetables around Nairobi. *International Journal of Humanities and Social Science, 1*(8), 198–207.

Jain, V., Pal, M., Raj, A., & Khetarpal, S. (2007). Photosynthesis and nutrient composition of spinach and fenugreek grown under elevated carbon dioxide concentration. *Biologia Plantarum, 51*(3), 559–562.

Keatinge, J. D. H., Waliyar, F., Jamnadas, R. H., Moustafa, A., Andrade, M., Drechsel, P., et al. (2010). Relearning old lessons for the future of food-by bread alone no longer: Diversifying diets with fruit and vegetables. *Crop Science, 50*, 51–62.

Kebede, S., Ngenoh, E., Faße, A., Karki, S., & Bokelmann, W. (2015). HORTINLEA household survey in Kenya 2014. Egerton, Kenya, Berlin, Hannover, Germany.

Lenné, J. M., Pink, D. A. C., Spence, N. J., Ward, A. F., Njuki, J., & Ota, M. (2005). The vegetable export system. A role model for local vegetable production in Kenya. *Outlook on Agriculture, 34*(4), 225–232.

Lloyd, S. J., Kovats, R. S., & Chalabi, Z. (2011). Climate change, crop yields, and undernutrition: Development of a model to quantify the impact of climate scenarios on child undernutrition. *Environmental Health Perspectives, 119*(12), 1817–1823.

Lotze-Campen, H. (2011). Climate change, population growth, and crop production: An overview. In S. S. Yadav, R. J. Redden, J. L. Hatfield, H. Lotze-Campen, & A. E. Hall (Eds.), *Crop adaptation to climate change*. New Jersey: Wiley.

Luoh, J. W., Begg, C. B., Symonds, R. C., Ledesma, D., & Yang, R. (2014). Nutritional yield of African indigenous vegetables in water-deficient and water-sufficient conditions. *Food and Nutrition Sciences, 5*(9), 812–822.

MAFAP. (2013). Review of food and agricultural policies in Kenya 2005–2011. MAFAP Country Report Series, Rome.

Masinde, P. W., & Stützel, H. (2005). Plant growth, water relations, and transpiration of spider plant [*Gynandropsis gynandra* (L.) Briq.] under water-limited conditions. *Journal of American Society Horticultural Science, 130*(3), 469–477.

Maundu, P., Ngugi, G. W., & Kabuye, C. H. S. (1999). *Traditional food plants of Kenya*. Nairobi: National Museum of Kenya.

Mayring, P. (2015). Qualitative Inhaltsanalyse, Grundlagen und Techniken. Beltz, Weinheim.

Met Office. (2011). Climate: Observations, projections and impacts, Exeter, UK.

Midmore, D. (2015). *Principles of tropical horticulture*. CABI: Queensland, Australia, Reading, UK.

Muhanji, G., Roothaert, R. L., Webo, C., & Stanley, M. (2011). African indigenous vegetable enterprises and market access for small-scale farmers in East Africa. *International Journal of Agricultural Sustainability, 9*(1), 194–202.

Müller, C., Waha, K., Bondeau, A., & Heinke, J. (2014). Hotspots of climate change impacts in Sub-Saharan Africa and implications for adaptation and development. *Global Change Biology, 20*(8), 2505–2517.

Muthomi, J., & Musyimi, D. M. (2009). Growth and responses of African nightshades (*Solanum scabrum* MILL) seedlings to water deficit. *ARPN Journal of Agricultural and Biological Science, 4*(5), 24–31.

Ngugi, I. K., Gitau, R., & Nyoro, J. (2007). Access to high value markets by smallholder farmers of African indigenous vegetables in Kenya. In *Regoverning markets innovative practice series*. IIED: London.

Niang, I., Ruppel, O. C., Abdrabo, M. A., Essel, A., Lennard, C., Padgham, J., et al. (2014). Africa. In Climate change 2014: Impacts, adaptation, and vulnerability. Part B: Regional Aspects. Contribution of Working Group II to the Fifth Assessment Report of the Intergovernmental Panel on Climate Change Cambridge University Press: 1199–1265, Cambridge, UK, and New York, NY, USA.

Noble, I. R., Huq, S., Anokhin, Y. A., Carmin, J., Goudou, D., Lansigan, F. P., et al. (2014). Adaptation needs and options. In C. B. Field, V. R. Barros, D. J. Dokken et al. (Eds.), *Climate change 2014: Impacts, adaptation, and vulnerability. Part A: Global and sectoral aspects* (Contribution of Working Group II to the Fifth assessment report of the intergovernmental panel on climate change) (pp. 833–868). Cambridge, UK, and New York, NY, USA: Cambridge University Press.

Nzeadibe, T. C., Egbule, C. L., Chukwuone, N. A., Agwu, A. E., & Agu, V. C. (2012). Indigenous innovations for climate change adaptation in the Niger Delta region of Nigeria. *Environment, Development and Sustainability, 14*(6), 901–914.

Okado, M. (2001). Background paper on Kenya off-season and specialty fresh vegetables and fruits. UNCTAD diversification and development of the horticultural sector in Africa. Regional workshop for horticultural economies in Africa, Kenya.

Oniang'o, R., Grum, M., & Obel-Lawson, E. (Eds.). (2008). Developing African leafy vegetables for improved nutrition. Regional workshop, December 6–9, 2005. Rural Outreach Program, Nairobi, Kenya.

Prota4u. (2015). Plant resources of tropical Africa. Wageningen University, Netherlands. http://www.prota4u.org/

Ramirez-Villegas, J., & Thornton, P. K. (2015). Climate change impacts on African crop production. CCAFS Working Paper, No. 119. CGIAR Research Program on Climate Change, Agriculture and Food Security (CCAFS), Copenhagen, Denmark.

Ruel, M. T. (2001). *Can food-based strategies help reduce vitamin A and iron deficiencies? A review of recent evidence*. Washington, DC, USA: IFPRI.

Sada, R., Shrestha, A., Shukla, A. K., & Melsen, L. A. (2014). People's experience and facts of changing climate: Impacts and responses. *International Journal of Climate Change Strategies and Management, 6*(1), 47–62.

Teka, O., Houessou, G. L., Oumorou, M., Vogt, J., & Sinsin, B. (2013). An assessment of climate variation risks on agricultural production: Perceptions and adaptation options in Benin. *International Journal of Climate Change Strategies and Management, 5*(2), 166–180.

UNEP. (2014). *The adaptation gap report 2014*. Nairobi, Kenya: United Nations Environment Programme.
Weinberger, K., & Msuya, J. (2004). *Indigenous vegetables in Tanzania: Significance and prospects*. Shanhua, Taiwan: Technical bulletin, AVRDC.
Wise, R. M., Fazey, I., Stafford Smith, M., Park, S. E., Eakin, H. C., Archer Van Garderen, E. R. M., et al. (2014). Reconceptualising adaptation to climate change as part of pathways of change and response. *Global Environmental Change, 28*, 325–336.
World Bank, & CIAT. (2015). Climate-smart agriculture in Kenya. CSA country profiles for Africa, Asia, and Latin America and the Caribbean series, Washington D.C., USA.
World Bank, FAO, IFAD. (2015). Gender in climate-smart agriculture, module 18 for the gender in agriculture, Washington, D.C., USA.
Yang R. Y., & Keding, G. B. (2009). Nutritional contributions of important African indigenous vegetables. In C. M. Shackelton, M. Pasquini, & A. W. Drescher (Eds.), *African indigenous vegetables in urban agriculture* (pp. 105–144). London, UK: Earthscan.

Enhancing Adaptation and Mitigation Activities Through Effective Climate Change Financing Policy in Ethiopia

Belay Simane and Neil Bird

1 Introduction

1.1 Climate Change in Ethiopia

There is considerable evidence that the climate of Ethiopia is changing and projections suggest that the rate of change will increase in the future. Mean annual temperature has risen by 1.3 °C between 1960 and 2006, an average rate of 0.28 °C per decade (Anon 2011). Daily temperature observations show increasing frequency of both hot days and hot nights. Climate models suggest that Ethiopia will see further warming in all seasons of between 0.7 and 2.3 °C by the 2020s and of between 1.4 and 2.9 °C by the 2050s (Conway 2011). An increase in rainfall variability is also predicted, with a rising frequency of both extreme flooding and droughts that could seriously affect agricultural production (McSweeney et al. 2008; World Bank 2010).

Drought and floods are predicted to become more severe under most climate change scenarios due to significant inter-annual climate variability, complex topography and associated local climate contrasts, erodible soils, and intense land pressure due to an increasing population and an economy that remains almost entirely dependent on smallholder, low-input agriculture (Simane 2013). From 1900 to 2009, Ethiopia suffered 12 extreme droughts and 47 major floods, killing 402,000 and 1957 people, and causing damages of USD 93 million and 16.5 million, respectively (Jeremy et al. 2003). The country faced three severe droughts in the 1990s as well as the 2000s, resulting in massive harvest failure and pro-

B. Simane (✉)
College of Development Studies, Addis Ababa University, Addis Ababa, Ethiopia
e-mail: simaneb@yahoo.com

N. Bird
Overseas Development Institute, 203 Blackfriars Road, London, UK
e-mail: n.bird@odi.org.uk

W. Leal Filho et al. (eds.), *Climate Change Adaptation in Africa*,
Climate Change Management, DOI 10.1007/978-3-319-49520-0_26

duction below subsistence level for many households. In 2015, close to 15 million people went hungry due to El Niño caused drought in Ethiopia.

1.2 The Climate Change Policy in Ethiopia

Ethiopia launched its Climate Resilient Green Economy (CRGE) strategy in 2011, building on the positive growth trajectory of the previous ten years whilst recognising the challenge to build a middle-income country that is both resilient to the impacts of climate change and is low-carbon (Anon 2010). This policy initiative is made up of two main components: climate resilience and green economy. The Green Economy (GE) strategy was launched in parallel with the CRGE vision in November 2011 and aims to support the country in reaching the middle income threshold by 2025 whilst keeping the country's growth carbon neutral. A number of elements and focus areas of this strategy have been identified and planning is underway. The Climate Resilience (CR) strategy has developed in a different way, by taking a sector focus. Both the Agriculture and Forestry Strategy and Water and Energy CR strategy documents were launched in 2015.

The Agriculture and Forestry Strategy focuses on agricultural crops, livestock, forestry, food security and disaster prevention (Anon 2015a). The most promising options have been selected to build resilience in the agriculture sector against the risks from current weather variability and future climate change. Total annual investment under this strategy in agriculture is estimated at approximately USD 1 billion, of which around 40% is government investment through the Ministry of Agriculture. Private sector investment currently stands at 20% and is expected to rise to over 40% by 2030.

The Water and Energy Strategy is designed to safeguard these important sectors, which are expected to contribute approximately USD 7.2 billion to the projected GDP growth over the next national planning period, the second Growth and Transformation Plan (GTPII) (Anon 2015b). Ethiopia's hydropower supply is largely dependent on rainfall, so assessing and combating the challenges related to rainfall variability, for example, will be critical to food security and livelihoods.

1.3 The Public Institutional Response

The CRGE is overseen by the CRGE Ministerial Steering Committee (an initiative under the Prime Minister's Office), the Ministry of Environment, Forest and Climate Change and the Ministry of Finance and Economic Cooperation. CRGE units have been established in key implementing line ministries and regions to translate the CRGE strategy into sectoral programmes and investment plans. Federal ministries and regional entities have been identified as national implementing entities, and will be responsible for implementing programmes and investment plans in partnership with non-state actors.

To build a middle-income climate resilient green economy by 2025, Ethiopia has established an innovative funding mechanism to support the implementation of the priorities set out in the CRGE strategy, the CRGE Facility (Anon 2013). Designed as a national funding mechanism, this is intended to make the administration of funds easier for the government to drive and manage international climate funds, donor funds and domestic funds in a coordinated manner. It is guided by the strategic directions set by the Environmental Council and the CRGE Ministerial Steering Committee. The Ministry of Finance and Economic Cooperation is responsible for the overall management of the Facility with the Ministry of Environment, Forest and Climate Change responsible for technical coordination. The Facility supports and incentivizes a programmatic approach to climate change activities that aims to minimize transaction costs and the duplication associated with a projectized approach. It provides a single engagement point where the Government, development partners, the private sector and civil society can engage and make decisions about climate change issues, thus enhancing national coordination and aid effectiveness.

2 The Study Methodology

This paper examines the policy effectiveness of the CRGE strategy to build a picture of the overall policy environment for climate change expenditure, from the formulation of climate change policy to its linkages to spending through national strategies and action plans.

We use an analytical framework of policy-related effectiveness principles, criteria and indicators (PCI) to assess the effectiveness of the CRGE (Bird et al. 2013). The PCI approach comprises principles (fundamental laws or truths, expressing a core concept), criteria (operational standards by which to judge the principles), and indicators (information to measure or describe observed trends). Four principles that are relevant to the effective implementation of climate change policy, as identified from the literature, are ease of implementation, legitimacy, coherence and transparency (Table 1).

Quantitative and qualitative data sets, collected from both primary and secondary sources, were used to quantify the application of each principle. Primary data were collected using a number of data collection techniques, included structured questionnaires, key informant interviews, and focus group discussions. We used a 5 scale Likert scale to measure respondents' attitudes towards the effectiveness of CRGE to particular indicators. A total of 38 experts participated in this study from all CRGE implementing sector ministries including NGOs.

Table 1 Policy-related effectiveness principles, criteria and indicators (PCI) used for this study

Principle	Criteria	Indicators
Climate change policy is designed for ease of implementation.	Policy objectives are clearly expressed	• Targeted objectives are listed in the policy documentation • Timelines to achieve the set policy objectives are articulated in the relevant policy documents • The method for mobilizing financial resources to implement the policy is contained within the policy statement
	Subsidiary instruments for implementation accompany the policies	• Subsidiary instruments to achieve specific policy objectives are identifiable within the policy documents • Timelines are in place to establish appropriate subsidiary instruments • Appropriate subsidiary instruments are legally gazetted
The legitimacy of climate change policies shall be recognized by stakeholders	Key stakeholders' interests are represented in policy-making processes	• Policy-making platforms exist, where key policy decisions are made (e.g. policy working groups, expert working groups, sector working groups) • Existing policy platforms provide for representation of key stakeholders from both government and civil society • Existing policy platforms provide opportunities for stakeholders to contribute to the policy-making process
	Policy-making is evidence-based	• The policy formulation process is preceded by, and benefits from, background analytical work • Policy think tanks and research institutions provide evidence-based analysis to support the policy process • Relevant policy documents contain explicit references to background analytical work and contributions from policy think tanks
Climate change policies shall be coherent with national development policies	Policy statements on climate change acknowledge national development goals	• Reference is made to national development in the national climate change policy
	Climate change actions are consistent with strategies and planning processes for national development	• Climate change strategy documents and national development goals refer to each other
Climate change policies shall promote transparency in climate finance delivery	Climate change policies provide for the establishment and operationalization of mechanisms and modalities to promote transparency	• Mechanisms and modalities exist to promote transparency of climate finance

Source Bird et al. (2013)

3 Results and Discussions

The effectiveness of the national policy processes associated with the CRGE are assessed based on the four principles of policy development described in Table 1. The reliability coefficient, measure of internal consistency, for the indicators show a relatively high coefficient of reliability (Cronbach's Alpha = 0.883). A reliability coefficient of 0.70 or higher is considered "acceptable" in most social science research situations.

3.1 Ease of Implementation

Based on the aggregate values for all indicators of ease of implementation, 60% of the respondents agreed that the CRGE has been designed for ease of implementation (Table 2).

Evidence of ease of implementation comes from the government having selected four initiatives to fast-track the implementation of the GE strategy: exploiting the country's vast hydropower potential; large-scale promotion of advanced rural cooking technologies; efficiency improvements to the livestock value chain; and reducing emissions from deforestation and forest degradation (REDD+). These initiatives represent a rational policy prioritization as they appear to have the best chance of promoting growth immediately, capturing large abatement potentials, and attracting international climate finance for their implementation.

The timelines to achieve the set policy objectives are articulated in the relevant policy documents. The GE Strategy aims to support the country in reaching the middle income threshold by 2025 whilst keeping the country's growth carbon neutral. It also preceded the CR strategy in order to capitalize on international financing opportunities and the relative simplicity of it preparation.

Mobilization of financial resources to implement the CRGE strategy has been facilitated through the establishment of the CRGE Facility. In addition to securing increased levels of climate finance, the CRGE Facility has also enhanced the coordination and targeting of its utilization by providing a single coherent system within which development partners, the private sector, civil society and other stakeholders can engage and determine how best to invest in relevant actions.

Thus it can be seen that since the development of the 2011 CRGE Vision, national climate change policy instruments have been designed in such a way that facilitates policy implementation through the successive iteration of more detailed strategies, plans, programmes and projects.

Table 2 Climate change policy is designed for ease of implementation

Criteria	Indicators	Understanding of experts (%)				
		Do not know	Disagree	Somewhat agree	Agree	Strongly agree
Objectives are clearly expressed	Targeted objectives are listed in the policy documentation	5.3	2.6	42.1	34.2	15.8
	Timelines to achieve the set policy objectives are articulated in the relevant policy documents	10.5	15.8	21.1	36.8	15.8
	The method for mobilizing financial resources to implement the policy is contained within the policy statement	15.8	21.1	23.7	34.2	5.3
Subsidiary instruments for implementation accompany the policies	Subsidiary instruments to achieve specific policy objectives are identifiable within the policy documents	21.1	18.4	28.9	26.3	5.3
	Timelines are in place to establish appropriate subsidiary instruments	50	21.1	15.8	10.5	2.6
	Appropriate subsidiary instruments are legally gazetted	44.7	13.2	23.7	15.8	2.6
Average		24.6	15.4	25.9	26.3	7.9

3.2 Legitimacy of the Policy

The CRGE strategy was reviewed in terms of whether key stakeholders' interests had been present in the policy making process and whether the policy was evidence-based. These criteria of legitimacy were recognized by 69% of the stakeholders (Table 3).

During the process of designing the CRGE Strategy, the government of Ethiopia used three different multi-stakeholder bodies. First, the Inter-Ministerial Committee (IMC) acted as the governing and decision making body for the CRGE initiative. Second, the Technical Committee of CRGE (TC) provided a platform to provide technical guidance that supported a Sub-technical Committee (STC) composed of experts from different ministers/sectors that aimed to help implement the CRGE at both the national and regional level. The leading institutions within these three entities that has allowed them to fulfil their primary governance functions are as follows:

- As Chair of the Inter-Ministerial Steering Committee, the Office of the Prime Minister provides overall guidance to the work conducted with respect to the CRGE and facilitates high-level decision making for the CRGE Facility;
- The Ministry of Environment, Forest and Climate Change (MEFCC) oversees the CRGE Task Force and thereby coordinates work conducted with respect to the Sectoral Reduction Mechanism (SRM) and the specification of the CRGE Strategy Framework; and
- The Ministry of Finance and Economic Cooperation (MoFEC) is responsible for hosting the CRGE Facility and for providing the necessary financial and programme management systems and expertise.

These newly established platforms provide for representation of key stakeholders from across all parts of the government administration, but less so for civil society groups. There are also national research institutions (e.g. the Ethiopian Development Research Institute) and other higher learning institutes that are tasked with providing evidence-based analysis to support the policy process. However, there is little evidence of such analysis having been completed so far.

3.3 Coherence with the National Development Policies

The climate change strategy is an integral part of the national development policy. The CRGE strategy document clearly acknowledges both the Growth and Transformation Plan (GTP) and the Environmental Policy of Ethiopia. The GTP, the main government policy instrument that guides the major economic and social development efforts of the country, sets out a goal for Ethiopia to achieve middle-income country status by 2025 through steady double-digit growth, whilst at the same time becoming carbon neutral. Boosting agricultural productivity,

Table 3 The legitimacy of climate change policies shall be recognized by stakeholders

Criteria	Indicators	Understanding of experts (%)				
		Do not know	Disagree	Somewhat agree	Agree	Strongly agree
Key stakeholders' interests are represented in policy-making processes	Policy-making platforms exist, where key policy decisions are made (e.g. policy working groups, expert working groups, sector working groups)	18.4	18.4	26.3	18.4	18.4
	Existing policy platforms provide for representation of key stakeholders from both government and civil society	18.5	15.8	23.7	23.7	15.8
	Existing policy platforms provide opportunities for stakeholders to contribute to the policy-making process	7.9	21.1	26.3	32.4	10.5
Policy-making is evidence-based	The policy formulation process is preceded by, and benefits from, background analytical work	15.8	18.4	34.2	21.1	10.5
	Policy think tanks and research institutions provide evidence-based analysis to support the policy process	18.7	15.8	28.9	26.3	7.9
	Relevant policy documents contain explicit references to background analytical work and contributions from policy think tanks	7.9	2.6	15.8	42.1	31.6
Average		14.5	15.4	25.9	27.3	15.8

Table 4 Coherence of CRGE with the national development policies

Criteria	Indicators	Understanding of experts (%)				
		Do not know	Disagree	Somewhat agree	Agree	Strongly agree
Policy statements on climate change acknowledge national development goals	Reference is made to national development in the national climate change policy	7.9	2.6	15.8	42.1	31.6
Climate change actions are consistent with strategies and planning processes for national development	Climate change strategy documents and national development goals refer to each other	13.1	5.3	23.7	31.6	26.3
Average		10.5	3.95	19.75	36.85	28.95

Table 5 Transparency in climate finance delivery

Criteria	Indicator	Understanding of experts (%)				
		Do not know	Disagree	Somewhat agree	Agree	Strongly agree
Establishment and operationalization of mechanisms and modalities to promote transparency	Mechanisms and modalities exist to promote transparency of climate finance	29	7.9	28.9	26.3	21.1

strengthening the industrial base and fostering export growth have been prioritized as vehicles for reaching this goal. The coherence of the CRGE with the national development plans was recognized by a large majority—86% of the stakeholders (Table 4).

3.4 Transparency in Climate Finance Delivery

The transparency of the CRGE Facility and SRM in climate finance delivery was not clearly understood by a sizable number—37% of the stakeholders (Table 5). The CRGE does not identify in explicit terms measures to ensure that the delivery of climate finance happens in an open and transparent manner. However, the mechanisms and modalities to promote transparency of climate finance are presented in the CRGE Facility and SRM. The Facility is a national institution, working with all stakeholders to support Ethiopia's climate change response. It is closely linked to the Ministry of Environment, Forest and Climate Change, the Office of the Prime Minister and the Ministry of Finance and Economic Cooperation. The core purpose of the CRGE Facility is to channel finance to the activities prioritised in the GE and CR strategies and plans.

The CRGE Facility was officially launched on September 2012 and it is responsible for attracting, allocating and channelling international climate finance. The Facility will look to leverage both public and private finance, from multilateral and bilateral sources. Ideally, climate finance will complement other forms of investment to bolster Ethiopia's core climate-compatible development activities (in areas such as food security, energy, infrastructure development and natural resources management). The government is also looking at the possibility of having a performance-based mechanism for allocating finance, which can be expected to enhance the transparency of climate finance delivery.

4 Conclusions and Recommendations

In terms of its strengths, climate change has clearly been mainstreamed within the national development planning process. The main policy instrument, the CRGE strategy, has been well designed for ease of implementation, with coherence across the two main elements of the strategy (mitigation and adaptation). The mitigation (or green economy) element began earlier and has been influential in informing the overall growth trajectory that aims to secure for Ethiopia middle income status by 2025 in a carbon neutral way. The adaptation (or climate resilience) part of the strategy now needs to be enhanced so as to secure the livelihoods of those most vulnerable to climate change. In that context, the fast tracking of the agricultural sector CR strategy demonstrates effective sector prioritisation.

The analysis recommends two areas where further effectiveness gains may be sought. The first concerns how best to secure the active participation of all stakeholders in the policy process so as to maximise the likelihood of active implementation of climate change programmes and projects. There is broad experience that demonstrates effective delivery of public programmes depend in large measure on early involvement of all affected groups. Creating additional space for non-government officials in the present policy platforms is worthy of consideration. Second, the present policy is silent on how it will promote transparency in climate finance delivery, which is a generally-held principle of public administration. The emphasis on performance-based approaches for the allocation of climate finance by the CRGE Facility represents an important opportunity in this regard.

References

Anon. (2005). Ethiopia: Building on progress: A plan for accelerated and sustained development to end poverty (PASDEP) (p. 133). Ministry of Finance and Economic Development (MoFED), The Federal Democratic Republic of Ethiopia.

Anon. (2010). Ethiopia's growth and transformation plan (GTP), 2011/12–2014/15, Addis Ababa. Ministry of Finance and Economic Development.

Anon. (2011). Ethiopia's climate resilient green economy (CRGE). Green Economy Strategy, Addis Ababa, Ethiopia (p. 188). Federal Democratic Republic of Ethiopia.

Anon. (2013). The sectoral reduction mechanism (SRM) framework. The Ministry of Environment and Forestry. Federal Democratic Republic of Ethiopia (Draft Document).

Anon. (2015a). Climate resilient strategy, agriculture and forestry. Ministry of Agriculture, Addis Ababa, Ethiopia (p. 70). Federal Democratic Republic of Ethiopia.

Anon. (2015b). Climate resilience strategy: Water and energy. Ministry of Water, Irrigation and Energy, Addis Ababa, Ethiopia (p. 64). Federal Democratic Republic of Ethiopia.

Bird, N., Tilley, H., Canales Trujillo, N., Tumushabe, G., Welham, B., & Yanda, P. (2013). Measuring the effectiveness of public climate finance delivery at the national level. London: Overseas Development Institute. Available at: http://www.odi.org.uk/publications/7342-measuring-effectiveness-public-climate-finance-delivery-national-domestic-level

Conway, D., & Schipper, E. L. F. (2011). Adaptation to climate change in Africa: Challenges and opportunities identified from Ethiopia. *Global Environmental Change* 21(1), 227–237.

Jeremy, F., Gars, J., Özdoğan, M., Simane, B., & Zaitchik, B. (2003). Weather and welfare in Ethiopia. Selected paper prepared for presentation at the Agricultural & Applied Economics Association's 2013 AAEA & CAES Joint Annual Meeting, Washington, DC, August 4–6, 2013.

McSweeney, C., New, M., & Lizcano, G. (2008). UNDP climate change country profiles—Ethiopia. Available at: http://country-profiles.geog.ox.ac.uk

MoFED. (2012). Ethiopia's climate resilient green economy (CRGE) facility terms of reference. Ministry of Finance and Economic Development. Federal Democratic Republic of Ethiopia, Addis Ababa, Ethiopia.

Simane, B. (2013). Climate change: Impacts and responses for carbon neutral and climate resilient development in Ethiopia. *Ethiopian Journal of Natural Resources, 13*(2), 82–115.

World Bank. (2010). *Economics of adaptation to climate change, Ethiopia*. Washington, DC: The World Bank Group.

Community Forest Management for Climate Change Mitigation and Adaptation in Ethiopia: Determinants of Community Participation

Desalegn Dawit and Belay Simane

1 Introduction

1.1 Background and Justification

Community-based forest management can significantly contribute to reduce forest emissions and increase forest carbon stocks, while maintaining other forest benefits (Bishaw et al. 2013). Forest-dependent communities are also at the center of climate change adaptation efforts, which must focus on strengthening people's adaptive capacity and resilience. Payments for ecosystem services (PES), in community forest management, are useful in preserving, acknowledging and rewarding good community forest management practices. However, such schemes require the establishment of transparent and fair benefit-sharing arrangements and exploration of possible aggregation mechanisms to reduce transaction costs. Active participation of communities in all aspects of forest management, taking into account people's needs, aspirations, rights, skills and knowledge, will contribute for the efficiency, sustainability and equity of forest-based measures to tackle climate change. According to the Climate Resilient Strategy for Agriculture and Forestry document (2015), forest sector is the second largest source of GHGs, currently accounting for 55 $MtCO_2e$ a year, and is estimated to increase to 90 $MtCO_2e$ by 2030 under a business as usual scenario. The main driver for this is deforestation caused by pressure for agricultural land and energy demand.

Africa had been left behind accounting only for less than 2.5% of the projects registered under Clean Development Mechanism (CDM) (Nussbaumer 2008)

D. Dawit (✉) · B. Simane
Centre for Environment and Development, College of Development Studies, Addis Ababa University, P.O.Box: 33715, Addis Ababa, Ethiopia
e-mail: dawitdesalegn6@gmail.com

B. Simane
e-mail: simaneb@yahoo.com; belay.simane@aau.edu.et

W. Leal Filho et al. (eds.), *Climate Change Adaptation in Africa*,
Climate Change Management, DOI 10.1007/978-3-319-49520-0_27

though recently showing progress of different projects in the continent. Among these, the Humbo CDM project is the first large-scale forestry CDM Project in Africa that registered with United Nations Framework Convention on Climate Change (UNFCCC) in order to sink 165,000 tons of carbon from atmosphere through its growth ten years growth time by providing more than US$700,000 to local communities from World Bank's BioCarbon Funds (Gondo 2012).

Ethiopia has a great potential to disseminate its scope for other parts of the country to scale up a noble results of the Projects through adopting, diversifying and managing over all ecosystem services in addition to forest ecosystem services to meet pressing goal of CDM through maintaining sustainable development by reducing the cost of complying with the provision of Kyoto Protocol as oppose to developed nations—where they expected to be prevented from unsafe interface with climate system as climate change mitigation (Nyambura and Nhamo 2014).

This would facilitate equitable and reasonable payments for ecosystem service for the entire world and especially for developing countries like Ethiopia to establish their development technologies environmentally friendly (Du Monceau and Brohé 2011). On the other hand, regulation of ecosystems motivation and economic incentives are very important in PES because we focus on the issues of natural resource and environmental economics so as to have efficient PES to force and strength ecosystem service into the market model. In the same way, conservation economics need to adjust prioritization of ecological sustainability of economic institutions to the physical characteristics of PES (Farley and Costanza 2010). This needs more concern of the local communities for their effective participation with regard to the sustainability of the existing forest ecosystem service, developing other ecosystem service like forest regeneration, and wetland management through controlled mass mobilization.

In general, forests provide a range of ecosystem services, which today have little direct cash-generating value but do have significant indirect economic value to people's livelihoods if the participation and management level is very high. Ill compensation system, knowledge gap, and socio-economic characteristics for the benefits from ecosystem services leads to forest degradation and deforestation. And this further generates sever environmental and social effects on the face of income from payments for ecosystem services. But if appropriately structured, that would result on preservation and regeneration of forest resources (De Gryze et al. 2009; Fahey et al. 2009) through effective participatory community forest management.

Hence, carbon sequestration in forest systems is rapidly becoming the primary ecosystem service for which a substantial market is emerging (Sala et al. 2002) while PES instruments primarily are developed to improve the efficiency of natural resource management and to realize environmental objectives, the effect on income redistribution is often an important side objective, especially in developing countries since those who provide environmental services often constitute poor groups in society (Farley and Costanza 2010) as in the case of Ethiopia too.

1.2 Participatory Forest Management in Ethiopia Using CDM to Sink Carbon

Participatory Forest Management (PFM) programs have been demonstrated to be successful in addressing deforestation and degradation. It has been used in Ethiopia since 1990s according to Forest Carbon Partnership Facility (FCPF) report in 2011 and Participatory Forest Management is an approach of global phenomenon (Ali et al. 2007). On the other hand, participatory forest management approach is important in changing the social capital of local households on top of addressing deforestation and degradations.

The local households clearly emphasis on their food security and income source security over forest management and protection as if these issues are keys for all other sources of capital (Ali et al. 2007). But study shows that participatory forest management has a great role in improving forest and livelihood diversification components in Ethiopia as compared to unmanaged adjacent forests (Mouton and Boshoff 2008; Edwards et al. 2010) to bring about holistic and efficient deal on ecosystem services payments.

There is an inherent pressure in PES schemes between the simultaneous objectives of local household participation in CDM project, benefit from the project and equitable distribution of benefit. In order to bring about effective and regular participation of the local households, payments need to be optimized and targeted to higher value land and other sources of incomes need to be considered on top of different determinants of socio-economic characteristics (Bishop and Landell-Mills 2002; Grieg-Gran et al. 2005). This involves higher transaction costs and the risk of creating inequities since targeted payments are more costly to manage, and higher value land tend to belong to wealthy land owners. On the other hand, in order to convey enough amount of benefit, PES schemes need to reduce transaction costs like travel time, forest monitoring, etc. (Meshack et al. 2006). This can be done through untargeted payments concentrating on large land users and reducing higher costs that spent by poor HHs. This may be done at the expense of smaller, often poor land users, and of decreasing active participation of the local HH in PES schemes.

PES schemes will be more equitable if untargeted payments are used. Equity also involves supporting numerous, small land users and consequently raising transaction costs (Bishop and Landell-Mills 2002). This shows that there is a striving trade-offs involved in developing PES schemes. The active role of local communities in sustainable forest management can secure the survival of forest ecosystems and enhances their environmental, sociocultural and economic functions. It can both maximize forests' contribution to climate change mitigation and help forests and forest-dependent people adapt to new conditions caused by climate change. As poverty is one of the drivers of deforestation in the rural Ethiopia, adaptation for and mitigation of climate change must be addressed in unison with the fight against poverty and actions towards sustainable forest management (Olsen 2007). Improved forest management practices for climate change mitigation and adaptation should be planned and implemented, as they are closely linked. The objective of this study was to assess the determinant factors of community participation community forest management practices designed for PES.

2 Methodology

2.1 Sampling Procedure

The data was collected with the randomly selected three farmers' cooperatives of the Humbo District. From these three cooperatives, 157 farmers/beneficiaries were selected by using probability proportional to size.

2.2 Econometric Model Estimation

The level of participation of the household either to involve in the Clean Development Mechanism (CDM) of community forest management or not is carried out by using probit model in order to see the major determining factors of the individuals on the study area. Probit model gives the probability of the local household either to participate or not on CDM community forest management as climate change adaptation locally and mitigation mechanism globally to make discreet choice of decision of individuals and furthermore to see the marginal effect of the change on the dummy variables. Not only this but also the Probit model estimation provides and enables us normalized mean, variance and rational choice behavior of the participants, respectively. Therefore, the equation for the study is as follows using the selected variable with their interpretation on Table 1;

$$\begin{aligned} RAPR_i = {} & b_0 + b_1 EDU + b_2 SEX + b_3 LASI + b_4 AGE + b_5 FASI \\ & + b_6 ATME + b_7 MAST + b_8 WTA + b_9 CPCDM + b_{10} BPCDM + b_{11} DIPR \\ & + b_{12} OSINC + b_{13} AGEC + b_{14} INVPUB + u_i \end{aligned}$$

where all the variables described on Table 1 with respect to expected effects.

3 Results and Discussion

3.1 Household's Participation with CDM Project and Free Riders Problem

A significant number of the local households are participating in the CDM project of forest management and ecosystem services for different reasons (Fig. 1).

Table 1 Description of determinant of rate of participation of an individual in CDM project on community forest management

Variable	Characterization	Expected effect
RAPR	Rate of participation of the head of HH with CDM project; regular participation (1), otherwise = 0	Dependent variable
$LNBEN_I$	Natural logarithm of net benefit	Dependent variable
EDU	Class of years of education of the individual; 0 = illiterate, 1 = primary, 2 = junior, 3 = secondary, 4 = preparatory	+/−
SEX	Sex of the respondent; male (1), Female (0);	+/−
LASI	Actual land size of the HH head in hectare	−/+
AGE	Age of the head of HH in years;	+/−
FASI	The total number of the family;	+/−
ATME	Attending on community meeting with the concern of forest management = 1, otherwise = 0	+
MAST	Marital status of the head of HH; 1 = married, 0 = otherwise	+/−
WTA	Willingness to accept the burden while making use of the forest for the future generation (1) and otherwise (0);	−/+
CPCDM	Cost of participation in terms of ETB[a]	−/+
BPCDM	Benefit of participation in terms of ETB	+/−
DIPR	Distance to the project site from home in km;	−
OINC	Having other source of income (1) and otherwise (0)	−
AGEC	Agro ecology of the HH resident; 1 = weynadega[b], 0 = Klla[c]	+/−
INVEPUB	Investing benefit on public goods (1), otherwise (0),	−/+
u_i	Random error term	

[a]Ethiopian Birr/currency
[b]Mid-highland
[c]Law land

From the total observation 26.11% households participated irregularly while the remaining 73.89% are participated on forest management regularly and contributed more. Irregular participants reason out different kind of problems for irregularity of participation and emphasis on their food security (Ali et al. 2007) and income source security rather than participation for forest management. This is due to knowledge gap of the local households and priority of food security. The most frequently observed problems. are distance of the project area from their home (43.9%), different socio economic and cultural problems of the family (31.2%), the low return of participation (17.2%), exclusion problem (4.5%) and problems other than these (3.2%).

The study shows that the level of participation in forest management is positive. Because, though there are some difficulties, local households sacrificed for the regeneration of the forest—for example the ratio of regular participation to irregular participation is greater than 1. This shows that the regular participants have covered 2.83 times of the irregular participants' position by using their full potential per year

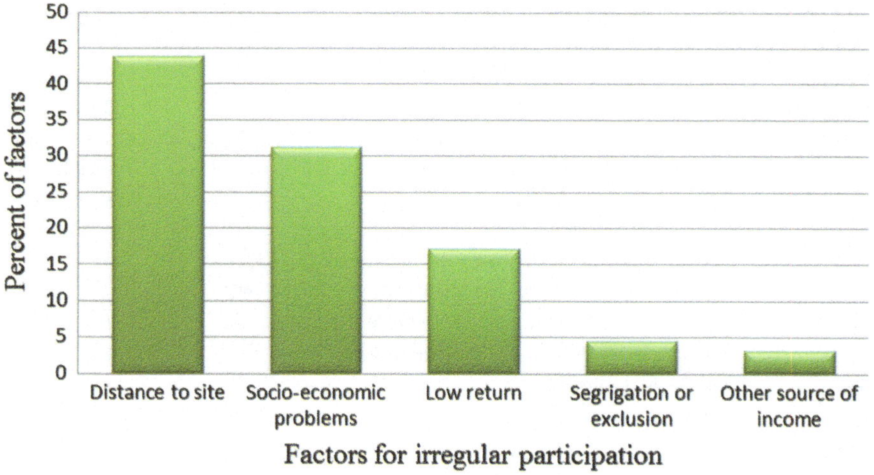

Fig. 1 Factors that influence local households for irregular participation

or regular participants done 2.83 times more than the irregular participants per year per work load. The implication is that irregular participants earn more annual net benefit at the expense of regular participants. This means that 26.11% are free riders to the resources without exact labor service provision to ecosystem services (forest management program).

According to the study, average net benefit distribution to the local individual with respect to participation level for irregular participant is greater than for regular participants per year per individual. The consequence shows that 26.11% respondents gain more than their fair share of labor (i.e. they pay less than their fair share of cost to the regenerated forest) and the remaining local households earn less for their fair share of labor and costs per year.

3.2 Determinants of Participation of the Local HH with the CDM Project

The level of participation with the CDM project on forest management adversely and positively determined with regard to education (EDU), gender/sex, agro-ecology (AGEC), attending on the general assembly of the members (ATME), cost of participation (CPCDM), investing income from carbon/Bio-Carbon fund/on public goods (INVEPUB), distance to the project site (DIPR) and other source of income (OSINC) of the HH head (Table 2).

Among the factors, gender discrimination (SEX), investing income from carbon/Bio-Carbon fund/on public goods (IBIP), distance to project site from residence (DIPR) and other source of income (OSINC) determined the state of

Table 2 Determinants of participation of the local HH with CDM project using Heckman's step one

Variables	Coefficient	Standard error	Marginal effect
EDU	0.062716	0.116038	0.018171
SEX(*)	−0.64747**	0.366524	−0.15979
LASI	0.077379	0.166963	0.02242
AGE	0.009463	0.016097	0.002742
ATME(*)	1.218667***	0.276269	0.411114
FASI	−0.02736	0.067303	−0.00793
MAST(*)	0.164569	0.493156	0.050036
WTA	−0.2594	0.270156	−0.07516
CPCDM	0.000267*	0.00014	7.73E−05
BPCDM	6.59E−06	0.000114	1.91E−06
DIPR	−0.10233*	0.097367	−0.02965
OSINC(*)	−0.31522*	0.301036	−0.09737
AGEC	0.076105*	0.375966	0.022051
INVEPUB(*)	−0.62528**	0.257605	−0.18259
_CONS	0.031991	1.325893	

Number of obs = 157
Censored obs = 41
Uncensored obs = 116
Wald chi^2 (24) = 77.60
LR chi^2 (14) = 36.65
Prob > chi^2 = 0.0008
Pseudo R^2 = 0.2033

Source Author's Computation using own Survey data 2012
(*) dF/dx is for discrete change of dummy variable from 0 to 1—which is marginal effect
*, **, *** indicates statistical significance at 10, 5, 1% probability level, respectively

participation with the project is negatively while the rest coefficient with star sign determined the level positively with their respective results in (Table 2).

Concerning marginal effects (Table 2), the highest amount of earnings or benefit goes to male headed HH as compared to female headed HHs. The marginal effect shows that a shift from female to male cause the rate of participation of female headed HH with CDM project to decline by 15.97%. This implies that the level of female headed HH participation less when they loss the required amount of net benefit per year from the total earnings. Female headed local households more busy and emphasis on their regular house work and subjected to less likely to participate on participatory forest management and ecosystem services.

The need of the local households on investing the public goods (INVEPUB) using income from Bio-carbon fund (income from carbon sequestration program) shows that the majority of the local HHs reluctant to invest on community wider projects. About 62.52% of the level of participation in forest management reduced by the reluctances of local HH to invest and demanding the fund for individual purpose.

Having one additional other source of income (OSINC) other than CDM forest regeneration program affected the level of participation of the local HH with the CDM project by reducing nearly 9.73% with negative correlation. This implies that there are other sources of income for HH that hinders them from going to forest management and at the same time their preference shift from CDM to other higher return/earnings. Where they delegate other family member on behalf of the HH head for forest management activity is also increased consistently.

Distance to project site from the residence "*DIPR*" using one additional kilometer impedes the HH head from having full effort investment of time on forest management activity approximately 2.96%. This may be convoyed by different social, cultural and economic challenges of the local household on top of the distance to the project.

As far regular meeting in the forest management decision and resource utilization "*ATME*" as concerned the local HH head participation level increased by almost 41.11%. This implies that the level of decision making process is crucial in participatory communal forest management program as compared to other explanatory variables whereas the level of education of the sample household portrays that having one additional step on educational cycle will result to increase the level of participation by 1.81% and this further improves benefits as though the return is not on the basis of fair labor share.

As result of dummy variable of gender, the level of participation of female headed household decreased by 15.97% as compared to male headed household since the marginal effect result of discrete change from female to male carried out. On the other hand, agro-ecology also maters the level of participation when we move from *Weyna Dega* to *Kola* and vice versa with positive implications of the progress of ecosystem services efficiency (Corbera et al. 2009). The implication is that peoples in *Kola* are participating more regularly than *Weyna Dega*. This is possibly due to the state of *Kola* region/area is more vulnerable to climate change hazard and the community understood the pain of recurrent drought and poverty. Also cost of participation affected its level positively other than net benefit.

4 Conclusion and Policy Implication

The level of participation determines the level of benefit of an individual from CDM and this further fixes the level of satisfaction to provide further ecosystem services in order to practice adaptive strategies and mitigation mechanism to climate change. Thus, participation significantly determined by distance of the site (where the residents reside from the forest), knowledge status which influences the level of ecosystem service efficiency, other source of income, family size, willingness to invest the fund on public goods and agro-ecology among different other factors and this further affected the level of adaptation and mitigation mechanisms of climate change in different ways.

Female headed HHs negatively affected for equal level of participation with their male headed counter parts on same activity of the project. Due to this reason, the income that they earn from project becomes declined along their side comparatively with respect to their labor share. On the other hand, attending on the general assembly meeting has a great importance in decision making process and increasing the level of earning from the forest to the local HHs as an adaptive strategy to climate change.

Most of the local HHs emphasized on other food security activities of short term and focusing on income source security rather than participation on forest management—that would sustain their agriculture system naturally and in a visible ways and concentrated on earnings of cash but not on the long term effect of the forest management. For example, majority of the local people neglected investing income from the forest carbon trade/sale on public goods (like on warehouse construction, flour mill installation, bridge construction, health facilities etc.) rather to use for their own purpose like as other source of incomes.

Due to these determining factors, the local people's earning becomes deceased satisfactorily by exacerbating bottlenecks on community forest management as climate change adaptation and mitigation mechanisms in the study area. This arbitrates to search for different policy implications to subdue the problem in its infant stage and find possible solutions. Policy makers need to set possible solutions for the barriers of participation of the local people from being participating on forest management as climate change mitigation mechanism to get more return and adaptation for the purpose of efficient payments for ecosystem services and boosting up their benefits on annual base with clear policy directions.

Furthermore, for some local HHs provide less labor and earn more benefit as compared to regular participant. Therefore, there should be a system that allows equal share of labor and benefit distribution mechanisms among the users so that this would facilitate on gearing the way forward of climate change adaptation. Because some households are free riders i.e. there should be equal share of benefit and cost among the participants, no one should pay the cost of others without earning more additional benefit. In addition, there is a knowledge gap among the local HHs vis-à-vis the project. Thus, critical awareness creation and pervasive turning point for training on merits and demerits of ecosystem services, climate change related problems should be provided to gear their perception on how to invest the fund from carbon trade and emphasizing on public goods benefit for all the local community.

References

Ali, T., et al. (2007). Impact of participatory forest management on financial assets of rural communities in Northwest Pakistan. *Ecological Economics, 63*(2), 588–593.

Bishaw, B., et al. (2013). *Farmers' strategies for adapting to and mitigating climate variability and change through agroforestry in Ethiopia and Kenya*. Corvallis, Oregon: Forestry Communications Group, Oregon State University.

Bishop, J., & Landell-Mills, N. (2002). Forest environmental services: An overview. *Selling forest environmental services—Market-based mechanisms for conservation and development, 2002*, 15–36.
Corbera, E., Soberanis, C. G., & Brown, K. (2009). Institutional dimensions of Payments for Ecosystem Services: An analysis of Mexico's carbon forestry programme. *Ecological economics, 68*(3), 743–761.
De Gryze, S., et al. (2009). Payment for ecosystem services: Developing forest carbon projects in Nepal. EnterpriseWorks/VITA, Terra Global Capital, USAID and ANSAB.
Du Monceau, T., & Brohé, A. (2011). Sustainable Development and Social Equity
Edwards, S., et al. (2010). Successes and challenges in ecological agriculture experiences from Tigray, Ethiopia. Tigray Project.
Fahey, T. J., et al. (2009). Forest carbon storage: Ecology, management, and policy. *Frontiers in Ecology and the Environment, 8*(5), 245–252.
Farley, J., & Costanza, R. (2010). Payments for ecosystem services: From local to global. *Ecological Economics, 69*(11), 2060–2068.
Gondo, P. C. (2012). *A review of forest financing in Africa*. Zimbabwe: Southern Alliance for Indigenous Resources (SAFIRE).
Grieg-Gran, M., et al. (2005). How can market mechanisms for forest environmental services help the poor? Preliminary lessons from Latin America. *World Development, 33*(9), 1511–1527.
Meshack, C. K., et al. (2006). Transaction costs of community-based forest management: Empirical evidence from Tanzania. *African Journal of Ecology, 44*(4), 468–477.
Mouton, J., & Boshoff, N. (2008). The Federal Democratic Republic of Ethiopia.
Nussbaumer, P. (2008). Clean development mechanism and sustainable development: A discussion on the link between the CDM and Sustainable Development, an analysis of the current status of the CDM portfolio, and an multicriteria evaluation of the effects of additional incentives in order to foster broad local Sustainable Development dividends from the CDM projects.
Nyambura, B., & Nhamo, G. (2014). CDM projects and their impact on sustainable development: A case study from Kenya.
Olsen, K. H. (2007). The clean development mechanism's contribution to sustainable development: A review of the literature. *Climatic Change, 84*(1), 59–73.
Sala, E., Aburto-Oropeza, O., Paredes, G., Parra, I., Barrera, J. C., & Dayton, P. K. (2002). A general model for designing networks of marine reserves. *Science, 298*(5600), 1991–1993.

Pastoralists and Farmers Coping and Adaptation Strategies to Climate Variability and Their Perceived Success in Ethiopia

Desalegn Yayeh Ayal, Muluneh Woldetisadik Abshare, Solomon Desta and Walter Leal Filho

1 Introduction

1.1 Background

Agriculture is the mainstay of pastoralists and farmers in developing countries like Ethiopia. Climate variability has been posing unprecedented challenges to agriculture, and to the old way of life of pastoralists and farmers (IPCC 2007). Moreover, agricultural adaptation efforts are compromised by low objective and subjective adaptive capacity (Ayal and Muluneh 2014).

Agriculturalists embrace adaptation measures either induced by government decisions or experience based autonomous initiative (IPCC 2007). Although people can make steadily growing adaptation concomitant with gradual climate variability, rigorous adaptation is a response made when traditional practices are no longer reliable to cope with the effects of climate extremes (Smit and Wandel 2006). This explains the distinction between adapting and coping. Coping simply refers to temporary mechanisms employed to survive in a given constraint while adapting signifies a drastic overhaul of systems to permanently deal with the pitfalls of climate variability (IPCC 2007). Pastoralists and farmers develop coping strategies

D.Y. Ayal (✉)
Addis Ababa University Center For Food Security Studies and MARIL
(Managing Risk for Improved Livelihood), 150129 Lebu, Addis Ababa, Ethiopia
e-mail: desalula@gmail.com

M.W. Abshare
Addis Ababa University Geography and Environmental Studies, Addis Ababa, Ethiopia

S. Desta
MARIL (Managing Risk for Improved Livelihood), Addis Ababa, Ethiopia

W. Leal Filho
Manchester Metropolitan University, Manchester, UK

W. Leal Filho et al. (eds.), *Climate Change Adaptation in Africa*,
Climate Change Management, DOI 10.1007/978-3-319-49520-0_28

through a gradual process of personal experience while adaptation is necessitated only when available coping mechanisms fail to serve their purposes especially in relation to drastic climate changes. Adaptation is crucial to those in agricultural sector for two main reasons:

i. it may help them to upkeep productivity, and secure income
ii. it may help them to ensure food security

Eventhough these two elements are very basic, it has to be understood that theoretically, pastoralists' and farmers' responses to the challenges of climate variability are a logical outcome of their perceptions towards their own ability to react to the challenges effectively. People decide to take action to climate related problems when they are confident that their action can change the situation for the better (Kuruppu and Liverman 2011). On the other hand, peoples' psychological disposition is in turn dependent on their asset bases such as natural, physical, human, social, financial and institutional resources, as well as culture, norms, beliefs and values (Ericksen et al. 2011; Smit and Pilifosova 2001). Hence, pastoralists and farmers perceptions of the success of their coping and adaptation practices has both subjective (internal or emotional) and objective (external or material) dimensions (Kuruppu and Liverman 2011; Smit and Pilifosova 2001). Pastoralists and farmers objective and subjective adaptive capacity should be balanced. Otherwise, it leads to cognitive biases i.e. underestimating and or overestimating the risk. Therefore, confidence in the success of coping and adaptation practices is the outcome of balanced resource bases and perceived adaptive capacity. Hence, a concerted effort to understand how pastoralists and farmers perceive their adaptive capacity and examine the actual outcome of their adaptation mechanisms is crucial to avert the hazards of climate variability. Accordingly, this study explores the objective and subjective foundations of pastoralists' and farmers' perceptions of the success of their adaptation practices.

Assessing current coping and adaptation practices helps us to understand pastoralists' and farmers' perceived and actual adaptive capacity so as to initiate appropriate measures that enhance their adaptive capacity, both to minimize the effect of climate related shocks (Adger et al. 2003a, b) and to benefit from opportunities. Even though it is difficult to neatly differentiate spontaneous adaptation responses (coping mechanisms) from planned adaptation mechanisms, this study sought to identify the matrix of the objective and subjective dimensions of coping and adaptation practices by pastoralists and farmers to climate extremes. By providing micro-level empirical evidence for the actual and perceived efficacy of adaptive responses, this study can inform policy makers to design effective adaptation strategies.

2 Description of the Study Areas and Methodology

2.1 Study Area

This study was conducted in the Yabello and Arero woredas of Borana zone of Oromia National Regional State and Gozamin and Enemy woredas[1] of East Gojjam of Amhara National Regional State, representing the lowland and the highland agro-ecology respectively. The lowland study site is situated (CCAFS learning site) between 4°35′–5°5′ N and 38°15′–38°40′ E, covering an area of 900 km. Population density is low; the population size of the two woredas is an estimated 150,493 and about 94% of those inhabits live in rural areas (CSA 2008). The area generally has a gentle plain topography except for minor interruptions by volcanic cone-shaped hills and mountains of less than 2200 m (Coppock 1994).

It has bi-modal rainfall where the main rainfall is received from March to May while the small rainfall occurs from September and November. The study sites suffer from highly variable rainfall and surface and ground water sources. Eellas and ponds are the main water sources, and they are unreliable. The mean annual maximum and minimum temperatures are 25.1 and 13.3 °C respectively. Livestock rearing remains the principal livelihood of the local population. However, in recent years, pastoral are increasingly practicing crop production (Coppock 1994).

Gozamin and Enemay woredas are located between 10°18′–10°40′ and 38°00′–38°22′ and 10°20′–10°40′ and 37°15′–37°45′ respectively. These woredas are inhabited by a population of 299,175. Less than 3% of the population lives in urban areas (CSA 2008). High population density has limited the size of landholding which seems to have reached the limits of further expansion (Tesfaye 2004a, b). The topography is characterized by rugged hills, mountains and gentle plain land. The woredas receives average annual rainfall of 1200 mm. It is drained by the Abay (Blue Nile), Muga, Yegudfin, Chemoga, Kulich and Degell rivers. Mixed farming is practiced in the area. The majority of land is allocated for cultivation (45.7%), following by human settlement (41.41%), grazing land (10.4%) and pasture (2.5%).

2.2 Research Methods

2.2.1 Study Site Selection and Sampling Methods

Yabello, Arero, Gozamin and Enemy woredas and then three rural kebele[2] in each lowland woredas and five kebeles in each highland woredas (totally sixteen) were selected using a purposive sampling method due to their physical accessibility. The research was conducted in Yabello Dikale, Dembelaseden and Abuno kebeles and

[1]Woreda is an administrative unit equivalent to a district.

[2]Kebele is the smallest administrative unit equivalent to a peasant association in the woreda.

in Arero Alona, Gedda and Selle kebeles in Gozamin Libanos, Denba, Yebona Erjina, Wonka and Enerata kebeles and in Enemay Yeser Eysus, Mahibre Birhan, Yetenbinaweyinam, Sekela and Dema kebeles. Accordingly stratified purposeful sampling was applied to select 12 experts, 32 development agents, 8 community key informants and 47 FGD participants. Stratified purposeful sampling allows researchers to capture major variations in pastoralists and farmers coping and adaptation practices and their perceived adaptation success (Patton 1990). Finally, stratified random sampling techniques on the basis of agro-ecological zones, sex and wealth status were used to select 300 sample households. Site and key informant selection processes were conducted in consultation with development agents and the local community.

2.2.2 Data Sources and Collection Methods

Qualitative and quantitative data were gathered from experts, development agents, and the community (key informants, FGD and survey participants). Data from various stakeholders were gathered using a questionnaire survey, PRA (FGD, overt observation, interview). By using a questionnaire survey, both quantitative and qualitative data on household local coping and adaptation strategies to climate variability and allied non-climate stressors were collected from sample households. Likert scale items that measure the respondents' opinions on success and perceptions of their coping and adaptation practice in the crop and livestock sector were employed. Using interview, data on objective and subjective adaptive capacity and challenges were collected. A total of four focus group discussion sessions with an average of 12 participants of male headed and female headed households were conducted i.e. one in each woreda. By using FGD, data on the nature and behavior of coping and adaptation strategies, the pastoralists' and farmers' perceptions of success were gathered. Overt observation was used to collect data on the seasonality of climate related shocks, farming activity, social networks, and the pressure on farming practices to cope with the impact of climate related problems.

2.2.3 Data Analysis

The study used both qualitative and quantitative data analysis method. Pastoralists' and farmers' coping and adaptation practices and the perceived success on crop and livestock production are the function of their objective and subjective adaptive capacity. Respondents' age group was classified using Erikson's psychosocial theory of age classification. Then, the last three age stages, namely early adult hood (20–30s years), middle adulthood (40–50s years) and late adulthood (60 years and above) were considered (Santrock 2011). A local wealth classification criterion was applied to group pastoralists and farmers according to their wealth status (see Ayal and Muluneh 2014). Descriptive statistics were also employed. Their perceived success in coping and adapting to climate shocks were analyzed using Multivariate Analysis of Variance (MANOVA). MANOVA helps to appreciate the role of a particular independent variable and interactional effect of interdependent factors on pastoralists and farmers' perception of the success of coping and adaptation practices. SPSS was used to analyze quantitative data. The qualitative data collected from interviews and focus group discussions and observation were analyzed thematically.

3 Results and Discussion

3.1 Pastoralists and Farmers Coping and Adaptation Strategies in the Crop and Livestock Sectors

Although all pastoralists and farmers in Ethiopia are involved in crop production and livestock rearing, they have different priorities depending on the nature of their habitation. In the highlands, farmers keep livestock mostly for the purpose of crop production (ploughing, transportation and manure) and to generate income from the sale of livestock and their products to cover agricultural and other expenses. For the lowland pastoralists, however, where crop production is in its infancy, livestock rearing has been the chief basis of their livelihood. From the varying emphases given to crop production and livestock rearing in the highlands and lowlands, various coping and adaptation strategies to climate extremes are borne. This is demonstrated in the Table 1.

Table 1 Crop sector adaptation measures to climate variability

Adaptation practice	Lowland (pastoralists)		Highland (farmers)	
	N = 29		N = 184	
	No. replied	%	No. replied	%
Grow different crop variety	5	16.6	103	56.0
Varying planting and harvesting dates	20	68.6	184	100.0
Use fast growing varieties	1	2.9	136	73.9
Use drought resistant varieties	0	0	30	16.3
Intercropping	1	4.9	44	23.9
Using fertilizer	3	8.8	184	100.0
Using pesticides	0	0	58	31.5
Using herbicides	1	2.9	44	23.9
Mulching	0	0	17	9.2
Increased use of irrigation	0	0	44	23.9
Water and soil conservation	11	37.3	175	95.1

Obviously, agriculturalists are not passive victims of the negative effects of climate variability. Table 1 shows that pastoralists and farmers responded to the problem in different and similar ways on crop production. As shown in Table 1 above the most widely practiced adaptation mechanism among all households is varying planting and harvesting dates, as confirmed by all farmers and 68.6% of pastoralists. Soil and water conservation is ranked the second most practiced adaptation strategy approved by 95.1 and 37.3% farmers and pastoralists respectively. The most remarkable difference between the farmers and pastoralists was in their use of fertilizer and planting fast growing crop varieties (see Table 1). Table 1 illustrates that farmers were practicing more types of adaptation strategies than pastoralists. The high disparity is likely due to the fact that crop production is not fundamental to the livelihood of pastoralists.

In the highlands, particularly in Libanos and Denba kebeles, FGD participants and key informants explained that farmers use the most productive and drought resistant *teff* variety—locally called *Mengisto*—which also enjoys a high market price. On the whole, with the exception of changing planting and harvesting dates and varying fertilizers, the rate of 'soft' and 'hard' agricultural technology adoption is very low. As indicated in Table 1, the use of drought resistant crop varieties, intercropping, pesticides, herbicides, mulching and irrigation as adaptation strategies is negligible, especially in the lowland. Although poverty would make chemicals fertilizers unaffordable, the infrequent practice of a zero-cost adaptation practice (intercropping) suggests that the problem is one of awareness as much as it is economic.

Understandably, the rare practice of mulching is attributable to the use of crop residues for animal feed. Likewise, poor irrigation practice could be explained by a shortage of irrigable land, the high cost of irrigation technology and a shortage of water during the dry season. It appears that the most frequently used adaptation strategies are those that cost very little or nothing, such as changing planting dates, crop rotation and soil and water conservation. This being said, however, the range of measures practiced in all agro-ecological zones would not qualify pastoralists and farmers as 'adaptive' to the impact of climate variability. Even the use of fertilizer may not have the desired effect since their ability to use the appropriate amount of fertilizer at the right time may be questionable. Besides, as Kassahun (2011) claimed, applying fertilizer helps only in times of good weather which means that, without other adaptation technologies, the use of fertilizer only exacerbates the adverse impact of climate variability. Table 2 shows the ccoping and adaptation strategies in the livestock sector

As indicated in Table 2, destocking is a strategy reported to have been practiced in all sites by a nearly equal percentage of respondents, averaging more than 66%. This seems to be a commendable practice since destocking reduces stress on available scarce resources while creating opportunities to increase the quality of livestock. Farmers practiced food storage, keeping drought resistant species and engaged in a cut and carry system more than pastoralists. On their part, pastoralists practiced bush clearing and the extended search for water and pasture more than farmers do. Enclosure is the most common adaptation strategy in the lowland: this

Table 2 Coping and adaptation strategies in the livestock sector

Adaptation practice	Lowland		Highland	
	N = 102		N = 184	
	No. replied	%	No. replied	%
Feed storage	28	27.5	128	69.6
Destocking	66	64.7	153	83.2
Keeping drought resistant species	20	19.6	83	45.1
Veterinary service	79	77.5	138	75.0
Enclosure/protected farmland	94	92.2	89	48.4
Extended search of water and pasture	72	70.6	26	14.1
Bush clearing	78	76.5	0	0.0
Growing feeder	22	21.6	78	42.4
Cut and carry	40	39.2	135	73.4

was practiced by 92% of pastoralists. Pastoralist key informants underline that area enclosure is a successful adaptation strategy in which villagers' collectively spare communal land to graze a small number of their selected stock, mostly for calves, sick and milk cows in time of feed shortage. It is governed by oral bylaw and administered by highly respected village elders. Thus, the idea that adaptation requires a society to act collectively (Puri 2007) against the challenges of food and water supply seems to be appreciated among pastoralists more than farmers. However, only 50% of farmers enclose a portion of their farmland to produce animal feed exclusively for the animals that need it most—ox or milk cows—so as to increase traction power and milk production. The findings indicate that the selection of a particular coping and adaptation practice reflects livelihood requirements and the permissible degree of action in a given environment.

FGD participants and key informants in pastoral areas underscored that their coping and adaptation strategies are threatened by frequent drought and other climate extremes. Due to frequent drought, their asset base is heavily eroded and their social bond is weakened. As Puri (2007) noted, social networks play an important role in alleviating individual members' problems and hence enhances objective and subjective adaptive capacity. In general, the experience of the study sites is a testament to the validity of the claim that successful adaptation requires livestock diversification, destocking and providing professional veterinary services (Nyanga et al. 2011).

3.2 Coping and Adaption Strategies to Climate Variability on and Beyond the Farm

People respond to climate variability and change differently following environmental dictates. Cultural codes also shape attitudes for or against adaptation of a

Table 3 Non-farm coping and adaptation strategies to climate variability

No.	Types of adaptation and coping practices	Lowland		Highland	
		N = 102		N = 184	
		No. of replied	%	No. of replied	%
1	Land contracting (renting, share cropping)	0	0	12	6.5
2	Borrowing from credit union	19	18.6	20	10.9
3	Borrowing from friends and or families	42	41.2	27	14.7
4	Selling of wood and tree	0	0	113	61.4
5	Selling of charcoal	9	8.8	21	11.4
6	Saving (crop seed, money etc.)	21	20.6	92	50.0
7	Free support (money, labor, material) from the clan or relatives and friends	47	46.1	0	0.0
8	Remittance from family elsewhere	12	11.8	61	33.2
9	Migration	8	7.8	28	15.2
10	Aid from government and NGOs	34	33.3	3	1.6
11	Avoidance, postponement or reduction of expenses during religious and public holidays or family occasion's for example, dowry, wedding, mourning feasts, etc.	93	91.2	104	56.5
12	Using media (TV, radio etc.) forecasting	18	17.6	41	22.3
13	Water harvesting technology	0	0	9	4.9
14	Diversification of livelihoods through non-farm activities: petty trading, hand craft	16	15.7	28	15.2
15	Using indigenous forecasting and or own experiences	102	100	174	94.6

certain technology (Thomsen et al. 2012; Fazey et al. 2011; Smit and Wandel 2006). Therefore, assessing pastoralists and farmers' coping and adaptation practices beyond agricultural activity could help to understand the available response options. Table 3 shows non-farm coping and adaptation strategies practiced by pastoralists and farmers in the face of changing climate.

As shown in the table, items from 1 to 11 are coping mechanisms while the other four are adaptation strategies. Reliance only on agricultural income is not sustainable especially due to the ever-rising cost of agricultural inputs, the shortage of farmland, the decline of rangeland potential and declining soil fertility. Pastoralists and farmers had found coping and adaption strategies outside the field of agriculture by way of reducing expenses and diversifying their sources of income. Of the coping and adaptation options illustrated in Table 3, land contracting, money borrowing, tree selling, charcoal production, remittance and aid could be regarded as maladaptive responses to climate variability related shocks. These manipulative behaviors could help to avert only current problems at the expense of the very natural asset bases. As highlighted by Mortimore and Adams (1999) these exploitive coping strategies will not sustain and be successful. Our argument is supported by Thomsen et al. (2012) who indicated that these options could only be used for the short term and had the potential to undermine future adaptive capacity and create dependency paths. Therefore, as reported by Barnett and O'Neill (2010) these response options could increase risks and should be excluded.

On the other hand, coping and adaptation options such as savings, income diversification and avoidance of extravagance during cultural and family events should be seen as accommodative responses and changing challenges to opportunities. Adaptive measures that are harmonious with existing ecological systems can be sustainable and self-regulating. Accommodative adaptive measures have the potential to enhance a community's resilience to climate variability shocks (Thomsen et al. 2012). Reducing the expenses of festivities and gifts during public holidays or family events, avoiding them altogether or postponing events are the most common coping mechanisms practiced by an average of about 69% of the respondents in all regions, with the highest (91.2%) being pastoralists. Similarly, wedding gifts for children are also reduced. The cutting of expenses is accompanied by the revitalization of saving and credit associations, as confirmed by an average of more than 28% of the respondents. Though as of yet insufficient, the emergent growth in savings and credit associations is a promising development to mitigate climate variability related problems. Thus, pastoralists and farmers who participate in such adaptation options had a higher adaptive capacity. As noted by FGD and key informants, defensive response options are mostly practiced by poor and female-headed households. In this regard, Adger et al. (2005), argued that for successful adaptation in the face of climate variability, effective, efficient, equitable and legitimate strategies are very important. Our result confirms the conclusion that climate variability hurts disadvantaged groups most and worsens their level of poverty.

A small but relatively important source of income for pastoralists is access to borrowing or free support from friends and or families. Surprisingly, free support from clan members and relatives is absent among farmers (see Table 3). Similarly financial support from NGOs is also negligible in sharp contrast to the case of pastoralists. Except for a few NGOs involved in watershed management and related interventions, there are no NGOs in the highlands distributing humanitarian and relief aid. However, free aid and support for pastoralists during bad times is blamed for the weakening of communal solidarity, which is a critical component of indigenous coping and adaptation strategies. The role of remittance obtained from family members living outside of the household to cope with climate related shocks is relatively more common among farmers than it is pastoralists. The eucalyptus tree is most important as it offers extra income to farmers from the sale of lumber and firewood. In all research sites, particularly in the highlands, the contribution of credit from institutions and friends or families is very low.

The variation among farmers and pastoralists could be explained by differences in their culture. The strength of social bond is very loose among farmers compared to that of pastoralists. Key informants asserted that, though weaker than they were in the past, there are strong socio-cultural institutions in the pastoralist area through which risks are shared and managed. Given dwindling asset holdings, communal risk sharing has declined in ways that victimizes the poor as they lack the necessary collateral to get loan in their locality. Overall, the institutional setting of all areas pointed that only actual income is a guarantee of life and property. It seems plausible to argue that pastoralists' and farmers' vulnerability to climate extremes and shocks is worsened by the precarious social and institutional landscape which has been eroded over time. Although borrowing is claimed to be an important coping mechanism elsewhere (Paavola 2008; Young et al. 2010), its role in this study areas is marginal. Therefore, credit and savings have little contribution in building pastoralists' and farmers' adaptive capacity to climate related risks. Cognate with our findings, Asres et al. (2013) argue that farmers are marginalized from credit services because of stringent loan requirements.

Table 3 also shows that 8.8 and 11.4% of pastoralists and farmers respectively are involved in charcoal production for sale, thereby endangering the ecology with irreversible damage. The trees cut to produce charcoal are immensely vital to balance the local ecosystem, and charcoal production is an exploitive response mechanism to climate variability and extremes. Even further, removing vegetation cover breeds resource based conflict and deepens poverty (Thomsen et al. 2012).

Migration as an alternative source of income is one of the least preferred coping mechanisms of pastoralists and farmers (see Table 3). Farmer key informants explained that mostly landless youth and poor farmers migrate temporarily in search of seasonal employment opportunities without abandoning their farmland for good. As learnt during individual and group interviews, the rate of seasonal migration has increased over time due to lack of alternative sources of income in their locality and the growing job opportunities created by the booming commercial agriculture and construction sectors. As Corbett (1988) observed, migration is a last adaptation option to the farming community lacking other sources of income. Migrants remit some money back home to support their families' purchase of agricultural inputs and other basic needs. In fact, seasonal migration in order to earn and remit money to family has become the norm rather than the exception in rural Ethiopia (Tesfaye 2004a, b).

The nature of migration varies between pastoralists and farmers. The poor female-headed pastoralists permanently migrate with their family to the nearby urban areas due to the persistent recurrence of drought. In contrast, poor male farmers migrate seasonally, leaving families at home. As reported by PACECMRP (2008) it might be wrong to reduce migration to the problems of climate extremes since other factors, such as bad governance, political exclusion and resource scarcity, could play an important role.

About 6.5% of respondent farmers practiced sharecropping and land renting. The absence of land contracting in either form could be explained by the fact that in the lowland there is no privately owned farmland to be rented. According to FGD and key informants, most female-headed and poor households contract their land due to financial constraints in order to buy oxen and other agricultural inputs such as fertilizer and improved crop seeds. Moreover, labor limitations and cultural influences are other factors which lead female-headed households to land contracting. The low rate of land rent is a necessary evil that farmers accept given their desperation to get some cash for their basic needs, particularly following bad harvest seasons. Therefore, this segment of the community would be more vulnerable to climate variability due to their comparative disadvantage caused and perpetrated by unequal asset backgrounds, dysfunctional institutions and cultural barriers. It seems clear that variation in the adaptive capacities of pastoralists and farmers reflects the cumulative effect of environmental, socio-cultural and institutional factors.

The diversification of income from petty trade and handcraft production as a coping mechanism to climate extremes is nearly the same between pastoralists and farmers. Key informants explained that landless youth farmers are involved in rural-urban petty trade in agricultural and industrial products while youth pastoralists operate in the livestock trade. This corresponds to the claim that highland Ethiopian agriculturalists cope with land degradation and the shortage of farmland through an increased participation in wage labor and petty trading (Tesfaye 2004a, b). Innovative adaptation mechanisms, such as fattening, dairy farming, the use of more productive exotic breeds, compost production etc., were practiced by a few better-off farmers. The application of compost could reduce the cost of chemical fertilizer and neutralizes the burning effect of chemical fertilizers on seedlings in times of rainfall interruption. However, poor farmers didn't participate in fattening and dairy farming due to initial capital problems.

The majority of pastoralists and farmers lack access to modern meteorological information. Nearly all respondents are relying on their own experiential judgments and indigenous weather information sources for their coping and adaptation decisions (see Table 3). Key informants reported that the reliability and acceptance of indigenous weather forecasts is decreasing over time due to repeated false predictions. Pastoralists and farmers accept and take proper action only when they trust the source of information (Frank et al. 2011). In support of the argument of Barrett et al. (2007), when pastoralists and farmers are unsure about the rainfall condition for the next season, they become more reluctant to invest in crop and livestock production than they might otherwise have been.

The result shows that pastoralists and farmers in the study areas practice on-farm and off-farm coping and adaptation strategies in the face of climate variability in crop and livestock production. Therefore, by way of endorsing previous findings, one may conclude that pastoralists and farmers make the necessary modifications of their livelihood at different levels (Adugna et al. 2013; Woldeamlak 2012). Adaptation practices are area-specific as they vary in different agro-ecologies (Temesgen et al. 2008). Both adaptation and maladaptation practices are employed by pastoralists and farmers to cope with climate variability in their livestock and crop production.

3.3 Measuring the Perceived Success of Coping and Adaptation Practices

A multivariate analysis on the impact of independent variables on the perceived success of coping and adaptation practices in crop and livestock production showed mixed results. Gender and agro-ecology differences show no statistically significant effect on the perceived success of coping and adaptation measures. Given unequal access to extension services and differences in actual adaptive capacity across gender and agro-ecology (Ayal and Muluneh 2014; Asres et al. 2013), the result might indicate optimistic bias among females and lowlanders about the success of their coping and adaptation practices. The discrepancy could be attributed to high uncertainty that distorts judgment (Grothmanna and Pat 2005). As Grothmanna and Pat (2005) pointed out, an unbalanced objective and subjective adaptive capacity induces cognitive biases so that people distort their actual adaptive capacity.

Household income differences showed a statistically significant, $F(5, 215) = 2.854$, $P < 0.05$, and $F(5, 215) = 9.666$, $P < 0.01$, impact on pastoralists' and farmers' perceptions of the success of their coping and adaptation measures in crop and livestock production respectively. The mean scores and standard deviations between income groups reveal that those with income levels greater than $3000 USD expressed a favorable outlook, $M = 2.5$, $SD = 2.138$ and $M = 4.437$, $SD = 2.731$, about the success of their coping and adaptation practices in crop and livestock production respectively. The opposite result was obtained for groups with income levels of $100–200 USD, whose perception of success for the crop and livestock sector stood at $M = 1.44$, $SD = 1.27$ and $M = 1.93$, $SD = 0.26$ respectively. Therefore, the better-off are confident in their coping and adaptation practice in the face of climate related shocks. This could be due to the better-off enjoying better access to extension services and facilities (Lefort 2010). The objective and subjective conditions justify that they had better confidence in the efficacy of their adaptation practices and vice versa. This implies that the better-off are likely to avoid risks of climate extremes in the future because they had the right mindset and the necessary resource base to undertake coping and adaptation practices.

Education showed a statistically significant impact on pastoralists' and farmers' perceived success of their coping and adaptation strategies ($F(4, 215) = 3.942$, $P < 0.01$) and ($F(4, 215) = 2.552$, $P < 0.05$) in crop and livestock production respectively. The mean and standard deviation ($M = 4.437$, $SD = 2.731$) for 9–10 graders shows lower success perception than those with education level above grade 10, both in the crop and livestock sectors. With a score of ($M = 1.734$, $SD = 1.596$ and ($M = 2.114$, $SD = 0.382$) the illiterate had the lowest perceived success. The result shows that a higher level of education makes pastoralists and farmers more confident in the success of their coping and adaptation practices. A plausible explanation for this difference could be that literacy allows for a better understanding of development agents' advice and recommendations to pastoralists and farmers. Further, education could also improve pastoralists' and farmers' access to information and the adoption of technology. Rational causal attribution to climate extremes could also give advantages to literates over illiterates, who would rather deem efforts to reverse the effects of climate extremes, which they believe is caused by God, futile. Illiteracy promotes divine causal attribution and nourishes cynical attitudes from which stems a maladaptive syndrome, a denial of the threat, wishful thinking and fatalism (Grothmanna and Pat 2005). This shows how education is a powerful force to rectify an erroneous world outlook and promote a rational causal attribution that is fundamental to nurture a spirit of self-worth and a realistic assessment of the value of one's adaptation measures.

Table 4 illustrates that age has ($F(1, 215) = 3.988$, $P < 0.05$) a statistically significant influence on how pastoralists and farmers perceive their coping and adaptation practices' success to climate variability in crop and livestock production. As the mean result shows, perceived success decreases with increases in age so that young age groups (20–40) had the strongest perception of success while the oldest (above 60) had the lowest perception of success. Grothmanna and Pat (2005) indicated that people lacking technology and knowhow have little self-confidence. Ayal and Muluneh (2014) observed that the youngest group had better access to education and agricultural extension services and facilities. Given that age had a strong effect on pastoralists' and farmers' perceptions of the adverse effect of climate variability (ibid), the opposite result about the relationship between age and the perceived success of adaptation strategies connotes that the correct perception of climate variability does not imply better perception of success in adaptation strategy. Therefore, arguably, better access for education and the extension of services would make one less amenable to religious influence, which would give an individual better causal attribution to climate extremes and a logical disposition to do something about it.

The interactive effect between age and sex ($F(2, 215) = 3.098$, $P < 0.05$), sex and income ($F(3, 215) = 6.270$, $P < 0.01$) and agro-ecology and level of education ($F(2, 215) = 6.713$, $P < 0.01$) has a statistically significant influence on pastoralists' and farmers' coping and adaptation practices' success. The statistically significant effect on perceived success that an absence of sex and agro-ecology has illustrates the danger of overemphasizing the effect of a single variable on perception.

4 Conclusion

Pastoralists and farmers are practicing mixed adaptation and maladaptation practices in the face of climate variability and extremes. Further, the behavioral response of many pastoralists and farmers proves that they are not adaptive enough —a point easily demonstrated by their inability to exist without humanitarian food aid in the wake of severe drought. Therefore, it is logical that instead of leaving them at the mercy of current inefficient "coping and adaptation strategies', the government should embark on the promotion of advanced and affordable technologies to deal with the impact of climate change and variability on agriculture.

Table 4 MANOVA results—the role of demographic, economic and environmental variables on the perceived success of coping and adaptation practices

Source	Dependent variable on perception of coping and adaptation practices successfulness:	SS	df	MS	F	Sig.
Corrected model	Crop production	368.794[a]	76	4.853	2.876	0.000
	Livestock production	32.417[b]	76	0.427	5.960	0.000
Sex	Crop production	2.631	1	2.631	1.559	0.214
	Livestock production	0.050	1	0.050	0.693	0.406
Age	Crop production	13.458	2	6.729	3.988	0.021
	Livestock production	0.135	2	0.067	0.940	0.039
HH_Income	Crop production	24.076	5	4.815	2.854	0.017
	Livestock production	3.459	5	0.692	9.666	0.000
Agro ecology	Crop production	2.197	1	2.197	1.302	0.256
	Livestock production	0.147	1	0.147	2.059	0.154
EduHHH	Crop production	26.605	4	6.651	3.942	0.005
	Livestock production	0.731	4	0.183	2.552	0.042
Sex * Age	Crop production	10.454	2	5.227	3.098	0.048
	Livestock production	0.108	2	0.054	0.755	0.472
Sex * HH_Income	Crop production	8.501	3	2.834	1.679	0.174
	Livestock production	1.346	3	0.449	6.270	0.001
Age * EduHHH	Crop production	40.579	7	5.797	3.436	0.002
	Livestock production	1.459	7	0.208	2.912	0.007
Agro-ecology * EduHHH	Crop production	2.348	2	1.174	0.696	0.500
	Livestock production	0.961	2	0.480	6.713	0.002
Age * HH_Income * Agro ecology	Crop production	5.087	3	1.696	1.005	0.393
	Livestock production	0.571	3	0.190	2.659	0.051
Age * HH_Income * EduHHH	Crop production	33.283	10	3.328	1.973	0.041
	Livestock production	0.891	10	0.089	1.245	0.268
Corrected total	Crop production	603.333	215			
	Livestock production	42.365	215			

[a]$R^2 = 0.611$ (Adjusted $R^2 = 0.399$)
[b]$R^2 = 0.765$ (Adjusted $R^2 = 0.637$)

Moreover, the government and NGOs initial intervention should target awareness creation and support for those who show efforts to adapt to climate variability. As Weinreich (1999) suggested, supporting those who practice different coping and adaptation practices would motivate them for higher success and, through demonstrative effect, initiate others to imitate adaptation activities. As proposed by Frank et al. (2011), involving influential pastoralists and farmers in information dissemination and technology demonstration efforts enhances the chances of success in realizing the desired behavioral change.

The perceived coping and adaptation success of the poor, female headed and illiterate could be explain by the practical challenges that they are facing to secure agricultural input and maladaptation practices, such as land contracting, charcoal production and migration. Whereas the pastoralists and female headed households didn't understand the scope of their actual asset base. Therefore, the experience in the study sites suggests that perceived success are determined by both objective conditions and subjective problems. Thus, the perceived success of pastoralists and farmers—and therefore their readiness to adopt technology—could be enhanced through preferential treatment that increases their access to financial and extension services. Providing them with custom-made technical assistance is necessary (Maddison 2007), especially since they have limited access to extension services and facilities.

References

Adger, N., Arnell, N., & Tompkins, E. (2005). Successful adaptation to climate change across scales. *Global Environmental Change, 15*(2), 77–86.

Adger, N., Huq, S., Brown, K., Conway, D., & Hulme, M. (2003a). Adaptation to climate change in the developing world. *Progress in Development Studies, 3*(3), 179–195.

Adger, W., Brown, K., Fairbrass, J., Jordan, A., Paavola, J., Rosendo, S., et al. (2003b). Governance for sustainability: Towards a 'thick' analysis of environmental decision-making. *Environment and Planning A, 35,* 1095–1110.

Adugna, T., Gazahgne, A., Mengistu, K., & Edrias, G. (2013). Adaptation to climate change and variability in eastern Ethiopia. *Journal of economics and sustainable development, 4*(6), 91–103.

Asres, E., Makoto, N., Kumi, Y., & Akira, I. (2013). Effect of agricultural extension program on smallholders' farm productivity: Evidence from three peasant associations in the highlands of Ethiopia. *Journal of Agricultural Science, 5*(8), 163–181.

Ayal, D., & Muluneh, A. (2014). *Smallholder farmers' vulnerability to climate variability in the highland and lowland of Ethiopia: Implications to adaptation strategies*. Doctoral Thesis, University of South Africa, Geography Department.

Barnett, J., & O'Neill, S. (2010). Maladaptation. *Global environmental Change Human and Policy Dimensions, 20*(2), 211–213.

Barrett, C., Barnett, B., Carter, M., Chantarat, S., Hansen, J., Mude, A., et al. (2007). *Poverty traps and climate and weather risk limitations and opportunities of index-based risk financing*. IRI Technical Report No. 07-03.

Coppock, L. (Ed.). (1994). *The Borana Plateau of Southern Ethiopia: Synthesis of pastoral research, development and change, 1980–1991*. ILCA (international livestock center for Africa), Addis Ababa, Ethiopia, 393 pp.

Corbett, J. (1988). Famine and household coping strategies. *World Development, 16*(9), 1099–1112.

CSA. (2008). *Summary and statistical report of the 2007 population and housing census*. Addis Ababa, Ethiopia: The Federal Republic of Ethiopia. CSA.

Ericksen, S., Aldunce, P., Bahinipati, C., Martins, R., Molefe, J., Nhemachena, C., et al. (2011). When not every response to climate change is a good one: Identifying principles for sustainable adaptation. *Climate and Development, 3*(1), 7–20.

Fazey, I., Pettorelli, N., Kenter, J., Wagatora, D., & Schuett, D. (2011). Maladaptive trajectories of change in Makira, Solomon Islands. *Global Environmental Change, 21*(4), 1275–1289.

Frank, E., Eakin, H., & Lo ′pez-Carr, D. (2011). Social identity, perception and motivation in adaptation to climate risk in the coffee sector of Chiapas, Mexico. *Global Environmental Change, 21*(1), 66–76.

Grothmanna, T., & Patt, A. (2005). Adaptive capacity and human cognition: The process of individual adaptation to climate change. *Global Environmental Change, 15*(3), 199–213.

IPCC. (2007). *Climate change 2007: Impacts, adaptation and vulnerability*. Contribution of Working Group II to the 4th Assessment Report of the IPCC, Cambridge, Cambridge University Press.

Kassahun, A. (2011, July). *The impact of climate variability on crop production in Ethiopia: Which crop is more vulnerable to rainfall variability*? The 9th International Conference of EEA/EEPRY Addis Ababa, Ethiopia.

Kuruppu, N., & Liverman, D. (2011). Mental preparation for climate adaptation: The role of cognition and culture in enhancing adaptive capacity of water management in Kiribati. *Global Environmental Change, 21*(2), 657–669.

Lefort, R. (2010). Powers—Mengist—And peasants in rural Ethiopia: The post-2005 interlude. *Journal of Modern African Studies, 48*(3), 435–460.

Maddison, D. (2007). *The perception of and adaptation to climate change in Africa*. Policy Research Working Paper 4308.

Mortimore, M., & Adams, W. (1999). *Working the Sahel: Environmental and society in Northern Nigeria*. London: Routledge.

Nyanga, H., Johnsen, F., & Aune, J. (2011). Smallholder farmers' perceptions of climate change and conservation agriculture: Evidence from Zambia. *Journal of Sustainable Development, 4* (4), 73–85.

Paavola, J. (2008). Livelihoods, vulnerability and adaptation to climate change in Morogoro, Tanzania. *Environmental Science and Policy, 11*(7), 642–654.

PACECMRP (Parliamentary assembly of Council of Europe Committee on Migration, Refugees and populations). (2008). Environmentally induced migration and displacement: *A 21st-century challenge,* December 23, 2008, Doc. 11785.

Patton, M. (1990). *Qualitative evaluation and research methods* (pp. 169–186). Beverly Hills, CA: Sage.

Puri, R. (2007). Responses to medium-term stability in climate: El Niño, droughts and coping mechanisms of foragers and farmers in Borneo. In R. Ellen (Ed.), *Modern crises and traditional strategies: Local ecological knowledge in Island Southeast Asia* (pp. 46–83). Berghahn Books: Oxford, UK.

Santrock, J. W. (2011). *Life-span development* (13th ed.). New York: McGraw-Hill Companies Inc.

Smit, B., & Wandel, J. (2006). Adaptation, adaptive capacity and vulnerability. *Global Environmental Change, 16*(3), 282–292.

Smit, I., & Pilifosova, O. (2001). Adaptation to climate change in the context of sustainable development and equity. In J. McCarthy, O. Canziani, N. Leary, D. Dokken, & K. White (Eds.), *Climate change 2001: Impacts, adaptation and vulnerability* (pp. 877–912). Cambridge: Cambridge University Press.

Temesgen, D., Hassen, R., Tekie, A., Mahmud, Y., & Claudia, R. (2008). *Analyzing the determinants of farmers' choice of adaptation methods and perceptions of climate change in the Nile Basin of Ethiopia*. IFPRI Discussion Paper 00798.

Tesfaye, T. (2004a). *The migration, environment, conflict nexus: A case study from Eastern Wollega*. Forum for Social Studies (FSS) studies on poverty No. 4.

Tesfaye, T. (2004b). *Natural resources scarcity and rural conflict: Case studies evidence on the correlates from Ethiopia*. Forum for Social Studies (FSS) studies on poverty No. 4.

Thomsen, D., Smith, T., & Keys, N. (2012). Adaptation or manipulation? Unpacking Climate Change Response Strategies. *Ecology and Society, 17*(3), 20. Available from: http://www.ecologyandsociety.org/vol17/iss3/art2 [Accessed September 27, 2014].

Weinreich, N. (1999). *Hands-on social marketing: A step-by-step guide*. CA: Sage Publications.
Woldeamlak, B. (2012). Climate change perceptions and adaptive responses of smallholder farmers in central highlands of Ethiopia. *International Journal of Environmental Studies, 69* (3), 507–523.
Young, G., Zavala, H., Wandel, J., Smit, B., Salas, S., Jimemez, E., et al. (2010). Vulnerability and adaptation in a dry land community of the Elqui Valley, Chile. *Climatic Change, 98*(1), 245–276.

Role of Sustainable Land Management (SLM) in Adapting to Climate Variability Through Agricultural Practices—Experiences from Ethiopian Highlands

Georg Deichert, Ashenafi Gedamu and Befikadu Nemomsa

1 Introduction

Economic growth has been a key challenge in developing countries including Ethiopia. Multiple determinants of growth had been identified including policy, institutional, geographical factors as well as resource endowments. There is increasing evidence that climate change is a growing and important factor in explaining poor economic performance of agrarian economies. The agriculture sector is to a large extent associated with climate change, as a major cause as well as farmers being highly affected by the impacts of this climate change. These climate signals are being felt as increased frequency of extreme weather events (such as droughts, storms and floods) or an increased variability in rainfall pattern (Tröger et al. 2011). Therefore it seems obvious that low-latitude economies with large shares of rain-fed and subsistence agriculture are especially vulnerable (Aragie 2013).

Ethiopia's highlands are the backbone for the nation's agriculture production which is dominated by smallholder mixed farming systems. Increasing population in combination with unsustainable land management and farming practices continue enhancing pressure on the natural resources of land and water. Unpredictable climate variability further contributes increasingly to low agricultural productivity and production. In response to these challenges soil and water conservation has to be combined with sustainable agriculture practices, whereby the latter need a strong

G. Deichert (✉)
GIZ Sustainable Land Management Programme, Rahem Building, Diaspora Square, Meganagna, P.O. Box 100009, Addis Ababa, Ethiopia
e-mail: georg.deichert@giz.de

A. Gedamu · B. Nemomsa
EU Global Climate Change Alliance (GCCA-E) Project, GIZ-SLM, Rahem Building, Diaspora Square, Meganagna, P.O. Box 100009, Addis Ababa, Ethiopia

W. Leal Filho et al. (eds.), *Climate Change Adaptation in Africa*,
Climate Change Management, DOI 10.1007/978-3-319-49520-0_29

focus on strengthening adaptation among smallholding farmers (Morras and Mungai 2015).

The Sustainable Land Management Program (SLMP), under the Government of Ethiopia's Ministry of Agriculture and Natural Resources (MoANR) addresses the challenges of agriculture production in the highlands through a three-stage watershed-management approach of (i) community mobilization, (ii) soil and water rehabilitation and (iii) putting rehabilitated land into productive use through agricultural production (Fig. 1).

Soil and water conservation (SWC) measures directly decrease soil erosion and rehabilitate degraded land thereby enhancing the availability of productive farmland. Therefore they are not directly climate-smart in terms of agriculture production, but rather a pre-condition for climate-smart agriculture production.

This has evolved a discussion with regard to what is called "Climate Smart Agriculture" (CSA). Climate smart agriculture is commonly understood to address three aspects simultaneously that are (i) an adaptation to climate change or climate variability impacts, (ii) a mitigation of Green House Gas emissions and (iii) a positive livelihood effect.

The FAO (2013a) defines that Climate-smart agriculture "sustainably increases productivity, resilience (adaptation), reduces/removes greenhouse gases (mitigation) while enhancing the achievements of national food security and development goals" (Fig. 2).

Accordingly "Climate Smart Agriculture" comprises practices and technologies that sustainably increase productivity and support smallholder farmers to adapt to climate change impacts and thereby strengthen the farmers' resilience towards hazards caused by climate variability. At the same time "CSA" measures are expected to reduce greenhouse gas (GHG) emissions and/or contribute to Carbon sequestration.

In this way climate-relevant adaptation measures are strongly linked to income generating agriculture production activities.

While this understanding is commonly accepted there has hardly been any undertaking so far to systematically screen soil and water conservation (SWC) and agriculture measures with regard to their climate relevance according to the above definition. As Corner-Dollof et al. (2015) put it decision-support frameworks are needed to identify best-bet CSA practices and programs for specific contexts. This paper describes and assesses various SWC and agriculture measures for their adaptation, mitigation and livelihood potential. The assessment is based on the piloting of measures under the GIZ/SLM/EU/GCCA-E[1] project in 34 selected micro-watersheds of the Blue Nile catchment from 2012–2015 (GIZ-SLM 2014).

[1]GIZ/SLM/EU/GCCA-E Project was an EU funded project under the Global Climate Change Alliance (GCCA) of the EU, implemented in more than 35 countries across the world. Global Climate Change Alliance—Ethiopia (GCCA-E) was implemented as a tripartite project of European Union—Ministry of Agriculture and Natural Resources—Deutsche Gesellschaft für Internationale Zusammenarbeit (2012–2015) with the main objective of piloting and testing climate—relevant agricultural interventions.

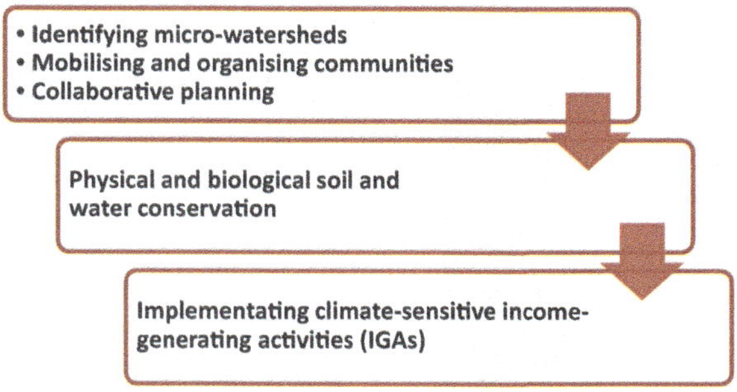

Fig. 1 Three stage watershed management approach

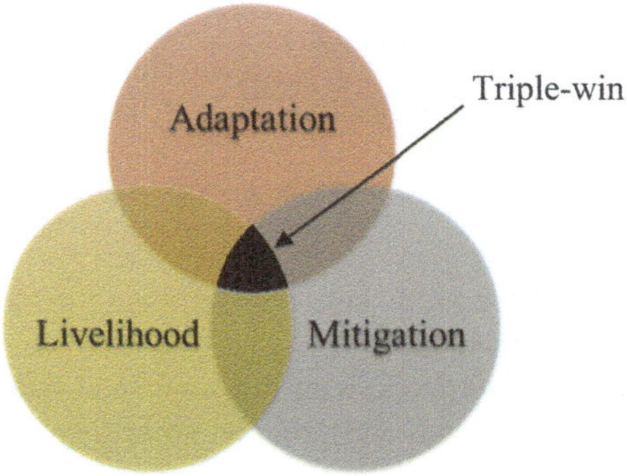

Fig. 2 Components of climate-smartness

2 Materials and Method

Looking at the many publications concerning "climate-smart agriculture (CSA)" production (Branca et al. 2011; FAO 2013a; Nysimi et al. 2014; Martins 2012; Word Bank 2014) one quickly comes to the conclusion that climate-smart agriculture is not an "either-or" that can be simplified into ones and zeros, but rather a more complex continuum between the two.

In order to get a more systematic classification of climate-smart agriculture versus non-climate smart agriculture the project developed a basket of climate relevant options of agricultural measures, named as a 'basket-of-options' (BoO) in the following. Basically the BoO describes a matrix listing agricultural measures or practices and rating them according to their adaptation, mitigation and livelihood enhancement potentials. The project team identified a number of sub-parameters for assessing the effects of agriculture intervention measures on adaptation, mitigation and livelihood respectively. These sub-parameters help to describe the direct effects of a measure on the three components of CSA (Fig. 3).

The BoO presents the rating for each measure with regard to the direct effect on the respective sub-parameter. The rating was done subjectively by the project team and can range from +3 to −3 for each of the respective sub-parameters. The ratings are being supplemented by a respective justification. As the ratings are based on

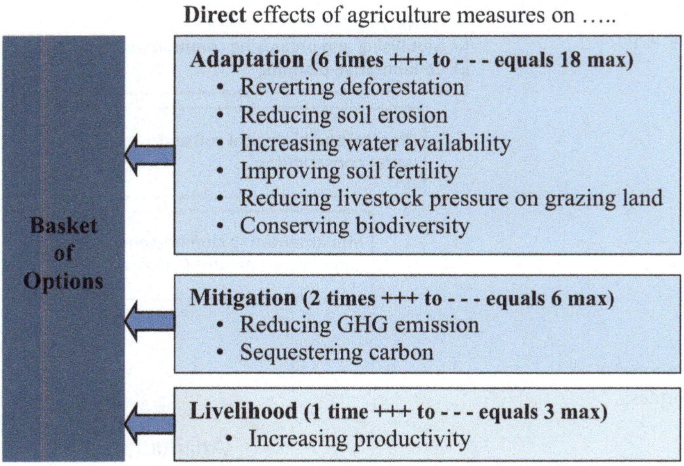

Fig. 3 The "Basket of Option" (BoO) and its sub-parameters

subjective judgments they are being reviewed with ongoing experiences gained in implementation.

In accordance with the watershed management approach measures in the BoO are considered and grouped according to following land use types:

- Degraded hill sides (forest land),
- Farm land,
- Homestead, and
- Grazing land.

Livestock interventions form an additional separate group. A particular single measure may appear in different land use types and even with a different rating, because the magnitude of expected effects might differ, for example tree planting in degraded hill sides versus tree planting in homesteads. Also the ratings do not always reflect a consistent consideration with regard to time aspect of a measure, e.g. the mitigation effect of planting trees is realized only after a number of years.

The following table provides an overall rating for climate-smartness of selected measures based on the respective ratings for adaptation, mitigation and livelihood. The maximum rating for adaptation, mitigation and livelihood would be $6 \times 3 = 18$, $2 \times 3 = 6$ and $1 \times 3 = 3$ respectively and totaling to $18 + 6 + 3 = 27$ (Table 1).

The total CSA rating being the summary of all 9 sub-parameter ratings automatically gives a very strong focus on adaptation because the adaptation component is made up of 6 sub-parameters compared to two and one sub-parameter for mitigation and livelihood respectively. Nevertheless if one wants to focus on measures with maximum effects on mitigation one could identify respective measures for a land use type from the BoO by selecting those with the highest mitigation rating.

Table 1 Selected measures and its ratings by order of total rating

Direct effects on →	Adaptation							Mitigation			Livelihood		CSA
Measures by land use type	Forest degradation	Soil degradation	Water availability	Soil fertility	Livestock pressure	Biodiversity	Subtotal	Reducing emission	Storing carbon	Subtotal	Increasing productivity	Subtotal	Total rating
a													
Degraded hillsides													
Planting trees	+++	++	++	+	NDR	++	10	+++	++	5	+	1	16
Forage production	+	++	+	+	++	+	8	+	+	2	++	2	12
Area enclosure	++	++	++	+	–	+	7	++	++	4	+	1	12
Beekeeping	+	NDR	NDR	NDR	NDR	++	3	–	NDR	−1	+++	3	5
Physical soil and water conservation (SWC)	NDR	++	++	–	NDR	–	2	NDR	NDR	0	+	1	3
Farm land													
Agroforestry	NDR	++	++	++	+	+	8	++	++	4	++	2	14
Applying compost	NDR	++	++	+++	NDR	++	9	–	++	0	++	2	*11*
Mulching	NDR	++	+++	+	NDR	+	7	+	+	2	+	1	10
Forage production	NDR	+	+	+	++	+	6	+	+	2	+	1	9
Conservation agriculture	NDR	++	+	++	+	+	7	+	0	1	+	1	9
Intercropping	NDR	++	++	+	NDR	+	6	NDR	NDR	0	++	2	8
Green manuring	NDR	+	++	++	NDR	+	6	0	+	1	+	1	8

(continued)

Table 1 (continued)

Direct effects on →	Adaptation							Mitigation			Livelihood		CSA
Measures by land use type	Forest degradation	Soil degradation	Water availability	Soil fertility	Livestock pressure	Biodiversity	Subtotal	Reducing emission	Storing carbon	Subtotal	Increasing productivity	Subtotal	Total rating
Using bio-fertilizer	NDR	NDR	NDR	+++	NDR	+	4	+	NDR	1	++	2	7
Applying lime on acidic soils	NDR	+	+	+++	NDR	+	7	–	NDR	−1	+	1	7
Crop residue management	NDR	++	+	+	–	+	4	+	+	2	+	1	7
Crop rotation	NDR	+	+	+	NDR	+	4	NDR	NDR	0	++	2	6
Planting with space/row planting	NDR	+	+	0	NDR	0	2	NDR	NDR	0	++	2	4
Changing crop varieties	NDR	0	NDR	+	NDR	–	0	0	0	0	++	2	2
b													
Homestead													
Multi-storey cropping	+	+	+++	++	+	++	10	+	++	3	++	2	15
Woodlot establishment	+++	++	–	+	0	++	8	++	++	4	+	1	13
Making compost	NDR	++	++	+++	NDR	+++	10	–	+	0	+++	3	13
Producing biogas	++	0	0	++	0	++	6	+++	+	4	++	2	12

(continued)

Table 1 (continued)

Direct effects on →	Adaptation							Mitigation			Livelihood		CSA
Measures by land use type	Forest degradation	Soil degradation	Water availability	Soil fertility	Livestock pressure	Biodiversity	Subtotal	Reducing emission	Storing carbon	Subtotal	Increasing productivity	Subtotal	Total rating
Using fuel saving stove	+++	NDR	NDR	NDR	NDR	++	5	+++	NDR	3	+++	3	11
Water harvesting and storage	NDR	+	+++	NDR	NDR	NDR	4	NDR	NDR	0	++	2	6
Production diversity of vegetables and fruit varieties	NDR	+	+	+	NDR	+	4	NDR	NDR	0	++	2	6
Communal grazing land (incl. pastures)													
Area enclosure	NDR	++	++	++	0	+	7	NDR	+	1	0	0	8
Controlled grazing	NDR	+	+	+	0	NDR	3	NDR	+	1	+	1	5
Livestock and poultry (irrespective of any particular land use type)													
Manure management	NDR	++	++	+++	NDR	+	8	+	+	2	++	2	12
Limiting the number of grazing livestock units (LU) on micro-watershed	NDR	+	NDR	NDR	+++	NDR	5	0	NDR	0	+	1	6
Fattening of sheep/goats/cattle	NDR	+	NDR	0	++	NDR	3	0	NDR	0	+	1	4

(continued)

Table 1 (continued)

Direct effects on →	Adaptation							Mitigation			Livelihood		CSA
Measures by land use type	Forest degradation	Soil degradation	Water availability	Soil fertility	Livestock pressure	Biodiversity	Subtotal	Reducing emission	Storing carbon	Subtotal	Increasing productivity	Subtotal	Total rating
Breed improvement	NDR	NDR	NDR	NDR	+	NDR	1	0	NDR	0	++	2	3
Improving market access	NDR	NDR	NDR	NDR	++	NDR	2	NDR	NDR	0	+	1	3
Poultry production (as a means of switching to less GHG Emitting animals)	NDR	1	NDR	NDR	NDR	NDR	1	−1	NDR	−1	2	2	2
Switching from large to small ruminants aiming at de-stocking	−	+	+	−	+	NDR	1	+	NDR	1	−	−1	1

NDR no direct relation

Table 2 Example of rating climate-smart measures and their justifications

Degraded hillsides	Direct effects on …	Planting trees (includes enrichment planting and buffer planting)	
		Rating	Justification
Adaptation	Forest degradation	+++	Improves forest quality through planting of diverse tree species, demarcate forest boundaries and reduces pressure on natural forest through increase of biomass
	Soil degradation	++	Improves soil cover (depending on litter raking), reduces run-off water and associated erosion, increases biomass and maintains natural drainage system
	Water availability	++	Enhances infiltration, maintains soil moisture and supports evenly distribution of water flows throughout the year (depending on the species type)
	Soil fertility	+	Continuously supply of soil OM and maintains natural nutrient cycle, though its utilization extracts biomass (nutrient)
	Livestock pressure	NDR	No direct relation
	Biodiversity	++	Maintains fauna and flora and other micro-organisms, protect the forest and facilitate the regeneration of native species
	Subtotal	10	
Mitigation	Reducing emission	+++	Reduces emission through enhancement of forest carbon stocks and reduction of forest/protected area encroachment, and serve as source of biomass energy
	Storing carbon	++	Reforestation of harvested areas through increased soil organic carbon stock, biomass carbon stock (depending on the utilization type)
	Subtotal	5	
Livelihood	Increasing productivity	+	Increases utilization potential of the forest, enhances ecosystem services and provide alternative income source
	Subtotal	1	
CSA	Total rating	16	

NDR no direct relation

Each rating is underpinned with a justification. These justifications are very important for understanding and possible reviewing of the ratings. The following two tables present the ratings with respective justifications for planting trees on degraded hillsides and applying compost on farm land respectively (Tables 2 and 3).

While the BoO serves as a tool to identify climate relevant agriculture interventions it also proposes parameters for measuring the impacts of the interventions. The different components of CSA require different methods of monitoring.

Table 3 Example of rating climate-smart measures and their justifications

	Direct effects on …	Applying compost	
		Rating	Justification
Adaptation	Forest degradation	NDR	No direct relation
	Soil degradation	++	The organic matter soil nutrients are better maintained
	Water availability	++	Enhances water holding capacity through improved soil structure
	Soil fertility	+++	Adds soil organic matter (SOM)
	Livestock pressure	NDR	No direct relation
	Biodiversity	++	Maintains and improves soil biota
	Subtotal	9	
Mitigation	Reducing emission	–	Increases GHG emissions if exposed
	Storing carbon	++	The absorption of compost directly increases soil organic matter
	Subtotal	0	
Livelihood	Increasing productivity	++	Directly increases crop yield depending on compost quality and amount applied
	Subtotal	2	
CSA	Total rating	11	

Effects of measures on livelihood are usually described by cost-benefit or gross margin analyses. There are standard procedures for calculating gross-margins for example, but there are also differences in the way gross margins are calculated for smallholder farming enterprises, especially with regard to the use of family labor.

Mitigation could be measured directly as the actual amount of Carbon sequestered and through directly measuring emissions. While direct measurement of GHG emissions and Carbon-sequestration is often costly, there are a number of internationally accepted formulas by which both emissions and sequestrations can be modelled (Brown 1997; Picard et al. 2012), e.g. the EX-ACT tool (FAO 2013b). These formulas require certain input data and estimate the emissions and sequestrations based on a number of assumptions.

Measuring adaptation poses the greatest challenge. Most reports on adaptation or resilience relate it to increased production or income. This, however, would confine climate-smart versus non-climate-smart agriculture measures to mitigation effects only. Adaptation often expressed as resilience also is strongly related to risks. This doesn't necessarily make measuring adaptation easier. Conducting vulnerability analyses is a widely accepted method to qualitatively assess resilience and risks.

3 Results

Comparing the degree of climate-smartness of single measures provide first ideas for identifying suitable climate-relevant interventions. The following graph visualizes the degree of climate-smartness for major interventions on farmlands (Fig. 4).

Minimum tillage and agroforestry are the strongest overall climate relevant measures followed by compost application, mulching and green manuring. Together they are also the strongest measures in terms of strengthening resilience. As can be seen from the relation to the total score of 18 + 6 + 3 = 27 also there is hardly any single measure that is very strong in all three components of adaptation, mitigation and livelihood. Therefore in order to overcome trade-offs between the three components within a single measure combination of measures within a land use type are being recommended. The combinations also allow strengthening specifically certain components more than other.

In the following some combinations with a strong adaptation focus are being proposed as example for different land use types. Apart from ratings the proposed combinations also consider the feasibility and possibility of adopting and upscaling. Therefore not more than 5 single measures are included in one combination. Furthermore the recommended combinations are based on one or two key interventions while additional measures are included as one of several alternatives, e.g. green manuring or mulching. Another consideration for recommended combination was that especially on farm land at least three out of five measures should be practiced that can be implemented by farmers without depending on external inputs.

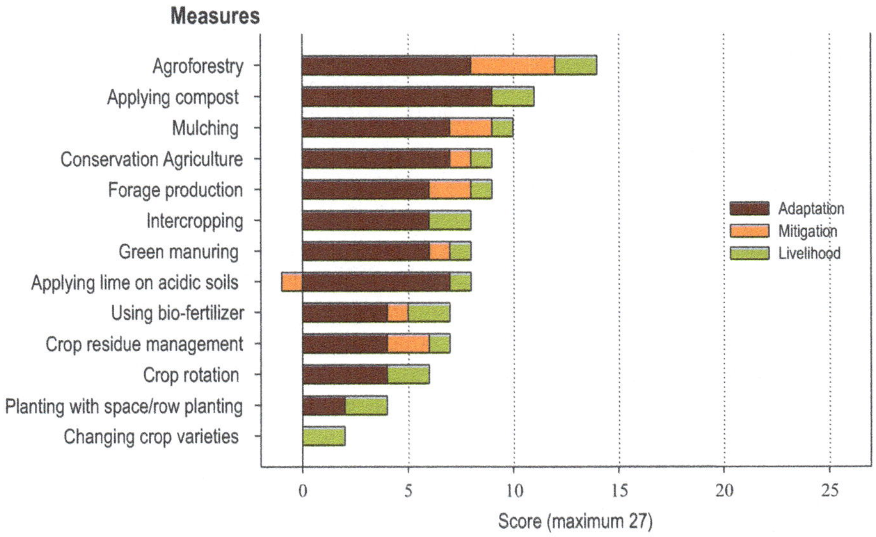

Fig. 4 Adaptation, mitigation and livelihood ratings for measures on farm land

Table 4 Combination of measures for degraded hillside

Measures … and direct effects on →	Adaptation	Mitigation	Livelihood	CSA
Area enclosure	7	4	1	12
Forage production	8	2	2	12
Physical soil and water conservation (SWC)	2	0	1	3
Planting trees	10	5	1	16
Beekeeping	3	−1	3	5
Total CSA rating of combination	30	10	8	48

Following combination is one recommended example for degraded hill sides (Table 4).

Watershed development starts from the top of the watershed. An area enclosure has been proven to be a highly effective measure not only for degraded hillsides. As benefits are not immediate additional measures should be combined. In this combination forage production together with physical SWC measures like terraces, trenches, etc. are recommended. The climate relevant effectiveness can be further enhanced with supported tree planting, which contributes a strong mitigation effect. Integrating bee keeping as a group enterprise boosts the income component for this CSA combination on degraded hillsides.

Identifying farmland based climate smart combination of measures provides many more options. A combination should contain at least three practices that do require no or little external inputs. Also there should be at least one measure with a strong direct effect on soil fertility, for example compost application, bio-fertilizer, or minimum tillage. Crop rotation, intercropping, green manuring, crop residue management or mulching can be added alternatively. Other optional measures like row planting, IPM, etc. can contribute significantly in strengthening resilience of crop production. The following table proposes an example of farmland based combination with a strong adaptation rating (Table 5).

The above combination does not include establishing soil and water conservation measures. Although they are often very useful and necessary on farm land as well, they do require a fair amount of investment. Change in crop variety is a popular means of climate-smart agriculture. However, unless the new variety has clearly defined properties with regard to negative climate signals such as drought resistance, shorter vegetation period, etc., the effect is visible through increased yields only. At the same time hybrid varieties tend to increase the dependency of farmers to external seed supply thereby reducing farmers' resilience.

Strengthening homestead based resilience focusses on diversity of food, cash crop, forage and animal production. Following is one example of a combination with a high rating of adaptation (Table 6).

A good combination should have at least ten different production types of fruits, vegetables, forage, spices, and animal varieties. Multi-storey cropping is a strong measure for effectively increasing diversity. Water harvesting and regular compost

Table 5 Combination of measures for sustainable crop production on farmland

Measures … and direct effects on →	Adaptation	Mitigation	Livelihood	CSA
Conservation agriculture	7	1	1	9
Applying compost	9	0	2	11
Agroforestry	8	4	2	14
Planting with space/row planting	2	0	2	4
Total CSA rating of combination	26	5	7	38

Table 6 Combination of measures for diversified homestead based production

Measures … and direct effects on →	Adaptation	Mitigation	Livelihood	CSA
Production diversity of vegetables and fruit varieties	4	0	2	6
Water harvesting and storage	4	0	2	6
Making compost	10	0	3	13
Poultry production	1	−1	2	2
Multi-storey cropping	10	3	2	15
Total CSA rating of combination	29	2	11	42

making should be a must in all homestead climate-smart combinations. They assure a high level of production of fruits and vegetables. Homestead based measures can also be combined with household energy measures, poultry or fish production, beekeeping or animal fattening.

Making livestock production climate smart poses specific challenges, as livestock is contributing significantly to GHG emissions and on the other hand animal keeping is very important for the livelihood of farming households in the highlands. Therefore simply reducing the number of animals cannot be the answer. One possible way is therefore to increase the productivity per animal without increasing the actual number of livestock units.

Different strategies are being applied for increasing productivity per head of cattle, sheep or goats, but there seem to be consensus, that increasing productivity has to start from the feeding rather than from the breeding. Therefore it is recommended that any "climate-smart" combination for livestock must have forage production in the center. Even more so as livestock based combinations are cross-cutting over different land use types.

The following table proposes one combination as an example for making livestock production climate-smart (Table 7).

Manure management and forage production are very strong measures to make livestock keeping climate-smart and therefore should be included in any livestock based combination. Although not very significant in the rating, improving market access can be an important strategy to make livestock keeping climate-smart by fostering a faster turn-over of marketable animals.

As the combinations are strongly focusing on adaptation the impact of the measures on adaptation are of prime interest. In an attempt to quantify the

Table 7 Combination of livestock based measures

Measures … and direct effects on →	Adaptation	Mitigation	Livelihood	CSA
Forage production	6	2	1	9
Improving market access aiming at de-stocking	2	0	1	3
Manure management	8	2	2	12
Limiting the number of grazing livestock units (LU) on micro-watershed level	5	0	1	6
Total CSA rating of combination	21	4	5	30

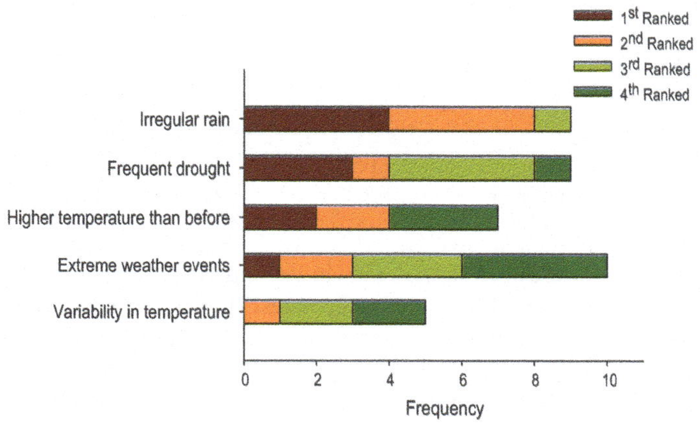

Fig. 5 Climate signals expressed by farmers

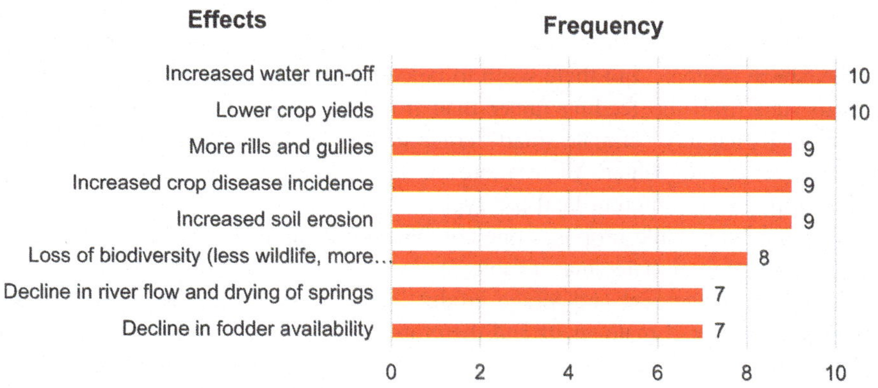

Fig. 6 Effects of climate signals

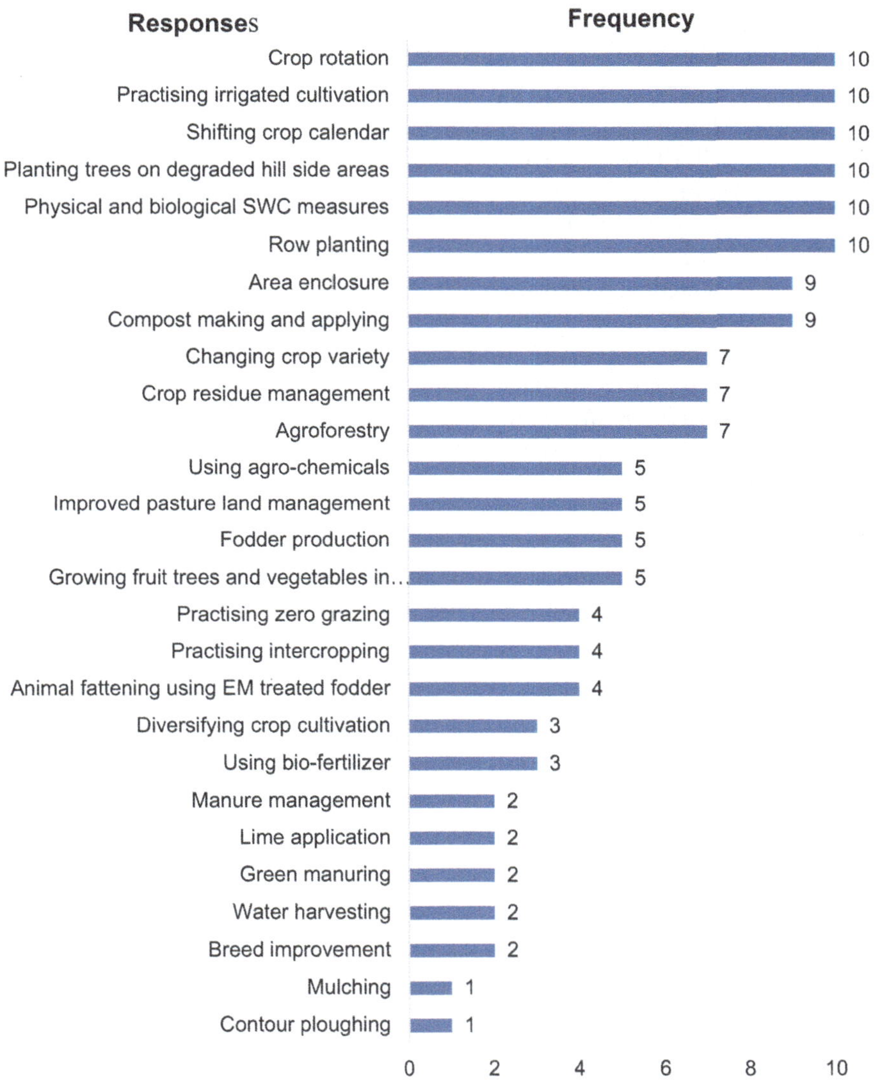

Fig. 7 Reponses to effects of climate signals

adaptation component of CSA the project conducted a 'lean' vulnerability analysis in 10 target micro-watersheds involving 126 male and 93 female beneficiaries. The focus group discussions addressed the following three questions:

- What are the climate signals that you experienced in the last 5–10 years and what is the order of their significance (ranking)?
- What are the effects of the climate signals on your life, environment and agriculture production, and to which signals do you relate them?

- Which responses did you undertake yourself or with project support to counteract the negative effects of climate signals?

Irregular rainfall was reported the most felt climate signal (9 out of 10). Frequent and recurrent drought situations were of second highest important signal (3 times 1st ranking). All 10 focus groups experienced extreme weather events, although with different levels of importance (Fig. 5).

Increased surface water run-off and lower crop yields were reported by all 10 micro-watershed focus groups, followed by increased soil erosion, more rills and gullies and increased crop disease incidences with 9 out of 10 groups respectively. Loss of biodiversity (8 out of 10), drying of springs and decline in fodder availability (each 7 out of 10) also were mentioned with considerable importance. Although reported only by 2 groups, the effect of people migrating away should not be underestimated in its importance (Fig. 6).

Six measures were mentioned by all farmers as responses to the effects perceived, including SWC and irrigation structures, but also adjusted farming practices like crop rotation and changes in crop calendar. Area closure and compost making ranked second with 9 times mentioning out of 10. Surprisingly in half of the cases the use of agro-chemicals was seen as response to negative climate effects (Fig. 7).

Preliminary data from an intensive data recording survey tend to coincide with the responses from the vulnerability analysis. The following table presents adoption figures from some of the farmland practices (Table 8).

The data show a large increase in practicing row planting and crop rotation over the conventional practices. The concepts of planting with space and crop rotation have been experienced to strengthen resilience with different crops and in different countries (SRI-Rice 2014). Also using compost appears to be a very well adopted measure to lower risk and strengthen resilience. On the other hand green manuring and mulching seem to be more difficult to be accepted by farmers. Although minimum tillage is one recommended practice, the number of ploughings was higher in the treatment plots compared to control plots. This is indicative of the strong believes in ploughing, but the results should also be analyzed by crop type, as Tef growing is normally practiced with higher number of ploughings.

Table 8 Scope of implementation of farm land based CSA recommendations in GCCA target micro-watersheds

Single measures	Unit	n	Treatment plot	n	Control plot
Compost application	Qt.	338	4251.5	128	1631.2
Row planting	%	598	90.8	269	36.1
Intercropping	%	601	41.4	601	13.3
Crop rotation	%	500	62.4	238	69.7
Green manuring	%	601	4.2	265	2.3
Mulching	%	601	2	601	0.3
Rounds of ploughings	No.	601	3.9	601	1.8

4 Discussion

Mitigating GHG emissions has been dominating the climate discussions since its beginning globally as well as for the agriculture sector which in most developing countries is dominated by smallholder farming systems. Meanwhile several reports and studies suggest that due to the past changes in the Carbon market and a virtual collapse of the Carbon price from over 20 USD/t to around 1 USD/t there is hardly any chance for smallholder farmers to gain any financial benefits from mitigation measures (Kossoy et al. 2015). Therefore the focus of the climate-relevant agriculture has shifted to adaptation already. Nevertheless mitigation remains an important component of CSA and needs continued to be considered and measured especially on national level.

Adaptation or resilience toward climate signals is of high importance for smallholder farmers. Although there are great challenges in quantifying adaptation, it should play a dominant role in climate relevant agriculture. According to Morras and Mungai (2015) CSA is often not discussed in the context of its socio-economic and cultural environment. Therefore the above concept contributes to this discussion by proposing a systematic, although subjective, way of rating "climate-smartness". By grouping agriculture interventions by land use types they can be very well integrated into a watershed development approach. It is further notable that in order to achieve significant resilience to climate signals, one should ideally combine several single measures. Combination of measures especially on farmland should also be aware of risk involved as well as level of dependency on external inputs. Both factors will strongly influence sustainable adoption of the measures. Upscaling of measures need to consider the land use type. Measures on communal land need different approaches than measures on individual farm land. Both will have different contributions to private and eco-system benefits.

5 Conclusion and Recommendation

A single measure very often doesn't have a strong potential in adaptation, mitigation and livelihood improvement. Therefore this paper recommends combinations of measures which show a strong triple-win potential in the combined effects. The proposed combinations can be of implementing guidance and allow sufficient flexibility for modification. This degree of flexibility is also necessary as the effects on adaptation, mitigation and livelihood are more often realized through the way "how" things are done rather than "what" is being done.

The flexibility of combining single measures also allows adjusting for respective stronger orientation to one or more of the 3 pillars. This is important in view of the changing framework, e.g. while few years back the focus was very much on the mitigation aspect, the changes in the international Carbon market moved the focus much more to the adaptation aspect.

On the other hand the BoO approach is limited by its subjective rating. Furthermore the ratings are constraint by the time factor as well as by a scaling factor, especially if absolute mitigation effects are of prime consideration.

Overall the BoO aims at providing a helpful decision making tool when designing projects and programs for upscaling climate relevant sustainable land management measures. By doing so the proposed "Promising and Good Practices" of climate relevant SWC and agriculture measures may also contribute to the targets for achieving Ethiopia's Climate Resilient Green Economy (CRGE).

Acknowledgements Special thanks to EU, GCCA project funder and MoANR at federal and regional offices and the farmers in the respective regions who participated in piloting and testing of climate relevant practices.

References

Aragie, E. (2013). *Climate change, growth, and poverty in Ethiopia* (Working Paper No. 3, Climate Change and African Political Stability). Austin, USA: University of Texas.

Branca, G., McCarthy, N., Lipper, L., & Jolejole, M. (2011). *Climate-smart agriculture: A synthesis of empirical evidence of food security and mitigation benefits from improved cropland management*. Rome, Italy: FAO.

Brown, S. B. (1997). *Estimating biomass and biomass change of tropical forest, a primer* (FAO Forestry paper 134). Rome: FAO.

Corner-Dolloff, C., et. al. (2015). *Towards a scalable framework for evaluating and prioritizing climate-smart agriculture practices and programs, CCAFS/CIAT presentation* (unpublished).

FAO. (2013a). *Climate-smart agriculture sourcebook*. Rome, Italy: FAO.

FAO. (2013b). *The EX-Ante carbon-balance tool (Ex-ACT)*. Rome, Italy: FAO.

GIZ-SLM. (2014). GCCA-E project: Piloting climate-smart agriculture interventions in the SLMP. Addis Abeba, Ethiopia: GIZ.

International Centre for Tropical Agriculture (CIAT). (2014). *Building an eco-efficient future: CIAT strategy 2014–2020, a synthesis*. Cali, Colombia: CIAT.

Kossoy, A., Peszko, G., Oppermann, K., Prytz, N., Klein, N., Blok, K., et al. (2015) *State and trends of carbon pricing 2015. State and trends of carbon pricing*. Washington, DC, USA

Martins, C. (2012). *Übersichtsstudie zu Climate-Smart Agriculture*. Berlin, Germany (unpublished).

Morras, E., & Mungai, C. (2015) How can climate—Smart agriculture enhance food security and community—Based Adaptation? In V. Atakos, M. Randey, C. Mungai, J. Kinyangi (Eds.), *The Climate—SmartAG Partner*. CCAFS East African Quarterly Newsletter, Nirobi, Kenya.

Nysimi, M., Amwata, D., Hove, L., Kinyangi, J., & Wamukoya, G. (2014). *Evidence of impact: Climate-smart agriculture in Africa* (CCAFS Working Paper No. 86). Copenhagen, Denmark.

Picard, N., Saint-André, L., & Henry, M. (2012). *Manual for building tree volume and biomass allometric equations: From field measurement to prediction* (p. 215). Montpellier: Food and Agricultural Organization of the United Nations, Rome, and Centre de Coopération Internationale en Recherche Agronomique pour le Développement.

SRI-Rice. (2014). *The system of crop intensification: Agroecological innovations to improve agricultural production, food security, and resilience to climate change*. Ithaca, New York: SRI International, Network and Resources Center (SRI-Rice), Cornell University.

Tröger, S., Grenzebach, H., Heide, F., Kuhnet, G., Lang, B., Peth, S., et al. (2011). *Failing seasons, ailing societies: Climate change and the meaning of adaptation in Ethiopia*. Addis Abeba, Ethiopia: HoA-REC and GIZ.

World Bank, CIAT, CATIE. (2014). *Climate-smart agriculture in Colombia, CSA country profiles for Latin America Series*. Washington DC, USA.

Silvopasture Using Indigenous Fodder Trees and Shrubs: The Underexploited Synergy Between Climate Change Adaptation and Mitigation in the Livestock Sector

Mulubrhan Balehegn

1 Introduction

In the dry lands of northern Ethiopia, the frequency of drought has increased greatly in recent decades (Gebrehiwot et al. 2011), albeit, against some predictions of increased rainfall in the Eastern Africa region (Christensen et al. 2007). Future climate predictions for the Sahel and East African region foresee increasingly intensive and erratic rain, exposing farmers and pastoralists to more droughts, floods and other climatic hazards (Giannini 2010). Droughts in arid and semi arid areas (ASALs) are not meteorological per se, but are results of intensive and erratic rainfall which, coupled with land degradation, ineffective land use systems, overgrazing and deforestation, make cultivation difficult and precarious. As a result, even though increased net primary productivity has been observed and predicted for the Sahel and East Africa (Doherty et al. 2010), in the absence of proper management, the increased moisture may not have any positive contribution towards increased productivity. In fact, whatever, the climatic predictions are, a vulnerability study in Ethiopian dry lands indicated that most of the regions, specially the arid and semiarid areas, are highly vulnerable to climate change related hazards and risks (Deressa et al. 2008).

Climate change, usually manifested in the form of extreme climate variability in the ASALs is one of the important constraints to agriculture in general and livestock production in particular, causing decrease in productivity by directly reducing feed availability and increasing the incidence of change induced livestock diseases and parasites (Jones and Thornton 2009). The livestock sector, however, is not only a victim of climate change, but also one of the main contributors. Livestock sector's

M. Balehegn (✉)
Department of Animal, Rangeland and Wildlife Sciences,
Mekelle University, Post Box 231, Mek'ele, Ethiopia
e-mail: mulubrhan.balehegn@mu.edu.et

W. Leal Filho et al. (eds.), *Climate Change Adaptation in Africa*,
Climate Change Management, DOI 10.1007/978-3-319-49520-0_30

contribution to global green house gases emission is estimated to be 51% (Goodland and Anhang 2009).

Therefore, the future of the livestock sector in the ASALs, both as an adaptable industry to the ensuing climate change, and as politically acceptable production system, from climate politics point of view, is dependent on its future ability to adapt to climate change and to be able to reduce its contribution to it. Within dryland rural settings of Ethiopia, the future of the livestock sector depends on the devising and promotion of climate resilient and adaptable livestock production systems which can maintain and increase livestock productivity under threats of rainfall shortage and variability (climate change).

1.1 Silvopastoralism: A Climate Resilient Alternative Livestock Production System

Silvopastoralism defined as an agroforestry system of livestock production where trees and shrubs, which provide diverse ecosystem services to humans are kept on pasturelands (Rigueiro-Rodríguez et al. 2011). Trees planted or naturally occurring on pasturelands, backyards, wastelands, farm-boundaries are used as a sole source of livestock feed, while at the same time providing other multipurpose ecosystem services and benefits (Gallo 2005).

Silvopastoralism has recently gained prominence as a sustainable and climate resilient livestock production system (Mosquera-Losada et al. 2011), although usually only theoretically (Dagang and Nair 2003). Silvopastoralism, compared to conventional pasture systems, has resulted in better livestock productivity (Schoeneberger 2009), and improved soil fertility (Moreno and Obrador 2007). Most importantly, silvopastoralism, as a tree based livestock production system provides an effective synergy between climate change adaptation and mitigation (Dube et al. 2011; Stavi and Lal 2013), by directly contributing to sequestration of Green House Gases (GHG), while at the same time buffering livelihoods against variability/recurrent drought as trees are less affected by variability in climate than grass pastures (Nair 2012).

Silvopastoralism contributes to climate change mitigation and adaptation in the following ways: first silvopastoralism is more adaptive to drought, than pastures because foliage production from trees and shrubs is less affected by varying precipitation, temperature and other climatic variables thus enabling farmers to sustain livestock production even at extreme weather condition (Papanastasis et al. 2008); second woody fodder species in a silvopastoral system increases the fertility of grazing lands there by providing suitable conditions for grasses, and improving micro-sites for grass growth (Moreno and Obrador 2007); thirdly livestock productivity is higher under a silvopastoral system compared to traditional pasture (Fig. 1) based systems, because fodder trees are of higher quality than many of the herbaceous grasses (Ibrahim et al. 2005), a fact that also enables reduction in the

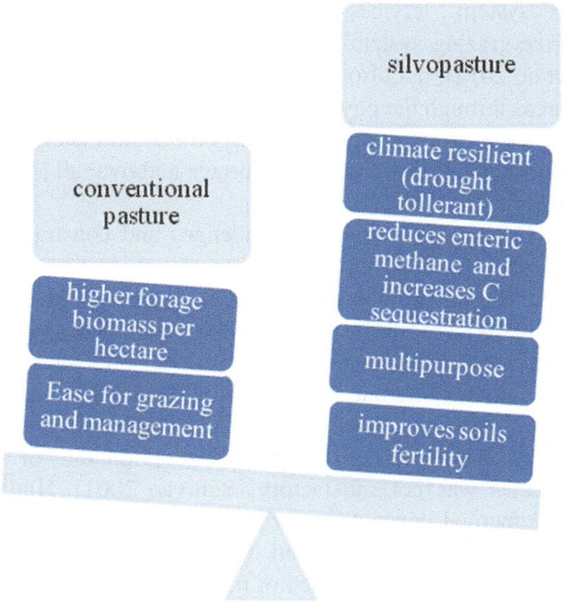

Fig. 1 Conceptual comparison of silvopasture with conventional pasture

amount of enteric methane and other GHG from livestock (Williams et al. 2011); fourth woody biomass increases the ecosystem carbon stock compared to herbaceous covered grazing lands (Mekuria et al. 2011), thereby making the system ecologically sustainable and politically acceptable.

Despite its proved benefits, however, silvopastoralism as a sustainable climate-resilient intensification of livestock sector has been overlooked in many livestock development programs. While work is underway in many places to develop varieties of staple food crops that can adapt to existing climatic changes (Legrève and Duveiller 2010), in contrast there is a serious dearth of information with regard to adaptable fodder trees. For instance, the Clean Development Mechanism (CDM) has spawned over 3000 climate change adaptation and mitigation projects but only 22 afforestation/reforestation projects, with none of those specifically focused on silvopastoralism (Lobovikov et al. 2012). Similarly, the Climate Change National Adaptation Program of Action (NAPA) of Ethiopia fails to mention the term, literally or conceptually (Tadege 2007).

1.2 Current State of Silvopastoral System Practices and the Need for Focusing on Indigenous Species

Traditional free-grazing systems have contributed to the steady degradation of land that has taken place in the highlands of Ethiopia for centuries (Taddese 2001), and has resulted in continuous decline in the availability of livestock feed and other ecosystem services from natural rangelands (Gebremedhin et al. 2004). Free-grazing system did not only result in increased land degradation (Meshesha et al. 2012), but also has limited the effectiveness of human endeavour in degraded areas through the physical destruction of soil and water conservation structures such as terraces, reforestation among others, ultimately causing a vicious circle of land degradation, livestock feed shortage and over all lower productivity of the livestock sector (Yisehak et al. 2013).

Cognizant of the feed challenges and constraints, and associated overgrazing induced land degradation, some strategies have been adopted and launched to alleviate the problems. Common strategies include; introduction of improved forage plants (Gebremedhin et al. 2003), supplementation with high nutritive value concentrates (Mengistu et al. 2006), and implementation of zero grazing (Gebreyohannes and Hailemariam 2011). Moreover, due to the inability of most of the introduced 'improved' species to adapt to local socio-ecological settings and shortage of land and water, the adoption rate of 'improved' introduced forage species was not satisfactory (Sullivan 2001). Higher cost and unavailability of commercial concentrate feeds have also obliterated the possibility of supplementation feeding (Jenkins and Miklyaev 2014). Similarly, due to problems of large numbers of livestock, lack of fodder bank, and cultural limitations, the endeavor to implement zero-grazing system in northern Ethiopia was not successful (Gebreyohannes and Hailemariam 2011).

Indigenous Fodder Trees and Shrubs (IFTS) are important part of animal feed and agro biodiversity in developing countries (Le Houèrou 2000). Attempts made to increase knowledge on IFTS have proved that indigenous browse species play a significant role in livestock production primarily by providing animals with feed resources rich in protein, energy, vitamins and minerals at a time when feed is scarce or of low quality (Osakwe and Drochner 2006). Most of the times, indigenous fodder trees and shrubs are appreciated for the higher content of crude protein (CP) they provide (Roothaert and Franzel 2001), and hence have been recommended as supplements, inclusions and replacements to commercial concentrates (Balehegn et al. 2014a).

Unlike many of the 'improved' introduced fodder trees, which are neither locally adaptable nor multipurpose (Balehegn and Eniang 2009a, b), most IFTS are multipurpose and provide benefits and services such as food, fiber, shade, soil improvement and conservation, timber, fire-wood and live fences across all of the agro-ecological zones of Africa. Indigenous browse species are also preferred over the 'improved' forage plants for being of low cost, available and accessible to local communities, adaptable to local environmental conditions, requiring little or no management input, and resistant to diseases and parasites (Inam-ur-Rahim et al. 2011). Therefore, incorporating IFTS into the agro-silvopastoral systems will not only help solve the livestock feed problem, but also contribute to environmental rehabilitation, stability, and improve livelihoods (Murgueitio et al. 2011). Despite their ecological and economic potential, however, there is still very limited research and development attention given to IFTS, compared to introduced 'improved' species (Le Houèrou 2000).

Silvopastoralism, a livestock production system where trees are planted on pasturelands, farmland boundaries and backyards etc., has recently gained prominence as a sustainable livestock production system (Mosquera-Losada et al. 2011), although usually only theoretically (Dagang and Nair 2003). Compared to conventional pasture systems, silvopastoralism, has resulted in better livestock productivity (Ibrahim et al. 2005; Schoeneberger 2009), and improved soil fertility (Moreno and Obrador 2007). It has also provided synergy between climate change adaptation and mitigation (Dube et al. 2011; Stavi and Lal 2013), by directly contributing to sequestration of Green House Gases (GHG), and being climate resilient (McAdam and McEvoy 2008; Nair 2012).

However, while work is underway in many places to develop varieties of staple food crops that can adapt to existing climatic changes (David and Christopher 2007; Legrève and Duveiller 2010), in contrast there is a serious dearth of information with regard to adaptable fodder trees. For instance, the Clean Development Mechanism (CDM) has spawned over 3000 climate change adaptation and mitigation projects but only 22 afforestation/reforestation projects, with none of those specifically focused on silvopastoralism (Lobovikov et al. 2012). Similarly, the Climate Change National Adaptation Program of Action (NAPA) of Ethiopia fails to mention the term, literally or conceptually (Tadege 2007).

It is therefore, based on the appreciation of the positive economic and ecological benefits of agro-silvopastoral system that this paper synthesizes the findings of a seven year study on ecological and livelihoods roles of *Ficus thonningii* (FT) silvopasture common in northern Ethiopia, while at the same time elucidating the ecological and livelihoods and climate change adaptation benefits of the silvopastoral system.

2 Methodology

The study was conducted among indigenous Tigrigna speaking people in Ahferom District central Tigray, Northern Ethiopia. Intensive field ecological and socio-economic survey was carried out using standard field equipment and materials. Individual interviews and focused group discussions were held with different members of the community representing men, women and elders. The interviews and socio-economic surveys focused on understanding the perceptions of the local people on climate change? and role of *Ficus thonningii* silvopasture in climate change adaptation, livelihood improvement and soil and water conservation (Balehegn et al. 2014b). To compare the value of *Ficus thonningii* relative to other local tree species, respondents were asked to rank ten local species according to different local criteria (Table 2). The average number of times a criteria was mentioned by each age category of the respondents is calculated by the number of people who mentioned it for any of the top ten tree species, divided by the 60 (total number of people in each age category). Therefore, if all respondents in a given age category mention the criteria for all the species, then the average number of times a

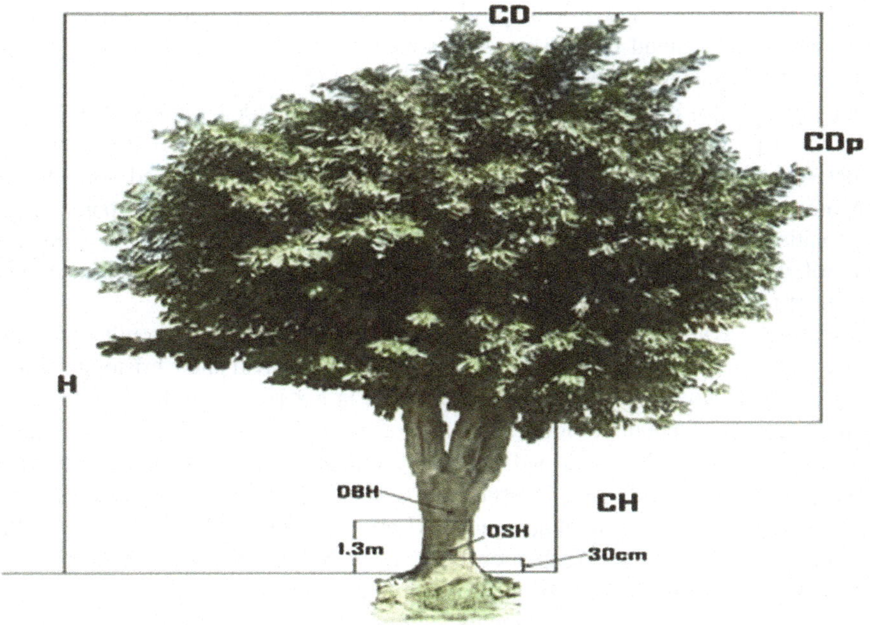

Fig. 2 Pictorial presentation of measured dendrometric parameters of *F. thonningii*

criteria is mentioned is equal to 60 * 10/60 = 10. Therefore, the maximum average number of times a criteria is mentioned by a given age category is 10 (i.e. if 60 of the respondents in a given age category mention it for every one of the 10 species), and the minimum is 0 (i.e. if none of the 60 respondents in a given age category mentioned it for any of the ten species)

To estimate the browse biomass production potential of *Ficus thonningii*, different tree dendrometric parameters, indicated in Fig. 2, were measured from 12 representative trees, four representing each of the three age groups (G1 > 50 years, G2 between 10–20 years, and G3 < 5 years). Measured tree dendrometric parameters including Height (*H*), Crown Height (*CH*) in meters and Diameter at Stump Height (*DSH*) and Diameter at Breast Height (*DBH*) in cm, Crown Diameter and estimated parameters including Crown Depth (*CDp*), Crown Hheight (*CH*), Crown Aarea (*CA*) and Crown Volume (*CV*) were used in a regression analysis to identify the best predictor of browse biomass (Balehegn et al. 2012).

To understand the effect of feeding *F. thonningii* fodder on animals, twenty four weaned central highland male goats of 7 ± 1.5 SD months of age and weighing (15 ± 1.86 SD) kg, were used in complete randomized block design with six goats in four dietary treatments involving partial replacement at different levels of commercial concentrates (Balehegn et al. 2014a).

Table 1 Local communities' perception and manifestation of climate change

S. no	Local manifestation of declining eco-climatic conditions	Percentage of respondents (n = 120)
1	Reduced crop yield	88.6
2	Reduced milk yield	74.3
3	Reduced forage options for livestock	71.75
4	Disappearance of wildlife species	64.2
5	Early drying up of perennial rivers	52.4
6	Increased soil degradation	50.2
7	Increased conflict with neighboring villages	32.4
8	Miscellaneous	12.3

3 Results

3.1 *Local Perception of Climate Change and Parameters Used for Selection of Tree Species for the Climate Resilient Silvopasture*

Local communities use different bio-physical indicators to express how they perceive climate change and its consequences (Table 1).

Moreover, respondents mentioned twenty local tree selection criteria falling under three categories namely: Animal based, plant based, and multipurpose (Table 2).

3.2 *Multipurpose Merits of* F. thonningii *and Trends in Its Use in a Silvopastoral System*

The proportion of *F. thonningii*, compared to other exotic fodder trees is increasing from time to time (Fig. 3).

3.3 *Indigenous Practices of Propagation and Use of* Ficus thonningii

Table 3 summarizes indigenous procedures, protocols and nursing practices involved at the different phases of propagation and growth of *F thonningii*.

Table 2 The average number of times fodder tree selection indigenous criteria was mentioned

Category of criteria	Fodder tree selection criteria	Average number of times a criteria was mentioned
Animal based criteria	Rumen fill	8.12[a]
	Health	4.21[b]
	Palatability	8.97[a]
	Milk production	3.21[b]
	Fattening	2.14[b]
Plant based criteria	Drought tolerance	10.00[a]
	Termite resistance	1.36[b]
	Early re-growth	8.76[a]
	Biomass productivity	9.45[a]
Multipurpose criteria	Food	6.23
	Fire wood	8.98[a]
	Charcoal	7.24
	Fencing	8.98[a]
	Timber and construction	3.22[b]
	Farm implements	8.72[a]
	Ethno medicine	8.94[a]
	Shade and shelter	3.24[b]
	Bark for making ropes	1.24[b]
	Sanitation (smoking utensils, tooth brush)	3.22
	Market value (cash)	1.2[b]

Values down the column with different super scripts are significantly different ($P < 0.05$)

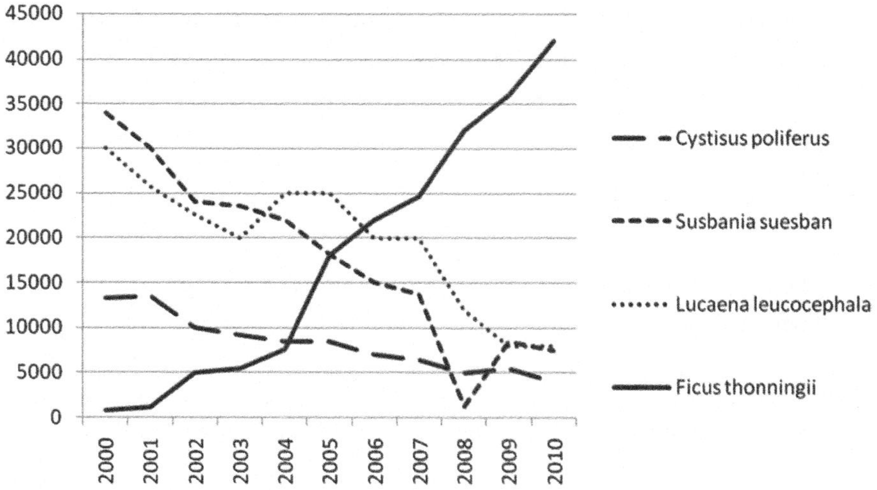

Fig. 3 A decade trend of the number of different fodder trees planted in the backyards of the farmers of Sefeo community (Balehegn et al. 2014b)

Table 3 Indigenous protocol and requirements for successful propagation of *Ficus thonningii*

Requirement	Measurement/indicator	Remark
Age of plant	>2 years	Depends on the size and maturity
Length and width of cutting	Width 10–30 cm, and length <4 m	Long cutting easily shaken by wind thus susceptible to death
Season for cutting	March–May and end of August	Soil should be wet, but not swampy
Care during cutting	Avoid peeling barks and shading cut part	All branches should be cut at once, as shading of cut part causes drying
Incubation time	About 1–2 months for those cut March–May, <1 month for those cut end of August	To make cuttings lose some moisture
Care during Incubation time	Covering the overall cutting with dung, or thorny branches to repel animals	Avoid peeling barks
Depth and width of planting pit	50 cm depth and 20 cm width	Adjusted to fit the dimensions of the cutting.
Special care in planting	Put a flat circular rock at the base of the pit in wet soil	Avoids rotting of new roots due to water logging
Watering	May not need watering	Survive on water stored in the stem watering is not bad though
Care after planting	Complete protection form animals by dipping the whole cutting in dung, attaching cutting to a tree or a stationary peg	Shaking hinders root development and causes subsequent dying of cuttings

This qualitative information on the propagation of *Ficus thonningii* was collected from group discussions and individual interviews (Balehegn et al. 2014b)

3.4 Ficus thonningii *Browse: Production and Its Nutritive Value*

3.4.1 Browse Biomass

The average values of browse biomass in dry matter for three age groups of trees are; 50.36 kg for G1, 5.96 kg for G2, and 0.914 kg for G3 (Table 4).

3.4.2 Nutritive Value of *Ficus thonningii* Foliage

The proximate nutritional value of *F. thonningii* foliage for this study is given in Table 5.

Table 4 Average predicted biomass for sampled trees of *F. thonningii* in Sefeo, Central Tigray, northern Ethiopia (Balehegn et al. 2012)

No	Age group	DSH (Giannini et al. 2008)	CV (m^3)	Average predicted DW (kg)$tree^{-1}year^{-1}$
1	G1 (20–100+ years)	62.098	77.404	50.356
2	G2 (5–20 years)	33.595	17.58	5.9611
3	G3 (<5 years)	12.65	6.1755	0.9136

Prediction formula: DW = 0.8470 * CV − 0.2202 * DSH − 1.5315

3.4.3 Effect of Feeding *Ficus thonningii* on Productivity of Animals

Replacement of local commercial concentrate by *F. thonningii* leaf meal at 50% level weight based has resulted in higher body weight gain compared to lower levels of replacements (Table 6) (Balehegn et al. 2014a).

4 Discussions

4.1 Local Perception of Climate Change and Parameters Used for Selection of Tree Species for Climate Resilient Silvopasture

Local communities' observations and perceptions of climate change are similar in principle to those observed in Kenya and Ethiopia (Ifejika Speranza et al. 2010; Balehegn and Tafere 2013), which also reflect people's focus on observable bio-physical changes when describing climate change focusing on drought. Similarly, answering to question on how they select forage plants in the study areas, 67% of respondents (Table 2) emphasized importance of characteristics of the plant to tolerate drought (plant based characteristics) than the ability of the plant to maintain animal productivity (animal based characteristics) (Balehegn et al. 2014b). The emphasis on drought tolerance imply that drought has become unavoidable part of the agrarian life in northern Ethiopia as noted by Gebrehiwot et al. (2011).

4.2 Multipurpose Merits of Ficus thonningii *and Trend in Its Use in a Silvopastoral System*

Ficus thonningii produces a large amount of nutritious foliage, which exceeds that of many conventional fodder trees (Berhe and Tanga 2013; Balehegn et al. 2012). Other ecological merits such as ability to withstand lopping; absence of allelopathic

Table 5 Nutritional value of *Ficus thonningii* foliage compared to common feed ingredients (Balehegn et al. 2014a)

Feed ingredients	DM	CP	Ash	OM	NDF	ADF	ADL	Tannin
Cotton seed cake	90.00 ± 12.3	28.00 ± 9	7.20 ± 3.0	94.23 ± 32.1	38.48 ± 15.0	39.00 ± 9.1	18.00 ± 3.4	–
Wheat bran	85 ± 12.4	11.00 ± 2.2	47.25 ± 13.0	96.24 ± 42.1	11.00 ± 2.2	50.20 ± 8.2	25.29 ± 18.2	–
Maize	85.40 ± 13.4	10.00 ± 3.4	13.00 ± 8.4	98.26 ± 42.2	12.00 ± 7.3	3.00 ± 0.5	1.00 ± 0.01	–
FT leaf	90.00 ± 9.4	18.00 ± 9.0	14.28 ± 9.0	88.00 ± 15.0	42.12 ± 7.3	36.00 ± 5.1	16.05 ± 2.3	0.06 ± 0.03
Wheat straw	90.00 ± 10.4	4.00 ± 2.0	7.10 ± 3.0	89.00 ± 32.1	81.00 ± 8.4	51.00 ± 7.3	8.42 ± 4.0	–

Values are Mean ± Std Dev, *DM* dry matter, *CP* crude protein, *OM* organic matter, *NDF* neutral detergent fiber, *ADF* acid detergent fiber, *ADL* acid detergent lignin, *FT Ficus thonningii*

Table 6 Body weight gain and feed conversion efficiency of experimental goats

	Treatments				SEM
	T1	T2	T3	T4	
Body weight gain					
Average initial body weight	14.50	14.08	15.00	15.00	2.00
Average final body weight	17.06^{a}	16.23^{b}	19.12^{ac}	17.00^{b}	2.10
Live weight gain (kg)	3.00^{a}	2.15^{b}	5.00^{ac}	2.25^{b}	1.60
ADG (g day^{-1})	29.00^{a}	24.00^{b}	50.37^{ac}	25.00^{b}	19.16
Feed conversion efficiency					
g ADG/g DM	11.00^{a}	8.00^{b}	13.03^{c}	6.37^{d}	5.41
g ADG/g CP	90.00^{a}	56.41^{b}	94.00^{a}	48.36^{c}	52.00

$^{a\text{–}d}$Values across rows with different superscripts are significantly different ($P < 0.05$)

effects to other plants mainly to the economically important shrub, *Ruhamnus prinoides* or '*Gesho*',[1] which grows underneath *F. thonningii*; easiness for propagation, and positive effect of soil fertility (Balehegn 2011; Berhe et al. 2013; Balehegn and Eniang 2009a; Tegegne 2008). All these ecological and livelihoods benefits make it a very opportune choice as a silvopastural fodder tree. As a result intensive use of *F. thonningii* for animal feeding, and thus its steady planting in silvopastoral system practices started following severe feed scarcity problems in the area caused by repeated droughts and over grazing which reached alarming levels between the years 1985 and 1991 (Balehegn and Eniang 2009a).

Appreciating its extensive benefits for livelihoods and adaptation to reoccurring drought, people have been extensively planting FT on a variety of land types, that the proportion of FT compared to other 'improved' introduced forage species (mainly *Cystisus proliferus*, *Sesbania sesban* and *Leucaena leucaocephala*) has been increasing from time to time, despite the orchestrated push by governmental and non-governmental agricultural and environmental agencies to introduce the latter (Fig. 2). Multipurpose drought tolerant plants such as *Ficus thonningii* and Bamboo will dominate the future of species for climate change adaptation, as they simultaneously help poor communities economically, while enabling them to adapt to climate change (Lobovikov et al. 2012). This tendency of people to develop a locally relevant system than just accepting what is recommended by outsiders is also a case confirming that local species are more adapted to local socio-ecological settings (German et al. 2006).

[1]A common local shrubby plant used by indigenous people to produce "*Swa*" or a traditional alcoholic beverage that yields considerable income to small scale producers. Owing to the economic value of *Ruhamnus prinoides* or '*Gesho*', the indigenous community would have preferred *F. thonningii* eliminated, if its existence had any negative impacts on the former.

4.3 *Indigenous Practices of Propagation and Use of* Ficus thonningii

Through experience, local communities in northern Ethiopia have developed and perfected indigenous procedures and protocols for successful propagation of the plant (Table 3). It takes between 2 and 5 years for a newly planted cutting to reach the stage of a matured tree ready for harvest of both cuttings and leaves. Time at first harvest seems to depend on the original size of the cutting used for propagation as it takes longer time for small cuttings to reach harvest size than it does to larger cuttings (Balehegn 2011). However, large cuttings are susceptible to damage by wind shaking and animals (Balehegn 2011; Balehegn and Eniang 2009a). Moreover, since the barks are usually watery at the cutting stage, they are highly relished by animals which nibble and harm the cuttings, thus farmers usually use either protective thorny fencing or rub cow dung on the young cuttings to protect them from animals (Balehegn and Eniang 2009a). Some of the local practices in the use and propagation of *F. thonningii* are also given in Fig. 4.

Fig. 4 Preparation, incubation and planting of *Ficus thonningii* cuttings (*upper row*), *Ficus thonningii* leaves used for animal feeding (*lower row*)

4.4 *Browse Production and Nutritive Value of* Ficus thonningii

4.4.1 Browse Production

The average values of dry matter biomass for each age group in (Table 4) translates into 222.2 ton/ha, 26.48 ton/ha and 4.05 ton/ha for G1, G2, and G3 respectively (Balehegn et al. 2012). These results indicate that *F. thonningii* produces by far higher biomass annually as compared to the average 31–87 Mg DM ha^{-1} for 26 native and exotic fodder species in similar non tropical drylands in west Africa (Larbi et al. 2009). This higher browse biomass production capacity of *F. thonningii* is also appreciated by farmers in the study area and other areas (Balehegn and Eniang 2009a; Mekoya et al. 2008).

4.5 *Nutritive Value of* Ficus thonningii *Foliage*

The reasonably higher nutritional values of the foliage of *F. thonningii* compared to the other common feed ingredients (Table 5) is the main reason for the higher palatability of *F. thonningii* by all species of livestock (Balehegn et al. 2014a). The crude protein (CP) of *F. thonningii* is similar to CP values of 21.5 (Bamikole and Ikhatua 2010). Moreover, unlike common fodder trees and shrubs, *FT* has very low tannin content of only 0.06% (Balehegn et al. 2014b). This value is comparable to the value of 0.04% reported by (Mekoya et al. 2008). This low tannin content coupled with reasonably higher protein content, could be a reason for its higher palatability to all farm animals reported in the silvopastoral system in northern Ethiopia (Tegegne 2008; Balehegn and Eniang 2009a; Berhe and Tanga 2013). Similarly, the NDF and ADF contents of FT reported in this study are also lower than the threshold safe level recommended by NRC (2007), again contributing to its nutritive value and subsequent acceptability by animals.

4.6 *Effect of Feeding* Ficus thonningii *on Productivity of Animals*

The higher body weight gain at the 50% level of replacement (Balehegn et al. 2014a) is comparable to the values of 37.6–54 g day^{-1} for naturally grazing and concentrate supplemented goats observed by Zewdu and Taye (2013). The larger body weight gain in the combined diets than concentrate only diet indicate that *F. thonningii* affected digestibility and assimilation of diets, as also observed in (Bamikole and Ikhatua 2010).

Increased body weight gain in the goats fed 50% *F. thonningii* leaf meal was also reflected in improved carcass components. Higher values in proximal thoracic limb, lumbar and abdominal region and the proportion of neck to hot carcass weight observed in goats fed treatments with 50 and 75% level of replacement of commercial concentrates by *F. thonningii* leaf meal (Balehegn et al. 2014a). This is expected because these two commercial cuts (proximal thoracic limb and lumbar and abdominal regions) are where initial growth in growing animals takes place (Mahgoub and Lu 1998). Aside from feed, nutritive and afore-listed multipurpose values of *F. thonningii* silvopasture has also resulted in improved habitat for wildlife species including *Cercopithecine* monkeys, frugivorous birds and insects the most significant of which is the endangered White billed starling (Balehegn and Eniang 2009b).

5 Conclusions and Implications

This review revealed an indigenous innovation for recurrent drought and climate change adaptation using *F. thonningii*—a drought resistant multipurpose indigenous fodder tree as a key species in an emerging silvopastoral system. This indigenous innovation silvopastoral system enabled herders in the dryland areas of northern Ethiopia to adapt to recurrent drought, while improving their livelihoods through contribution to much sought after nutritious animal fodder, improvement of soil fertility and maintenance of ecological integrity.

Looking at the potential of *Ficus thonningii* towards adapting and sustaining livelihoods of farmers living in fragile ecosystems, it should be adopted in future adaptation programs. Being indigenous to 33 other Sub-saharan African countries with similar agro-ecological settings with northern Ethiopia, *F. thonningii* silvopasture has a potential for similar use in these other countries, but most importantly these studies provide an insight for the use of indigenous tree species in the agro-silvopastoral system in future local climate-change adaptation and mitigation programs.

References

Balehegn, M. (2011). *Forage and multipurpose uses of Ficus thonningii in Northern Ethiopia: Nutritive value, biomass estimation, and impact on soil fertility*. Saarbrücken, Germany: LAP Lambert Academic Publishing.

Balehegn, M., Eik, L., & Tesfay, Y. (2014a). Replacing commercial concentrate by Ficus thonningii improved productivity of goats in Ethiopia. *Tropical Animal Health and Production*, 1–6.

Balehegn, M., Eik, L. O., & Tesfay, Y. (2014b). Silvopastoral system based on Ficus thonningii: an adaptation to climate change in northern Ethiopia. *African Journal of Range & Forage Science*, 1–9.

Balehegn, M., & Eniang, E. (2009a). Assessing indigenous knowledge for evaluation, propagation and conservation of indigenous multipurpose fodder trees towards enhancing climate change adaptation in Northern Ethiopia. In J. Parotta, O.-Y. Cobbinah, & D. Jacob (Eds.), *Traditional forest-related knowledge and sustainable forest management in Africa: Papers from the conference held in, 15–17 October 2008*. Accra, Ghana.

Balehegn, M., & Eniang, E. A. (2009b). Assessing indigenous knowledge for evaluation, propagation and conservation of indigenous multipurpose fodder trees towards enhancing climate change: adaptation in Northern Ethiopia. Traditional forest-related knowledge and sustainable forest management in Africa. *IUFRO World Series, 23*, 39–46.

Balehegn, M., Eniang, E. A., & Hassen, A. (2012). Estimation of browse biomass of Ficus thonningii, an indigenous multipurpose fodder tree in northern Ethiopia. *African Journal of Range & Forage Science, 29*, 25.

Balehegn, M., & Tafere, K. (2013). Gendered impacts and adaptation mechanisms to climate change among Afar Pastoralists in North Eastern Ethiopia. In Mulinge & Getu (Eds.), *Impacts of climate change and variability on pastoralist women in Sub-Saharan Africa*. Addis Ababa, Ethiopia: Organization of Social Sciences Research in Eastern and Southern Africa (OSSREA).

Bamikole, M. A., & Ikhatua, U. J. (2010). Nutritional evaluation of *Ficus thonningii-Panicum maximum* mixtures in West African dwarf goats. *Nutrition & Food Science, 40*, 280–288.

Berhe, D. H., Anjulo, A., Abdelkadir, A., & Edwards, S. (2013). Evaluation of the effect of Ficus thonningii (blume) on soil physicochemical properties in Ahferom district of Tigray, Ethiopia. *Journal of Soil Science and Environmental Management, 4*, 35–45.

Berhe, D. H., & Tanga, A. A. (2013). Nutritional evaluation of Ficus thonningii Blume leaves as ruminant livestock feed in the Ahferom district of Tigray, Ethiopia. *African Journal of Range & Forage Science*, 1–6.

Christensen, J. H., Hewitson, B., Busuioc, A., Chen, A., Gao, X., Held, R., et al. (2007). *Regional climate projections. Climate Change, 2007: The Physical Science Basis*. Contribution of Working group I to the Fourth Assessment Report of the Intergovernmental Panel on Climate Change, University Press, Cambridge, Chapter 11, pp. 847–940.

Dagang, A. K., & Nair, P. K. R. (2003). Silvopastoral research and adoption in Central America: Recent findings and recommendations for future directions. *Agroforestry Systems, 59*, 149–155.

David, B. L. & Christopher, B. F. (2007). Global scale climate—crop yield relationships and the impacts of recent warming. *Environmental Research Letters*, 2, 014002.

Deressa, T., Hassan, R. M., & Ringler, C. (2008). Measuring Ethiopian farmers' vulnerability to climate change across regional states. *International Food Policy Research Institute*.

Doherty, R. M., Sitch, S., Smith, B., Lewis, S. L., & Thornton, P. K. (2010). Implications of future climate and atmospheric CO_2 content for regional biogeochemistry, biogeography and ecosystem services across East Africa. *Global Change Biology, 16*, 617–640.

Dube, F., Thevathasan, N., Zagal, E., Gordon, A., Stolpe, N., & Espinosa, M. (2011). Carbon sequestration potential of silvopastoral and other land use systems in the Chilean Patagonia. In B. M. Kumar & P. K. R. Nair (Eds.), *Carbon sequestration potential of agroforestry systems*. Netherlands: Springer.

Gallo, L. (2005). Silvopastoralism with native tree species in Uruguay. In M. Mosquera-Losada, A. Rigueiro-Rodríguez, & J. Mcadam (Eds.), *Silvopastoralism and sustainable land management: Proceedings of an International Congress on Silvopastoralism and Sustainable Management Held in Lugo, Spain, in April 2004*. 2005 Lugo, Spain. CABI, p. 49.

Gebrehiwot, T., van der Veen, A., & Maathuis, B. (2011). Spatial and temporal assessment of drought in the Northern highlands of Ethiopia. *International Journal of Applied Earth Observation and Geoinformation, 13*, 309.

Gebremedhin, B., Ahmed, M. & Ehui, S. (2003). Determinants of adoption of improved forage technologies in crop-livestock mixed systems: Evidence from the highlands of Ethiopia.

Gebremedhin, B., Pender, J. & Tesfay, G. (2004). Collective action for grazing land management in crop–livestock mixed systems in the highlands of northern Ethiopia. *Agricultural Systems, 82*, 273–290.

Gebreyohannes, G. & Hailemariam, G. (2011). Challenges, Opportunities and available good practices related to zero-grazing in Tigray and Eastern and Western Hararghe zones of Oromia National Regional States of Ethiopia. Dryland Coordination Group (DCG).

German, L. A., Kidane, B., & Shemdoe, R. (2006). Social and environmental trade-offs in tree species selection: a methodology for identifying niche incompatibilities in agroforestry. *Environment, Development and Sustainability, 8,* 535–552.

Giannini, A. (2010). Mechanisms of climate change in the semiarid African Sahel: The local view. *Journal of Climate, 23,* 743.

Giannini, A., Biasutti, M., Held, I., & Sobel, A. (2008). A global perspective on African climate. *Climatic Change, 90,* 359–383.

Goodland, R., & Anhang, J. (2009). Livestock and climate change: What if the key actors in climate change are... cows, pigs, and chickens? *Livestock and climate change: what if the key actors in climate change are.... cows, pigs, and chickens?*

Ibrahim, M., Villanueva, C., & Mora, J. (2005). Traditional and improved silvopastoral systems and their importance in sustainability of livestock farms. In M. Mosquera-Losada & A. Rigueiro-Rodriguez (Eds.), *Proceedings of an International Congress on Silvopastoralism and Sustainable Land Management held in Lugo, Spain, in April 2004.* CAB International.

Ifejika Speranza, C., Kiteme, B., Ambenje, P., Wiesmann, U., & Makali, S. (2010). Indigenous knowledge related to climate variability and change: Insights from droughts in semi-arid areas of former Makueni District, Kenya. *Climatic Change, 100,* 295–315.

Inam-ur-rahim, Maselli, D., Rueff, H. & Wiesmann, U. (2011). Indigenous fodder trees can increase grazing accessibility for landless and mobile pastoralists in northern Pakistan. *Pastoralism: Research, Policy and Practice,* 1, 2

Jenkins, G. P. & Miklyaev, M. (2014). Cost-Benefit analysis of small ruminants fattening with feed concentrates in the highlands of Ethiopia. JDI Executive Programs.

Jones, P. G., & Thornton, P. K. (2009). Croppers to livestock keepers: Livelihood transitions to 2050 in Africa due to climate change. *Environmental Science & Policy, 12,* 427–437.

Larbi, A., Khatib-Salkini, A., Bolus Jamal, P., & Iniguez, L. (2009). Shrub yield and fodder quality variations in a non-tropical dryland environment in West Asia. *Agroforestry Systems, 75,* 147.

Le Houèrou, H. N. (2000). Utilization of fodder trees and shrubs in the arid and semiarid zones of West Asia and North Africa. *Arid Soil Research and Rehabilitation, 14,* 101–135.

Legrève, A., & Duveiller, E. (2010). Preventing potential disease and pest epidemics under a changing climate. In M. Reynolds (Ed.), *Climate change and crop production.* Wallingford, Cambridge: CABI.

Lobovikov, M., Schoene, D., & Yping, L. (2012). Bamboo in climate change and rural livelihoods. *Mitigation and Adaptation Strategies for Global Change, 17,* 261.

Mahgoub, O., & Lu, C. D. (1998). Growth, body composition and carcass tissue distribution in goats of large and small sizes. *Small Ruminant Research, 27,* 267–278.

Mcadam, J. & Mcevoy, P. (2008). The potential for silvopastoralism to enhance biodiversity on grassland farms in Ireland. *Agroforestry in Europe.* Springer.

Mekoya, A., Oosting, S. J., Fernandez-Rivera, S., & van der Zijpp, A. J. (2008). Multipurpose fodder trees in the Ethiopian highlands: Farmers' preference and relationship of indigenous knowledge of feed value with laboratory indicators. *Agricultural Systems, 96,* 184–194.

Mengistu, A. (2006). Contry pasture/forage profile Ethiopia. Availalbe at http://www.fao.org/ag/agp/agpc/doc/counprof/pdf%20files/ethiopia-english.pdf

Meshesha, D. T., Tsunekawa, A. & Tsubo, M. (2012). Continuing land degradation: Cause—effect in Ethiopia's Central Rift Valley. *Land degradation & development, 23,* 130–143.

Moreno, G., & Obrador, J. (2007). Effects of trees and understorey management on soil fertility and nutritional status of holm oaks in Spanish dehesas. *Nutrient Cycling in Agroecosystems, 78,* 253–264.

Mosquera-Losada, M. R., Freese, D., & Rigueiro-Rodríguez, A. (2011). Carbon sequestration in European agroforestry systems. In B. M. Kumar & P. K. R. Nair (Eds.), *Carbon sequestration potential of agroforestry systems.* London, Netherlands: Springer.

Murgueitio, E., Calle, Z., Uribe, F., Calle, A. & Solorio, B. (2011). Native trees and shrubs for the productive rehabilitation of tropical cattle ranching lands. *Forest Ecology and Management, 261*, 1654–1663

Nair, P. K. R. (2012). Climate change mitigation: A low-hanging fruit of agroforestry. In P. K. R Nair, & D. Garrity, (Eds.), *Agroforestry—The future of global land use*. Netherlands: Springer.

NRC. (2007). *Nutrient requirements of small ruminants: sheep, goats, cervids, and new world camelids: National Research Council*. Committee on Nutrient Requirements of Small Ruminants.

Osakwe, I. I. & DROCHNER, W. (2006). Nutritive value of Morinda lucida and its fermentation parameters in West African dwarf (WAD) sheep when fed as supplement to grass hay. *Small Ruminant Research, 64*, 107–115.

Papanastasis, V. P., Yiakoulaki, M. D., Decandia, M., & Dini-Papanastasi, O. (2008). Integrating woody species into livestock feeding in the Mediterranean areas of Europe. *Animal Feed Science and Technology, 140,* 1–17.

Rigueiro-Rodríguez, A., Rois-Díaz, M., & Mosquera-Losada, M. R. (2011). Integrating silvopastoralism and biodiversity conservation. In E. Lichtfouse (Ed.), *Biodiversity, biofuels, agroforestry and conservation agriculture*. Netherlands: Springer.

Rootheart, R. & Franzel, S. (2001). Farmers' preferences and use of local fodder trees and shrubs in Kenya. *Agroforestry Systems, 52*, 239–252.

Schoeneberger, M. M. (2009). Agroforestry: Working trees for sequestering carbon on agricultural lands. *Agroforestry Systems, 75,* 27–37.

Stavi, I., & Lal, R. (2013). Agroforestry and biochar to offset climate change: A review. *Agronomy for Sustainable Development, 33,* 81–96.

Sullivan, G. A. (2001). Forages: Looking at the Potential and the Pathways toward Adoption in the Ethiopian Farming Systems.

Taddese, G. (2001). Land degradation: A challenge to Ethiopia. *Environmental Management, 27,* 815–824.

Tadege, A. (2007). *Climate change National Adaptation Programme of Action (NAPA) of Ethiopia*. Addis Ababa: National Meteorological Agency (NMA).

Tegegne, E. D. (2008). *Importance of Ficus thonningii Blume in soil fertility improvement and animal nutrition in Gondar Zuria, Ethiopia*, na.

Williams, C. M., Eun, J. S., Macadam, J. W., Young, A. J., Fellner, V., & Min, B. R. (2011). Effects of forage legumes containing condensed tannins on methane and ammonia production in continuous cultures of mixed ruminal microorganisms. *Animal Feed Science and Technology, 166–167,* 364–372.

YIisehak, K., Belay, D., Taye, T. & Janssens, G. P. J. (2013). Impact of soil erosion associated factors on available feed resources for free-ranging cattle at three altitude regions: Measurements and perceptions. *Journal of Arid Environments, 98*, 70–78.

Zewdu, S., & Taye, M. (2013). Growth performance and carcass characteristics of central highland goats in Sekota District, Ethiopia. *Agricultural Advances, 2,* 250–258.

The Need for Transformation: Local Perception of Climate Change, Vulnerability and Adaptation Versus 'Humanitarian' Response in Afar Region, Ethiopia

Teklehaymanot Geremeskel and Mesele Abera

1 Introduction

Climate change is a growing challenge for many communities in the world. It is particularly a critical problem for communities in the rural developing world (Adger et al. 2003). Frequent droughts, floods, storms, and other related problems are some of the major indicators of a changing climate. Ethiopia as one of the developing countries also faces some of the worst effects of climate change.

There have been several interventions by national states and other actors in response to such disasters. Interventions in the past particularly in pastoral regions of Ethiopia however mainly focused on alleviating immediate consequences of disasters and not long term adaptation. This focus on emergency aid by donors has been criticized for several reasons. Thus, there are calls for more investments towards adaptation and reduction of vulnerability than just emergency responses to climate change disasters in the developing world (Mirza 2003).

The vagueness of climate change effects is one of the most problematic concepts in the climate change literature. Many scholars, policy makers and practitioners alike consider climate change as a linear process in which a system bounces back to normality after the disaster. However, our climate is always under uncertain future which full of disruptions and accidents some of which are irreversible. Social-ecological systems are of a non-equilibrium state and changes are less predictable and more uncertain. Policies and programs which are based on the equilibrium view of social-ecological systems more often than not tend to focus on instantaneous responses during disasters and not on supporting long term change in the structural factors leading to vulnerability in the first place.

T. Geremeskel (✉) · M. Abera
Institute of Environment, Gender and Development Studies, Mekelle University, P. O. Box 231, Mekelle, Ethiopia
e-mail: gebremeskel8@yahoo.com

W. Leal Filho et al. (eds.), *Climate Change Adaptation in Africa*,
Climate Change Management, DOI 10.1007/978-3-319-49520-0_31

Using a rights-based approach to development provides a more convenient framework to study people differences in their perception of climate change risk and adaptation pathways and anticipatory actions they take (Tanner et al. 2015).

Very limited number studies are carried out about climate change adaptation in pastoral areas and particularly in Ethiopia. The role of humanitarian aid in bringing long-term adaptation of pastoralists is particularly less explored despite its tremendous importance. Examining the perception of local actors about climate change, vulnerability and adaptation is very important in understanding their interests in the decision making process about adaptation to climate change.

The overall purpose of this study was therefore to critically examine the ways in which climate change, vulnerability, adaptation and the contribution of humanitarian aid response towards long term adaption are perceived by different local actors in pastoral communities in Afar region, Ethiopia. An ethnographic fieldwork which involved key informant interviews (with local government and NGOs representatives), local level individual interviews with different members of communities, focus group discussions and observation by researchers was carried out in three districts of Afar region between 2013 and 2015.

2 Climate Change Adaptation

Adaptation to climate change is defined as an "*Adjustment in natural or human systems in response to actual or expected climatic stimuli or their effects, which moderates harm or exploits beneficial opportunities*" (IPCC 2007). According to Pelling (2011), there are three forms of adaptation depending on how vulnerability is viewed by decision makers. The first is *resilience* which requires changes that can allow existing functions and practices to persist without questioning underlying power relations and root causes of vulnerability in a society. The assumption is that climate change related disturbances are evident but stability is a possibility after each disturbance. It is based on the belief of a system's ability to bounce back to equilibrium.

The second form is *transition* which refers to the incremental changes in the system to absorb opportunities and mitigate challenges brought by a changing climate. Such forms of adaptation generally assume prevailing governance regimes as conducive for adaptation and hence discourage responses to proximate causes of vulnerability (Pelling 2011; Wise et al. 2014). *Transformation* is the third form of adaptation. This type of adaptation can be considered relevant and achieved if vulnerability is assumed to be an outcome of wider social processes shaping how people view themselves and others, their relationship with the environment and role in social processes. The outcome of transformation is changes in the fundamental social values and turning against vested interests and status quo (Pelling 2011, p. 136).

According to Tanner et al. (2015), the concept of resilience requires attention if the aim is to address the limits to adaptation and the development needs of the poorest sections of societies. Mainstream views of resilience tend to focus on persistence of existing system which contributed to the original destruction over its transformational restructuring (p. 24).

Aid, be it development or humanitarian should hence support adaptation actions leading towards social, economic and cultural transformation of a society in the long term. However, the question in Ethiopia is whether or not current humanitarian and development aid is supporting long term transformations and adaptation to changes.

Eriksen et al. (2011) stated that sustainable adaptation responses should fundamentally be guided by four basic principles. These are (1) recognize the context of vulnerability, including multiple stressors, (2) acknowledge that different values and interests affect adaptation outcomes, (3) integrate local knowledge into adaptation responses and (4) consider potential feedbacks between local and global processes (p. 16).

Participatory assessment of vulnerability is hence important in understanding the extent of vulnerability and adaptation ways relevant to the problem locals face (Smit and Wandel 2006).

2.1 Drivers of Adaptation Decision Making

Adaptation decision making involves different stakeholders such as community groups, individuals, local and national governments, NGOs, business and industry groups. Decisions by such an array of stakeholders can be thus be influenced by different drivers. Hence, it is important to critically look into the mindsets and motivations of all decision makers and stakeholders concerned (Maani 2013).

2.2 Detrimental Impact of Small and Repeated Disasters

Rentschler (2013) brought a new perspective into how critical the impact of small and repeated disasters is on the lives of the poorest groups. The impacts of such small and repeated disasters are mostly not recognized by decision makers because their direct effect is never visible and is mainly long-term. This is because there are always deep-seated problems with data about smaller and repeated weather events as these are in most cases poorly monitored. Diagram 1 shows how small and repeated disasters impact the resilience capabilities of poor households and go unrecognized by policy makers. Different decision making actors mainly pay attention to major disasters with greater direct impacts.

The long-term welfare level attainable by the poor Rentschler (2013) argue are not only affected by the severity of disasters but also by their frequency (p. 15).

Diagram 1 Poverty traps: the vicious cycle of perpetuated poverty and vulnerability (Rentschler 2013)

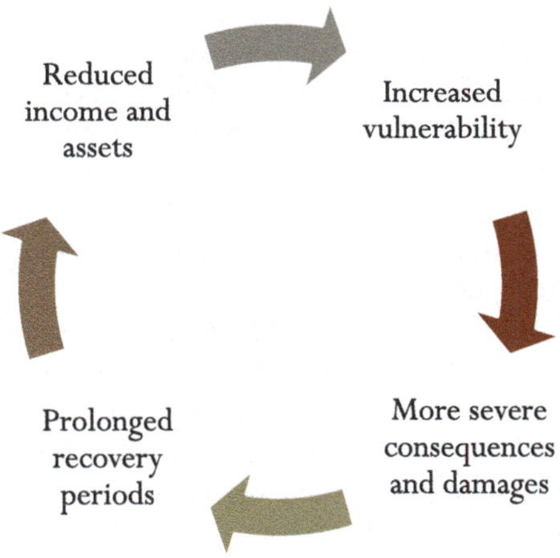

Management of risks due to such events requires flexibility of systems to be able to build resilience to the various levels of their impacts. Policy makers need to act in a forward looking, well coordinated and integrated manners. This is however not an easy task for governments and aid actors in low income societies and countries.

It is also important to strengthen the institutional capacity and improve coordination of different agencies and levels of government. Another important way to make adaptation efforts effective is to choose measures which the highest long and short term impact.

3 Different Actors' Perception of Climate Change and Related Concepts in Afar, Ethiopia

Climate change, vulnerability and adaptation can be differently understood by different actors. There are different humanitarian and development interventions with an attempt to reduce the livelihood effects of climate change. Humanitarian actors in particular provide support for communities during times of disaster. Emergency food, shelter and medical support are some of the most common humanitarian interventions.

With ever growing climate change related problems, there is a need for alterations in the way humanitarian aid is provided. Mainstream ways of humanitarian aid which focus on addressing immediate effects of climate change have now to shift towards supporting long term adaptation. The question however is if current ways of intervention particularly in Ethiopia are working towards achieving this goal-the goal of long term adaptation.

How do different actors perceive climate change and related concepts such as climate change vulnerability, adaptation and aid? (Map 1).

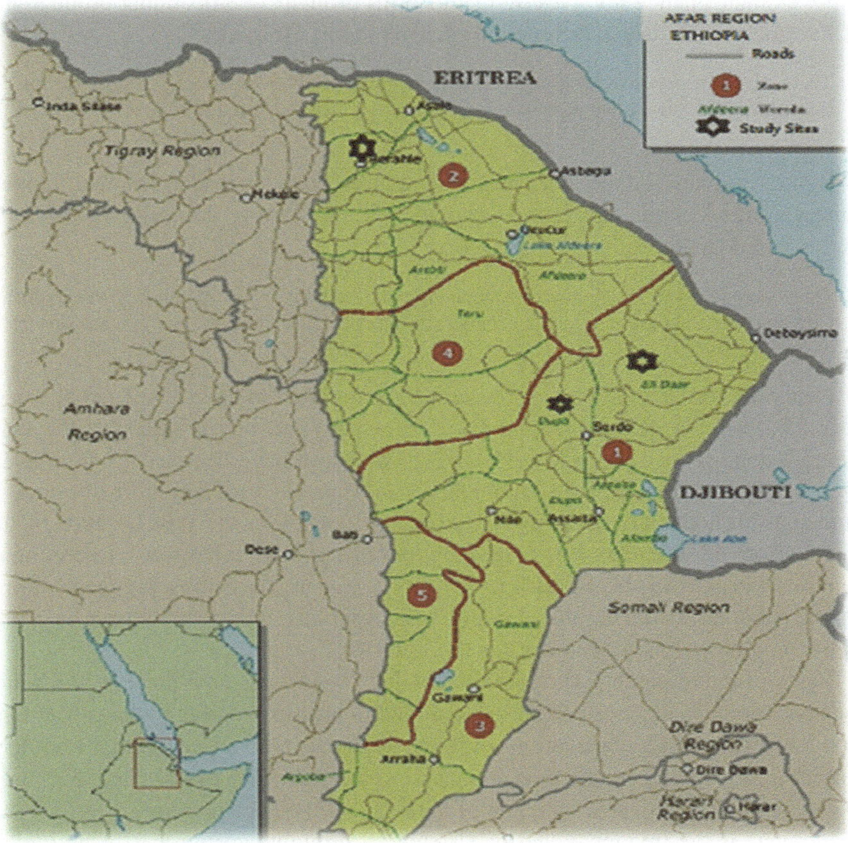

Map 1 Study area (Afar region): study sites are marked with a *black star* sign

4 Local Actors' Perception of Climate Change

Climate change according to many interviewees and discussion participants is real in Afar. They believed that climate is alarmingly changing and mentioned several observable indicators of change. Raising temperature, declining rainfall, frequent storms, declining forest cover and transformation of the tree/plant species have been identified as the major indicators of a changing climate. Decline of livestock productivity, different health problems, and malnutrition are some of the consequences of climate change.

According to one of the local interviewees, there are two types of winds in the area. One of the winds brings rain with good productivity and the other reduces productivity of the forests and crops. These winds have always been there in balance throughout history. The difference is, now there is more of the later type of wind. Climate change involves frequent storms drought, and declining productivity of livestock which cause destruction of livelihoods. He said;

> There are two types of winds. One of them reduces productivity by limiting growth–storms. Some plants/forests are not producing/ giving fruits which they used to- due to this type of wind.

Similarly, many of the local respondents used changes in the seasonal temperature variations as indicators of a changing climate. One of the respondents in Eli De'ar woreda said;

> There were seasonal variations in temperature in the past. Six cooler months and six warm. Now the six cooler months are also getting warm which affects the productivity and health of our livestock.

Generally, the following changes have been reported to have happened in recent years. Forests are declining, rainfall is decreasing and this is affecting the temperature conditions, livestock productivity/ownership have declined from an average of 200 to current high 3–4 animals per households, production of livestock is limited to in-house and in-farm instead of pastures as in the past because of the shortage or/and absence of pasture, abortion/miscarriage in livestock is becoming common because of the unseasonal fertility (i.e. animals getting pregnant during harsher seasons and facing miscarriage), even if it rains, it does not change anything as the area has already been dry for the last 10–15 years.

Many of the research participants stated that 10–15 years ago the area had better forest coverage, and grass variety has changed to grasses inedible by livestock. One of the most common observation made by the local interviewees is that rain does not make any difference this time compared to the past when grasses and trees used to regenerate fast with small amount of rain. One of the explanations given for these phenomena is the frequency and persistency of drought causing loss of biological regenerating capability of plants. In the past, drought used to happen every 5–10 years and for a shorter time causing less destruction.

Tree species which used to give nutritious fruits for people are getting extinct or their productivity is declining faster than ever. According to locals, these trees are being substituted by trees of no importance for human and livestock use. The invasion of large amount of lands in central and southern Afar by "*Prosopis-juliflora*" is one of such instances.

5 Climate Change and Livelihoods

Climate change has multitude effect on the lives of communities in Afar. First, communities historically used to depend on livestock production as a livelihood base. This sector is currently suffering from serious shortage of water due to frequent droughts. Declining pasture, decreasing livestock size, shortage of food and lower household income are the outcomes of climate change. Changes in rainfall

Fig. 1 Salt traders use camels as transportation in Berahle area (*Photo* Teklehaymanot G. 2015)

and temperature patterns are affecting pastoralist lives. Second, in areas such as Berahle, most men used to engage in salt trade using camels as means of transportation. This livelihood activity is declining and only few can go into salt trade now because they do not have camels to carry salt out of the salt mines (Fig. 1).

Pastoralist livelihoods are also changing towards a more sedentary life based on external aid, remittance from international migration, retail trade, and other non-pastoralist employments. Communities which used to be dependent on pastoralism and who claim to have never expected any external supports are now almost completely dependent on government food aid. They are left with very limited traditional pastoralist assets to sustain livelihoods. Some households with better human capital (i.e. households with young and working family members) migrate elsewhere to earn income. Many these young people migrate to the Arab states and cities within and outside the region for non-pastoralist employments.

6 Vulnerability

6.1 Who Are the Most Vulnerable?

Rentschler (2013) stated that poor sections of societies are disproportionately impacted by climate change disasters.

Based on the reflections of local people, though all members of the community suffered in the worst drought prone times, the elderly, children and women are considered as the most vulnerable sections of communities for several major reasons. First, while the young and men have the opportunity to move to better of areas in drought times and travel elsewhere and work for non-pastoralist employments, these three sections of the society are incapable of moving to other places; the elderly are weak and unhealthy, the children are dependent on their mother and the women/mothers are responsible to take care of the children and elderly. So, since they cannot move, they suffer the most when worst conditions happen. Second, the elderly and children are weak and prone to the epidemic diseases and weather changes which happen most in these times. Third, especially on the part of the elderly is the changes in their dietary conditions which are new and further exacerbate their health conditions in general.

In addition to the aforementioned points, women are also responsible for most of the pastoralist works. Hence, declining pastoral livelihood means reduced options for women to secure income for their households. One of such major challenges in Afar is the growing shortage of water. Climate change significantly affects distance of water points from home, and as distance to water access points become longer and longer with increasing frequency of droughts, it is getting harder and harder for women who are responsible for collecting water and child care. In many places, it takes women almost a whole day to bring water home, and when they arrive there are tasks such as child care and whole other domestic duties waiting for them. According to some unpublished researches, Afar women work for 18 h of a day.[1] Women take most responsibilities in a household including activities such as collecting water, livestock herding, child care, and all of the other domestic/home tasks. Men are responsible for collection of fuel wood and they share child care with the women.

According to Mr. Mekonen, Afar Disaster Resilience and Sustainable Livelihood Program (DRSLP) coordinator, there is a saying in the region that,

> Every activity in Afar is women's activity, so if any developmental work is done, it absolutely helps the women.

Furthermore, because of the culture which demands that they stay at home, women have no options for employment other than pastoralism which is hard hit by climate change. With exacerbating effect, climate change by reducing livestock productivity which used to be the source of food, more affects the nutrition of women and children compared to men who have the freedom to go elsewhere. Of course, the worst of all are the women who are bread winners of their family, though they share similar responsibilities like the other women, they must also fill the contributions of husbands other women have. Women only wait for what the men bring them. Ahmed Sule, the administrator of Eli'De'ar Woreda believe that the situation is getting worse for women. He said;

[1]Interview with Mr. Mekonen, coordinator of the Afar DRSLP (Disaster Reduction and Sustainable Livelihood program).

Fig. 2 Women are responsible for collecting water travelling long distances in Afar (*Photo* Teklehaymanot G.)

> In the past, despite many problems, women used to get enough nutrition and better income which they could not at this time because of climate change.

However, albeit these main challenges, some respondents believe that climate change and social dynamism related to it have also brought some advantages for women and children. Women traditionally were expected to stay at home and cook for men. Because of the changes, men could not manage to earn sufficient income for their households. As a result women are getting chances to be involved in more outdoor economic activities thereby getting more freedom. Similarly, there is a general tendency among households to send their children to school which was not possible in the past. Children, especially girls, in the past were involved in helping their parents in pastoralist activities (Figs. 2 and 3).

7 Effectiveness of Current Aid Strategies

There are different views regarding the effectiveness of current aid strategies in Afar region. According to our observation and official statistical evidences, while there are several aid agencies in Afar their distribution across woredas/districts is uneven. Most aid projects are spatially concentrated. There are very few programs which are widely distributed throughout the region. Productive Safety Net Program (PSNP) is one of such major aid programs operating across all Woredas in Afar.

Fig. 3 Children at school in Kora 'Kebelle', Berahle Woreda (*Photo* by Teklehaymanot G.)

PSNP provides two forms of aid targeting different sections of communities. The first form is the public work program which targets people who are poorest of the poor but are capable of working. People under this type are involved in a form of food-for-work program in which they contribute labor and earn food rations in return. On the other hand, PSNP provides free food aid (relief) to people who are poor and unable to work because of special reasons such as age (children and elderly), disability, and gender (pregnant women).

There are also aid agencies and government programs which provide emergency and development support. Emergency aid in Afar in most cases focus on providing food, medicine and other support with the aim of addressing immediate needs during times of disaster. It is given to people who face acute problems during disasters. These programs mostly supply supplementary foods for children with severe malnutrition in times of emergency such as drought and flood disasters.

Other major problems among humanitarian and development agencies include lack of coordination and lack of participation and consultation of local people in the works of the different organizations. According to the representative of Afar regional Bureau of Finance and Economic Development, one of the major challenges is the lack of coordination among humanitarian and development actors despite the fact that there is good NGO presence in Afar compared to other regions. The impact of their works is however low because they tend to concentrate in few areas doing duplicated activities. Awash 7, which is the closest city to Addis Ababa has 38 NGOs providing aid to communities while there are woredas such as Berahle and Eli De'ar with almost no NGO. They are mostly concentrated in areas close to Addis Ababa or urban areas in the region such as Samara (the regional capital) and Dubti.

Currently, there are efforts to coordinate and monitor humanitarian and development agencies to bring an organized support for sustainable improvement of livelihood of pastoralists in the region. This initiative is led by the Unit for Coordination of Charities and Societies under the regional Bureau of Finance and Economic Development. The following box is an account of a meeting that the research team was invited to attend in which the Unit for Coordination of Charities and Societies was trying to push a new charity which wanted to start working in Awash 7 city to other areas with lower NGO presence.

Box 1: Account of a meeting the research team attended during field visit to Samara (Afar region's capital)

Two NGOs with similar activities wanted to work in Awash woreda particularly Awash 7 town: the coordination unit under the Bureau of Finance and Economic Development (BoFED) believed that the project is a duplication of activities which has been done by another charity which is working on children. Both projects despite being planned by two different organizations have the same activities and are planned to be implemented in one (the same area) place. The unit in BoFED suggested that the new aid agency should move to other places/areas with no such aid.

In addition to technical comments to the proposal, BoFED encouraged the organizations' to fill what the government wants to but couldn't. According, to BoFED there are 38 NGOs operating in Awash Woreda and they don't want to move even to other accessible woredas along the Samara road, albeit the peripheral woredas that demand such support at most. They are all concentrated in areas close to Addis Ababa. Of these 38 NGOs, 8 work on children related activities. The two currently contending Charity NGOs also want to start operating in the same area which is unfair given the fact that there are many areas where children do not have such opportunities while aid resources are being misused in some areas. Ali Mohammed, the head of the coordination unit said:

"What we desire is a fair distribution of resources. We do not want some people to suffer while others have more resources than what they basically need." "ሃብት የሚጣፍጠው ለታለመለት ዓላማ ሲውል ብቻ ነው" [Wealth is meaningful only when it serves its intended purpose].

Representative of the NGO said: we have very limited budget and we can logistically not be able to work in far areas, we want to use stations/churches close to Awash 7 Kilo, if you let us start with the small budget in a closer area we want to work, otherwise there is no way we can continue. I have no option than to move away with my money feeling sad that I couldn't help.

Ali Mohammed: we don't want to add more NGOs in places where we see duplication of projects. We can't violet government laws to obey donor interests and put all NGO activities in small areas and let the other parts of Afar stay the way they are. The government's interest is to have NGO activities distributed evenly. As a solution, I suppose we can write support letter to your donors explaining why we wanted to shift your area of operation. We don't want to reject incoming resources/wealth but we want it to be put in the right places for the right beneficiaries. Finally, the representative of the NGO asked for the support letter to be sent to his donor explaining the situation.

According to the above account, the regional government/BoFED—Unit for Charities and Societies Coordination is in a difficult position to decide about coordinating distribution of aid agencies across the region. What makes it more difficult is the fact that the Disaster Prevention Bureau (i.e. previously responsible for the task of NGO coordination) had no clear documentation of which NGOs are working on different issues and areas. Aid agencies mostly follow donor interests when deciding where and what to work on and not local needs and community/government expectations. A 2015 report by IGAD (Inter-governmental Authority on Development) also showed that coordination among projects funded by different donors is a critical problem in the IGAD region which includes Ethiopia, Djibouti, Kenya, Uganda, Somalia and Sudan.

The researchers observed striking differences in aid distribution among Woredas during field works for this study. Such difference is visible between Berahle and Woredas around Samara. While there are more aid options in Samara area, there are very limited aid programs if any other than PSNP in areas such as Berahle and Eli De'ar. However, people in and around Samara seem to have more option in terms of livelihood diversification towards irrigated agriculture and trade than in Berahle which has none other than an almost non-existent pastoralism. Contrary to this fact, (i.e. Berahle despite being dry and with fewer resources to diversify livelihood into) gets only PSNP and very limited other emergency aid schemes.

Current systems of aid delivery are generally considered ineffective by locals in addressing long-term adaptation needs for many reasons. First, the amount of aid provided is quite small. The impact of climate change on the livelihoods of communities is immense and solving the problem requires more resources. This is particularly true for PSNP which combines the aims of addressing immediate needs and long term adaptation but fails to do so because of its limited budget. PSNP aid covers six months of a year and beneficiaries have no way to sustain lives the rest of the year. It is thus difficult to imagine adaptation considering the level of support provided through the PSNP program and the magnitude of changes in the livelihood situation as a result of climate change.

Second, public work activities under such aid programs do not consider the biophysical features of different areas. PSNP program focuses on rehabilitating degraded areas using public work as one of its major aims. Many of our local interviewees think that there is no point in building soil and water conservation structures where there is no rain. Some community members admitted that they participate in public work programs not because they believe in the intended purposes but to earn food to address their immediate household needs.

8 Conception of Adaptation to Climate Change

There have been different views about climate change adaptation in Afar among different actors. Some argued that adaptation actions should bring slower changes not to affect the cultural and natural on which the lives of the Afar have been built. On the other hand many members of pastoral communities believed that there is no way pastoral lives could be rebuilt and argued for new strategies to be incorporated. There is a wide-ranging predisposition among local people to underestimate their capability to plan their own adaptation pathways. One of their major reasons is the lack of choice due to the collapse of their traditional livelihood system (i.e. pastoralism) and lack of knowledge to find new strategy under a changing circumstance. They feel like they do not know what to do in the new context created by climate change after the collapse of the pastoralist livelihood. According to participants of this study, there is a need for external help be it from the government or any actors in identifying appropriate pathways and leading the people through the transition process. One of our respondents said "*we are only hoping that the government will find us a way forward. We have no idea about what we can do about this [climate change] problem.*"

This complete dependence on the government and external agencies possibly emanates from four major reasons. First, local (indigenous) knowledge of communities is based on pastoralist life style and people have no idea about what to do under emerging circumstances that they never experienced before. Second, the aid system so far has followed a top-down approach. Despite PSNP program's attempt to change this by involving community members in decision making, local participation is still quite restricted resulting in a bigger dependency on decisions made by external actors.

Similar arguments have also been made by some government and non-government actors interviewed for this study.

> We cannot transform to agriculture overnight as our people do not know anything other than pastoralism. You cannot just think and then do it. The people should decide what they want to do.

Third, the Afars are predominantly Muslim and they have a tradition of externalizing everything be it good or bad to their God. When it comes to climate change, they think it is brought by God and they could not do anything about it.

Third, local community members stated that the projects which could solve their problems are capital intensive but PSNP and other aid programs support labor intensive projects and people have no idea about how they can build sustainable living with such projects. Local people tend to conclude that what are needed in the area are not labor intensive projects but rather capital and technology intensive. Hence, they have no idea about what they can do by themselves with limited outside support.

However, many of there search participants have also put forward measures to promote adaptation to climate change. One of the suggestions is building of water infrastructures to increase access. The major problem in Afar is water and solving this could significantly reduce the challenges in many ways. First, access to safe drinking water for people could improve health and in so doing increasing productivity in the society. It can also reduce women's burden and enable them to engage in other more productive activities. Second, there are some areas with great potential for agriculture. Building water infrastructure in such areas can make agriculture an important alternative pathway. And finally, one of the major causes of the decline in livestock production is the lack of access to water for livestock. Providing access to clean water for the livestock could improve livelihoods and facilitate adaptation. Local people suggested that, aid instead of making people build SWC (Soil and Water Conservation) structures everywhere, should provide support towards building irrigation schemes and water harvesting techniques to be able to start irrigated agriculture.

Monetary assistance to engage in trade and production of commercial products was also one of the possible support mechanisms put forward by the locals. Establishing small and micro-scale enterprises and supporting them to involve in livestock production, bee farming and similar other business can be very effective. One of the research participants said; "*if they* [*the government and aid agencies*] *give us animals they* [*animals*] *will die…if they give us money we could start business such as coffee houses* [*shops*] *and retail trade and make profit to payback the debt.*" This suggests that there is a greater need to move out of traditional livelihood strategies. In relation to this, some suggested for flexibility of aid delivery. In one of our group discussions the respondents agreed that everything could not be solved by simply giving out money or livestock. People who want to continue their pastoral livelihoods should be supported to do so and people who want to shift their livelihoods away from pastoralism should be provided with aid that supports their desired pathways. 'One size fits all' strategy by PSNP program lacks the flexibility to specific conditions of aid receiving communities.

Some interviewees also suggested distribution of livestock as good way to restart/recover pastoralist livelihood. Such suggestions were brought by people in areas where there still is better potential for pastoralism. However, there have been critical arguments against it. Some argued that even if the government gives them livestock to restart pastoralism it would not be a sustainable solution as the animals will die shortly because of frequent drought. The patterns of climate change disasters according to the second group tend to be more frequent and people have less capability to recover from each disaster before the next one. Droughts are recently coming every second year and pastoralists have little capacity and very short time to recover before every disaster.

Strikingly, despite so many international critics against the government of Ethiopia, resettlement to more fertile irrigable areas is seen as an alternative pathway by many of the local respondents. They opted for resettlement to urban centers and areas of potential for irrigation within and outside the region.

9 Aid Agencies' Views of Climate Change

9.1 How Is Climate Change Seen by Different Aid Actors?

There are different views regarding climate change depending on how it is perceived by different actors. External actors (i.e. governmental and non-governmental aid agencies) defined climate change in their activity areas in Afar using different indicators such as rainfall and temperature variability, forest degradation and reduced productivity of livestock.

Climate change, according to the Productive Safety Net Program (PSNP)'s Berahle representative, is the changes in rainfall and temperature patterns. Raising temperature, declining amount of annual rainfall, and unpredictable storms are characteristic features of a changing climate. Other aid agencies also see climate change in terms of variations in temperature and frequency of droughts.

According to some of the NGO representatives interviewed, adaptation is seen as changes in housing and other structures to be able to make adjustments to changing temperatures and rainfall patterns. Representative of African Humanitarian Action (AHA) which is a regional humanitarian agency working on an Eritrean refugee camp suggested changing housing materials from traditional to use of modern and efficient ones to reduce deforestation and protect extreme temperatures.

PSNP representatives in different levels also believed that it is not possible to bring adaptation with current strategies employed by external humanitarian and development agencies. PSNP's Berhale representative said;

> Since PSNP is the only visible project that works on climate adaptation related activities, it is not possible to achieve the goal of adaptation with only PSNP efforts. It is important to design and implement other climate smart initiatives/projects to be able to meet the goal of climate change adaptation and resilience.

He argued that PSNP despite its importance in relieving current food shortages, its budget as well as the proportion of the targeted population are far below what is expected by the people to solve their existing problems, albeit bringing climate smart communities with adaptive and resilient capacities. On the one hand, due to shortage of funds, the program does not cover all members of the community, though almost all are poor and vulnerable to the problems of climate change and related hurdle.

Second, though the project aims at public works that intend towards long-term climate change adaptation efforts such as terrace building, water shade management etc., due to the small amount of fund appropriated for each locality, the public works could not bring the desirable results both through the eye of the community and the project. The program's activities could not bring sustainable improvements of livelihoods given the magnitude of the problem and the extent of support provided. Public work programs in PSNP demand that 80% of the resources needed to build the public works should be a labor contribution made by the community members and this significantly limits the selection/design of activities to small scale structures with a very low long-term impacts on adaptation.

One of the senior experts we interviewed in the regional PSNP office said:

> PSNP public work intervention programs are labor intensive, but the need in the region is for capital intensive projects. Although public works programs are planned by communities, the plans are constrained by the PSNP's goals of public works based on 80% of community labor and not their actual and future needs.

Community needs are much more than what PSNP provides. Additional development works are needed to build long term adaptive capacities of communities. Since water is the major problem, PSNP public work programs focus on addressing this problem. However, the activities are limited to community labor and digging underground water using human labor is difficult. One of the Kebelle leaders interviewed for this study said;

> …water is a big problem. It couldn't be found by digging using only human labor. To find water, we need more technological support.

The 'community committees' who are supposed to consult and select public works according to the priorities of the community tend to emphasize on the above mentioned short-term tasks than long-term based large projects mainly because of the limitation of adequate funds to run such huge projects. One member of these committees forwarded that:

> In all our consultative meetings with the community, the priorities of the people are always the same every year - to build big water harvesting projects, building big dams in nearby rivers or digging deep wells - so as to get rid of their number one problem, acute shortage of water. But, due to the limitations of fund, we tend to select the practical public works.

Furthermore, there is no community based planning among NGOs in Afar. Even though, their motto is 'working with the communities' their plans are imposed from top and this is a big problem. Donors influence most of the decision about the type of aid and where it should be provided (PCU 2015).

10 Conclusion and Policy Implications

10.1 Conclusion

The aim of this study was to look into how climate change, vulnerability and adaptation are seen from the perspective of local actors and how humanitarian intervention responses addressed local concerns during decision making about climate change adaptation. Analysis of the views of local community members shows that climate change is indeed a reality with many indicators. Households are facing everyday challenges as a result of the climatic changes. Livelihoods have changed from a once entirely reliant on pastoralism to a currently significantly dependent on external aid. Children, elderly and women are seen as the most vulnerable sections of communities for reasons related to mobility and assets ownership.

There are several governmental and non-governmental aid programs addressing immediate humanitarian and long term adaptation needs. Despite the numbers of intervention programs, aid system so far has been criticized by local people because of its limited scope and focus on solving immediate challenges than supporting long term adaptation. PSNP aid despite good intentions to support long term adaptation ended up doing less because of the limited nature of its focus and resources.

Generally, while what is needed in Afar is transformation, current aid is merely focused on shock responses aiming at reducing the consequences of immediate disasters. It seems it is not possible to bring long-term adaptation or build resilience with the current aid system. Households in Afar could no longer mostly recover to their pre-disaster level of income before every next disaster and they find themselves in the vicious cycle of a poverty trap. They have low coping capacities, face more severe consequences, need longer recovery periods and persistently reduced incomes which perpetuate spiral poverty in the future.

However, despite a detailed analysis of local level perceptions of the Afars we have made, this paper would have been richer by including perspectives of different policy makers other than those at local level. Further study of the views of policy makers about the issues studied here would be of great help in understanding the differences in the perspectives between local and external actors. Such analysis can bring us a clearer picture of the actors in play in the policy making process regarding adaptation to climate change. This can enable us identify the links between perception and the role of power in policy making.

10.2 Policy Recommendations

If aid agents want people to adapt to climate change, they have to work towards enabling them to make structural transformation of their socio-ecological systems. First, people in a condition such as Afar should be provided with resources beyond satisfying their basic necessities. Local communities need bigger technological support to build resilient livelihoods. Hence, providing basic food aid to people in a context like Afar and dreaming big changes is a daydream given the dramatic changes in climate and alarmingly deteriorating livelihoods. No matter how cleaver local people are in terms of their knowledge of adaptation to climate change, they cannot make a living out of an almost non-existent livelihood base. Bigger changes are needed to transform livelihoods. Suggestions made by locals such as resettlements, irrigation based farming and bigger financial and capacity building programs could be of great help.

Second, policy makers need to take into account small but frequent disasters as well as slow onset changes with irreversible effects during decision making processes because the cumulative impact of such changes could be as severe as these from major disasters but get less attention. Attention of policy makers and aid actors are mostly directed towards immediate and large-scale disasters limiting their interventions to only emergency responses. Such considerations enable policy makers design strategic measures with highest long-term adaptation impact.

Third, policy makers and intervention actors should devise mechanisms to support existing local dynamics. Migration both local and international, branching out to agriculture and trade and other activities are some of the strategies followed by local Afars in their endeavor to adapt to climate change. Aid programs should thus support such dynamism as part of their efforts to bring transformation. Finally, coordination of aid actors is needed to channel aid towards more effective strategies of addressing long-term adaptation concerns through measures with the highest possible impacts (Figs. 4 and 5).

Fig. 4 Camels are used as major transportation in the Salt trade

Fig. 5 Goats used to be the mainstay of households

References

Adger, W. N., Huq, S., Brown, K., Conway, D., & Hulmea, M. (2003). Adaptation to climate change in the developing world. *Progress in Development Studies, 3*(3), 179–195.

Eriksen, S., Aldunce, P., Bahiniparti, C. S., Marins, R. D. A., Molefe, J. I., Nhemachena, C., et al. (2011). When not every response to climate change is a good one: Identifying principles for sustainable adaptation. *Climate and Development, 3*(2011), 7–20.

IPCC. (2007). *Climate change 2007: Working group II: Impacts, adaptation and vulnerability*.

Maani, K. (2013). *Decision-making for climate change adaptation: A systems thinking approach* (p. 67). Gold Coast: National Climate Change Adaptation Research Facility.

Mirza, M. M. Q. (2003). Climate change and extreme weather events: can developing countries adapt? *Climate Policy, 3*(3), 233–248. doi:10.1016/S1469-3062(03)00052-4

PCU. (2015). *IDDRSI programming report 4th IDDRSI platform steering committee meeting*. Addis Ababa, Ethiopia, March 25–26, 2015. Addis Ababa.

Pelling, M. (2011). *Adaptation to climate change: From resilience to transformation*. London: Routledge.

Rentschler, J. E. (2013). *Why resilience matters—The poverty impacts of disasters*. The World Bank: Washington, DC.

Smit, B., & Wandel, J. (2006). Adaptation, adaptive capacity and vulnerability. *Global Environmental Change, 16*(3), 282–292. doi:10.1016/j.gloenvcha.2006.03.008

Tanner, T., Lewis, D., Wrathall, D., Bronen, R., Cradock-Henry, N., Huq, S., et al. (2015). Livelihood resilience in the face of climate change. *Nature Climate Change, 5*(1), 23–26. doi:10.1038/nclimate2431

Wise, R. M., Fazey, I., Stafford Smith, M., Park, S. E., Eakin, H. C., Archer Van Garderen, E. R. M., et al. (2014). Reconceptualising adaptation to climate change as part of pathways of change and response. *Global Environmental Change, 28*, 325–336. doi:10.1016/j.gloenvcha.2013.12.002

Uptake of Resilient Crop Interventions to Manage Risks Through Climate-Smart Villages Approach in Nyando, Western Kenya

John W. Recha, Maren Radeny, James Kinyangi and Philip Kimeli

1 Introduction

The eastern Africa region has been described as a major "poverty and hunger hotspot" (Sanchez et al. 2005). Attaining food security and sustainable agriculture remain a daunting challenge for smallholder farmers (Rarieya and Fortun 2010), especially in light of climate change. East Africa's climate is warmer than it was a century ago and models project that this warming will continue (Christensen et al. 2007). In Kenya, climate change has affected precipitation and contributed to the high frequency of extreme events such as droughts and floods (Kiunga 2015). These in turn affect farming systems through changes in length of the rainy and growing seasons, soil erosion, increased pests and diseases (Rarieya 2007).

The Nyando basin around Lake Victoria in western Kenya is one of the highest populated rural localities in east Africa (Kinyangi et al. 2015) with a population density exceeding 400 persons per square kilometer. The basin is rich in agricultural resources, such as rainfall and land for production of cereals, legumes, root crops, and vegetables, as well as livestock. About 40% of the landscape is degraded (Macoloo et al. 2013), and about 81% of the families experience 1–2 hunger months in a year. Meanwhile 17% of the families experience 3–4 hunger months per year, whereby they are unable to produce food from their own farm (Mango et al. 2011). The main source of livelihood is mixed farming involving keeping livestock and growing crops. However, crop productivity levels are low, for example the average maize yields are 100 kg/ha (Ndwiga et al. 2013).

Sustainably increasing agricultural productivity in Nyando is central to the future of food security, and there is need for strategies that enhance climate-smart agriculture as the first step for sustainable agriculture (Lipper et al. 2015). To address

J.W. Recha (✉) · M. Radeny · J. Kinyangi · P. Kimeli
International Livestock Research Institute (ILRI),
P.O. Box 30709–00100, Nairobi, Kenya
e-mail: j.recha@cgiar.org

W. Leal Filho et al. (eds.), *Climate Change Adaptation in Africa*,
Climate Change Management, DOI 10.1007/978-3-319-49520-0_32

this challenge, the Consultative Group for International Agricultural Research (CGIAR) Research Program on Climate Change, Agriculture and Food Security (CCAFS), in collaboration with Kenyan research and development organizations, partnered with rural communities to develop Climate-Smart Villages as models of local actions that ensure food security, promote adaptation and build resilience to climatic stresses (Aggarwal et al. 2013). Researchers, local partners, farmers' groups and policy makers collaborate to select the most appropriate technological and institutional interventions based on global knowledge and local conditions to enhance productivity, increase incomes, achieve climate resilience and enable climate mitigation. Climate information is an important component of Climate-Smart Villages and consideration is always given to integrating village developmental and adaptation plans together with local knowledge and institutions into the project. There is no fixed package of interventions or a one-size-fits-all approach. The emphasis is on tailoring a portfolio of interventions in different research sites that complement one another and that suit the local conditions (Aggarwal et al. 2013).

There are very few studies that show a relationship between changes in agricultural practices that respond to climate related risks at the household level (Kristjanson et al. 2012). It is important to establish if households that are changing their farming practices to address the risks related to changing climate can have higher agricultural productivity. The study explores the changes in farming practices made by households which are arising from the partnership. These can be useful in generating evidence for policy makers to decide the types of investments that could be promoted for helping farming households deal with a changing climate.

2 Methods

2.1 Study Area

The study was done in the Nyando river basin of Western Kenya. The site, measuring 10 by 10 km, was identified by CCAFS based on climate related risks like changing rainfall patterns, high frequency of drought, and flood events. Its temperature is relatively stable during the year, with an average annual temperature of 25 °C. It has an annual rainfall of about 1400 mm, with a bimodal rainfall pattern consisting of a long rainy season in March to May and short rainy season in September to November (Tobella 2009; Verchot et al. 2007). The changing climate has led to erratic rainfall that is characterized by late onset and early cessation, with more frequent extreme events (Macoloo et al. 2013). The driest months are January and February (Hoshino et al. 2006). CCAFS is developing Climate-Smart Villages model in Nyando through a partnership that is facilitating the testing of a portfolio of climate-smart agriculture interventions, allowing farming households to make progressive changes to crops and cropping patterns. Figure 1 shows the site in western Kenya. The red numbers show the villages with sampled households.

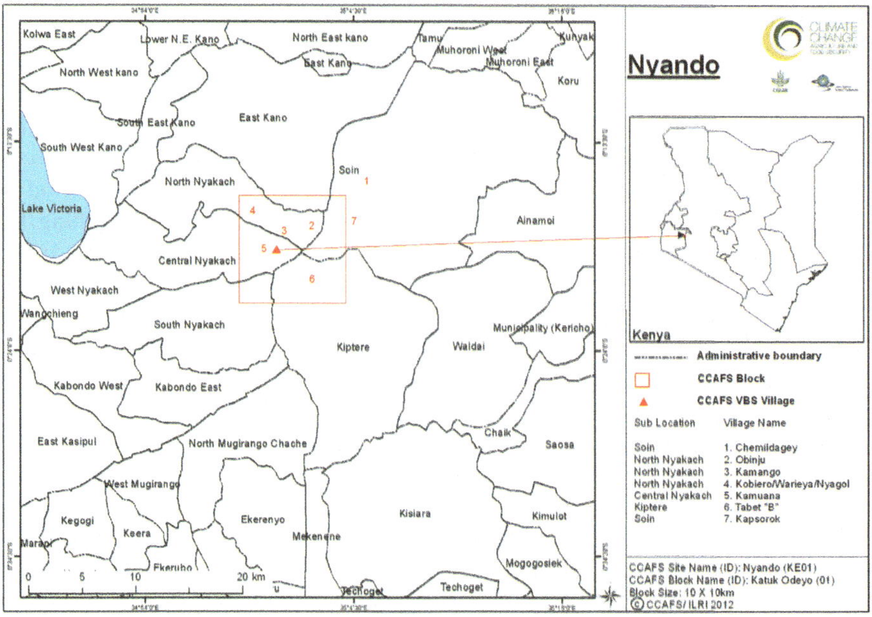

Fig. 1 The study site. *Source* Mango et al. (2011)

2.2 *Data Collection and Analysis*

The study made use of a household-level baseline survey designed by the CCAFS team that was implemented in early 2011 (Mango et al. 2011). One of the objectives of this survey was to develop simple, comparable household level indicators, for which changes can be evaluated with time of food security, household assets, diversity in on-farm agricultural production, adaptation, and farming practices (Kristjanson et al. 2012). The household-level baseline survey involved 139 randomly selected farmers from seven villages within the 10 by 10 km area in the year 2011 (Mango et al. 2011). A portfolio of crop related interventions that include the use of terraces for soil and water conservation, use of intercropping method, use of improved crop varieties, and the planting of new crops were introduced thereafter. Annual monitoring and data collection was done in the seven villages targeting the previous year's households and additional ones for four consecutive years (2012–2015) using questionnaires. The number of farmers was 196 in the year 2012, 314 in the year 2013, 370 in the year 2014, and 364 in the year 2015. MS-Excel was used to generate graphs, and data analysis was done using R-package (The Comprehensive R Archive Network; https://cran.r-project.org/). Confidence intervals and estimated proportions were calculated using *proptest* package in R-package (Dalgaard 2002).

3 Results

The surveys were done to assess changes in farming practices made by households in Nyando. The soil and water conservation measures introduced include establishment of water-harvesting pans, ploughing across contours, and use of terraces. Use of terraces was the main method taken up by farmers during the period of study. Terraces specifically reduce the velocity of water runoff by breaking the length of the slope that runoff has available. There was a gradual increase in the proportion of households using the terraces from 0.132 to 20.264 within the five year period. As shown in Table 1, the proportional increase between years 2011–2012, 2012–2013, 2013–2014, and 2014–2015 were not significant (Alpha = 0.05). The significant difference (Alpha = 0.05) was between the years 2011 and 2015, as well as 2012 and 2015 (Table 1).

The intercropping method was increasingly used by farmers over the five year period. The various crops intercropped include maize, beans, cowpeas, green grams, sorghum, indigenous vegetables like *Crotolaria* species, and sweet potatoes. Some of the crops were seeded at the same time (mixed intercropping), while others were seeded at different times (relay intercropping). There was no significant increase (Alpha = 0.05) in households doing intercropping between the years 2011 and 2012 (Table 2). However, this increase was significant (Alpha = 0.05) between 2012 and subsequent years (Table 2).

Table 1 Use of terraces for soil and water conservation

Use of terraces				
Year	N	Proportion	Lower 95% CI	Upper 95% CI
2011	139	0.132^{a}	0.0757	0.1883
2012	196	0.1429^{a}	0.0939	0.1919
2013	314	0.2006^{ab}	0.1563	0.2449
2014	370	0.2297^{ab}	0.1868	0.2726
2015	364	0.264^{b}	0.2187	0.3093

Proportions with different letters are significantly different (Alpha = 0.05)

Table 2 Use of intercropping method

Use of intercropping				
Year	N	Proportion	Lower 95% CI	Upper 95% CI
2011	139	0.341^{a}	0.2622	0.4198
2012	196	0.4081^{a}	0.3393	0.4769
2013	314	0.5669^{b}	0.5121	0.6217
2014	370	0.6216^{b}	0.5722	0.6710
2015	364	0.64^{b}	0.5917	0.6903

Proportions with different letters are significantly different (Alpha = 0.05)

In addition, other changes that farmers made included the use of improved crop varieties such as hybrid maize, hybrid sorghum, beans, green grams, cowpeas, and sweet potatoes. The proportion of households using improved crop varieties increased gradually (Fig. 2), but this was not significant (Alpha = 0.05) for the years 2011, 2012 and 2013. However, there was a significant increase (Alpha = 0.05) from the first three years (2011–2013) to the year 2014 and 2015 (Fig. 2).

There was also a significant increase (Alpha = 0.05) in number of households adopting at least three more new crops between 2011 and 2012, and between 2013 and 2014 (Table 3). The crops include cowpeas, pigeon peas, groundnuts, green grams, cassava, sweet potatoes, sorghum, finger millet, bananas, butternuts, water melons, kales, cabbages, collards, onions, tomatoes, and indigenous vegetables. There was a significant drop (Alpha = 0.05) in number of farmers adopting new crop types in 2015 from the year 2014 (Table 3).

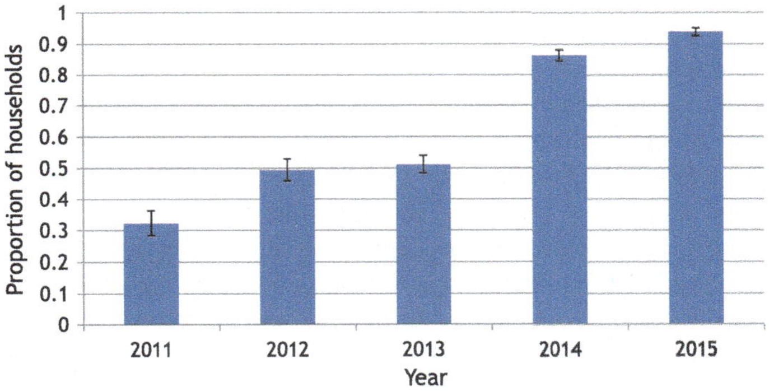

Fig. 2 Use of improved crop varieties (Alpha = 0.05)

Table 3 Households adopting three or more new crops

Adopting three or more crops				
Year	N	Proportion	Lower 95% CI	Upper 95% CI
2011	139	0.317[a]	0.2396	0.3944
2012	196	0.484[b]	0.4140	0.5540
2013	314	0.44[ab]	0.3851	0.4949
2014	370	0.899[c]	0.8683	0.9297
2015	364	0.23[a]	0.1868	0.2732

Proportions with different letters are significantly different (Alpha = 0.05)

4 Discussion

Climatic conditions, specifically precipitation and frequency of extreme rainfall events are some of the factors influencing the rate of soil erosion by water (Zhang et al. 2008). Various types of soil erosion including sheet, rill and gulley erosion may be triggered by rainfall (Askoy and Kavvas 2005; Valentin et al. 2005), leading to land degradation (Beskow et al. 2009). The Nyando farmers are increasingly using the terracing technique for collecting surface runoff water thus increasing infiltration and controlling water erosion, to help transform the landscape (Zuazo et al. 2005). The main types of terraces used by the Nyando farmers are stone-wall terraces and *Fanya-juu* terraces. Stone wall terraces have stone bunds deployed along the slope, where sediment deposition occurs to create the terrace (Nyssen et al. 2000). *Fanya-juu* terraces are made by digging the trenches along contour lines, and excavated soil is thrown uphill. The efficiency of terraces in limiting the soil erosion rate comes about due to reducing the volume and speed of rain surface runoff because the amount of lost soil is directly related to surface water flow (Zuazo et al. 2005).

The Nyando farmers have been incrementally practicing intercropping. This crop intervention spreads the risk of crop failure because the crops have different patterns of growth, and are affected by different pests and diseases. The food insecurity risk is reduced, and there could be a chance of getting higher yield from the two crops grown as an intercrop with sufficient rainfall (Li et al. 2003). The introduction of new crops and growing of improved varieties were done concurrently in Nyando. There was however a reduction in number of households planting new crops between the year 2014 and 2015. This reduction could be attributed to an already high number of households having embraced the new crop types in previous years, thereby exhausting the basket of options of the new crops available. The major driving forces for crop diversification in Nyando included balancing household food demand and increasing income. Specifically, the improved crop varieties had capabilities of tolerating water stress during dry periods, had faster growth rates, and could tolerate some pests and diseases (Mba et al. 2012).

5 Conclusions and Recommendations

To deal with climate related risks, Nyando households are diversifying crop choices. Early results show a shift to use of improved agronomic practices and high diversification. Households are now adopting more than three crops, greatly expanding on-farm choices for resilient varieties. However, there is need to have climate information services to guide farmers in decision making on crop types and varieties.

Acknowledgments This research was funded by the CGIAR Research program on Climate Change, Agriculture and Food Security (CCAFS). The authors appreciate the support of Kenya Agricultural and Livestock Organization (KALRO), Maseno University, Kisumu and Kericho County Departments of Agriculture, Livestock and Fisheries, ViAgroforestry and Community Based Organizations, and all local farmers.

References

Aggarwal, P., Zougmoré, R., & Kinyangi, J. (2013). Climate-smart villages: A community approach to sustainable agricultural development. Copenhagen, Denmark: CGIAR Research Program on Climate Change, Agriculture and Food Security (CCAFS). http://hdl.handle.net/10568/33322

Askoy, H., & Kavvas, M. L. (2005). A review of hillslope and watershed scale erosion and sediment transport models. *Catena, 64,* 247–271.

Beskow, S., Mello, C. R., Norton, L. D., Curi, N., Viola, M. R., & Avanzi, J. C. (2009). Soil erosion prediction in the Grande River Basin, Brazil using distributed modeling. *Catena, 79,* 49–59.

Christensen, J. H., Hewitson, B., Busuioc, A., Chen, A., Gao, X., Held, I., et al. (2007). Regional climate projections. In S. Solomon et al. (Eds.), Climate change 2007: The physical science basis. Contribution of Working Group 1 to the Fourth Assessment Report of the Intergovernmental Panel on Climate Change. Cambridge: Cambridge University Press.

Dalgaard, P. (2002). *Introductory statistics with R* (p. 267). New York: Springer.

Hoshino, M., Kadohira, M., Yamamoto, K., Yoshida, H., Sugitani, K., Katsurada, Y., et al. (2006). Soil erosion and conservation in Western Kenya. In *Research project report*. Nagoya, Japan: Nagoya University.

Kinyangi, J., Recha, J., Kimeli, P., & Atakos, V. (2015). Climate smart villages and the hope of food security in Kenya. CCAFS info note. CGIAR Research Program on Climate Change, Agriculture and Food Security (CCAFS). Copenhagen, Denmark. http://hdl.handle.net/10568/65144

Kiunga, S.K. (2015). An Assessment of the Phenomena of Climate Change in Kenya. *Asian Journal of Multidisciplinary Studies. 3*(8).

Kristjanson, P., Neufeldt, H., Gassner, A., Mango, J., Kyazza, F. B., Desta, S., et al. (2012). Are food insecure smallholder households making changes in their farming practices? *Evidence from East Africa. Food Security, 4*(3), 381–397. doi:10.1007/s12571-012-0194-z

Li, L., Zhang, F., Li, X., Christie, P., Sun, J., Yang, S., et al. (2003). Interspecific facilitation of nutrient uptake by intercropped maize and faba bean. *Nutrient Cycling in Agroecosystems, 65,* 61–71. doi:10.1023/A:1021885032241

Lipper, L., Thornton, P., Campbell, B. M., Baedeker, T., Braimoh, A., Bwalya, M., et al. (2015). Climate-smart agriculture for food security. *Nature Climate Change. 5.* doi:10.1038/NCLIMATE2437

Macoloo, C., Recha, J., Radeny, M., & Kinyangi, J. (2013). Empowering a local community to address climate risks and food insecurity in Lower Nyando, Kenya. Case Study prepared for Hunger Nutrition Climate Justice (2013). A New Dialogue: Putting People at the Heart of Global Development. Dublin, Ireland: Irish Aid. http://cgspace.cgiar.org/handle/10568/27889

Mango, J., Mideva, A., Osanya, W., & Odhiambo, A. (2011). Summary of baseline household survey results: Lower Nyando, Kenya. CGIAR Research Program on Climate Change, Agriculture and Food Security (CCAFS). Copenhagen, Denmark. http://cgspace.cgiar.org/handle/10568/27889

Mba, C., Guimaraes, E. P., & Ghosh, G. (2012). Re-orienting crop improvement for the changing climatic conditions of the 21st century. *Agriculture & Food Security, 1,* 7.

Ndwiga, J., Pittchar, J., Musyoka, P., Nyagol, D., Marechera, G., Omanya, G., et al. (2013). Integrated *Striga* management in Africa poject. constraints and opportunities of maize production in Western Kenya: A baseline assessment of *striga* extent, severity, and control technologies. Integrated *Striga* Management in Africa (ISMA). p. 34.

Nyssen, J., Poesen, J., Haile, M., Moeyersons, J., & Deckers, J. (2000). Tillage erosion on slopes with soil conservation structures in the Ethiopian highlands. *Soil and Tillage Research, 57,* 115–127.

Rarieya, M. (2007). *Environmental degradation, food security, and climate change: an STS perspective on sustainable development in Western Kenya.* Rensselaer Polytechnic Institute: Publication No. 3286339.

Rarieya, M., & Fortun, K. (2010). Food security and seasonal climate information: Kenyan challenges. *Sustainability Science, 5,* 99–114. doi:10.1007/s11625-009-0099-8.

Sanchez, P. A., Swaminathan, M. S., Dobie, P., & Yuksel, N. (2005). *Halving hunger: It can be done*. London: Earthscan.

Tobella, B.A. (2009). Water infiltration in the Nyando River basin, Kenya. Masters Thesis. Swedish University of Agricultural Sciences.

Valentin, C., Poesenm, J., & Li, Y. (2005). Gully erosion: Impacts, factors and control. *Catena, 63,* 132–153.

Verchot, L., van Straaten, O., & Zomer, R. (2007). Pre-feasibility report Nyando River Watershed, Kenya. Biophysical Characterization, Land Suitability and Project Scenario Analysis.

Zhang, K. L., Shu, A. P., Xu, X. L., Yang, Q. K., & Yu, B. (2008). Soil erodibility and its estimation for agricultural soils in China. *Journal of Arid Environments, 72,* 1002–1011.

Zuazo, V. H. D., Ruiz, J. A., Raya, A. M., & Tarifa, D. F. (2005). Impact of erosion in taluses of subtropical orchard terraces. *Agriculture, Ecosystems & Environment, 107,* 199–210.

Urban Heat Island Effect of Addis Ababa City: Implications of Urban Green Spaces for Climate Change Adaptation

Ermias Teferi and Hiwot Abraha

1 Introduction

Change in land use and land cover (LULC), mainly conversion of natural environments into impervious surfaces, has become a major environmental concern worldwide (Weng 2001). In Ethiopia, the rapid population growth together with the fast economic growth has accelerated unplanned growth of urban or built-up areas. Urbanization causes drastic changes in vegetation cover, hydrological regimes and local scale climates. The most obvious climatic impact of urbanization is increase in land surface temperature (LST) in urban areas relative to the surrounding rural areas (Zhou and Wang 2011; Pongracz et al. 2006), and this phenomenon is referred to as the Urban Heat Island (UHI) effect. Elevated temperatures (i.e. heatwaves) can affect air quality (IPCC 2013) and human health (WHO 2013). Heatwaves can have undesired effects on urban population. The effect of heatwaves on human health is exemplified by the deaths of 70,000 Europeans in 2003 (Robine et al. 2008). Thus, the increasing frequency and intensity of heatwaves is becoming an important health concern for policymakers (WHO 2013).

The LST is an important factor controlling many physical, chemical and biological processes of the Earth. Thus, knowledge of the LST is necessary for many environmental studies and management activities of the Earth's surface resources (Li and Becker 1993). It is one of the key parameters in the physics of land surface processes on regional and global scales, combining the results of all surface–atmosphere interactions and energy fluxes between the atmosphere and the ground (Mannstein 1987).

E. Teferi (✉)
Center for Environment and Development Studies,
Addis Ababa University, Addis Ababa, Ethiopia
e-mail: ermias52003@yahoo.com

H. Abraha
Addis Ababa Environmental Protection Authority, Addis Ababa, Ethiopia

W. Leal Filho et al. (eds.), *Climate Change Adaptation in Africa*,
Climate Change Management, DOI 10.1007/978-3-319-49520-0_33

Changes in LST due to urbanization can be assessed by comparing historical point data from meteorological stations located within urban areas and in the surrounding rural areas where such data exist. In the absence of such records, satellite remote sensing data can provide useful information to evaluate land surface temperature changes in response to urbanization. Previous studies have demonstrated that land surface temperature product retrieved from thermal infrared (TIR) sensors can be used to monitor the UHI effect. For example, Pongracz et al. (2006) used LST time series data derived from Moderate Resolution Imaging Spectroradiometer (MODIS) to determine UHI intensities over ten most populated cities of Hungary. Hung et al. (2006) analyzed the UHI effect in 18 Asian mega cities using MODIS LST products acquired between 2001 and 2003. They examined spatial patterns of UHI effects for each city in a diurnal cycle and seasonal variations. They found that both the magnitude and extent of UHI effect were positively correlated with population size of the cities, indicating the significant impact of urban expansion and population growth on local and global climates. Amiri et al. (2009) examined the relationship between land cover changes and LST using Landsat images for urban area of Iran. However, in Ethiopia, no study has so far been conducted, on the Changes in LST due to LUCC.

The impacts of LUCC on thermal environment change can be investigated in two ways. The first approach involves the comparison of the LST of categorical land cover data (Chen et al. 2006; Xiao and Weng 2007). The second approach involves the analysis of LST based on remote sensing indices such as the Normalized Difference Vegetation Index (NDVI) (Yuan and Bauer 2007) and the Normalized Difference Built-up Index (NDBI) (Zha et al. 2003). The NDVI is a proxy for greenness of an area (Chen and Brutsaert 1998). The NDBI is an indicator of built-up and barren land.

The objective of this study is to appraise the urban heat island effect of Addis Ababa, which is the capital and largest city of Ethiopia. The specific objectives were to (i) determine extent and rate of expansion of Addis Ababa city since 1986, (ii) investigate the urban heat island effect of Addis Ababa city using remote sensing and GIS technologies, and (iii) suggest some interventions that could reduce the urban heat island effect in the city. The paper is divided into four sections. The following section (Sect. 2) presents descriptions of the study area and data and methods of the study. Section 3 presents results and discussion, and the final section (Sect. 4) concludes the study.

2 Materials and Methods

2.1 Study Site: Addis Ababa City

Addis Ababa city, the capital of Ethiopia, is located between 8° 50′N–9° 5′N and 38° 38′E–38° 52′E. The city lies at the foot of an isolated mountain called Entoto, and elevation varies from 2015 to 3150 m a.s.l (Fig. 1). The high elevation

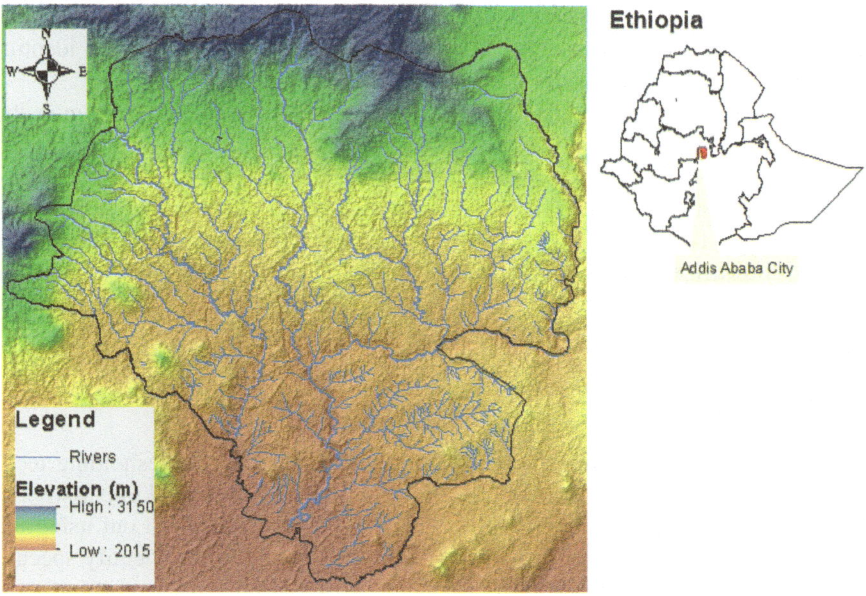

Fig. 1 Location map of the study area

moderates temperatures year-round, and the city's position near the equator means that temperatures are relatively constant from month to month. The city has increasingly expanded southward, westward and eastward. Expansion to the north is limited because of topography. Based on the 2007 census results, Addis Ababa has a total population of 2,738,248, consisting of 1,304,518 men and 1,433,730 women. The city is fully urban, with no rural dwellers within the city's administrative boundaries. For the capital city 662,728 households were counted living in 628,984 housing units, which results in an average of 4.2 persons to a household. Addis Ababa contains 22.9% of all urban dwellers in Ethiopia and for about 3.7% of the total population of the country.

2.2 Data Used and Image Pre-processing

The study used topographic map of the area at a scale of 1:50,000 dated 1984 and Landsat images of years 1986 (TM) and 2011 (TM). Scenes were required to be of the same phenological cycle (dry season) and have little or no cloud cover. All Landsat images were accessed free of charge from U.S. Geological Survey (USGS) Center for Earth Resources Observation and Science (EROS) (via http://glovis.usgs.gov/). All scenes supplied by the EROS Data Center had already been geo-referenced to the Universal Transverse Mercator (UTM) map projection (Zone 37), WGS 84 datum and ellipsoid. Re-projection to the local level projection system was made (UTM, map projection; Clarke 1880, Spheroid, and Adindan Datum). High resolution SPOT-5 imagery (5 m) was also used for visual feature identification and collection of inaccessible ground control points.

In order to eliminate the effects of atmospheric scattering and absorption in the image and to increase the accuracy of land cover classification, the original Digital Number (DN) values were converted to reflectance at the surface of the earth by using FLASH atmospheric correction. This procedure is divided into two stages: (i) converting DN values to spectral radiance and (ii) transferring the sensor detected radiance into surface reflectance.

2.3 *Image Classification*

A hybrid image classification based on the combined use of a thresholding technique, and supervised/unsupervised classification approach was used to classify the 1986 and 2011 images. The unsupervised classification was carried out using the Iterative Self-Organizing Data Analysis (ISODATA) algorithm to identify spectral clusters in the images. The ISODATA method uses minimum spectral distance to assign a cluster for each candidate pixel in image classification (Ball and Hall 1965). Based on the results of the unsupervised classifications, training sites were chosen from the images. For each image, spectral signatures for the training sites were carefully chosen and examined and also using the spectral clusters ground truth were collected to associate the spectral classes. A maximum likelihood classifier (MLC) was then employed for the image classification. MLC is one of the well-known parametric classifiers used for supervised image classification (Foody et al. 1992). The advantage of the MLC as a parametric classifier is that it takes into account the variance–covariance within the class distributions. Two remote sensing indices were computed to assist land cover types identification. Normalized Difference Vegetation Index (NDVI) was computed using Eq. 1 for the purpose of estimating emissivity. In Eq. 1, ρ_4 represents reflectance of the near infrared band of TM and ρ_3 represents the red band of TM image.

$$NDVI = \frac{\rho_4 - \rho_3}{\rho_4 + \rho_3} \tag{1}$$

To improve the accuracy of bare land identification Zhao and Chen (2005) introduced Normalized Difference Bareness Index (NDBaI). NDBaI was used to extract the different types of bare lands: primary bare land (no vegetation), bare land associated with grassland and cultivated soils. Band 5 and Band 6 are the Short Wave Infrared (SWIR) and Thermal Infrared (TIR) bands Landsat TM, respectively.

$$NDBaI = \frac{Band\,5 - Band\,6}{Band\,5 + Band\,6} \tag{2}$$

2.4 Land Surface Temperature Retrieval

To obtain a reasonably good quality of LST, four steps may be required (Fig. 2): (1) conversion to at-sensor spectral radiance; (2) land surface emissivity estimation; (3) atmospheric correction; and (4) LST retrieval.

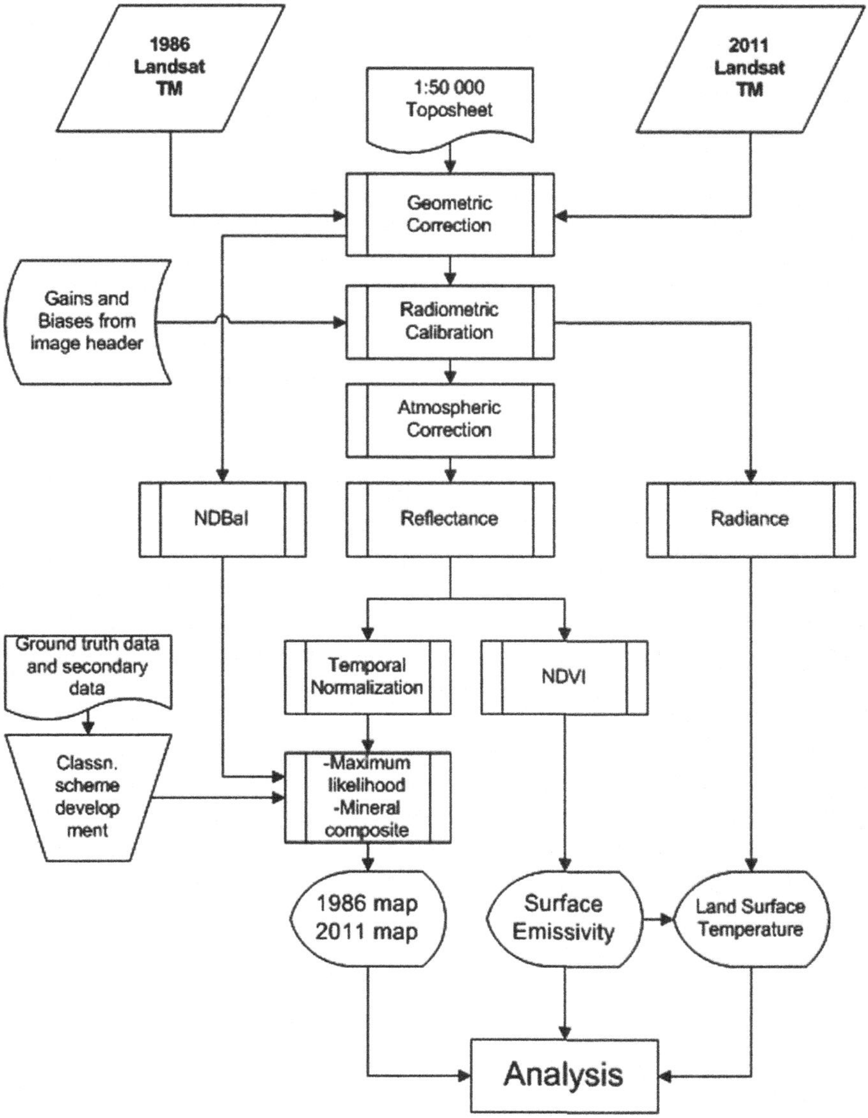

Fig. 2 Methodological flowchart

Conversion to at-sensor spectral radiance: At-sensor spectral radiance (L_{sen}) was computed for the Landsat 5-TM quantized calibrated pixel values in DNs (Q_{cal}). The conversion from Q_{cal} to at-sensor spectral radiance (L_{sen}) is performed using sensor calibration parameters published by Chander et al. (2009) and in image header file by following equation:

$$\mathbf{L}sen = \mathbf{G_{rescale}} \times \mathbf{Q_{cal}} + \mathbf{B_{rescale}} \quad (3)$$

where

L_{sen} = Spectral radiance at the sensor's aperture $[W/(m^2 sr \mu m)]$

Q_{cal} = Quantized calibrated pixel value $[DN]$

$G_{rescale}$ = Band 6 rescaling gain factor $[(W/(m^2 sr \mu m))/DN]$ (0.05518).

$B_{rescale}$ = Band 6 rescaling bias factor $[W/(m^2 sr \mu m)]$ (1.2378). *Land surface emissivity estimation:* Emissivity for ground objects from passive sensor data has been estimated using different techniques. Among other techniques, NDVI methods are easy to apply (Van de Griend and Owe 1993). A slight modification of the first one was used in this paper to make the equation fit to the local condition of the study area (Eq. 4). To derive emissivity image the above equation was written in the spatial modeler of ERDAS.

$$\varepsilon = \begin{cases} 0.99, NDVI \leq 0.01 \\ 1.0094 + 0.047 * \ln(NDVI), otherwise \end{cases} \quad (4)$$

Atmospheric correction: The impacts of the atmosphere and the emitted ground are unavoidably involved in the sensor-observed radiance. Thus, correction is necessary for retrieving true LST from Landsat TM6 data. An atmospheric correction was applied to thermal band of Landsat TM using local values for several meteorological parameters. An online atmospheric correction parameter calculator (http://atmcorr.gsfc.nasa.gov/) was used to calculate the atmospheric-correction parameters (τ, L↑, and L↓) required in the atmospheric radiative transfer equation. However the calculator will not work to correct scenes earlier than 2000. To overcome this problem similar parameters with that of the 2011 imagery were used to correct for the earlier scene (e.g. 1986). Since the scenes are from the same time of year, it is possible to assume a similar atmospheric condition.

The radiance acquired by a thermal band of the remote sensor can be described quantitatively with the Radiative Transfer Equation (RTE) as follows:

$$B(LST) = \frac{L_{sen} - L\uparrow}{\varepsilon\tau} - \frac{1-\varepsilon}{\varepsilon} L\downarrow \quad (5)$$

From Eq. 5, LST can be obtained using the inverted Planck function (Eq. 6). Therefore, the LST obtained is corrected for atmospheric and emissivity effects.

$$LST = \frac{k_2}{\ln\left(\frac{k_1}{B(LST)} + 1\right)} \quad (6)$$

where B(LST) is the blackbody radiance given by the Planck's law in W/m^2/sr/um; $L_{atm}\uparrow$ is the upwelling atmospheric radiance in W/m^2/sr/um (0.79); $L_{atm}\downarrow$ is the downwelling atmospheric radiance in (1.32); τ is the total atmospheric transmissivity between the surface and the sensor (0.89); ε is the land surface emissivity; LST is the land surface temperature in Kelvin (K); k_1 and k_2 are the calibration constants for Landsat TM with $k_1 = 607.76$ W/m^2/sr/um and $k_2 = 1260.56$ W/m^2/sr/um

3 Results and Discussion

3.1 Land Use and Land Cover Change

Figure 3 depicts the classified maps for 1986 and 2011. The overall accuracy of the land use/cover maps of 1986 and 2011 were 89.2 and 91.5% respectively. The Kappa coefficients for the 1986 and 2011 maps were 0.81 and 0.84, respectively. Applying the methods of Congalton and Green (2009) the above results indicate strong agreement between the ground truth and the classified classes. Furthermore, the maps met the minimum accuracy requirements to be used for the subsequent post-classification operations such as change detection (Anderson et al. 1976). Table 1 summarizes the producer's and User's accuracy figure for the 1986 and 2011 image classification.

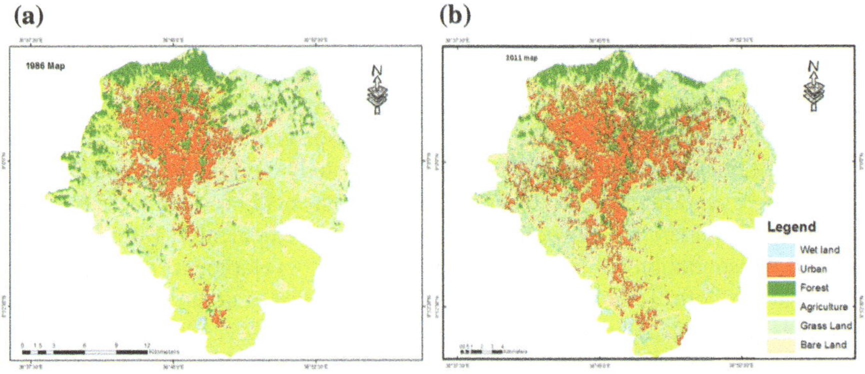

Fig. 3 Land use and land cover maps of **a** 1986, and **b** 2011

Table 1 Producer's and user's accuracy for individual LULC

Land cover classes	1986 accuracy (%)		2011 accuracy (%)	
	Producer	User	Producer	User
Urban/built up	83.33	71.43	75.09	76.09
Forest	84.62	73.3	83.33	71.43
Agriculture	94.59	92.11	97.87	92.00
Grassland	67.67	88.89	60.09	85.71
Wetland	78.00	79.00	77.78	87.50
Bare land	75.00	85.71	80.00	80.00

Table 2 Summary of land use/cover transitions (%)

	Total 1986	Total 2011	Gain	Loss	Total change	Swap	Absolute value of net change
Urban	19.20	34.45	18.71	3.46	22.17	6.92	15.25
Forest	11.48	9.63	4.19	6.04	10.23	8.39	1.84
Cropland	40.62	37.51	12.85	15.96	28.81	25.69	3.12
Grassland	23.30	14.95	8.13	16.49	24.61	16.25	8.36
Barren land	5.40	3.46	2.51	4.44	6.96	5.02	1.93
Total	100	100	46.39	46.39	46.39	31.14	15.25

The proportion of gain, loss, swap and net change of each land use/cover for the 1986 and 2011 are presented in Table 2. There has been a considerable change (~46% of the landscape) in land use/cover in the study area during the 24-years period. Urban class has shown a growth of about ~79%. Urban class experienced the highest gain in about 19% of the landscape, whereas grassland experienced the highest loss in about 17% of the landscape, followed by cropland in about 16% of the landscape. Losses in grassland and cropland are most likely due to urban expansion. The change attributable to quantity (net change) is highest for urban (about 15% of total change for urban); whereas the change attributable to location (swap) is highest for cropland (25% of total change for cropland). Swap land change dynamics accounted for 31% of total landscape change.

Most of the new developments during this study period, took place in the suburbs as organized clusters for accommodating especially residential expansions, industrial, commercial, condominium, emerging settlements, warehouses, or external transportation facilities, in addition to rapid developments on the outskirts of the old city core. Table 3 shows the spatial occurrence of urban expansion within sub-cities. The highest urban expansion occurred in Bole (25.12 km^2), Akaki Kality (21.63 km^2), Nifassilk Lafto (16.56 km^2) and Kolfe Keranyo (15.64 km^2) sub-cities. Massive urban sprawl in eastern and southern part of Addis Ababa can be attributable to rural urbanization, which is a common phenomenon in the post-*derg* regime. Dramatic urban development was observed in the early 2000s. The reconstruction, expansion and upgrading of Bole airport to international status;

Table 3 Areal extent of urban expansion by sub-cities

Sub-cities	Areal extent of urban expansion (km^2)	Contributions of			
		Cropland	Grassland	Barren land	Forest
Addis Ketema	0.39	–	–	–	–
Akaki Kality	21.63	16.54	4.16	0.72	0.2
Arada	0.78	–	–	–	–
Bole	25.12	12.5	9.87	2.13	0.6
Gulele	2.34	–	–	–	–
Kirkos	0.96	–	–	–	–
Kolfe Keranyo	15.64	6.48	5.93	0.82	2.4
Lideta	0.99	–	–	–	–
Nifassilk Lafto	16.56	6.84	7.42	1.3	0.98
Yeka	12.81	4.12	5.63	1.4	1.6

buildings of different factory in industrial area; establishment of private College of Education, Office buildings, hotels, and recreational facilities played greater role in shaping and added attribute to a modern Addis Ababa city. Ring road was constructed to provide the foundation for future expansion.

In the late 2000, Addis Ababa witnessed yet another significant expansion in terms of both land area and population. This is evident in areas like Ayat, Lebu, Summit, Asko, Jemo, and other areas where various stages of land development like condominiums, single and multistory residential and other types of constructions have been taking place. Furthermore, a number of renewals, upgrading or slum improvement strategies undertaken such as redevelopment Lideta area, Arat Killo area, Merkato area, and etc. Moreover, the land use and land cover maps derived from the Landsat images show that the urban and built-up area increased by 15.25% in the 25-year period (1986–2011). Generally, the city has shown rapid Vertical and horizontal expansion both planned and unplanned.

3.2 Changes in Land Surface Temperature in Response to LULC Change

Figure 4 depicts the LST maps for 1986 and 2011. It is evident from the map that there is a thermal gradient as progressed from the central part to the countryside. The lowest temperature in the built-up areas (290.16 to 300.95 K) appeared in: (i) Northern part foot of Entoto hill and areas along Gojam and Ambo road. (ii) Eastern part around Kotebe (iii) Southeastern parts around Adwa Park and in the outskirts of city except southern parts along Deberezeit road. The standard deviation of the 2011 LST is greater than that in 1986 (Table 4), indicating that the surfaces experienced relatively considerable variation in land surface temperature during these periods. The central parts showed a high temperature of over

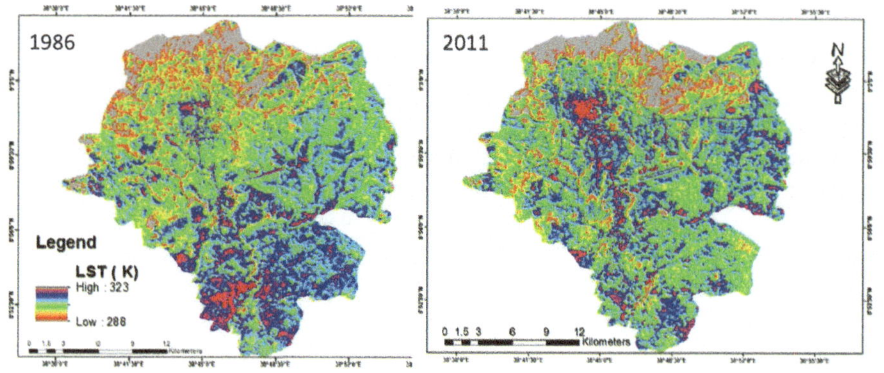

Fig. 4 Derived land surface temperature (LST)

Table 4 Land Surface Temperature (LST) averaged over each land cover classes

Land use and land cover	LST 1986 (K)	LST 2011 (K)	*dT* (K)
Urban/built-up	303.4 ± 2.88	304.42 ± 3.01	1.02
Forest	296.51 ± 3.3	295 ± 2.99	−1.51
Agriculture	302.93 ± 2.69	302.98 ± 2.92	0.05
Grass land	302.79 ± 2.28	303.81 ± 2.46	1.02
Wet land	301.53 ± 1.45	302.48 ± 1.68	0.95
Bare land	303.3 ± 2.27	304.25 ± 2.64	0.95

dT is the Mean temperature difference between 1986 and 2011

323.23 K, while in the outskirts, a lower temperature of 288.28 K exist. As areas in cities develop, more vegetation is lost, and more surfaces are paved or covered with buildings. The change in ground cover results in less shade and moisture to keep urban areas cool. Thus, built up areas contribute to elevated surface and air temperatures, while green spaces moderate temperatures by providing shade, thus helping reduce the risk of heat-related illnesses for city dwellers (IPCC 2013; WHO 2013; Wolch et al. 2014).

The distinctive land surface temperature patterns are associated with the thermal characteristics of land cover classes. To better understand the impact of urban development on land surface temperatures, the thermal signature of each land cover type was obtained by overlaying a land surface temperature image with a land use and land cover map of the same year. The average value of land surface temperature by land cover types is summarized in Table 4. Of all the LULC categories, urban and built-up had the highest LST for both years. This implied that urban development, due to the replacing of natural vegetation with non-evaporating, non-transpiring surfaces, such as stone, metal, and concrete, did have an effect on raising LST. The lowest LST in 1986 was observed in forest, followed by wet land, grass land, agriculture, and bare land. The average LSTs for urban/built-up area are 303.4 and 304.42 K in 1986 and 2011, respectively. Forest land exhibited a

Table 5 Land surface temperature based on land cover change categories

Land use and land cover change categories	*dT* (s.d.) (K)	Adjusted *dT* (K)
Unchanged urban	1.54 (2.91)	0.00
Forest to built-up	1.21 (2.54)	3.32
Agriculture to built-up	1.97 (2.99)	1.65
Grass land to built-up	2.42 (2.84)	1.27
Bare land to built-up	1.24 (2.53)	0.29

decreasing trend in LST from 296.51 K in 1986 to 295 K 2011. Bare land possessed the highest LST next to built-up areas among all the LULC types. In general, the LST of the city shows an increasing trend in the years under study due to changes in land cover and land use.

To investigate the effect of landscape transformation on LST change, LST difference was calculated by subtracting LST in 1986 from LST in 2011 for each land cover change category (Table 5). LST increased about 3.32, 1.65, 1.27 and 0.29 K for areas where forests, agricultural lands, grass lands and bare lands converted into built-up areas, respectively.

3.3 The Urban Heat Island Effect at Addis Ababa

The growth and expansion of Addis Ababa has resulted in change in land surface temperature. Since 1986, the urban built up area has expanded dramatically. Figure 5 shows the urban-rural temperature difference between the urban center and its surrounding areas. Transect #1 (Red line in Fig. 4) shows the maximum LST of the urban center is 315 K (41.85 °C) and the LST of the surrounding rural areas is 300 K (26.85 °C). Thus, the urban-rural temperature difference between the urban center and its surrounding areas can reach a maximum of 15 °C. The land surface temperature in the urban center was a mean of 5 K warmer than that in surrounding rural areas. The results show that UHI effect existed significantly in the city of Addis Ababa. The increased temperature is a reflection of urban expansion during the 25-year period considered. In addition to urban expansion, there have been significant changes in the old parts of the city. A large area of land in these parts has been re-developed by the government or private investors for residential, commercial and industrial developments. The conversion of the surrounding forest and agricultural land into built-up land has contributed to the increased land surface temperature. The government has relocated many residents to the outskirts of the city in order to rebuild the city. The new houses and infrastructure were frequently located in high-quality agricultural land or forestland. This relocation has affected agricultural area and increased local temperatures.

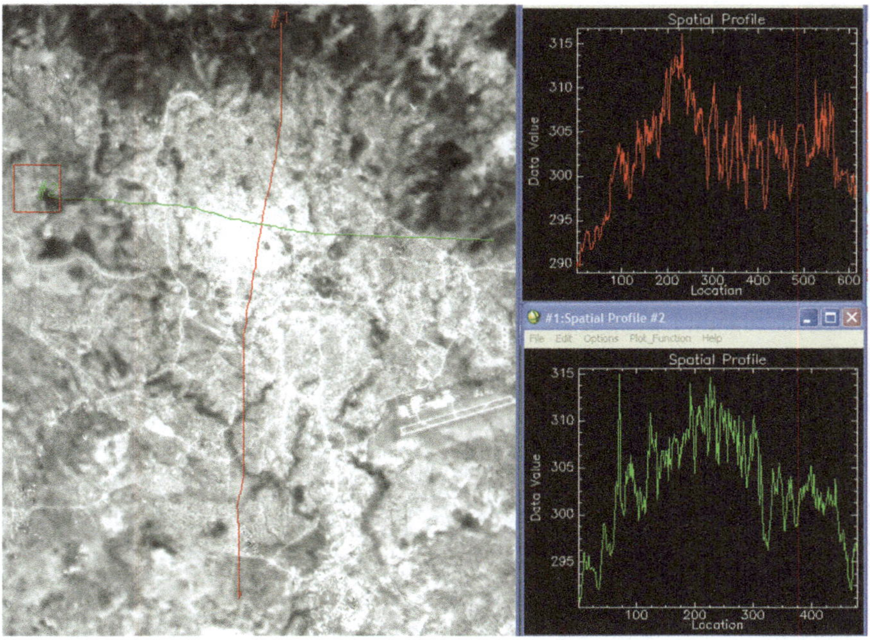

Fig. 5 Land surface temperature difference between Addis Ababa urban center and its rural surrounding

3.4 *Urban Greening to Adapt Addis Ababa City to Climate Change*

The relationship between LST and NDVI clearly shows that vegetation has great impact on reducing UHI effect (Fig. 6). The strong, negative correlation between LST and NDVI implies that the higher biomass a land cover has, the lower the land surface temperature. Because of this relationship between LST and NDVI, changes in land use/cover have an indirect impact on surface temperatures through NDVI. Thus, it is very crucial to promote and support urban greening, such as new planting in the public areas and green infrastructure. In many existing urban areas, it is not feasible to create large new greenspaces. In that case, the creative use of the green infrastructure such as street tree planting is one of the most promising opportunities for adaptation.

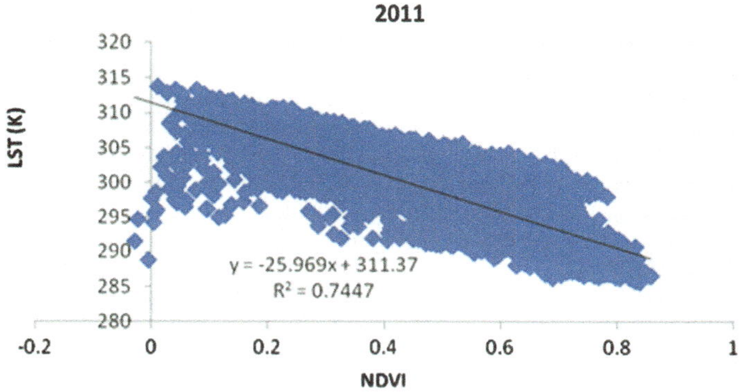

Fig. 6 The relationship between Land Surface Temperature (LST) and Normalized Vegetation Index (NDVI)

4 Conclusion

This study has examined LULC changes in the capital of Ethiopia, Addis Ababa from 1986 to 2011. The results indicate that urban/built-up areas expanded dramatically, while agricultural land and forest declined. Barren land increased, mainly in the boundary areas between forest and dry croplands, especially in steeply sloping areas. The observed changes in LULC were largely attributed to population pressure on the land, a rapidly growing infrastructure and poor land use planning. Changes in LULC were accompanied by changes in LST. Moreover, temperature differences between the urban/built-up and the surrounding rural areas significantly widened. The study assessed the UHI spatial patterns and temporal variations in the Addis Ababa city. The urban-rural temperature differences between the urban core and its surrounding areas of Addis Ababa show a maximum difference of 15 K. This indicates the existence of intensified urban heat island effect in Addis Ababa city. The results of this study suggest that an increase in urban green space could significantly ameliorate the rising temperatures associated with climate change and the UHI effect. Thus, development of UHI effect reduction strategies such as the creation of greenways, community gardens, and street gardens is very important.

References

Anderson, J. R. (1976). A land use and land cover classification system for use with remote sensor data (Vol. 964). US Government Printing Office.

Amiri, R., Weng, Q., Alimohammadi, A., & Alavipanah, S. K. (2009). Spatial–temporal dynamics of land surface temperature in relation to fractional vegetation cover and land use/cover in the Tabriz urban area. *Iran Remote Sensing of Environment, 113*(12), 2606–2617.

Ball, G. H., & Hall, D. J. (1965). *ISODATA, a novel method of data analysis and pattern classification*. Melno Park, CA: Stanford Research Institute.

Chander, G., Markham, B. L., & Helder, D. L. (2009). Summary of current radiometric calibration coefficients for landsat MSS, TM, ETM+, and EO-1 ALI sensors. *Remote Sensing of Environment, 113,* 893–903.

Chen, D., & Brutsaert, W. (1998). Satellite-sensed distribution and spatial patterns of vegetation parameters over a tallgrass prairie. *Journal of the Atmospheric Sciences, 55*(7), 1225–1238.

Chen, X. L., Zhao, H. M., Li, P. X., & Yin, Z. Y. (2006). Remote sensing image-based analysis of the relationship between urban heat island and land use/cover changes. *Remote Sensing of Environment, 104*(2), 133–146. doi:10.1016/j.rse.2005.11.016.

Congalton, R. G., & Green, K. (2009). Assessing the accuracy of remotely sensed data: Principles and practices. CRC Press.

Foody, G. M., Campbell, N. A., Trodd, N. M., & Wood, T. F. (1992). Derivation and applications of probabilistic measures of class membership from the maximum-likelihood classification. *Photogrammetric Engineering and Remote Sensing, 58*(9), 1335–1341.

Hung, T., Uchihama, D., Ochi, S., & Yasuoka, Y. (2006). Assessment with satellite data of the urban heat island effects in Asian mega cities. *International Journal of Applied Earth Observation and Geoinformation, 8*(1), 34–48.

IPCC, Climate Change. (2013). The physical science basis—working group I contribution to the fifth assessment report of the intergovernmental panel on climate change. New York: Cambridge University Press.

Li, Z. L., & Becker, F. (1993). Feasibility of land surface temperature and emissivity determination from AVHRR data. *Remote Sensing of Environment, 43,* 67–85.

Mannstein, H. (1987). Surface energy budget, surface temperature and thermal inertia. In *Remote Sensing Applications in Meteorology and Climatology,* (pp. 391–410). Springer Netherlands.

Pongrácz, R., & Bartholy, J. (2006). Dezső, Zs. Remotely sensed thermal information applied to urban climate analysis. *Advances in Space Research, 37,* 2191–2196.

Robine, J.-M., Cheung, S. L. K., Le Roy, S., Van Oyen, H., Griffiths, C., Michel, J.-P., et al. (2008). Death toll exceeded 70,000 in Europe during the summer of 2003. *Comptes Rendus Biologies, 331*(2), 171–178.

Van de Griend, A., & Owe, M. (1993). On the relationship between thermal emissivity and the normalized difference vegetation index for natural surfaces. *International Journal of Remote Sensing, 14,* 1119–1131.

Weng, Q., & Lo, C. P. (2001). Spatial analysis of urban growth impacts on vegetative greenness with Landsat TM data. *Geocarto International, 16*(4), 17–25.

WHO. (2013). Protecting health from climate change: vulnerability and adaptation assessment. Geneva, Switzerland: World Health Organization.

Wolch, J. R., Byrne, J., & Newell, J. P. (2014). Urban green space, public health, and environmental justice: The challenge of making cities 'just green enough'. *Landscape and Urban Planning, 125,* 234–244.

Xiao, H., & Weng, Q. (2007). The impact of land use and land cover changes on land surface temperature in a karst area of China. *Journal of Environmental Management, 85*(1), 245–257.

Yuan, F., & Bauer, M. E. (2007). Comparison of impervious surface area and normalized difference vegetation index as indicators of surface urban heat island effects in Landsat imagery. *Remote Sensing of Environment, 106*(3), 375–386.

Zha, Y., Gao, J., & Ni, S. (2003). Use of normalized difference built-up index in automatically mapping urban areas from TM imagery. *International Journal of Remote Sensing, 24*(3), 583–594.

Zhao, H. M., & Chen, X. L. (2005). Use of normalized difference bareness index in quickly mapping bare areas from TM/ETM+. *Geoscience and Remote Sensing Symposium, 3*(25–29), 1666−1668.

Zhou, X., & Wang, Y. C. (2011). Dynamics of land surface temperature in response to land-use/cover change. *Geographical Research, 49*(1), 23–36.

Climate Change Adaptation Activities for Agricultural Development in Ethiopia: A Review of Potentials

Esubalew Abate, Negussie Semie and Berhanu Ayenew

1 Introduction

Climate change is weather phenomenon that occurs when the average long-term weather patterns of a region are altered for an extended period of time, typically decade or longer. Climate change basically refers to long-term changes or trends in the average climate (such as annual average temperature, rainfall or precipitation) or trends in climate extremes such as the frequency of intense rainfall events, patterns of winds, and more frequent and severe extreme storms and experienced by people as individual weather events that fluctuates on annual, seasonal and decade basis. In other words, climate change can be distinguished as the change in atmospheric composition over an identified period of time. Therefore, for this review purpose, climate change can be defined as 'a change in the climate which is attributed directly or indirectly to human activities that alter the composition of the global atmosphere in addition to natural variability observed over comparable time periods' [IPCC (2007) cited in Abegaz and Wims 2015].

The two terms (Global Warming and Climate Change) are used in the scientific literature more frequently to refer to the occurrence of physical phenomena in the earth's atmospheric temperature. However, though they are used interchangeably in scientific writings, there is distinction between the two. Climate change is a long term changes in average global weather conditions which occurs because of changes to Earth's environment. On the other hand, global warning refers to long-term rising in average global temperature.

E. Abate (✉) · N. Semie · B. Ayenew
Institute of Development and Policy Research (IDPR),
Addis Ababa University, Addis Ababa, Ethiopia
e-mail: esubalewabate@gmail.com; esubalew.abate@aau.edu.et

W. Leal Filho et al. (eds.), *Climate Change Adaptation in Africa*,
Climate Change Management, DOI 10.1007/978-3-319-49520-0_34

2 Causes of Climate Change

In many scientific writings, it is common to find an argument that excessive concentration of greenhouse gases in the atmosphere as a result of different human activities is major cause of change in earth's climate system. Human activities such as burning of fossil fuels, industrial production, and to some extent agricultural practice have the potential to increase greenhouse gas levels that traps more heat in the atmosphere and drives global warming and climate change. Intergovernmental Panel on Climate Change (IPCC), in its 4th assessment report, concludes that climate change is on the way resulting from past, current, and future greenhouse gas emissions with its potential adverse impacts on socio-economic development of nations (NMA 2007). The causes of change in earth's climate can be distinguished into natural and human or anthropogenic factors. Since in most cases the effect of natural factor is not commonly considered as significant and it will not be addressed here. Moreover, important causes of climate change which are associated to human activities are greenhouse gases (water vapor, carbon dioxide, methane, nitrous oxide, chlorofluorocarbons (CFCs)), aerosols and land-use change (Hope 2009; Nelson et al. 2009 cited in Abegaz and Wims 2015; www.wmo.int).

Since the time of industrial revolution, socio-economic activities have been the major source of an increase in the amount of greenhouse gasses in the atmosphere. The increased amount of gases, which absorb heat, directly leads to more heat being retained in the atmosphere and thus an increase in global average surface temperatures. Greenhouse gases and aerosols affect climate by altering incoming solar radiation and out-going infrared (thermal) radiation that are part of Earth's energy balance. Moreover, land-use changes which results from cutting down forests to create farmland) have led to changes in the amount of sunlight reflected from the ground back into space (the surface albedo) (www.wmo.int). It is true that about half of the land use changes are estimated to have occurred during the industrial era, much of it due to replacement of forests by agricultural cropping and grazing lands. The cumulative effect of these human activities would increase in frequency of extreme weather events, sea levels rise, recurrent flood, and change in precipitation patterns (Zerihun 2012). The highest proportion of change in the earth's climatic condition is largely associated with the emergence of the industrial revolution. This change in natural climate system would have significant implication for the socio-economic activities of men particularity on the potential of agricultural productivity that is highly dependent on changes in temperature and rainfall patterns (*ibid.*).

Ethiopia's topography consists of 45% high plateau mountainous ranges with elevation of 1500 m above sea level where 90% of the population resides and the surrounding lowland areas with elevation less than 1500 m which mostly populated by pastoralists (Robinson et al. 2013). The topography of Ethiopia is traditionally divided into three agro-ecological zones 'Kolla (warm semiarid), <1500 m above sea level; Woinadega (cool sub-humid temperate zone), 1500–2400 m above sea level; and Dega (cool and humid zone), mostly >2400 m above sea level' (*ibid.*).

However, Ethiopia is a country characterized by large heterogeneity in terms of agro-ecology even within a small administrative unit (Di Falco et al. 2011). Thus, the above traditional classification of the agro-ecology of Ethiopia may be highly crude to provide a better understanding of the country.

3 Ethiopia's Vulnerability to Climate Change

Among the agro-ecological classification of the country which is being used for agricultural planning and development is the one which was developed by Hurni in 1998 as part of Soil Conservation Research Project. According to this classification the agro-ecologies of Ethiopia fall into six categories that include Bereha (desert, below 500 m a.s.l.), Kolla (lowlands, 500–1500 m a.s.l.), Weynadega (midlands, 1500–2300 m a.s.l.) and Dega (highlands, 2300–3200 m.a.s.l.), Highdega (3200–3700 m a.s.l.) and Wurch, an area with 3700 m a.s.l. (*ibid.*). The implication is that different agro-ecological zones of the country have different levels of susceptibility to change in climatic conditions. For instance, Robinson et al. (2013) divided the country into five agro-ecological zones based on the spatial variation of climate change risk as drought prone, humid lowland moisture reliable, moisture reliable-cereals, moisture reliable-*enset* and pastoralist.

The level of climate change related risks for different administrative regions of Ethiopia and vulnerability to climate change impact is not uniform across the administrative regions of the country. For instance, according Deressa et al. (2008), over the past century, Amhara region has suffered most from droughts and floods, with Oromia and Somali regions following closely behind, whereas Beneshangul Gumuz and Afar regions have experienced the lowest number of droughts and floods. They also argue that the vulnerability to climate change as measured by net effect of sensitivity, exposure, and adaptive capacity shows great variability across regions in Ethiopia. Based on this information, Afar, Somali, Oromia, and Tigray regions are relatively more vulnerable to climate change than the other regions.

Furthermore, with regard to the cause of their vulnerability, they found that vulnerability in Afar and Somali is attributed to their low level of rural service provision and infrastructure development whereas, in Oromia and Tigray, it is because of higher frequencies of droughts and floods, lower access to technology, fewer institutions, and lack of infrastructure (*Ibid.*). The intensity of vulnerability is considerably high for the four regions including Afar, Ormia, Somali and Tigray. This indicates that ex ante risk management and ex post shock-coping abilities of the households in different region and agro-ecologies cannot be addressed using uniform intervention measures.

Another study by Gutu et al. (2012) showed contrary result to the general expectations that farmers living in lowland areas are less vulnerable to the adverse impact of climate change than those living in the highland areas and the report was based on their vulnerability index calculation for each of the agro-ecological areas with reference to factors such as social vulnerability (like age, gender, education, etc.), economic vulnerability (livestock ownership, crop diversity, non-farm income, landholding size, etc.) and environmental vulnerability variables (slope and fertility of farmland, natural hazards, rainfall, temperature etc.).

4 Trends in Climate of Ethiopia

During the past four decades, Ethiopia experienced significant change in patterns of rainfall, temperatures, and extreme disasters (Emerta 2013) all of which are important climate trend indicators that affect economic activities in agrarian economies unless some adaptation and/or mitigation measures are put in place. This section presents the details of experienced trends of climatic conditions in Ethiopian along with their implication for the overall performance of the agriculture sector.

4.1 Temperature

According to Emerta (2013), over the past four decades, 'the average annual temperature in Ethiopia has been increasing by 0.37 °C every ten years, which is slightly lower than the average global temperature rise with the highest temperature being observed during the 1990s'. He argues that the rise in temperature was profound in the dry and hotspot areas particularly in the northern, northeastern, and eastern parts of the country; and the lowland areas, the more vulnerable to flooding during the time of highest precipitation in the highland areas.

Furthermore, the IPCC's projection for Ethiopia indicates that the mean annual temperature will increase in the range between of 0.9 and 1.1 °C by 2030 and in the range of 1.7–2.1 °C by 2050 (*ibid.*). This kind of trend in mean annual temperature will obviously pose a threat to Ethiopian economy in general and to agriculture in particular as seasonal temperature increases have potential to decease yields by impacting the growth of some crops like teff, barley, and wheat, which are important sources of staple food in the country (Emerta 2013). The rise in mean annual temperature will, of course, has not only an impact on crops but also on livestock and on the pests and diseases they are exposed to.

4.2 Rainfall

The distribution of mean rainfall in Ethiopia exhibits a great variation across diverse agro ecological zones (Zerihun 2012). 'Mean annual rainfall ranges from about 2000 mm over some areas in the south west to less than 250 mm over the Afar lowlands in the northeast and Ogaden in the southeast' (*ibid.*). Ethiopia experienced a decreasing trend of rainfall pattern during the past decades and the 'precipitation has shown a general decreasing trend since the 1990s' (Emerta 2013). Extreme weather events such as drought, floods, heavy rains, strong winds, frost, heat waves (high temperatures), etc. have become common and costly in recent decades (Emerta 2013; NMA 2007). In the Ethiopian context, though the historical and social impacts of such extreme weather conditions are not well documented, the country has experienced recurrent drought and flooding over the last two decades that have resulted in loss of life and property.

5 Impact of Climate Change on Agriculture

Climate change due to natural variability or as a byproduct of human activities (such as the burning of fossil fuels to produce energy, deforestation, industrial processes, and some agricultural practices) is the major challenge of development worldwide (Dawit et al., n.d). Climate change is posing huge challenge on socio-economic activities in developing economies including Ethiopia. Most of the African countries whose economies depend on weather-sensitive agriculture and with low level of adaptive capacity are highly vulnerable to climate change (Di Falco et al. 2011). These authors argue that 'the vulnerability been demonstrated by the devastating effects of recent flooding and the various prolonged droughts of the twentieth century'.

The relationship between climate change and agriculture is quite complex and diverse (Abegaz and Wims 2015). According to Bosello and Zhang (2005) cited in (Abegaz and Wims 2015), the interdependency between climate change and agriculture evolves dynamically over time, often cover a large spatial and temporal scale, and is still surrounded by large uncertainties. A growing number of empirical reports show that climate change has direct and obvious effect on agricultural performance and thereby on the food security situation, health of the people and animals. In fact, there are tangible empirical evidences showing 'climate change is decreasing the productivity of crops across the globe, thus increasing the risk of food shortages in developing countries where agricultural systems are low-tech and malnutrition is common' (Moreland and Smith 2012). Furthermore, agriculture and allied activities are highly dependent on climate patterns and variability. This implies that climate change will have a significant impact on agricultural condition and related food supply and food security which in turn obstruct national or regional economic progress and poverty alleviation.

Ethiopia's economy is 'heavily dependent on rain-fed agriculture, and its geographical location and topography in combination with low adaptive capacity entail a high vulnerability to adverse impacts of climate change' (Robinson et al. 2013). Though Ethiopia has experienced vibrant and sustained economic growth over the past years, it has been historically affected by climate extremes (particularly droughts) that result in income swings; and is expected to be exposed to even more pronounced and frequent climate shocks in the future (Robinson et al. 2013). According to Bezabih et al. (2011) and Zhai et al. (2009) cited in Dawit et al. (n.d), the impact of climate change in Ethiopia is on increase as its economy is dependent on subsistence agriculture, implying devastating effect on the performance of the sector and on the whole economy. As a result frequent droughts, floods, heat waves and other extreme weather conditions are common problems affecting the practice of agriculture in Ethiopia.

Furthermore, Agriculture is a dominant source of economy in Ethiopia. For instance, during 2006/07 budget year, agriculture accounted for 46% of GDP followed by industry and services, produced 80 percent of exports, employed 80% percent of the labor force, and is a means of survival for the nation's poor (You and Ringler 2010). Since agriculture is the main source of export earning, supply of raw materials to domestic agro-industries and a source of livelihood for the majority of the Ethiopians, and any negative shock in the agricultural sector will have a devastating impact for the whole economy of the country (Dawit et al., n.d).

Agriculture constitutes subdivisions of crop and livestock production, fishery, and forestry that are most susceptible to the impact of climate change (Temesgen et al. 2014). Moreover, the cause of agricultural underproduction in Ethiopia is mainly recurrent drought which often causes famine (Temesgen and Rashid 2009). Available data over the last three decades showed us that the contribution of the agricultural sector to the GDP of Ethiopia is linked with the amount and distribution of the annual rainfall; that 'years of drought and famine (1984/1985, 1994/1995, 2000/2001) are associated with very low contributions, whereas years of good climate/rain fall (1982/83, 1990/91) are associated with better contributions' (*ibid.*).

5.1 Crop Production/Food Security/and Climate Change

There is an increasing consensus that climate change is threatening crop production and food security globally making the supply of sufficient foods for increasing population challenging while ensuring sustainability of the already stressed environment. During the past three decades, Ethiopia experienced countless droughts, five of which resulted in famine. This situation continues even today as the country is currently facing serious drought in northern, north east, east and south east of the country where more than 10 million people are in need of food aid.

There are indications that, change in the atmospheric composition has already caused a significant impact on food security, water resource availability and human health in Ethiopia. Unfolding climate change causes shift in average climatic variables and intensifies weather variability which exposes Ethiopia's agriculture to higher levels of risks which in turn jeopardizes economic growth, food security and poverty reduction (You and Ringler 2010).

Heavy dependence of Ethiopia's agriculture on rainfall makes crop production vulnerable to climate change and variability. There are an extensive research works that have been done to investigate the impact of climate change on the crop production and productivity [see Deressa 2007; Deressa et al. 2008 and Yesuf et al. 2008 cited in Zenebe et al. (2011)]. For instance, a study by Zerihun (2012) investigated the impacts of climate change on crop yields (for Teff, Maize and Wheat) in Ethiopia over a period of 28 years. He concluded that there is direct correlation between pattern of rainfalls and crop productivity in Ethiopia and the negative impact of future climate entails serious damage on production of some crops such as teff and wheat whereas it will result in the increase of the productivity of such crops as maize by 2050 (*ibid.*).

Similarly, Temesgen and Rashid (2009) showed, from their study of the effect of climate change on net crop revenue as a proxy for crop productivity, that marginal increase in temperature during summer and winter would significantly reduce crop net revenue per hectare whereas an increase in precipitation during spring would significantly increase net crop revenue per hectare. They further projected that between 2050 and 2100, the impact of climate would be significant that smallholder farmers would experience a reduction in net crop income per hectare of agricultural landholding. On the other hand, a study by Dawit et al. (n.d) revealed that CO_2 emission negatively affects agricultural productivity and household welfare and that the emission of such gas into the atmosphere leads to decrease in crops production.

5.2 Livestock Production and Climate Change Impacts

Ethiopia is home to about 52.1 million heads of cattle, 24.2 million sheep, 22.6 million goats and 44.9 million poultry and it is the continent's top livestock producer and exporter (Temesgen et al. 2014). Livestock production is a major source of livelihood for the population residing in the arid and semi-arid agro-ecological areas of Ethiopia. It is the major source of pastoralists' socio-economic capital as it forms the main source of income and social security for agro-pastoralist families. Thus, livestock production plays a crucial role in ensuring food security and the alleviation of the problem of hunger at household, community and even regional and national levels in Ethiopia. 'Livestock are extremely important as they serve a wide variety of functions in society from social to subsistence purposes' [Behnke 2010 and Kassahun 2008 cited in Kombal 2011). Livestock sector forms an integral part of the farming systems in the Ethiopian economy and serves as source of many social and economic values such as food, draught power, fuel, cash income, security and investment in both the highlands and the lowlands/pastoral farming systems (Temesgen et al. 2014).

Nonetheless, livestock productions like cropping are greatly vary across regions and agro-ecologies and dependent on and vulnerable to climate change and variability. 'Livestock productivity will be impacted by increased temperature with higher-yielding breeds more likely to be negatively affected than more-robust local breeds' (IFAD 2011). Climate change can affects dairy, meat and wool production mainly by impacting grassland and/or rangeland productivity (*ibid.*)

The distributions as well as the availability of pasture and water are closely related to the distribution and pattern of rainfall. Studies to-date claim that changes in the patterns of rainfall and ranges of temperature can affect feed availability, productivity of grazing land, and incidence of diseases that directly affect the livestock productivity and hence food security in areas where people rely on livestock production as livelihood source. With regards to this Temesgen et al. (2014) reported that, changes in climatic factors such as temperature, precipitation and the frequency and severity of extreme events like droughts directly affected livestock yields through their impact on availability of feed and water, grazing lands, and disease incidence. For instance, an empirical study by Zelalem et al. (2009) cited in Temesgen et al. (2014) reported four major effects of climate change on livestock production in Borana pastoralists of Oromia Regional State in Ethiopia. According to these authors, feed shortage, water shortage, reduced productivity, and decreased mature weight and/or longer time to reach mature weight are the negative effects of climate change observed among the Borana pastoralist communities.

5.3 Fisheries

According to FAO (2007) long-term climate change has important feedback loops to global ocean circulation patterns, sea level rise and sea temperature rise which causes fish to inhabit different depth ranges and changes in ocean salinity all of which affect the biological properties and distribution of fish species and it was further reported that FAO is giving priority to an Ecosystem Approach to fisheries which requires addressing impacts of the wider environment in order to manage fishery resources and the ecosystems on which they depend. Though Ethiopia is a landlocked country and has no access of large water bodies (ocean and sea); it is imperative to investigate the potential impact of climate change on Ethiopians' inland water body fishery resource.

5.4 Forestry

Thirty percent of the total land surface is covered by forests and only 34% of forests are intensely managed for wood production worldwide. Forests can help human societies to adapt to climate change and adverse climate change impacts (FAO 2007). In addition to benefits such as the provision of wood and non wood forest products, restoration of soil fertility, and the conservation of biological diversity, trees and forests improve the microclimate by buffering winds, regulating the water table, providing shade to crops and animals, and stabilizing coastal areas. They thus contribute to sustainable agricultural production and food security (*ibid*).

6 Adaptation Strategies to Climate Change

The effect of climate change is a reality that has a profound impact on the efforts of development in emerging economies, especially in Sub-Saharan Africa where Ethiopia is located, with frequent and intensified climate related disasters like recurrent drought, flood and erratic rainfall. Sub-Saharan African agriculture, thus, is extremely vulnerable to climate change and extreme weather conditions are posing a negative pressure on the performance of agriculture. Changes in frequency and severity of droughts and floods could pose challenges for farmers and ranchers. Climate change affects the production of crops by increasing the chance of occurrence for hazardous events such as floods, insect outbreaks, hailstorm, alien weeds, disease and pests, and droughts all of which have a negative effect on the likelihood of crop production (Gutu et al. 2012).

Ethiopia is facing unprecedented climatic changes that are having long range impacts on socio-economic activities in general and the agricultural sector in particular and hence on its food security. Rainfall variation which manifest itself in the form of too little or too much rain has been a challenging for subsistence farmers in Ethiopia as it impacts the production of crops and/or livestock. As a result the country is not able to address persistent food insecurity; reduce the problem of poverty and still remains as one of the top food aid receivers in the world. Cognizant of the structural nature of the problem, during the past ten years, the government made a radical shift in policy from ad hoc to systematic approach in addressing drought related stocks such as those experienced in 2002 (FDRE 2002).

7 Adaptation Activities to Climate Change in Agriculture

There are two possible measures that nations can consider for addressing the challenge of climate change: Adaptation and Mitigation. The theme of this review document is to dwell on the adaptation to climate change with special emphasis on adaptation activities to climate change in agriculture. According to FAO (2007) the impacts of climate change among others are either biophysical like physiological effects on crops, pasture, forests and livestock (quantity, quality); changes in land, soil and water resources (quantity, quality); increased weed and pest challenges; shifts in spatial and temporal distribution of impacts; sea level rise, changes to ocean salinity; sea temperature rise causing fish to inhabit different ranges or socio-economic which includes decline in yields and production; reduced marginal GDP from agriculture; fluctuations in world market prices; changes in geographical distribution of trade regimes; increased number of people at risk of hunger and food insecurity; migration and civil unrest.

Adaptation refers to all those responses to climate change that may be used to reduce vulnerability or to actions designed to take advantage of new opportunities that may arise as a result of climate change (Burton 1996), while "Adaptive capacity is the ability of a system to adjust to climate change, including climate variability and extremes, to moderate potential damages, to take advantage of opportunities, or to cope with the consequences" (IPCC 2001).

The major classes of climate change adaptation in agriculture are considering seasonal changes and sowing dates; using different adaptive variety or species; improved water supply and irrigation system; using other inputs (fertilizer, tillage methods, grain drying, other field operations); introducing new crop varieties and adaptive livestock breeds; forest fire management, promotion of agroforestry, adaptive management with suitable species and silvicultural practices (Reilly and Schimmelpfennig (1999) and FAO (2005).

According to Stern (2006 cited in The AEA group 2007) various types of adaptation can be distinguished, including anticipatory and reactive adaptation, private and public adaptation, and autonomous adaptation and policy-driven adaptation. Easterling (1996 cited in FAO 2007) described the two main types of climate change adaptations. The first one is autonomous, which is a short-term adjustment where the reaction of farmers to changing precipitation patterns is observed, meaning the farmers change crops or use different harvesting and planting/sowing dates. Stern (2006 cited in The AEA group 2007) further described autonomous adaptation as actions "taken 'naturally' by private actors, such as individuals, households, businesses in response to actual or expected climate change, without the active intervention of policy".

Moreover, the agricultural sector is one in which autonomous adaptation is a particularly important category because farmers have traditionally adapted their methods in response to felt changes. The planned or policy-driven climate change adaptations are also sometimes called long-term adaptation measures which are major structural changes to overcome adversity, with conscious policy options or response strategies, multisectoral in nature, and aimed at altering the adaptive capacity of the agricultural system or facilitating specific adaptations (Easterling 1996 cited in FAO 2007). Changes in land-use to maximize yield under new conditions; application of new technologies; new land management techniques; and water-use efficiency related techniques are among the planned climate change adaptations (FAO 2007).

According to Stern (2006 cited in The AEA group 2007) policy driven adaptation is "the result of a deliberate policy decision" and it is therefore associated with public agencies, either in that they set policies to encourage and inform adaptation or they take direct action themselves, such as public investment. Accordingly, HM Government UK (2006) categorized planned adaptations into two main groups: "building adaptive capacity" and "taking adaptive action". These are complementary forms of action, with adaptive actions usually coming after adaptive capacity has been built up. Building adaptive capacity involves ensuring that the scientific, technical and socio-economic evidence, the skills, the governmental and non-governmental partnerships, the policies and the resources are in place to enable adaptation to be undertaken (Defra 2005). In many cases, efforts to build adaptive capacity may be best made at a sectoral level, but even within individual organization, or for an individual farmer, a certain amount of capacity building (e.g. awareness-raising, education) is initially required as the foundation for the next step of taking actual adaptive action. Efforts to take adaptive action will necessarily be location and context specific, as they require a deliberate change of practice, whether in management, process or infrastructure (The AEA group 2007).

A study by The AEA group, categorized three types of adaptation options in the agriculture sector which includes: management, technical/equipment and infrastructural. The type of measure will largely determine the extent to which farmers can adopt them without additional assistance. Farmers should be able to carry out some changes in management measures without getting support from others. This could be true for technical measures, while infrastructural measures may require significant capital investment. Accordingly, (1) Management measures are like the choice of crop variety and pesticide which farmers make every year. These decisions could be based on information taken from a number of sources: agrochemical industry publications and representatives, government extension service advisers, discussions with other farmers, etc. and market forces that have driven historic innovation and adaptation may continue to drive the adoption of new measures; (2) The Technical/equipment measures differ from management measures arbitrarily, as technical understanding is needed to implement management decisions. However, the introduction of new crops or livestock, together with the agrochemicals needed is technical as the husbandry requirements could be new to the farmers in a given area. The introduction of improved irrigation equipment could also be considered as technical. Mean while, technical Advice may be required from government agencies and extensive breeding and testing programmes may be necessary to identify cultivars and breeds appropriate to changing local conditions. (3) Infrastructural measures may vary greatly in scale and expense, and require an element of capital investment and rainwater harvesting could be a good example which requires adding guttering to the roofs of buildings and collecting water in an earth-banked reservoir.

8 Conclusion and Recommendations

The purpose of this review work is to consolidate up to date internationally available scientific information regarding the climate change adaptation of agricultural activities for learning and possible utilization of the information by policy makers, academicians, farmers, businessmen and the like in Ethiopia and elsewhere. Accordingly, the following concluding remarks are presented.

Agriculture is the backbone of Ethiopia's economy and it is vulnerable to climate change. Moreover, the major causes of climate change are the results of human activities (greenhouse gases, land-use change, etc.). Furthermore, the cumulative effect of these human activities increased the frequency of extreme weather events like sea levels rise, recurrent flood, and change in precipitation patterns.

Like in the other parts of the world, in Ethiopia the vulnerability to climate change impact is not uniform across the administrative regions of the country. Over the past century, it was learned that the risk management and post shock-coping abilities of the farm households in different regions and agro-ecologies was not uniform depending on their coping strategies and resilience which implies the need for well thought and location specific climate change adaptation intervention measures.

Available literature evidences indicate that the change in the atmospheric composition has started causing negative impact on food security, water resource availability and human health in Ethiopia. It is apparent that there is direct correlation between pattern of rainfalls and crop and livestock productivity in Ethiopia. Some reports are showing that the negative impact of future climate will result in serious damage on production of some C-3 crops such as Barley and wheat and may result in increase in the productivity of C-4 crops like maize, sorghum and sugarcane. Accordingly, livestock productivity will be also impacted by increased temperature with higher-yielding breeds more likely to be negatively affected than more-robust local breeds and climate change can affect dairy, meat and wool production mainly by impacting on grassland and/or rangeland productivity.

This information is necessary for Ethiopia as a country, which is mainly relying on agriculture, to plan and to take appropriate actions to adapt the agricultural activities in line with changing climate. Available scientific information indicates that the major classes of climate change adaptation in agriculture include considering seasonal changes and sowing dates; using different adaptive crop variety or livestock species; improved water supply and irrigation system; using other inputs (fertilizer, tillage methods, grain drying, other field operations); introducing new crop varieties and adaptive livestock breeds; forest fire management, promotion of agroforestry, adaptive management with suitable species and silvicultural practices.

As a whole, there are two main types of climate change adaptations (autonomous and planned) in agriculture. Autonomous adaptation are actions taken naturally by private actors or farmers, such as individuals, households, businesses in response to actual or expected climate changes, while planned or policy-driven climate change adaptations which are sometimes called long-term adaptation measures are major structural changes to overcome adversity, with conscious policy options or response strategies. They are multisectoral in nature, and aimed at altering the adaptive capacity of the agricultural system or facilitating specific adaptations.

Furthermore, planned adaptations are also classified into two main groups (building adaptive capacity and taking adaptive action). These are complementary forms of actions, where adaptive actions usually come after adaptive capacity has been built up. Moreover, adaptive actions are location and context specific, as they require a deliberate change of practice in management (choice of crop variety, livestock breed and pesticide/insecticide, etc.), technical (introduction of new crops or livestock, together with the agrochemicals, etc.) or infrastructure (which requires an element of capital investment like rainwater harvesting).

In conclusion, the purpose of this review work is to consolidate up to date scientific information regarding the climate change adaptation agricultural activities for learning and possible utilization of the information by policy makers, academicians, farmers, businessmen and the like in Ethiopia and elsewhere.

Finally, it is recommended that, as the threat of climate change on agriculture is on increase, Ethiopia in line with the green economy policy need to give due attention and focus on location and context specific planned climate change adaptation actions at national level and to support and motivate autonomous climate change adaptation actions being carried out by individual farmers and pastoralists and local communities. Both autonomous and planned climate change adaptation actions will bring about synergetic positive development impact on Ethiopia's agriculture.

Acknowledgements The authors wish to express their sincere gratitude to the Institute of Development and Policy Research (IDPR) of Addis Ababa University for encouraging the authors and thanks to all colleagues for their insightful review of the manuscript.

References

Abegaz, D. M., & Wims, P. (2015). Extension agents' awareness of climate change in Ethiopia. *The Journal of Agricultural Education and Extension, 21*(5), 479–495. doi:10.1080/1389224X.2014.946936

Bezabih, M., Chambwera, M., & Stage, J. (2011). Climate change and total factor productivity in the Tanzanian economy. *Climate Policy, 11*(6), 1289–1302.

Burton, I. (1996). The growth of adaptation capacity: Practice and policy. In J. Smith, N. Bhatti, G. Menzhulin, R. Benioff, M. I. Budyko, M. Campos et al. (Eds.), *Adapting to climate change: An international perspective* (pp. 55–67). New York, NY, USA: Springer.

Dawit, W., Zerayehu, S., & Tsegaye, G. (n.d.). *CO_2 emissions impact on agriculture productivity and household welfare in Ethiopia: Computable general equilibrium analysis.*

Defra. (2005). *Review of Defra's climate change impacts and adaptation (agriculture) R&D programme.*

Deressa, T. T., & Hassan, R. M. (2009). Economic impact of climate change on crop production in Ethiopia: Evidence from cross-section measures. *Journal of African Economies, 18*(40), 529–554. doi:10.1093/jae/ejp002. Online date 15 March 2009.

Deressa, T. T., Hassan, R. M., & Ringler, C. (2008). *Measuring Ethiopian farmers' vulnerability to climate change across regional states*, IFPRI Discussion Paper No. 806, http://www.ifpri.org/pubs/dp/ifpridp00806.asp. (Washington, DC: IFPRI, 2008).

Di Falco, S., Yesuf, M., Kohlin, G., & Ringler, C. (2011). Estimating the impact of climate change on agriculture in low-income Countries: Household level evidence from the Nile Basin, Ethiopia. *Environmental and Resource Economics*. doi:10.1007/s10640-011-9538-y

Easterling, W. E. (1996). Adapting North American agriculture to climate change in review. *Agricultural and Forest Meteorology, 80*(1), 1–54.

Emerta, A. (2013). *Climate change, growth, and poverty in Ethiopia. Climate change and Political Stability (CCPS)*. The Robert S. Strauss Center for International Security and Law. Working Paper No. 3.

FAO. (2007). Adaptation to climate change in agriculture, forestry and fisheries: Perspective, framework and priorities. In *International working Group on Climate change*. Viale delle Terme di Caracalla - 00100 Rome, Italy.

FAO (Food and Agriculture Organization of the United Nations). (2005). Impact of climate change, pests and diseases on food security and poverty reduction. In *Special event background document for the 31st session of the Committee on World Food Security*. Rome, May 23–26, 2005.

FDRE (Federal Democratic Republic of Ethiopia). (2002). *Food security strategy*. March 2002, Addis Ababa, Ethiopia.

Gutu, T., Bezabih, E., & Mengistu, K. (2012). Analysis of vulnerability and resilience to climate change induced shocks in North Shewa, Ethiopia. *Agricultural Sciences, 3*(6), 871–888.

HM Government. (2006). *Climate change: The UK programme*. HMSO. http://www.wmo.int/pages/themes/climate/causes_of_climate_change.php#a. Web site visited on January 28, 2016.

Hope Sr, K. R. (2009). Climate change and poverty in Africa. *International Journal of Sustainable Development & World Ecology, 16*(6), 451–461. doi:10.1080/13504500903354424.

IFAD (International Fund for Agricultural Development). (2011). *Climate-smart smallholder agriculture: What's different?* IFAD occasional paper 3.

IPCC (Intergovernmental Panel on Climate Change). (2001). Climate change 2001: The scientific basis. In J. T. Houghton, Y. Ding, D. J. Griggs, M. Noguer, P. J. van der Linden, X. Dai, et al. (Eds.), *Contribution of working group I to the third assessment report of the intergovernmental panel on climate change*. Cambridge: Cambridge University Press, 881 pp.

Kombal, T. (2011). Livestock production systems and their environmental implications in Ethiopia. In *Environmental policy review 2011: Research highlights*.

Moreland, S., & Smith, E. (2012). *Climate change, food security, and population: Pilot-testing the model in Ethiopia*. SR-12-69.

Nelson, G. C., Rosegrant, M. W., Koo, J., Robertson, R. R., Sulser, T., Zhu, T., et al., (2009). *Climate change: Impact on agriculture and costs of adaptation*. Intl Food Policy Res Inst. http://www.ifpri.org/sites/default/files/publications/pr21.pdf. Accessed January 25, 2016.

NMA(National Meteorological Agency). (2007). *Climate change national adaptation programme of action (Napa) of Ethiopia*. Report output for a project entitled "Preparation of National Adaptation Programme of Action for Ethiopia" that was supported by the GEF through the UNDP. Addis Ababa, Ethiopia.

Reilly, J. M., & Schimmelpfenning, D. (1999). Agricultural impact assessment, vulnerability, and the scope for adaptation. *Climatic Change, 43,* 745–788.

Robinson, S., Strzepek, K., & Cervigni, R. (2013). *The cost of adapting to climate change in Ethiopia: Sector-wise and macro-economic estimates*. ESSP Working Paper 53.

Temesgen, G., Aleme, A., & Mulata. H. (2014). Climate change and livestock production in Ethiopia. *Academia Journal of Environmental Sciences,* 2(4), 059–062. July 2014. doi:10.15413/ajes.2014.0108

The AEA group (AEA Energy & Environment and Universidad de Politécnica de Madrid). (2007). *Adaptation to climate change in the agricultural sector*. Report to European Commission Directorate—General for Agriculture and Rural Development. Issue Number 1. December 2007.

You, G. J.-Y., & Ringler, C. (2010). *Hydro-economic modeling of climate change impacts in Ethiopia*. IFPRI Discussion Paper No. 960. Washington, DC: International Food Policy Research Institute.

Zelalem Y., Aynalem H., & Emmanuelle, G. (2009). Effect of climate change on livestock production and livelihood of pastoralists in selected pastoral areas of Borana, Ethiopia. In Climate change, livestock and people: Challenges, opportunities, and the way forward. Zelalem Yilma, Aynalem Haile (eds). Proceedings of the 17th annual conference of the Ethiopian Society of Animal Production (ESAP) held in Addis Ababa, Ethiopia, September 24 to 26, 2009. 3–21.

Zenebe, G., Stage, J., Alemu, M., & Atlaw, A. (2011). *Climate change and the Ethiopian economy: A computable general equilibrium analysis*. Environment for development. Discussion Paper Series, EfD DP 11-09.

Zhai, F., Lin, T., & Byambadorj, E. (2009). A general equilibrium analysis of the impact of climate change on agriculture in the People's Republic of China. *Asian Development Review, 26*(1), 206–225.

Zerihun, G. (2012). *An analysis of the impacts of climate change on crop yield and yield variability in Ethiopia*. PhD Candidate in Economics and Management. Graduate School of Social Sciences, University of Trento.

Local Climate Change Perceptions and Adaptation Strategies in East Gojjam Zone, Northwestern Ethiopia: Anthropological Approach

Takele Merid, Guday Emirie and Belay Simane

1 Introduction

There is ample evidence that climate in Ethiopia is changing and projections suggest that the rate of change will increase in the future (Simane 2013). Climate change is currently adversely affecting food security and livelihoods of millions of people. Agricultural production and food security in many African regions will likely be severely compromised by climate change and climate variability. This situation is further worsened by its poor state of economic development and low adaptive capacity. Extreme poverty, frequent natural disasters such as droughts and floods and heavy dependence of agriculture on rainfall further increases Ethiopia's vulnerability.

In anthropology, "adaptation" is long been the subject of study, but in relation with environmental change, it has become popular among anthropologists in the 1950s with the emergence of cultural ecology approach, which deals with social and cultural adaptation to environmental change (Sutton and Anderson 2010). Since very recently, anthropologists such as Fiske et al. (2014) define adaptation in the context of climate change. According to these scholars, "climate change adaptation"

This paper is part of an on-going dissertation work, titled "Gender Aspects of Livelihood Strategies in the Face of Changing Environment: the Case of Selected Communities in East Gojjam Zone, Northwestern Ethiopia".

T. Merid (✉)
Institute of Ethiopian Studies, Addis Ababa University, Addis Ababa, Ethiopia
e-mail: chochomerid@yahoo.com

G. Emirie
Department of Social Anthropology, College of Social Sciences, Addis Ababa University, Addis Ababa, Ethiopia

B. Simane
College of Development Studies, Addis Ababa University, Addis Ababa, Ethiopia

W. Leal Filho et al. (eds.), *Climate Change Adaptation in Africa*,
Climate Change Management, DOI 10.1007/978-3-319-49520-0_35

refers to the process of improving chances for survival in a given environment. Sutton and Anderson (2010) indicate that environment is always in a state of change, thus people should adapt to those changes. Peoples' adaptation always starts with the environment which they inhabit. Adaptation is, therefore, a way to respond to environmental change, be it economic, social, behavioral or physical adaptation. In these adaptations, "culture is a way in which groups of people can adapt to environment, through collective behavior or technology" (Sutton and Anderson 2010: 12). According to Eller (2009: 151), environment is "a particular combination of physical factors-climate, water and food supplies, natural resources…and adaptation is the process by which humans fit themselves into cooperate with their surroundings". Despite the slight differences between scholars and that of institutional definitions, in this study, adaptation to climate change incorporates cultural adaptation including changes in land use, farm-based activities and agricultural schedule. It also incorporates process of socioeconomic adaptation in which households respond to climate change impacts.

Studies that focus on climate change adaptation in East Gojjam zone are few. The reviewed studies focus on the Blue Nile Watershed areas. For instance, Simane et al. (2012, 2013), Zaitchik et al. (2012), Simane and Tedla (2011) and Deressa et al. (2009) are concerned with changing climate in the study areas. According to Deressa et al. (2009), in the Blue Nile Watershed areas, local climate change has brought weather extremes that resulted in frequent drought, flood and hail storms, which severely affect peoples' ways of lives. The authors then presented socioeconomic vulnerability to climate change in the Nile basin. This study also shows that the population in the Nile basin is susceptible to climate change. Moreover, according to Zaitchik et al. (2012), in the "Abay highlands", "climate variability" and "local climate contrasts" (2012: 435) are the major events that lead people to vulnerability. These calamities were accompanied by population pressure on physical environment that make the population highly vulnerable to negative effects of climate change. The changes led the study people to unfavorable condition in that through time their livelihood sources are devastated. However, perceptions of local people about local climate and physical environment changes are not given due attention. To overcome the problems, as Simane et al. (2012: 611) suggested, it is important to deal with understanding of peoples' "culture, historical settings" and study diverse income generating activities.

Study by Bewket and Alemu (2011) shows that there is a direct relationship between climate change and agriculture of the rural population of East Gojjam. According to the authors, climate change is manifested through rainfall variability and/or temperature increase. These negatively affect agricultural practices. However, to understand local adaptation strategies, it is important to know local perception of changes of local climate. Additionally, Bewket (2007) studied vulnerability of agricultural activities to rainfall variability in East Gojjam of Amhara region. He concluded that, the detail of the relationship between rainfall variability and fluctuations of agricultural production at rural level is not adequately studied.

Understanding about local climate change and adaptive strategies has become an important way to address and solve socioeconomic problems. As climate and physical environment are the only resources people exploit, their changes are the most constraining factors of development. Anthropological study that focuses in the study area is hardly found in this case and this paper fills this lacuna. The overall aim of this paper is, therefore, to provide the local perception of climate change, indicators of change and adaptive strategies that the study people used to overcome challenges of climate change.

2 Methodology

2.1 Study Area and the People

East Gojjam is located in the northwestern part of Ethiopia, situated at 37°42′36″ to 37°58′24″E longitude and 10°33′06″ to 10°50′24″N latitude. It incorporates four conventional agro-ecologies: *wurč*[1] (alpine) *däga* (highland), *wäina-däga* (midland), and *qolla* (lowland). In its agro-ecology, majority of East Gojjam zone is *däga,* followed by *wäina-däga.*[2] Topographically, the zone lies in altitudinal range of 800–4100 m above sea level. The highest point of the zone is the massive of Choqe Mountain, which has 4100 m of height. The lowest point is Abay Gorge, locally called Abay *bäräha*, which has an elevation of 800 m above sea level.

This study was conducted in three *worädas* of East Gojjam zone namely, Sinan, Gozamen and Dejen, from which three research sites (*qäbäles*) were selected to represent three agro-ecological settings. Gedamawit is selected from Sinan *woräda* to represent highland agro-ecology. In this area, households cultivate crops include varieties of potato and barley. They also cultivate oats, wheat, *əngədo* (*Avena* species), *awaquñ* (triticale), beans and peas. Livestock production is another livelihood activity that they depend on. Common types are sheep, cattle, horse and poultry. Enerata is selected from Gozamen *woräda* to represent midland agro-ecology. In this area, barley, wheat and *ṫef* are widely cultivated crops. In addition, maize, barley, *awaquñ*, and *əngədo* are important crops for food as well as cash sources. Onion, garlic, potato, tomato, pepper and carrot are sources of income for several households. Kurar is selected from Dejen *woräda* and represents lowland agro-ecology. In Kurar, households produce especially, sorghum *ṫef* and sesame as well as follow their livestock. Kurar is found between 800 and 1200 m above sea level. Annual average rainfall is between 250 and 800 ml; while annual temperature ranges between 20 and 28 °C.

---

[1]As key to transliteration system for local terms, see the attached appendix, at the end of the paper (Appendix I).

[2]According to data received from East Gojjam zone Agricultural and Rural Development Office, in East Gojjam zone, *wurč* is (2%) *däga* (45%), *wäina-däga* (37%), and *qolla* (16%).

2.2 Study Methods

To conduct this research, the principal researcher spent the total of 13 months in the field. During his stay in the field, he used different data collection methods. In order to address objectives of this study, both primary and secondary data were used. Primary data were gathered through qualitative and quantitative approaches. Qualitative methods[3] consist of observation,[4] key informant (KI)[5] and in-depth (IDI)[6] interviews as well as Focus Group Discussions (FGD).[7] Qualitative data were triangulated across each other to increase the validity and reliability of information gathered from different sources. To support evidences collected through qualitative methods, household survey was conducted on 341 households. Of the 341, 331 questionnaires were correctly filled by respondents. The 331 questionnaires were entered into the Statistical Package for Social Sciences (SPSS) and analyzed. Result of the household survey is presented in tabular form where it supports the qualitative analysis. To understand themes of this study, secondary data sources of both published and unpublished literatures were reviewed. Moreover, archival materials were also reviewed to substantiate data gathered through qualitative and quantitative methods.

3 Results and Discussions

3.1 Local Perceptions of Climate Change

Climate change is locally known as *yäwäqt mäqäyayär*, which is to say, “change related to the usual seasons’ pattern”. Study participants[8] perceive local climate change consists of three major variables: rainfall, temperature and wind. Rainfall deviation from its usual pattern of timing is characterized as absence or delay, early or late, last long or short. Change of rainfall pattern is also variability of rainfall intensity, which can be high or low or normal, with or without hail. Secondly, *yäwäqt mäqäyayär* is also related to local temperature increase, which takes soil moisture out. Thirdly, local climate change is about the advent of unexpected,

[3]Profiles of the qualitative data sources area attached at the end of this paper (Appendix II).

[4]This was conducted during the ethnographic fieldwork undertaken between March 1, 2013 and February 28, 2014.

[5]There were 13 key informants who provided relevant information throughout the study.

[6]There were a total of 52 informants (interviewees) from all the agro-ecologies in which 20 from the highland (Gedamawit), 16 were from Enerata (midland) and 16 were from Kurar (lowland) areas.

[7]There were a total of nine focus group discussions organized in the three agro-ecologies.

[8]All the FGDs agree that there are local climate and physical environmental changes.

undesirable and unusual wind that often devastates crops. Local seasonal variation depends on local climate, which in turn determines households' working schedules.

By considering the onset and withdrawal of rainfall, study participants, for instance, KI2 and KI4 explain about rainfall variability. Sometimes, the rain is accompanied by *bäräd* (hail). In another time, it is followed by soil erosion and land degradation. This is due to the erosive rain and erodible soil nature of the area (Zaitchik et al. 2012). Thus, rainfall variability is considered as the major cause and occurrence of unusual seasonal pattern. Unusual seasonal pattern is also against the schedule of agricultural activities, which, in turn, adversely impacting the overall amount of grain produced. Deviation of rainfall also negatively affects quality and quantity of livestock.

There are three major seasonal farm-work schedules that households depend on, locally known as *wäqtoč* (seasons) to perform their agricultural activities. These working schedules depend on the beginning and end of rainfall and situation of one season determines the other. Households locally predict and adjust their livelihood activities based on consistency of the seasons. One of the seasons is the *bälg*, also known as *yäbäga zənab* (dry season's rain) in which households expect a simple and short duration of rain. This is usually from about mid-February or early March to April (in highland and midland) and roughly from April to May (in lowland). The *bälg* rain helps soil to get adequate moisture. Thus, households are expected to prepare croplands, fill some land with potato (in highland and midland) and continue to work on the land and plant sorghum in the next month (in lowland). As a result, households consider *bälg* as the season of hope and happiness if the rain comes on time as expected and it is a desperate season if otherwise. An informant states, however, due to shortage of rain, if they are not successful in planting crops during *bälg*, implies a bad situation in the months that follow. Their current experience shows that *bälg* rain is being diverted from its normal pattern. Moreover, informants explain that, they know whether rain is adequate or not, simply by observing duration of the rain and by looking at the inside part of the soil. For instance, if the inside part of the soil is dry, they understand that rain is inadequate so that seeds cannot properly grow.

The *bälg* season is followed by *kərämt* (the long rainy season). This is usually from June[9] to October in lowland and extended to November and December in highland and midland areas. *Kərämt* starts with a relatively heavy rain that should be followed by simple rain important to shower crops. Due to its longer duration, *kərämt* is considered as a peak season in which households actively engage in farm activities and often encounter with labor shortage. It is also a season of hardworking that needs endurance and patience of household heads. As a temporary means of relief for the shortage of food and cash in the highland and midland areas, potato plays an irreplaceable role while *boqolo eśät* (ripen maize) is a temporary means of

[9]Due to climate change, sometimes the *kərämt* rain comes in May and/or in July, deviated from its normal time. It also ends in unusual manner, either in August (too early) or in December and January (too late).

relief in the lowland area. However, this success depends on the reliability of rain. During this season, if the rain does not come on time, as an informant expresses, *qänum čäläma näw* (meaning, the day time is also dark).[10]

In the study area, *kərämt* is considered as a tough season in which most farming households are challenged by shortage of food and cash. For shortage of food and cash, there are several reasons that FGD2, FGD5 and FGD8 explain. One of the reasons is, households have less chance to get alternative sources of income and cash. Another reason, May is a month of wedding in which households spend a lot of grain and other properties to prepare feast in the form of *aqolquay* (hosting wedding feast for members in kinship). To host the banquet on a wedding ceremony, households use grain in the stock, which often leads them to shortage of food and cash in the following months. Additionally, in the low and medium income households, during *kərämt* season, there is a shortage of food and cash because farming households plant various seeds and they lack extra seed reserved at home. Thus, they use any grain for seed purpose that they reserved at home. Subsistence nature of their agriculture, which they call as *käəj wädaf*, meaning, "hand to mouth" is another reason explained as a factor to households' food shortage. This, in turn, is because, they produce grain only once in a year, the product is small and not enough to feed household members throughout the year. In the survey households were asked whether the lands they own were enough or not to the sustenance of their households. Accordingly, 59% of the respondents said that the land they currently owned is not enough to sustain their households. Whereas, 39.5% said that the land is enough while 1.5% responded that the land they owned is more than enough to the sustenance of their households.

Kərämt is followed by what is locally called *mähär wäqt* (a harvesting season). It is usually between December and January in highland and midland areas; and from mid October to December in lowland area. During *mähär*, rain is not favored and should be very light otherwise. Just like *kərämt*, this season is a peak working season, since households engage in harvesting their crops. However, their success and/or failure to harvest crops depend on situation of the rain during *kərämt*. Success of households also depends on their strength to work hard during the previous two seasons: *bälg* and *kərämt*. As KI9 and KI12 indicate, in the past, the three seasons (*bälg*, *kərämt* and *mähär*) have distinctive timeframes that they used to follow and prepare themselves in order to act accordingly. Now, they are unable to predict all these.

Participants in FGD1 and FGD4 perceive that, the cause of climate change is God's punishment for their sins and for their misbehaviors.[11] But it does not mean that all of them are cursed for their wrong actions. One of the participants in FGD1,

---

[10]In the study area, darkness is a symbol of any unfortunate event and desperation. At dark time, there is no work at all, because there is no electricity that gives light and enables people to move from place to place for work.

[11]These are, cutting indigenous trees, encroaching grazing lands for personal use, extravagancy, neglect fasting and praying, improper use of agricultural products, drunkenness, theft, deceiving, not helping the needy, etc.

said, the issue is, *lä-hatan yämäta lä-šadqan yətärfal*, this is to say, "God treated both sinners and the righteous persons together". This shows that while few individuals devastate the resource, majority is negatively affected. They believe that they do not have any alternative other than adapting to local climate change and its negative consequences. The above statement is in line with a study done by Adem and Amsalu (2012) who explained that communities in the southern lowlands of Ethiopia perceive that climate change is what God brought to them. They explain that in most cases they do not have capacities to adapt to climate change impact that affects their livelihoods. As a solution, they suggest to work on rehabilitation of their physical environment.

3.2 Indicators of Climate Change

There are indicators of local climate change that study participants explain. One of the indicators is locally known as *dərq* (drought). FGD1, FGD4 and FGD6, defined drought in various ways. First, drought is related to the absence and delay of rain from its usual seasonal pattern, no matter how long the absence of the rain is. Then, drought is related to loss of soil moisture or wetness of their land due to reduction in the amount of rain and increase in local temperature that led local crops being unable to grow. Study participants remember that since in the past twenty years, in addition to major droughts, minor droughts have also occurred in their locality each year or at least within two years of interval.

For example, they remember and narrate about three major *dərq* occurred in their areas. The 1983–84 *dərq* was followed by famine, known in the area as *tərgäte* (meaning, drought that trampled the people). It occurred due to the absence of the *bälg* and *kərämt* rains. Secondly, the 1991–92 *dərq* was followed by famine, known in the area as *mančəloś* (meaning unbearable), which they attribute the problem to the failure of the *bälg* and *kərämt* rains. The third is the 2001/2 *dərq* and starvation, locally known as *adära* (this is when someone is in a situation of begging the drought not to harm him/her). It mainly occurred in the highland and midland areas due to shortage of *kərämt* rain, which led to the total crop failure. Focus group participants also indicate that, since in 2002, rainfall fails within a short interval so that crops fail to grow. Though not explained in detail, occurrence of frequent drought in the Blue Nile area is explained by (Deressa et al. 2009). The overall effect of *dərq* is decline in the amount of crop production.

IDI5, IDI17, IDI30 and IDI42, conclude that drought eventually devastated what they have accumulated, particularly, the natural and social asset they have. To understand the effects of rainfall variability on grain productivity, households were asked to rank five years productivity situations. The summary of survey result is indicated in Tables 1 and 2.

Table 1 Households experiences of yearly productivity variation by agro-ecology

Average ranking on comparison of yearly productivity							
Production year		2013/14	2012/13	2011/12	2010/11	2009/10	2008/09
Agro-ecology							
Highland	Mean	2.4	4.7	3	3.2	3.4	4.4
	Rank	1	6	2	3	4	5
Midland	Mean	1.4	2.5	4	3.8	4.6	4.9
	Rank	1	2	4	3	5	6
Lowland	Mean	3.5	3.5	3	3.5	3.6	4
	Rank	2	4	1	3	5	6
Overall	Mean	2.3	3.6	3.3	3.5	3.8	4.5
	Rank	1	4	2	3	5	6

Source Household Survey by the Researcher, December 25, 2014–February 17, 2015

Table 2 Experiences of yearly productivity variation by household type

Agro-ecology	Household type	Average ranking on comparison of yearly productivity						
		Year	2013/14	2012/13	2011/12	2010/11	2009/10	2008/09
Highland	FHHH	Mean	2.7	4.7	3.9	3.8	2.4	3.7
		Rank	2	6	5	4	1	3
	MHHH	Mean	2.4	4.7	2.7	3.1	3.6	4.6
		Rank	1	6	2	3	4	5
Midland	FHHH	Mean	2.3	3.1	3.1	4.5	4.1	4
		Rank	1	2	2	6	5	4
	MHHH	Mean	1.2	2.3	4.2	3.5	4.7	5.2
		Rank	1	2	4	3	5	6
Lowland	FHHH	Mean	3.5	3.7	2.7	3.7	3.9	3.6
		Rank	2	4	1	5	6	3
	MHHH	Mean	3.5	3.5	3	3.5	3.5	4
		Rank	5	4	1	2	3	6

Source Household Survey by the Researcher, December 25, 2014–February 17, 2015

Table 1 presents the ranking of yearly productivity over five years, between 2008/9 and 2013/14. Respondents show that, 2013/14 was a relatively productive year and ranked first. Then, they ranked 2011/12 the second favorable year. This is because, in 2011/12, respondents produced in a better way next to the year 2013/14. The year 2012/13 was considered as a relatively unfavorable year, which is ranked fourth. This result goes with statements of the study participants (for instance, IDI16, IDI18, IDI25 and IDI38), who explain that in 2013/14 and 2011/12, rain was consistent from *bälg* to *kərämt* and it rained on time as desired. They also argue that, particularly in addition to the good condition of rain in 2013/14, they have started to benefit from what is locally known as *yätäfasä ləmat* (watershed development)[12] in their localities. It is also possible to understand that respondents remembered 2008/09 and 2009/10 as bad years respectively, since they produced low amount of grain in quintals. In these years, as the above study participants indicate, they witnessed drought at the beginning of *bälg*. Then, during *kərämt*, the intensity of rain was very high and accompanied by hail that devastated crops. As a result, productivity was considerably negligible.

From Table 1, it is possible to understand that, specifically, in the year 2012/13, compared with midland and lowland areas, in highland, productivity was ranked last. This was due to the hail that devastated their crops such as potato and led households to harvest small amount of potato and grain (IDI6). In both midland and lowland areas, year 2008/09 was ranked last because households earned small amount of product. 2011/12, 2012/13 and 2013/14 were the 2nd productive years for the highland, midland and lowland areas respectively. From the evidences provided, we can understand that there is a deviation of rainfall from its usual patterns (i.e. highly variable in its intensity and accompanied by the two weather extremes such as hail and drought affected the livelihood strategies of the people.

Table 2, supports perceptions and experiences of Female Household Heads (FHHHs) and Male Household Heads (MHHHs) about productivity over the past consecutive five years (2008/9 and 2013/14). The result shows that both types of households have similarities and differences. For instance, in the highland area, except the production year of 2012/13, in all the other years, MHHHs and FHHHs had different rank in production, which shows their different experiences towards the change in the local climate conditions. In the midland area, FHHHs and MHHHs had similar experiences in the production years of 2013/14 and 2012/13 that in both years, MHHHs and FHHHs earned a relatively better product of grain. In the rest of the years, the two groups of respondents had differences in their experiences to the yearly grain production. In the lowland area, MHHHs and FHHHs have similar views in the years of 2011/12 and 2012/13 that both household types ranked their level of production in fourth and first, respectively. That is, in 2011/12, both male and female household heads ranked their productivity forth and the year 2012/13 ranked first. This similarity in rank shows that FHHHs and MHHHs experienced the effect of climate change in similar ways.

Furthermore, women and men differently experience negative effects of local climate change. For instance, IDI1 and IDI2 describe that in highland area, potato is the most important crop that households use as a source of cash and consumption. Households used to plant potato twice a year at *bälg* and at the end of *kərämt* seasons. Potato during *bälg* is produced for cash, locally known as *yäzär mägźa* (to purchase seeds), while potato produced during *kərämt* is mostly for consumption, which is locally considered as *yä-kərämt mäśägagäria* (a means of transition to the next season). The most appropriate season to plant potato, however, is *bälg*, which is becoming difficult because of the short duration and/or absence of rain. Thus,

[12]Watershed management has been started in 2003 as the national environmental rehabilitation program.

now potato is often planted once in a year. If potato is not planted during *bälg*, households do not earn income from the sale of potato. This situation would lead them to a risk of starvation and purchasing seed because they cannot substitute potato with other grain, which they produce once in a year. Households also use potato as an indicator and predictor of upcoming seasons of their livelihoods and forecast whether the future will be better or not. In the lowland, a crop important to predict situation of upcoming season is sorghum, which they have to plant during *bälg* season. According to Bewket (2007), the relationship between climate variability and agricultural production is not adequately studied. However, Simane et al. (2014) indicate that, due to climate change in highland of the study area, livelihoods of the communities are disturbed. This in turn, shows the disturbance of agricultural activity, the mainstay of most of the population of the study area.

Increase in temperature is another indicator of local climate change. For instance, households were able to compare the past and present temperature by indicating that their area was covered with *ćəgag* (fog). They also indicated that they used to get sunlight only rarely and now it is almost the opposite of past temperature. In midland and lowland areas, study participants complain about the unusual high temperature and express it as *märetu nädädä* (meaning, the land is burnt) because of the increasing heat from sunlight. However, most of them were unable to explain the time when local temperature has begun to rise.

Women and men have different views while they explain about temperature change. Men perceive temperature increment both positively and negatively. As they explain, increase in temperature negatively affected their physical environment, which in turn, affected their agriculture. For instance, the high temperature increased evaporation and took the soil moisture and nutrients then changed the soil into unproductive status.

There is positive aspect of temperature increase that informants explained. For instance, in the highland area, contrary to the past, now, they are able to work several hours in a day because the day is sunny and the air is relatively calm. An informant perceives that, *muqätuma ləbs näw*, meaning, "the increase in temperature serves as a cloth". Since highland area had a very cold temperature, the people used to spend money to wear several clothes at a moment. Increase in temperature reduced the cold weather of the area so that they are not forced to buy as many clothes as they used to.

In highland and midland areas, according to women informants, they benefited from temperature increase. For example, in their local market, there is a high demand for local drinks such as *aräqe* (local liquor) and *ṫälla* (local beer). In order to brew the drinks, ingredients such as *geśo*[13] (dried hops) and *bəqəl*[14] (malt) have become important cash sources, especially for women who make the drinks and supply the ingredients. Preparing local drinks and the ingredients are exclusively women's domain. In the past, women used to supply the ingredients once in two

[13]Species of *Rhamnus Prinoides*.

[14]They prepare malt from barley and sometimes from maize.

weeks or in ten days interval. This was mainly due to the low temperature in the area did not help dry hops and malt as fast as needed. Thus, they were unable to get cash when they need. Now, due to increase in the local temperature, women supply the items twice a week and have begun to earn a relatively more income than they used to. Women produce hops on the plot of land at homesteads and they were often supported by their children to treat the plant through watering and harvesting the weed.

In highland, another indicator of local climate change is the introduction of maize. Informants stated that since recently, due to increase in the temperature of the area, maize, which has never been grown in the highland, has started to grow. This is locally called *yägubəl mädäläya* (a means to pacify female children). Maize is particularly important during *kərämt* season because, at this time, shortage of food is chronic and maize ripens in the mid of July. This means that, during rainy season, most households encounter with food shortage and children cannot tolerate the hunger. After the introduction of maize, they enjoyed maize in the form of *bäqolo eśät* (ripen maize) at home and those who produce it are temporarily relieved their children from the *kərämt* hunger. From maize, they not only get grain but also benefit from the cane and corncob that serve households as sources of fuel energy. In addition, in the past, highland weather was not suitable to rear goats. Now, there are households that own goats due to local temperature increase. However, households have not started to fully produce the new products for different reasons. Firstly, they fear that new crops may fail to give adequate product so that they fear it would be unnecessary cost if crops fail to give yield. Secondly, households do not have sufficient land to produce new crop types. Third, most households do not get adequate technical support from agricultural extension workers. As climate change indicator, FGD1, FGD4 and FGD7 also mention prevalence of untimely and unexpected wind that disturbs their livelihood activities. They call the wind *əngəda näfas*[15] (unpredictable wind). Wind unusually comes during harvesting season, which has considerable negative effects on the ripen crops such as *ṫef*, barley, sorghum and sesame. Through time, especially during harvesting season, the wind gets pervasive and devastates crops by dispersing the seed, which causes locally called *bəkənät* (waste) thus, eventually reduces the amount of expected grain yield.

In addition, IDI6, IDI7 and IDI21 explain the occurrence of human, animal and crop diseases as different from in the past. They perceive, diseases are the cumulative effects of local climate change. Most of the diseases are prevailed especially in the past two decades. There are also diseases locally called *näbbar bäśəta* (diseases existed in the past) that were common in the study areas, but persist today in different forms. The prevalence of diseases in different ways makes the subsistence nature of peoples' livelihoods to further deteriorate that in turn, affect their

[15]The term *əngəda* literally means "guest" and *näfas* means "wind". This is to compare the untimely advent of wind with the guest who unexpectedly arrive at his/her relatives. The study people consider both events as undesirable.

economy and their lives in general. For instance, as human diseases, study participants in the highland area suspect the prevalence of *gunfan* (common cold) and *täqmaṫ* (diarrhea) different from in the past. They indicate that these diseases were easily cured and once treated, for long they never appear again. Now, they recurrently prevail and are not easily cured. They relate these situations with temperature increase in their locality.

According to KI4 and KI7, in the past, highland and midland areas were known for persons who lived more than 100 years. As KI4 expresses, "until recent past, individuals used to die natural death and without being contracted by any diseases, due to the cold and healthy climate condition of the area. Now, we notice that friends and relatives are dying very early, live not more than 60–70 years". In midland area too, as informants explain, in their area, prevalence of *nədad* (malaria) is not common. They also here, in the past, mosquito was found only in the lowland area. In the midland area, nowadays, the number of persons affected by this disease is increasing. Moreover, in the past ten years, new respiratory diseases occurred in all the study agro-ecologies. Deterioration of health conditions of the people is attributed to local climate change.[16]

With regard to crop diseases, FGD3, FGD6 and FGD9 indicate that *wag*, sometimes called *bića-wag* (rust that affects crops) *magəd* (a pest that burns crops) and *arämamo* (a disease that spoils and kills crops before they are cultivated) are related to local climate change that attack their crops. Informants perceive that, in the past twenty years unexpected wind brought new types of *arämamo* from other areas. Focus Group participants also explain that controlling the problem is beyond their capacity unless they receive support from government. In the lowland area, a worm that devastates their crops is known as *tämč* (armyworm). This worm used to occur within ten or so years of intervals, mainly because of the unusual rain pattern. Since very recently, the armyworm occurs in every one-year interval. Thus, it has become a challenge to properly harvest their crops. KI3 and KI12 indicate, the occurrence of armyworm within a short period of interval is the consequence of climate change, especially the problem of rainfall timing and intensity.

In addition to the above, prevalence of what is locally known as *qorṫəm* (a pest that cuts the crop) is a very recent crop disease occurred in the highland and midland areas. Though it affects all the crops, it particularly devastates a crop locally called *yädabbo sənde* (wheat used to make bread). Due to this, women were unable to provide bread for their children. An informant explains that *yäləj gənəbar gäräfäñ*. This is to say that "I could not fulfill my children's need of bread, thus the adverse condition is painful". Others, locally called *nədəft*[17] and *qurba*[18] are, examples that informants mentioned as newly prevailed animal diseases in their

[16]Sinan *Woräda* Health Office, April 21, 2014; Gozamen *Woräda* Health Office, June 8, 2014 and Dejen *Woräda* Health Office, October 28, 2014.

[17]Broken-windedness (disease of equines transmitted by fly bites).

[18]Blister, pustule, malign bubo, which attacks hands or feet; sudden illness causing diarrhea.

localities, which is related to local climate change. The overall result is deterioration of their incomes and unhealthy lives.

3.3 Adaptive Strategies of the People to Climate Change

Study participants, despite they claim that they lack capacity to adapt to climate change effects, there are strategies that they use. An informant states, that "*əsu yamätawən əsu yəmäləsäwal*", this means, "it is only God that can give us a solution." Thus, they believe that, to reduce negative effects related to local climate change, their intention is to pray to God.[19] That is, to respond to local climate change, the people need support of God in the first place. What is expected of them is confessing their past misbehaviors in order to get mercy from God. Informants, argue that it is not the practice of observing saints that brought climate change and poverty to their areas. Rather, it is their misbehavior that resulted in the overall adverse impacts. Celebrating religious festivities and gathering on Sundays are not only religious in their objectives, but also the ways they strengthen their social networks and the way they control individuals who misbehave on the local environment. Therefore, after they restarted their religious celebrations and meetings on every Sundays, according to informants, God helped them to have positive attitude towards their physical environment. They also started discussing about their local environmental situation, which enabled them to work on environmental rehabilitation. For instance, currently every one, women and men participate in the practice of watershed management.

Shifting the time of farm operation is another way of responding to negative climate change effects. In the past households had, at least a one-month break time between the end of harvesting and the beginning of land preparation. They use the one-month break time to renew their social networks. Now, this is not possible and they are tight with farm works throughout the year and without interrupting their farm tasks. Thus, farming households make time adjustment and prepare farmland as early as possible immediately after they harvest. They also make adjustment of planting and harvesting months in order to cope up with rainfall variability. In the lowland, according to informants, during planting season, for instance, if *bälg* rain delays from its normal time, they often prefer to wait for the coming of rain. That is, in lowland area, households prefer to plant crops only when the rain comes.

One of the reasons for flexibility in their working plan is that, especially, in lowland area, there is no/limited amount of moisture that enables the seed to survive for a relatively longer time without rain and moisture. Moreover, to support their farming activities, there is no subsidy that households receive if grains fail to grow. Providing seed in subsidy form is important when households lack extra seed at

[19]This means, they need to confess their wrong acts through fasting, respect saints, celebrate religious festivals, avoid theft, etc.

home that they replant if the first seedling fails to grow. To minimize any risk of crop failure, farmers prefer to wait for the rain. To overcome this problem, they suggest locally known as *mäsəno* (irrigation) from the rivers. Households also fear planting seed due to the debts that they receive in the form of seed and other inputs. This is because, they often do not pay back unless they adequately produce from the seed they plant. In this regard, an informant clearly states that, *märet mərt baysät mänəgəst ədahən mäč yətäwələhal*, this is to say, "even if production fails, the government does not forgive the debt". Therefore, households often use their comparative advantages as a way to cope up with the problems of local climate change. One of this is abstaining from receiving farm inputs in debt form.

In highland and midland areas, it is indicated that even if there is a risk of delay or/and absence of *bälg* rain, households opt to plant seed instead of waiting for rain and postpone seed planting. For example, if land for potato is well-prepared, even if *bälg* rain delays, potato seed can tolerate few days without being spoiled. However, if there is no *bälg* rain at all, instead, households are forced to dismiss the potato and substitute with another crop. This is because, potato seed may be spoiled if it does not get adequate rain within a month or two after it is planted. Besides, potato may not give yield at all if there is no rain within five to six weeks after it is planted. It should also be ripen before heavy rain starts in June and July, thus, it should be planted in the early days of *bälg* rain. By the rain of June and July potato may not properly give yield because it would be devastated by heavy rain. The same is true in lowland area, where crops that do not tolerate heavy rain should be planted in the early days of the *bälg* so that they would ripe before the heavy rain begins.

In order to minimize or even to avoid risk of crop failure, farming households prefer to apply short duration and stronger crop varieties that tolerate both temperature and shortage of rainfall. The problem is that, it is only better-off and medium income households that have access to improved seeds. The assumption is that, these households can pay back the debt. An informant asserts that "better-off households have the capacity to purchase improved seed in cash while price of the seed is not affordable for most of medium and low income households". In addition to improved seed, most of the medium and low income households are denied access to drought tolerant seeds and the seeds are mostly for better-off households.

In response to the failure of local agricultural offices to provide new improved varieties of seeds, households indicate that they diversify crops on their plot of lands.[20] Crop diversification is to avert the risk of crop failure in that the assumption is, if one crop fails to grow, the other may succeed. Moreover, crop diversification helps to maintain soil fertility. In highland area, where there is a relatively high land scarcity, households plant crops in the form of row cropping. In this practice, they plant two or three types of crops simultaneously on a given plot of land.[21] In

[20]This is in the form of intercropping (mixed) and row-cropping (not mixed crops but side-by-side on the same plot).

[21]In the highland, for instance, bean and wheat or bean and barley; whereas in the lowland, the mixings of *ṭef* and sunflower; sesame and sorghum are common ways of intercropping.

lowland, diversification is in the form of intercropping and row cropping. In highland and midland areas, intercropping is favored only on *yäguaro märet* (backyard) that they plant varieties of crops such as hops (Rhamnus Prinoides), bean and wheat in the form of intercropping.

According to informants, during harvesting, to protect their grain products from unexpected and untimely rain and wind, households mobilize domestic labor, which has become difficult. Thus, currently, better-off households often hire daily laborers, in addition to the labor they mobilize from their households and their kinship members. As a result, better-off households are often on the safe side and they are able to increase their chance to harvest their crops without being damaged by local climate change. Moreover, in lowland area, it is common to observe men and women in group of 5–10. They are fellow kin groups and enter into contracts with better-off households to harvest crops on the field. The price of the labor during the fieldwork (March 2013–2014) was is 200–300 ETB per 0.25 ha, i.e., one *ṫəmad* of crop land and this is known as *quṙt* (in cash). Here, the poor, unhealthy and female household heads are vulnerable to risks of climate change effects. This is due to lack of capacity to mobilize others' labor and because they often look for the support of others to harvest their crops.

Despite their efforts, households could not escape from the negative consequences of local climate change. The change is continuously hitting most of them and their farm activities. This, in turn, devastates, especially their physical environmental resources. The change in physical environment decreases land productivity and eventually lowers households' incomes. To explain the challenges that households face with local climate change, an informant clearly explains, "*bänqərt lay joro dägəf honäbən*", this is to mean, "this is a double challenge". It is to indicate that, the challenge is not only from climate change effects, but also from the physical environmental degradation. The statement shows communities' lack of adaptive capacity. However, generally, as study by Simane (2013) shows, there is a need to work on the sustainability of how communities can adapt to climate change impacts.

4 Conclusion and Recommendations

In the past twenty years or so, farming households in the study areas experienced significant change in local climate. Climate change is the change in rainfall situation. This includes deviation of rainfall from its usual seasonal patterns, variability of rain intensity and the general decrease in the amount of rainfall. They also perceive climate change is about temperature increase in the area and its negative as well as positive effects on the physical environment. These two changes are resulted in frequent drought. Local climate change is also about the advent of untimely and unexpected wind. The study communities believe that, it is difficult to reverse climate change, but they try to adapt to the adverse impacts of the change. One of the adaptation strategies is to confess their misbehaviors towards nature. As a result, they have started to celebrate weekly and monthly religious observance that they were told to stop and discuss about their environment on their religious get-together. Religious observance strengthens their solidarity and social network, thus, easy to control individuals who misbehave on physical environment. This also encourages them to participate in the watershed management and rehabilitate their local environment. They also respond by extending their farm activities throughout the year without any break. This, in turn, negatively affected their social networks that they used to have during a one-month break, immediately after harvesting time. Furthermore, in order to overcome the problems attributed to rainfall variability, in the lowland, people tend to delay the time of planting their seeds; whereas, in the highland and midland, they plant their seed, even if the rain delays. Local climate change increased peoples' activities with low return. Though utilization of farm input has its own problem, people use it as a means to respond to local climate change. In this regard, better-off households benefitted from using farm input. To get additional incomes, female household heads engaged in petty trade and selling their labors in the nearby towns. This, in turn, negatively affected women, in that, tasks at home also remain for women, after they spend their time outside of home. Male household heads participate in planting eucalyptus tree on their cultivable land. Men also migrate to towns to work as laborers that enable them supplement their income from farming activities. Migration of men to towns led women to act as household heads in the absence of their husbands. This situation makes women more burdened. Overall, due to local climate change, women are the most negatively affected group. In the process of responding to climate change also, women have become more burdened and vulnerable to other socioeconomic problems. Therefore, poor households such as female headed households and unhealthy household heads are the most vulnerable social groups. These households need to be supported in the process of their adaptation to climate change. For instance, their farming activities need to be subsidized. This could be in terms of providing farm inputs and improved seed varieties that better-off households can easily access to. It is important if new technologies such as small scale machineries are introduced in the study area that enable communities to harvest their crops before it is damaged by undesirable rain and unexpected rainfall and wind. In addition, information about climate variability and agricultural extension services should be equally addressed to all social groups without discrimination. This may enable households to harvest their grain on time. Then, there should be further studies that deal with prevailing new crops, animals' and human diseases and their relationships with local climate change.

Appendix I: Key to Transliteration System for Local Terms

I. The seven sounds of the Ethiopic alphabet are represented as follows:

Order	Sounds	Example Amharic	English translation
First	በ = B(b)ä	*bäqällä*	Grown
Second	ቡ = B(b)u	*buqayya*	Shooting grain
Third	ቢ = B(b)i	*billawwa*	Knife
Fourth	ባ = B(b)a	*bahərzaf*	Eucalyptus
Fifth	ቤ = B(b)e	*bet*	House
Sixth	ብ = B(b)ə	*bəddər*	Loan
Seventh	ቦ = B(b)o	*boräbor*	Eroded land

II. Regarding the 6th form in the above list, it must be noted that "i" will be suffixed to the letter only if the letter is vocalized or stressed, otherwise, it will not be required at all. As a general rule also, the "i" is not required when the 6th form is the last letter of a word. Examples,

መምህር	Mämihir	A teacher
ብድር	Biddir	Loan

III. Palatalized sounds are represented as follows:

Sounds	Represented by	Amharic example	English translation
ሸ	ś	*bäśita*	Disease
ቸ	č	*čəggər*	Problem
ኘ	ñ	*moññ*	Foolish
ዠ	ź	*mazäźa*	An order
ጀ	j	*ənjära*	A type of food

IV. Glottalized sounds are represented as follows

Sounds	Represented by	Amharic example	English translation
ቀ	q	*bəqəl*	Malt
ጠ	ṫ	*ṫälla*	Local beer
ጨ	ć	*ćəgag*	Fog
ጸ፣ ፀ	š	*šäbäl*	Holy water
ጰ	ṕ	*ṫäräṕieza*	A table
ፐ	p	*police*	Police
ከ	k	*kəlkəl*	Forbidden
ተ	t	*täqaṫala*	Burnt

V. Gemination should always indicated by doubling

Käbbädä
Taddässä

Appendix II Profiles of Study Participants

I. Key Informants

Key informant	KI2	F	52	Widowed	March 15–May 14, 2013	Gedamawit *qäbäle*
Key informant	KI3	M	70	Married	March 15–May 14, 2013	Gedamawit *qäbäle*
Key informant	KI4	M	60	Married	March 15–May 14, 2013	Gedamawit *qäbäle*
Key informant	KI7	M	42	Married	May 15–July 15, 2013	Enerata *qäbäle*
Key informant	KI8	M	62	Married	May 15–July 15, 2013	Enerata *qäbäle*
Key informant	KI9	M	48	Divorced	May 15–July 15, 2013	Enerata *qäbäle*
Key informant	KI10	F	40	Divorced	September 15–October 20, 2013	Kurar *qäbäle*
Key informant	KI12	M	56	Married	September 15–October 20, 2013	Kurar *qäbäle*

II. In-depth Interview

Information source	Code	Sex	Age	Marital status	Date of interview	Place of interview
In-depth interview	IDI1	F	45	Divorced	March 16, 2013	Gedamawit
In-depth interview	IDI2	M	52	Married	March 16, 2013	Gedamawit
In-depth interview	IDI5	M	58	Married	March 17, 2013	Gedamawit
In-depth interview	IDI6	M	36	Married	March 18, 2013	Gedamawit
In-depth interview	IDI7	M	63	Married	March 18, 2013	Gedamawit
In-depth interview	IDI12	M	55	Married	March 24, 2013	Gedamawit
In-depth interview	IDI14	M	60	Married	March 24, 2013	Gedamawit
In-depth interview	IDI16	M	49	Married	April 2, 2013	Gedamawit
In-depth interview	IDI17	F	49	Widowed	April 2, 2013	Gedamawit
In-depth interview	IDI18	M	42	Married	April 3, 2013	Gedamawit

(continued)

(continued)

Information source	Code	Sex	Age	Marital status	Date of interview	Place of interview
In-depth interview	IDI21	M	56	Married	May 17, 2013	Enerata
In-depth interview	IDI25	M	40	Married	May 18, 2013	Enerata
In-depth interview	IDI30	M	60	Married	May 22, 2013	Enerata
In-depth interview	IDI38	M	49	Married	September 22, 2013	Kurar
In-depth interview	IDI42	M	51	Married	September 26, 2013	Kurar

III. Focus Group Discussion Participants

Code	Sex	No. participants	Age	Date of interview	Place of interview
FGD1	Men	9	18–65	February 2, 2014	Gedamawit *qäbäle*
FGD2	Women	7	20–50	February 3, 2014	Gedamawit *qäbäle*
FGD3	Men and women	8	18–50	February 4, 2014	Gedamawit *qäbäle*
FGD4	Men	10	20–60	February 10, 2014	Enerata *qäbäle*
FGD5	Women	6	20–48	February 11, 2014	Enerata *qäbäle*
FGD6	Men and women	7	20–50	February 12, 2014	Enerata *qäbäle*
FGD7	Men	10	20–52	February 22, 2014	Kurar *qäbäle*
FGD8	Women	8	18–40	February 23, 2014	Kurar *qäbäle*
FGD9	Men and women	9	20–45	February 24, 2014	Kurar *qäbäle*

References

Adem, A., & Amsalu, A. (2012). Climate change in the southern lowlands of Ethiopia: Local level evidences, impacts and adaptation responses. *Ethiopian Journal of Development Research, 34* (1), 1–36.

Bewket, W. (2007). Rainfall variability and agricultural vulnerability in the Amhara Region, Ethiopia. *Ethiopian Journal of Development Research, 29*(1), 1–34.

Bewket, W., & Alemu, D. (2011). Farmers' perceptions of climate change and its agricultural impacts in the Abay and Baro-Akobo River Basins, Ethiopia. *Ethiopian Journal of Development Research, 33*(1), 1–28.

Deressa, T., Hassan, R. M., & Ringler, C. (2009). Assessing household vulnerability to climate change: The case of farmers in the Nile basin of Ethiopia. *International Food Policy Research Institute Discussion Paper 00935*. Environment and Production Technology Division.

Eller, J. (2009). *Cultural anthropology: Global forces, local lives*. New York and London: Routledge.

Fiske, S. J., Crate, S. A., Crumley, C. L., Galvin, K., Lazrus, H., Lucero, L., et al. (2014). Changing the atmosphere: Anthropology and climate change. In *Final report of the AAA global climate change task force* (137 pp). December 2014. Arlington, VA: American Anthropological Association.

Simane, B. (2013). The sustainability of community-based adaptation in the Choke Mountain Watershed, Blue Nile Highlands, Ethiopia. *Sustainability, 5*, 1–24.

Simane, B., & Tedla, S. (2011). Hibretesebin na Siri'ate Mihidarin Yamakele Yeteqenaje Yelimat Inqisiqase: Ye Choqe Terara Yilma Yagegim Hizbum Yibeltsg. In B. Semani (Ed.), (2011) *Ye'ayer Nibret Lewiten Yemiquaquam Arenguade Economi Be Choqe Sir'ate Mihidar Masfen* (pp. 8–15). Addis Ababa: Addis Ababa University Press.

Simane, B., Zaitchik, B. F., & Foltz, J. D. (2014). Agro-ecosystem specific climate vulnerability analysis: Application of the livelihood vulnerability index to a tropical highland region. In *Mitigation adaptation strategies to global change*, Springerlink.com. Accessed July 16, 2015.

Simane, B., Zaitchik, B. F., & Mesfin, D. (2012). Building climate change resilience in the Blue Nile/Abay Highlands: A framework for action. *International Journal of Environmental Research and Public Health, 9*, 610–631.

Simane, B., Zaitchik, B. F., & Ozdogan, M. (2013). Agro-ecosystem analysis of the Choke Mountain Watersheds, Ethiopia. *Sustainability, 5*, 592–616.

Sutton, M., & Anderson, E. N. (2010). *Introduction to cultural ecology* (2nd ed.). New York: Altamira Press.

Zaitchik, B., et al. (2012). Building climate resilience in the Blue Nile/Abay Highlands: A role for earth system sciences. *International Journal of Environmental Research and Public Health, 9*, 435–461.

Uptake of Climate-Smart Agriculture Through a Gendered Intersectionality Lens: Experiences from Western Kenya

C. Mungai, M. Opondo, G. Outa, V. Nelson, M. Nyasimi and P. Kimeli

1 Introduction

1.1 Climate Change, Agriculture and Food Security Issues in Kenya

The impacts of climate change and variability are expected to significantly alter food production patterns and shift Africa's farming systems and create new challenges for food security and economic development (IPCC 2014). Effects of climate change and variability are already being felt at national and local level. For example, in recent years Kenya has had its share of climate-related impacts: prolonged droughts; frost in some of the productive agricultural areas; hailstorms; extreme flooding; receding lake levels; drying of rivers and wetlands among others leading to large economic losses and adversely impacting food security and the livelihoods of communities (GoK 2013; Waithaka et al. 2013). The number of Kenyans requiring food assistance rose from 650,000 in 2007 to almost 3.8 million in 2009/2010 (GoK 2013). More recently, unfavorable 2013 October-to-December short rains and 2014 March-to-May long rains have resulted in deteriorating food

C. Mungai · M. Nyasimi · P. Kimeli
CGIAR Research Program on Climate Change,
Agriculture and Food Security (CCAFS)/International
Livestock Research Institute (ILRI), Nairobi, Kenya

C. Mungai (✉)
The Climate Impact Research Capacity and Leadership Enhancement
in Sub-Saharan Africa Programme (CIRCLE), London, UK
e-mail: c.mungai@cgiar.org

M. Opondo · G. Outa
Institute of Climate Change and Adaptation, University of Nairobi, Nairobi, Kenya

V. Nelson
Natural Resources Institute, University of Greenwich, London, UK

W. Leal Filho et al. (eds.), *Climate Change Adaptation in Africa*,
Climate Change Management, DOI 10.1007/978-3-319-49520-0_36

security in parts of northern Kenya (USAID 2014). Although food insecurity has relatively improved since 2011, nearly 1.1 million people required emergency food assistance as of August 2015 (USAID 2015). The improvement in food assistance can be attributed by humanitarian and government efforts to address food insecurity. To address food security challenges, the government has partnered with researchers, donors and non-governmental organizations to develop technologies and practices to cope with climate change and variability effects in the agricultural sector (ASDS 2010).

Some factors which influence the uptake of climate change adaptation and mitigation strategies in the agricultural sector include regional and national agricultural policies, economic conditions, levels of education and the availability of information, land tenure systems, and the preferences of individual farmers, which are conditioned by societal and community-based norms (Waithaka et al. 2013). In addition, adoption is also determined by labor requirements and abilities to apply new technologies and practices (Harvey et al. 2014). Physiological, behavioral and socio-institutional factors also influence how communities react to climate related stress (Gifford 2011). Consequently, measures which support local innovation processes and governance systems are needed to help stakeholders in the agriculture sector cope with the changes in emerging and projected weather patterns. It is increasingly argued that climate-smart agriculture (CSA) offers unique opportunities to improve food security and enhance adaptation to climate change (AGRA 2014; FAO 2012; Neufeldt et al. 2013). CSA is defined as agricultural practices that "*sustainably increase productivity, enhance resilience, reduce/remove greenhouse gas emissions, and enhance the achievement of national food security and development goals*" (FAO 2010). The CSA approach also identifies the importance of creating an enabling environment through the alignment of policy frameworks, financial investments and institutional arrangements to support transformation in the agriculture sector (FAO 2013).

A recent study by the Food and Agriculture Organization of the United Nations (FAO) on policies, programs, projects and activities on CSA in Kenya shows that some of the CSA practices being undertaken by farmers include improved livestock breeds, crop diversification, mixed cropping, tree planting, irrigation and diversification of livelihoods and enhancing early warning systems with drought monitoring and seasonal forecasts (Osumba and Rioux 2015). As a way forward, the study calls for a more in depth understanding of the political, economic, environmental and socio-cultural factors that might influence the uptake of CSA practices at local scales. Participatory social learning processes link community knowledge with that of science and can therefore be used to address the uncertainties inherent in agricultural contexts and also with changing climate.

1.2 Gender, Intersectionality and Climate Smart Agriculture

Individuals and communities experience the impacts of climate change differently depending on their position in society, as determined by gender, race, class, ethnicity, religion, and age, among other factors (Beuchelt and Badstue 2013; IPCC 2014; Nelson et al. 2002; Ray-Bennett 2009; Skinner 2011). According to CARE (2009) gender is a social construct that defines what it means to be a man or woman, boy or girl in a given society and includes specific roles, status and expectations with households, communities and culture. This means that policy and research should not focus only on gender differences but try to understand a person's position in society based on factors such as age, race, ethnicity, religion as well as the power dynamics that these imply (Denton 2002; FAO 2007; Nelson et al. 2002; Nelson and Stathers 2009; Kaijser and Kronsell 2014). This approach to understanding the impacts of climate change enables researchers, policy makers, development workers and other stakeholders to understand the social and cultural dimensions of climate change, and to therefore structure policies, projects, and research in a manner that acknowledges these complexities and accounts for different local priorities and needs (AGRA 2014).

Intersectionality is an analytical tool for studying, understanding and responding to the ways in which gender intersects (or interacts) with other identities and how these intersections contribute to unique experiences of oppression and privilege (Symington 2004). Intersectionality acknowledges that factors such as race and skin color, caste, age, ethnicity, language, ancestry, sexual orientation, religion, socio-economic class, ability, culture, geographic location and so on combine to determine one's social status (Symington 2004). Intersecting inequalities mean that whereas men in most societies enjoy the benefits of male privilege, they may share with women in their lives similar experiences of indignity, subordination and insecurity as a result of discrimination or social and economic oppression (Carr and Thompson 2014; Demetriades and Esplen 2010). Consequently, researchers need to find spaces within gender and climate change frameworks to acknowledge and communicate the vulnerabilities that men also experience.

The aim of an intersectional approach therefore, is to make visible not only similarities and differences between the sexes, but also to recognize how the diversity that stems from other social categories affect communities in different ways (Niskanen 2011). Indeed, Kaijser and Kronsell 2014 emphasize the need of undertaking intersectionality studies in the field of climate change because how individuals relate to climate change depends on their positions in context-specific power structures based on social categorizations. Although complex due to the uncertainties of exploring climate change impacts on agriculture as a sector and on the associated gender roles, an explicit intersectionality lens may ensure a more sophisticated analysis or a reminder to unpack different dimensions of identity and social differentiation. Focusing on the agricultural sector, the aim of this context-specific participatory research is to highlight local realities, knowledge and coping strategies of different groups of women and men in the face of accelerating climate change (Demetriades and Esplen 2010).

2 Methods

2.1 Study Area

This study was undertaken in the Lower Nyando Basin in western Kenya, and more specifically in Nyakach sub-County in Kisumu County and Soin- Sigowet sub-county in Kericho County. Nyakach sub-County has a population of 133,041 while Sigowet/Soin has a population of 113,312. The male to female ratio is 1:1.01 (KNBS 2013). The population density exceeds 400 persons per square kilometer making it one of the highest populated rural localities in east Africa (Kinyangi et al. 2015). The study site is inhabited by two distinct ethnic groups; the Luo in Kisumu and Kalenjin in Kericho. Both communities are already grappling with reduced agricultural production and food insecurity as a result of increased variability in climate patterns and extreme events such as droughts and floods (Förch et al. 2013). For example, a baseline study undertaken in 2011 shows that for 3–4 months, about 17% of households are unable to meet their food needs. This has resulted in malnutrition, estimated at 45% among under-fives (Mango et al. 2011). This is an area where over half of the population lives below the poverty line,[1] and human health issues such as high HIV prevalence are widespread (GoK 2013, 2014). Most people rely on their own small (on average less than one hectare) mixed crop-livestock farms for their livelihoods (Mango et al. 2011). The CGIAR Research Program on Climate Change, Agriculture and Food Security (CCAFS) has established Climate-Smart Villages which are research sites where scientists collaboratively work with farmers and development partners to reduce local climate risks and vulnerabilities and increase resilience in agricultural systems. The goal of the climate-smart village (CSV) approach is to respond to climate variability, reduce periodic hunger, ensure food security and enhance household incomes (Aggarwal et al. 2013). This is achieved through the participatory application of climate-smart agricultural practices and technologies (Kinyangi et al. 2015) (Fig. 1).

[1]Poverty line is defined as the level of personal or family income below which one is classified as poor according to governmental standards. At first the World Bank defined this a 1$ per day but as differences in the cost of living across the world evolve, the global poverty line has to be periodically updated to reflect these changes. In 2008 $1.25 was the global line. As of October 2016, the new global line was updated to $1.90. See http://www.worldbank.org/en/topic/poverty/brief/global-poverty-line-faq.

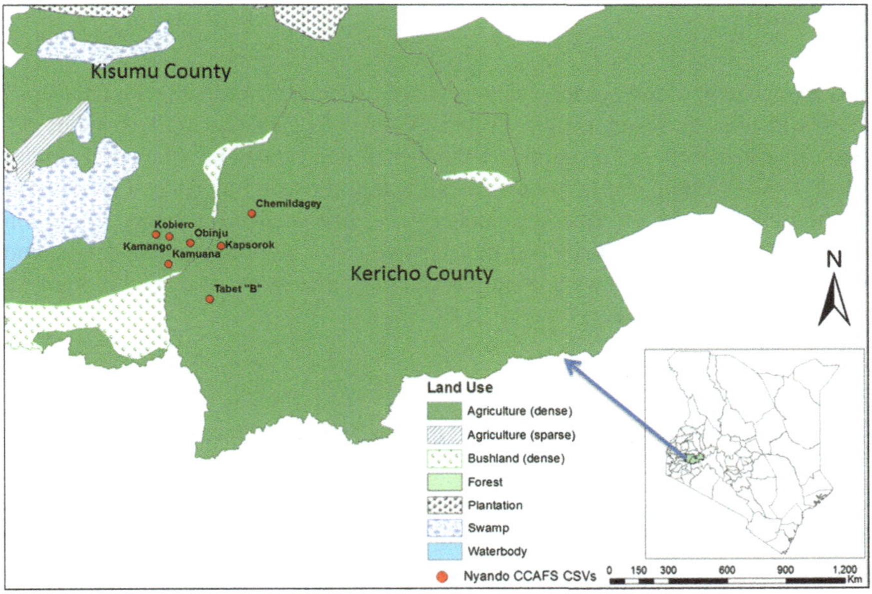

Fig. 1 Map showing CCAFS climate-smart villages in Western Kenya

2.2 Survey and Focus Group Discussions

CCAFS and partners have been working with farmers in the study area since 2011. This study sought to examine the uptake of CSA practices and technologies across the 139 households in the seven CSVs involved in the participatory action research processes since the project was initiated. Building on previous household baseline surveys (Kristjanson et al. 2012; Mango et al. 2011) and gender studies (Bernier et al. 2015; Jost et al. 2014; Twyman et al. 2014), this study applied a mixed method research design for an in-depth analysis of the uptake of CSA interventions among smallholder farmers. Using a gendered intersectionality lens, the study explored how factors such as age, ethnicity, education and marital status intersect with gender to influence the uptake of CSA technologies and practices. Data for the study was collected from farmers and households participating in the CSA related initiatives in the study area.

This paper draws on results from a quantitative survey undertaken with 51 farmers and from 4 focus group discussions (FDGs) held with 33 farmers (19 males and 14 females). In Nyakach the FGD comprised of 17 participants (9 females and 8 males) while in Soin-Sigowet 16 farmers (5 females and 11 males) participated in the FGD. The participants for the study were selected randomly drawing from CCAFS household baseline data for the study area.

3 Results

Fifty five percent of the households interviewed were male-headed, with an average household size of seven people (Table 1). The dependency ratio is about 0.9, suggesting that there are a significant number of dependants i.e. old age and those below 15 years compared to those who fall within the working group. Over 70% of the households had at least a member who had attained primary education. One-quarter of the households had members who had completed secondary school education, with very few households having members who had completed post-secondary education.

3.1 Uptake of Climate-Smart Agriculture Practices

Farmers in the study area have been exposed to a wide range of CSA practices which focus on crops, livestock and sustainable land management. These CSA practices have been tested together with local government, research institutions and non-governmental organizations working in the area. Table 2 shows some of the CSA interventions which have been focused on in the study area. These range from improved high yielding varieties (HYVs), improved stress tolerant varieties (STVs), agroforestry, destocking, water harvesting, and greenhouse production among others. Results from the FGDs suggest that there were different factors which hindered or supported the uptake of CSA interventions in the study area. For instance, some of the key barriers hindering the uptake of crop technologies include lack of labor to undertake farm activities, lack of starting capital and unavailability of planting materials. Constraints hindering uptake of livestock technologies was lack of space to construct shelter for the small ruminants on the small holdings and unavailability of funds to purchase fodder. Lack of labor was cited as a key constraint to the uptake of sustainable land management technologies and practices. Some of the factors which support CSA uptake include observation of success stories from fellow farmers, availability of information such as agro-advisories and availability of funds. The quantitative survey results indicated that about 85% of the respondents were willing to adopt CSA interventions if constraining factors are resolved.

3.1.1 Gender Differentiated Uptake of CSA Interventions

The data shows that despite over 90% of both female and male respondents having some awareness of the different interventions (Table 2), there was no significant difference in terms of gender. During the FGDs it emerged that some interventions being implemented by research and non-governmental institutions were gender

Table 1 Demographic characteristics of the survey participants

Demographic characteristics	Percent of households (n = 51)
Sex of respondent (%)	
Male	54.9
Female	45.1
Marital status (%)	
Single without partner	5.9
Married living together	74.5
Married living apart	5.9
Divorced/separated/widowed	13.7
Highest level of formal education of any household member (%)	
None	13.7
Primary	64.7
Secondary	15.7
Tertiary (post-secondary training)	5.9
Average household size	
Number of persons in a household	7
Number of young dependents (<15 years)	3.22
Number of people of working age (15–64 years)	3.53
Number of old dependents (>60 years)	0.25
Dependency ratio[a]	0.98

[a]Dependency Ratio is a measure of the portion of a population which is composed of dependents (people who are too young or too old to work)—expressed as Dependency Ratio = (<15 + >64)/(>15 + <64) * 100. The ratio is equal to the number of individuals aged below 15 or above 64 divided by the number of individuals aged 15–64, can be expressed as a percentage. This indicator gives insight into the number of people of non-working age compared to the number of those of working age. A high ratio means those of working age—and the overall economy—face a greater burden in supporting the aging population

specific. According to the farmers, deliberate attempts were made to ensure the participation of men and women as well as youth in the study area.

3.1.2 Uptake of CSA Interventions Based on Ethnicity

Using the ethnicity lens to examine the uptake of interventions, we found that there was a higher uptake of CSA interventions amongst the Luo community compared to the Kalenjin (Fig. 2). Although these communities neighbor one another, their experiences of climate change—and respective adaptive capacities to deal with its effects on their farms—varied significantly (Deverux and McKune 2014). They thus present an interesting case study of the role that ethnicity and the accompanying gender dynamics play in the adoption of climate adaptation strategies (Deverux and McKune 2014; Twyman et al. 2014). For instance 41.2% of the Luos practiced destocking compared to 29.4% of the Kalenjin. This could be attributed to the fact

Table 2 Uptake of CSA practices using a gender lens

CSA practices	% HH aware of and using CSA practices and technologies		
	Male	Female	Total n (51)
Improved high yielding varieties (HYVs)	53.1	46.9	96.1
Scientific weather forecasting	52.2	47.8	92
Efficient fertilizer use	52.3	47.7	86.3
Cover cropping	50	50	82.4
Improved stress tolerant varieties (STVs)	54.8	45.2	82.4
Manure management	43.9	56.1	80.4
Terraces/bunds	51.3	48.7	78
Agroforestry	56.8	43.2	72.5
Integrated pest management	45.9	54.1	72.5
Destocking	61.1	38.9	70.6
Crop residue mulching	51.5	48.5	64.7
Water harvesting	51.6	48.4	62
Composting	55.2	44.8	58
Improved feed management	61.5	38.5	51
Adoption of resilient livestock breeds	60.9	39.1	46
Traditional weather forecasts	38.1	61.9	41.2
Tree nurseries	63.2	36.8	37.3
No/minimum tillage	41.2	58.8	34
Green house production	40	60	19.6

that land sizes are larger amongst the Kalenjin giving them an "allowance" to retain more livestock.

Results from the FGDs further illustrate how cultural practices and norms based on ethnic groups influence the uptake of interventions. Among the Kalenjin it emerged that women during menstruation are believed to possess evil powers such as witchcraft and hence are barred from undertaking all farm activities including weeding, harvesting, milking of cows etc. It is believed that the land and livestock will have poor yields. Amongst the Luo community some of the cultural practices mentioned include, in case of harvesting, the mother-in-law has to harvest first before the daughter in law regardless of whether her (the daughter in law) crops are ready. This can lead to increased conflict and/or loss of yields the daughter in law plants faster maturing crops. To illustrate this, a female FGD participant reported that:

> I was happy to learn about early maturing crops as a way of ensuring I feed my family. However, I was worried about applying this because my mother in law is not keen to take on the new practices and this may interfere with my crop when it comes to harvesting because I have to wait for her to harvest first.

Basically, all farming practices have to be initiated by the eldest family member in the homestead, in most cases this will be the parents. This means that the children

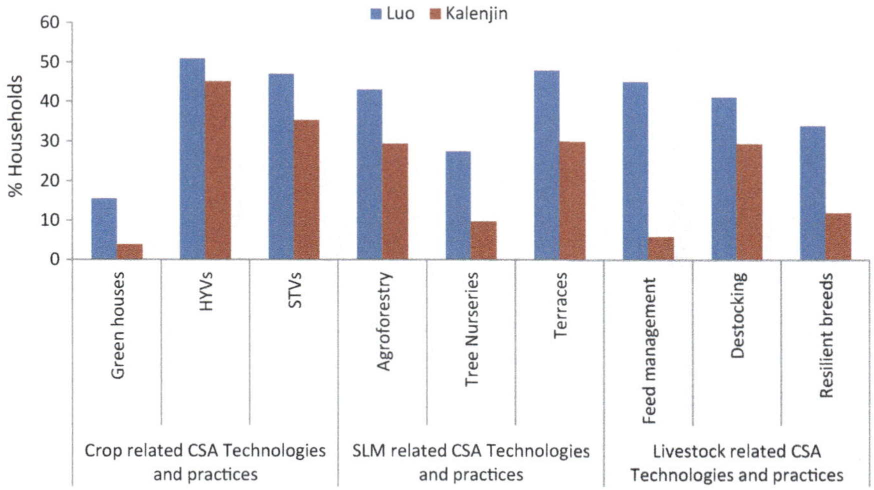

Fig. 2 Uptake of CSA interventions using the ethnicity lens

have to wait for the parents even though they would like to practice early planting as may be advised based on agro-advisories. In the case of land preparation and sowing, it emerged that these activities have to be initiated by the man and this is followed by a sexual encounter. If the man is not around, the woman has to wait leading to delayed farm activities which may affect the adoption rate of CSA interventions.

3.1.3 Uptake of CSA Interventions Based on Education

Figure 3 shows that a higher percentage of CSA uptake takes place amongst farmers with primary and secondary level education. This is attributed to the fact that information is more accessible and can be easily applied by farmers with some level of formal education. Looking at crops, for example, there was a high uptake (62.7%) of improved high yielding varieties (HYVs) amongst farmers with primary level education. Reduced uptake amongst those with post-secondary levels of education could be attributed to the fact that at this level, most individuals tend to migrate to urban areas in search of jobs hence do not undertake farming. While results indicate some level of uptake (11.8% in HYVs for instance) amongst farmers with no education levels, these findings emphasize the need to address literacy levels amongst smallholder farmers in order to facilitate the increased uptake of CSA interventions.

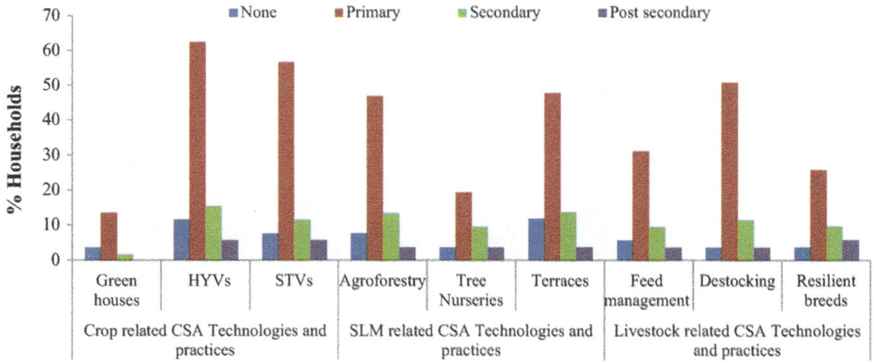

Fig. 3 Uptake of CSA interventions using the education lens

3.1.4 Uptake of CSA Interventions Based on Age

Using age as a lens we find that the highest uptake of interventions takes place amongst farmers aged between 41 and 50 years of age. This is the most productive age during which farmers are able to undertake labor intensive activities such as terracing. We also found few farmers aged below 30 participating in CSA activities in the study area. We find a seemingly high uptake amongst farmers aged above 50 years of age. During the FGDs, it emerged that while some young people are interested in taking up agri-business, a larger percentage prefer to move to urban areas. Labor intensive activities such as terracing are also taken up by farmers between 41 and 50 age bracket (Fig. 4).

3.1.5 Uptake of Interventions Based on Marital Status

Analysis using the marital lens shows a high uptake amongst married couples living together (Fig. 5). This can be attributed to the fact that husbands and wives can share responsibilities as well as information. This is especially critical where information is shared in spaces or through channels which are not accessible to either the husband or wife such as female only investment groups. Married couples living together displayed a high level of uptake of HYVs (70.6%) compared to their single counterparts and those divorced or widowed (5.9 and 13.7% respectively). This is attributed to access to land whereby for instance women are not allowed to inherit and/or own land. During the FGDs it emerged that women are very innovative in looking for alternative sources of income, an essential livelihood diversification factor for successful CSA uptake. For instance, one of the FGD participants had the following to say about his wife:

> Women are very innovative when it comes to looking for money, my wife makes baskets which she sells in the market and comes home with money which we use to buy drought tolerant seeds.

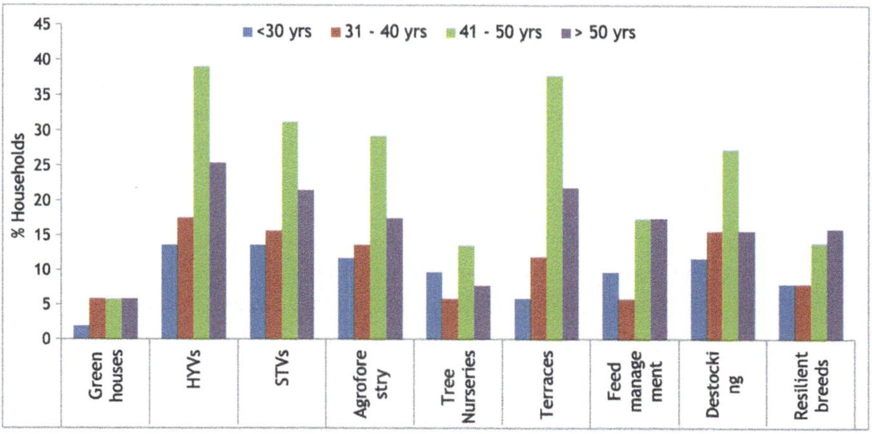

Fig. 4 Uptake of CSA interventions using age as a lens

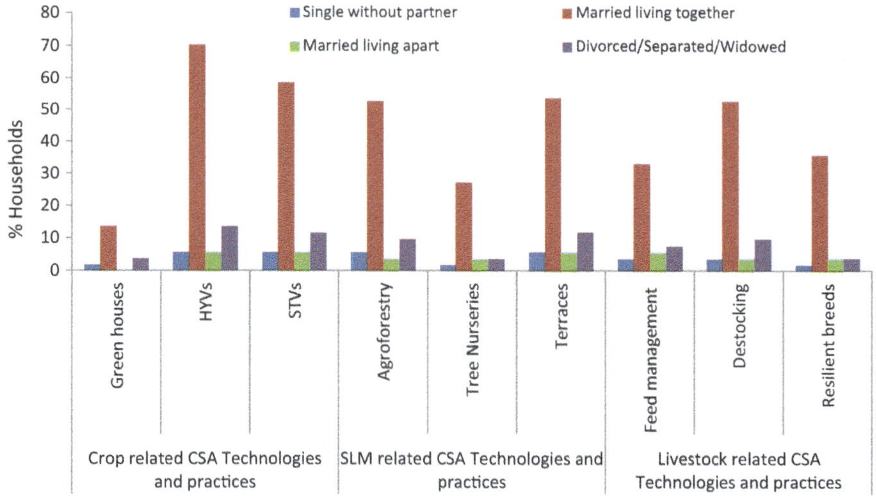

Fig. 5 Uptake of CSA practices using marital status as a lens

4 Discussion

This study demonstrates how the application of a gendered intersectionality lens reveals fundamental factors which influence the uptake of CSA interventions introduced by CCAFS and partners in the study area. Overall we find that there is no significant difference in the uptake of CSA practices based on gender. This is in tandem with a study undertaken in the study area by Bernier et al. (2015 p. 48) which found that "*gender does not seem to be a factor that necessarily prevents or impedes the adoption of CSA amongst those who know about the technologies.*"

However, using lenses such as marital status, age, ethnicity and education we find that these factors play a role in influencing the uptake of CSA practices.

Marital status, for example helps us to understand the vulnerability and adaptive capacity of men and women within the context of marriage, or as singles, divorcees or as widows/widowers. A recent study by Van Aelst and Holvoet (2016) concludes that women's vulnerability and adaptation capacity is highly influenced by her marital status. However, if the man leaves the marriage for instance through separation or as a result of divorce, his capacity to adapt may not necessarily worsen. In our study we find that there is an increased uptake of CSA amongst married couples. This can be attributed to factors such as easy access to land through the husband, sharing of house hold duties and responsibilities, sharing of information and income.

While ethnicity still influences how decisions are made amongst farming communities, it is important to note that religion and formal education are starting to erode some of the social and cultural beliefs held by the communities. This may play a positive role in addressing some of the challenges farmers face such as making decisions on whether to adopt fast maturing crops.

Our study shows that formal education (for both female and male farmers) is a critical element influencing the uptake of CSA interventions in the study area. Using the age lens we find that there is a high uptake among the youth. Young people are increasingly viewing agriculture as a source of income and because of increased literacy levels and access to information through technologies such as mobile phones and internet, they are able to adopt promising CSA interventions. A member of the youth group involved in the green house project supported this:

> When I finished my secondary school education, I went to live with my aunt in Nairobi in the hope of finding a job. However, after one year, I did not get anything stable, so I moved back to the village. This was an excellent decision because I got involved in CCAFS activities focusing on greenhouse production and I am now encouraged to explore agriculture as a business.

Our results show that while gender analysis is an important tool for supporting the design of gender sensitive policies and projects, an intersectionality lens further unravels underlying factors which need to be considered to develop policies and projects that yield equitable benefits for communities.

5 Conclusion and Recommendation

This study reveals that while gender analysis plays an important role in developing gender sensitive response strategies in the agricultural sector, further analysis using an intersectionality lens demonstrates the need to develop targeted solutions to address climate change in agriculture. It is anticipated that knowledge generated from an intersectional analysis can be used to develop tools which are better tailored to implementing gender equality policies (Niskanen 2011). Using a gendered intersectionality lens strengthens the argument for targeted interventions which focus on local needs and priorities while recognizing local contexts as informed by social, cultural and economic factors. This study reveals that farmers, regardless of gender, are willing to adopt climate smart technologies and practices. However, factors such as ethnicity, education, age and marital status determine the levels of uptake of CSA technologies and practices.

Kenya has developed a National Climate-Smart Agriculture Program 2015–2030 which acknowledges that climate-smart agriculture will play a key role in increasing productivity, enhancing adaptation and resilience of Kenya's farming systems while reducing emissions in order to achieve sustainable development and reduce poverty (GOK n.d). While this creates a framework within which to fundraise for and implement CSA interventions in Kenya, it is important to draw on emerging lessons to develop projects and programmes which will enhance the equitable uptake of CSA technologies and practices in Kenya. This study is therefore timely in informing the process of implementing CSA related policies, programmes and projects in the country.

6 Acknowledgements

This paper is part of the first author's post-masters research undertaken through funding from the Department for International Development (DfID) under the Climate Impact Research Capacity and Leadership Enhancement (CIRCLE) programme.

References

AGRA. (2014). *Africa agriculture status report: Climate change and smallholder agriculture in sub-Saharan Africa*. Nairobi: Kenya Alliance for Green Revolution in Africa.

Agriculture Sector Development Strategy 2010–2020. (2010). Government of Kenya.

Aggarwal, P., Zougmoré, R., & Kinyangi, J. (2013). Climate-smart villages: A community approach to sustainable agricultural development. Copenhagen, Denmark: CGIAR Research Program on Climate Change, Agriculture and Food Security (CCAFS), https://cgspace.cgiar.org/bitstream/handle/10568/33322/CCAFSClimate-SmartVillages2013.pdf, accessed January 12, 2016.

Bernier, Q., Meinzen-Dick, R., Kristjanson, P., Haglund, E., Kovarik, C., Bryan, E., & Sivestri, S. (2015). *Gender and Institutional Aspects of Climate-Smart Agricultural Practices: Evidence from Kenya*.

Beuchelt, T. D., & Badstue, L. (2013). Gender, nutrition-and climate-smart food production: Opportunities and trade-offs. *Food Security, 5*(5), 709–721.

Carr, E. R., & Thompson, M. C. (2014). Gender and climate change adaptation in agrarian settings: Current thinking, new directions, and research frontiers. *Geography Compass, 8*(3), 182–197.

Demetriades, J., & Esplen, E. (2010). The gender dimensions of poverty and climate change adaptation. *Social Dimensions of Climate Change*, 133.

Denton, F. (2002). Climate change vulnerability, impacts and adaptation: Why does gender matter? *Gender & Development, 10*(2), 10–20.

Deverux, T., & McKune, S. (2014). *Gender, power and climate information in Nyando, Kenya*. https://ccafs.cgiar.org/blog/gender-power-and-climate-information-nyando-kenya#.VqslI_mqpBc, accessed January 29, 2016.

FAO. (2007). *People-centered climate change adaptation: Integrating gender issues—Awareness of climate change from a gendered perspective*. Rome: FAO. ftp://fao.org/docrep/fao/010/a1395e/a1395e00.pdf, accessed on January 29, 2016.

FAO. (2012). *Gender and climate change research in Agriculture and food security for rural development training guide: Food and Agriculture Network*.

FAO. (2013). *Climate smart agriculture source book*. Rome: Food and Agriculture Organization of the United Nations. http://fao.org/docrep/018/i3325e/i3325e.pdf, accessed January 27, 2016.

Förch, W., Sijmons, K., Mutie, I., Kiplimo, J., Cramer, L., Kristjanson, P., et al. (2013). *Core sites in the CCAFS regions: East Africa, West Africa and South Asia version 3: Copenhagen*. Denmark: CCAFS.

Gifford, R. (2011). The dragons of inaction: Psychological barriers that limit climate change mitigation and adaptation. *American Psychologist, 66*(4), 290.

GoK. (2013). *National climate change action plan 2013–2017*. Government of Kenya.

GoK. (2014). *Kenya HIV County Profiles*. Government of Kenya.

GoK. (n.d.). Kenya Climate-Smart Agriculture Program 2015-2030, http://canafrica.com/wp-content/uploads/2015/06/Kenya-CSA-Program-.pdf, accessed January 11, 2016.

Harvey, C. A., Chacon, M., Donatti, C. I., Garen, E., Hannah, L., Andrade, A., et al. (2014). Climate-smart landscapes: Opportunities and challenges for integrating adaptation and mitigation in tropical agriculture. *Conservation Letters, 7*(2), 77–90.

Jost, C., Ferdous, N., & Spicer, T. (2014). *Gender and Inclusion Toolbox: Participatory Research in Climate Change and Agriculture*.

Kaijser, A., & Kronsell, A. (2014). Climate change through the lens of intersectionality. *Environmental politics, 23*(3), 417–433.

Kenya National Bureau of Statistics. (2013). *2009 Kenya Population and Housing Census Analytical Reports*. KNBS 2013.

Kinyangi, J., Recha, J., Kimeli, P., & Atakos, V. (2015). *Climate-smart villages and the hope of food secure households*.

Kristjanson, P., Neufeldt, H., Gassner, A., Mango, J., Kyazze, F. B., Desta, S., et al. (2012). Are food insecure smallholder households making changes in their farming practices? *Evidence from East Africa. Food Security, 4*(3), 381–397.

Mango, J., Mideva, A., Osanya, W., & Odhiambo, A. (2011). *Summary of baseline household survey results: Lower Nyando, Kenya*.

Nelson, V., Meadows, K., Cannon, T., Morton, J., & Martin, A. (2002). Uncertain predictions, invisible impacts, and the need to mainstream gender in climate change adaptations. *Gender & Development, 10*(2), 51–59.

Nelson, V., & Stathers, T. (2009). Resilience, power, culture, and climate: a case study from semi-arid Tanzania, and new research directions. *Gender & Development, 17*(1), 81–94. doi:10.1080/13552070802696946.

Neufeldt, H., Jahn, M., Campbell, B. M., Beddington, J. R., DeClerck, F., De Pinto, A., et al. (2013). Beyond climate-smart agriculture: toward safe operating spaces for global food systems. *Agric Food Secur, 2*(12), 10–1186.

Niskanen, K. (2011). Gender and power in the Nordic countries. *Gender and Power in the Nordic Countries. Oslo: NIKK*, 11–54.

Osumba, J., & Rioux, J. (2015). Scoping study on climate-smart agriculture in Kenya: Smallholder integrated crop-livestock farming systems. In *Mitigation of Climate Change in Agriculture (MICCA) Programme Background report 8*. Rome: Food and Agriculture Organization.

Ray-Bennett, N. S. (2009). Multiple disasters and policy responses in pre-and post-independence Orissa, India. *Disasters, 33*(2), 274–290.

Skinner, E. (2011). *Gender and climate change: Overview report*: Bridge.

Symington, A. (2004). *Intersectionality: A tool for gender and economic justice*.
Twyman, J., Green, M., Bernier, Q., Kristjanson, P., Russo, S., Tall, A., & McKune, S. (2014). *Adaptation actions in Africa: Evidence that gender matters*.
USAID. (2014). *Kenya food insecurity fact sheet*. https://www.usaid.gov/sites/default/files/documents/1866/kenya_fi_fs01_09-30-2014.pdf, accessed May 23, 2016.
USAID. (2015). *Kenya food insecurity fact sheet*. https://www.usaid.gov/sites/default/files/documents/1866/kenya_fi_fs01_09-30-2015.pdf, accessed May 23, 2016.
Van Aelst, K., & Holvoet, N. (2016). Intersections of gender and marital status in accessing climate change adaptation: Evidence from rural Tanzania. *World Development, 79*, 40–50.
Waithaka, M., Nelson, G. C., Thomas, T. S., & Kyotalimye, M. (2013). *East African agriculture and climate change: A comprehensive analysis*. Intl Food Policy Res Inst.

Carbon Isotope Discrimination as a Surrogate of Grain Yield in Drought Stressed Triticale

Lawrence Munjonji, Kingsley Kwabena Ayisi, Bram Vandewalle, Geert Haesaert and Pascal Boeckx

1 Introduction

Water availability in semi-arid regions is increasingly becoming threatened by erratic rains, frequent droughts, rising temperatures and high evapotranspiration rates leading to over-reliance on irrigation to meet food demand. Currently, about 80% of world fresh water used by humans is used for irrigation (Morison et al. 2008) and in many dry areas such levels of consumption are unsustainable (Condon et al. 2004) as water resources are also under increased pressure from other uses. To feed the projected nine billion people by 2050, crop management, particularly of cereals, must adapt to climate change and variability through the use of varieties that use water efficiently (Barnabás et al. 2008).

Improving cereal water use efficiency (WUE) has for a long time been one of the main targets of crop research particularly in arid and semi-arid environments, with the aim of finding sustainable ways of increasing crop productivity while reducing water loses (Foley et al. 2011). Crop WUE plays an important role in the exchange of water between the biosphere and the atmosphere and thus affects the global water cycle (Seibt et al. 2008). However, one of the major bottlenecks in cereal breeding to produce "more crop per drop" is the lack of or the evaluation of appropriate traits (Araus et al. 2008). Direct measurement of WUE under field conditions remains a

L. Munjonji (✉) · B. Vandewalle · P. Boeckx
Isotope Bioscience Laboratory—ISOFYS, Faculty of Bioscience Engineering,
Ghent University, Ghent, Belgium
e-mail: Lawrence.Munjonji@Ugent.be

L. Munjonji · K.K. Ayisi
Risk and Vulnerability Science Centre/VLIR-IUC Programme,
University of Limpopo, Sovenga, South Africa

G. Haesaert
Department of Applied Bioscience, Faculty of Bioscience Engineering,
Ghent University, Ghent, Belgium

W. Leal Filho et al. (eds.), *Climate Change Adaptation in Africa*,
Climate Change Management, DOI 10.1007/978-3-319-49520-0_37

big challenge due to the large amount of work involved (Tardieu 2013) and thus have stalled the use of WUE trait in crop improvement programs. There is therefore need to identify reliable proxies of WUE that can be measured quickly; are correlated to yield and that can also provide the highest repeatability and heritability.

Stable isotope ratios of plant material are a powerful tool in ecological research as they indicate key environmental and physiological processes (Barnard et al. 2012). Carbon isotope composition ($\delta^{13}C$) or the discrimination value ($\Delta^{13}C$) have frequently been used as a time-integrated measure of the intrinsic water-use (Barbour et al. 2011; Cabrera-Bosquet et al. 2009; Farquhar et al. 1989). Intrinsic water use efficiency ($WUE_{intrinsic}$) is the ratio of photosynthetic rate relative to stomatal conductance (*A/gs*). Stomatal conductance (*gs*) plays an important role in the trade-off between water conservation and carbon assimilation as it controls both CO_2 uptake and water loss (Araya et al. 2010). Changes in stomatal conductance result in changes in leaf $\delta^{13}C$ and in turn crop WUE (Farquhar and Richards 1984). However, according to Seibt et al. (2008), the relationship between $\delta^{13}C$ and WUE is not direct due to the influence of external biotic and abiotic factors on the ratio of intercellular CO_2 to that of the atmosphere (*Ci/Ca*) the primary determinant of discrimination against ^{13}C–CO_2 in leaves. It has been argued that WUE at leaf level depends on evaporative demand, which does not directly affect $\delta^{13}C$ (Seibt et al. 2008). Thus, WUE and $\delta^{13}C$ can vary independently of one another, making the use of $\delta^{13}C$ as a proxy for WUE questionable (Seibt et al. 2008).

In this study; the use of $\Delta^{13}C$ as a proxy for $WUE_{intrinsic}$ was tested together with its correlation to grain yield through empirical studies using four field grown spring triticale varieties under four soil moisture levels. The utility of $\Delta^{13}C$ to infer sources of carbon assimilate to grain filling was explored.

2 Materials and Methods

2.1 Study Site and Experimental Design

The study was carried out at the University of Limpopo experimental farm, Syferkuil (23°50′ S; 029°41′ E), Limpopo Province, South Africa, during two winter seasons; June to October in 2013 and July to November in 2014.

The experimental design was a randomized complete block design (RCBD) in factorial arrangement of moisture levels and genotypes as treatment factors. The four moisture levels were as follows:

- Well-watered (WW): 25% soil moisture depletion before recharging to field capacity (FC);
- Moderately well-watered (MW): 50% soil moisture depletion before recharging to FC;
- Moderate stress (MS): 75% soil moisture depletion before recharging to 50% of FC;

- Severe stress (SS): moisture was allowed to dry out from detectable first node, with 40 mm supplementary irrigation being applied at milk stage to avoid complete failure of the crop.

These four moisture levels were selected to enable the assessment of triticale performance in Limpopo, from unlimited water supply (WW) to water limited conditions (SS). The moisture levels were applied after crop establishment corresponding to Zadoks Growth Stage 31 (GS31) (Zadoks et al. 1974). Zadoks et al. (1974) describes GS31 as the growth stage when the plant is at stem elongation with one detectable node. Soil moisture was measured every other day using Diviner 2000 (Campbell Scientific, Australia) from access tubes installed at the center of each plot. Soil moisture was measured for the whole profile at 10 cm intervals.

A preliminary study was carried out with eight genotypes in 2012 to select genotypes that were most suitable to the growing conditions at Syferkuil. Triticale is an uncommon crop in the Limpopo province hence there was a need for a pre-valuation of its suitability. From the preliminary study, the best four medium maturity genotypes from Plant Breeding Unit at Stellenbosch University, South Africa; Agbeacon, Bacchus, Rex, and US2007 were selected based on above-ground biomass yield as affected by soil moisture. Triticale was planted in rows, 25 cm apart using a tractor drawn planter at a density of 200 plants m^{-2}. The plot sizes were 10 m × 10 m and irrigation was by sprinklers (Rain Bird, USA), fitted with water meters (M20, Arad, Israel) to record amount of water applied. The soil was classified as Chromic Luvisol (Hypereutric) (WRB 2014). The soil depth ranged from 60 to 70 cm, with sandy loams and sandy clay loams overlaying sandy clays and available water capacity of 80 mm. Fertilization was adapted to local practice: Nitrogen (N) (12.5 kg ha^{-1}), Phosphorus (21 kg P ha^{-1}) and Potassium (12 kg K ha^{-1}) where applied as mixed fertilizer [NPK; 2:3:2 (22%)] before planting. Another dose of N (50 kg N ha^{-1}) was applied in the form of urea at the booting stage.

2.2 *Agronomic Measurements*

Above-ground biomass was measured 19 weeks after emergence after physiological maturity (GS92) together with grain yield. Sampling was done from the middle rows of each experimental unit by cutting the plants at 10 cm above-ground and oven drying them to constant weight at 65 °C.

2.3 *Leaf Gas Exchanges*

Flag leaf gas exchange measurements were carried out using LCi-SD Ultra Compact Photosynthesis System (ADC BioScientific, UK). The flag leaf was selected for these measurements because it is considered to be the main source of assimilates for grain filling. The measurements were taken three times each season on clear sunny days between 11h00 and 13h00. The specific data collected included: photosynthetic rate (*A*), stomatal conductance (*gs*), transpiration (*E*), intercellular CO_2 concentration (*Ci*), and atmospheric CO_2 (*Ca*).

2.4 Isotope Analyses

Flag leaves for isotope analyses were sampled at GS71 and GS92 and grain samples at GS92. Samples were dried at 65 °C to constant weight and ground to a fine texture using a ZM200 mill (Retsch, Germany). The $^{13}C/^{12}C$ isotope ratio (R_{sample}) of both leaf and grain samples were analyzed using Automated Nitrogen Carbon Analyser—Solid and Liquids (ANCA-SL, SerCon, UK) interfaced with an Isotope Ratio Mass Spectrometer (IRMS) (20–20, SerCon, UK). The isotope composition was reported as $\delta^{13}C$ in ‰ using Vienna Pee Dee Belemnite (V-PDB) as international standard ($R_{standard}$) and calculated using Eq. 1.

$$\delta^{13}C_{sample} = \left(\frac{R_{sample}}{R_{standard}} - 1\right) \times 1000 \tag{1}$$

The ^{13}C discrimination ($\Delta^{13}C$) was then calculated from $\delta^{13}C_{sample}$ using Eq. 2.

$$\Delta^{13}C = \frac{\delta^{13}C_{air} - \delta^{13}C_{sample}}{1 + (\delta^{13}C_{sample}/1000)} \tag{2}$$

Where $\delta^{13}C_{air}$ and $\delta^{13}C_{sample}$ is the carbon isotope compositions of air and plant sample (leaves or grain), respectively. $\delta^{13}C_{air}$ was put at −8.15‰ (CDIAC 2015).

2.5 WUE Determination

Integrated WUE ($WUE_{biomass}$) was calculated as the ratio of aboveground dry biomass to total amount of water used and also as the ratio of grain yield against total amount of water used (WUE_{grain}). At leaf level, Intrinsic WUE ($WUE_{intrinsic}$) and instantaneous WUE (WUE_{inst}) were calculated as follows: $WUE_{inst} = A/E$ and $WUE_{intrinsic} = A/gs$.

2.6 Statistical Analyses

A two-way analysis of variance (ANOVA) was performed to calculate the effects of moisture level and genotype on the studied parameters. Post Hoc multiple comparisons for observed means was done using Tukey and different means were denoted by *, ** or *** for significance levels $P \leq 0.05$, $P \leq 0.01$ and $P \leq 0.001$, respectively. All data were analyzed using the SPSS 20 statistical package (SPSS, USA).

3 Results

3.1 Weather Conditions at the Study Site

Weather conditions experienced in the two growing seasons are shown in Fig. 1. Figure 1a shows the daily maximum and minimum temperatures recorded on station at Syferkuil in 2013 and 2014 growing seasons using at automatic weather station. Daily maximum temperatures during the reproductive months (September to November, shown by double arrow on Fig. 1a) of triticale reached as high as 35 °C in both years. The same Fig. 1a also shows optimum temperature range for triticale (12–25 °C), represented by lines at 25 and 12 °C. Mean temperatures for both years are shown on Fig. 1b together with total monthly rainfall for the two seasons.

3.2 Grain Yield, Above-Ground Biomass and Integrated WUE

Table 1 shows the effect of moisture levels on grain yield, total above ground dry biomass, $WUE_{biomass}$ and WUE_{grain}. Moisture level significantly influenced grain

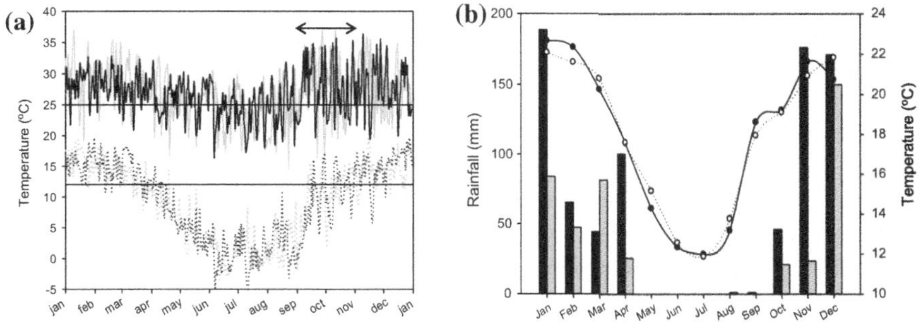

Fig. 1 **a** Daily maximum and minimum temperatures in the two seasons (2013 maximum *solid grey*; 2013 minimum *dashed grey*; 2014 maximum *black solid*; and 2014 minimum *dashed black*). The *double arrow* shows the reproductive growth period. Figure 1b mean monthly temperatures for Syferkuil in 2013 (*black circles* and *line*) and 2014 (*open circles* and *dashed line*). Figure 1b also shows monthly total rainfall for 2013 (*black bars*) and 2014 (*gray bars*)

Table 1 Grain yield, total dry biomass, $WUE_{biomass}$ and WUE_{grain} under four moisture levels in 2013 and 2014

Year	Irrigation	Grain yield	Total dry biomass	$WUE_{biomass}$	WUE_{grain}
		t ha^{-1}		kg ha^{-1} mm^{-1}	
2013	Well-watered	3.5^{a}	13.6^{a}	31.0^{b}	8.1^{b}
	Moderate water	3.9^{a}	13.6^{a}	40.9^{a}	11.7^{a}
	Medium stress	2.4^{b}	9.6^{b}	29.8^{b}	7.3^{b}
	Severe stress	0.8^{c}	6.9^{c}	29.7^{b}	3.4^{c}
		***	***	*	***
2014	Well-watered	4.9^{a}	15.0^{a}	39.4^{a}	11.7^{a}
	Moderate water	4.0^{ab}	12.2^{b}	34.0^{a}	10.1^{a}
	Medium stress	3.1^{bc}	9.5^{c}	32.5^{b}	9.5^{b}
	Severe stress	1.8^{c}	6.8^{d}	27.6^{b}	7.4^{b}
		***	***	*	*

Significance levels: *$P < 0.05$, **$P < 0.01$, ***$P < 0.001$. Different letters in the same column indicate significant differences

yield and above-ground biomass in both seasons (Table 1). Triticale performed better under non-limited water supply compared to water limited conditions with yields ranging from 0.8 t ha^{-1} under SS to 3.5 t ha^{-1} under WW in 2013 and 1.8–4.9 t ha^{-1} in 2014. Differences were also observed for $WUE_{biomass}$ and for WUE_{grain}. The general trend observed was a decreasing WUE with decreasing soil moisture. In 2013, WUE was significantly higher under MW compared to other moisture levels while in 2014, WUE was significantly higher under WW and MW compared to MS and SS. Neither genotypic differences nor the interaction of genotype and moisture levels were observed.

3.3 *Effect of Moisture Level on $\Delta^{13}C$*

Moisture level strongly influenced both flag leaf $\Delta^{13}C$ and grain $\Delta^{13}C$ (Table 2). Triticale discriminated more against ^{13}C under well-watered conditions compared to water limited conditions as evidenced by higher $\Delta^{13}C$ values under WW and MW at the two sampling stages. A general decrease in $\Delta^{13}C$ was observed from WW to SS in both leaf and grain samples over the two study seasons. In 2013 at GS71, all moisture levels had significantly different $\Delta^{13}C$ values, but flag leaves sampled later in the season at GS92 showed no differences between WW and MW. SS resulted in the lowest $\Delta^{13}C$ values at both stages. Similar to the 2013 season, SS plants produced the lowest $\Delta^{13}C$ in 2014. However, WW and MW did not differ at both stages. Across the two seasons and sampling stages, flag leaf $\Delta^{13}C$ ranged from 17 to 20.6‰. A T-test performed on the 2013 data showed that ^{13}C discrimination values were on average 0.65‰ higher ($P < 0.001$) at GS71 compared to GS92 while in 2014 no differences ($P > 0.05$) were observed but still GS71 had

Table 2 Flag leaf $\Delta^{13}C$ measured at early milk stage (GS71) and harvest maturity (GS92) under four moisture levels in the two growing seasons

Year	Irrigation	$\Delta^{13}C$ (‰)		
		GS71	GS92	Grain
2013	Well-watered	20.6^{a}	20.0^{a}	n.a.
	Moderate water	20.0^{b}	19.7^{a}	n.a.
	Medium stress	19.1^{c}	18.6^{b}	n.a.
	Severe stress	18.1^{d}	17.0^{c}	n.a.
	LSD	***	***	
2014	Well-watered	19.0^{a}	19.0^{a}	17.4^{a}
	Moderate water	18.7^{ab}	18.5^{ab}	16.7^{a}
	Medium stress	18.4^{b}	18.3^{b}	15.2^{b}
	Severe stress	17.9^{c}	17.7^{c}	14.9^{b}
	LSD	**	***	***

Significance levels: $^{*}P < 0.05$, $^{**}P < 0.01$, $^{***}P < 0.001$, n.a. means not available. Different letters in the same column indicate significant differences

slightly higher values than GS92. In addition, T-test performed on 2014 data between flag leaf $\Delta^{13}C$ at GS71 and grain $\Delta^{13}C$ showed that; under SS, flag leaf $\Delta^{13}C$ values were on average 2.9‰ (2.2–3.6‰) higher than grain $\Delta^{13}C$ while under WW, the average difference in $\Delta^{13}C$ values between the two was 1.7‰ (0.71–2.8‰). Significant differences were also observed between flag leaf $\Delta^{13}C$ and grain $\Delta^{13}C$ under MW (1.9‰) and MS (3.3‰).

3.4 *Relationship Between Different Traits*

Grain yield was significant and positively correlated to flag leaf $\Delta^{13}C$ under SS (0.54**) and MS (0.34*) (Fig. 2a) but the correlations were not significant under WW and MW. Similar results were observed with grain $\Delta^{13}C$ (Fig. 2b). The correlation coefficient values were higher with grain $\Delta^{13}C$ than flag leaf $\Delta^{13}C$. WUE_{grain} showed a significant positive correlation with flag leaf $\Delta^{13}C$ under SS (0.49**) but was negatively correlated to flag leaf $\Delta^{13}C$ under WW (0.53**) (Fig. 2c). WUE_{grain} was positively correlated to grain $\Delta^{13}C$ under SS (0.64*) and MS (0.63*) and showed non-significant negative correlations to MW and WW (Fig. 2d). The correlations between $WUE_{intrinsic}$ and flag leaf $\Delta^{13}C$ and between $WUE_{intrinsic}$ and grain $\Delta^{13}C$ were not significant when the data was separated according to moisture levels. However, when data for all moisture levels was combined, negative correlations were observed (Fig. 2e, f) with a stronger correlation being observed in 2013 (0.31***) compared to 2014 (0.06*). Other important significant correlations observed in this study were between $WUE_{intrinsic}$ and WUE_{grain} as well as between $WUE_{intrinsic}$ and grain yield. $WUE_{intrinsic}$ was negatively related to both grain yield and WUE_{grain} and was significant in both seasons for grain yield but was only significant in 2013 for WUE_{grain}.

Fig. 2 Correlation plots: grain yield and flag leaf $\Delta^{13}C$ (**a**); grain yield and grain $\Delta^{13}C$ (**b**); WUE_{grain} and flag leaf $\Delta^{13}C$ (**c**); WUE_{grain} and grain $\Delta^{13}C$ (**d**); $WUE_{intrinsic}$ and leaf $\Delta^{13}C$ (**e**); $WUE_{intrinsic}$ and grain $\Delta^{13}C$ (**f**). r_{SS}, r_{MS}, r_{MW} and r_{WW} are correlation co-efficiencies for respectively: severe stress, medium stress, moderately well-watered and well-watered. r_{13} and r_{14} are correlation co-efficient for growing season 2013 and 2014 respectively. *Lines* were fitted for significant correlations only

4 Discussion

4.1 Grain Yield, Biomass Accumulation and Integrated WUE

Grain yield, aboveground biomass, $WUE_{biomass}$ and WUE_{grain} responded to moisture levels in the two study seasons. Higher grain yield and aboveground biomass observed under high moisture levels was expected since improved water status result in higher levels of transpiration and therefore higher rates of plant growth (Cabrera-Bosquet et al. 2009). Low grain yield and biomass were observed under SS and MS due to water stress' limitation to photosynthesis (Tezara et al. 1999) through reduction of *gs*. These results confirm that crop yields will continue to be negatively affected by decreasing water availability in future due to climate change. There is therefore need for water use efficient crops that can produce good yields under limited water supply. This is particularly important in Africa where the population is expected to double by 2050 (United Nations 2015) hence the need to improve crop productivity even under harsh conditions.

$WUE_{biomass}$ and WUE_{grain} decreased with decreasing moisture level which was rather contrary to the findings of some studies on sugar beet were decreasing soil moisture resulted in increased integrated WUE (Bloch et al. 2006). Our results are however not uncommon as they corroborate with the findings of other studies (Erice et al. 2011). Under water stress photosynthesis is inhibited through decreased ribulose bisphosphate supply (Tezara et al. 1999) thereby retarding biomass accumulation, and in turn WUE. Furthermore, high water loss via evaporation during the early growth stages could be a plausible reason for the observed low WUE particularly under SS where irrigation was stopped at GS31.

4.2 Carbon Isotope Discrimination in Grain and Flag Leaves

The observed influence of soil moisture on $\Delta^{13}C$ is consistent with findings of other studies (Cabrera-Bosquet et al. 2009; Barbour et al. 2011; Cabrera-Bosquet et al. 2011; Erice et al. 2011; Cernusak et al. 2013; Wang et al. 2013). According to Davies and Gowing (1999), plants are known to be very sensitive to small differences in soil moisture and they respond by regulating their *gs*, which in turn affects

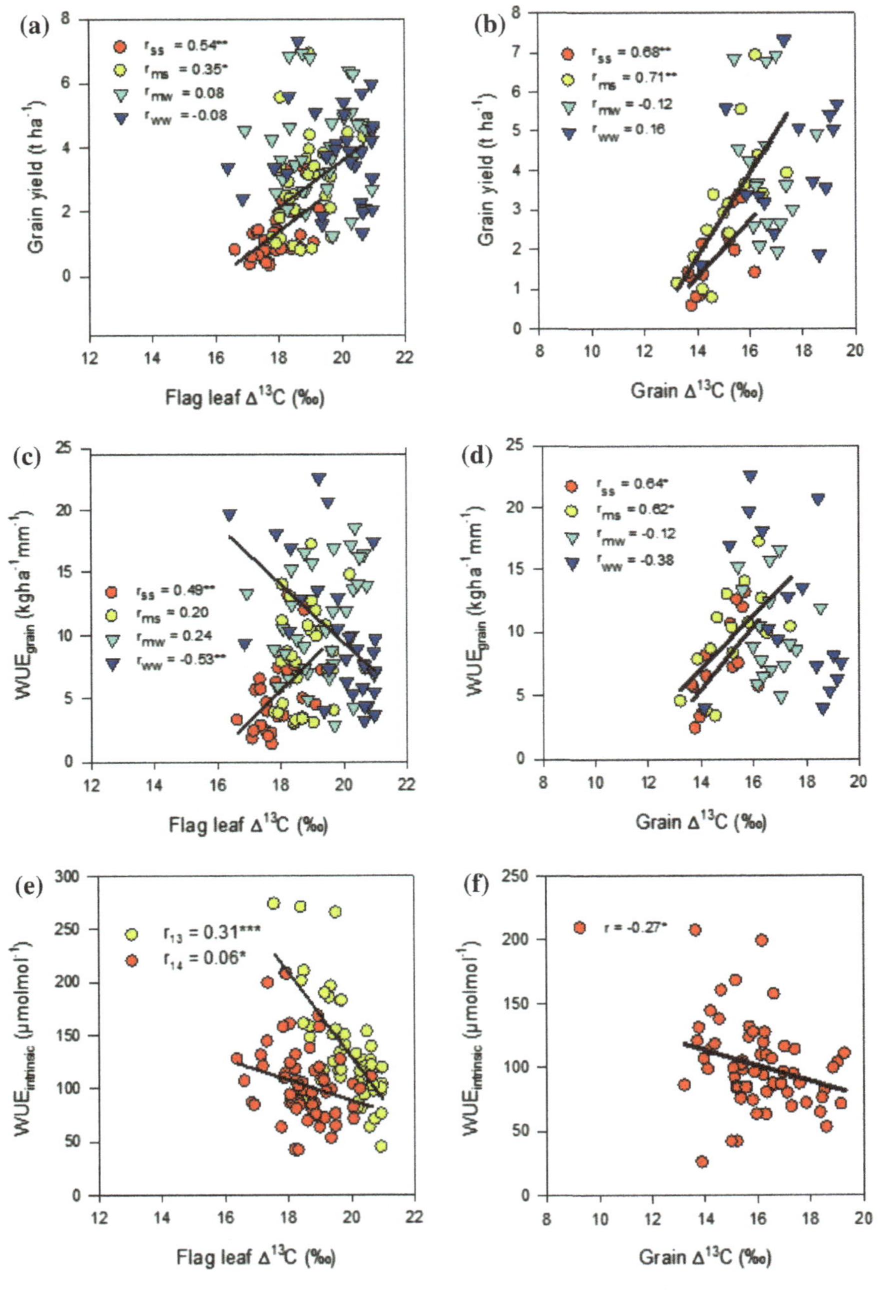

^{13}C enrichment in leaves. In C_3 plants, discrimination by Rubisco has a huge effect on $\Delta^{13}C$ variation as controlled by *gs* (CO_2 supply side). Hence, the variation in $\Delta^{13}C$ observed in flag leaf leaves under the four moisture levels arose because of the differences in the level of discrimination by Rubisco as affected by CO_2 supply. Higher discrimination values observed under WW conditions compared to SS resulted from the high CO_2 supply (high *gs*) compared to the restricted CO_2 supply (low *gs*) under SS. According to Farquhar and Richards (1984), higher *gs* promotes discrimination by Rubisco which has a discrimination factor of about 30‰.

$\Delta^{13}C$ was measured in flag leaves at GS71 and GS92 and in 2014, extra measurements were also done on grain samples. Measurements at GS71 indicate early response of triticale to moisture stress while later measurements are an integration of the entire season. Though both growth stages were not far apart, the results showed that earlier measurements were less enriched in ^{13}C than those measured later in the season. In 2013 it was also observed that early season measurements were more efficient in separating moisture levels than later measurements (Table 2). The lower $\Delta^{13}C$ values in grain samples and in flag leaves measured at GS92 compared to leaf $\Delta^{13}C$ at GS71, can be attributed to the increase in evaporative demand occurring later in the growing season (Smedley et al. 1991; Condon and Richards 1992).

In several studies (Anyia et al. 2007; Cabrera-Bosquet et al. 2011), grain $\Delta^{13}C$ has been found to be lower than leaf $\Delta^{13}C$, which is consistent with the results of this study. Using isotope signatures alone, our results and those of the above mentioned studies seem to suggest that most of grain filling assimilates come from post anthesis photosynthesis and this is regardless of the moisture level, as evidenced by significantly higher flag leaf $\Delta^{13}C$ values compared to grain $\Delta^{13}C$ under all four moisture levels. If the suggestion is true, it is contrary to the common belief that assimilates for grain filling under drought conditions are exported from stored pre-anthesis assimilates (Blum 2011; Xue et al. 2014). If the majority of assimilates were exported from pre-anthesis stages, grain $\Delta^{13}C$ would be expected to be higher or similar to flag leaf $\Delta^{13}C$ measured under WW at GS71 because the assimilates would have been produced in the absence of water stress. In fact, the difference between flag leaf $\Delta^{13}C$ and grain $\Delta^{13}C$ was even greater under severe stress (2.9‰) suggesting less contribution of pre-anthesis assimilates.

It is known that other leaves and organs contribute to assimilate production but their discrimination levels would still be expected to be in the same range of flag leaves as the growing conditions were the same. The difference in $\Delta^{13}C$ between the flag leaves and grain was just too wide to suggest major contribution of pre-anthesis assimilates. This therefore means that most of the assimilates for grain filling in triticale particularly under SS were probably from other sources such as ear photosynthesis. Ear photosynthesis is a reasonable source since empirical evidence has also shown higher tolerance of ear photosynthesis to water stress compared to flag leaf (Tambussi et al. 2005) hence ear photosynthesis would be expected to contribute more to grain filling than flag leaves. In addition, transport of assimilates by the phloem from the source (leaves or stems) to the sink (grain) is more likely to be hindered under water stress. In the study of Araus et al. (1993), ear $\Delta^{13}C$ closely resembled grain $\Delta^{13}C$ more than flag leaf $\Delta^{13}C$ and in a more recent study which determined contributions of different organs to grain filling in durum wheat, it was also found, using $\Delta^{13}C$, that ears contributed more to grain filling than flag leaves (Merah and Monneveux 2015).

The study therefore shows that triticale adaptation to drought conditions may be dependent on post anthesis photosynthesis, particularly ear photosynthesis. Hence, triticale/small grain breeders may need to focus on ear photosynthesis tolerance to water stress for improved yield in future drier climates.

4.3 *Relationship of $\Delta^{13}C$ and Grain Yield, Grain Yield WUE, Intrinsic WUE*

Most breeding programs target high grain yields or traits that are highly correlated to it. Under water limiting conditions, WUE_{grain} and $WUE_{biomass}$ would probably be the most relevent traits, as they indicate higher yield potentials under drought. However, they are difficult to directly measure in the field (Tardieu 2013) hence the need for surrogate traits like $\Delta^{13}C$. Several studies with cereals, particularly with wheat have shown positive correlations between grain yield and $\Delta^{13}C$ under drought conditions (Monneveux et al. 2005; Wahbi and Shaaban 2011). In Australia some wheat cultivars have been selected for high biomass and grain yield using $\Delta^{13}C$ as a selection tool (Rebetzke et al. 2002). Accordingly, our results concur with such findings, as $\Delta^{13}C$ was positively related to grain yield under SS and MS (Fig. 2a and b). These positive correlations were observed both for flag leaf $\Delta^{13}C$ and grain $\Delta^{13}C$ under SS and MS but were non-significant under MW and WW. These results show that $\Delta^{13}C$ is a useful selection trait for grain yield under water limited conditions.

Also, the relationship between flag leaf $\Delta^{13}C$ and WUE_{grain} was significantly positive under SS (0.49*) and so was the relationship between grain $\Delta^{13}C$ and WUE_{grain} (Fig. 2). These results further confirm the usefulness of $\Delta^{13}C$ as a selection tool under drought conditions. The observed positive correlation suggests that high yielding genotypes have a higher stomatal conductance (i.e. high discrimination) than low yielding genotypes under dry conditions. The lack of a significant relationship between grain yield and $\Delta^{13}C$ under well-watered conditions is congruent with the findings of Monneveux et al. (2005) and thus, other traits should be used for indirect cultivar selection for higher grain yield under well-watered conditions.

5 Conclusions

The study showed that $\Delta^{13}C$ could be useful as a triticale grain yield predictor under drought conditions. In addition, $\Delta^{13}C$ also offers potential as a proxy of intrinsic WUE and could be used for breeding towards improved drought tolerance in triticale. In essence, the study also revealed the importance of post anthesis photosynthesis to grain filling in triticale.

Acknowledgements This work was financially supported by VLIR-IUC, an Inter University Cooperation with University of Limpopo (UL), South Africa under Project 6 on Food Security. The authors would also like to acknowledge support from the Risk and Vulnerability Science Centre at UL. The following people are also acknowledged for their contribution to the work: Katja Van Nieuland, Willem Botes, Louis Eloff and Jimmy Motloutsi.

References

Anyia, A., Slaski, J., Nyachiro, J., Archambault, D., & Juskiw, P. (2007). Relationship of carbon isotope discrimination to water use efficiency and productivity of barley under field and greenhouse conditions. *Journal of Agronomy and Crop Science, 193*(5), 313–323.

Araus, J., Brown, H., Febrero, A., Bort, J., & Serret, M. (1993). Ear photosynthesis, carbon isotope discrimination and the contribution of respiratory CO_2 to differences in grain mass in durum wheat. *Plant, Cell and Environment, 16*(4), 383–392.

Araus, J. L., Slafer, G. A., Royo, C., & Serret, M. D. (2008). Breeding for yield potential and stress adaptation in cereals. *Critical Reviews in Plant Science, 27*(6), 377–412.

Araya, Y. N., Silvertown, J., Gowing, D. J., McConway, K., Linder, P., & Midgley, G. (2010). Variation in delta 13C among species and sexes in the family Restionaceae along a fine-scale hydrological gradient. *Austral Ecology, 35*(7), 818–824.

Barbour, M. M., Tcherkez, G., Bickford, C. P., Mauve, C., Lamothe, M., Sinton, S., et al. (2011). delta ^{13}C of leaf-respired CO_2 reflects intrinsic water-use efficiency in barley. *Plant, Cell and Environment, 34*(5), 792–799.

Barnabás, B., Jäger, K., & Fehér, A. (2008). The effect of drought and heat stress on reproductive processes in cereals. *Plant, Cell and Environment, 31*(1), 11–38.

Barnard, H., Brooks, J., & Bond, B. (2012). Applying the dual-isotope conceptual model to interpret physiological trends under uncontrolled conditions. *Tree Physiology, 32*(10), 1183–1198.

Bloch, D., Hoffmann, C. M., & Märländer, B. (2006). Impact of water supply on photosynthesis, water use and carbon isotope discrimination of sugar beet genotypes. *European Journal of Agronomy, 24*(3), 218–225.

Blum, A. (2011). *Plant breeding for water-limited environments*. New York: Springer Science + Business Media.

Cabrera-Bosquet, L., Albrizio, R., Nogues, S., & Araus, J. L. (2011). Dual Delta 13C/delta 18O response to water and nitrogen availability and its relationship with yield in field-grown durum wheat. *Plant, Cell and Environment, 34*(3), 418–433.

Cabrera-Bosquet, L., Molero, G., Nogues, S., & Araus, J. L. (2009). Water and nitrogen conditions affect the relationships of Delta C-13 and Delta O-18 to gas exchange and growth in durum wheat. *Journal of Experimental Botany, 60*(6), 1633–1644.

CDIAC. 2015. *Monthly atmospheric 13C/12C isotopic ratios for 11 SIO Stations*. Carbon Dioxide Information Analysis Center (CDIAC).

Cernusak, L. A., Ubierna, N., Winter, K., Holtum, J. A., Marshall, J. D., & Farquhar, G. D. (2013). Environmental and physiological determinants of carbon isotope discrimination in terrestrial plants. *New Phytologist, 200*(4), 950–965.

Condon, A., & Richards, R. (1992). Broad sense heritability and genotype × environment interaction for carbon isotope discrimination in field-grown wheat. *Crop and Pasture Science, 43*(5), 921–934.

Condon, A., Richards, R., Rebetzke, G., & Farquhar, G. (2004). Breeding for high water-use efficiency. *Journal of Experimental Botany, 55*(407), 2447–2460.

Davies, W., & Gowing, D. (1999). *Plant responses to small pertubations in soil water status. Physiological plant ecology* (pp. 67–89). Oxford, UK: Blackwell.

Erice, G., Louahlia, S., Irigoyen, J. J., Sánchez-Díaz, M., Alami, I. T., & Avice, J.-C. (2011). Water use efficiency, transpiration and net CO_2 exchange of four alfalfa genotypes submitted to progressive drought and subsequent recovery. *Environmental and Experimental Botany, 72*(2), 123–130.

Farquhar, G., & Richards, R. (1984). Isotopic composition of plant carbon correlates with water-use efficiency of wheat genotypes. *Functional Plant Biology, 11*(6), 539–552.

Farquhar, G. D., Ehleringer, J. R., & Hubick, K. T. (1989). Carbon isotope discrimination and photosynthesis. *Annual Review of Plant Biology, 40*(1), 503–537.

Foley, J. A., Ramankutty, N., Brauman, K. A., Cassidy, E. S., Gerber, J. S., Johnston, M., et al. (2011). Solutions for a cultivated planet. *Nature, 478*(7369), 337–342.

Merah, O., & Monneveux, P. (2015). Contribution of different organs to grain filling in durum wheat under mediterranean conditions I. Contribution of post-anthesis photosynthesis and remobilization. *In Journal of Agronomy and Crop Science, 201*(5), 344–352.

Monneveux, P., Reynolds, M. P., Trethowan, R., González-Santoyo, H., Peña, R. J., & Zapata, F. (2005). Relationship between grain yield and carbon isotope discrimination in bread wheat under four water regimes. *European Journal of Agronomy, 22*(2), 231–242.

Morison, J. I. L., Baker, N. R., Mullineaux, P. M., & Davies, W. J. (2008). Improving water use in crop production. *In Philosophical Transactions of the Royal Society B: Biological Sciences, 363*(1491), 639–658.

Rebetzke, G., Condon, A., Richards, R., & Farquhar, G. (2002). Selection for reduced carbon isotope discrimination increases aerial biomass and grain yield of rainfed bread wheat. *Crop Science, 42*(3), 739–745.

Seibt, U., Rajabi, A., Griffiths, H., & Berry, J. A. (2008). Carbon isotopes and water use efficiency: Sense and sensitivity. *Oecologia, 155*(3), 441–454.

Smedley, M. P., Dawson, T. E., Comstock, J. P., Donovan, L. A., Sherrill, D. E., Cook, C. S., et al. (1991). Seasonal carbon isotope discrimination in a grassland community. *Oecologia, 85*(3), 314–320.

Tambussi, E., Nogués, S., & Araus, J. (2005). Ear of durum wheat under water stress: Water relations and photosynthetic metabolism. *Planta, 221*(3), 446–458.

Tardieu, F. (2013). Plant response to environmental conditions: assessing potential production, water demand, and negative effects of water deficit. *In Frontiers in physiology, 4,* 17.

Tezara, W., Mitchell, V. J., Driscoll, S. D., & Lawlor, D. W. (1999). Water stress inhibits plant photosynthesis by decreasing coupling factor and ATP. *Nature, 401*(6756), 914–917.

United Nations, U. 2015. *World Population Prospects: The 2015 Revision, Key Findings and Advance Tables*. Department of Economic and Social Affairs.

Wahbi, A., & Shaaban, A. S. A. (2011). Relationship between carbon isotope discrimination (Delta), yield and water use efficiency of durum wheat in Northern Syria. *Agricultural Water Management, 98*(12), 1856–1866.

Wang, Y., Zhang, X., Liu, X., Zhang, X., Shao, L., Sun, H., et al. (2013). The effects of nitrogen supply and water regime on instantaneous WUE, time-integrated WUE and carbon isotope discrimination in winter wheat. *Field Crops Research, 144,* 236–244.

WRB. 2014. *World reference base for soil resources 2014*. International soil classification system for naming soils and creating legends for soil maps. World Soil Resources Reports.

Xue, Q., Rudd, J., Liu, S., Jessup, K., Devkota, R., & Mahano, J. (2014). Yield determination and water-use efficiency of wheat under water-limited conditions in the US Southern high plains. *Crop Science, 54*(1), 34–47.

Zadoks, J. C., Chang, T. T., & Konzak, C. F. (1974). A decimal code for the growth stages of cereals. *Weed Research, 14*(6), 415–421.

Enhancing Farmers' Resilience and Adaptive Capacity Through Access to Usable Weather Information: A Case Study from Uganda

Margaret Barihaihi and Jessica Mwanzia

1 Background

The Africa Climate Change Resilience Alliance (ACCRA) project was established in 2009 with the goal of improving poor and vulnerable communities' ability to adapt to climate change. ACCRA is a consortium comprised of Oxfam GB, Save the Children International, CARE International, World Vision International and Overseas development Agency ACCRA seeks to achieve improved national decision making in order to reduce vulnerability and to support community agency and resilience to climate change. ACCRA works in Ethiopia, Uganda and Mozambique supporting governments, civil society and the wider development community to integrate climate change adaptation and resilience into their policy and practice. Africa is most vulnerable to climate variability and change amidst other development stressors like high levels of poverty, poor infrastructure among others. Meteorologically related natural disasters are increasingly retarding the pace of the socio-economic development of Africa countries (Integrated African Strategy on Meteorology 2012). The agriculture sector which remains a backbone for local and national economies is highly vulnerable to erratic weather and climate change. Most farmers in Africa are increasingly susceptible to extreme weather, with about 70% of people living in sub-Saharan Africa dependent on rain-fed subsistence agriculture (IPCC 2015 5th Assessment Report; Hellmuth et al. 2007). Planting seasons are changing, rainfall is becoming less reliable, temperatures are rising, weather is becoming more extreme, and as a result, the fight against hunger is becoming much harder.

M. Barihaihi (✉)
Africa Climate Change Resilience Alliance (ACCRA), Oxfam, Great Britain, UK
e-mail: mbarihaihi1@oxfam.org.uk

J. Mwanzia
Pan Africa Programme, Oxfam, Great Britain, UK
e-mail: JMwanzia@oxfam.org.uk

W. Leal Filho et al. (eds.), *Climate Change Adaptation in Africa*,
Climate Change Management, DOI 10.1007/978-3-319-49520-0_38

In Uganda, the UN's Food and Agriculture Organization determined that the drop in the growth of the Ugandan economy from 6.6% in 2004–05 to 5.3% in 2005–06 was largely attributed to the variability of the weather, specifically, its impact on agriculture (National Planning Authority—Uganda National Development Plan 2010). More frequent extreme weather events including droughts, floods and landslides over the past several decades have been experienced. Recently, CDKN report (2015) on Economic Assessment of the Impact of Climate Change in Uganda shows among others, that some major food crops like cassava, potato will face a production decline of around 40% by 2050. Other crops such as millet, sorghum and pigeon peas will reduce by 10%. Cash crops like coffee which contributes about 18% of total exports, might reduce between 50 and 75% by 2050, as a result of a combination of yield reductions and (more importantly) loss of areas where coffee can be grown.

The fifth Assessment IPCC (2015) report, projects increased severity and frequency in droughts, floods, tropical cyclones in Africa. Therefore information about these impacts of climate variability and change is needed by communities and resource managers to adapt and prepare for larger fluctuations as global climate change becomes more evident. Empirical studies among African farmers have shown that usable climate forecasts can help them reduce their vulnerability to drought and climate extremes, while also allowing them to maximize opportunities when favorable rainfall conditions are predicted (Patt et al. 2005; Phillips et al. 2001; Roncoli et al. 2009). Seasonal rainfall forecasts are particularly suited for rain-fed farming systems (Klopper et al. 2006). However, a number of challenges are highlighted in these studies associated with getting weather information to farmers.

Africa has the least developed land-based observation network of all continents (Integrated African Strategy on Meteorology 2012). For example Uganda does not have adequate weather equipment. Most districts have weather stations but many of them are either not functional or lack staff to record and report the data. This makes it difficult for Uganda National Meteorological Authority and other weather distributors to collect and disseminate accurate weather information. The forecasting methods used by national meteorologist are more suitable for regional and global scenarios and sometimes fall short in providing accurate information about changes at local level which the rural farmers need most. This explains why rural communities tend to trust and prefer using the Indigenous knowledge forecasts over the modern forecasts.

Traditionally, local communities and farmers in Africa and elsewhere have used traditional knowledge to understand weather and climate patterns in order to make decisions about crop and irrigation cycles (Makwara 2013). This knowledge has been gained through many decades of experience, and has been passed on from previous generations. The knowledge is adapted to local conditions and needs. Indigenous knowledge provides valuable insights on how communities have interacted with their local environment. However, increasing variability in climate has reduced farmers' confidence in traditional knowledge and has led them to seek out scientific weather forecasts.

Women are more vulnerable to climate change than men since they comprise 70% of the world's poor and in sub-Saharan Africa women grow up to 80–90% of the food in the region Food and Agriculture Organization of the United Nations (FAO). It is therefore critical for farming communities especially women in Africa to have information about the weather during the cropping season so that they can plan better to prevent or avert impending disaster or loss of crops and livelihood. However, women don't always get the information. Most often the information and the dissemination channels are not gender sensitive and some categories of the communities are left out. This is why ACCRA attempts to ensure equal access to information and guidance, taking gender-based access barriers into account.

The World Meteorological Organisation (WMO) has overtime established institutional initiatives, especially the regional Climate Outlook Forums (COFs) that produce seasonal climate forecast (SCFs) (Patt et al. 2007; Roncoli et al. 2009). For example, Uganda is part of the Greater Horn of Africa Climate Outlook Forum (GHACOF). Regional bodies are supported by the Global Framework for Climate Services. Despite efforts and institutional advancements to promote access of weather forecasts by African farmers, most countries have not experienced significant benefits from using them. Climate forecasts and projections are supposedly providing an indication of what the future climate may look like. The information guides the people whose livelihoods are affected by changes in climate to design forward-looking and flexible plans that are adapted to a range of climate possibilities. However, people must understand the forecasts and their implications together with information on the uncertainty and probability of whether they will happen.

As weather and climate patterns and behaviour do not recognise political boundaries, enhanced cooperation among African countries, is required to effectively meet government and societal needs through weather and climate information and services.

The 2011 ACCRA research (Levine et al. 2011) aimed at understanding how development interventions have been able to increase adaptive capacity to climate change in Ethiopia, Mozambique and Uganda. One of the key research findings was that rural communities in these countries were not receiving seasonal forecasts, and if they did, they came late, always in English with difficult scientific terminologies. For example in Uganda, for centuries UNMA was the sole responsible agency for the production and dissemination of seasonal forecasts. Channels for dissemination of the final products generated centrally were limited to emails sent to government ministries and districts. This meant that the majority of rural farmers didn't have access to the forecast. Even those who got it was with no advisory messages to interpret the highly technical and bulky information, the chances of rural communities, understanding and using the outlook to plan farming activities for the coming season were thin. Specifically, women were not accessing the information due to the fact that the channels of dissemination and the timing were not favorable. The common channels which were being used were through internet and FM radios which women have no access and control. Consequently, the majority farmers did not have much interest in scientific meteorological services and relied almost fully on their experience and indigenous knowledge for decision making. And yet even though farmers turned to indigenous knowledge for their decision-making they also reported that they were no longer reliable due to disappearance of certain insects, birds, plants which constitute key indicators for traditional forecasting. Farmers were also experiencing more intense and frequent rains.

Overall, the research found that farmers—who were most vulnerable to unpredictable extreme weather events—were not able to access weather information. In that scenario, accurate, reliable and timely weather forecasts, combined with efficient early warning systems and other innovative decision-making tools like the ACCRA "Forward Looking and Decision Making Game" would be needed to enhance farmers' adaptive capacity. The Challenges of climate change and importance of accurate climate information is evidenced by the following summary from a farmer in Ogur Sub County in Otuke district in Uganda "*Traditionally, we predicted the weather by looking for signs; strong winds and dust followed by leaves drying off meant the dry season was about to begin whereas appearance of mist, frogs and changes in the sky indicated the arrival of the rainy season. But the seasons are no longer predictable. Everybody in this area has noticed climate change; the rain delays and as a consequence the level of agricultural production has decreased. It is very important for us to know the scientific forecast so that we can plan accordingly. If the forecast predicts drought then we can store food and prepare our gardens early*".

2 The Weather Forecast and Dissemination Model

In 2012, ACCRA stepped in by bringing together experts from government ministries, agencies and civil society to work together with UNMA to improve meteorological services in Uganda. The 2011 ACCRA study found that for decades, the UNMA used bulky and complicated terminologies to issue seasonal forecasts which were only published in English, which is only spoken by a minority in rural population. Furthermore, the channels for dissemination of the final products generated at national level were limited to emails sent to government ministries and districts. Due to unreliable internet connections, lack of defined distribution procedures as well as a non-existent dissemination budget, the information seldom reached end users, especially rural communities. If they did, the forecasts came at the end of the season therefore not supporting farming decision making.

The Uganda's climate and weather forecast model was initiated to address these problems. The model follows a process of producing a simplified and easily understandable seasonal weather forecast that is translated into local languages with sector specific advisories. The experiences and lessons learned are illustrated below.

2.1 The Process

The model follows five steps as outlined below

Step 1 **Issue of the technical seasonal climate forecast at regional level**
UNMA produces the seasonal weather forecast as part of the Greater Horn Regional Climate Outlook Forum where the seasonal forecast is agreed upon, by consensus with other countries. This is done three times annually including March–May; June–July; and September–December.

Step 2 **National technical expert meeting at national level**
UNMA in collaboration with ACCRA convenes experts to discuss and simplify the scientific seasonal forecast with sector-based advisories. The experts are drawn from different ministries and departments, civil society organisations, research institutions and private enterprises—representing various sectors including agriculture, health, disaster management, water, environment, energy and wild life among others. The last season's performance is also reviewed.

Step 3 **National press releases**
A press release is issued by the Minister of Water and Environment immediately after the national technical expert meeting and published in the national newspaper, New Vision.

Step 4 **Translations into local languages**
The forecasts are translated into local languages by specialized translators who have been trained in meteorological terminology. The translators work with UNMA's communications department to clarify any uncertain issues. The translations takes three to four days in total and the various versions are produced in electronic text, soft copies, and prerecorded audio CDs. ACCRA started off supporting 4 languages in 2012, and increased to 10. By the end of 2015, other donors like GIZ supported more 15 languages to make 25 in total with more pledges to cover the 52 languages in the country.

Step 5 **Dissemination process and products diversification**
Diversified dissemination channels and forecast products have been introduced to ensure all categories of the community are reached. Also the list of stakeholders to disseminate is greatly diversified in addition to Government structures. Non-government actors include ACCRA consortium members (World Vision, Oxfam, Save the Children, CARE) and other Civil Society Organizations (CSOs)' networks like Climate Action Network Uganda CAN-U, PELUM, Lutheran World Federation among

others and Academia like the Makerere University Centre for Climate Change Research and Innovations (MUCCRI).
Each distribution channel disseminates the appropriate forecast product(s) as follows;

1. Emails targeting the government ministries and departments, CSOs and networks, local governments, civil aviation, and selected media houses.
2. Local FM radio stations using pre-recorded CDs and printed translated copies targeting community members especially men, private sector representatives and the general public.
3. Primary and Secondary schools using printed materials that are simple and child friendly targeting school going children and teachers.
4. Churches, mosques, markets, community meetings, and regional meetings using printed versions of the translated forecast targeting women, people living in remote inaccessible areas and leaders.
5. Mobile phones through short messages (SMS) targeting the entire public (phone owners).
6. Print media through newspapers targeting urban communities and decision makers.

Step 6 **User feedback surveys**
User feedback surveys are conducted jointly by ACCRA, UNMA and the Office of the Prime Minister to enable UNMA's technical teams to capture access and utilization levels of the seasonal forecast information and how it influences the decisions of rural farmers including women farmers.

3 Gender Responsive Dissemination Channels

In Uganda, women are more involved in agriculture than men. Whereas men supervise the process, sell the produce and manage the household economy, women do the bulk of work; they cultivate the gardens, buy seeds, plant, weed, harvest and process the produce. All of which are directly affected by weather conditions. The community feedback surveys with gender disaggregated group discussions carried out by UNMA and ACCRA staff in selected implementation areas indicated that men and women have different preferences when it comes to receiving information. This could be attributed to their gender roles, norms and literacy level.

Women, for example, proposed that the forecasts should be disseminated to local community groups, markets and churches. They regarded these as accessible spaces. The commonly used channel, the FM radio stations turned out to be the least appropriate tool for reaching out to women. This was because first, many women either do not own radios or do not have time to listen to them due to their busy schedule of food production, care taking and performing domestic chores. Secondly, since men own the radios, they often move around with them or control which programmes to listen to based on their personal interests. Communication channels that are specifically designed to reach women become essential to enhance women access and utilize weather information.

Men, on the other hand, preferred radios, newspapers and local council meetings. This shows that knowledge and information delivery channels need to be analysed in relation to gender dynamics. Based on the results of the focus group discussions, the national dissemination strategy for the seasonal forecasts has been modified to accommodate the different needs of rural men and women.

4 Impacts of the Model

4.1 Farmers' Feedback on the Impact of Timely and Simplified Seasonal Weather Forecasts

Overall, in districts where there have been focused efforts on explaining the scientific forecast and the advisories to men and women farmers, the implementation rate has been considerably higher and as a direct result, fewer farmers are exposed to food insecurity. The testimonies below indicate the changes that farmers are experiencing because they now have access to timely and usable forecasts.

A woman farmer from Kitgum district testifies: *"For the first time we have received the weather forecasts in the local language, which has not happened before. When I heard the programme on the radio, I was excited and kept listening every day. I do not know how to write, so I instructed my son to write the important points. It has helped me to plan and so far the forecast has been accurate".*

A woman farmer in who is participant in a focus group discussion testifies: *"I have been very keen on following up the seasonal forecast given on radio or on phones. I usually note down what the forecast is predicting and use this information for my farming activities. The last seasonal forecast of June, to August 2013 advised farmers to plant in August and I followed the advice. All my crops did well and I have no problems with food scarcity. Those who waited and planted late because they thought the rains were not enough lost all their crops".*

A woman farmer from Ogur Sub-county, Otuke District *testifies: "I receive the information at community meetings from FAPAD (Facilitation for Peace and Development—a civil society organisation) and the forecasts are also posted at the sub-county headquarter. I have planted eucalyptus trees to prevent flooding around the house and raised the veranda to prevent rain water from getting inside the hut. I have grown cassava, invested in poultry, and fast maturing crops. I did it because the forecasts predicted enough rain and cassava is resistant to drought. We have also dug drainage channels in the garden to prevent the crop from drowning".*

A woman farmer from Ogur, Sub-county, Otuke District *testifies: "To cope with the negative effects of climate change, we have learned about the importance of planting trees, constructing trenches for the running water, building a veranda for the house to flood-proof it, and to grow climate resistant and fast growing crops".*

A male farmer from ***Bundibugyo District testifies:*** *"Because of unpredictable rains, I used to plant greens in the swamp. But this season we were advised to plant near home because the rains would be sufficient. I established a garden near my house and I have had a bumper harvest; we have had enough to eat and I sold the surplus which has brought extra income to the family".*

4.2 Impacts at Institutional Level

First, the model has significantly improved coordination among experts at national level. For the first time UNMA started inviting key experts from line-ministries such as agriculture, water, climate change department, energy, disaster management, health among others and representatives from the media, research and academia to discuss the seasonal weather forecast and its implications and to develop advisories for the rural population on how to cope with the forecasted seasons climatic conditions. Second, the initiative has reduced misinterpretation of the seasonal scientific weather forecast. By distributing timely and usable weather and climate forecasts and by explaining the importance of this information, rural farmers can make informed decisions and improve seasonal planning. Thirdly, the model has contributed to raise the profile and credibility of UNMA among rural community members. Farmers are increasingly gaining confidence in scientific meteorological services and they use their newly gained knowledge to improve their farming.

As reported by Deus Bamanya Director of Applied Meteorology, Data, and Climate Services, Uganda National Meteorology Authority (UNMA), "the early successes of the improved seasonal forecast model encouraged the government of Uganda to set aside a dedicated budget, so that UNMA is now able to meet the costs of producing the climate forecasts as well as improving the model".

Finally, the model has generated a demand for translated weather forecasts, particular at lower levels of governments such as districts. Some districts have responded by introducing notice boards which have been termed 'weather boards' for displaying the information for the public to use.

4.3 Challenges with the Model

- It is still evident that rural communities prefer to use both IK and scientific forecasts due to lack of trust. The modern forecasts are not 100% accurate, so farmers tend to adhere to what they are used to. How to bring together traditional knowledge and modern science while respecting both sets of values and building on their respective strengths remains a challenge.
- Like many other developing countries, Uganda does not have adequate weather equipment. This makes it difficult for UNMA and other weather distributors to collect and disseminate accurate weather information.
- Another challenge is to ensure timely dissemination of the weather forecast. Uganda relies on regional and global circulation models through the GHACOF regional forum and most times, the final products are disseminated in the middle of the season. This has not been helpful for farmers to make appropriate decisions. This gap forces farmers to seek for information from various sources on radios which is put together by internet, newspapers, etc. and the quality of the weather information is a challenge.

5 Key Lessons and Policy Suggestions—How UNMA Can Sustain the Model

- For sustainability purposes, Government structures ought to be used systematically to boost dissemination effectiveness, starting from ministries to local communities. Non-government actors should supplement government efforts in a coordinated manner. Focal persons in ministries, agencies and departments should be nominated to form a permanent team mandated to provide expert support to the existing technical committee responsible for downscaling and developing advisories.

- UNMA needs to establish a Monitoring and Evaluation system as part of existing government structures to account for who disseminates where, how is the forecast supporting early warning and how it is causing changes in planning and decision making for farming activities.
- Indigenous knowledge on forecasting can no longer be ignored and kept isolated from scientific predictions. A study by UNMA and ACCRA (2015) on indigenous forecasting, in addition to user feedback, revealed that traditional methods strongly influence farmers' decisions. For this reason UNMA and ACCRA are currently strategizing and fundraising to pilot integration of science and traditional weather forecasting. ACCRA's alliance member CARE International is spearheading such approach in Otuke district.
- Adequate funding should be deliberately allocated by UNMA and various ministries and trickle down to local government to aid effective dissemination and monitoring and evaluation of weather forecasts utilization. More funding should also serve in restoring weather monitoring equipment countrywide to improve on the data quality.
- Long term strategic Public Private Partnerships should be part of the dissemination model for meaningful scale up and sustainability. For example UNMA should engage telecommunication companies to exploit the widespread use of mobile cellphones countrywide. A few short-term projects have been piloted with UNMA and GRAMEEN in Kasese District; and with World Vision in Soroti and Kabong Districts but the two projects closed with limited impact.
- UNMA needs to train different actors in Government, CSOs, Media, Academia and Private sector on the importance of the weather and climate information, linking the costs and benefits to development. Inadequate understanding of climate change and forecasts is still prevailing at all levels. Increased knowledge will interest them to disseminate and evaluate the performance of the seasonal forecasts for rural communities.
- Policy makers should be deliberately involved and trained by UNMA to increase awareness and appreciation to ease discussions around additional financing to UNMA for efficient and effective delivery of meteorological services in the country.

References

ACCRA. (2013). *The climate forecast model*. Available online: http://community.eldis.org/.5c3edc47/mlclimate-forecast-model-100914-en.pdf

CDKN—Climate Change Department. (2015). *Economic assessment of the impacts of climate change in Uganda study report*. Available at http://www.ccu.go.ug/index.php/projects/cdkn and www.cdkn.org

Hellmuth, M., Patt, A., & Ogallo, L. (2007). Learning from 10 years of climate outlook forums in Africa. *Science, 318*, 49–50.

Klopper, E., Vogel, C. H., & Landman, W. A. (2006). Seasonal climate forecasts—Potential agricultural-risk management tools? *Climatic Change, 76,* 73–90.

IPCC. (2015). Working group II to the fifth assessment report of the intergovernmental panel on climate change: Impacts, adaptation, and vulnerability and Part B: Regional aspects. In V. R. Barros, C. B. Field, D. J. Dokken, M. D. Mastrandrea, K. J. Mach, T. E. Bilir, et al. (Eds.) (pp. 1199–1265). Cambridge, United Kingdom and New York, NY, USA: Cambridge University Press. https://ipcc-wg2.gov/AR5/images/uploads/WGIIAR5-Chap22_FINAL.pdf

Levine, S., Ludi, E., & Jones, L. (2011). *Rethinking support for adaptive capacity to climate change: The role of development interventions*. London: ACCRA, ODI.

Makwara, C. E. (2013). *Indigenous knowledge systems and modern weather forecasting: Exploring the linkages*. Curriculum Studies Department, Great Zimbabwe University.

National Planning Authority. (2010, April). *The National Development Plan 2010/11–2014/15*. http://www.npa.ug/docs/NDP2.pdf

Patt, A., Ogallo, L., & Hellmuth, M. (2007). Learning from 10 years of climate outlook forums in Africa. *Science, 2007*(318), 49–50.

Patt, A., Suarez, P., & Gwata, C. (2005). Effects of seasonal climate forecasts and participatory workshops among subsistence farmers in Zimbabwe. *Proceedings of National Academy of Sciences USA, 102,* 12623–12628.

Phillips, J., Makaudze, E., & Unganai, L. (2001). Current and potential use of climate forecasts for resource-poor farmers in Zimbabwe. In C. Rosenzweig (Ed.), *Impacts of El Nino and climate variability in agriculture* (pp. 87–100). Participatory climate forecast dissemination in Burkina Faso (West Africa). *Climatic Change 2009, 92,* 433–460. Madison, Wisconsin: American Society of Agronomy Special Publication.

Roncoli, C., Jost, C., Kirshen, P., Sanon, M., Ingram, K. T., Woodin, M., et al. (2009). From accessing to assessing forecasts: An end-to-end study of.

World Meteorological Organization (WMO). *Intergovernmental Board on Climate Services*. Year? Available online: http://www.wmo.int/pages/publications/meteoworld/archive/july13/documents/meteoworld.pdf

Strengthening Farmer Adaptive Capacity Through Farms of the Future Approach in Nyando, Western Kenya

Philip Kimeli, Mary Nyasimi and Maren Radeny

1 Introduction

Climate change and variability has rapidly emerged as a leading global challenge, causing serious threats to sustainable development with adverse impacts on environment, human health, food security, economic activities, natural resources and physical infrastructure (IPCC 2007; Huq and Jessica 2006). In East Africa, the impacts of climate change are exacerbated by other compounding factors such as widespread poverty, declining soil fertility, inequitable land distribution, increasing human population, over dependence on rain-fed agriculture and poor market access (Jayne et al. 2014; Addae-Korankye 2014; IPCC 2007). Future climate projections show an increase in rainfall and with some seasons experiencing intensive droughts for East Africa (IPCC 2014). These changes in rainfall patterns and other extreme weather events are likely to lead to increased crop failures, pest and disease outbreaks, and water scarcity altering the functioning of agricultural landscapes in overwhelming and often destructive ways (IPCC 2007). While farmers in the region are already adapting to changing climate, the changes tend to be marginal rather than transformational and appear to be limited to farming practices that are fairly easy to undertake without major disruptions to the farming system (Kristjanson et al. 2012). Comprehensive agricultural adaptation strategies are therefore, needed to meet the food and income needs of current and future generations (Thornton et al. 2006).

In order to strengthen adaptive capacity and encourage transformative changes, farmers need to understand what their future climate is likely to be. Model based assessment of climate change impact on agricultural production is the first step for informing the development of site specific adaptation options that reduce the risks

P. Kimeli (✉) · M. Nyasimi · M. Radeny
CGIAR Research Program on Climate Change, Agriculture and Food Security (CCAFS), International Livestock Research Institute (ILRI), P.O. Box 30709, Nairobi 00100, Kenya
e-mail: p.kimeli@cgiar.org

W. Leal Filho et al. (eds.), *Climate Change Adaptation in Africa*,
Climate Change Management, DOI 10.1007/978-3-319-49520-0_39

associated with climate change. Studies have shown that only 30% of the world climates are expected to be completely novel under climate change, implying that 70% of expected future climates already exist somewhere else on the globe (Williams et al. 2007). The spatial and temporal variability in climate can be used as a means of having a real experiment of what the future holds for a specific site. The CGIAR Research Program on Climate Change, Agriculture and Food Security (CCAFS) has developed a Climate Analogue tool that can be used to connect sites with statistically similar climates across space and/or time (Ramirez et al. 2011). Spatial analogue refers to an area whose climate today appears to be similar to the future projected climate of another location and addresses the following questions:

- Where will I be able to find areas with a climate "possibly" similar in the future to the current climate of my site?
- Where can I find a place that currently has a climate that looks like the one expected for my site in the future?
- Where can I find the present climate of my site in the world currently?

2 The Farms of the Future Approach

Through the farms of the future (FotF) approach, the CCAFS climate analogue tool can be used to connect farmers to their possible future climates through farmer-to-farmer learning exchanges. In East Africa, this has already been piloted in Lushoto, Tanzania (Nelson et al. 2012), and recently implemented in Nyando, Kenya. The Farms of the Future approach focusses on two main objectives: First, to build on farmer-to-farmer exchanges to analogue sites as a valuable option to improve adaptive capacity and support knowledge transfer, second, to improve understanding of local practices and available tools for enabling change, as well as cultural, economic, or institutional obstacles to such adaptive change.

The FotF approach involves four key steps: First, using CCAFS climate-analogue tool to identify climate analogue sites, including the potential social, economic, environmental similarities and learning opportunities, second, facilitating learning between farmers and other agricultural innovation system (AIS) stakeholders, third, documenting the learning experience by farmers and other AIS stakeholders using photos and video cameras and lastly sharing of the lessons learnt by farmers and the other AIS stakeholders. The aim of this study was to strengthen the adaptive capacity of farmers and other Agriculture Innovation System (AIS) stakeholders from Nyando, Kenya through farmer exchange visits to their analogue sites in Kenya. This paper shares the learning experiences of Nyando farmers' across the different analogues sites they visited in Kenya.

3 Materials and Methods

3.1 Study Area

The study was conducted in Nyando basin in western Kenya, where CCAFS has established Climate-Smart Villages (CSVs) (see Fig. 1). CSVs are sites where researchers, local partners and farmers work together to evaluate and maximize synergies across a portfolio of climate-smart agricultural interventions (Aggarwal et al. 2013). The aim is to sustainably increase agricultural productivity and income, reduce local climate risks and vulnerabilities, build resilience to climate change, reduce greenhouse gas emissions and enhance achievement of national food security and development goals. This is achieved through the participatory application of climate-smart agricultural practices and technologies (Kinyangi et al. 2015). Currently there are over 2000 farmers involved across seven villages in Nyando CSV.

Nyando is a rich agricultural flood plain around the large Lake Victoria, characterized by humid to sub-humid climate (Yamane et al. 2015; Förch et al. 2011). Subsistence farming is primarily mixed rain fed crop-livestock farming systems, with small farm sizes (averaging less than 1 ha) and low agricultural potential due to low and erratic rainfall (Kristjanson et al. 2012). The population density is more than 400 persons per square kilometer making it one of the highest populated rural localities in East Africa (Kinyangi et al. 2015). Other characteristics include high levels of poverty and massive environmental degradation, including declining tree cover, serious soil erosion and declining soil fertility (Place et al. 2007; Were et al. 2006; Krishna et al. 2004; Odada et al. 2004). A baseline survey in 2011 showed that 81% of the families experience 1–2 hunger months (a period when they are unable to produce from their own farm source) in a year, while 17% of the families experience 3–4 hunger months (Mango et al. 2011). Agriculture remains the main source of livelihood. Crops grown include maize and sorghum in mixtures with legumes such as beans and cow peas. Livestock reared include zebu cattle, poultry, sheep and goats.

3.2 Identifying Nyando Climate Spatial Analogues Sites

The analogues approach is a novel way of supporting policy recommendations with on-the-ground empirical testing. The CCAFS climate-analogue tool was used to identify possible spatial analogue sites for Nyando. In essence, the tool locates a site whose climate today is similar to the potential future climate of Nyando—where in the world can we find today the potential future climate of Nyando. Using rainfall and temperature variables, socio-economic factors and varying the different seasons

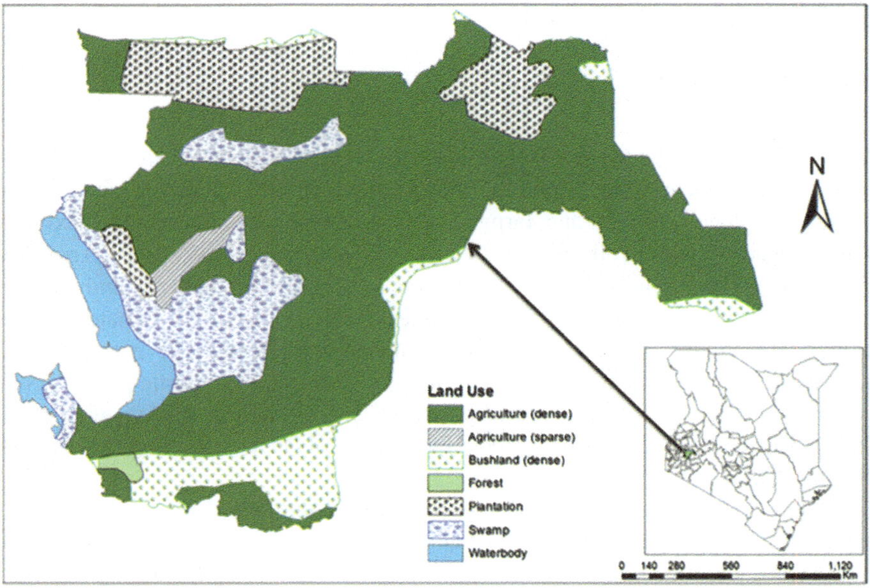

Fig. 1 Map of the study area (Nyakach and Sigowet sub-counties)

(March–May (MAM), October–December (OND) and January–December (all year)), several analogues maps were generated (Fig. 7). The socio economic factors used included market access, population density, poverty levels, and livelihood strategies derived from secondary data sources (see Mango et al. 2011 for example). A planning workshop was held for AIS stakeholders and other partners[1] to identify the best analogue map that is representative of Nyando's potential future climate. Participants identified Fig. 8 as the most plausible future climate of Nyando showing the 15% best analogues to which 80% of the models agreed. Using Fig. 8, and in a participatory process, the participants identified the most suitable sites (Fig. 2) that were mainly semi-arid areas of western Kenya.

This formed the farmer learning journey itinerary starting from Nyando CSVs and ending at Nandi (Fig. 2). After the planning workshop, a field meeting with farmers and the AIS stakeholders was held to select farmers to participate in the learning journey. A total of 16 farmers (7 men and 9 women) of different ages, gender, social-economic status and involvement in ongoing work by CCAFS and its partners were selected by fellow farmers to participate. Two farmers (male and female) were trained on the use of video as a participatory learning tool (Fig. 3).

[1]This included the district agricultural extension officer, community development officers, land use planners and field officers, and representatives from various NGOs and research organizations.

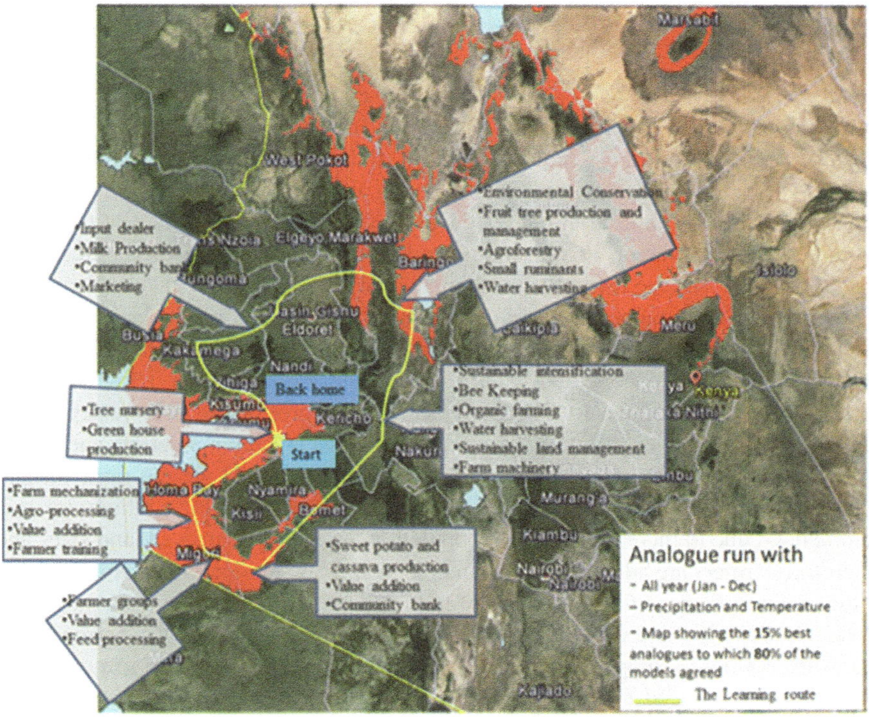

Fig. 2 A map showing the learning journey

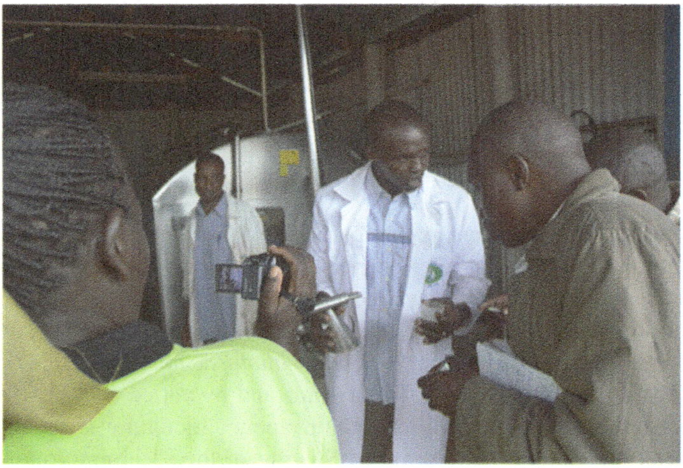

Fig. 3 Farmers using the video cameras to document the lessons

4 Results and Discussion

The 16 farmers and five key agricultural stakeholders from Nyando in Western Kenya took part in the 8-day learning journey to several learning sites including Katito, Homabay, Migori, Kuria, Nakuru, Baringo and Nandi in Rift Valley (Fig. 2). The AIS stakeholders accompanying the farmers were drawn from different sectors including community development, agricultural extension officers, community development organisations, and agricultural and livestock officers.

During the 8-days learning journey, the farmers and AIS stakeholders were exposed to a wide range of ongoing community-adaptation and risk-management strategies and innovations, both institutional and technological. The institutional innovations included collective action to manage climate risks. These included pulling financial resources to form savings and loans groups and community banks, establishment of tree nurseries, groundnut processing and marketing and sweet potato value addition. The technological strategies included simple farm mechanization, animal feed establishment and management technology which are recommended due to increase in human population leading to decreased land sizes and therefore farmers are adapting to the small land sizes by rearing goats which require little space and forage for production of milk and meat. The farmers were trained on fodder production and conversion of grass into silage using 30 PVC barrens each carrying 300 kg of silage that can feed 6 dairy cows for a day. Due to population increase there is need for zero grazing which requires silage and hay for livestock feed (Fig. 4). Other technological strategies included tree nurseries, soil and water conservation, water harvesting, root crops value addition, bee keeping, fruit trees establishment (Fig. 5) and management and small ruminants' management. The key highlights of the learning journey are summarized below.

4.1 The Learning Journey Enabled Farmers to Explore Future Scenarios

By observing different current scenarios across the learning route and later discussing with their peers what might happen in Nyando in the future, farmers drew comparisons about the resources available and climate differences they experienced on the tour and explored diverse scenarios. This created heated debate amongst the farmers on what might happen in Nyando '*without action being taken*' and what would happen '*with action*' (of different kinds and by varied actors). This process enabled farmers to think through and identify new agricultural ideas and agree to implement in the coming rainy seasons. Sharing of their learning journey experiences with other farmers, the rest of the community was inspiring and motivating to change, encouraging forward-thinking as adapt to climate change.

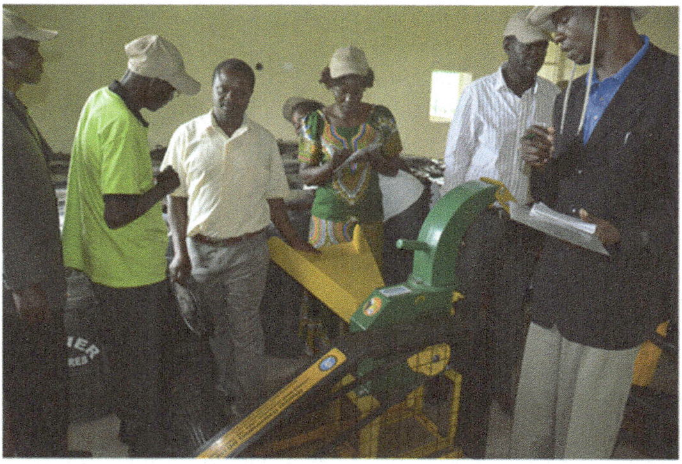

Fig. 4 Farmers learn how to make silage using PVC barrens

Fig. 5 Farmers learn how to establish grafted mangoes in KALRO—Perkerra

4.2 Motivation to Act, Articulate Demand for Technical and Financial Support

Farmers were motivated to act following the learning journey. There was significant interest in establishing a SACCO and community bank amongst the participants and to improve their farming and environmental management practices. According to a male participant, "*The important things which impressed us, first is the formation of the SACCO that started as a group activity and later transformed into a community bank. In addition, planting of agroforestry and fruits that can generate income, value addition of groundnuts and sweet potatoes, and bee keeping were initiatives that we can start doing in Nyando*".

Interacting with other farmers during the learning journey, the farmers learned that they could demand for extension and agro-advisory services. For example, farmers learned that a women group (called Moto Moto based in Kuria), requested IFAD and County government of Migori for high yielding sweet potato varieties that are adapted to the local climate. The group also demanded for support from the extension services for better agronomic practices. Later on, the group was able to initiate an innovation fund with initial support from IFAD and County government. Through this experience, Nyando farmers learned that they can involve their County government and NGOs working in their area for technical and financial support. According to a female participant "*Adaptation requires diversified assets and institutions*".

A key focus of the learning journey was to identify a broad range of learning opportunities that not only encompass particular adaptation, mitigation and risk management technologies, but also other institutional innovations. Technology transfer is highlighted as one of the solutions to agricultural adaptation, but due to associated uncertainties, which require localized processes of innovation can be an unlikely solution to rural farmers. However, institutions offering capacity building in adaptive agriculture are crucial. Further, while the endowment of a particular natural resources is relevant (e.g. the quality of soils, type of naturally occurring vegetation) and more importantly how resources are accessed, utilized and managed impacts adaptation. According to a male farmer, "*the farmers we have interacted with during the learning journey had diversified assets and income sources which is supported by institutions including the county government and project stakeholders. This is a challenge to us to diversify greatly our agricultural activities e.g. crops diversification and small ruminants, and with support from our partners in terms of capacity building will make Nyando a food secure region.*"

4.3 Effective Environmental Conservation Requires Learning and Adaptation

During evaluation of the learning tour, most of the farmers and agricultural stakeholders noted the importance of environmental conservation. They identified specific practices to test in Nyando, and also discussed the relationship between conservation, land availability, and diversification of income sources. A farmer from Kapsorok who had not participated in the learning journey, but who viewed the video said, that the changing climatic conditions have affected their ability to grow maize and sugarcane, and commented that the tree nurseries and fruit trees for income and environmental conservation could be a potential adaptation strategy.

The environmental management observed at Rural Organization for Just Environment (ROJE) in Katito, Kisumu County will have a positive impact, noting that there are knowledge gaps in Nyando regarding which trees should be planted and where, how to establish small groups to manage tree nurseries.

4.4 Access to Finance for Risk Management

Access to finance is an important element in adaptation and risk management to changing climate. The changes farmers proposed to make will require seed funding or access to credit. The visit to Masaba community bank and Lel Chego farmer groups, for example, included presentations from a successful savings and credit (SACCOs) group—a scheme which has enabled them to invest more in environmental conservation practices and expand their agricultural production. The links between income, livelihoods, markets, agriculture and environmental conservation were noted as well, with reduced youth out-migration being attributed to more secure and diversified livelihoods, environmental conservation and the SACCOs scheme. According to a Nyando community based organization (CBO) member, "*Over the years, our CBO is well established and members obtain credit to purchase inputs and other agricultural activities, from what we have learnt in Masaba and Nandi, we can register as a SACCO and ultimately as a community bank, this would benefit the entire community to save and borrow loans thus improving the community livelihoods*".

4.5 Social and Cultural Practices Hinder Adaptation

Exposure to different ethnic groups during the learning journey ensured that the Nyando farmers learned about other cultures and how farmers are changing their cultural practices as the climate is changing. Access to and control over resources is critical for *equitable* adaptation and development. Positive collaborations were noted in various learning locations, as being important for securing livelihoods, but also as a means to support specific groups. For example, young people were given employment by the SACCOs scheme in Masaba community bank. In such a brief initiative, it is limited how far entrenched gender norms which underpin discrimination can be challenged, but giving women and men farmers' equal status in the learning journey and training both groups in using the videos is an important demonstration that the voices of women and men are important.

This visit also sparked discussions involving the district community development and women farmers, regarding successful women's poultry keeping and agroforestry groups in Nyando and about how to successfully get organized. The learning journey did not necessarily challenge prevailing gender norms about appropriate roles in farming for women and for men—more time would be needed to achieve this and potentially a more central focus given in the learning journey. Indeed, learning journeys could be organized that place these issues at their heart—as opposed to climate change. It is likely to be easier to break out of fixed views on gender roles by meeting individuals and visiting communities where different gender relations are in evidence and by visiting positive role models and successful groups to see what can be achieved, with benefits brought to both women and men.

4.6 Adaptive Capacity Requires Innovation

The ability and willingness to innovate is included as a key part of the adaptive capacity framework. The learning journey did seem to encourage a willingness to innovate at least as reflected in the discussions of the participants. However, the extent to which they have the capability to innovate depends to some extent on the resources they have with which to take more risks and the support provided (e.g. access to information or seedlings). Thus, this aspect of adaptive capacity is easier to assess after a longer period and should consider the contextual factors that will support or constrain the participating farmers', other AIS stakeholders and their local communities over time.

In terms of cropping and livelihood diversification, crop diversification and enterprise diversification were among the lessons learnt. Farmers expressed interest in several of the trials, for example, a cassava variety (Fig. 6) and groundnuts, but

Fig. 6 Farmers learn how to produce high yielding cassava varieties

more time was needed for the farmers to explore the crops and varieties available. A seed fair might have been a useful means for the visiting farmers to access the wide range of diversity available. A detailed explanation of the benefits of bee keeping as a means of livelihood diversification attracted interest amongst a number of male farmers. Reflection and learning can also be stimulated by visiting and observing failed innovations, as much as successful ones. The visit to Baraka College exposed the farmers to various integrated systems and organic farming comprising of rabbit keeping, dairy goat keeping, bee keeping, and general farm machinery. A visit to *Moto Moto* women group (comprising of mainly old women) exposed the participants to many challenges one undergoes to succeed, they shared their testimony of previous failures in their projects and the current stage of success.

4.7 Collective Action and Changes in Decision-Making to Build Resilience

The learning journey encouraged farmers to explore more systematically and collectively the potential future scenarios they face, and with a focus on the changing climate. However, while there is greater willingness to act according to the learning journey participants, leadership, good governance and collective action, and external support are needed. It is very important to involve agricultural system stakeholders beyond the farmer level in order to also build understanding, willingness to act and to support the creation of new or improved linkages across scales between farmers and these stakeholders. Many of the participating farmers referred to the learning journey as being well run in the sense that links were made between farmers and AIS stakeholders are important and mostly the participants were treated as having equally important contributions to make. Participatory processes have in the past been criticized for failing to adequately take into account the identity and status of facilitators and external actors in externally driven processes. The increased uncertainties created by a changing climate, requires coordinated action that reaches across scales, but also means that extension workers and those providing advisory services will have to take on different roles, moving from teachers, to facilitators of farmer's own learning and experimentation.

5 Key Challenges

According to a male farmer, "despite participating in the learning journey, there are several challenges that we face as a community". First, we do not have training on some of the activities and enterprises we learnt on the journey. Second, that some of the adaptation activities require initial high investment capital, for instance, bee keeping demands for intensive training as well as high initial investment for equipment such as hives, equipment for harvesting and, processing and storage of honey.

Other challenges included inherent uncertainties associated climate modeling and projections, especially the difference in upland and lowland areas of Nyando villages, where major changes in altitude over short distances complicate the situation. Preparatory communication with the different hosts in the learning sites is important to ensure clarity on the purpose of the visit. Social and cultural barriers such as gender inequalities often restrict women's mobility and participation in learning journeys such as these and may hinder innovation and change from the learning process. Finally, on such a long journey, it is not always possible to stick precisely to the study-tour itinerary.

6 Key Lessons and Recommendations

Some key lessons and recommendations emerged from the learning journey:

- The climate-analogue tool is most useful as a learning tool, rather than a predictive tool, and can be used to explore future scenarios and spark learning for adaptive action. However, understanding climate modeling can be challenging for people who are not climate scientists and more resources are needed to enable learning (e.g. at sub-county level).
- There are other obstacles to farmers' action including structural challenges, lack of access to farm inputs, capital and information. The FotF approach is thus likely to be most effective when embedded within an overall participatory adaptation process, so that support can be given to the participants, their communities and wider agricultural system stakeholders to act on insights and to innovate.
- Farmers need space to explore future horizons and potential challenges and opportunities, and can effectively learn from their peers. This learning is not necessarily about technology transfer, but also about institutional change and developing localized solutions. It is critically important to engage actors from across the agricultural sector, because of the uncertainties posed by climate change, the need for more flexibility in responses (e.g. from agricultural advisors) and the potential scale of the challenges ahead.
- The participatory video documentation, where the farmers used video cameras to document the learning process, enabled them to share their learning experience with other farmers in their own communities who did not participate in the learning journey. This is an important dissemination tool whereby farmers are able to document and share with others.
- The learning journey increased awareness of the participants of the changing climate and environment, and the need to act, as well as specific practical plans such as the formation of a community based organizations (CBOs) by the farmers who participated in the learning journey. With the necessary support,

they anticipate to develop this into a fully-fledged Savings and Credit Cooperative (SACCO).

7 Conclusions

The Farms of the Future approach is a learning process that can enable farmers begin to read the world differently. It can help them to think critically about their future and encourage them to act. Significant opportunities exist for more farmer-to-farmer (and stakeholder-to-stakeholder) shared learning on adaptation. The learning journey to spatial analogue learning sites can enable farmers and agricultural stakeholders to better envision how their site-specific agricultural future might look and can facilitate exchange of knowledge through which strategies, farming technologies and information can be shared. Strengthening adaptive capacity will also require institutional, structural and policy changes. To assess the outcomes of the pilot requires a follow-up visit to explore whether the learning journey sparked new thinking and practical action and to assess its contribution to an overall participatory action research process, including identification of the social and cultural barriers to adaptation.

Acknowledgements The study team hereby offers grateful thanks to the communities and various government departments, NGOs and CBOs that hosted our extensive field journey. Most of all we extend our thanks to the farmers from Kapsorok and Kamwana villages. Further, we would like to thank the facilitators Wilson Aore, Reuben Chirchir, Jared Akuku and CCAFS field technicians; David Musuya and Wilson Okila for all their hard work in organizing the exchange visit.

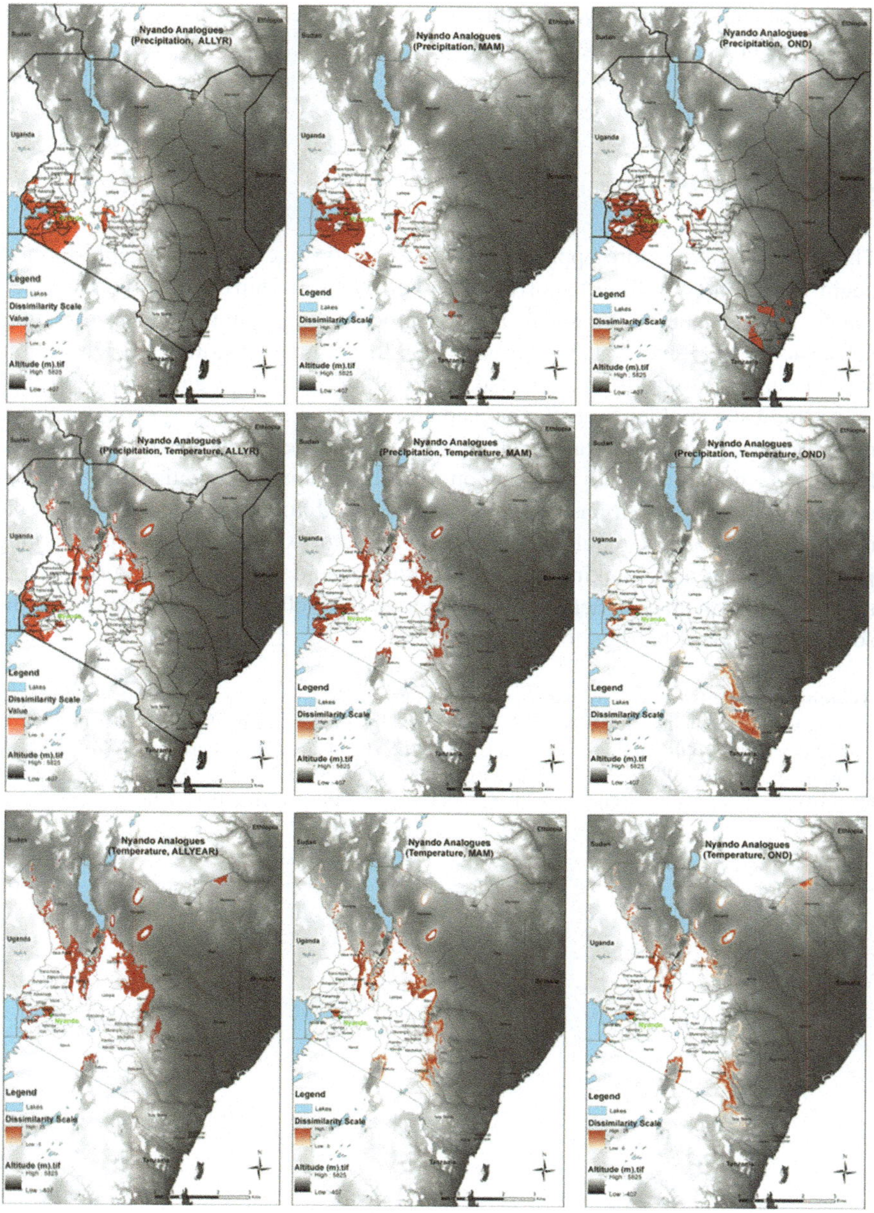

Fig. 7 Combined Nyando analogue maps generated by varying all input variables (precipitation, temperature and seasons i.e. MAM, OND, AllYR)

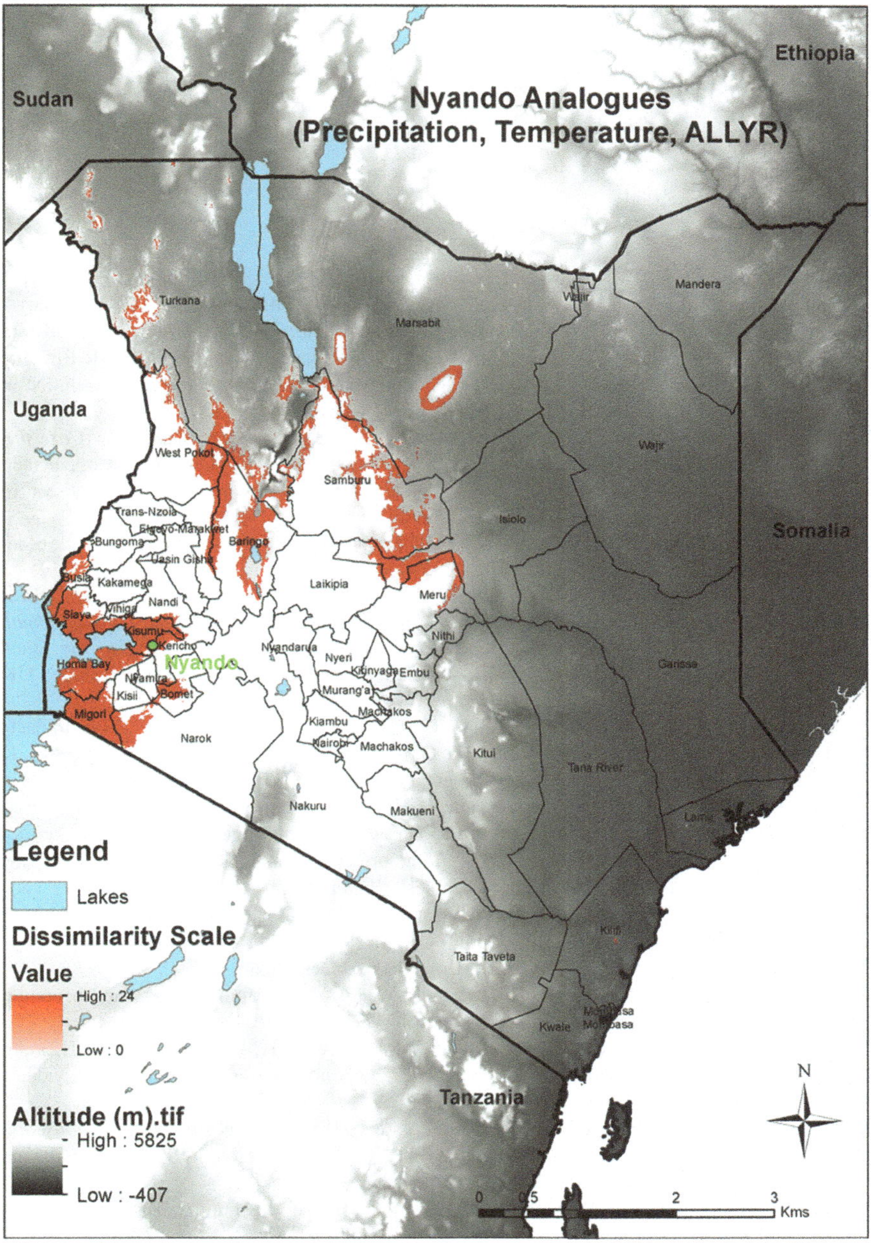

Fig. 8 Selected analogue map showing most plausible future climate of Nyando

Appendix

See Figs. 7 and 8.

References

Addae-Korankye, A. (2014). Causes of poverty in Africa: A review of literature. *American International Journal of Social Science*. ISSN 2325-4149(Print), ISSN 2325-4165(Online).

Aggarwal, P., Zougmoré, R., & Kinyangi, J. (2013). *Climate-smart villages: A community approach to sustainable agricultural development*. Copenhagen, Denmark: CGIAR Research Program on Climate Change, Agriculture and Food Security (CCAFS). Available online at: www.ccafs.cgiar.org

Förch, W., Sijmons, K., Mutie, I., Kiplimo, J., Cramer, L., Kristjanson, P., et al. (2011). *Core sites in the CCAFS regions: East Africa, West Africa and South Asia*, Version 3. Copenhagen, Denmark: CGIAR Research Program on Climate Change, Agriculture and Food Security (CCAFS). Available online at: www.ccafs.cgiar.org

Huq, S., & Jessica, A. (2006). Working paper: The possibilities for combining mitigation and adaptation in vulnerable communities in the developing world.

Intergovernmental Panel on Climate Change (IPCC). (2007). Summary for policymakers. In S. Solomon, D. Quin, M. Manning, Z. Chen, M. Marquis, K. B. Averyt, et al. (Eds.), *Climate change 2007: The physical science basis. Contribution of working group I to the fourth assessment report of the Intergovernmental Panel on Climate Change*. Cambridge, UK: Cambridge University Press.

Intergovernmental Panel on Climate Change (IPCC). (2014). Climate change 2014: Impacts, adaptation and vulnerability. In V. R. Barros, C. B. Field, D. J. Dokken, M. D. Mastrandrea, K. J. Mach, T. E. Bilir, et al. (Eds.), *Part B: Regional aspects. Contribution of working group II to the fifth assessment report of the Intergovernmental Panel on Climate Change* (688 pp.). Cambridge, United Kingdom and New York, NY, USA: Cambridge University Press.

Jayne, T. S., Jordan, C., & Derek, D. H. (2014). Land pressures, the evolution of farming systems, and development strategies in Africa: A synthesis. *Food Policy, 48*, 1–17.

Kinyangi, J., Recha, J., Kimeli, P., & Atakos, V. (2015). Climate smart villages and the hope of food security in Kenya. CCAFS Info Note. CGIAR Research Program on Climate Change, Agriculture and Food Security (CCAFS), Copenhagen, Denmark.

Krishna, A., Kristjanson, P., Radeny, M., & Nindo, W. (2004). Escaping poverty and becoming poor in 20 Kenyan villages. *Journal of Human Development, 5*(2), 211–226.

Kristjanson, P., Neufeldt, H., Gassner, A., Mango, J., Kyazze, B. F., Desta, S., et al. (2012). Are food insecure smallholder households making changes in their farming practices? Evidence from East Africa. *Food Security, 4*(3), 381–397.

Mango, J., Mideva, A., Osanya, W., & Odhiambo, A. (2011). *Summary of baseline household survey results: Lower Nyando, Kenya*. CGIAR Research Program on Climate Change, Agriculture and Food Security (CCAFS), Copenhagen, Denmark.

Nelson, V., Lamboll, R., & Nathaniels, N. (2012). Farms of the Future, Tanzania. CCAFS Technical Report.

Odada, E., Olago, D., Kulindwa, K., Ntiba, M., & Wandiga, S. (2004). Mitigation of environmental problems in Lake Victoria, East Africa: Causal chain and policy options analyses. *Ambio, 33*(1–2), 13–23.

Place, F., Adato, M., & Hebinck, P. (2007). Understanding rural poverty and investment in agriculture: An assessment of integrated quantitative and qualitative research in western Kenya. *World Development, 35*(2), 312–325.

Ramírez-Villegas, J., Lau, C., Köhler, A. K., Signer, J., Jarvis, A., Arnell, N., et al. (2011). *Climate analogues: Finding tomorrow's agriculture today*. Working Paper 12. Cali: CCAFS. www.ccafs.cgiar.org/resources/working-papers

Thornton, P. K., Jones, P. G., Owiyo, T. M., Kruska, R. L., Herrero, M., Kristjanson, P., et al. (2006). *Mapping climate vulnerability and poverty in Africa* (200 p). Nairobi (Kenya): ILRI.

Were, E., Swallow, B., & Joy, J. (2006). *Water, women, and local social organization in the Western Kenya highlands*. CAPRI Working Paper No. 51. Washington, D.C.: International Food Policy Research Institute.

Williams, J. W., Jackson, S. T., & Kutzbach, J. E. (2007). Projected distributions of novel and disappearing climates by 2100 AD. *Proceedings of the National Academy of Sciences, 104*(14), 5738–5742.

Yamane, Y., Asanuma, S., & Umenura, K. (2015). Influence of livestock farming on vegetation in a degraded soil area on the East Coast of Lake Victoria in Western Kenya: A case study of Jimo East Sub-Location in Nyando Sub-County. *Journal of Environmental Protection, 6*, 824–836. doi:10.4236/jep.2015.68075.

Coffee Shade Tree Management: An Adaptation Option for Climate Change Impact for Small Scale Coffee Growers in South-West Ethiopia

Weyessa Garedew, Binyam Tesfaw Hailu, Fikre Lemessa, Petri Pellikka and Fabrice Pinard

1 Introduction

Coffee (*Coffea arabica* L.) is one of the most important cash crops that have been contributing a lion's share to Ethiopia's economy. It is a crop that is grown within a large variation of shade cover (Lin 2007). It varies from traditional forest coffee (i.e. rustic coffee), which is planted under a forest canopy to intensive coffee agriculture, which has little or no shade cover. In Ethiopia, coffee is generally grown under four types of production systems, namely: forest, semi-forest, garden and plantation coffee (Weldetsadik and Kebede 2000; CFC 2004). The major coffee production system in the highland of south-western Ethiopia is referred to as 'semi-forest coffee' (Labouisse et al. 2008); although in many of the forest, wild coffee (forest coffee) is still found sparsely scattered in remote parts of the forest (Samnegard et al. 2014). The semi-forest coffee is grown in the understory of natural moist afro-montane forests with low annual management. Such areas are widespread in the margin of the large contiguous forests. Coffee is also grown isolated from the contiguous forests in forest patches surrounded by an open agricultural matrix of

W. Garedew (✉) · F. Lemessa
Jimma University College of Agriculture and Veterinary Medicine,
P.O. Box 307, Jimma, Ethiopia
e-mail: woyessa.garedew@ju.edu.et

B.T. Hailu
Department of Earth Sciences, Addis Ababa University,
P.O. Box 1176, Addis Ababa, Ethiopia

B.T. Hailu · P. Pellikka
Department of Geosciences and Geography,
University of Helsinki, Helsinki, Finland

F. Pinard
CIRAD-ICIPE, UR 106 Bioagresseurs,
TA-A 106/D, Campus de Baillarguet, 34398 Montpellier, France

W. Leal Filho et al. (eds.), *Climate Change Adaptation in Africa*,
Climate Change Management, DOI 10.1007/978-3-319-49520-0_40

open crops and shaded by trees (Gove et al. 2008; Samnegard et al. 2014). Because of forest fragmentation, such a setting is typical in coffee producing areas of southwestern Ethiopia.

The exposure of forest environment to external climatic conditions, due to forest fragmentation and the creation of forest edges, could reduce the ability of a forest to buffer its internal microclimate from more extreme macroclimate conditions (Ewers and Banks-Leite 2013). The microclimate could differ, between coffee grown under shade in the margin of the large contiguous forests and small forest patches surrounded by an open agricultural matrix of open crops. It could be assumed that such a contrast could both directly or indirectly affect the internal microclimate condition of each coffee plot. The land-cover change provides an additional major factor that alters climate, through changes in the physical properties of the land surface (Pielke et al. 2002). The land use/cover types surrounding the coffee plots might have an impact on the internal microclimate variability of each coffee agro-ecosystem. As in agricultural landscapes, trees and shrubs occur on farms in different spatial and temporal arrangements with crops and outside farms in communal lands.

Depending on the quality of the surrounding matrix, it can be expected that these processes of fragmentation has caused significant changes on microclimate of fragmented coffee plots. Moreover, the contrast between coffee grown under shade and in the open in terms of microclimate variability were not done extensively in Ethiopia. Therefore, the objectives of this study were to characterize the influence of the land use/cover types on the internal microclimate of coffee plots and compare the microclimate variability under shade and in the open in south-west Ethiopia.

2 Materials and Methods

2.1 Description of the Study Site

The experiment was carried out in Ethiopia, Oromia Regional State, Jimma Zone, Ageyo-Setema research site (08°03″96′–08°04″19′N, 36°32″84′–36°47″04′E; altitudes ranging from 1505 to 2124 m above sea level) from June 2012 to 2015 (Table 1). The coffee plots were selected from two districts/woredas of Jimma Zone: Gumay and Setema (Fig. 1). It is located around 100 km in the northwest of Jimma town. The local rainfall pattern of the study site is nearly mono-modal, with a main rainy season from June to August, inducing a single coffee crop harvest season from October to December. The topography of the area is undulating landscape that consist of a mosaic of crop land, pasture, forest fragments managed for coffee production and isolated farmsteads, and patches of exotic timber tree species.

Table 1 Mean daily temperature, relative humidity and wetness duration among the different coffee classes during wet season at Ageyo-Setema study site in 2012 and 2013

Coffee class	Year					
	2012			2013		
	Temperature (°C)	Relative humidity (%)	Wetness duration (h/day)	Temperature (°C)	Relative humidity (%)	Wetness duration (h/day)
1	17.73[a]	89.90[c]	0.15[c]	17.78[a]	89.74[c]	0.21[c]
2	16.97[b]	90.81[b]	0.18[b]	16.84[b]	91.23[b]	0.25[b]
3	16.52[c]	92.53[a]	0.23[a]	16.58[c]	92.89[a]	0.27[a]

Means sharing the same letter in the column do not differ significantly at 5% probability level using HSD test

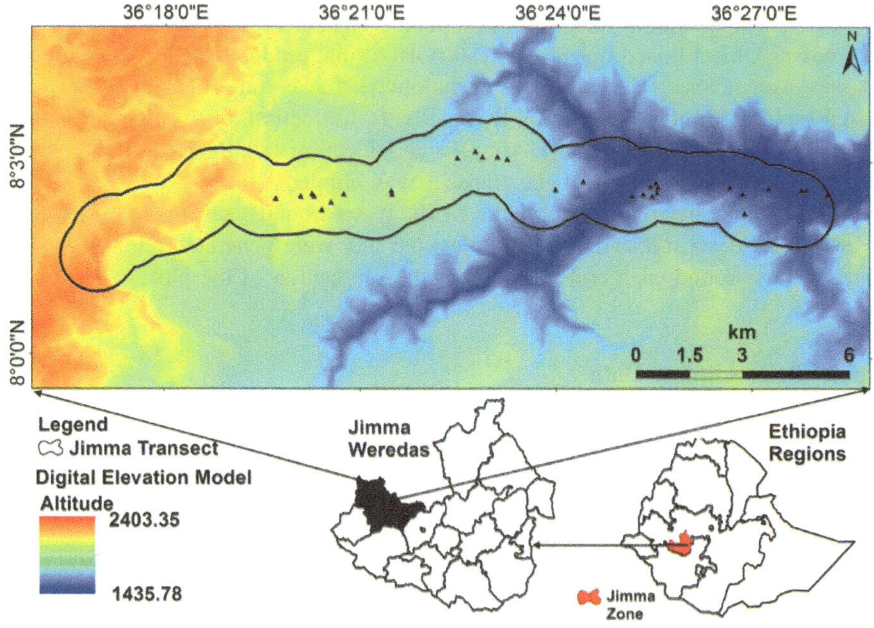

Fig. 1 Map of the study area: Ageyo-Setema research site (*triangle shaped dots* represent coffee plots of the present investigation)

2.2 *Experimental Plot Selection*

A total of 30 coffee plots of 20 m × 20 m, smallholder coffee farms were selected along Ageyo-Setema altitudinal transect (Fig. 1). The plots were selected along the transect considering the level of fragmentation. The coffee produced in the study area is exclusively *Coffea arabica* and the coffee trees used for the study were more than 25 years old local landraces (variety). Coffee field management namely slashing of the understory shrubs and weeding is done once or twice per year, one or two weeks before harvesting the berries. Farmers do not fertilize nor apply any other chemical input on their coffee farms. The dominant shade trees used in the selected coffee plots are *Albizia gummifera, Cordia africana, Millettia ferruginea, Croton macrostachyus, Acacia abyssinica, Ficus vasta, Ehretia cymosa, Dracaena steudneri* and *Vernonia amygdalina.*

2.3 Data Collection

2.3.1 Land Use Land Cover

The Land Use Land Cover (LULC) of the Ageyo-Setema transect were identified from aerial and satellite images as described by Hailu et al. (2014). The classification was Object Based Image Analysis (OBIA) that produces eight classes. These classes were Closed Herbaceous Vegetation (pasture), Indigenous Forest, Small Sized Field of Graminoid Crop(s), River, Roads, Extraction Sites, Urban Area, and Exotic Forest (Hailu et al. 2014). After the classification and digitization, the LULC (at 2 m × 2 m resolution) of the transect was clipped with 50, 100 and 200 m radius of each coffee plots to know the LULC types around each coffee plot. Hence, a total of 24 descriptors (LULC variables) per plot were generated (eight LULC at three radius around each coffee plot) for characterization of the plot.

2.3.2 Microclimate Data

Data loggers, that is, Maxim iButton, were installed in each coffee plot from June 2012 to 2014 to record the temperature (°C) and relative humidity (%) every hour with specified accuracy of ±0.05 °C and 0.0625% resolutions, both in coffee plots and in the open areas. But wetness duration (h) was calculated from relative humidity; a value recorded when relative humidity become 100%. The loggers were hung on coffee trees at the center of each coffee plot and single tree in the open on average at 2 m above the ground and protected from direct sun. The collected data were downloaded from data loggers to a laptop computer using a cable on average at two months interval. The hourly collected temperature and relative humidity data were aggregated on daily, monthly and season basis during the analysis.

2.4 Statistical Analysis

The clipped eight LULC variables generated at each scale was used for characterization of the 30 coffee plots based on the number of pixels they generated around each coffee plot. Preliminary analysis was done on the 24 LULC variables generated before using all the descriptors for further analysis. After preliminary analysis road, urban, extraction site and river variables were found to be less contributors to the variation at all the scales considered and were excluded from further analysis.

Indigenous trees, exotic trees, crop land and pasture land were only used to understand the strategies of farmers in using open land. To avoid multi-colinearity problem between these LULC variables, Principal Component Analysis (PCA) was run at each scale separately. This step was important to identify whether there was correlation between the four selected LULC variables or not to use them for further analysis. In all the scale considered, crop and pasture land were negatively correlated to indigenous and exotic trees respectively; and were then excluded from the analysis while explaining the relationship between LULC with microclimate of the transect as well as each coffee plots along the altitudinal gradient (see result part).

On the other hand, indigenous and exotic trees were independent to each other at each scale; and were used for further analysis. Consequently, total trees (indigenous + exotic trees) were added together at each scale (T50, T100 and T200). However, T50, T100 and T200 were highly correlated to each other. Using these three variables, PCA was further run to construct two independent synthetic variables (Trees1 and Trees2) by considering the first two principal components (PC). Trees1 and Trees2 were created depending on the two axis (PC). Trees1 represent the global tree canopy density around each plot (more contribution from T100) while trees2 represents the tree canopy density close to the plot (contributed more by T50) and at the periphery of the plot (contributed more by T200). Then, hierarchal cluster analysis based on "mcquitty" method was employed to classify the 30 coffee plots into different groups using Trees1 and Trees2 variables. The number of clusters was determined by running Multivariate Analysis of Variance (MANOVA) test as well as our personal observation of the actual coffee plots used for the study.

To determine the variation of microclimate among different coffee classes particularly during wet and dry season, daily mean, minimum and maximum temperature (°C) and relative humidity (%) was calculated from logger data over two years; 2012 and 2013 for wet season (June to August) and 2013 and 2014 for dry season (February to April). Daily sum of wetness duration was calculated by summing up the total wetness duration over the specified period and divided by 24 h to get daily sum of the wetness over the transect. For the wet season, data were collected and summarized over 70 consecutive days from June 22 to August 30 in 2012 and over 91 consecutive days from first of June to August 30 in 2013. For the dry season, data were compiled over 85 consecutive days from February 5 to April 30 of 2013 and 2014. These specific months were selected because they represent the wet and dry season. January was not included in the dry season because of non availability of the data. Furthermore, the temperature under shade and open area was also compared.

Finally, analysis of variance (ANOVA) was run to test whether there was significant difference between the different classes of coffee plots with respect to daily mean temperature and relative humidity for both seasons and years separately. Tukey's 'Honestly Significant Difference' (HSD) method was calculated and used to identify significantly different means. All statistical analysis was performed using R software version 2.15.1 (R Development Core Team 2012).

3 Results

3.1 Characterization of Coffee Plots Using LULC

The clipped LULC analysis indicated that the coffee plots have a great variety of LULC in their surrounding areas (Fig. 2). It includes the proportions of area covered with trees (both exotic and indigenous), crop lands, pasture lands, urban, roads and river areas. Some of the plots are composed of small areas of trees and large areas of cropland. For example, plot 20 has 177 pixels of trees, which encompasses 1106.25 m^2 area and 4848 pixels of cropland, which is 30,300 m^2 areas at 100 m scale (Fig. 2). On the other hand, some plots have large areas of trees and small areas of cropland. For example, plot 13 has 4230 pixels of trees (26,437 m^2 area), 525 pixels of cropland (3281 m^2 area) and 265 pixels of pasture (1656 m^2 area). In addition, there are plots with almost equal proportion of cropland and trees. For

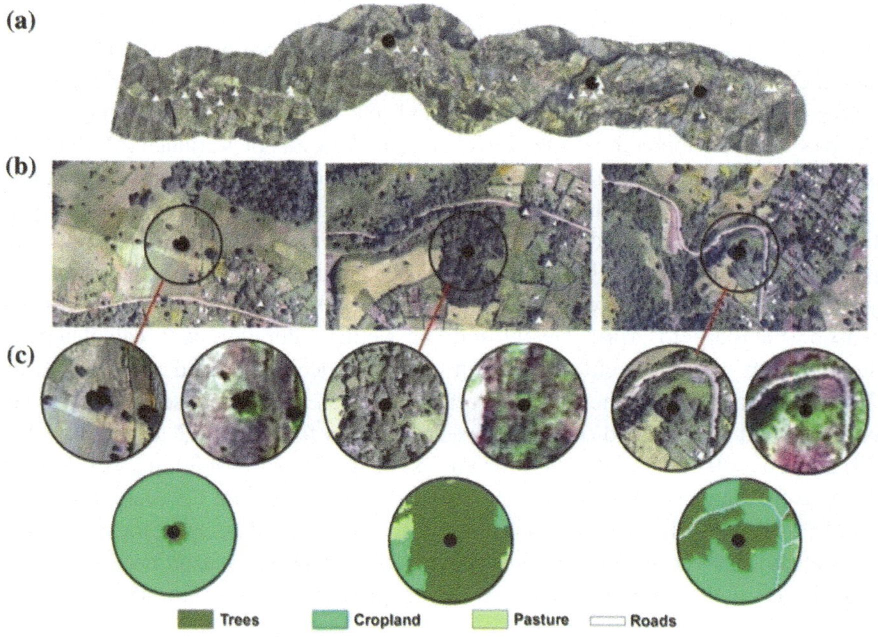

Fig. 2 **a** The 30 plots including plot 20, 13 and 6 with *black dots* distributed on Jimma transect on aerial image, **b** closer view of the three plots with 100 m radius circles, and **c** plot 20 (*left*), 13 (*middle*) and 6 (*right*) with 100 m radius on aerial image with 0.5 m resolution, Spot 5 satellite image with 2.5 m resolution with RGB:432. and LULC classes respectively

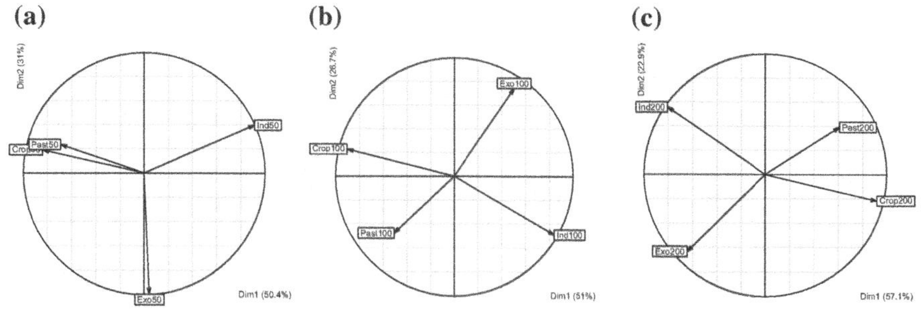

Fig. 3 PCA analysis on LULC variables at 50 m (**a**), 100 m (**b**) and 200 m (**c**) scale on indigenous (Ind) and exotic (Exo) tree canopy coverage, crop land (crop) and pasture land (past)

example, plot 6 has 1963 pixels of trees (12,268.75 m^2 area), 2808 pixels of cropland (17,550 m^2 area), 62 pixels of pasture (388 m^2) and 186 pixels of road (1163 m^2). Generally, plot 13, 30, and 1 have the highest tree cover in the 50, 100 and 200 m radius scale respectively. However, plot 8, 20 and 15 had the lowest tree cover in the three respective scales respectively. Plot 20 had the highest crop cover in the 50 and 100 m radius scale but in the 200 m radius scale, plot 4 had the highest crop cover. Plot 13, 27 and 30 had the lowest crop cover in 50, 100 and 200 m radius scale respectively (data not presented).

The PCA analysis at 50 m radius around each coffee plot (Fig. 3a) indicated that pasture and crop lands were positively correlated and prevented indigenous trees. There was strong negative correlation ($r = -0.71$) between crop land and indigenous trees. The analysis at 100 m radius (Fig. 3b), on the other hand, gave a more precise pattern; the presence of cropped land strongly opposes the maintenance of indigenous trees ($r = -0.79$) while the presence of exotic tree meant the absence of pasture land ($r = -0.30$). In addition, the analysis at 200 m scale (Fig. 3c) produced a more or less similar trend to the scale of 100 m. For example, cropped land was negatively correlated with indigenous trees ($r = -0.80$). Moreover, the first two principal components (PC) explained 81.4, 76.7 and 78.0% of the total variation at scale of 50, 100 and 200 m radius respectively (Fig. 3).

Furthermore, at all scales, the presence of indigenous trees was independent to exotic tree (tree plantation); their correlations were very small. Consequently, total trees (indigenous + exotic trees) were added together at each scale (T50, T100 and T200) (Fig. 4b). However, there was strong correlation between exotic trees at 50, 100 and 200 m radius (Fig. 4a). For example, there was strong positive correlation ($r = 0.97$) between exotic trees at 100 and 200 m. Similarly, there was significant association between indigenous trees observed at the three scales (Fig. 4a). For instance, indigenous tree at 100 m radius was highly correlated ($r = 0.80$) with indigenous trees at 200 m. Therefore, the total trees at the three scales were de-correlated to produce a set of synthetic, linearly uncorrelated two independent variables; Trees1 and Trees2 using the two axis (Fig. 4b). Trees1 represent the global tree canopy density around each plot on the first axis (Dim1) accounting for

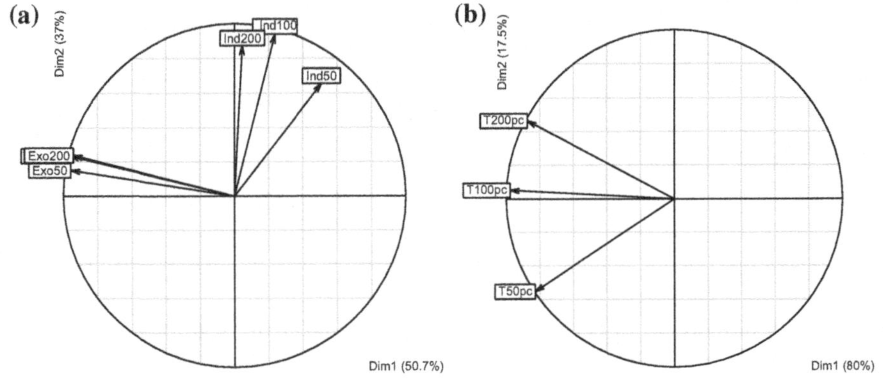

Fig. 4 PCA analysis using indigenous (Ind) and exotic (Exo) trees at 50, 100 and 200 m radius (**a**) and total tree density (T50, T100 and T200) (**b**)

80.01% of the variation while Trees2 represents the tree canopy density close to the plot and at the periphery of the plot on the second axis (Dim2) explaining 17.47% of the total variation (Fig. 4b). The two synthetic uncorrelated variables explained 97.48% of the total variation.

3.2 Cluster Analysis

Cluster analysis using Trees1 and Trees2 variables indicated that the 30 coffee plots were classified into various groups (Fig. 5). The dendrogram showed that there were two big categories of plots at global level; one group of the plots was characterized by high tree density in their surrounding area while the other group was the isolated coffee plots. Multivariate Analysis of Variance (MANOVA) test indicated that only three coffee classes were detected. This test was also supported with our personal observation of the actual coffee plots observed in the field. The MANOVA test showed significant difference when the coffee plots were grouped into three classes at maximum. This classification was also clearly indicated on the dendrogram. The first groups of the coffee plots (coffee class1) were characterized as isolated coffee plots with single/few shade trees; the second group (coffee class2) were characterized as coffee plots within a planted area surrounded with high density of patches of shade trees and the last group of coffee plots (coffee class3) were characterized as coffee plots surrounded by large contiguous forests (Fig. 5). Majority of the coffee plots used for this study were grouped under coffee class2 and class3 (13 coffee plots each) while only few coffee plots (four coffee plots) were categorized under coffee class1.

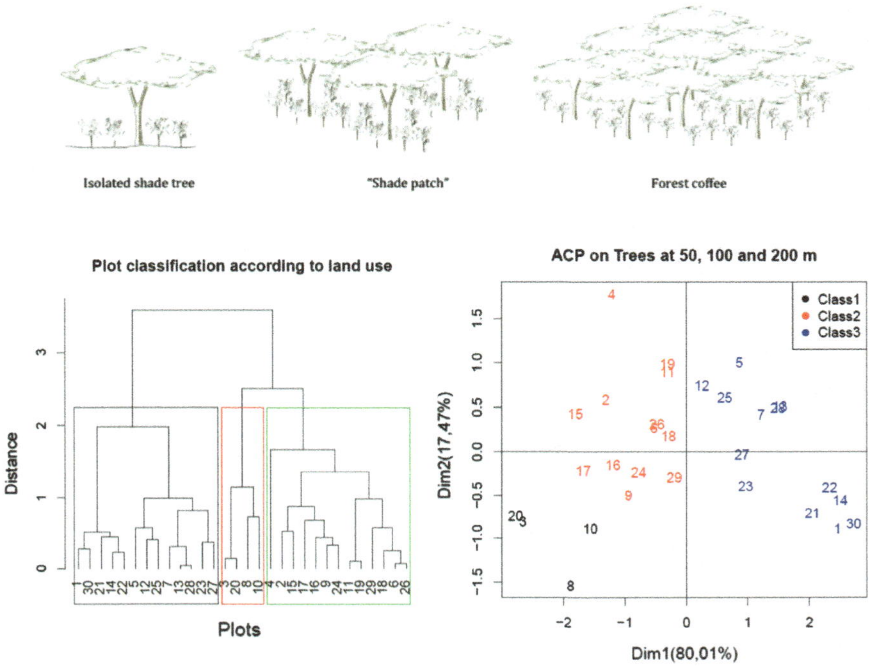

Fig. 5 Classification of coffee plots using Trees1 and Trees2 variables (LULC)

3.3 Climate Condition Among Coffee Classes

The analysis of variance for mean temperature, in both seasons, indicated that there was highly significant ($P < 2e-16$) difference between different coffee classes (Tables 1 and 2). Coffee class1, isolated coffee plots (characterized by low tree density), had significantly higher mean temperature compared to others. On the other hand, coffee class3, characterized by high tree density, gets cooler in both seasons than coffee class2 and class1, which have relatively less tree density (Tables 1 and 2).

Moreover, the analysis of variance indicated that coffee class3, characterized by high tree density, were found to be more humid compared to others, in both years of wet and dry seasons (Tables 1 and Table 2). During the wet season, there was a mean difference of 2.63 and 3.15% between the isolated coffee plots and coffee plots with contiguous forest in 2012 and 2013 respectively (Table 1). But, the difference was relatively small during the dry season as the difference was only 2.35 and 1.76% in 2013 and 2014 respectively (Table 2).

Table 2 Daily mean temperature, relative humidity and wetness duration among coffee classes during dry season at Ageyo-Setema study site in 2013 and 2014

Coffee class	Year					
	2013			2014		
	Temperature (°C)	Relative humidity (%)	Wetness duration (h/day)	Temperature (°C)	Relative humidity (%)	Wetness duration (h/day)
1	22.60[a]	57.29[c]	0.012[b]	20.76[a]	67.65[c]	0.050[c]
2	21.90[b]	58.14[b]	0.014[b]	20.24[b]	68.41[b]	0.097[a]
3	21.57[c]	59.64[a]	0.023[a]	19.75[c]	69.41[a]	0.066[b]

Means sharing the same letter in the column do not differ significantly at 5% probability level using HSD test

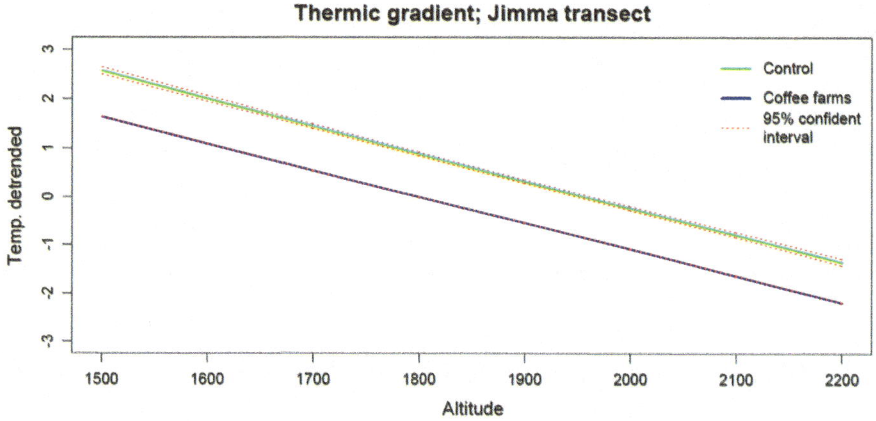

Fig. 6 Mean temperature between control (*open area*) and coffee farms (*under shade*) along altitudinal gradient in Jimma area

Furthermore, the analysis of variance for wetness duration during the wet and dry season showed there was highly significant difference ($P < 2e-16$) between the different coffee classes (Tables 1 and 2). The result indicated that there was more wetness duration on coffee class3 both in 2012 and 2013 of wet season compared to the other classes (Table 1). During the dry season, coffee class3 had more wetness duration in 2013 while coffee class2 had more wetness duration in 2014 compared to the others (Table 2).

The comparison of temperature between coffee plots and in the open indicated that there was higher mean temperature under open conditions compared to shaded coffee plot (Fig. 6). A difference of 1 °C was observed between the two contrasting conditions.

4 Discussion

The results of the study indicated that the coffee plots in the study area have a great variety of LULC in their surrounding matrix depending on the level of fragmentation. The existence of different LULC around the coffee plots created a significant impact on the internal microclimate of the coffee plots. Some of the plots are composed of small areas of trees and large areas of cropland and vice versa. Such LULC types might be created due to deforestation for agricultural activities and other ventures such as resettlement. Because of these activities, the coffee produced in the study area is under fragmented land with patchy forests. Farmers use open land either for pasture land or trees plantation (exotic trees) in a competitive manner. On the other hand, the competition for land between cropping and indigenous forests further indicated that farmers are involved in deforestation of indigenous forest to use the land for agriculture crop production. The existence of many patchy coffee plots (fragmented coffee plots) within planted areas in the study area could be used as an indicator for the practice of deforestation. However, the existence of indigenous trees has no impact on the presence of exotic trees, indicating that farmers do not deforest the indigenous forest for tree plantation (exotic trees).

The LULC classified the 30 coffee plots into 3 major groups. The analysis of variance showed significant difference between the three coffee classes for mean temperature, relative humidity and wetness duration during the wet and dry seasons. During the wet seasons (both years) coffee class1, characterized by lower tree density, had a higher mean temperature compared to the other coffee classes. A similar trend of mean temperature variation within the different coffee classes was observed during the dry seasons (both in 2013 and 2014). Hardwick et al. (2015) also reported that human modification of forest results in a change in climate within the forest. Generally, during in the wet season, there was a maximum temperature difference of 1.21 °C among the coffee classes while in the dry season it was 1.03 °C. Such a difference has a practical implication with the implementation of climate change adaptation options. This could also explain that the moderating effect of high density canopy on below canopy microclimate was greater during the wet season compared to the dry season. This might be due to shedding of leaves during the dry season from shade trees; the difference in shade level between the coffee classes was minimal.

On the other hand, the higher tree density in coffee class3 resulted in a higher relative humidity and wetness duration in 2012 and 2013 of the wet season. This might be because of more transpiration by leaves and evaporation from the soil and plant surfaces which add water to the air while cooling and thus lowering the water holding capacity of the air. This resulted in increased relative humidity and wetness duration below the canopy. During the dry season, similar trends of relative humidity and wetness duration among the different coffee classes were observed except for wetness duration of 2014. The highest wetness duration was recorded from coffee class2 in 2014 of the dry season which is difficult to explain. Hence, forest canopy creates a specific understory microclimate that differs from the surrounding local climate. This alteration of local climate is the result of a complex interplay of several stand characteristics and physiographic settings (Arx et al. 2012).

A mean temperature difference of about 1 °C between open and under shade conditions indicates trees have been playing a buffering effect on temperature fluctuation. Along the gradient, the variation was similar showing a possibility of developing shade trees management strategy as an adaption option to climate change impact on coffee at all locations. Therefore, promoting shade tree management as an adaptation option is recommended for small scale coffee growers in Ethiopia.

5 Conclusion

Shade modifies the microclimate of coffee plots. Shade management should be targeted in the strategy to adapt or mitigate climate change in coffee producing country. A better understanding of the landscape structure impact on local climate, which may be used to design and implement landscape management strategies favouring the global performance (production, sustainability) of shade coffee agro-systems.

Acknowledgement The authors would like to thank Ministry of Foreign Affairs of Finland for funding through CHIESA project and Ageyo-Setema coffee growing farmers for allowing us to implement the activity on their coffee farms. The authors also would like to thank Jimma University College of Agriculture and Veterinary Medicine for logistic support.

References

Arx, G. V., Dobbertin, M., & Rebetez, M. (2012). Spatio-temporal effects of forest canopy on understory microclimate in a long-term experiment in Switzerland. *Agricultural and Forest Meteorology, 166–167,* 144–155.

CFC (Common Fund for Commodities). (2004). *Improving coffee quality in east and central Africa through enhanced processing practices* (pp. 10–11) (A (CFC/ICO/22) Project for Rwanda and Ethiopia, Final Appraisal Report). The Netherlands, Amsterdam.

Ewers, R. M., & Banks-Leite, C. (2013). Fragmentation impairs the microclimate buffering effect of tropical forests. *PLoS ONE, 8*(3), e58093. doi:10.1371/journal.pone.0058093

Gove, A. D., Hylander, K., Nemomisa, S., & Shimelis, A. (2008). Ethiopian coffee cultivation-implications for bird conservation and environmental certification. *Conservation Letter, 1,* 208–216.

Hailu, B. T., Maeda, E. E., Hurskainen, P., & Pellikka, P. K. E. (2014). Object-based image analysis for distinguishing indigenous and exotic forests in coffee production areas of Ethiopia. *Applied Geomatics, 6,* 207–214.

Hardwick, S. R., Toumi, R., Pfeifer, M., Turner, E. C., Nilus, R., & Ewers, R. M. (2015). The relationship between leaf area index and microclimate in tropical forest and oil palm plantation: Forest disturbance drives changes in microclimate. *Agricultural and Forest Meteorology, 201,* 187–195.

Labouisse, J. P., BEllachew, B., Kotecha, S., & Bertrand, B. (2008). Current status of coffee (*Coffea arabica* L.) genetic resources in Ethiopia: Implications for conservation. *Genetic Resources and Crop Evolution, 55,* 1079–1093.

Lin, B. B. (2007). Agroforestry management as an adaptive strategy against potential microclimate extremes in coffee agriculture. *Agricultural and Forest Meteorology, 144,* 85–94.

Pielke, R. A., Marland, G., Betts, R. A., Chase, T. N., Eastman, J. L., Niles, J. O., et al. (2002). The influence of land-use change and landscape dynamics on the climate system: Relevance to climate-change policy beyond the radiative effect of greenhouse gases. *Philosophical Transactions of the Royal Society of London A: Mathematical, Physical and Engineering Sciences, 360,* 1705–1719.

Samnegard, U., Peter, A. H., Nemomissa, S., & Hylander, K. (2014). Local and regional variation in local frequence of multiple coffee pests across a mosaic landscape in *Coffea arabica's* native range. *Biotropica, 46,* 276–284.

R Development Core Team. (2012). *R: A Language and Environment for Statistical Computing.*

Weldetsadik, W., & Kebede, K. (2000). Coffee production systems in Ethiopia. *Proceedings of the workshop on the control of coffee berry disease in Ethiopia.* Addis Ababa, August 13–15, 1999.

Characterizing Long Term Rainfall Data for Estimating Climate Risk in Semi-arid Zimbabwe

M. Moyo, P. Dorward and P. Craufurd

1 Introduction

Most of the potential benefits that could be realised from the use of a range of forecasts, including the SCF are yet to be achieved, partly due to failure to fully communicate agriculturally relevant climate information to users (Hansen et al. 2011; Stern and Cooper 2011). There is a mismatch between the farmers' needs and the forecasts, in terms of relevance and their scale (Patt and Gwata 2002; Manatsa and Gadzirayi 2010). There is therefore a need to research into ways of improving dissemination of the weather related information as suggested in various studies (Patt and Gwata 2002; Manatsa and Gadzirayi 2010; Hansen et al. 2011; Stern and Cooper 2011).

Despite the existence and dissemination of SCF amongst many communities, there is still much more information that could be provided by national meteorological services to augment and improve the seasonal forecast information (Hansen et al. 2011). Simple rainfall analyses using long term rainfall records could be able to assist farmers in terms of proving valuable information in relation to climate risk management (Stern and Cooper 2011). This type of analyses, when used with SCF will hopefully be useful to farmers, aiding them in making ex-ante agricultural decisions as well as going a long way in addressing some of the constraints to using

M. Moyo (✉)
International Crop Research Institute for the Semi-Arid Tropics (ICRISAT), P.O. Box 776, Bulawayo, Zimbabwe
e-mail: m.moyo@cgiar.org

P. Dorward
School of Agriculture Policy and Development, University of Reading, Box 236, RG6 6AT Reading, UK

P. Craufurd
The International Maize and Wheat Improvement Center (CIMMYT)
ICRAF House, United Nations Avenue, Gigiri, P.O. Box 1041, Nairobi, Kenya

W. Leal Filho et al. (eds.), *Climate Change Adaptation in Africa*,
Climate Change Management, DOI 10.1007/978-3-319-49520-0_41

SCF (Mupangwa et al. 2011; Stern and Cooper 2011). A number of studies in the 1980s (for example, Stern et al. 1982; Sivakumar 1988) have described the importance accessing long-term daily rainfall records to enable "weather-within-climate" analyses that can be tailored to the needs of different groups of users to be done. Mupangwa et al. (2011) notes that the analysis of daily rainfall data in most semi-arid regions shows that the crop moisture related problems are associated with intra-seasonal dry spells during critical stages of crop growth rather than cumulative rainfall.

The importance of increasing temperatures in terms of climate change is however acknowledged; but this paper focuses mainly on rainfall data as under rain fed agriculture, season to season variability in rainfall and possible changes in the pattern and in the variability are likely to be of more immediate concern to farmers (Stern and Cooper 2011). Crops primarily respond to daily climate or sequences of daily climate and in particular daily rainfall is a key parameter in rain-fed agriculture (Stern et al. 1981; Cooper et al. 2006).

The main objectives of the study included (i) using daily rainfall values to examine the risk or probability of getting a number of weather events that would be useful for farmers such as (a) annual rainfall, (b) rainfall onset and cessation, (c) number of rain days, and (d) risks of dry spells during the growing season in different seasons (i.e. in El Nino, Ordinary and La Nina seasons) and (ii) using Markov Modelling to quantify the chance of rain.

2 Methodology

Long term rainfall data from Hwange District (1963–2009) was used for this study. The principal station used was Victoria Falls Airport (17.56° latitude and 25.50° longitude). The dataset had three years of data missing, i.e. from 1981 to 1984. The methods of Stern and Cooper (2011) were used to analyse the rainfall data. The software package used was Instat Statistical programme (University of Reading 2008). The rainfall data were categorised in El Nino, La Nina and Ordinary years (Table 1) based on information from the International Research Institute for Climate and Society (IRI), (http://portal.iri.columbia.edu/).

2.1 *The Weather-Induced Risk Analyses*

2.1.1 Number of Rain Days

A way of exploring rainfall data in terms of its distribution and amount is to look at the number of rain days. In this study, a rainy day is regarded as a day with measurable rain; i.e. a day yielding 2.95 mm or more. Meteorologically, Mupangwa et al. (2011) notes that 0.85 mm over 1 or 2 days can be classified as a

Table 1 The different season types in the study period

La Nina	Ordinary	El Nino
1970	1964	1963
1973	1966–1969	1965
1975	1971	1972
1984	1974	1982
1988	1976–1981	1986–1987
1998–1999	1983	1991
2007	1985	1994
	1989–1990	1997
	1992–1993	2002
	1995–1996	2006
	2000–2001	2009
	2003–2005	
	2008	

Source International Research Institute for Climate and Society (IRI) (http://portal.iri.columbia.edu/)

rainy day, but that agronomically the threshold value is too low (Mupangwa et al. 2011; Stern, personal communication). Stern and Cooper (2011) indicate that setting or defining the threshold for rain is usually a complication in rainfall data analysis. This is mainly due to that the smallest amounts recorded are 0.1 mm, and in some countries in the early years, the lower limit was 0.01 inches. Below this value, days could be recorded as having trace rainfall. The ideal would be to record all non-zero values, i.e. to set the limit as 'trace and above'. The daily data, from which the annual totals are calculated, contains a mixture of zero values (dry days) plus those with rain.

2.1.2 Onset of the Rains

The start of the season in this study was modified from the ones defined by Mupangwa et al. (2011) and Stern et al. (2003) and it was defined as the first day after 1 November when the rainfall accumulated over 1 or 2 days is at least 20 mm. Further to this definition, to avoid a false start to the season through long dry spells, a condition that this day (start of the season) should not be followed by 10 consecutive dry days within 21 days of the start date, was set.

2.1.3 Dry Spells

The dry spells were defined as any spells within the season that had 7 days or more without rainfall (less than 2.95 mm) in the season after the 1st of November (Day 124) to the end of the rains which technically occurs by 30 April of each season.

2.1.4 Risks of Dry Spells or Replanting 10, 12 and 15 Days After Planting

The risks of dry spells of 10, 12 and 15 days between 1 November and 15 December (the planting window or possible planting dates) were determined.

2.1.5 Date of the End of the Rains

This was taken as the last day with 10 mm by the end of April, i.e. any day that has 10 mm or more before the end of April. This approach defines a single date for the latest possible end of rains date each year. Hence the dataset gave a set of 44 values in Hwange (1963–2009, with 3 years missing data).

2.1.6 Length of Rains

This was derived as the number of days between the end of the rain and the start of the rains.

2.1.7 Using the El Nino Factor

As already indicated, the data were categorized in El Nino, La Nina and Ordinary years based on information from the International Research Institute for Climate and Society (IRI), (http://portal.iri.columbia.edu/). The weather-induced risk analyses for different events were then conducted and risks of the different weather within climate events in the different El Nino, La Nina or Ordinary seasons were established. For each of the analyses, a test of significance (i.e. one way analysis of variance with the Y variate being the weather aspect of interest for example, number of rain days by factor (El Nino, Ordinary or La Nina) was done to find out if there were significant differences in the El Nino, Ordinary or La Nina seasons.

2.1.8 A Modelling Approach to Rainfall Analyses

The long term daily rainfall was also fitted to simple Markov chain models as outlined by Stern et al. (1982, 2003) and Stern and Cooper (2011). For this study Markov chain models of order one and two were considered. Markov modelling intends to further analyse the rainfall through using a more 'sensitive' and precise method of analysis of the rainfall data that would therefore have a chance of detecting smaller changes in the pattern of rainfall and can therefore be of use in augmenting the current SCF information.

3 Results

3.1 Risk Analysis for Total Rainfall Data

The risk of receiving rainfall in a year of more than 800 mm is low, i.e. 28%, meaning that in only 3 out of 10 years can expect rainfall above 800 mm. The risk of receiving rainfall that is below the mean is relatively low, i.e. 43%, meaning any 6 out of 10 years could receive above 625 mm. The risk of receiving rainfall lower than 400 mm is even lower, at 16%. The risk of receiving rainfall less than 200 mm is non-existent (Table 2).

3.1.1 Risks Associated with Annual Rainfall in El Nino, Normal and La Nina Seasons

There are more Ordinary seasons than El Nino seasons and the least are La Nina seasons (Table 3). However, there are no statistically significant differences in rainfall amounts received in the different season types; El Nino, Ordinary and La Nina years ($p > 0.05$).

3.1.2 Risk Analysis for the Annual Rainfall Total for El Nino, Normal and La Nina Seasons

The risk of receiving rainfall amount that is less than the mean amount of 625.4 mm significantly decreases in La Nina seasons compared to El Nino and Ordinary

Table 2 General descriptive statistics of total annual rainfall and associated risks

Rainfall summary statistics (Rainfall in mm)	
Minimum	211
Maximum	1043
Mean rainfall	625
Std. deviation	212
Count ≤ 200	0
Count ≤ 400	7
Count ≤ 600	19
Count ≤ 800	32

Table 3 Mean rainfall amounts in the different season types

Season type	Number of seasons	Mean rainfall (mm)	Standard deviation
El Nino	11	598.4	193.9
Ordinary	26	630.7	225.7
La Nina	7	648	213

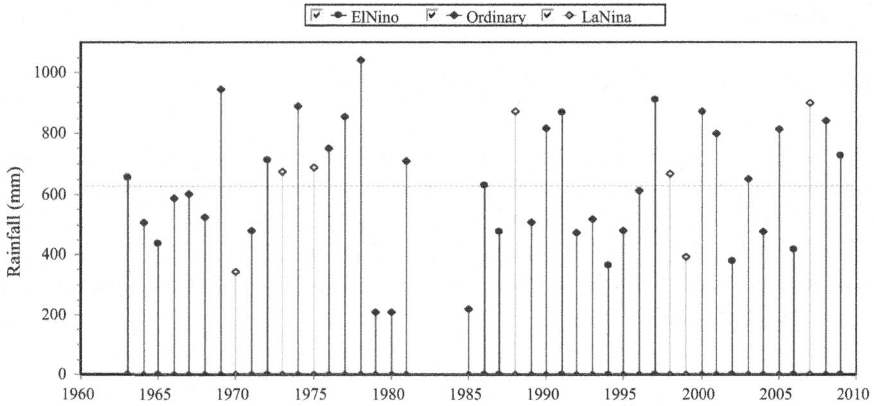

Fig. 1 Occurrence of different season types in comparison to the long term rainfall mean

seasons (Fig. 1). There is a 50% chance that during El Nino seasons rainfall could be below the long term average of 625.4 mm. The risk analysis for the different season types implies that farmers in Hwange may need to take precautionary measures in their farming practices (especially in El Nino and Ordinary seasons that have a high risk of receiving below the mean annual rainfall). However, there is still need for further investigating the in the within season rainfall distribution, which is more important to the farmers than the annual rainfall amount.

3.2 *Analysing Number of Rain Days in Different Season Types*

The mean number of rain days since 1963–2009 was found to be 36 days (±14.6 days). There are no significant differences in rain days' count for the different season types ($p > 0.05$); 34 days with rain in El Nino seasons; 38 days of rain in Ordinary seasons and 35 days of rain in La Nina seasons. There is about 50% probability of El Nino seasons having less than the mean of 36 rain days. However the risk of less than the mean of 36 rain days in Ordinary and La Nina seasons is relatively low (3 in 10 years or 30%) (Fig. 2).

3.3 *Start of Season in Different Season Types*

Analysis of the start of the season is very important in semi-arid areas, where seed inputs are expensive and farmers do not really afford replanting. Data analysis revealed that the mean start of the season in Hwange is Day 154 (1 December)

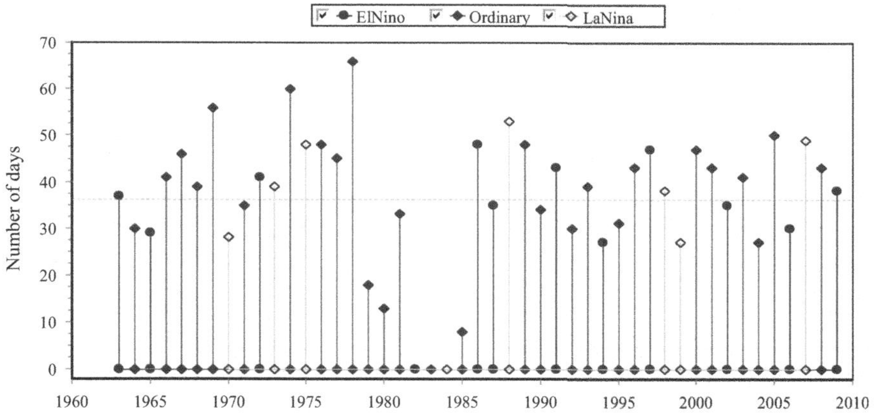

Fig. 2 Number of rain days in the different season types (mean is 36 days)

(±25 days). The earliest start of the season was found to be Day 124 (1 November) and the latest was found to be Day 245 (1 March). There were no significant differences ($p > 0.05$) in the start of seasons for the different season types (El Nino years' mean start day of the season was Day 151–28 November; Ordinary years' was Day 153–30 November; and La Nina was Day 163–10 December).

3.3.1 Risk Analysis of Start of Season Date in El Nino, Normal and La Nina Seasons

Six out of the 11 El Nino seasons had their start of the season dates on earlier dates than the mean date of Day 154 (1 December) (Fig. 3). This means that if farmers in Hwange have to decide when to plant, they could be informed that there is about 50% chance that the season starts before 1 December and they would be advised to stagger the sowing dates (before and after 1 December) in El Nino seasons. Approximately 60% of the Ordinary seasons had their season onset before 1 December hence farmers could be advised to plant before the 1st of December. In La Nina seasons, the advice could be that farmers plant after the 1st of December as the probability that the start of the season is going to be then is high (70%).

3.4 Dry Spells and Their Occurrence in Different Season Types

Dry spells were analysed from 1 November and these were defined as any spells within the season that had 7 consecutive days or more without rainfall after the 1st of November of each season. The mean frequency of dry spells in Hwange is

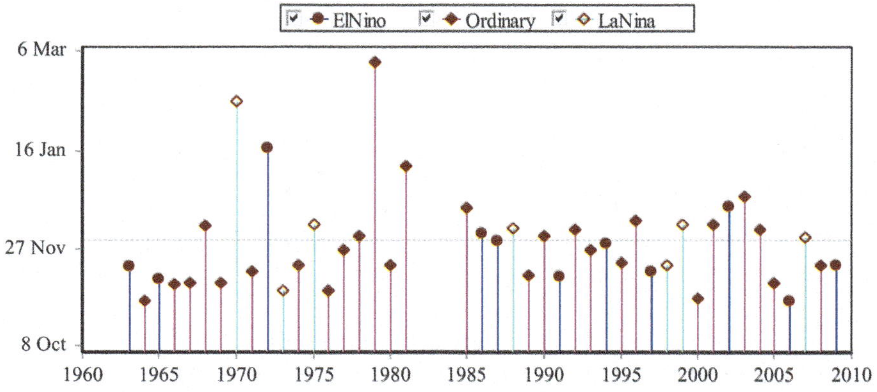

Fig. 3 Dates of the onset of seasons for the different season types (mean is 1 December)

15.4 (±1.5). During El Nino years, the mean frequency of dry spells was found to be 7.7, for Ordinary years, it was 6.5 and in La Nina seasons, it was 6.1. Although there are no statistically significant differences in dry spells in different season types, there is a higher chance (82%) of getting more dry spells in El Nino seasons, i.e. 9 out of the 11 seasons, compared to 46% in Ordinary years and 43% in La Nina seasons.

3.4.1 Risk of Dry Spells Longer Than 10, 12 and 15 Days

Figure 4 gives the proportion of years that had a dry spell longer than 10, 12 or 15 days during the 21 days following planting dates, ranging from the 1st of November to the 15th December, conditional on the initial day being rainy. The top curve show that for a crop planted on 1 November, the risk of a dry spell of 10 days or more, in early November is very high at 60%. The risk of a dry spell of 15-days or more is relatively less, about 30%, or one year in three seasons can farmers encounter a dry spell of more than 15 days in early November. By planting at the end of November, the risk of a 10-day dry spell has considerably decreased to 30%, and shortly after the risk has reached a plateau in Hwange. In terms of extension advice, someone wishing to minimize their risk of dry spells could be advised to wait if they were considering planting in early November, and they could plant in mid-November because the risk does not decrease further, but the chance of a damaging dry spell later in the season might increase. Also, the growing of drought tolerant crops that can withstand a 15 day dry spell after date of planting is recommended. Because of the high risk of dry spells in early November and December there is a high chance of having to replant in the semi-arid areas.

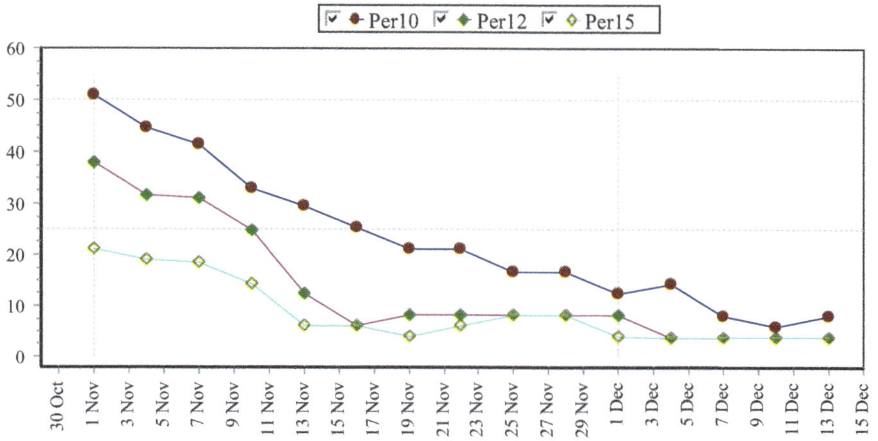

Fig. 4 Risk of dry spells (10, 12 and 15 days)

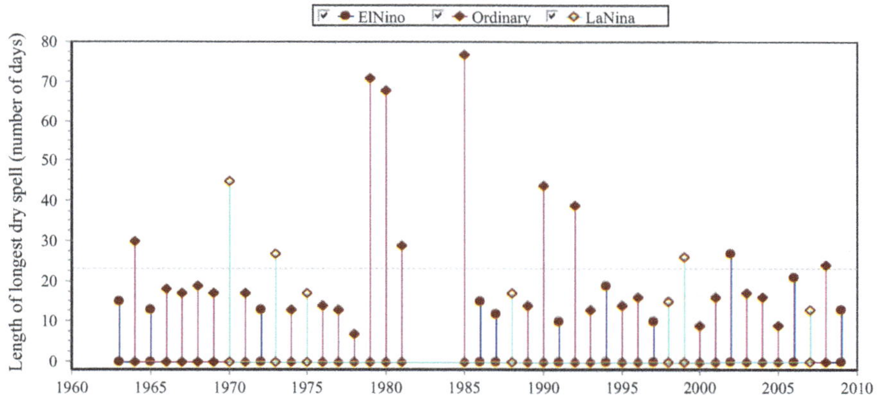

Fig. 5 Length of longest dry spell in different season types (mean is 23 days)

3.4.2 Mean Length of the Longest Dry Spells in El Nino, Normal and La Nina Seasons

The study finds that the length of the longest dry spells is statistically different for the different season types in Hwange district ($p < 0.05$), being longest in Ordinary seasons (26 days) compared to El Nino (15 days) and La Nina seasons (23 days). The risk of very long dry spells in El Nino seasons is not very high (10%), although moisture conservation strategies are necessary (Fig. 5).

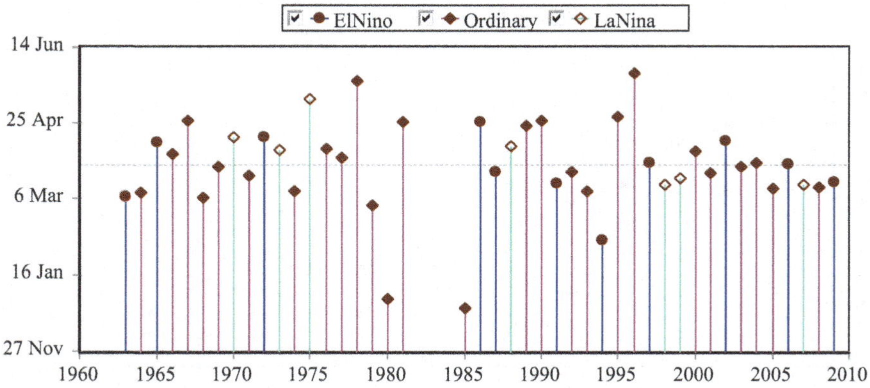

Fig. 6 End of rains in the different season types (mean is 28 March)

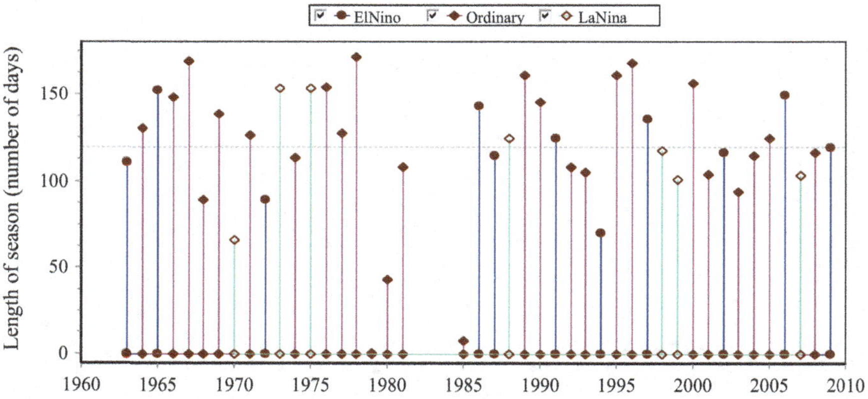

Fig. 7 Season length in El Nino years (mean is 119 days)

3.5 End of the Rains and Its Occurrence in Different Season Types

The mean last day of the season is Day 272; the 28th of March. The latest day of the end of the season recorded was Day 333 (28th May) and the earliest was Day 179 (26th December). There are no statistically significant differences in the end of season dates in different season types ($p > 0.05$). In the El Nino and Ordinary seasons, the mean end of rains is on Day 271 or 27th March, with the mean end of rains in La Nina seasons being on Day 279 (4th April). However, there is a high risk that the rainfall cessation dates across the different season types occur before the mean date of the 28th March (Day 272) (~60% in El Nino and Ordinary seasons and ~40% in La Nina seasons) (Fig. 6).

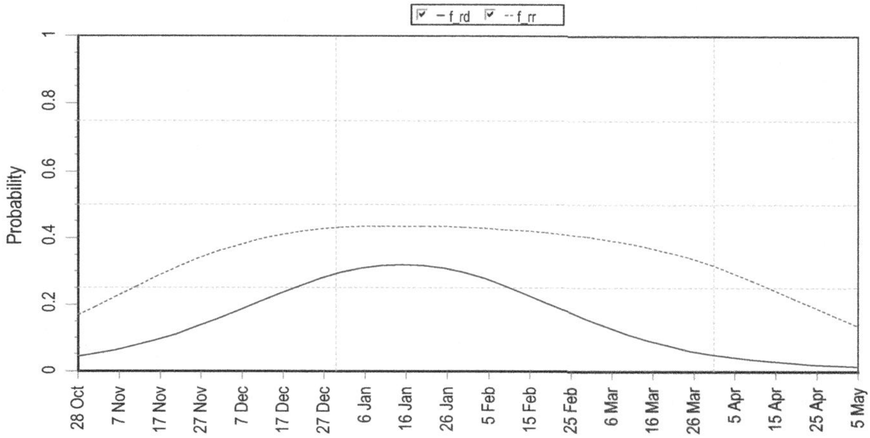

Fig. 8 First order chance of rain

3.6 Length of the Season (Season Length in El Nino, Normal and La Nina Seasons)

The mean season length was found to be 119 days (±38). There are no significant differences in the season lengths for the different season types ($p > 0.05$); 120 days in El Nino years, 114 and 107 days in Ordinary and La Nina years respectively. There is a high risk that the season length across the different season types is less than the mean length of 119 days (Figs. 7 and 8).

3.7 A Modelling Approach

The results in the Markov modelling of the chance of rain in Hwange in the First-order-Markov chain are presented in Fig. 8. The top curve in Fig. 8 is the chance of rain when the previous day also had rain (f_rr; representing the chance of rain when the previous day had rain). This is therefore the chance that a rainy spell continues for a further day. In January this is about 0.45, i.e. about 45% of rainy days continued and had rain on the next day. The chance of rain after a rain day earlier in the season (in November) is low, at 0.25, meaning only 25% of the days had rain on the next day. The bottom curve in Fig. 8 is the chance of a rainy day if yesterday had no rain, i.e. the chance of rain after a single dry day (f_rd; representing the chance of rain when the previous day had rain). The results indicate that the chance of a rainy day after a dry day is low during the peak of the season, in January, at 0.25, i.e. 25% of the days have rain after a dry day.

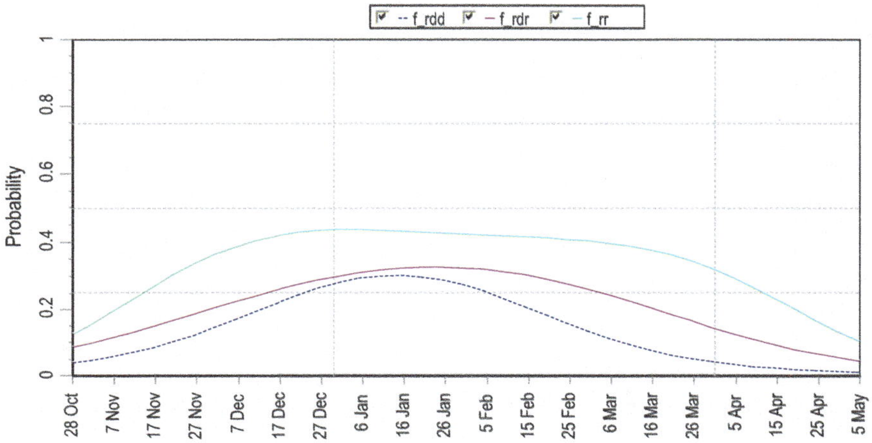

Fig. 9 Second order chance of rain

3.7.1 Second-Order Markov Chains

The top curve in Fig. 9 is the chance of rain when the previous two days also had rain (f_rr). This is therefore the chance that a rainy spell continues for a further day. The chance of rain after 2 rain days is highest in December, and this is over 0.5, i.e. about half of the rainy days continued and had rain on the next day. The middle curve in Fig. 9 shows the chance of rain after a dry spell is in between rain days (f_rdr). The chance of rains when a dry day is sandwiched between rain days is lower than when the previous 2 days were raining. Lastly, the bottom curve in Fig. 9 shows the chance of rain after a dry spell of two or more days. This is therefore the chance that a dry spell (of longer than one day) is broken. This is a smaller probability than the chance of a rain spell continuing, rising to about 0.2 or 20% in January. The chance of rain 'returning' is greater after just a single dry day, than if a dry spell has been in experienced for two days or more.

4 Discussion

The analysis of the daily rainfall data was done so as to investigate how it could help farmers (together with the SCF). This type of analyses, when used with SCF will hopefully be useful to farmers, aiding them in making ex-ante agricultural decisions such as what crop varieties to plant, when to plant, weed and apply fertilisers, as well as going a long way in addressing some of the constraints to using SCF.

The information on season length is vital for farmers so that they can choose crops and varieties that mature within the season, whilst moisture is still available.

The rainfall onset and cessation dates are of use to farmers as they help determine the planting time and early planting is usually encouraged to ensure successful establishment and early survival of the crop, as well as ensuring that harvesting problems do not occur. It is of importance that farmers get to know the onset dates of any given season as these help farmers plan accordingly, to secure inputs on time so that they are not "surprised" by the rains. Since the characterization of the rainfall finds that it is possible to estimate risk for El Nino, Ordinary and La Nina seasons, these could be used together with the SCF to inform farmers of when the season is likely to start.

Crop and livestock management practices are encouraged in the semi-arid areas, irrespective of season types. Staggering of planting irrespective of season types (El Nino, Ordinary or La Nina) is advisable to farmers in the semi-arid areas due to the high risk associated with the start of the rainy season. It would also be wise for farmers in the semi-arid areas to diversify and grow a variety of cereal crops to spread the risk associated with the rainfall distribution in these areas. However as Stern and Cooper (2011) rightly point out, the start of the season is also influenced by several factors such as the farmers' degree of risk aversion, the frequency and amounts of early rainfall events as well as the texture of their soil which will determine how deep any sequence of rainfall events will penetrate and be stored in the soil.

Although the risk of getting a 15 day dry spell was found to be low, it is recommended that farmers grow crops (or varieties) that could withstand a 15 day dry spell. The growing of drought tolerant crops is one decision that farmers could take or the use of simple moisture conservation measures such as soil surface mulching could also help increase the risk of successful crop establishment. Due to the high risk of encountering 10 days long dry spells, the farmers would be advised to implement soil moisture conservation techniques irrespective of season quality. There is also considerable advantage to planting early in semi-arid areas, but this opportunity has often been offset because early planting might have a higher risk of being followed by a long dry spell resulting in seedling death and the need to replant. Because of the high risk of dry spells in early November and December there is a high chance of having to replant in the semi-arid areas and information on risks of dry spells at the start of the season could really assist poor smallholders. The economic cost of replanting is high, so if the planting window has been characterized and is known, farmers could make informed decisions on when to plant. The choice of crop could also be determined by knowledge on the frequency of dry spells and the risks of encountering dry spells after the onset of the rains.

Further rainfall analysis using Markov modelling helps in detecting smaller changes in the pattern of rainfall. Such type of analysis will be beneficial to farmers so that they can make decisions especially on when to plant, apply dressing fertilizer and when to weed. Cooper et al. (2006) have indicated that farmers may be risk averse in using fertility amendments, mainly because of fear that if they do not have information on how rains would fare might lead to inappropriate application

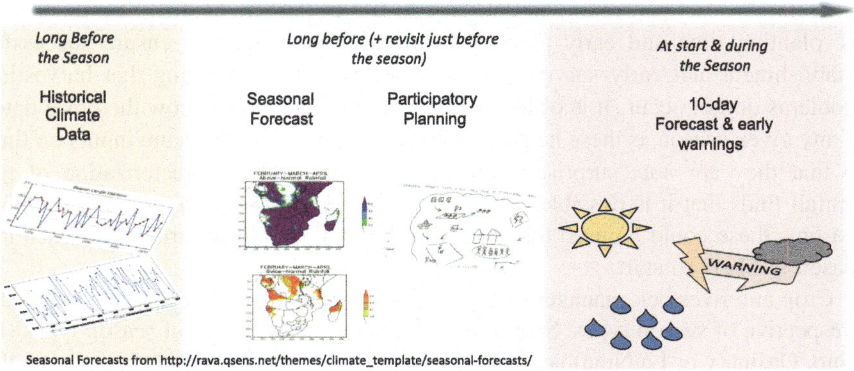

Fig. 10 A schematic plan for introducing the use of weather within climate information with SCF and weather related information to smallholders. *Source* Adapted from CCAFS (2012)

rates and times. However if the chance of rain in a given season has been determined, it makes it possible for better decision making in terms of when farmers could apply the fertility amendments and how much could be applied.

5 Conclusions and Recommendations

The results from the study do help in answering some of the farmers' concerns in terms of rainfall distribution within the seasons, and helps in showing risks associated with different season types. There are difficult risks in the semi-arid areas of Zimbabwe in terms of dry spells, and since the amount of rainfall cannot be influenced, technologies that enhance water use efficiency could also be one of the major areas of research that should be integrated into the semi-arid farmers' existing strategies to cope with climate variability and ultimately change.

Lastly, one key question that also needs to be answered is how best the characterized climate information could be introduced to the farmers to help them make crop management decisions. Answers to this question lie within suggestions that facilitating smallholders to make better plans and decisions and utilise climate and weather information could be done through some participatory methods (CCAFS 2012). Enhanced communication of climate-related information is an option that could assist in adaptation strategies and timely decision-making by farmers. Packaging SCF with historic climate data as well as bringing in the shorter range forecasts, together with the experience of the season as it develops is a way in which value could be added to climate information dissemination (Fig. 10).

Acknowledgements The authors gratefully acknowledge the funding provided to ICRISAT through the CGIAR Program on Climate Change, Agriculture and Food Security (CCAFS) for this work.

References

Climate Change Agriculture and Food Security (CCAFS). (2012). *Identifying good practice in the provision of climate services for farmers in Africa and South Asia*. Background paper for the workshop: Scaling up Climate Services for Farmers in Africa and South Asia. Senegal. December 10–12, 2012. http://ccafs.cgiar.org/node/1766. Accessed December 13, 2014.

Cooper, P. J. M., Dimes, J., Rao, K. P. C., Shapiro, B., Shiferaw, B., & Twomlow, S. J. (2006). Coping better with current climatic variability in the rainfed farming systems of sub-Saharan Africa: A dress rehearsal for adapting to future climate change? ICRISAT Global Theme on Agro-Ecosystems Report No. 27 (p. 24).

Hansen, J., Mason, S. J., Sun, L., & Tall, A. (2011). Review of seasonal climate forecasting for agriculture in sub-Saharan Africa. *Experimental Agriculture, 47*(2), 205–240.

Manatsa, D., & Gadzirayi, C. (2010). Tailored seasonal climate forecast system for small holder farmers in Chiredzi district. Coping with drought and climate change project. www.ema-cwd.co.zw/index.php?option=com. Accessed January 25, 2013.

Mupangwa, W., Walker, S., & Twomlow, S. (2011). Start, end and dry spells of the growing season in semi-arid southern Zimbabwe. *Journal of Arid Environments, 75*(11), 1097–1104.

Patt, A. G., & Gwata, C. (2002). Effective seasonal climate forecasts applications: Examining constraints for subsistence farmers in Zimbabwe. *Global Environmental Change and Human Policy Dimensions, 12,* 185–195.

Sivakumar, M. V. K. (1988). Predicting rainy season potential from the onset of rains in Southern Sahelian and Sudanian climatic zones of West Africa. *Agricultural Forestry Meteorology, 42,* 295–305.

Stern, R. D., & Cooper, P. J. M. (2011). Assessing climate risk and climate change using rainfall data—A case study from Zambia. *Experimental Agriculture, 47*(2), 241–266.

Stern, R. D., Dennett, M. D., & Dale, I. C. (1982). Analysing daily rainfall measurements to give agronomically useful results. 1—Direct methods. *Experimental Agriculture, 18,* 223–236.

Stern, R. D., Dennett, M. D., & Garbutt, D. J. (1981). The start of the rains in West Africa. *Journal of Climatology, 1,* 59–68.

Stern, R. D., Knock, J., Rijks, D., & Dale, I. (2003). *Instat climatic guide* (398 pp.) http://www.reading.ac.uk/ssc/software/instat/climatic.pdf. Accessed September 26, 2013.

University of Reading. (2008). *Instat—An interactive statistical package*. UK: Statistical Services Centre, University of Reading.

Supporting Farmer Innovation to Enhance Resilience in the Face of Climate Change in Farming Systems in Machakos and Kitui Counties, Kenya

Eunice Karanja, Geoffrey Kamau, Chris Macoloo, Makonge Righa, Laurens van Veldhuizen and Ann Waters-Bayer

1 Introduction

Africa's population is expected to rise to about 1.3 billion by 2020 and a major challenge is how to match food supply with overall population growth and rising urban demand. The faster growth of urban population has serious implications on demand for food, since most urban dwellers depend on rural communities for food production. However, this could decline, as most rural-to-urban migrants are young people, leaving farming to the elderly. This challenge is further compounded by lack of well-adapted agricultural technologies (Lynam et al. 2005). Other challenges include unreliable rainfall and new severe pests and diseases such as the large grain borer and the maize necrotic disease, all of which have been aggravated by the adverse effects of climate change. Agriculture is the backbone of Kenya's economy, accounting for 32% of the GDP and 65% of the export earnings. (GOK 2010). About 80% of the country is made

E. Karanja (✉)
Prolinnova, Nairobi, Kenya
e-mail: ekaranja12@gmail.com

G. Kamau
Kenya Agricultural and Livestock Research Organization, Nairobi, Kenya
e-mail: gmkamau_1@yahoo.com

C. Macoloo · M. Righa
World Neighbors, Nairobi, Kenya
e-mail: cmacoloo@wn.org

M. Righa
e-mail: jrigha@wn.org

L. van Veldhuizen · A. Waters-Bayer
Prolinnova International Support Team, Leusden, Netherlands
e-mail: L.v.Veldhuizen@kit.nl

A. Waters-Bayer
e-mail: waters-bayer@web.de

W. Leal Filho et al. (eds.), *Climate Change Adaptation in Africa*,
Climate Change Management, DOI 10.1007/978-3-319-49520-0_42

up of arid and semi-arid land (ASAL). Unfavourable changes in the climate have impacted negatively on the resource capability of ASAL areas, thus impeding their expected contribution to farmer livelihoods. Farmers are poorly equipped to handle the challenges emanating from these effects, thus rendering the current food production methods incapable of meeting today's and tomorrow's growing needs, if there is not continuous adaptation of technology and innovation (UN 2013). Apart from this, small-scale farmers generally have poor access to external agricultural inputs and to markets for their products, which further impacts negatively on their production and livelihoods. This has made it necessary for communities of small-scale farmers to innovate and adapt how they use and manage agricultural resources, so as to sustain their environment and gain economic and social welfare viability.

There have been enormous strides in agricultural research and development (ARD) worldwide that have seen various agricultural technologies and practices developed to enhance agricultural production and improve livelihoods of small-scale farmers. In addition, ARD approaches have also evolved over the years, as agricultural researchers continuously ask themselves how they can be more effective in working with farmers to develop better technologies and farming practices. In conventional ARD however, the interaction between researchers and farmers still tends to follow a top-down approach whereby broad-based "solutions" to challenges are presented to farmers who are not directly involved in the decision-making processes about developing these "solutions". Many of the top-down approaches that have been applied have failed to take into account the facts that small-scale farming practices are very specific to particular environmental, sociocultural and economic conditions and that agricultural innovations have to fit into the local context (Waters-Bayer et al. 2005). They also fail to address the important potential of small-scale farmers as innovators in the process of developing locally adapted innovations (Frank 2012). This failure can be attributed to the colonial legacy in many countries in the Global South, including economic reliance on commercial export agriculture for tax revenue. Much public-sector research has focused on commercial export agriculture and has disregarded the voices and needs of small-scale and subsistence farmers.

This situation has been made worse by the dominance of the aforementioned top-down or linear model of technology generation and transfer with its assumptions about fixed roles assigned to the various actors involved. In the linear model, agricultural professionals in research and extension determine what constitutes a (presumed) improved technology for farmers to use. Very little or no attention is paid to local innovations, despite their potential to improve the livelihoods of farmers and to benefit the wider community.

Since the advent of farming, farmers have been engaging in innovation to improve their production and adapt to various changes (Reij and Waters-Bayer 2001). However, most of the formal researchers and extensionists are not aware of the many initiatives by farmers to address the challenges they face. The focus is often on offering advice to farmers or developing solutions, often without involving the farmers, resulting in technologies that hardly address the farmers' problems. There is need to bridge this gap in order to facilitate learning and sharing between farmers and formal researchers and to ensure participation of farmers and other stakeholders such as extensionists in developing and improving agricultural technologies and techniques.

Prolinnova–Kenya (PK) is a multi-stakeholder platform in Kenya, which is part of the international Prolinnova network (www.prolinnova.net) and is promoting farmer-led research and development. PK, together with three other country platforms in Eastern Africa, seeks to respond to the above-mentioned challenges through a project called Combining Local Innovative Capacity with Scientific Research (CLIC–SR). This aims to strengthen community resilience to change, including climate change, by building on the inherent innovative capacities of local people. PK is enhancing farmers' own creativity and innovativeness by supporting them in their efforts to develop new and better ways of doing things, using low-cost locally available resources, on their own initiative and in good contact with other relevant actors in the agricultural innovation system. This refers to Participatory Innovation Development (PID), in which scientists and development agents jointly work with farmers to help them further develop, adapt and test local ideas and initiatives, integrating local and scientific knowledge (Waters-Bayer et al. 2004).

Furthermore, the CLIC–SR project applies an approach that puts farmers at the centre of decision-making in ARD, namely a farmer co-managed funding mechanism known as the Local Innovation Support Fund (LISF). This is an approach to promoting innovation that leads to more fundamental outcomes than only new and site-appropriate technologies. It also strengthens local capacities to continue to innovate and adapt to changing conditions. Through LISFs, ARD funds are channelled to farmer innovators to try out or further develop new ideas based on their own ideas and needs, to engage in sharing and learning about farmer innovation, to document and disseminate their innovations and, thus, to accelerate local innovation processes (Prolinnova 2012; Triomphe et al. 2012).

The LISF approach recognises that poor farmers are resourceful. Indeed, because of their poverty, farmers are forced to be innovative in order to survive. They are resourceful through their innovative use of locally available resources in their attempts to address new challenges, including the effects of climate change, even under extremely difficult circumstances. Such farmers, if given the chance, are capable of engaging with formal researchers in contributing to the development of sustainable agricultural and NRM practices. Furthermore, LISF helps identify local innovators and innovations, encourages participatory documentation, and facilitates knowledge sharing among farmers and other stakeholders. The innovations that result from LISF-supported activities of farmers have great potential in improving food security, NRM and livelihoods.

2 Area Characterization

Machakos County (1.5177°S, 37.2634°E) covers an area of 6208 km^2 with most of it being semi-arid and of hilly terrain at an attitude of 1000–2100 m above sea level. The county experiences two rainfall seasons, and subsistence agriculture is practised with maize and drought-resistant crops such as sorghum and millet being grown. Productivity is, however, constrained by dependence on rainfed agriculture and limited use of external inputs, among other factors (GOK 2013b).

Kitui County (1.3751° S, 37.9952° E) covers an area of 30,496 km^2, of which 14,137 km^2 are regarded as arable land. The county is mostly semi-arid and experiences a temperature range between 14 to 34 °C and an average rainfall of 900 mm annually. Small-scale farming is practised, where maize, millet, sorghum, green grams, cowpeas and pigeon peas are grown (GOK 2013a).

Returns on farming have decreased over the years because of unreliable weather conditions, pests and diseases, post-harvest losses and soil exhaustion, exacerbated by climate change. Out of necessity, farmers have been forced to adapt their practices in agriculture and NRM. They innovate through tacit knowledge they have acquired over the years through experience, insights, informal experimentation on a trial-and-error basis, as well as drawing on explicit knowledge from various interactions with other stakeholders.

3 Operation and Structure of the LISF

The LISF is accessible to individual farmers as well as to farmer groups and is managed at the local level within operational structures known as Local Steering Committees (LSCs), while the National Steering Committee (NSC), consisting of representatives from research institutions, civil society organizations and universities, is responsible for overall administration at national level.

The LSC is composed of farmer representatives, local administrative and government officers, and people from local non-governmental organisations and research institutions. Generally, farmers are the key office bearers and have the largest representation in the LSC. This committee is responsible for raising awareness about the LISF at grassroots level, calling for innovation proposals, vetting and approving the proposals, disbursing funds to farmers and then monitoring the implementation of the activities supported by the funds. The process of selecting the proposed innovations, experiments and learning activities to be funded begins with a call for proposals by the LSCs. The committee announces the call through various channels, including public meetings, churches, informal community gatherings and personal communication within their networks to ensure that all eligible innovators in their respective areas receive information about the call.

Proposals are received within a specific period of time and screened by the LSCs using criteria they have developed to decide which proposals will be funded. The proposals are then recorded in an LISF register and feedback on the decision of the LSC is provided to all applicants. The LSC disburses funds to successful applicants and is responsible for monitoring the use of the funds and the progress of the experimentation on the innovations or the other activities funded. The committee, with guidance from the NSC, is also responsible for analysis and documentation of the farmers' experiences and lessons learnt, for the purpose of knowledge sharing with other farmers (Kamau et al. 2012). This provides opportunities for learning and

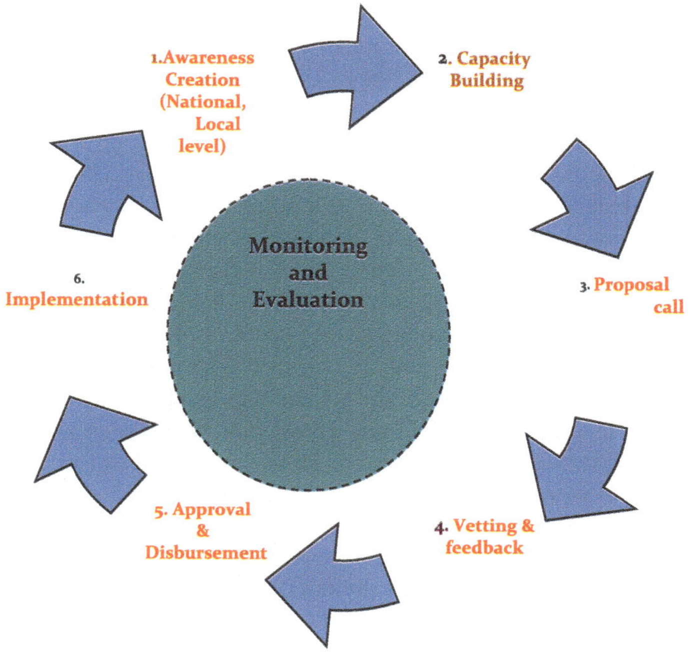

Fig. 1 LISF cycle

uptake of innovations and of own experimentation and adaptation by other farmers and other stakeholders. The cycle followed in the LISF process is shown in Fig. 1.

Farmers in Machakos and Kitui semi-arid Counties have been supported through locally managed LISFs to improve innovations related to crop, livestock and NRM, thus contributing to building community resilience. Five of these innovations are examined in this chapter.

4 Cases of Innovation

4.1 Case 1: Raising Finger Millet in Nurseries

4.1.1 *Context*: *Finger Millet Cultivation in Machakos*

Finger millet (*Elusine indica*) used to be a common cereal crop in the semi-arid areas of Machakos. Its popularity was mainly due to its high nutritive value and hence it was used as a weaning food for babies and for feeding invalids. The major challenges to its production have been the high cost of seed and the high labour demand associated with bird scaring and post-harvest processing. In Machakos, these challenges were exacerbated by the adverse effects of climate change, as rainfall became intermittent and less reliable, thus resulting in farmers receiving minimal returns despite investing heavily in labour. The farmers have conventionally been broadcasting the finger millet seeds but, upon germination, a major threat to the seedlings was arthropod pests and birds that destroyed the crop in the very young stages. This forced the farmers to over-seed and hence waste a lot of seed in an effort to compensate for the damage.

Faced with these challenges, farmer Simon Masila experimented on his own initiative with establishing finger millet in a nursery bed. He watered the nursery and thereafter closely tended the millet seedlings, which he later transplanted in the field. Despite minimal rainfall, the finger millet flourished compared to seeds that his neighbours broadcast directly in the fields, as their crops failed completely. Other general observations that he made were that transplanted finger millet was of better quality and gave higher yields than millet sown directly in the field. The intense tillering of the single finger millet seedlings produced up to 25 tillers per stand. Through this approach, the farmer saved about 60% on seeds and also gave his plants a head start at the onset of rains. This innovation therefore addressed the challenge of seed wastage as well as adverse environmental conditions arising out of climate change, namely less reliable start of the wet season and often shorter wet seasons. This farmer also discovered how to reduce bird attack on his crop by planting other types of finger millet preferred by birds, e.g. pearl millet, at the margins of his farm so that the attack on the main millet crop was reduced.

This farmer sought a grant from the LISF and was granted KSh 20,000 (USD 200) to support his further experimentation. The LISF-supported experiment attracted the attention of formal research, and Simon and neighbouring farmers are now working with researchers from the Kenya Agricultural and Livestock Research Organization (KALRO) in testing different seed densities and transplanting regimes with a view to coming up with the best mechanism for use by farmers in the region.

In comparison, nursery-grown finger millet on a plot of 150 by 50 m produces approximately 200 kg as compared to 10 kg or less produced through broadcasting. On account of the success of this approach, 120 farmers in the area have picked up the nursery-and-transplanting idea and have successfully applied it in other dryland crops such as watermelons, pumpkins, maize and sweet potatoes. The plants are established in polythene bags in the nurseries and, at the onset of rains, they are transplanted into the field, thus giving them a head start.

4.2 *Case 2: Wall Terraces for Water Harvesting*

4.2.1 *Context*: *Low Rainfall in Kalama, Machakos*

The amount of rainfall received in Kalama area of Machakos County has been diminishing over the years (Jaetzold et al. 2003). Because of reduced moisture, crop production has become a big challenge. In addition, termites have been infesting the farms and sometimes have destroyed every green plant in a farm. Faced with this challenge and having returned to the farm after a career in the hotel industry, farmer Daniel Kivyuvi tried to figure out how to cultivate crops on his recently inherited

3.5 acre farm located next to a path leading to a dry riverbed. During the occasional rainfall events, he observed the water running down the slope and creating a gully along the footpath. He then came up with an idea that seemed strange to his neighbours and the agricultural extension officers. On his land, he made "*fanyajuu*", terraces that are constructed by digging ditches and throwing up the soil to reinforce the upper side of the ditches so as to form wall-like bunds following the contours. The main costs involved are one's own labour in doing the digging. He then opened an inlet at one corner of his farm and let the water flow into the farm through the uppermost trench and he guided the water in a zigzag pattern throughout his farm. After the rains, the water collecting in the trenches would remain there for some time and thus provide the much-needed moisture to his crops. He would then grow crops like arrowroot, passion fruits, pawpaws, and mangoes in the trench and on the sides of the trench. He noticed that, because of the higher moisture levels, the termites no longer attacked his crops. He applied for a grant of KSh 30,000 (USD 300) from the LISF to continue his experiments.

During his experimentation, he observed that water was getting exhausted quickly from the structures. He therefore added another structure in the form of pits lined with cement-soaked sisal bags. This ensured that seepage of water out of the pit was kept to a minimum. Loss of water through evaporation remained a challenge. To deal with this, he constructed a roof over the pits and thatched it with local grasses. He is now experimenting with a live "thatch" in the form of climbing passion-fruit plants. He has been using the water collected for production and this has ensured that the farm remains productive all year round in this fairly arid area. He now produces approximately 1.3 t of maize, 100 kg of passion fruits, 40 kg of pawpaws and 100 kg of mangoes per acre. Fifteen farmers in his vicinity have now adopted the innovation.

4.3 Case 3: Combining Rock-Hyrax Manure with Farmyard Manure

4.3.1 *Context: Hillside Farming*

The rock-strewn hill masses of Machakos provide an ideal habitat for many animals, among which are rock hyraxes (*Procavia capensis*), which live in colonies of several scores. These are small hare-like rodents that feed on grass and other soft broad-leaved plants. The animals accumulate a lot of droppings in their hideouts and also in the area where they feed. During the rains, these droppings are washed down to the foothill of the farms and eventually to the seasonal streams that form after the rains.

Farmer Phillip Kilaki observed pockets of crops doing very well in his farm and, on further examination, found that these were sites where rock-hyrax manure had been deposited. He then experimented with an idea that enhanced the collection of the manure, using cut-off drains that he constructed on his farm. This ensured that no rock-hyrax manure escaped from his farm. He used the manure to grow mainly bananas, oranges and mangoes, which all performed very well. He applied for a grant from the LISF of KSh 30,000 (USD 300) to do further experimentation with mixing farmyard manure from domestic livestock with the rock-hyrax manure. He discovered that this enhanced the effectiveness of the manure in increasing soil fertility. He realised that this supplementation of the droppings of the rock hyrax (which are locally and readily available) with farmyard manure (which is generally limited because most small-scale farmers have only few animals) substantially increases soil fertility due to increased quality of manure, thereby retaining the soil moisture level and ultimately increases yields. Rock-hyrax manure collection has now become a commercial activity after other farmers saw the benefits of the manure. This manure is sold either mixed with manure from domestic livestock or in pure form. Crop production has improved and farmers are now protecting the rock hyraxes almost like domestic animals.

4.4 Case 4: Cultivation of Local Grasses as Fodder For Goats

4.4.1 Context: Challenge in Feeding Goats During the Dry Season

Farmers in Kitui County keep goats as a source of income and food but a major challenge has been the dry season extending up to seven months in a year, which disrupts goat production on account of feed shortages. Certain grass species that are very palatable to goats in the area have been declining. A farmer group therefore decided to cultivate these grasses during the rainy season for feeding in the dry season. They submitted a proposal for support from the LISF and received KSh 25,000 (USD 250). This allowed them to try to grow the grass and bale it as hay for dry-season feeding.

In order to increase the quality of feed, the farmer group of about 30 members led by woman farmer Julia Simon mixed the grass with other palatable tree leaves and baled these together with grass. This mixture provided more nutritious feed that allowed farmers to keep more productive breeds of goats.

To speed up the baling process, the farmers abandoned the conventional baling box and used a pit dug in the ground into which they placed two strings on the length and breadth of the hole and after placing the grass in the hole, they compacted it with their feet and then tied the hay bale up. This innovation in growing grasses and baling hay solved the problem of shortage of goat feed during the dry season and also earned the group income through sale of hay to neighbouring farmers.

4.5 *Case 5: Sexing of Chicken Eggs*

4.5.1 *Context: Challenge of Market Demand*

Because climate change is negatively affecting their cropping activities, many farmers are seeking alternative sources of livelihoods in agriculture, including activities they hope will be more resilient to climate change. One example of such diversification as a form of adaptation to climate change is an increased attention to rearing local chickens. This has now become a common activity in almost all homesteads in Makueni and Kitui Counties. The local chickens are fairly resilient under the prevailing circumstances and provide both food and income for the farm families. Some of the income generated from selling chickens is used to purchase food and farm inputs when other farm-based sources of income fail. A major challenge for small-scale farmers has been their inability to supply large numbers of chickens to traders and local hotels. This challenge is more pronounced when the traders are seeking specifically cocks or hens only.

When farmer Christine Kilonzi found herself in such a predicament, she decided to experiment with selecting eggs that would hatch into cocks or hens only. She first tried selecting the eggs based on colour, later based on size and eventually based on egg shape—and she succeeded with the last-mentioned. She then applied for an LISF grant and received KSh 30,000 (USD 300) to expand her experiments. Today, she is able to select male and female eggs with a success rate of about 80%. By being able to predetermine the sex of chickens before they hatch and raise only those needed by the market, she has been able to improve her income, as she can supply chickens according to the needs of her clients. She now sells one-week-old chicks at KSh 300 (USD 30). She has used her skills to assist the poultry unit of the national research institute to sex eggs. She is currently working on developing a device that can be used to determine the various shapes and hence accurately apply this egg-sexing technique widely. This is being done through joint experimentation with formal researchers in testing different prototypes.

5 Discussion and Lessons Learnt

(i) *Specific needs of farmers*

Climate change has posed challenges that have tasked farmers to come up with innovative experiments and thereby come up with solutions. A major difference between the farmers' experimentation and that of the research scientists is that the farmers' experiments are tailored to specific and felt needs, whereas formal researchers' experiments often address general needs. This is illustrated by all five cases where farmers are addressing local challenges but, in the process, this is also leading to spinoffs that are beneficial for others.

(ii) *Assessment methods*

The methods of assessment also differ between the farmers and the research scientists in that the farmers broadly observe what happens with reference not only to the specific parameter being tested but also to any other effects on the farm. Hence, while farmer Simon was concerned about survival of millet seedlings, he also realised that he was able to save on seeds. The same applied to farmer Daniel's experience, where he not only collected water but also eliminated a termite menace. Farmer Julia sought to domesticate grass and also managed to come up with a way of baling that was more practical than the conventional one. The same can be said of farmer Phillip, who collected rock-hyrax manure and, in the process, mixed it with farmyard manure and enhanced its value so that what started off as a manure-collection exercise has led to efforts to protect hyraxes. Farmer Christine's challenge in egg sexing has led to a solution that is slowly gaining popularity in the region. All these examples confirm what Richards (1985) stated about farmers' experimentation methods: that farmers experiment in time while formal researchers experiment in space (Richards 1994, 1989). Others have referred to farmers "tinkering" to arrive at workable solutions (Biggs 1989).

(iii) *Dynamic farmer experimentation: responding to changes*

A further observation is the dynamic nature of the experimentation, where what may seem to be most challenging may quickly change, as demonstrated by Daniel's water-harvesting. From successful water-harvesting came the new challenge of water loss through seepage and evaporation, which then had to be addressed by the farmer innovator. The same can be said of the grass domestication, where new baling techniques became necessary, as well as in the collection of the rock-hyrax manure. This confirms Drinkwater's (1994) assertion that agricultural reality bubbles with change, disorder and multiple social processes in the same time and place, with perhaps different elements moving in different directions at the same time and hence the need to be up to date with new innovations (Drinkwater 1994). This is an opportunity that research institutions and the extension services usually do not pay attention to, because of either their resource limitations or the institutional and policy environment. The farmers therefore continue to be ahead of the formal researchers with respect to responses to changing environments.

One challenge in farmer innovation could be regarded as the specificity of local innovations, which—almost by definition, as the innovations are developed to suit local conditions—cannot always be widely applied. It is, however, noteworthy that the principles espoused by initiatives such as the LISF and others contribute to supporting farmers to experiment on their innovations in their own self-determined ways, thereby not only establishing the viability of their innovations but also giving recognition to their capacity to innovate. Through LISFs, there has been a general acceptance by stakeholders in research and extension that farmers are innovators in their own right, and this gives them more influence within the ARD sector. Furthermore, the funds have enhanced knowledge sharing and uptake of innovations also among non-grantees, as other farmers realize that they can build on what local farmer innovators are doing. This is illustrated by farmers who are raising watermelons and other such crops using transplanting techniques after observing the farmer innovator's work on transplanting from finger millet nurseries. This approach is expected to further promote the scaling up of innovations and of farmer-led experimentation processes by other farmers, formal researchers and extensionists.

6 Conclusion

In acknowledging the LISF as one of the mechanisms that should be adopted in the face of climate change to promote resilience through strengthening and supporting farmer innovativeness, a major challenge that has confronted the approach is acceptance by formal researchers and extension agents. However, their attitudes continue to change gradually over time. It is expected that the approach will eventually be institutionalised and become part and parcel of the curricula in training institutions to equip future research and extension agents. The realization of the capacity of farmers to come up with solutions that can be combined with formal research technologies for enhanced outcomes in the farms ought to be supported to provide local solutions to local challenges and to enhance local capacities to continue to adapt to change.

References

Biggs, S. D. (1989). *Resource poor farmer participation in research: A synthesis of experiences from nine agricultural research systems.* OFCOR Comparative Study Paper No. 3. The Hague: International Service for National Agricultural Research (ISNAR).

Drinkwater, M. (1994). Knowledge, consciousness and prejudice: Adaptive agricultural research in Zambia. In I. Scoones, J. Thompson, & R. Chambers (Eds.), *Beyond farmer first: Rural people's knowledge, agricultural research and extension practice* (pp. 133–139). London: Intermediate Technology Publications.

Frank M. (2012). *Local innovation in the development of smallholder agriculture: a literature review and case study in the Department of Cochabamba, Bolivia.* Witzenhausen: University of Kassel/German Institute for Tropical and Subtropical Agriculture (DITSL).

Government of Kenya. (2010). *Agricultural sector development strategy 2010–2020.* Nairobi.

Government of Kenya. (2013). Kitui County development profile.

Government of Kenya. (2013). Machakos County development profile.

Jaetzold, R., Schmidt, H., Hornetz, B., & Shisanya, C. (2007). *Farm management handbook of Kenya: natural conditions and farm management information,* Vol. *II, Part C: Eastern Kenya* (2nd ed., pp. 37–107). Nairobi: Ministry of Agriculture in corporation with Germany Agency for Technical Cooperation (GTZ).

Kamau, G. M., Kirigua, V., Righa, M., & Ng'ang'a, T. (2012). *Farmer Access to Innovation Resources: experiences with piloting local innovation support funds in Kenya.* Leusden: Prolinnova International Secretariat, ETC Foundation.

Lynam, J. K., Ndiritu, C., & Mbabu, A. N. (2005). *Transforming agricultural research in Africa: Lessons from Kenya.* Michigan State University Press.

Prolinnova. (2012). *Farmer access to innovation resources: Findings & lessons learnt on facilitating local innovation support funds.* Leusden: Prolinnova International Secretariat, ETC Foundation.

Reij, C. P., & Waters-Bayer, A. (2001). *Farmer innovation in Africa: A source of inspiration for agricultural development.* London: Earthscan.

Richards P. (1985). *Indigenous agricultural revolution.* London: Hutchinson/Boulder CO: Westview Press.

Richards, P. (1989). Agriculture as performance. In R. Chambers, A. Pacey, & L. Thrupp (Eds.), *Farmer first: farmer innovation and agricultural research* (pp. 39–43). London: Intermediate Technology Publications.

Richards, P. (1994). Local knowledge formation and validation: the case of rice in Central Sierra Leone. In I. Scoones, J. Thompson, & R. Chambers (Eds.), *Beyond farmer first: rural people's knowledge, agricultural research and extension practice* (pp. 165–170). London: Intermediate Technology Publications.

Riomphe, B., Wongtschowski, M., Lugg, D., & Veldhuizen, L. V. (2012). Module 5: Innovative activity profile 4: Providing farmers with direct access to innovation funds. *Agricultural innovation systems: an investment sourcebook* (Module 7, pp. 117–126), Washington, DC: World Bank.

UN. (2013). *World economic and social survey 2013: Sustainable development challenges.* New York: Department of Economic Affairs, United Nations Secretariat.

Waters-Bayer, A., van Veldhuizen, L., Wettasinha, C., & Wongtschowski, M. (2004). Developing partnerships to promote local innovation. *Journal for Agricultural Education and Extension, 10*(3), 143–150.

Water-Bayer, A., van Veldhuizen, L., Wongtschowski, M., & Killough, S. (2005). Innovation support funds for farmer-led research and development. In *Indigenous Knowledge (IK) Notes* 85. Washington DC: World Bank.

Towards a Methodological Approach to Document and Analyze Local Knowledge of Climate Change: With Evidence from Rift Valley and Blue Nile Basins, Ethiopia

Abate Mekuriaw

1 Introduction

It is widely acknowledged that poor and indigenous communities are among the most vulnerable and the first to confront the impacts of climate change due to their reliance on natural resources (IPCC 2007; Salick and Byg 2007; Green and Raygorodetsky 2010). Nevertheless, the pendulum of policy and scientific discussion on climate change has been mainly swinging among scientists and government delegates at higher levels. Besides, these discussions are largely drenched with controversy, skepticism and inaction (Cobb 2011). Such state of the discourse has two clear implications to the vulnerable people around the developing world. First, since the discourse is mainly apprehended with controversies and embedded economic and political interests, world collective action is more likely to delay for quite some time to come, and hence local communities (such as farming communities) will continue to suffer from the impacts. Secondly, as climate change discourse generally remains very weak at engaging local (indigenous) knowledge in an ongoing scientific discussion and decision making, policies that eventually emanate from the discourse tend to be top-down and thus less likely responsive to local climate problems both in terms of policy ingredients and time frame. To lessen these effects of top-down approach and thus utilize local knowledge as invaluable source to design community or context based adaptation strategies, local knowledge

This article largely borrows from Bizuneh (2013).

A. Mekuriaw (✉)
College of Development Studies, Addis Ababa University, Addis Ababa, Ethiopia
e-mail: abatemek@gmail.com

W. Leal Filho et al. (eds.), *Climate Change Adaptation in Africa*,
Climate Change Management, DOI 10.1007/978-3-319-49520-0_43

has been in the lens of many researchers in the last few years (World Bank 1998; Nyong et al. 2007; Macchi et al. 2008; Green and Raygorodetsky 2010; Bizuneh 2013).

It has been acknowledged that local knowledge plays a key role in solving environmental and development problems (World Bank 1998; Nyong et al. 2007). Despite this, the global discourse of climate change fails to formally engage local knowledge into scientific and policy discussions. Among other things, traditionally perceived inferiority of local knowledge as compared to scientific mode of inquiry has hampered the role of local knowledge in the discourse. As such, climate change discourse is based purely on scientific inquiry. Such approach, according to Jasanoff (2010) and Cobb (2011) detaches knowledge from meaning, given that meanings emerge from embedded experiences and interactions with the social and natural environment. Jasanoff (2010:235) argues that "science is not the only, not even the primary, medium through which people experience climate". Senses supplemented with experience play an important role in understanding climate, and epistemic claims of climate change are most trusted when they engage practices and local values (Jasanoff 2010). Therefore, it would be benefiting for climate change discourse if it uses multiple epistemic sources that are composed of social and environmental interactions (Jasanoff 2010; Cobb 2011). In this regard, local knowledge, as a knowledge body founded on experience and social and environmental interaction, could play a pivotal role to enrich the discourse, complement scientific evidence and fill data paucity. Most importantly, it provides effective tools for managing climate change impacts at local levels and designing policies that are participatory and responsive to local needs by reversing or balancing the top-down approach of decision making.

With this understanding, some studies stress the importance of incorporating local knowledge into scientific inquiry and formal mitigation and adaptation strategies (Nyong et al. 2007; Macchi et al. 2008; Kelman and West 2009; Green and Raygorodetsky 2010; Cobb 2011; Raygorodetsky 2011). But, much work remains to be done as regard to tracking, documenting and systematically integrating local knowledge with scientific knowledge (Gagnon and Berteaux 2009). At the outset, methodology to document or analyze local knowledge is a prerequisite for incorporation with the scientific knowledge. However, methodology of documentation is lacking in the context of climate change. Thus, this study attempts to address this prerequisite by introducing a methodological approach and documents and critically analyzes this knowledge system among subsistence farmers in the highlands[1] of Ethiopia.

[1]Very commonly, Ethiopia's landmass is classified into two as highlands (land area above 1500 m asl) and lowlands. Highlands are home to 90% of the country's population (Kloos and Adugna 1989).

2 Methods

The scientific study of climate change apparently involves four major steps to document relatively comprehensive knowledge on the subject matter. The first step involves detection of climate change. The next step goes with attribution of the changes. The third step entails the study of impacts. The fourth step centers on response mechanisms. This study principally adopted the analogy of these scientific steps with the introduction of different local techniques in documenting and analyzing indigenous ways of understanding climate change.

The study was carried out among subsistence farmers in five Weredas[2] (Yilmana Densa, Fogera, Arsi Negele, Shashemene and Shalla) in the Blue Nile and Rift Valley Lakes Basins of Ethiopia. Farmers were selected through simple random sampling technique. The study employed mixed methods approach where both qualitative and quantitative data were collected concurrently and results from both approaches were compared and contrasted.

2.1 *Data Source and Collection Instruments*

Primary data was collected through questionnaire, interview and focus group discussion. Two hundred fifty heads of farming households participated in filling the semi structured questionnaire which seeks socioeconomic and environmental information. Two focus group discussions in each of the Basins were carried out with farmers to explore historical accounts of the climate and agricultural practices in the respective areas. Elderly and experienced farmers, with a belief that they have deep knowledge on local climate and ecology, were selected purposefully by the help of agricultural development agents and Kebele[3] administrators. Interview was also conducted with farmers and experts. Expert interviews mainly included agricultural development agents (DAs) and in some cases environmental and health experts at regional and/or Wereda levels. An in-depth interview with women and men farmers, which were selected based on agro ecological knowledge, was also conducted. On the other hand, secondary data was collected from Central Statistics Agency (CSA), agricultural and health offices at various levels, and Werede and Kebele administrations. Precipitation and temperature data for the nearest weather stations to the respective residential areas of farmers was obtained from National Meteorological Agency of Ethiopia (NMA).

[2]Administrative unit equivalent to district.

[3]Lowest administrative unit in Wereda.

2.2 Method of Data Analysis

Content and thematic analysis for the qualitative data and descriptive and relevant statistical methods for the quantitative data were used. Recordings from focus group discussions and interviews were transcribed verbatim. From the transcription, texts and descriptions were summarized and categorized under the pertaining climate variable and themes (the four steps indicated above, i.e., detection, attribution, impact and response). The interpretation was performed based on the categories and themes. In the analysis, direct quotations and instances were also used to supplement the interpretation. Quantitative data, with the aim of triangulating the subjective assessment of farmers, was analyzed descriptively and graphically along with simple regression method.

3 Results and Discussion

3.1 Farmers' Detection of Climate Change

Detection of climate change through the lens of local knowledge was approached through three methods. The first method is detection based on farmers' perception. The second is detection through farmers' characterization of the changes and signals of the climate over several decades. The third approach draws from concrete evidences from farmers' daily life. Each of these is discussed as follows.

3.1.1 Local Detection Through Perception

Farmers were asked about the trend of precipitation and temperature in the past two decades. Twenty years was taken as a reference for two reasons. The first reason pertains to meteorological recording which is more limited but relatively good in the last two decades at the weather stations. The second and the most important reason goes with the memories of the farmers. Since farmers were asked about perception that involves long years of insights, assisting their memories through some unique events could be helpful. In this case, the year 1991, a year the change of government took place in the country, was taken as a reference from when farmers assess the trend of precipitation and temperature.

Accordingly, 96.4% of the households noticed changes in the trend of precipitation. Out of which, 82.8% perceived a decreasing trend on the amount of precipitation, and only 13.6% noted the contrary (see Fig. 1). As regard to temperature, the great majority (95.6%) of the households perceived a rise in temperature and only 3.6% indicated the reverse (Fig. 2).

Fig. 1 Perception of farmers about the trend of precipitation in the last twenty years

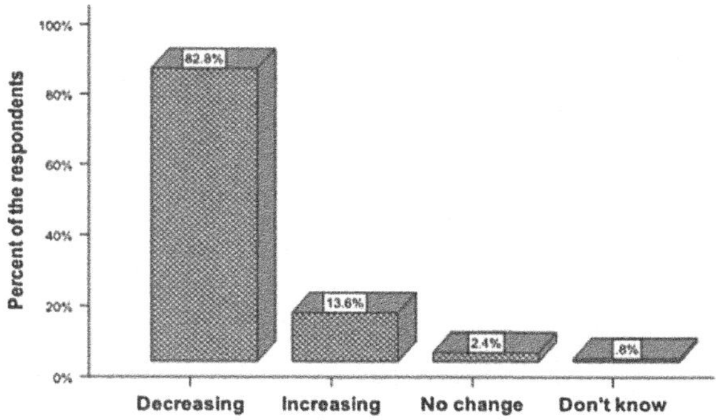

Fig. 2 Perception of farmers about the trend of temperature in the last twenty years

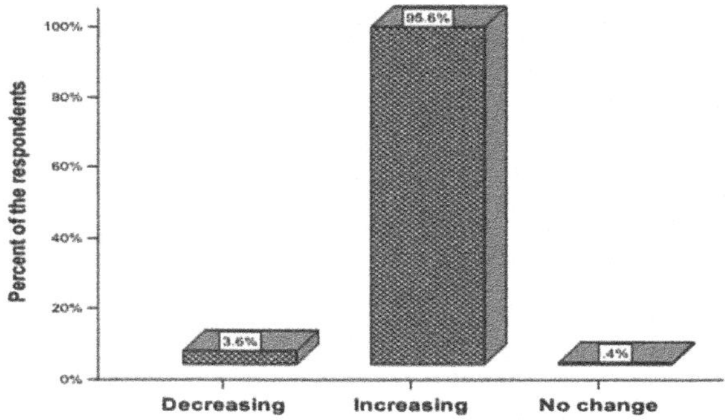

This subjective assessment of farmers was checked against meteorological recordings over the period of 1990–2009[4] at nearby weather stations (Figs. 3, 4, 5, 6 and 7). Accordingly, precipitation in all of the five weather stations is declining. Time (in years) explains 5.8–35.74% of the variation. The decline varies from 8.88 to 28.62 mm per year, the lowest decline being in Adet and the highest in Arsi Negele stations. The decline is statistically significant at 10% and 1% levels at Alaba Kulito and Shashemene stations, respectively. Assuming the negative sign in the trend of precipitation in all of the five weather stations, farmers' perception appears to be in accordance with meteorological recordings. Particularly, the statistical significance of the trend in Alaba Kulito and Shashemene weather stations strengthens the subjective assessment of farmers.

Maximum and minimum temperature in all of the stations visibly increased except minimum temperature at Arsi Negele station. The rising trend of maximum temperature is statistically significant in four of the stations (Adet, Alaba Kulito and Arsi Negele stations at 1% level and at 5% in Wereta station). As regard to

[4]Initially data from 1991 to 2010 was thought. But data for the year 2010 was not ready at the NMA at the time of data collection. Thus, meteorological record from 1990 to 2009 was considered. In the case of Arsi Negele, 12 years (from 1998 to 2009) were considered due to unavailability of data from July 1992 to December 1997.

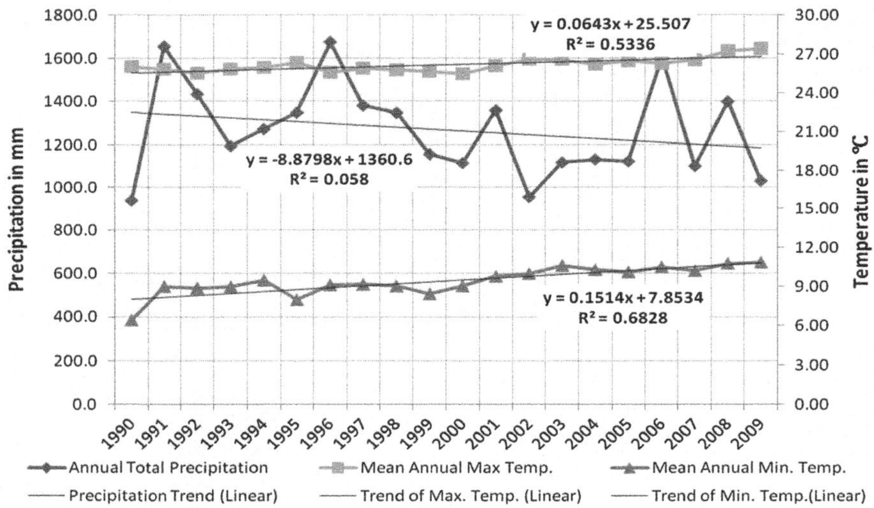

Fig. 3 Trend of precipitation and temperature in Adet (all the figures of precipitation and temperature at the weather stations are constructed based on raw data of National Meteorological Agency of Ethiopia)

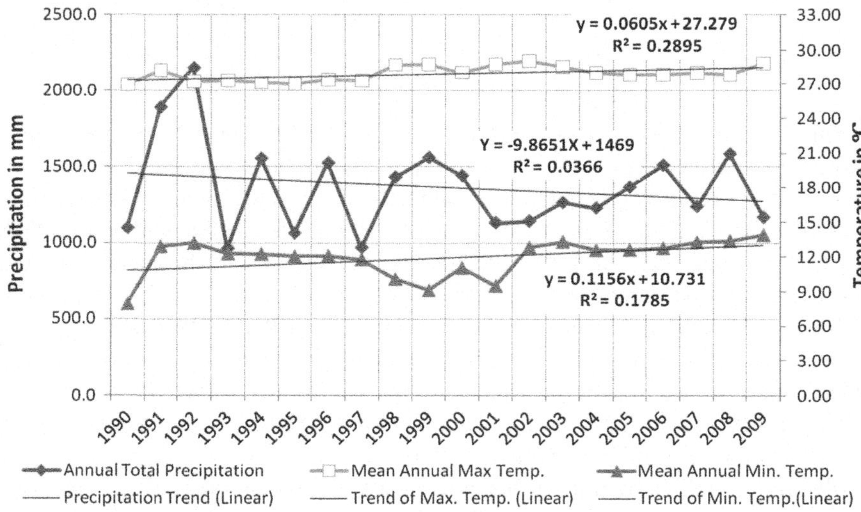

Fig. 4 Trend of precipitation and Temperature in Wereta

minimum temperature, the rising trend is significant in four of the stations, namely Adet, Wereta, Alaba Kulito and Shashemene. It is evident that the rising trends of both maximum and minimum temperature are significant at Adet, Wereta and Alaba Kulito stations. At Shashemene station, although maximum temperature is statistically insignificant, both minimum and maximum temperature are indeed rising.

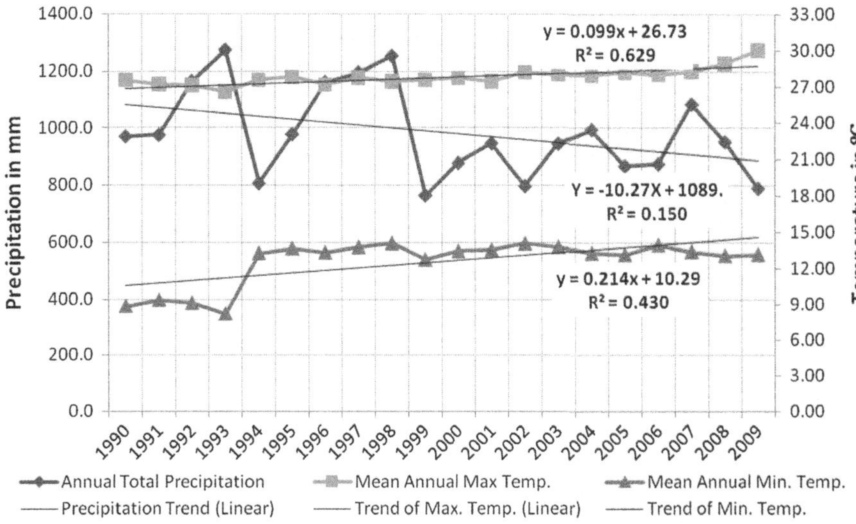

Fig. 5 Trend of Precipitation and temperature in Alaba Kulito

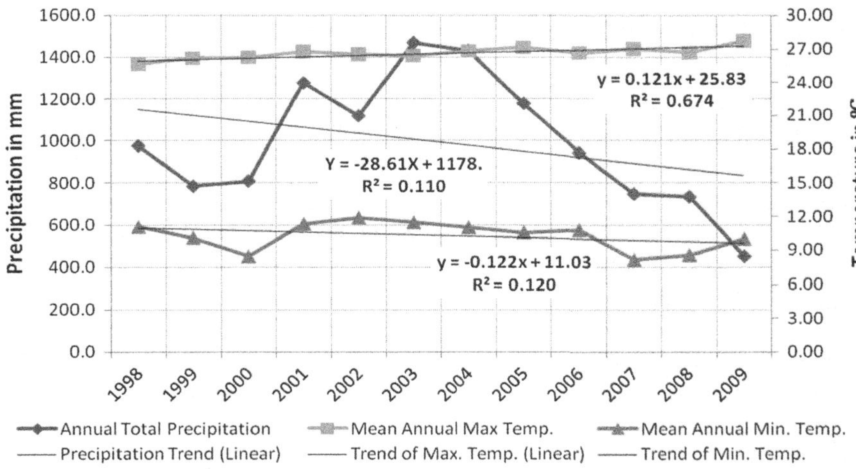

Fig. 6 Trend of precipitation and temperature in Arsi Negele

The exception is only Arsi Negele, where opposing trend is observed. At this station, minimum temperature has been declining insignificantly whereas maximum temperature has been rising significantly. Such significant rise of maximum temperature presumably outweighs the insignificant trend of the minimum temperature. Accordingly, the perception of farmers coincides with temperature recordings of the weather stations.

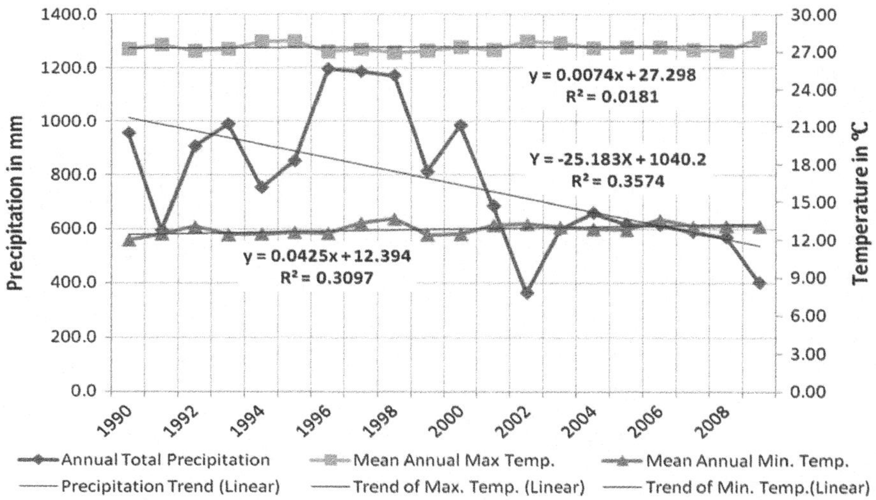

Fig. 7 Trend of precipitation and temperature in Shashemene

3.2 *Local Detection Based on Observation and Characterization of the Long Term Climate*

Household heads' perception about the trends of precipitation and temperature is one aspect to detect and document changes in the local climate. However, simple trend analysis hideouts the variability of the climate. This subsection, thus, builds on local detection of climate change based on characterization of long-term changes as learnt from interviews and focus group discussions. At the outset, it has to be noted that farmers did not seek to differentiate between climate variability and change, i.e., they appear to have no interest to differentiate or have no awareness on the distinctions. As a result, they interchangeably use '*ye ayir lewit* or *ye ayer mezabat*', and '*ye ayer huneta melewawet* or *mekeyayer*'[5] in expressing changes and fluctuations in the climate in general, which in the academic world the first more referring to climate change and the latter to climate variability. Nevertheless, their expression of the changes on specific ingredients of the climate, i.e., precipitation and temperature, is clear and explicit.

3.2.1 Rainfall

As a major climate variable that has an upper hand in their livelihoods, farmers described rainfall performance in various forms of manifestations comparing it with the old times. Old time here refers to at least 3 to 4 decades back, and the farmers

[5]In Amharic language.

also denote it as their childhood time and/or their fathers' time. Farmers typically portrayed the features of rainfall in indicator words such as onset, cessation, duration, drought, flood, intensity, distribution, cloud, timing and amount. For analytic purpose, such expressions were condensed into four categories of precipitation as presented below.

(a) ***The timing of rainfall (onset and cessation of rainfall)***

According to farmers, fluctuation on the timing of rainfall from time to time is increasing. Onset and cessation timings were relatively stable in older times and are tending to be more unpredictable these days, in some years being early and in some other years being late. They further explained that seasons that receive rainfall in the right time of onset and cessation are dwindling these days as compared to the old times. They were more inclined to pronounce an unusual lateness on the start and early withdrawal in a season.

(b) ***The duration and pattern of rainfall***

Characteristics such as interruption, lack of continuity and shortness of rainfall at a time and season are the main reference terms that farmers used to describe the duration and pattern of rainfall. They said that as compared to the old times, present day rains quite often lack continuity and tend to interrupt at any point in a season. In older times, the rain was generous with rare interruptions in a season and stayed as long as 6 months, and sometimes even more than that. This has now diminished significantly up to 4 months and sometimes even less than that. The farmers further stressed that nowadays the rain is intermittent and tends to downpour in a short period of time which is usually followed by interruptions for days, or weeks to the worst. After disappearing for days or weeks, it then pours down again.

(c) ***The amount of rainfall***

The amount of rainfall in terms of intensity is another major characteristic of rainfall that was critically viewed by farmers. Generally, farmers have developed the sense of viewing the rainfall being short in duration and more intense than they had known it in their childhood. There is also clear consensus among farmers about the overall diminishing tendency of rainfall amount over time. Such perception replicates the views of household heads discussed above.

(d) ***Occurrence of extreme events: drought and flood***

Farmers maintained that they are aware of flooding in the old times. But, they marked out the flooding of recent years as more frequent and strong. In fact, this might not be surprising given the increasing trend of flood in Ethiopia in the recent decades. Formerly, the country was well known of drought, but not as such pronounced with widespread flooding. For instance, floods which destroyed property and claimed lives in 1997 in several parts of the country were the worst in 40 preceding years (Reuters, 12/5/1997 in Wolde-Georgis 2000). Other recurrent floods had then followed in the North, South and Eastern parts of the country to the level that could be characterized as a national catastrophe (Ayalew 2007).

As regard to drought, there is also consensus among farmers that its frequency has been seemingly growing. But their expression is not strictly a meteorological drought but of agricultural drought, where intermittent droughts sustain in between periods of rainfall events. In the words of farmers, such occurrences of drought are getting worse from time to time and are seriously impeding farming.

3.2.2 Temperature

Farmers described the changes in temperature consistently in terms of increment of air temperature, warmness of days and strength of the sun. In these measurements, the air tends to be drier, days tend to be very hot and the sun feels quite harsh as compared to their childhood time. In these expressions, there is similar expression among farmers in all the study areas.

In a nutshell, it has been clearly indicated by farmers that the climate has changed in ways discussed above. Experts' (agricultural development agents) views were found to strengthen farmers' accounts. Alike farmers, experts clearly believed in the rise of temperature. With respect to rainfall, they characterized it being more variable from year to year, tending to be less predictable, heavy, inconsistent in its onset and cessation, sometimes inadequate for crop production, and sometimes coupled with interruptions at any point in the season (Fig. 9).

3.3 Detection Through Farmers' Concrete Experiences

So far, farmers' detection knowledge was analyzed based on perception and long-time observation. Both perceived and observed events and associated interpretation might be influenced by culture and recent climate events. To minimize these drawbacks, one further step but simple technique was introduced to scrutinize farmers' detection knowledge. This step involved examination of concert experiences of farmers.

One simple technique to see the changes of temperature is to examine the longevity of fermentation and preservation trends of local staple foods and beverages in older times and the present. This technique is vivid as it is very easy to assess, involves day-to-day activities and is meaningful to farmers. Since women are culturally tied with food and beverage preparation, they were particularly interviewed tacitly without making them aware of climate change as the study's intent. Accordingly, they reported marked changes currently in the length of food preservation and fermentation practices as compared to 2–3 decades ago. Table 1 below shows the reported changes in the respective regions of the study.

As seen in the tables, there is a clear change in fermentation and preservation longevity of local foods and beverages. Such changes could be due to technological

Table 1 Longevity of food and beverages' fermentation and preservation

a. Blue Nile Basin Villages

Food/ Beverage Item Longevity(fermentation and staying fresh)	Unit of Measurement	Time Frame	1	2	3	4	5	6	7	8	9	10	11	12
Tella (local beer) malt fermentation	Day	Old Time						*10 to 12*						
		Currently						*3 to 5*						
Tella drink (staying fresh)	Day	Old Time						*4 to 5*						
		Currently						*2 to 3*						
Tella malt (staying fresh)	Month	Old Time							*Up to 6*					
		Currently							*Up to 2*					
Injera (flat bread) dough fermentation	Day	Old Time								*A maximum of 7*				
		Currently								*A maximum of 4*				
Baked injera (staying fresh)	Day	Old Time								*A maximum of 7*				
		Currently								*A maximum of 3*				

b. Rift Valley Lakes Basin Villages

Food/Beverage Item Longevity(fermentation and staying fresh)	Unit of Measurement	Time Frame	1	2	3	4	5	6	7	8	9	10	11	12
Keneto (local beer) malt fermentation	Day	Old Time								*3 to 7*				
		Currently								*1 to 3*				
Keneto drink (staying fresh)	Day	Old Time									*6 to 8*			
		Currently									*2 to 4*			
Injera dough fermentation	Day	Old Time				*2 to 3*								
		Currently				*Within 1 day but also up to 2 days*								
Baked injera (staying fresh)	Day	Old Time					*Up to 4*							
		Currently					*Up to 2days*							
Baked bread (staying fresh)	Day	Old Time											*Up to 10*	
		Currently								*Up to 7*				

and/or climate change. One of the advantages of interviewing women without being aware of the researcher's intention lays in exploring potentially underlying causes (climatic or non-climatic) behind the changes. To identify this, the women were further asked to point out the reason (s) behind the changes. Their answers were unswervingly '*muketu new*', which means 'it is warmness of the air temperature'.

In a nutshell, farmers' perception, observation and concrete experience shows that the climate has changed in the ways discussed above. If the changes are detected in such manner through the lens of local knowledge, the next question should then be to explore to what factors does this knowledge attribute the changes observed in the climate?

3.4 Local Attribution of Observed Changes in the Local Climate

Understanding the cause of climate change is essential in designing any form of response mechanisms to offset the impacts. Similarly, farmers take the root causes into consideration while attempting to implement adaptation strategies. For instance, if they believe that the cause is a supernatural force, this perception by itself might limit their counteractions, and thus they might resort to cultural and ritual practices that might have nothing to do with concrete adaptation actions. Hence, bringing forth the major reasons that farmers associate with the perceived and described changes in the climate is very important to deeply understand the response mechanisms of these communities. Accordingly, farmers identified the following as major causes behind the changes.

(a) ***Population increment***

The farmers clearly stated that due to burgeoning population, communal grazing sites, tree and grass lands, swampy plains, moisture places and marginal lands by now are either settled or used for agriculture or degraded due to over grazing and over utilization. These areas, on the accounts of farmers, were vital in maintaining the balance of air temperature. A historical account by an elderly farmer, aged 76, clearly connects how population pressure acts as a source in aggravating the changes in the local climate.

> In the old time, there were forests within the distance of 1 to 2 km from our village. The air was fresh. There was also good rain at that time. Now, those trees and the forest are already cleared. We, ourselves, cleared the forest foremost because we became many.

Population growth and expansion of agricultural land in Ethiopia clearly supports the account of farmers. The 2007 Census of the country put the national population growth at an average annual rate of 2.6% between 1994 and 2007 (CSA 2008). This growth rate marks Ethiopia being among the nations with bourgeoning population in the world. Given the great majority of the population living in rural areas (83.9%) and its growth proportion (2.26%), demand for new land openings to meet food requirements would be immense. Figure 8 shows how cultivated land area in the country has been increasing in the last 17 years among private peasant holdings in rural sedentary areas.

Land coverage of grains in 2010/11 has increased by 69.86% from that of 1996/97. In these years, land coverage of grain has expanded averagely 355, 600 hectares per year. No doubt that the expansion of cultivated land in rural areas

Fig. 8 Land coverage by crops

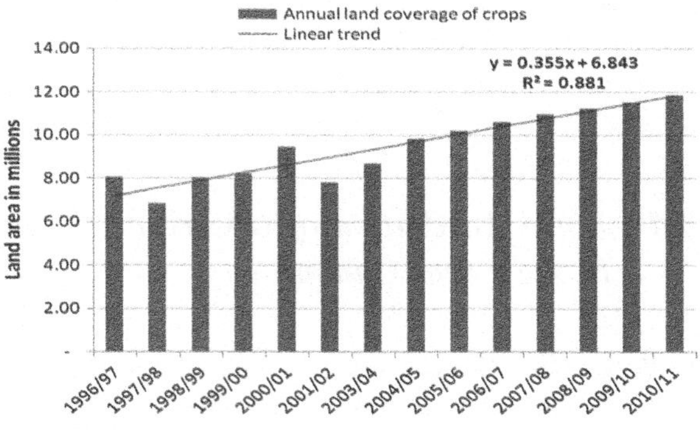

comes from clearing and reclaiming tree and grass lands, grazing sites, moisture places and marginal lands. This is what farmers interviewed in this study attempted to connect growing population with the pressure it creates on land and its resources, which in turn plays an important role in affecting the local climate.

(b) ***Deforestation***

In older times, as reported by farmers, hills, stream sides, river banks and several fields were covered by trees, and there were considerable forest and bush lands. Due to increased demand for more land and firewood, trees and forests (which are instrumental in attracting moisture and stabilize the temperature, as described by farmers themselves) had vanished. In connection to this, a farmer aged 43 said the following:

> I am now 43. I no more see those trees which I used to see in my childhood. Even I do not find the types of trees I used to see just as near as 15 years ago. Those trees are not in place now. They are not found even in the gorges and the fields. We cut them. It is due to such act of ours that the climate has been changing, and the temperature has been rising.

Lack of protection to the environment (inability to plant trees as replacements for the cuttings and taking care of the existing ones) was also mentioned as a contributing factor along with deforestation. In a focus group discussion, a farmer plainly stated: "*I could say that it is we who are responsible for the changes that occurred to the climate. It is we that contributed for the rise of the temperature*".

(c) ***God's act***

Although farmers blame their actions as a root cause for changes in the climate, still they maintain God's will as taking the upper hand. This articulation has been observed in all of the focus group discussions with the exception of one focus group discussion, where farmers' own actions were predominantly blamed. Interestingly in that focus group discussion, the participants were found to be relatively literate ones (at least read and write) and composed of the younger generation. In another aspect, there was also a tendency among farmers to attribute extreme cases such as drought, hailstones and flood to God, and other gradual manifestations (a good example is hotness of days, lack of moisture in the air) largely to human action.

Further question was posed to farmers to understand as to how they associate the changes with God. Their views could be classified into two classes of thought and could be termed in this study as: the *notion of punishment* and *lack of stewardship*. As per their explanation, when people fail to obey God's words and commands, He punishes them through drought, flood, thunderstorms, hail, strong winds and plague of worms. The *notion of lack of stewardship*, on the other hand, links the changes with people's failure to protect the nature that God has given to them. There is a sentiment among farmers that these days people are becoming more egocentric and demanding, and living is also becoming more expensive. Consequently, everybody tries to make the most out of nature to fulfill his/her demands while paying little or no due regard to nature. Such breach of human stewardship of nature has led them to the changes that they have faced now. The notion of stewardship, in fact,

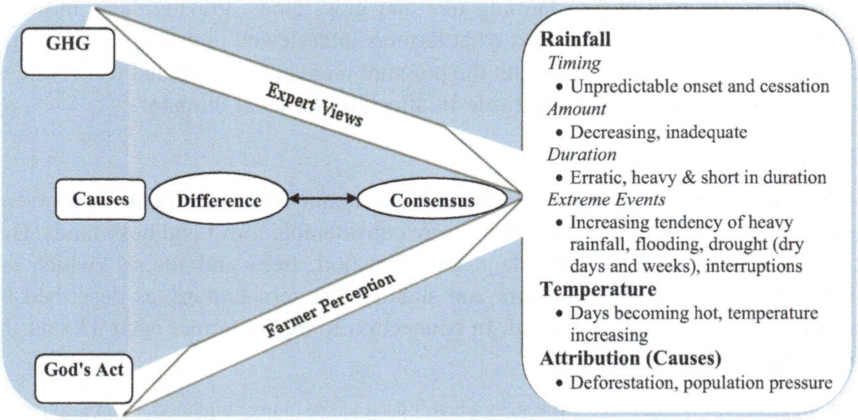

Fig. 9 Farmers' and experts' description and attribution of changes in the local climate

is a cultural explanation of human induced causes described above, such as deforestation.

Generally, from the discussion so far, it is evident that farmers relate the changes in the climate with human activities (deforestation and population growth) and cultural belief (God). While emphasizing on the Devine power, the cultural belief also relates the changes to human activity (in the form of lack of stewardship), which still shows the importance of anthropogenic causes. But, does this attribution by farmers relate with that of experts?

As depicted in Fig. 9 above, both farmers and experts share similar views in characterizing rainfall and temperature. However, in attributing the changes, they entertain both similarities and differences. They have similar ideas in the sense that they both believe in anthropogenic causes. But the scope of anthropogenic cause is limited to local actions when it comes to farmers and goes beyond that for experts. While maintaining the role of local human actions, experts prominently link it with greenhouse gas emissions. Major difference among farmers and experts, however, is marked in the cultural aspect of the cause. For farmers, God has place in the materialization of the change, the point where experts depart from farmers. Such difference is not surprising given experts' education level and proximity to the climate change discourse and its information outlet, and conversely the role of culture among farmers and their detachment from the discourse.

In general, farmers do come in conformity with the scientific discourse that asserts the changes as largely of anthropogenic. In fact, farmers do carry also the belief of God's will. Still in this line of belief, they emphasize lack of stewardship as a cause, which strengthens the human role in the observed changes.

3.5 *Local Account of the Impacts*

Farmers are increasingly concerned with impacts that are related to availability of water, agricultural production and health. In addition to these major problems, they also mention additional problems that seem to threaten their psychosocial establishments. Here under is a brief presentation of these problems.

3.5.1 Impact on Availability of Water

According to farmers, streams and rivers that used to flow the whole year are gradually receding and some villages are living under water stress and residents travel several hours (as much as 2–3 h) to get drinking water. For example, a river called Shina in Adet Zuria Kebele and Divdiv Stream in Mosobe Kebele (Yilma ena Densa Wereda) are now said to recede nearly to dryness after January until the next rain season. But formerly both of these water bodies were said to flow the whole year. Because of this, water for domestic usage and irrigation has been getting scarce from time to time. With respect to this, a farmer in a focus group discussion gave the following account:

> For instance, until 1995 I was using irrigation from a small stream in my village. But today, let alone irrigation, drinking water has become scarce. Formerly, there was water in the gorges but no more these days. At present, we struggle to protect the remaining drinking water bodies by keeping a fence around them. However, we feel that our protection now is like "a dog barking after the hyena has gotten away".[6]

However, it is hard solely to link these accounts with climate change given an increasing practice of irrigation throughout the country. As part of the government policy to enhance food self sufficiency, farmers are encouraged to use rivers and streams for small-scale irrigation schemes. The farmers involved in this study also affirmed that small scale irrigation schemes are increasing alongside rivers and streams. Some rivers and streams are said to be used excessively by upstream farmers, and others being utilized beyond the carrying capacity so that valleys and gorges at downstream end up dewatered.

3.5.2 Impact on Health

One of the new phenomena that emerged in recent decades in one of the study areas is Malaria. Malaria, largely influenced by temperature, rainfall and humidity, is well known disease in Ethiopia which is predominantly associated in areas less than 2000 m asl (MOH 2008). But, in recent years, the impermeability of the highlands, which were considered to be free of the disease due to their altitude, seems to be breaking.

---

[6]An Ethiopian proverb.

An important example of farmers' account in Blue Nile Basin (Adet Zuria and Mosebe Kebele's of Yilmana Densa Wereda) represents the incursion of malaria into the highlands. Until the 1980s, malaria was not known to these areas, being highlands with an altitude over 2000 m above sea level and someone had the chance of contracting the disease only when he/she had visited lowlands or malaria prone areas. Here is an experience of a farmer in Adit Zuria Kebele, aged 46.

> In 1978, I was seriously sick. There was no clinic in this area at that time. So, no one was having the knowledge of my sickness, as no one knew clearly the symptoms. While I was seriously sick, luckily a relative from a far area came to visit us, and the first thing he asked after looking to me was whether I had been somewhere else. My family told him that I was in Bahir Dar[7] sometime ago. Then he told them that the disease was nidad,[8] and it was after this that I went for a treatment.

In these Kebeles, at the mid and end of 1990s, several people were reported to have died by malaria outbreak. Farmers, for instance, witnessed that they had counted 141 cases of deaths in 1999 in Adet Zuria Kebele alone. They further stressed that, at that time, they had buried up to nine people per day. They recall that period as a malaria epidemic ever to hit their place. Although these days there is an aggressive malaria prevention program by the government, NGOs and communities at large, and consequently rare deaths cases, malaria has got a prominent place as one of the top diseases in the area. For instance, the available 3 years disease recordings at Yilmana Densa Wereda Health Bureau shows malaria ranking first among the list of 20 top diseases in the years 1999, 2000 and 2002. Similarly, the household data collected through questionnaire for this study showed that averagely 4 in an average household size of 6 in Yilmana Densa have a history of malaria infection with only 4 percent of the households reporting no history of infection among the household members. Surprisingly, nearly two decades ago this area was believed to have been malaria free, and now about 95% of the households have at least one family member with a history of malaria infection. Given the relationship between temperature and malaria, the rise of temperature certainly takes the blame for the incursion of the disease in the area. In the other four Weredas (Fogera, Shashemene, Arsi Negele and Shala), although malaria is not a new disease, farmers pinpointed seasonal climate having an upper hand influence over the spread of the disease.

3.5.3 Impact on Crop Production/Farming

In stating the impacts, farmers' grave concern is not quite the gradual decrement of rainfall, but its irregularity and distribution. Irregularity in the timing (late onset and early cessation) and distribution over time (mainly interruptions at the middle) are

[7]Capital of Amhara National Regional State, and is a malarious area.

[8]Name referring to malaria in older times, when malaria was regarded as a lowland disease. Presently, that name is replaced by nationwide name 'weba'.

said to have significant and sometimes unbearable effects on crop production. Besides impacting the whole farming communities, these climate features pose extra burden on specific sections of farming communities, mainly female-headed households who lack male labor for traction of land, sowing and harvesting. Due to cultural and traditional norms in Ethiopia, women do not plough with oxen. Consequently, they are forced to rent out their land in the form of sharecropping to male farmers. When the timing and distribution of the rain tends to be inconsistent, these households are said to be badly hit by the effects. At the time of irregularities of rainfall (mainly on onset and cessation), the tenants prioritize their own land in farm activities as they will be forced to compete with the irregularity of the rainfall. It is later that they come to attend the rented land. In such cases, the rented land might lack timely cultivation and farm management decisions. Because of this, production in such lands are said to suffer the most.

3.5.4 Psychosocial Impact

Psychosocial impact here refers to feelings of confusion and tearing of social and cultural establishments. Due to unpredictability and fluctuations in the onset of rainfall, farmers especially get confused on the timing of planting, i.e., they are more uncertain of when to sow crops due to increasing unpredictability of the onset of the rain.

The other impact is related to tearing of the social fabrics of helping each other. There has been a long tradition among communities to help the disadvantaged groups (widows, disabled and old people without a working family member) by giving priority in land preparation, ploughing, sowing and reaping in each farming season. In the current era, as farmers are in predicament with the behavior of rainfall, the disadvantaged groups are less likely to enjoy the statuesque of the social assistance. Even if they get, the assistance comes late after farmers have attended their own farms. This has a negative effect on crop productivity of these groups especially if they fail to get timely assistance in the critical periods of crop planting. Here is an account put by an elderly man:

> In our tradition unless we first prepare and attend the land of widows, disabled and the old, we fear God that He will halt the rain at all. These days, there is less regard to this custom as farmers are forced to compete with the behavior of the rain.

These groups of the community are socially disadvantaged who have limited or no access to productive labor, among other things, which in turn jeopardizes food security of the group. Hence, they are forced to fall back on the assistance from the community or in broader terms, on social capital. Social capital (in the form of social trust, norms and networks) is quite extensively regarded as an important instrument towards facilitating collective action in responding to disaster and development challenges (Dynes 2002; Pelling 2011). Particularly, social capital in the form of norms is considered as a powerful means to subjugate self-interest to the needs of the community (Dynes 2002). Similarly, in the climate change literature, social capital is considered as a social apparatus to mitigate impacts and facilitate individual or collective adaptation to climate change and extreme events (Adger 2001, 2003; Pelling 2011). Nevertheless, social capital itself could be broken when it faces immense pressure from climate risks. Farmers' experience in this study typically attests this situation when survival is at a grave risk, social capital might not stand to its expectations.

3.5.5 Impacts on Infestation of Insects and Weed

Farmers also described the impacts through infestation of insects and weeds. As compared to the old days, insects invading crops have increased following the rise of temperature which created a favorable condition for breeding and infestation. Farmers said that without the use of chemicals, it is almost impossible to store grain at present. In regard to weeds, there is a general tendency among farmers to attest the spread as compared to the olds times. Weeds that were typical to lowlands are already observed in the highlands, particularly, such behavior was pronounced in Yilmana Densa Wereda.

3.6 Response to the Observed Changes in the Local Climate

Detection, attribution and impacts of climate change are documented exclusively based on local knowledge. What is now remaining is assessing adaptive responses of farmers to the perceived changes. Understanding adaptive responses of farmers to climate change represents serious challenges for researchers (Gbetibouo 2009). Among the challenges, isolating climate induced response from non-climate motives that farmers face in actual world is a major one (Maddison 2006; Nhemachena and Hassan 2007; Gbetibouo 2009). Despite this, it became a trend for researchers to document adaptation by asking farmers what adaption strategies they had adopted as a response to climate change (Maddison 2006; Nhemachena and Hassan 2007; Deressa et al. 2008; Gbetibouo 2009). Normally, such approach might lead farmers to respond in a certain way and hence influence the authenticity of farmers' responses in several ways. First, it might lead farmers to answer in such a way that they believe what the interviewer wants to hear about. Second, they might assume themselves as if they were supposed to undertake adaptation strategies and might respond positively as if they were undertaking while not in actual cases. Third, they might also tag adaptation strategies that they had adopted as a response to non-climate motives as if they were performing as a response to climate change. In this case, non-climate responses might be regarded as climate driven and thus might be wrongly included into analysis.

To minimize the possibility of such bias, if not avoid completely, an attempt is made in this study to address this concern by following a technique that extracts adaptive strategies systematically from farmers. By consulting the literature and local experts, a list of adaptation strategies believed to be practiced in the localities as a response to climate variability and change were entered into the questionnaire. First, respondents were cross-checked against each of the strategies as to whether they were performing (or not) while without mentioning about climate variability and change. Later, they were asked as to whether they have performed adaptive strategy (strategies) as a response to perceived changes in the local climate. It was assumed that while a farmer is performing an adaptation strategy and if he/she says that he/she is not adapting to climate variability and change, this might be an indication that the farmers' implementation of the strategy is most probably for non-climatic reason(s).

In general, the majority of households (65.2%) have employed adaptive strategies in contrast to 34.4% who have done nothing. Among the strategies employed by the households, changing of planting dates and using different crop varieties are found to be the most used ones followed by increased use of fertilizer and planting trees. On the other end of the spectrum, increased use of irrigation, diversification into non-farming activities and water harvesting are found to be the least used strategies. But, it has to be noted that most households employed multiple strategies rather than limiting with one specific strategy. This may show the probability that no one adaptive strategy is adequate to withstand climate predicaments that farmers are facing (Table 2).

However based on the outlined technique, nearly 11% of the households (27 in number) said that they did not perform any adaptive strategy as a response to climate variability and change while they have actually carried out one or some of the adaptive strategies listed in the above table. This result may lead us to raise some critical issues, and even in some cases to new openings for further research. First, it shows the likelihood of implicit learning of adaptation where farmers might not even notice that they are adapting to climate variability and change while actually implementing some of the adaptation strategies. Second, farmers who are seemingly carrying out adaptive strategies but who claim they are not responding to climate stimuli might be carrying out adaptation strategies for other non-climate

Table 2 Adaptive strategies carried out by the households

No.	Adaptation strategy	No.	Percentage
1	Using different crop varieties	111	65.2
2	Changing planting dates	128	
3	Adopting drought resistant crops	52	
4	Increased use of soil and water conservation techniques	55	
5	Diversification into non-farming activities	19	
6	Water harvesting	11	
7	Planting trees	75	
8	Increased use of irrigation	29	
9	Increased use of fertilizer	77	
10	Done nothing	86	34.4

motives such as market, policy, productivity or simple imitation from neighbors. Nevertheless, the adaptive strategies that these farmers carry out (for whatever reason) will continue to act against the effects of climate variability and change. In this case, how do these farmers adaptive behavior be considered in the face of climate change, i.e. should we consider them as adapters or non-adapters? Third, since there is lack of adaptation strategies meant specifically to respond to climate change, singling out climate driven adaptive behavior of subsistence farmers is very difficult, and thus it warrants clearly acknowledging non-climate driven adaptation along with that of climate driven in studying adaptive behavior of farmers. Addressing these issues is very important in any attempt to understand adaptive behavior specifically targeting climate change. As a matter of fact, addressing these issues certainly requires further research and follow-up studies.

4 Conclusion and Outlook

In this article, a simple but sensible methodology was introduced and an attempt was made to document local knowledge of climate change. While this methodology, as a starting exercise, shades light in the documentation and analysis of local knowledge in other regions, it has to be noted that modification and contextualization to fit specific localities is necessary.

As regard to the knowledge system, local knowledge has demonstrated pronounced changes in the local climate. Through cross examination (wherever possible), this knowledge system is found to be in conformity with objective data. It offered worthwhile information on detection, attribution and impacts of climate change in the localities. It provided several instances of the changes observed and detected in simple ways what science strives to detect through various rigorous processes and sophisticated technologies. It supported its claims through concrete examples of detection mechanisms. It also offered an important lesson as regard to attribution, one of the major controversial topics in the science of climate change. In fact, it attributes the changes both to anthropogenic sources and God's will. The former cause brings it in line with what is said largely in the climate change discourse. The later cause takes it back to religious (cultural) beliefs. Even though local knowledge believes in God's will and regards the changes as signs of punishment for disobeying Him, it interestingly puts weight on man's infringement of nature's stewardship given to him by God, the creator. This notion of lack of stewardship (as termed in this study) although appears to come literally under God's will, it is a cultural form of attributing the causes to human activities. Therefore, local knowledge directly or indirectly supports the idea of anthropogenic causes.

When it comes to the impacts, this knowledge system demonstrated the impacts observed on the availability of water, health, farming, and infestation of insects and weeds. In addition to these impacts, it has also revealed some psychosocial impacts that are rarely reported in the literature. The psychosocial impacts basically refer to confusion on timing of planting and tearing the social fabrics of helping each other. Besides these psychosocial findings, the other general impacts revealed by the knowledge system are found to confirm assertions of the science of climate change but with unique details and contexts.

Given the insights revealed in this study, both science and policy dialogue need to consult and exploit local knowledge. In this case, even if local knowledge is limited in scope and space (the respective localities), several cases from localities around the world could serve as living laboratories to study the climate and complement the science through experiential knowledge and tangible evidences. Finally, care has to be taken in scrutinizing adaptive behavior. Since singling out adaptation to climate change among subsistence farmers is a difficult task, climate driven adaptive behavior should be understood with a mix of non-climate drivers of adaptation.

Acknowledgements The author wishes to express his heartfelt gratitude to the Catholic Academic Exchange Service (KAAD), and Institute of Development Research and Development Policy (IEE) and RUB Research School of Ruhr University Bochum, Germany, for directly or indirectly providing financial support to carry-out this research as part of his PhD study.

References

Adger, W. N. (2001). Social capital and climate change. *Tyndall Centre Working Paper* No. 8, Tyndall Centre for Climate Change Research and CSERGE, University of East Anglia, Norwich.

Adger, W. N. (2003). Social capital, collective action, and adaptation to climate change. *Economic Geography, 79*(4), 387–404.

Ayalew, S. M. (2007). *Flood forecasting and early warning system (FFEWS) an alternative technology for flood management system and damage reduction in Ethiopia: A concept note.* LARS: Catchment and Lake Research. 2007.

Bizuneh, A. M. (2013). *Climate Variability and Change in the Rift Valley and Blue Nile Basin, Ethiopia: Local Knowledge, Impacts and Adaptation.* UAMR Studies on Development and Global Governance, Vol. 62, Logos Verlag Berlin, ISBN 978-3-8325-3524-7.

Cobb, A. (2011). *Incorporating indigenous knowledge systems into climate change discourse. Colorado Conference on Earth System Governance: Crossing Boundaries and Building Bridges.* Colorado State University. http://cc2011.earthsystemgovernance.org/pdf/2011Colora_0130.pdf.

CSA (2008). *Summary and statistical report of the 2007 population and housing census results: Population size by age and sex.* Addis Ababa: FDRE Population Census Commission.

Deressa, T., Hassan, R. M., Alemu, T., Yesuf, M., & Ringler, C. (2008). Analyzing the determinants of farmers' choice of adaptation methods and perceptions of climate change in the Nile Basin of Ethiopia. *IFPRI Discussion Paper 00798.*

Dynes, R. R. (2002). The importance of social capital in disaster response. In *Preliminary paper, No. 327*, Disaster Research Center, University of Delaware, Newark.

Gagnon, C. A., & Berteaux, D. (2009). Integrating traditional ecological knowledge and ecological science: a question of scale. *Ecology and Society, 14*(2): 19. http://www.ecologyandsociety.org/vol14/iss2/art19/

Gbetibou, G. A. (2009). Understanding farmers' perception and adaptation to climate change and variability. The Case of Limpopo Basin, South Africa. Environment and Production Technology Division, *IFPRI Discussion Paper 00849.*

Green, D., & Raygorodetsky, G. (2010). Indigenous knowledge of a changing climate. *Climatic Change, 100,* 239–242.

IPCC (2007). *Climate change 2007: Impacts, adaptation and vulnerability. Contribution of Working Group II to the Fourth Assessment Report of the Intergovernmental Panel on Climate Change.* Cambridge: Cambridge University press.

Jasanoff, S. (2010). A new climate for society. *Theory, Culture and Society, 27*(2–3), 233–253.

Kelman, I., & West, J. J. (2009). Climate change and Small Island Developing States: A critical review. *Ecological and Environmental Anthropology, 5*(1).

Kloos, H., & Adugna, A. (1989). The Ethiopian population: Growth and distribution. *The Geographical Journal, 155*(1), 33–51.

Macchi, M., Oviedo, G., Gotheil, S., Cross, K., Boedhihartono, A., Wolfangel, C., & Howell, M. (2008). *Indigenous and traditional peoples and climate change. Issues Paper.* IUCN.

Maddison, D. (2006). The perception of and adaptation to climate change in Africa. In *Policy Research Working Paper 4308,* Development Research Group, The World Bank, Washington DC.

MOH. (2008). *Ethiopian national malaria indicator survey 2007.* Ethiopia: Addis Ababa.

Nhemachena, C., & Hassan, R. (2007). Micro-level analysis of farmers' adaptation to climate change in Southern Africa. *IFPRI Discussion Paper 00714.*

Nyong, A., Adesina, F., & Elasha, B. O. (2007). The value of indigenous knowledge in climate change mitigation and adaptation strategies in the African Sahel. *Mitigation Adaptation Strategies for Global Change, 12,* 787–797.

Pelling, M. (2011). Climate change and social capital. *Discussion paper 5.* A Report commissioned as part of the UK Government's Foresight Project on the International Dimensions of Climate Change, Government Office for Science. London.

Raygorodetsky, G. (2011). *Why traditional knowledge holds the key to climate change.* United Nations University. http://unu.edu/publications/articles/why-traditional-knowledge-holds-the-key-to-climate-change.html

Salick, J., & Byg, A. (2007). *Indigenous peoples and climate change.* Oxford: Tyndall Centre for Climate Change Research, Tyndall Centre Publication.

Wolde-Georgis, T. (2000). *The case of Ethiopia reducing the impacts of environmental emergencies through early warning and preparedness: The case of the 1997–98 El Niño.* http://archive.unu.edu/env/govern/ElNIno/CountryReports/pdf/ethiopia.pdf

World Bank (1998). *Indigenous knowledge for development: A framework for action.* Washington DC: The World Bank, Knowledge and Learning Center, Africa Region.

Adaptation Opportunities to Climate Variability and Potential Effects on Sustainable Development. The Case of Nigeria's Niger Delta Region

Chika Ogbonna, Eike Albrecht and René Schönfelder

1 Introduction

Climate variability and climate change have become major environmental challenges facing developing countries in the 21st century. The Nigeria's Second National Communication under the United Nations Framework Convention on Climate Change (FME 2014) affirms that the Nigeria nation in general faces climate-related impacts such as severe floods, windstorms, heat waves, ocean surges, and a wide range of climate extremes that have negative impact on the country's socio-economic activities. Projections show that a rise of 0.5 m in sea level could result in the loss of 35% coastline of the highly-productive delta (INDC 2015). The area faces extensive climate risks ranging from storm surges, increasing risks of flooding, variable but extreme rainfall and heat waves. However, besides seasonal flooding, the 2012 flood incident in the coastal Niger Delta was devastating in the area. On the other hand, following the global climate temperature trend, observed climate shows that temperature in Nigeria have been increasing in the last five decades and it has become more significant since the 1980s and this is also the case in the Niger Delta (FME 2014). The Fifth Assessment Report of the International Panel on Climate Change (IPCC) concludes that "climate variability and climate change is a threat to sustainable development" (IPCC 2014a).

C. Ogbonna · E. Albrecht
Brandenburg University of Technology Cottbus Senftenberg,
Senftenberg, Germany
e-mail: ogbonchi@b-tu.de

E. Albrecht
e-mail: albrecht@b-tu.de

R. Schönfelder (✉)
Institute for Software Engineering and Programming Languages,
University of Lübeck, Lübeck, Germany
e-mail: rene.schoenfelder@isp.uni-luebeck.de

W. Leal Filho et al. (eds.), *Climate Change Adaptation in Africa*,
Climate Change Management, DOI 10.1007/978-3-319-49520-0_44

The concept of sustainable development started in Stockholm, Sweden during the 1972 United Nations Conference on Human Environment. The World Summit on Sustainable Development held in Johannesburg, South Africa from 2 to 4 September 2002 reaffirms the commitment to Sustainable development. Sustainable development is a process of development that seeks to achieve, in a balanced manner, the three pillars: social development, economic growth and environmental protection. However, the concept of sustainable development was popularized by the Brundtland report published by the World Commission on Environment and Development (WCED) in 1987, also known as "Our Common Future" (UN 2010). The report defined sustainable development as "development which meets the needs of the present without compromising the ability of future generations to meet their own needs" (WCED 1987). While Agenda 21 adopted by more than 178 governments at the 1992 Earth Summit brought a comprehensive plan to the table, the 2012 Rio+20 conference asked countries to renew their commitment to sustainable development (IPCC 2014a).

1.1 Rationale and Significance of Research

The findings from the Working Group II of Fifth Assessment Report (AR5) of the United Nations Intergovernmental Panel on Climate Change and the Nigeria Intended National Determined Contribution (INDC) to the Paris Agreement (COP21) emphasised on the adverse effects of current and future climate impacts on sustainable development. In particular, the Working Group II of the AR5 states that climate change poses a moderate threat to current sustainable development and severe threat to future sustainable development (IPCC 2014a). On the other hand, the Nigeria Intended National Determined Contribution made specific reference to increasing coastal vulnerability in the Niger Delta due to global warming induced accelerated sea level rise. To this end, the paper is a contribution to adaptation in the Niger Delta. It examines the links between current and future impacts of climate variability on future sustainable development initiatives in the area. The paper attempts to illustrate the direct implications of the results of the study on the 2030 Agenda for sustainable development. Therefore, this paper provides perspectives on adaptation opportunities essential to reducing climate risks and building a resilient future to achieve and reinforce sustainable development goals.

The rationale for this study is based on the evidence of climate and weather related impacts in the Niger Delta reported in literature pointing on how current and potential challenges of flooding and coastal erosion due to accelerated rise in sea level, change in rainfall pattern and increasing heat waves pose problems on the human environment in the area. However, there is presently a dearth of information on how such current and expected effects could jeopardize achieving sustainable development in the area. It has therefore become necessary and urgent to examine the links between observed and future climate variability and sustainable development options. Hence, the relevance of discussing the potential effects of climate variability and its connection to sustainable development is due to the obvious fact that climate change and climate variability is anticipated to pose major threat to future sustainable development in Nigeria; particularly in the Niger Delta. In essence, such climate impacts make it hard to achieve sustainable development. This study, provides insight on how climate associated events endanger current and future sustainable development and presents viable adaptation measures that need to be considered in order to reinforce development initiatives in the region.

2 Current and Future Impacts of Climate Variables

Climate variability and climate change is a big challenge to equitable and sustainable development in Nigeria where 55% of the population lives below the poverty level, on less than $1.90 per day (World Bank 2009; Bayelsa State Government 2011; Oxfam 2016). The Nigerian nation faces climate-related impacts such as severe floods, windstorms, heat waves, ocean surges and a wide range of climate extremes which have negatively impacted socio-economic activities (FME 2014). In the coastal Niger Delta region, flooding is the most pressing climatic and weather associated challenges, having severe impacts in and around coastal settlements. For example, the flood incident that happened in 2006 as reported by Douglas et al. (2008 cited in IPCC 2014a). This extreme 10-h rainfall drove 10,000 residents out of their homes and caused widespread traffic chaos in Port Harcourt City (Satterthwaite et al. 2007). This unprecedented flooding, submerged houses, paralyzed economic activities and displaced some residents of the Mgbuoba, Diobu and Nkpolu communities (Zabbey 2007). Similarly, heavy rains between July and October 2012 combined with rising water levels resulting from runoffs contributed to the flooding of human settlements located downstream of the Kainji, Shiroro, and Jabba dams (PDNA 2013). According to the National Emergency Management Agency (NEMA), 363 people were killed, 5851 injured, 3891, 314 affected, due to the resulting floods in the entire country including the coastal Niger Delta Communities (PDNA 2013). Weighing the potentiality of communicable diseases as a result of the 2012 flood disaster, malaria was found to be of immediate and long-term concern, while cholera/typhoid/shigellosis and acute lower respiratory tract infections were of immediate concern (PDNA 2013). Moreover, flood and erosion pose a threat to city infrastructures such as roads and affect freshwater resources, thereby threatening the well being, lives and properties of residents in the area.

The problem is exacerbated by the fact that the area serves as the hub of Nigeria's oil and natural gas production. Nigeria is one the world's leading oil producers with 2.2 million barrels of oil being produced in the Niger Delta daily (Niger Delta Biodiversity Project Report 2013). There could be a severe deleterious impact on the environment and the Nigerian economy if oil and gas infrastructures and movement of people, goods and services are affected. Moreover, the South-South region of Nigeria, where the core Niger Delta is located is the most vulnerable of the three zones in the South. This could be as a result of the challenges of coastal flooding and erosion combined with the challenges related to petroleum exploration in this part of the country (FME 2014). According to the projection of the United Kingdom's Department for International Development (DFID), there could be a possible rise in sea level from 1990 levels to 0.3 m by 2020 and 1 m by 2050, and a rise in temperature of up to 3.2 °C by 2050 under a high climate change scenario. The figure is based on IPCC's climate change assumptions, latest research findings and results of a consultation exercise in Nigeria (DFID 2009; FME 2014).

Following the global climate temperature trend, observed climate shows that there has been a steady increase in temperature in Nigeria over the last five decades and it has become more significant since the 1980 s and this is also the case in the Niger Delta (FME 2014). For example, a climate scenario over Nigeria shows an increase in the future occurrence of extreme temperature over its agro-climate zones in a B1 and A2 scenarios. The findings in the B1 scenario indicate that by the middle of the century, the number of days with a temperature greater than 38 °C would increase by 3 days in the Mangrove and 5 days in the Rainforest zone where the Niger Delta is situated, while under the A2 scenario, the increase would be 7 and 23 days respectively. Similarly, by the middle of the century, it is expected that the number of days with heat waves would increase by 23 days in the Mangrove and 32 days in the Rainforest zone under the B1 scenario and 39 and 51 days respectively under the A2 scenario (FME 2014).

On the other hand, population is one of the socio-economic issues that predisposes the region to the adverse effects of climate variability and climate change. According to the Niger Delta Biodiversity Project Report, the Niger Delta is one of most densely populated regions in Sub-Saharan Africa, with a population of nearly 33 million people in the region in 2006 (>265 people per km^2). The region's population continues to rise and is expected to be over 45 million people by 2020 as projected by German Society for International Cooperation (GIZ) and reported in the Niger Delta Development Master Plan of the Niger Delta Development Commission (NDDC 2006). The increasing population could accelerate urbanization, thereby straining public infrastructure beyond capacity.

2.1 The Case Study: The Case of Nigeria's Niger Delta Region

The Niger Delta region is located on the Atlantic Coast of Southern Nigeria, where the river Niger splits into numerous distributaries. It is one of the largest delta regions in the world and the rich part of Nigeria in terms of natural resources (Ogbonna 2014). For example, it has large oil and gas deposits and biodiversity. The region covers an area of about 70,000 km^2, starting from a coastal zone of swamps, which extends Northwards towards the rainforest that gradually merges

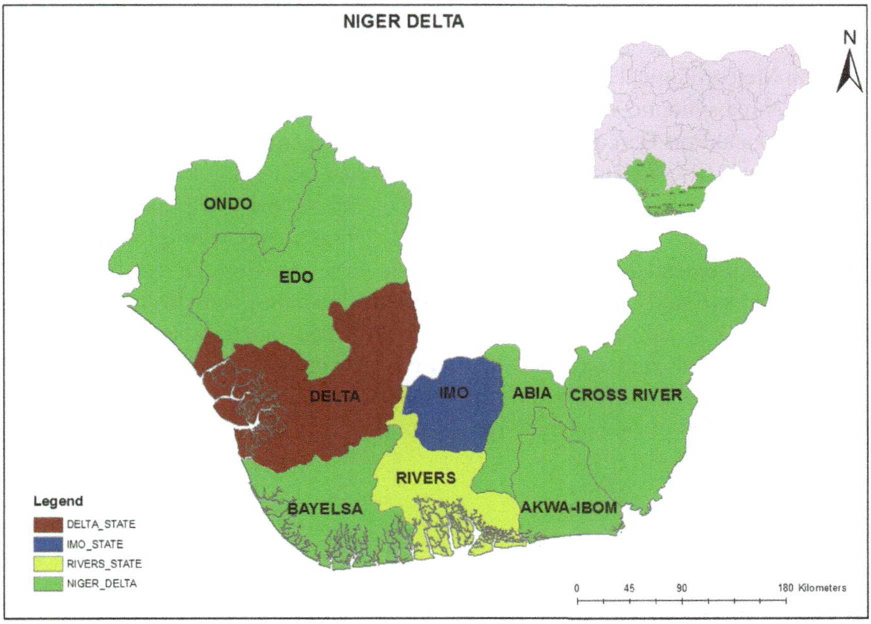

Fig. 1 Map of Nigeria

with woodlands and savanna grasslands in central Nigeria (UNEP 2011). The region has nine member states, which include: Abia, Akwa-Ibom, Cross River, Bayelsa, Edo, Delta, Imo, Rivers and Ondo. The region has about 185 local government areas (LGAs) and over 800 ethnic groups (Omeri et al. 2014). The population of the Niger Delta region is above 31 million according to the most recent national population census (NPC 2006). According to Twumasi and Meren (2006), the projected population of the region could reach above 45 million in the year 2020. The Niger Delta is located on latitudes from 4°25′ N to 6°00′ N and longitudes from 5°00′ E to 7°5′ E. Figure 1 shows a map of Nigeria indicating the location of Niger Delta region and the nine States of the region. Three focal States for this study include Delta, Imo, and Rivers States.

3 Materials and Methods

The study aims to analyze the trends of observed climate variability in the 3 selected States and provides a reflection on how these climate variables (monthly maximum temperature and monthly rainfall) could affect current and future development initiatives in the area. The observational data relates to climate variation within a 30-year period ranging from 1980 to 2010. The minimum rainfall data and maximum temperature data used in trend analysis was collected from the Nigerian Meteorological Agency (NIMET), Lagos. The climate data were compiled from the three synoptic stations in the region namely: Warri (Delta State); Owerri, (Imo State) and Port Harcourt (Rivers State). The temperature and rainfall data were imported into 'R' statistical software and plotted using ggplot2. The monthly maximum temperature is averaged for an indication of heat waves and the monthly rainfall is added up to get the annual rainfall. These values were plotted in a line chart and then extended by a linear regression curve showing the trend. The study employed a case study approach and incorporated existing facts from publications through a desk-based analysis. The study also relied on the analysis of various secondary materials and scientific literature, relevant journals, government policy papers, research articles, and books that reveal evidence of climate vulnerability and weather associated events in the Niger Delta region.

4 Results and Discussion

4.1 Trend Analysis of Temperature Data

As can be seen on the result (Fig. 1), the average monthly maximum temperature at Warri, Owerri and Port Harcourt showed an increasing trend in temperature patterns. In Warri (Delta State), the lowest temperature of 30.3 °C was recorded in 1985 while the highest temperature of 32.3 °C was observed in 1994. Results also showed that from 1980 to 2010, the average temperature in Warri, Delta State increased by 0.14 °C per decade, whereas the average monthly maximum temperature in Owerri, Imo State over the 30-year period indicated a gradual increase in temperature. The lowest maximum temperature in Owerri was recorded in 1985 at 31.4 °C, which rose again rapidly beyond 32 °C around 1987, and gradually fluctuated until the highest temperature of 33.10 °C was attained by the end of 1998. The linear trend indicates a progressive increase in temperature with time recorded in years. The temperature trend analysis also showed that the average temperature over the years increased by about 0.21 °C per decade in Owerri. On the other hand, the monthly maximum temperature at Port Harcourt, Rivers State varied between 1980 and 2010. The lowest temperature recorded within this period was in 1992 with a record of 30.25 °C, which rose and fluctuated within 31 °C and 31.75 °C between 1994 and 2005. After 2005, average temperature increased again and fluctuated until the highest was reached in 2010 at 32.05 °C. Overall, temperature did not increase extensively in Port Harcourt in comparison with Owerri and Warri as the trend analysis indicated a slight rise in temperature as the years progressed. Results also show that the monthly maximum temperature in Port Harcourt increased at a rate of 0.042 °C per decade.

4.2 Trend Analysis of Rainfall Data

The graph (Fig. 2) indicates that Delta State (Warri) experienced a record high of rainfall of about 3400 mm in 1995, whereas a record low of rainfall of about 2400 mm occurred in 1981. Observed rainfall data in Port Harcourt, Rivers State shows a wide variation in rainfall. There was a sharp decline in rainfall pattern recorded in 1982 at 2200 mm with an all-time record low of 2000 mm in the years 2000 and 2008, while a peak in rainfall pattern of 2800 mm was observed in 2007. In Imo State, there was an average rainfall decline to 1550 mm and 1650 mm in 1983 and 1998 respectively, a sharp increase to 2,300 mm in 2007 and ultimately, a decline to 2200 mm in 2010 (Fig. 3).

The entire results correspond to the existing research on the Niger Delta region (BNRCC 2011; Emaziye et al. 2012; Okorie et al. 2012) and with the global predictions of the IPCC (2007, 2014b). It also strengthens confidence that temperature anomalies in the States follow the same trends of global warming. High values in average monthly maximum temperature clearly depict the intensity of maximum temperature (heat waves) all through the years. The results also corroborated the research conducted by Okorie et al. (2012) in Owerri, Imo State which found that "an increase in surface temperature and torrential rainfall causes flash floods and flood disasters, increasing frequency and intensity of extreme

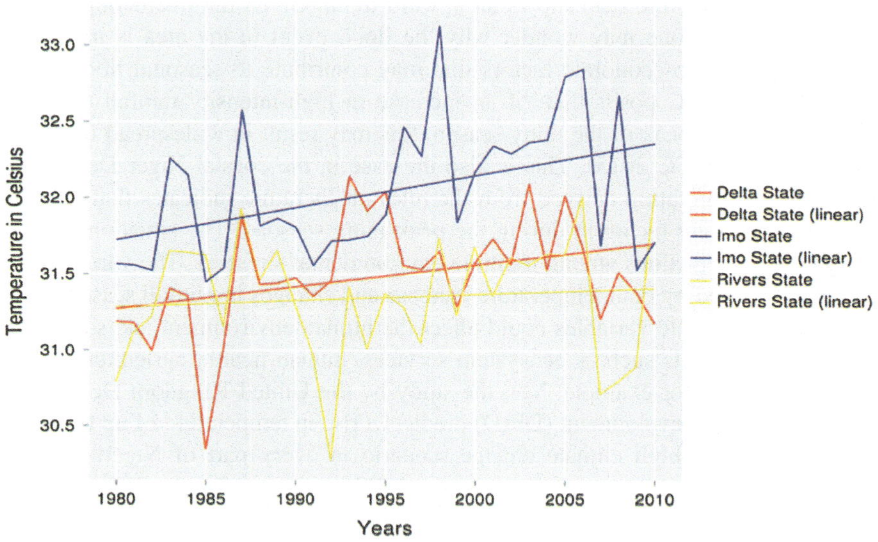

Fig. 2 Maximum temperature and yearly average from Owerri (Imo State), Warri (Delta State) and Port Harcourt (Rivers State) synoptic stations respectively

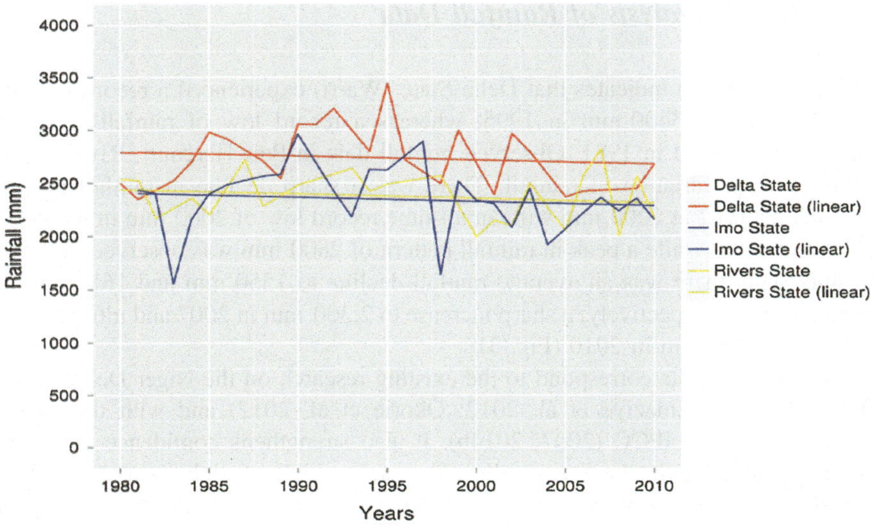

Fig. 3 Average annual maximum rainfall for Owerri (Imo State), Port Harcourt (Rivers State) and Warri (Delta State)

weather events such as thunderstorms and unpredicted rainfall patterns, as well as other climate related disasters." However, while rainfall data show fluctuating variations, temperature data shows an upward trend. Given the fluctuating rainfall pattern in Fig. 2, one may wonder why the flood event in the area is increasing besides other socio-economic factors that may contribute to seasonal flooding. In this vein, the IPCC posits that "if an increase in high-intensity rainfall events is concurrent to the peak of the rainy season, this may result in widespread flooding" (Sylla cited in IPCC 2015). This is also the case in the coastal Niger Delta communities as information derived from the relevant literature indicates that increase in flood events are recurrent during the peak rainy seasons. The situation is compounded by interactions among multiple environmental stressors. The variability in climate portrayed by both temperature increase and changes in rainfall seasons casts light on how climate variables could affect the human environment and sources of people's livelihoods such as ecosystem services, human health, agricultural activities, and fishing for example. Yet, the study by the United Kingdom Department for International Development (DFID) predicts a rise in temperature of up to 3.2 °C by 2050 under a high climate change scenario in every part of Nigeria; it also predicts a significant sea level rise from the 1990 levels to 0.3 m by 2020 and 1 m by 2050 along the Nigerian coasts (DFID 2009; FME 2014). However, considering the fact that the coastal Niger Delta settlement is a low-lying area the future predictions of rise in sea levels call for action.

5 Implications for Sustainable Development

The concepts of sustainable development and adaptation to CC are interlinked in two ways. First, adaptation policies and programs play a role directly or indirectly in achieving some national and regional sustainable development initiatives. Secondly, a climate robust adaptation reduces society's vulnerability to CC by strengthening adaptive capacity and people's resilience to the effects of CC, thereby addressing the challenges of sustainable development goals. The link between climate change adaptation and sustainable development has also been recently recognized within the United Nations Framework Convention on Climate Change (UNFCCC).

It is essential that sources of human livelihood such as water, agriculture, infrastructure and health be protected while considering the changing conditions in sustainable development agendas in order to maintain and improve people's well-being. Deyal et al. (2006) argues that sustainable development will be a reality if adaptation to CC is internalized for citizens, sectors, and stakeholders. Given the numerous environmental challenges associated with climate change affecting development, there is a need to increase resilience through the provision of robust climate change adaptation strategies in the region. Particularly, now that adaptation to climate risks has emerged as a relevant development policy.

Climate variability and climate change pose potential threats to various sectors of development in the Niger Delta and thus to all the prospective sustainable development agenda. Events associated with climate variability are major threats to the sustainable development of Nigeria, including the Niger Delta (FME 2014). The Niger Delta coastal and marine environment which stretches for 150 km in the Delta and 25 km East of the Delta is endangered by extreme climate-related events which add to the existing environmental degradation in the area. The number of days with extreme temperature has become higher than the historical average in the region. For instance, in Asaba, Delta State Capital, 26 days of extremely high temperature were recorded in the area in 2013 (TACC 2013).

The above analyses show evidence of observed climate variability and its potential risk to people and communities in the area and draw on the Nigeria's Intended National Determined Contribution to the Paris Agreement which states that "climate change poses a significant threat to the achievement of development goals, especially those related to eliminating poverty and hunger and promoting environmental sustainability" (INDC 2015). In other words, achieving sustainable development entails dealing with climate vulnerability (IPCC 2014c). Figure 4 summarizes the current effects of climate and weather related events on sustainable development goals. Apparently, the increasing trend in temperature and irregular, yet decreasing trend in rainfall pattern, portends extensive negative environmental and socioeconomic implications unless developmental policies consider resilient and robust adaptation actions. For instance, the following three new United Nations' proposed sustainable development goals (SDGs) among others set to be

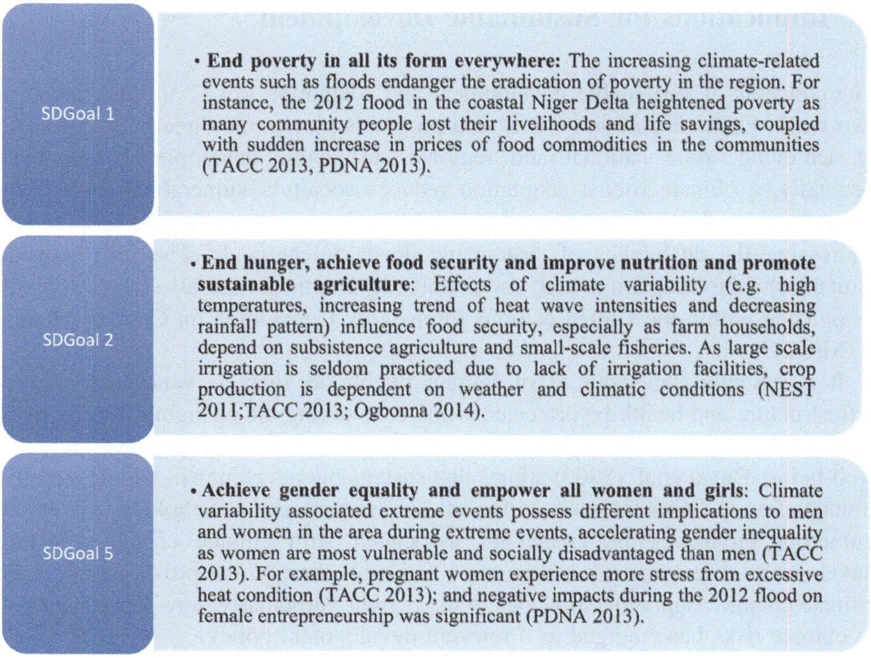

Fig. 4 Effects of Climate related events on Sustainable Development. *Source* NEST (2011), TACC (2013), PDNA (2013) and Ogbonna (2014)

achieved by 2030, are potentially threatened by current and anticipated climate and weather related impacts in the region:

5.1 SDGoal 1: End Poverty in All Its Form Everywhere

The increasing climate-related events such as floods endanger the eradication of poverty in the region. For instance, the 2012 flood in the coastal Niger Delta heightened poverty as many community people lost their livelihoods and their life savings (TACC 2013, p. 41). In fact, the Niger Delta low-lying coastal areas are particularly vulnerable to sea-level rises, which constitute threats to communities livelihoods, resources and economy. Although the result shows a declining rainfall pattern, there remains an increasing uncertainty in the timing and amount of rainfall. Extreme rainfall during a peak of raining season has contributed to coastal erosion and flooding. Flooding and coastal erosion adds to existing environmental stressors thereby threatening community livelihoods. The high intensity of rainfall escalates runoffs and erosion. Coastal erosion and flooding have become more prevalent in the coastal Niger Delta and this is worrisome as the study of the United Kingdom's Department for International Development (DFID) predicts a significant sea level rise from the 1990 levels to 0.3 m by 2020 and 1 m by 2050 along the Nigerian coast.

5.2 SDGoal 2: End Hunger, Achieve Food Security and Improve Nutrition and Promote Sustainable Agriculture

Effects of climate variability (e.g. extreme temperatures and decreasing/fluctuating rainfall pattern etc.) influence food insecurity, especially as the local population farm households, depend on subsistence agriculture and small-scale fisheries. Hence irrigation is not commonly practiced due to inadequate irrigation facilities, food availability dependents on weather and climatic conditions (TACC 2013). For example, the study, by Building Nigeria's Response to Climate Change (2011), showed that during extreme weather and climate events in the area, food expenditure is usually higher than normal (BNRCC 2011, p. 60); such environmental stressors aggravate challenges to improving nutrition and food security. The irregularity in rainfall distribution also influences farmers' decisions with respect to crop planting and harvesting. Farming and fishing are the mainstay of rural coastal communities in the Niger Delta. Consequently, decreasing rainfall and increasing temperature foretells severe impacts on sustainable agriculture, crop production, sustainable livelihoods, household sustainability and economic growth in these coastal communities.

5.3 SDGoal 5: Achieve Gender Equality and Empower All Women and Girls

Climate variability, climate change and weather associated extreme events accelerate gender inequality as women are most vulnerable and socially disadvantaged. The study, by the DFID predicts a rise in temperature of up to 3.2 °C by 2050 under a high climate change scenario in every part of Nigeria (DFID 2009; FME 2014). As a result of the current and potential increase in temperature, urban and rural areas of the region are faced with the challenge of increasing heat waves. Such current and future weather and climate impacts would have more effects on women and girls than men and boys. This is apparent as the aftermath of flood effects in the coastal Niger Delta in 2012 induced stronger gender inequality as household resources were exhausted and the poor were made poorer with women being worse off (TACC 2013). These heat waves add to the degradation of the mangrove forest and pollution of water bodies as a result of gas flaring, which tremendously affected fishing and farming activities; thereby leading to high level of poverty for women whose main source of livelihood depend directly on fishery and agriculture (Omeire et al. 2014).

6 Conclusion and Policy Recommendations

Section 20 of the Nigerian 1999 constitution requires that "the state shall protect and improve the environment and safeguard the water, air and land, forest and wildlife of Nigeria". The United Nations Framework Convention on Climate Change provides the basis to adapt to the impacts of climate change; Nigeria ratified the Convention on the 29 August 1994 (Albrecht and Ogbonna 2015). Article 4.1 of the Convention is crucial for undertaking adaptation and paragraph 1(e) of the Article asked all parties to the Convention to cooperate in preparing for adaptation in the areas of coastal management and agriculture with particular reference to Africa. Drawing upon the above study, climate variability combined with other environmental stressors exerts potential pressure on current and future sustainable development initiatives in the Niger Delta. Based on the result the variability and change in climate and weather events may be gradual but the current impacts associated with such changes are of concern.

However, the limitations and constraints of this research lie on available climate data, whereas climate and weather variables vary in specific locations. Climate data (maximum temperature and monthly rainfall) were obtained from synoptic stations predominantly located in urban areas and major business districts (Owerri, Port Harcourt, Warri). Furthermore, the direct consequence of gas flaring and oil spills add to the conundrum of environmental degradation in the area, coupled with potential climate and weather-related effects on sustainable development. In this regard, there is a need to delineate such environmental challenges through an in-depth environmental assessment while updating baseline data for robust environmental strategies necessary to reinforce sustainable development and build future resilience to take off.

On the other hand, adaptation provides opportunities to accelerate development and reinforce current and future sustainable development initiatives. Adaptation will enhance social, economic and environmentally sustainable management of the region's natural resources. In this regard, adaptation measures that need to be addressed to reinforce current and future sustainable development goals include, but not limited to the following are:

6.1 Application of Precautionary Principle

Application of the precautionary principle is imperative to reduce climate risks, protect coastal settlements, infrastructures, and people's livelihoods in the Niger Delta. The principle 15 of the Rio Declaration on Environment and Development

(UN 1992) which states that "…if there are threats of serious or irreversible environmental damage, lack of full scientific certainty should not be used as a reason for postponing measures to prevent environmental degradation." The states and local governments should employ the precautionary principle in reducing the risks posed by adverse impacts of climatic variability and climate change, as well as natural related hazards and disasters. The precautionary principle is imperative considering the uncertainty of climate projections.

6.2 Provision of Appropriate Legal and Policy Frameworks

Appropriate legal and policy frameworks, which will mean bringing together all necessary legal and policy instruments that can enhance adaptation implementation. This will accelerate policy initiatives and allow for robust implementation of adaptation options, thereby reducing communities' vulnerability and improve both individual and institutional adaptive capacities. Provision of legal and policy frameworks by authorities in the Niger Delta region can encourage designation of disaster prone areas and can also make it mandatory for local governments to implement adaptation options, for example, the provision of warning alerts in vulnerable coastal areas. Legal frameworks are imperative for effective adaptation to be carried out, notably without legal strategies in place, policies might be difficult to implement no matter how robust such policies are.

6.3 Provision of Public Access to Climate Information

Provision of accurate weather and climate predictions are imperative for adaptation. This entails making climate information available to the public. Accurate weather and climate predictions are not panaceas to extreme weather events but will enhance proactive adaptation actions and improve self-protective behaviors. Such effects may undermine conservation and sustainable development efforts in the area. On the other hand, new and modern information services should be provided to cover the rural areas so that NIMET can carry out its required functions effectively. Reliability and usefulness of weather and climate forecasts underpin decisions for land use planning in the face of adverse impacts of predicted climate variability and change. Considering the fact that climate information largely depend on specific locations; there is a need for NIMET to establish more synoptic stations, particularly in the suburbs and rural communities so as to have accurate observational data to inform planning.

6.4 Scaling up Research that Would Improve Traditional Agricultural Practices

It has become urgent for the government and relevant institutions to scale up research that could help discover new approaches to agricultural practices. For example, improved crop species that could be resilient to extreme changes in temperatures have become imperative. Furthermore, since irrigation is seldom practiced on large scale, the government should encourage mechanized irrigation systems such as the sprinkler irrigation to encourage agricultural production as agriculture is the mainstay of the local economy besides oil and natural gas production. There is a need for agricultural risk assessments, as it has become crucial to sustainable agricultural production in the area. Such could help avert influences on hunger, food security and poverty.

References

Albrecht, E., & Ogbonna, C. U. (2015). Strategic environmental assessment as a tool to integrate climate change adaptation: A perspective for Nigeria. In W. Leal Filho (Ed.), *Handbook of climate change adaptation* (pp. 1239–1260). Berlin: Springer.

Bayelsa State Government. (2011). *Bayelsa State climate change policy* (pp. 38). Bayelsa state Government Ministry of Environment state secretariat complex, Yenegoa. Nigeria: Ministry of Environment Bayelsa State.

BNRCC. (2011). *National adaptation strategy and plan of action on climate change for Nigeria (NASPA-CNN)* (p. 82). Federal Ministry of Environment special climate change unit. Building Nigeria's response to climate change (BNRCC) Project, Ibadan, Nigeria.

Department for International Development (DFID). (2009). *Impact of climate change on Nigeria's economy*. Abuja, Nigeria: United Kingdom Department for International Development (DFID).

Deyal, A., Bouwer, L. M., Aerts, J. C. J. H., Ahmad, Q. K., Van Schalk H., Hoyte, P. A., et al. (2006). Experts address the question: How can we maximize the impact of adaptation measures to support sustainable development? A United Nations Sustainable Development Journal. *Natural Resources Forum, 30*(4), 328–330.

Douglas, I., Alam, K., Maghenda, M., Mcdonnell, Y., Mclean, L., & Campbell, J. (2008). Unjust waters: Climate change, flooding and the urban poor in Africa. International Institute for Environment and Development (IIED). *Environment and Urbanization, 20*(1), 187–205.

Emaziye, P. O., Okoh, R. N., & Ike, P. C. (2012). A Critical analysis of climate change factors and its projected future values in delta state Nigeria. *Asian Journal Agriculture Rural Developement, 2*(2), 206–212.

Federal Ministry of Environment (FME). (2014). *Nigeria's second national communication on climate change* (p. 132). Abuja, Nigeria: Federal Ministry of Environment of the Federal Republic of Nigeria.

Intended Nationally Determined Contribution (INDC). (2015). *Nigeria's intended nationally determined contribution*. The Federal Government of Nigeria, available at: http://www4.unfccc.int/submissions/indc/Submission%20Pages/submissions.aspx. Accessed May 26, 2016.

IPCC. (2007). Climate change 2007: The Physical science basis. In *Contribution of working group I to the fourth assessment report of the intergovernmental panel on climate change* (p. 996) [S. D. Solomon, M. Manning, Z. Chen, M. Marquis, K. B. Averyt, M. Tignor and H. L. Miller (Eds.)]. Cambridge, United Kingdom and New York, NY, USA: Cambridge University Press.

IPCC. (2014a). Climate change 2014: Impacts, adaptation, and vulnerability. Part A: Global and sectoral aspects. In *Contribution of working group II to the fifth assessment report of the intergovernmental panel on climate change* (pp. 1132) [C. B. Field, V. R. Barros, D. J. Dokken, K. J. Mach, M. D. Mastrandrea, T. E. Bilir, M. Chatterjee, K. L. Ebi, Y. O. Estrada, R. C. Genova, B. Girma, E. S. Kissel, A. N. Levy, S. MacCracken, P. R. Mastrandrea, and L. L. White (Eds.)]. Cambridge, United Kingdom and New York, NY, USA: Cambridge University Press.

IPCC. (2014b). Climate change 2014: Synthesis report. In *Contribution of working groups I, II and III to the fifth assessment report of the intergovernmental panel on climate change* (pp. 151) [Core Writing Team, R. K. Pachauri and L. A. Meyer (Eds.)]. Geneva, Switzerland: IPCC.

IPCC. (2014c). Climate change 2014: Mitigation of climate change. In *Contribution of working Group III to the Fifth Assessment Report of the intergovernmental panel on climate change* [O. Edenhofer, R. Pichs-Madruga, Y. Sokona, E. Farahani, S. Kadner, K. Seyboth, A. Adler, I. Baum, S. Brunner, P. Eickemeier, B. Kriemann, J. Savolainen, S. Schlömer, C. von Stechow, T. Zwickel and J.C. Minx (Eds.)]. Cambridge, United Kingdom and New York, NY, USA: Cambridge University Press.

NDDC. (2006). *Niger delta regional development master plan: Our challenge–their future* (p. 258). Niger Delta Development Commission (NDDC). Port Harcourt: Federal Republic on Nigeria Publication, NDDC.

NEST. (2011). *Reports of research projects on impacts and adaptation. building nigeria's response to climate change (BNRCC)* (p. 330). Ibadan, Nigeria: Nigerian Environmental Study/Action Team (NEST).

Niger Delta Biodiversity Project Report. (2013). *United Nations Development Program (UNDP)/ Global Environment Facility (GEF) (2013–2016) Niger Delta Biodiversity Project*. PRODOC. Available at: http://www.undp.org/content/dam/undp/documents/projects/NGA/Niger%20Delta%20BiodiversityProdoc.pdf. Accessed May 20, 2016.

Nigeria Population Commission (NPC). (2006). *Nigeria Population Statistics* (No. 2, Vol. 96). Federal Republic of Nigeria Official Gazette. Abuja: Federal Republic of Nigeria.

Ogbonna, C. U. (2014). Adaptation to climate change in developing countries: A need in the Niger Delta region of Nigeria. In E. Albrecht, M. Schmidt, M. Mißler-Behr, & S. P. N. Spyra (Eds.), *Implementing adaptation strategies by legal, economic and planning instruments on climate change* (pp. 165–187). Heidelberg: Springer.

Okorie, F. C., Okeke, I., Nnaji, A., Chibo, C., & Pat-Mbano, E. (2012). Evidence of climate variability in Imo State of Southern Eastern Nigeria. *Journal of Earth Science Engineering, 2,* 544–553.

Omeri, E. U., Aveuya, A. A., Muoneme-Obi, C. T., Adolphus, G., Ufomba, A., & Omerire, C. A. (2014). Between the devil and the deep blue sea: Niger delta women and the burden of gas flaring. *European Scientific Journal, 10*(26), 151–162.

Oxfam. (2016). *An Economy for the 1%: How privilege and power in the economy drive extreme inequality and how this can be stopped.* Available at: https://www.oxfam.de/system/files/bp210-economy-one-percent-tax-havens-180116-en.pdf. Accessed June 06, 2016.

Post-Disaster Needs Assessment (PDNA). (2013). Nigeria post-disaster needs assessment. *A Report by the Federal Government of Nigeria, with Technical Support from the European Union, United Nation, World Bank, and other Partners, Nigeria Post-Disaster Need Assessment 2012 floods* (p. 154). Federal Government of Nigeria Abuja.

Satterthwaite, D., Huq, S., Pelling, M., Reid, H., & Lankao, P. R. (2007). *Climate change and cities: Adapting to climate change in urban areas: The possibilities and constraints in low-and-middle-income nations.* Human Settlement Discussion Paper Series (p. 112). International Institute for Environment and Development (IIED), London.

Sylla, M. B. (2015). *Regional climate change over West Africa: Trends, shift of climate zones and timing of the wet extremes* [In T. F. Stocker, D. Qin, G.-K. Platter & M. Tignor (Eds.), IPCC. (2015). Workshop Report of the Intergovernmental Panel on Climate Change Workshop on Regional Climate Projections and their Use in Impacts and Risk Analysis Studies, pp. 154–155]. IPCC Working Group I Technical Support Unit, University of Bern, Bern, Switzerland.

Territorial Approach to Climate Change (TACC). (2013). *Biophysical and socio-economic assessment of the nexus of environmental degradation and climate change* (p. 234). Nigeria: Delta State.

Twumasi, Y. A., & Merem, E. C. (2006). GIS and remote sensing applications in the assessment of change within a coastal environment in the Niger Delta of Nigeria. *Environmental Research and Public Health, 3*(1), 98–106.

United Nations (UN). (2010). *Sustainable development: From Brundtland to Rio 2012*. Background Paper Prepared for Consideration by The High Level Panel On Global Sustainability at Its First Meeting (p. 26). September 19, 2010. Geneva: International Institute for Sustainable Development (IISD).

United Nations Environmental Program (UNEP). (2011). *Environmental assessment of Ogoniland* (p. 257). Nairobi, Kenya: United Nations Environmental Program.

World Bank. (2009). *Data find 53.5% of the Nigerian population, or 83 million people living below $1.90/day Povcalnet, 2009 data, 2011 PPP.* Available at: http://povertydata.worldbank.org/poverty/country/NGA, most recent World Bank. Accessed June 06, 2016.

World Commission on Environment and Development: Our Common Future (WCED). (1987). *Report of the World Commission on Environment and Development: Our Common Future*. Available at http://www.un-documents.net/our-common-future.pdf. Accessed May 15, 2016.

Zabbey, N. (2007). *Climate change and flooding: Fate of Riverine communities in the Niger Delta*. Paper delivered at an interactive roundtable organized by Stakeholder Democracy Network (SDN) in commemoration of World Environmental Day, SDN Conference Room, Port Harcourt, Nigeria, June 5, 2007.

The Association of Monetary, Multidimensional and Traditional Poverty with Climate Change Adaptive Capacities in Northern Benin

Frédéric Kosmowski and Richard Lalou

Climate change is expected to adversely affect most countries and social groups. A change in local climatic conditions is expected for several indicators, including increased temperatures, changes in rainfall amounts and patterns, and increased frequency and severity of drought and flood events (IPCC 2013; Deme et al. 2015). The COP21 conference in December 2015 has reasserted that even if the temperature increase is kept below 2 °C, the impacts of climate change are likely to require some degree of adaptation. Understanding adaptation and adaptive capacities of climate-sensitive households is therefore crucial for policy-makers.

Past IPCC reports have reinforced the view that socially and economically disadvantaged and marginalized people are disproportionately affected by climate change. Smit and Pilifosova (2001) considered economic resources—such as economic assets, capital resources, financial means, wealth, or poverty—as a determinant of a household's adaptive capacities. Benefiting from the extension of scholarship in the climate change field, the list of factors driving adaptive capacities was broadened in following reports. Adaptive capacity was considered not only to be influenced by economic development and technology, but also by social factors such as human capital and governance structures (Adger et al. 2007). The recent AR5 reaffirms these views while stating that significant work remains in understanding context-specific and inter-related determinants of household adaptive capacities (Klein et al. 2014).

Indeed, a positive relationship between wealth and adaptive capacity is plausible: poor households are constrained by a lack of adequate financial resources.

F. Kosmowski (✉)
CGIAR Standing Panel on Impact Assessment, ILRI, Gurd Sholla Area,
P.O. Box 5689, Addis Ababa, Ethiopia
e-mail: frederic.kosmowski@fao.org

R. Lalou
LPED UMR D 151, Institut de Recherche pour le Développement,
Aix Marseille University, 13331 Marseille, France

W. Leal Filho et al. (eds.), *Climate Change Adaptation in Africa*,
Climate Change Management, DOI 10.1007/978-3-319-49520-0_45

Capital-intensive adaptation measures such as irrigation systems, improved crop varieties, climate-risk insurance and diversification of farm operations may not be affordable. Given the potential threat to food security, economically disadvantaged households may also demonstrate higher risk aversion in the face of innovation (Yesuf and Bluffstone 2007). However, taking into account the diversity of poverty outcomes that may influence adaptive capacities, the relationship becomes more complex.

Recent literature on the determinants of adaptations and adaptive capacities in sub-Saharan Africa provide evidence about the importance of factors related to economic resources. Evidence exists that farm income can have a positive and significant effect on several adaptation practices (Bryan et al. 2009; Silvestri et al. 2012; Gebrehiwot and van der Veen 2013). Access to credit has also been found positively related in several sub-Saharan contexts (Bryan et al. 2009; Hisali et al. 2011; Fosu-Mensah et al. 2012; Okonya et al. 2013), although not in all (Cavatassi et al. 2011; Wood et al. 2014). Using total land ownership as a proxy for traditional wealth, no association was found with the adoption of modern maize varieties in Ghana (Doss and Morris 2001) and Malawi (Fisher and Snapp 2014). Livestock ownership appears as a positive factor related to adaptation in Marenya and Barrett (2007), Bryan et al. (2009), Gebrehiwot and van der Veen (2013) and Comoé and Siegrist (2015). However, other studies demonstrate conflicting evidence (Cavatassi et al. 2011; Silvestri et al. 2012; Fisher et al. 2015).

Several lessons emerge from this overview of the literature. First, it is clear that most, if not all research on the determinants of household adaptation and adaptive capacity have conceptualized poverty through the economic lens. Wealth proxies measured through income or assets do not take into account the multidimensional nature of poverty, an important component of the Sustainable Development Goals. Second, conflicting findings in part of the existing literature suggest that different wealth proxies may capture different aspects of poverty and therefore highlight different linkages with household adaptive capacities. However, too few studies have investigated the overlaps between the heterogeneity of wealth proxies found in the climate change literature. In Ethiopia, Tache and Sjaastad (2010) found that income and expenditure indicators produce poverty rates that are significantly higher than traditional measures based on herd size. Arguably, the heterogeneity of wealth proxies used in the literature makes any definitive statement about the poverty-adaptive capacity nexus hasty. Third, there is a consensus that different wealth proxies are associated with some adaptation practices, but not with all. Although particular adaptations practices are manifestations of adaptive capacities (Adger et al. 2007), the total number of changes made by farmers over a given period arguably provides a better indication of potential adaptive capacities. Interestingly, recent works relying on such a proxy bring further conflicting evidence. Using rich datasets, Kristjanson et al. (2012) found no evidence of an association of access to credit and land ownership with the number of crop-related changes. Wood et al. (2014) observed that at the regional level, wealth appears to be significantly related to the likelihood of crop-related changes. However, when moving to the multi-region level, lower asset farmers are the most likely to change their on-farm practices. The authors explain this result by the vital necessity for these farmers to make low-cost adjustments to their cropping strategies in the face of vulnerability. Wealthier households, by contrast, have better access to off-farm incomes to mitigate risk.

In this article, we explore the effect of poverty on adaptive capacities in northern Benin. We hypothesize that the relationship between poverty and climate change adaptive capacity is likely to depend on the aspects of poverty captured by each measure. Little research has put into perspective conventional poverty indices in the context of sub-Saharan Africa; and even fewer have investigated households' multiple privations. Indeed, several other factors may enhance household adaptive capacity and modify the assumed linear relationship with poverty. It is therefore crucial to control the relationship of interest with additional variables related to the farm, household, social institutions and perceptions of climate change. Moreover, poverty and its manifestations may often interact with other predictors—thus giving a blurred and misleading picture. Consequently, exploring potential interactions is an important step for analysis. The aim of this study is to fill a gap in the literature and shed light onto the poverty-adaptive capacity nexus using a case study in northern Benin. This analysis is exploratory and has been designed to test the hypothesis of an association between three poverty indices and two proxies of adaptive capacities at the household level, or simply rule it out.

1 Data and Methods

(a) Study area

Located in western Africa, Benin is dependent on rainfed agriculture, which is a source of income for 45% of the population (World Bank 2010). In the Djougou district, in the north of the country, small-scale agriculture with hand tools is practiced by eight inhabitants out of ten in a context of land abundance. The area is prone to food insecurity and malnutrition (WFP 2011). The most important crops grown are maize, yam and sorghum. Although the Yoa ethnic group is dominant, the area is populated by several ethnic groups, including Ditamaris, Lokpas, Baribas and Peuhls. The rainfall pattern in Djougou is unimodal, with an average annual rainfall of 1200 mm over the last decades (Mahé et al. 2012) in a rainy season lasting from May to October. The Djougou district has experienced about a 1.2 °C rise in minimum temperatures over the past five decades. The tendency is less clear for maximum temperatures, where the increase is however about 1 °C. Rainfall experienced a quasi-linear decrease during the 1950s and the 1960s. Since the early 1980s, the regional climate seems to have recovered and no significant change in rainfall amount occurred since the 1990s (Kosmowski et al. 2015). The area has, however, observed high year-to-year rainfall variability and the 2007 season, particularly, delivered relatively few rains.

The analyses presented here are based on cross-sectional household survey data collected in July 2013 through exhaustive sampling. Two transect belts were used to survey households located along the north and south roads of Djougou town, as shown in Fig. 1. The household questionnaire collected information on household characteristics, land access, food security, expenditures and assets. In each household, up to two farmers who grew staple crops during the 2012 agricultural season were randomly selected to answer an individual questionnaire. When eligible, the household head was automatically selected. The individual questionnaire comprised questions on demographic characteristics, agricultural practises and innovations, coping strategies, climate change perception and social institutions. Close supervision by the study's author ensured collection of high quality data. Households recently settled in the area were dropped from the dataset, resulting in a total sample of 1006 households and 1136 farmers.

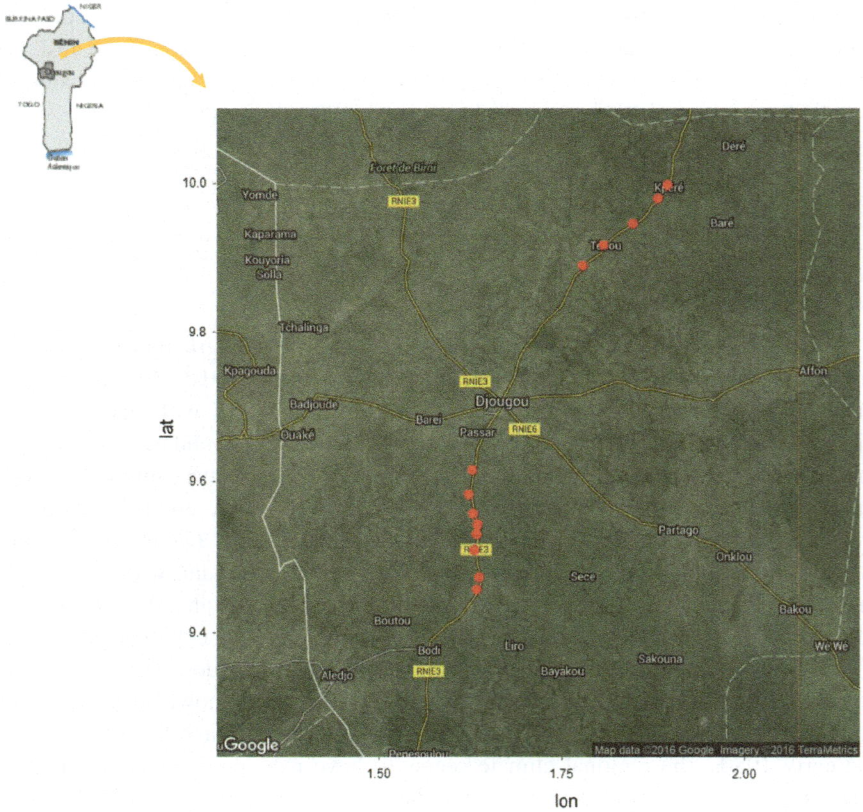

Fig. 1 Research sites locations

(b) **Poverty measures**

The general consensus in the literature is that there is no single measure that accurately captures the complexity of poverty (Stiglitz et al. 2009). In an attempt to explore different aspects of the phenomenon, three alternative indices of poverty were computed and used in a continuous form. Measures of monetary poverty have been implemented in official poverty statistics for most countries in the world since the major contribution of Ravallion et al. (1991). In the context of developing countries, consumption is considered a better indicator of poverty measurement than income (Banerjee and Duflo 2007). We constructed consumption aggregates from the survey responses using the methodology outlined by Deaton and Zaidi (2002) and Beegle et al. (2012). The monetary poverty index is formed by three sub-aggregates: food expenditures, non-food expenditures and durable goods. The survey used a seven-day consumption recall for 32 food items, which were adapted to the local context. For items that were consumed but not purchased, the item's price from the nearest market at the time of data collection was used to assess its value. Non-food expenditures were collected through a 30-day recall for 13 items. Expenditures related to education, repairing and ceremonies were estimated on a yearly basis and added to this sub-aggregate. The value of services that the household received from all the durable goods was estimated for the items for which the information was collected. Finally, total household expenditures were deflated using an adult equivalent scale.

Recent years have seen a renewed interest among scholars and policy makers for a measure of poverty that can better reflect Sen's capability approach (Stiglitz et al. 2009; Alkire and Santos 2014; Olsson et al. 2014). These efforts have led to the creation of a multidimensional poverty index (Alkire and Santos 2014), published in the Human Development Report (UNDP 2013) and monitored by the Sustainable Development Goals. Indeed, affordable quality services, such as water, health and education are frequently not provided through the market and are not guaranteed to improve with rising incomes. Similar to the methodology presented in Alkire and Santos (2014), a multidimensional poverty index was computed. Overall, the index is composed of ten indicators gathered into three dimensions of equal weight: health, education and standard of living. We used households' dietary diversity, a relevant proxy of an individual's nutritional status (Moursi et al. 2008; Headey and Ecker 2013) to account for privations in nutrition. The multidimensional poverty index computed is robust to a plausible range of alternative weights. The index ranges from 0 to 100, with 0 indicating a complete lack of privations, and 100 suggesting privations in all indicators.

Monetary and multidimensional poverty indices are conventional measures monitored by the Sustainable Development Goals to facilitate macro-social comparisons. However, it is recognized that a measure of poverty should take into account the social and cultural aspects of poverty and that these aspects are context-specific (Townsend 1993). In rural sub-Saharan societies, components of traditional wealth include agricultural land and livestock (Swinton 1988; Garenne 2015). It is therefore not surprising that livestock ownership is used as a wealth proxy in many studies (Silvestri et al. 2012; Gebrehiwot and van der Veen 2013; Comoé and Siegrist 2015). In order to construct an index of traditional poverty, total land area owned and number of livestock per capita were calculated for each household. Livestock was separated into two categories: cattle and small ruminants (sheep and goats). The index was computed by aggregating the three categories.

(c) **Measures of adaptive capacities**

The concept of adaptive capacity is complex and several pieces of past research have used the current status as a suitable proxy for identifying future adaptive capacity (Below et al. 2011; Kristjanson et al. 2012; Panda et al. 2013; Wood et al. 2014). Whether or not adaptive capacity is drawn upon to bring about adaptation depends on several uncertainties (Vincent 2007) and these measures of adaptive capacity generally capture an adaptation potential rather than a real capacity to adapt to future climate change. These limitations notwithstanding, adaptive capacity is defined as the ability of a farmer to make changes and to respond to climate-related shocks in a way that mitigates risk. Two proxies of potential adaptive capacities are used in this research: crop-related changes introduced in the last 10 years and perceived coping strategies in case of crop failure. It is argued that farmers applying a high number of innovations, or perceiving a high number of feasible strategies in case of crop failure, can better respond to future climate change impacts. On the one hand, innovation is an essential element of adaptation (Adger et al. 2007). The development of new management techniques and the adoption of new agricultural technologies represent a trade-off between risk and opportunity. Changes in farming practices demonstrate a capacity to experiment and modify a given production system. Thus, considering whether farmers made crop-related changes over the last decade shed light into their ability to make future changes in response to risk. On the other hand, it has been shown that cognitive barriers can impede a household's adaptive capacities (Grothman and Patt 2005; Weber 2010; Dow et al. 2013). Grothman and Patt (2005) found that adaptive actions are also determined by perceived abilities to adapt. In response to climate change and variability, several coping strategies are used by rural households (Dercon and Christiansen 2011; Traerup and Mertz 2011; Hisali et al. 2011). Given the increase in drought frequency and intensity expected over the following century, the number of perceived coping strategies in case of crop failure is an interesting proxy to explore. Since more innovative households are likely to better respond to future droughts, and indeed perceive themselves in a better capacity to do so, we assume a positive relationship between both proxies.

(d) **Statistical analysis**

Generalized linear models (GLM) are usually recommended for the analysis of count data (Nelder and Wedderburn 1972). To analyze the relationship between the three poverty indices and potential adaptive capacities, we used a generalized linear model of the following form:

$$Y_{i \sim binomial\ (\mu_i)}$$

$$\mu_i = \log_e \frac{\mu_i}{1 - \mu_i}(\propto + \beta_1 X_{i1} + \beta_2 X_{i2} + \beta_3 X_{i3} + \cdots + \beta_k X_{ik})$$

where μ_i is a count variable that can take values between 1 and 11 for crop-related changes and 0 and 9 for perceived coping strategies. The dependent variables are not normally distributed and we used a binomial estimator with a quasi-maximum likelihood for model estimation (Wooldridge 2010). This assumes that both dependent variables are binomially distributed, such as $E[Y_i] = \mu$ where $0 \leq \mu_1 \leq 1$. Thus, Y_i represents the proportion of success on a limited number of n_i binary trials. X_i corresponds to the dependent variable score for a farm household x_i. The logit link-function is represented by $\log_e \frac{\mu_i}{1-\mu_i}$ and $\propto + \beta_1 X_{i1} + \beta_2 X_{i2} + \cdots + \beta_k X_{ik}$ is a linear model.

Since there is a suspicion that the relationship between poverty and adaptive capacities suffers from endogeneity in the form of simultaneity, we choose household distance from population centers as an instrumental variable. Household distance from population centers is likely not to affect adaptive capacities but is a good predictor of poverty: distance can increase transport costs, limit market opportunities and thus qualifies as an exogenous determinant of poverty. We tested the assumption that distance from population centers is a strong instrument by calculating the F statistic. According to Staiger and Stock (1997), the weak instrument hypothesis is rejected if an F test is greater than 10. For the crop-related changes model, F statistics delivered values >35 for all poverty measures. Concerning the perceived coping strategies model, values were 18.5, 6.6 and 15.7 for monetary, multidimensional and traditional poverty measures respectively. We then tested for endogeneity using a Hausman specification test for the consistency of the three-stage least squares estimation (Hausman 1978). The tests indicate that there is no endogeneity bias for monetary poverty, while endogeneity is an issue for multidimensional and traditional poverty measures.

The independent variables were based on a review of the existing literature (Grothman and Patt 2005; Pelling and High 2005; Bryan et al. 2009; Ouédraogo et al. 2010; Weber 2010; Silvestri et al. 2012; Fosu-Mensah et al. 2012). Prior to including the independent variables in the regression analysis, we tested them for collinearity. Table 1 lists the variables included in the empirical models.

All independent variables were standardized (the mean subtracted from individual values divided by standard deviation). The coefficient estimate states by what percentage the adaptive capacity proxies change when a poverty index changes by one unit. Statistical analysis was performed in R version 3.1.2 (R Development Core Team 2015).

Table 1 Description of independent variables used in empirical models

Independent variable	Type
Household size	Numeric
Gender	Categorical (male, female)
Age	Numeric
Years of education	Numeric
Main occupation	Categorical (agriculture, other)
Monetary poverty index	Numeric
Multidimensional poverty index	Numeric
Traditional poverty index	Numeric
Total land cultivated	Numeric
Market orientation	Numeric
Soil erosion	Categorical (yes, no)
Past crop failures	Numeric
Village size	Categorical (<100, >100)
Membership in farmer group	Categorical (yes, no)
Access to agricultural extension services	Categorical (yes, no)
Access to credit	Categorical (yes, no)
Participation in village ceremonies	Categorical (yes, no)
Trust between residents	Categorical (yes, no)
Perception of temperature change	Categorical (yes, no)
Perception of annual rainfall change	Categorical (yes, no)
Perception of change on the onset	Categorical (yes, no)
Distance from population center	Numeric

2 Results

About half (56%) of the households surveyed are extremely poor in monetary terms. Daily median household consumption is about 531 FCFA (or 1.07$), with food expenditures accounting for 62% of household budgets. A vast majority of households (86%) are considered multidimensionally poor by international standards, meaning that they are deprived in more than 30% of the weighted indicators. Mean land area owned per capita is 13.6 ha, which is large when compared to other sub-Saharan contexts. However, nearly two thirds (64%) of households do not own any livestock. For livestock owners, the mean number per capita is 2.8 for cattle and 3.1 for small ruminants. A significant lack of overlap was found between the three poverty measures. The Spearman's r correlations among the three poverty measures were not correlated, and were all significant at the $p < 0.01$ level. This result, further explored in Kosmowski (2016), suggests that the three indices differ in the underlying aspect of poverty they attempt to capture.

Figure 2 shows the percentage of farmers that have introduced crop-related changes in the last ten years. Tree planting was mentioned as the most widely adopted innovation. Farmers generally favour cashew trees, declaring that it can generate income after only three years. Eucalyptus and teak trees are also planted and generally cut after a few years for selling or construction purposes. The second innovation used in our study area is to change planting dates. The practice, used by farmers to mitigate risk in response to year-to-year variability, has been reported in several other sub-Saharan contexts (Fosu-Mensah et al. 2012; Bryan et al. 2009; Yegbemey et al. 2013). Following the same purpose, sequential seeding is also practiced, but to a lesser extent. Expanding the areas of maize and yam grown, the two main staple crops, was practiced by one in five households. However, a lack of access to labour was reported as a major constraint. Farmers are also practicing soil management methods such as inter-cropping and crop rotations. These practices help to combat low fertility by increasing soil nutrients. Farmers in several sub-Saharan contexts reported switching crop species (Bryan et al. 2009). In the surveyed area, the last decade has seen a major increase in soybean adoption. In some villages surveyed, half of farmers are now growing soybean. If the crop represents a commercial opportunity, it also has a high nutritional content. Development projects have promoted the making and selling of soybean cheese that is now replacing meat in several households. Farmers in the Djougou district have also introduced early maturing varieties in their farming systems. Modern maize varieties, such as EV DT 97 STR W and 2000 syn. EE W provide drought tolerance as well as early maturation. The introduction of early maturing varieties of yam (Yanouha, Morkonnoudje, Coutonouma or Kpataga), generally imported from surrounding areas, was also reported by farmers. Nevertheless, geographic and economic barriers persist and limit farmers' adoption of early maturing varieties. Over the last decade, farmers introduced a mean number of 2.5 crop-related changes.

Farmers were also asked whether they would be able to implement various coping strategies in case of future crop failure. The survey identified using past savings, reliance on help from relatives, selling assets, reducing consumption,

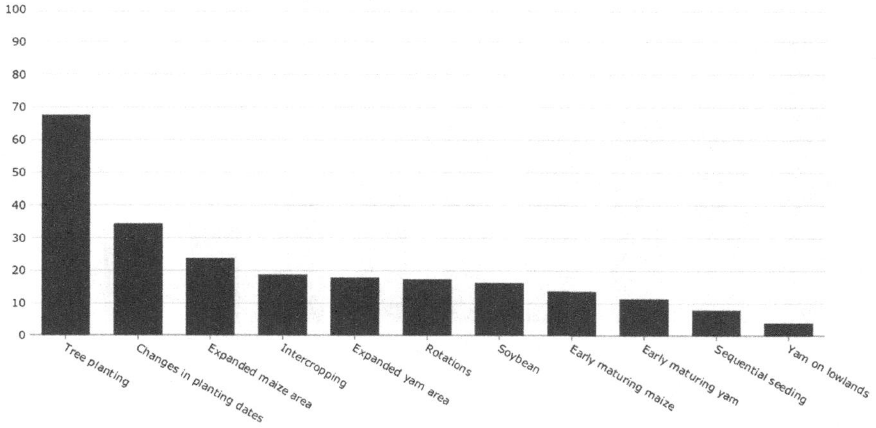

Fig. 2 Crop-related changes introduced in last 10 years (% of households)

borrowing and wage employment as potential coping strategies. Other strategies related to agricultural production include expanding areas and changing crop varieties. Results presented in Fig. 3 suggest that farmers perceive some coping strategies as more feasible than others. A large number of households consider relying on saving as a feasible strategy. Around 60% of households also reported that reliance on help from relatives, selling assets, reducing consumption and expanding farm area would be possible options. One third of households consider borrowing as a potential strategy and 27% think they would be able to reduce food expenditures. Overall, households perceive a mean number of 5.1 coping strategies as feasible in case of crop failure. Our assumption that more innovative households perceive a higher range of coping strategies in case of crop failure is supported by the data and a positive relationship exists between both proxies of adaptive capacities ($t = 2.25$; $p = 0.05$).

The analysis is articulated in two parts. For both proxies of adaptive capacities, we first study their association with monetary, multidimensional and traditional poverty indices. We then perform an analysis of the factors influencing households' adaptive capacities by including variables related to the farm, household, social institutions and perceptions of climate change in the regression models. Interactions between poverty indices and other covariates are explored, as well as the form of the relationship.

(a) Association between poverty measures and crop-related changes

The association between poverty measures and crop-related changes was first examined by entering all three indices as independent variables into a binomial generalized linear model (Table 2). We observe that monetary poverty is positively associated with crop-related changes while multidimensional poverty appears negatively associated. One additional change in household consumption is associated with a 5% increase in crop-related changes while one additional change in

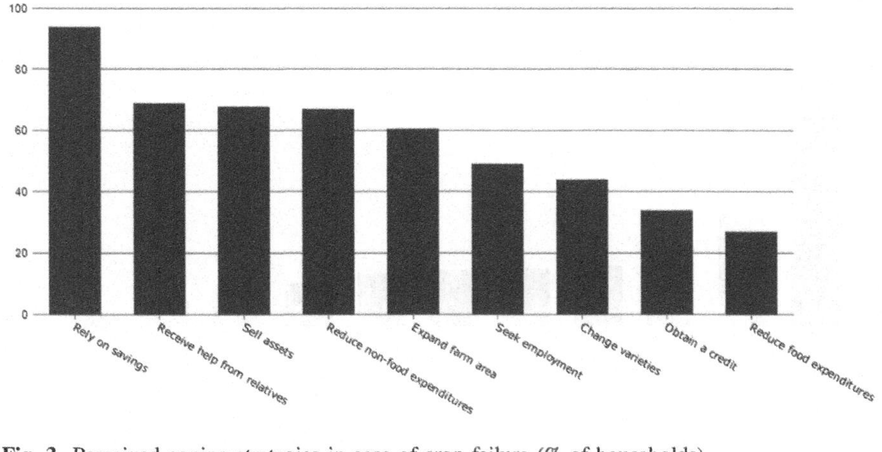

Fig. 3 Perceived coping strategies in case of crop failure (% of households)

household privations is associated with a 11% decrease. These associations are robust to alternative modelling choices such as Poisson and negative binomial.

Figure 4 provides a representation of the significant relationships observed between poverty indexes and crop-related changes. It is clear from Fig. 4a that the relationship of consumption expenditures with crop-related changes is curvilinear, not linear. We observe that monetary poor households are less prone to innovation than other households and that the advantages associated with economic resources accrue mainly from not being extremely poor. On the middle range, the curve displays a relative stability. Wealthier households distinguish themselves from the other by their higher number of crop-related changes, although a lesser number of observations forces us to be cautious. In the case of household privations (Fig. 4b), the curve is almost flat, but however reflect the fact that multiply deprived households are slightly less prone to innovation than other households.

Household changes in farming practices can be associated with other factors, which could bias the estimated results. Consequently, we re-estimated the regression models including factors related to the farm, household, social institutions and perceptions of climate change. As stated earlier, our analysis of cross-sectional data can only estimate associative relationships between these parameters and household adaptive capacities. The usual sources of statistical endogeneity, and particularly simultaneity, are likely to compromise the causal identification of the multidimensional and traditional poverty measures presented in Table 3. The results confirm the positive association of the monetary index with crop-related changes. While there was evidence of a relationship between multidimensional poverty and crop-related changes at a bivariate level, controlling for other factors resulted in the relationship disappearing. Female-headed household are significantly less innovative. A negative relationship is observed between age and crop-related changes, suggesting that younger farmers are making more changes in their farming practices than older ones. Other significant factors include the surface cultivated, market orientation, past crop failures and farmer's participation in social institutions. An exploration of potential interactions between the poverty indices and other covariates did not deliver additional information on the determinants of a household's crop-related changes.

Table 2 Association between monetary, multidimensional and traditional poverty indices and crop-related changes introduced in last 10 years

Independent variables	Estimate	Pr (>|z|)
Monetary poverty index	0.05 (0.02)	0.02**
Multidimensional poverty index	−0.11 (0.02)	<0.0001****
Traditional poverty index	−0.01 (0.02)	0.63
Distance from population center	0.00	<0.0001****
(Intercept)	−1.57 (0.04)	<0.0001****

Model was estimated with a binomial estimator. Coefficient estimates are shown with SEs in parentheses

, ** Statistically significant at the 0.05 and 0.001 level respectively

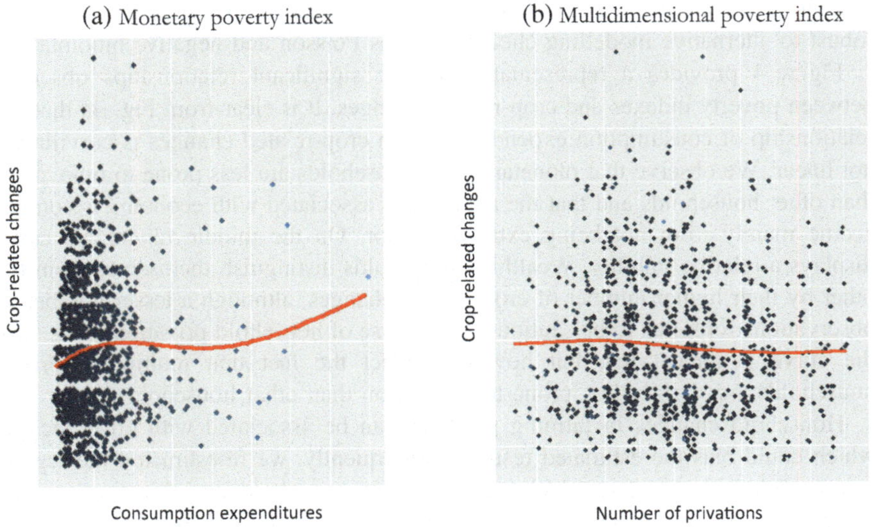

Fig. 4 The association of household consumption expenditures and household privations with crop-related changes introduced in last 10 years

(b) Association between poverty measures and perceived coping strategies

We then analyzed the relationship of the three poverty indices with households' perceived coping strategies. Table 4 shows results of regression models, where the number of perceived coping strategies is used as the dependent variable and the three poverty indices as independent variables.

Among the three poverty indices, multidimensional poverty is the only index to be significantly associated with perceived coping strategies. The model produces a negative coefficient of $\beta = -0.12$ for the multidimensional poverty index, suggesting that one additional change in privations is associated with a 12% decrease in perceived coping strategies. This association is robust to alternative modelling choices such as Poisson and negative binomial. The illustrative curve of the relationship between the multidimensional poverty index and perceived coping strategies is presented in Fig. 5. Two stages can be distinguished. The stability of the curve firstly suggests an absence of relationship for households facing little or no privations. At the second stage, perceived coping strategies diminish as a household's privations increase and the curve's slope decreases more sharply as we approach multiply deprived households.

We then analyzed the association between poverty and adaptive capacity by including additional independent variables into the regression model. After control of additional factors related to the farm, household, social institutions and perceptions of change, the association of multidimensional poverty with perceived coping strategies is confirmed ($\beta = -0.10$; $p = <0.0001$). In addition, social institutions emerge as an important cohesive force in fostering adaptive capacities:

Table 3 GLM model of the factors affecting the number of crop-related changes introduced in the last 10 years and the number of perceived coping strategies in case of crop failure

	Crop-related changes		Perceived coping strategies	
	Estimate	Pr (>|z|)	Estimate	Pr (>|z|)
Household				
Household size	0.03 (0.02)	0.23	0.01 (0.02)	0.56
Gender (Ref = Female)	−0.44 (0.09)	<0.0001****	−0.20 (0.08)	0.01**
Age	−0.09 (0.02)	<0.0001****	−0.01 (0.02)	0.50
Years of education	0.06 (0.02)	0.01**	−0.04 (0.02)	0.09*
Main occupation (Ref = Other)	0.02 (0.05)	0.69	−0.08 (0.05)	0.08*
Monetary poverty index	0.03 (0. 02)	0.15	0.00 (0.02)	0.98
Multidimensional poverty index	−0.06 (0.03)	0.03**	−0.10 (0.02)	<0.0001****
Traditional poverty index	−0.01 (0.02)	0.57	0.00 (0.02)	0.7
Farm				
Market orientation	0.10 (0.03)	<0.0001****	0.05 (0.02)	0.06*
Soil erosion	0.04 (0.07)	0.59	0.03 (0.06)	0.63
Past crop failures	0.04 (0.02)	0.09*	−0.02 (0.02)	0.36
Social institutions				
Village size (Ref = Small village)	−0.11 (0.05)	0.04**	−0.17 (0.05)	<0.0001****
Membership in farmer group	0.22 (0.05)	<0.0001****	0.14 (0.05)	0.003***
Access to agricultural extension services	0.31 (0.07)	<0.0001****	−0.03 (0.06)	0.66
Access to credit	0.05 (0.06)	0.35	−0.12 (0.05)	0.02**
Participation in village ceremonies	0.26 (0.05)	<0.0001****	0.18 (0.04)	<0.0001****
Number of friends	0.02 (0.02)	0.47	−0.12 (0.02)	<0.0001***
Trust between residents	0.17 (0.08)	0.04**	0.37 (0.07)	<0.0001****
Perceptions of climate change				
Perception of temperature change	0.02 (0.06)	0.77		

(continued)

Table 3 (continued)

	Crop-related changes		Perceived coping strategies	
	Estimate	Pr (>|z|)	Estimate	Pr (>|z|)
Perception of annual rainfall change	0.13 (0.12)	0.28		
Perception of change on the onset	0.12 (0.06)	0.06*		
Instrumental variable				
Distance from population center	0.00 (0.00)	<0.0001****	0.00 (0.00)	0.14
(Intercept)	−2.61 (0.17)	<0.0001****	−0.23 (0.11)	0.03**

All independent variables were standardized for comparison
*, **, ***, **** Statistically significant at the 0.1, 0.05, 0.01 and 0.001 level respectively

Table 4 Association between monetary, multidimensional, traditional poverty indices and perceived coping strategies in case of crop failure

Independent variables	Estimate	Pr (>|z|)
Monetary poverty index	−0.03 (0.02)	0.14
Multidimensional poverty index	−0.12 (0.02)	<0.0001****
Traditional poverty index	−0.03 (0.02)	0.20
Distance from population center	0.00	<0.0001****
(Intercept)	0.35 (0.02)	<0.0001****

Model was estimated with a binomial estimator. Coefficient estimates are shown with SEs in parentheses
**** Statistically significant at the 0.001 level

participation in village ceremonies, trust and solidarity between farmers are strongly associated with perceived coping strategies.

3 Discussion

The question of how to increase adaptive capacity of the poorest is of high relevance for research and policy. Poverty and adaptive capacity are tricky concepts to define and complex linkages may exist between them. In this article, we examined the association between poverty and adaptive capacity using a case study in northern Benin. The results challenge the widespread assumption that poverty readily correlates with adaptive capacities at the household level. What we are seeing is that not all poverty indices are associated with potential adaptive capacities and hat these relationships are not linear.

Consistent with prior findings, this research showed that poverty indices capture different aspects of poverty (Ruggeri-Laderchi et al. 2003; Tache and Sjaastad

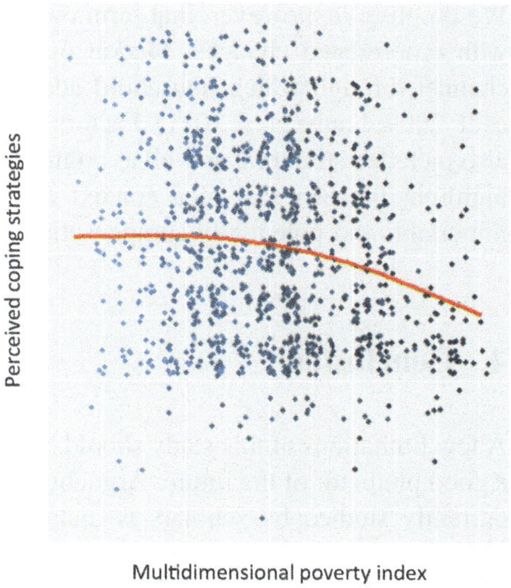

Fig. 5 The association of the multidimensional poverty index with perceived coping strategies

2010). Caution is warranted and using more than one indicator of poverty is advisable. Since monetary and multidimensional poverty reduction are important parts of the Sustainable Development Goals agenda, the use of these conventional indicators should be encouraged in future studies. It is also important for policy makers to be aware of which aspects of poverty each indicator portrays.

It should be acknowledged that the context of the study area shows a high level of homogeneity in terms of household's economic status. Thus, agricultural innovations and coping strategies in northern Benin are reflective of household's financial constraints and can be characterized as low-cost adaptations. Adaptive capacities of the wealthy, who represent a small proportion of our sample, may indeed be of very different nature. Our results suggest that multiple deprivations are negatively associated with both crop-related changes and perceived coping strategies. We also found that an improvement of economic status (through monetary or traditional asset growth) is associated with increased innovations, but only for the poorest households. Results reveal that these relationships are not linear. This suggests that we cannot assume that rich households will demonstrate higher adaptive capacities, nor can we assume that households' adaptive capacities will increase as they become wealthier, or less deprived. Further, as projections using a linear or non-linear approach may result in substantially different outputs, accounting for the non-linear nature of the association between poverty and adaptive capacities is crucial when assessing future climate change impacts.

However, we did not find evidence of an association of traditional poverty with both proxies of adaptive capacities. This is surprising, since land and livestock ownership, used as a proxies for wealth, has been found to be positively associated with agricultural innovation in various contexts (Maddison 2007; Bryan et al. 2009; Gebrehiwot and van der Veen 2013; Okonya et al. 2013; Fisher et al. 2015).

We do observe, however, that farm commercial orientation is positively associated with crop-related changes. Market opportunities may thus represent an effective channel through which household adaptive capacities can be stimulated (Barbier et al. 2009; Laube et al. 2011). Factors associated with increased adaptive capacities also included structural as well as cognitive measures of social capital. These results highlight the fact that in a context of rural poverty, social institutions play an important, and potentially compensating role for climate change adaptation.

4 Conclusion

A few limitations of this study should be acknowledged. First, the past is not always a good predictor of the future. Arguably, the way the concept of adaptive capacity is currently studied by scholars is dictated by empirical measures, rather than the opposite. Studies based on longitudinal datasets should explore the robustness of adaptive capacity proxies. Second, cross-sectional surveys based on case studies have a number of caveats. Results should be interpreted as context-specific and are only representative of the sample itself. With this in mind, the analysis does rely on a large, exhaustive sample, thus providing better accuracy than other sampling methods. Third, the issue of endogeneity, through simultaneity, measurement error, and unobserved heterogeneity is one that we should be aware of. Therefore, the interpretation of our results is one of statistical association and not necessarily causality. Whether and to what extent poverty measures can lead to lower adaptive capacities is an important avenue for future research that should clarify the causality behind the estimated associations. Indeed, the direction of the causality has important policy implications. If poverty limits adaptive capacities, targeting poor households is essential to foster innovation and coping capacities. Safety nets should also be encouraged to cope with the impacts of climate change-related events. On the contrary, if households' adaptive capacities affect poverty, a stronger emphasis should be put on capacity building, by promoting climate-smart adaptations through agricultural extension services. The possibility of a bidirectional causality should not be ruled out and stringent experimental studies are needed to assess the direction of causality.

This paper hopes to contribute to the existing literature by demonstrating that a purely economic view of the relationship between poverty and adaptive capacities, most often relying on a single poverty measure, is insufficient. Since an overestimation of the economic and technologic factors may not give justice to the complexity of the sub-Saharan context, the common discourse about the poverty-adaptive capacity nexus certainly needs some adjustments. The development of a conceptual framework taking into account the linkages between different indices of poverty and adaptive capacity as well as the interactions between these indices and other determinants of adaptive capacities is worth pursuing (Magnan 2010). Most likely, these linkages are different in different contexts. A clear understanding of the relationship linking poverty and households' adaptive capacity remains a crucial step for the design of development programmes that can achieve the goal of eradicating poverty under a changing sky.

Acknowledgements This work is part of the ESCAPE project (ANR-10-CEPL-005) funded by the French National Research Agency (ANR). We would like to thank the participants of the Symposium on Climate Change Adaptation in Africa which took place in Addis Ababa in February 2016.

References

Adger, W. N., Agrawala, S., Mirza, M. M. Q., Conde, C., O'Brien, K., Pulhin, J., et al. (2007). In M. L. Parry, O. F. Canziani, J. P. Palutikof, P. J. van der Linden, & C. E. Hanson (Eds.), *Assessment of adaptation practices, options, constraints and capacity. Climate Change 2007: Impacts, adaptation and vulnerability*. Contribution of Working Group II to the Fourth Assessment Report of the Intergovernmental Panel on Climate Change. Cambridge University Press, Cambridge, UK.

Alkire, S., & Santos, M. E. (2014). Measuring acute poverty in the developing world: Robustness and scope of the multidimensional poverty index. *World Development, 59,* 251–274.

Banerjee, A., & Duflo, E. (2007). The economic lives of the poor. *Journal of Economic Perspectives, 21*(1), 141–168.

Barbier, B., Yacouba, A., Karambiri, H., Zorome, M., & Some, B. (2009). Human vulnerability to climate variability in the Sahel: Farmers' adaptation strategies in Northern Burkina Faso. *Environmental Management, 43,* 790–803.

Beegle, K., De Weerdt, J., Friedman, J., & Gibson, J. (2012). Methods of household consumption measurement through surveys: Experimental results from Tanzania. *Journal of Development Economics, 98*(1), 3–18.

Below, T. B., Mutabazi, K. D., Kirschke, D., Franke, C., Sieber, S., Siebert, R., et al. (2011). Can farmers' adaptation to climate change be explained by socio-economic household-level variables? *Global Environmental Change, 22,* 223–235.

Bryan, E., Deressa, T. T., Gbetibouo, G. A., & Ringler, C. (2009). Adaptation to climate change in Ethiopia and South Africa: Options and constraints. *Environmental Science & Policy, 12*(4), 413–426.

Cavatassi, R., Lipper, L., & Narloch, U. (2011). Modern variety adoption and risk management in drought prone areas: Insights from the sorghum farmers of eastern Ethiopia. *Agricultural Economics, 42,* 279–292.

Comoé, H., & Siegrist, M. (2015). Relevant drivers of farmers' decision behavior regarding their adaptation to climate change: A case study of two regions in Côte d'Ivoire. *Mitigation and Adaptation Strategies for Global Change, 20,* 179–199.

Deaton, A., & Zaidi, S. (2002). *Guidelines for constructing consumption aggregates for welfare analysis*. LSMS Working Paper #135, The World Bank, Washington D.C., USA.

Deme, A., Galle, S., & Hourdin, F. (2015). «Les projections du climat en Afrique de l'Ouest: évidences et incertitudes». In B. Sultan, R. Lalou, M. Amadou Sanni, A. Oumarou, & A. Soumere (Eds.), Les sociétés rurales face aux changements climatiques et environnementaux en Afrique de l'Ouest, Editions IRD, Paris, France.

Dercon, S., & Christiaensen, L. (2011). Consumption risk, technology adoption and poverty traps: Evidence from Ethiopia. *Journal of Development Economics, 96*(2), 159–173.

Doss, C., & Morris, M. (2001). How does gender affect the adoption of agricultural innovations? The case of improved maize technology in Ghana. *Agricultural Economics, 25,* 27–39.

Dow, K., Klein, R. J. T., Midgley, G. F., Berkhout, F., Preston, B. L., Midgley, G., et al. (2013). Limits to adaptation. *Nature Climate Change, 3,* 305–307.

Fisher, M., Abate, T., Lunduka, R. W., Asnake, W., Alemayehu, Y., & Madulu, Y. A. (2015). Drought tolerant maize for farmer adaptation to drought in sub-Saharan Africa: Determinants of adoption in eastern and southern Africa. *Climatic Change, 133*, 283–299.

Fisher, M., & Snapp, S. (2014). Smallholder farmers' perceptions of drought risk and adoption of modern maize in Southern Malawi. *Experimental Agriculture, 50*(4), 533–548.

Fosu-Mensah, B. Y., Vlek, P. L. G., & Maccarthy, D. S. (2012). Farmers' perception and adaptation to climate change: A case study of Sekyedumase district in Ghana. *Environment, Development and Sustainability, 14*(4), 495–505.

Garenne, M. (2015). Traditional wealth, modern goods, and demographic behavior in rural Senegal. *World Development, 72*, 267–276.

Gebrehiwot, T., & van der Veen, A. (2013). Farm level adaptation to climate change: The case of farmers in the Ethiopian highlands. *Environmental Management, 52*, 29–44.

Grothmann, T., & Patt, A. (2005). Adaptive capacity and human cognition: The process of individual adaptation to climate change. *Global Environmental Change, 15*(3), 199–213.

Hausman, J. A. (1978). A generalized specification test. *Economics Letters, 8*, 239–245.

Headey, D., & Ecker, O. (2013). Rethinking the measurement of food security: From first principles to best practice. *Food Security, 5*, 327–343.

Hisali, E., Birungi, P., & Buyinza, F. (2011). Adaptation to climate change in Uganda: Evidence from micro level data. *Global Environmental Change, 21*(4), 1245–1261.

IPCC. (2013). Summary for policymakers in climate change 2013: The physical science basis. Contribution of Working Group I to the Fifth Assessment Report of the Intergovernmental Panel on Climate Change. In T. F. Stocker, et al. (Eds.), Cambridge, United Kingdom and New York, NY, USA: Cambridge University Press.

Klein, R. J. T., Midgley, G. F., Preston, B. L., Alam, M., Berkhout, F. G. H., Dow, K., et al. (2014). *Adaptation opportunities, constraints, and limits, in Climate Change 2014: Impacts, adaptation, and vulnerability*. Part A: Global and Sectoral Aspects. Contribution of Working Group II to the fifth assessment report of the intergovernmental panel on climate change. In C. B. Field, et al. (Eds.), Cambridge, United Kingdom and New York, NY, USA: Cambridge University Press.

Kosmowski, F. (2016). «Les Objectifs du développement durable convergent-ils dans l'identification des ménages pauvres? Une comparaison empirique au nord Bénin». *Mondes en développement, 174*, 44. doi:10.3917/med.174.0000

Kosmowski, F., Lalou, R., Sultan, B., Ndiaye, O., Muller, B., Galle, S., et al. (2015). «Changements climatiques: observations scientifiques et perceptions des populations dans trois contextes ouest-africains». In B. Sultan, R. Lalou, M. Amadou Sanni, A. Oumarou, & A. Soumere (Eds.), Les sociétés rurales face aux changements climatiques et environnementaux en Afrique de l'Ouest, Editions IRD, Paris, France. doi:10.13140/RG.2.1.1383.6880

Kristjanson, P., Neufeldt, H., Gassner, A., Sayula, G., Thiede, B., Förch, W., et al. (2012). Are food insecure smallholder households making changes in their farming practices? Evidence from East Africa. *Food Security, 4*, 381–397.

Laube, W., Schraven, B., & Awo, M. (2011). Smallholder adaptation to climate change: Dynamics and limits in Northern Ghana. *Climatic Change, 111*, 753–774.

Maddison, D. (2007). *The perception of and adaptation to climate change in Africa*. CEEPA discussion paper no. 10. Centre for Environmental Economics and Policy in Africa, University of Pretoria, South Africa.

Magnan, A. (2010). *For a better understanding of adaptive capacity to climate change: A conceptual framework*. IDDRI Analyses 2, Paris, France.

Mahé, G., Rouche, N., Dieulin, C., Boyer, J.-F., Ibrahim, B., Cres, A., et al. (2012). *Carte des pluies annuelles en Afrique*. Paris, France: Service Cartographie de l'IRD.

Marenya, P., & Barrett, C. B. (2007). Household-level determinants of adoption of improved natural resources management practices among smallholder farmers in western Kenya. *Food Policy, 32*, 515–536.

Moursi, M., Arimond, M., Dewey, K., Treche, R. M., & Delpeuch, F. (2008). Dietary diversity is a good predictor of the micronutrient density of the diet of 6- to 23-month-old in Madagascar. *Journal of Nutrition, 138*, 2448–2453.
Nelder, A. J. A., & Wedderburn, R. W. M. (1972). Generalized linear models. *Journal of the Royal Statistical Society, 135*(3), 370–384.
Okonya, J. S., Syndikus, K., & Kroschel, J. (2013). Farmers' perception of and coping strategies to climate change: Evidence from six agro-ecological zones of Uganda. *Journal of Agricultural Science, 5*(8), 252–263.
Olsson, L., Opondo, M., & Tschakert, P. (2014). Climate change 2014: Livelihoods and poverty, Contribution of working group II to the fifth assessment report of the intergovernmental panel on Climate Change. In C. B. Field, et al. (Eds.). Cambridge, United Kingdom and New York, NY, USA: Cambridge University Press.
Ouédraogo, M., Dembele, Y., & Some, L. (2010). Perceptions et stratégies d'adaptation aux changements des précipitations: cas des paysans du Burkina Faso. *Sécheresse, 21*(2), 87–96.
Panda, A., Sharma, U., Ninan, K. N., & Patt, A. (2013). Adaptive capacity contributing to improved agricultural productivity at the household level: Empirical findings highlighting the importance of crop insurance. *Global Environmental Change, 23*(4), 782–790.
Pelling, M., & High, C. (2005). Understanding adaptation: What can social capital offer assessments of adaptive capacity? *Global Environmental Change Part A, 15*(4), 308–319.
R Development Core Team. (2015). *R: A language and environment for statistical computing*. R Foundation for Statistical Computing, Vienna, Austria.
Ravallion, M., Datt, G., & Van de walle, D. (1991). Quantifying absolute poverty in the developing world. *Review of Income and Wealth, 37*(4), 345–361.
Ruggeri-Laderchi, C., Saith, R., & Steward, F. (2003). Does it matter that we don't agree on the definition of poverty? A comparison of four approaches. *Oxford Development Studies, 31*(3), 243–274.
Silvestri, S., Bryan, E., Ringler, C., Herrero, M., & Okoba, B. (2012). Climate change perception and adaptation of agro-pastoral communities in Kenya. *Regional Environmental Change, 12* (4), 791–802.
Smit, B., & Pilifosova, O. (2001). Adaptation to climate change in the context of sustainable development and equity in climate change 2001: Impacts, adaptation and vulnerability. In J. J. McCarthy, et al. (Eds.) (pp. 877–912). Cambridge, UK: Cambridge University Press.
Staiger, D., & Stock, J. (1997). Instrumental variables regression with weak instruments. *Econometrica, 65*(3), 557–586.
Stiglitz, J., Sen, A., & Fitoussi, J.-P. (2009). *Rapport de la Commission sur la mesure des performances économiques et du progrès social*. Paris, France: Éditions Odile Jacob.
Swinton, S. M. (1988). Drought survival tactics of subsistence farmers in Niger. *Human Ecology, 16*(2), 123–144.
Tache, B., & Sjaastad, E. (2010). Pastoralists' conceptions of poverty: An analysis of traditional and conventional indicators from Borana, Ethiopia. *World Development, 38*(8), 1168–1178.
Townsend, P. (1993). *The international analysis of poverty*. London, UK: Routledge.
Trærup, S. L. M., & Mertz, O. (2011). Rainfall variability and household coping strategies in northern Tanzania: A motivation for district-level strategies. *Regional Environmental Change, 11*(3), 471–481.
UNDP. (2013). *Human development report*. UN Plaza, New York, USA: United Nations Development Program.
Vincent, K. (2007). Uncertainty in adaptive capacity and the importance of scale. *Global Environmental Change, 17*, 12–24.
Weber, E. U. (2010). What shapes perceptions of climate change? *Psychology, 1*, 332–342.
WFP. (2011). *Analyse Globale de la Vulnérabilité, de la Sécurité Alimentaire et de la Nutrition (AGVSAN)*. Rome, Italy: Programme Alimentaire Mondial.
Wood, S. A., Jina, A. S., Jain, M., Kristjanson, P., & Defries, R. S. (2014). Smallholder farmer cropping decisions related to climate variability across multiple regions. *Global Environmental Change, 25*, 163–172.

Wooldridge, J. M. (2010). *Econometric analysis of cross section and panel data* (2nd ed.). Cambridge, Massachusetts, USA: The MIT Press.
World Bank. (2010). Available at http://data.worldbank.org/indicator/SL.AGR.EMPL.ZS/countries/BJ?display=graph [Accessed January 16, 2016].
Yegbemey, R. N., Yabi, J. A., Tovignan, S. D., Gantoli, G., & Haroll, S. E. (2013). Farmers' decisions to adapt to climate change under various property rights: A case study of maize farming in northern Benin (West Africa). *Land Use Policy, 34,* 168–175.
Yesuf, M., & Bluffstone, R. (2007). *Risk aversion in low income countries.* IFPRI Discussion Paper 00715, Washington, USA.

Vulnerability and Adaptation Strategies of Fishers to Climate Change: Effects on Livelihoods in Fishing Communities in Lagos State, Nigeria

Temitope A. Adewale, B. Tosan Fregene and Ibidun O. Adelekan

1 Introduction

Fishers are a highly vulnerable group of people owing to the nature of their job, which is laden with a lot of physical, environmental and social risks. This is further accentuated by the fact that most fisheries are highly variable by nature and subject to environmental change such as climate change (World Fish Center 2009). In addition, the African coast including Nigeria's coast, is among the global coasts identified to be highly vulnerable to climate change and sea-level rise (French et al. 1995; Turner et al. 1995; Boko et al. 2007; IPCC 2007a; Nicholls et al. 2007). There has been a lot of variation in the weather parameters such as rainfall and temperature. Local impacts of global warming, including sea-level rise and changing seasonal patterns of rainfall observed as decrease or virtual disappearance of the short dry season and reduced number of raindays, have been experienced in Nigeria's coastal zone (Nwilo 1997; Adelekan 2010; Chineke et al. 2010 cited in Adelekan and Fregene 2014). Frequent storm surges have also increased the incidence of coastal flooding.

Climate change has been documented to affect fishery production along many pathways. Fish reproduction, growth and migration patterns are all affected by temperature, rainfall and hydrology (Ficke et al. 2007). Changes in these parameters will shift patterns of species abundance and availability. According to Medugu et al. (2014) recent occurrences in some parts of Nigeria show that changes in precipitation

T.A. Adewale (✉)
Department of Agricultural Education, Michael Otedola College of Primary Education, P.M.B 1028, Noforija, Epe, Lagos, Nigeria
e-mail: adewaleta7@gmail.com

B.T. Fregene
Department of Aquaculture and Fisheries Management, University of Ibadan, Ibadan, Nigeria

I.O. Adelekan
Department of Geography, University of Ibadan, Ibadan, Nigeria

W. Leal Filho et al. (eds.), *Climate Change Adaptation in Africa*,
Climate Change Management, DOI 10.1007/978-3-319-49520-0_46

have affected seasonal flooding patterns that influenced inland fish production in some areas. Drier dry seasons also may threaten stocks of both wild and cultured fish. These changes in fishery production are likely to have the greatest impact on people living in the rural riverine and coastal communities whose primary livelihood activity is fishing. These people are often poorer and more marginal than those who own land and have other primary sources of income. Saltwater intrusion caused by rising sea levels may threaten freshwater fisheries while creating opportunities for catching and cultivating high-value brackish or marine species (World Fish Centre 2007).

Few positive impacts of climate change on fisheries have been reported. However most of the impacts are overwhelmingly negative (IPCC 2007b). Primary productivity is predicted to decline at lower latitudes where the majority of the world's small-scale fisheries are located, thereby reducing productivity of the fisheries (FAO 2008). Increased frequency of extreme weather events will affect the safety of fishers; damage homes, services and infrastructure, and many coastal ecosystems particularly in coastal areas (IPCC 2007b). Johnston et al. (2009) states that the effects of climate change on fisheries will harm the least equipped to cope hence the need to device strategies to mitigate them. Brugere (2015) notes that not only will livelihoods and national economies need to cope with immediate changes and trade-offs imposed by climate change, but they will also need to evolve in a way that allows them to develop positive adaptation mechanisms. Local people especially those in developing countries, are also to be worst hit by the change in climate, it becomes imperative to investigate their perception on the phenomenon and examine their adaptation strategies to the perceived change (Medugu et al. 2014).

Several adaptation measures to the impacts of climate change on fisheries and aquaculture and the fishing communities that depend on the resources for livelihoods have been suggested (World Fish Center 2007; Halls 2009; Williams and Rota 2012). Adoption of coping strategies to the impacts of climate change is crucial for the artisanal fisheries system and the rural fishing communities. Though they contribute very little to global warming but inevitably suffer more for the negative effects. However, the appropriateness and effectiveness of these measures would depend on the characteristics and vulnerability of the fisheries and the fishing communities. The paper therefore aims to assess factors that predispose fishers in the fishing communities of Lagos state to vulnerability of climate change. Various adaptive strategies used by the fishers to cope with the impact of climate variability and change were also identified.

2 Materials and Methods

2.1 The Study Area

The study area was Lagos State, located in the south-western part of Nigeria, with Ikeja as the state capital. Geographically, Lagos State lies between latitude 6°20′

North to 6°40′ North and long 2°45′ East to 4°20′ East. It is bounded in the north and east by Ogun State, in the South by the Guinea coast of the Atlantic Ocean and in the west by Republic of Benin. It occupies an area of 3345 km^2 mass (Osei et al. 2006) with 786.94 km^2 of it being lagoons and creeks in Lagos Island, Ibeju-Lekki, Ikorodu, Badagry and Epe Local Government Area (LGAs). The vegetation of the state ranges from mangrove to rainforest. The state is blessed with many lagoons, inland waters, creeks and coastlines. The people are engaged in various occupations but fishing activities is very prominent. Some of the fishes caught include *Sphyraena spp, Oreochromis niloticus, Sardinella spp Polydactylus quadrifilis* and *Gymnarchus niloticus*.

Lagos State (Fig. 1) consists of 20 LGAs, which are grouped into three Agricultural Development Programme (ADP) zones which are far-eastern, western and eastern zones. The far-eastern zone (Epe, Ibeju-Lekki, Etiosa and Lagos Island) has a wide coverage of the coastal areas with many lagoons and creeks, followed by the western zone (Badagry, Ojo, Amuwo odofin, Alimosho, Agege, Ifako ijaye, Ikeja, Apapa, Surulere, Oshodi and Mushin). The eastern zone (Lagos mainland, Ikorodu, Ajeromi-ifelodun, Kosofe and Somolu) consists mostly the mainland areas which are landlocked; hence, fishing activities are not as predominant as it is in other zones.

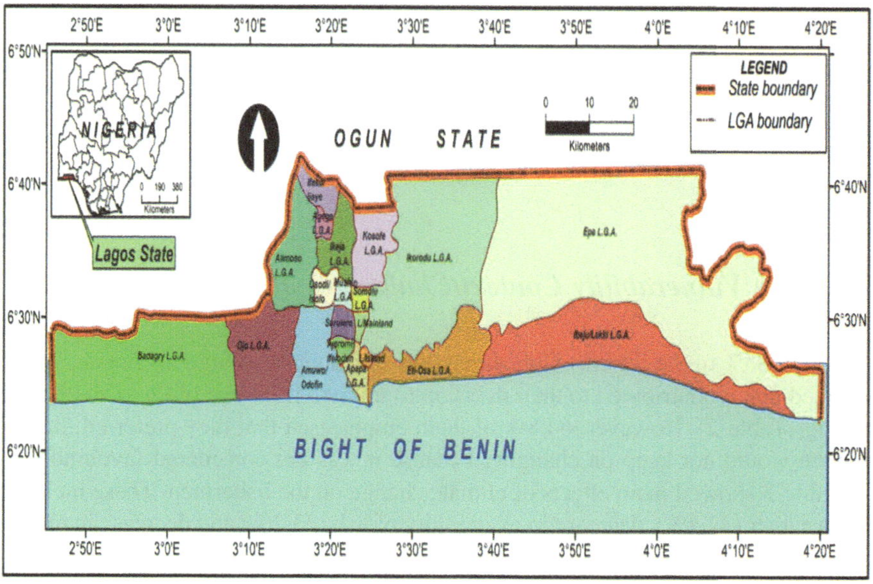

Fig. 1 Lagos state

2.2 *Sampling Technique and Sources of Data*

Samples of fishing households were selected from all the ADP zones and blocks. A two stage stratified sampling technique was used for selecting 247 fishers for the study (Table 1). In the first stage, communities adjacent to the coast, lagoons and rivers where fishing activities are predominant, were selected. Five percent of the fishing households were selected based on the list of registered fishermen with the ADP in the final stage.

The primary data were collected through observation, key informant interview, focus group discussion and collection of field notes. Copies of a structured questionnaire were administered to the fishers. Extension officers were trained as enumerators to collect primary data from the fishers. The questionnaire was designed to elicit responses on the forms of vulnerability of the fishermen, the effect of climate change on fishing as well as the adaptation strategies adopted by the fishermen. Secondary data on registered fishermen and their communities were collected from the office of the Lagos State Agricultural Development Programme (LSADP) and the Ministry of Agriculture and Cooperatives.

2.3 *Data Analysis*

The data collected were analysed using frequency count and simple percentage with graphical illustration.

3 Results

3.1 *The Vulnerability Context/Challenges of the Fishers*

A total of 85% of the fishers claimed that they had noticed change in climate, 74.5% affirmed that it contributed to their decision to migrate from one place to another for fishing (Table 2). However, 64.8% of them emphasised that they preferred that the climate would not keep on changing because it was not considered favourable.

Table 3 showed main effects of climate change on the fishermen. These included bad weather (34.4%), decrease in availability of fish (33.2%) and decrease in quality

Table 1 Selection of sample according to ADP structure

ADP zone	Blocks	No. of fishermen sampled
1. Far-Eastern zone	6	105
2. Eastern zone	4	62
3. Western zone	6	85
Total	16	247

Table 2 Perception of respondents on climate change

Perception	Yes	No	No response	Total
Have you really noticed any change in climate?	210 (85.0)	37 (15.0)	–	247 (100.0)
Does the climate change have anything to do with your decision to migrate?	184 (74.5)	63 (25.5)	–	247 (100.0)
Would you prefer that the climate keep on changing?	36 (14.6)	160 (64.8)	51 (20.6)	247 (100.0)

Table 3 Effects of climate change on the fishing communities

Which of these effect of climate change affects your community	Frequency	Percentage
Risk to life due to bad weather	85	34.4
Decrease in availability of fish	82	33.2
Decrease in quality of fish	30	12.1
Having to land fish further from landing sites	26	10.5
Less stable livelihood	24	9.7
Total	247	100.0

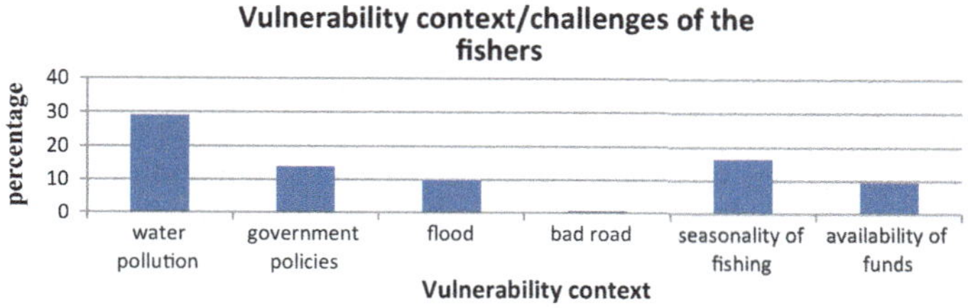

Fig. 2 Factors influencing fishing activities of fishermen

of fish (12.1%). The effects are usually caused by extreme temperatures, which make fish to migrate from their natural habitat or zone in the water bodies to more conducive places as well as lead to quick spoilage. Others were risk to life due to and having to land fish further from landing sites (10.5) and less stable livelihood as a result of fluctuations in fishing period (9.7%).

Figure 2 reveals some of the factors that affect fishing activities. These include water pollution, government policies, flood, bad roads, seasonality in fishing, availability of funds and change in climate. Water pollution, climate change, seasonality in fishing and government policies were the more predominant factors that affected the fishermen.

Fishermen encountered several forms of vulnerabilities. These includes injury, financial handicap, health breakdown susceptibility to STDs and HIV/AIDS, hostility by leaders and members of host communities for migrants, loss of fishing

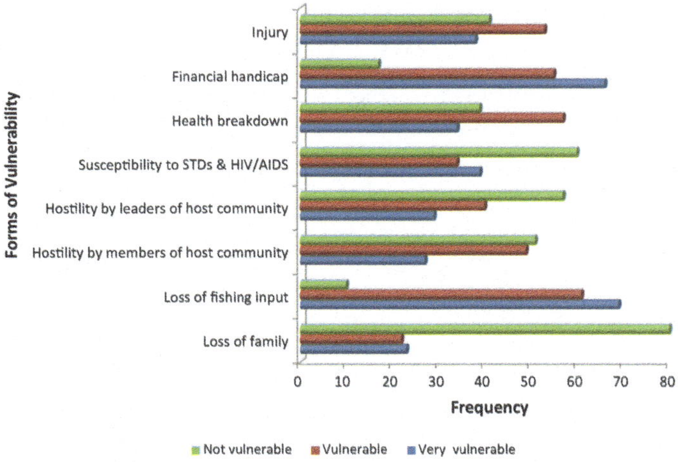

Fig. 3 Vulnerability of fishers in the study area

input and loss of family members (Fig. 3). Financial handicap and loss of fishing input make the fishers very vulnerable. Some of these factors may not be directly caused by climate change but the climate change phenomenon can lead to their occurrence. One factor could also make the fishers to be predisposed to the others. Injury, loss of family member and fishing input can be caused by a storm surge on the sea which leads to financial handicap and breakdown of health. Loss of fishing input and financial handicap could make the fisher decide to migrate to search for a means of livelihood and sustenance in another fishing community. This may make him face hostility by leaders and members of the host community. If he decides to go without his wife and members of his household, he may be involved in illicit relationships making him susceptible to sexually transmitted diseases (STDs) and HIV/AIDS. These are aspects where the fishermen require assistance, which could be provided by both government and non-governmental organisations.

Other areas of vulnerability include the interest rate on loans procured by the fishers, which is usually on the high side and the loans are not easily accessible due to demands for collateral which the fishers may not be able to meet. The information gathered from the fishers revealed that the average interest rate within the period of 2009–2012 ranged from 8.83% to as high as 18.47%. The cooperative societies also demanded for high interest rates especially when the payback period is more than a year. To mitigate this situation, they resorted to borrow from money lenders. These could be friends or relatives that were able to tolerate them for a longer period of time. Also 85% of the fishermen were not involved in any insurance policy, which ought to serve as a palliative in the time of disaster. Those that did insurance included, health insurance (2.4%), property insurance (6.1%), business insurance (0.8%) and life insurance (0.4%).

Table 4 reveals type of social amenities accessible to fishers. The highest was access to primary school education (29.8), while the least was access to the internet

Table 4 Social amenities accessible by the fishers

Social amenities	Frequency	Percentage
Good roads	64	10.6
Health care	97	16.1
Cyber cafe	20	3.3
Electricity	128	21.2
Primary school	181	29.8
Internet facilities	2	0.3
Pipe borne water	9	1.5
Secondary schools	90	14.9
Tertiary school	13	2.3

(0.3%) which is a very important source where they can get access to information on weather forecasting and ways to improve their fishing as well as management of the water bodies as a common resource. Other amenities included healthcare facilities (16.1%), electricity (21.2%) and secondary education (14.9%).

3.2 Adaptation Strategies

Table 5 shows some of the adaptive strategies the fishermen felt they could adopt to prepare for possible future climate change. These included getting modern fishing equipment, sourcing for necessary information that would help them to be better prepared, weather observations, alternative businesses and saving to have enough funds. The fishermen emphasised that they wanted government's intervention. Other points noted were awareness and information on climate change (39.3%), financial assistance (27.1%), provision of aids to assist those that have been affected (17.8%), provision of inputs (9.3%) as well as the need to curb or minimise ocean dredging (6.5%).

Flooding is one of the most frequently occurring disasters in fishing communities. Migration was one of the adaptive strategies used by fishermen to cope with some of the adverse effects of climate change. Others desired r alternative employments to survive. Majority (96.8%) participated in various contributions to prepare for future financial survival in case of accidents. Financial contributions included *esusu/ajo* (18.6%), cooperative contributions (19.8%), personal savings (45.4%), buying shares (6.1%) and bank savings (6.9%).

Table 5 Adaptive Strategies of Fishers

	Frequency
Expected assistance from government	
Provision of housing for flood victims	13 (5.3%)
Provision of fishing tools	19 (7.7%)
Financial provision for affected fishermen	20 (8.1%)
Extension lectures on how to cope	148 (59.9)
None	47 (19.0%)
How do you think you can prepare for possible future climate change	
Acquiring more fishing inputs and modern equipment	93 (37.7%)
Awareness and information	80 (32.4%)
Weather observation	37 (14.9%)
Alternative business	24 (9.7%)
Funds	13 (5.3)
Suggest ways government can help fishermen to cope better	
Awareness and information	97 (39.3%)
Financial assistance	67 (27.1%)
Provision of aids	44 (17.8%)
Provision of inputs	23 (9.3%)
Stop ocean dredging	16 (6.5%)
How to cope with flood situations and accommodation	
Migrate with family members to another favorable location	106 (43.0%)
Move family to stay with some relations and fisher go elsewhere for fishing	33 (13.4%)
I move my family to government settlement and go elsewhere for fishing	8 (3.2%)
I move with my family to government settlements and do other works to survive	9 (3.6%)
Never experienced such situation before	91 (36.8%)
How to cope with extreme temperature situations and fishing	
Migrate to where there is fish	186 (75.3%)
Stay in my location and survive doing other jobs until situation change	31 (12.6%)
Remain idle throughout surviving on others goodwill donations	4 (1.6%)
Never experienced such	18 (7.3%)
Others	8 (3.2%)
How you prepare for future financial survival in case of accidents	
Esusu/Ajo	46 (18.6%)
Cooperative contributions	49 (19.8%)
Personal savings	112 (45.4%)
Buying shares	15 (6.1%)
Bank saving	17 (6.9%)
No attempt	8 (3.2%)

4 Discussion

International Panel on Climate Change (IPCC) which examined implications of projected climate for fishery production concluded that the sector is vulnerable to and has potential to be strongly impacted by climate change (Bates et al. 2006). The forms of vulnerability caused by climate change included less stable livelihood, decrease in availability of fish, decrease in quality of fish, risk to life due to bad weather and having to land fish further from landing sites. Others included injury, financial handicap, health breakdown, susceptibility to STDs and HIV/AIDS, hostility by leaders and members of the host community for migrants, loss of fishing input and loss of family members. Involvement in insurance policies by the fishers were at a very low ebb, showing the need to educate fishers on the need for insurance. Stanley (2012) stresses the need to encourage insurance programmes to mitigate risk. Insurance companies covering property development in coastal areas could assess the potential for sea level rise, increased storm severity, flooding and other climate change impacts on fishers and incorporate appropriate measures in their documents. It would also be necessary to take a cursory look at the terms and conditions offered to fishers, which could have deterred them from taking up the policies.

According to Adelekan and Fregene (2014), over-dependence on fisheries resources and limited diversification of livelihoods are key factors affecting low adaptive capacity and therefore, vulnerability. A key to successful adaptation is diverse livelihoods which helps to ensure that if one economic option temporarily closes, people can resort to other options for making a living (World Fish Center 2009). Fishers need to be encouraged to take up alternative businesses, such as aquaculture, to help reduce the adverse effects of vulnerability. Indigenous people are not only keen observers of climate change but also actively trying to adapt to the changing conditions (Salick and Byg 2007; Macchi et al. 2008), the fishers sampled affirmed that awareness and information would enable them to be better prepared for the event of future climate change. Government institutions, like Nigerian Meteorological Agency (NIMET) should work on more accurate predictions and warning systems. Timely and quality information that will assist in disaster aversion and emergency management strategies to minimize losses should be incorporated.

5 Conclusion

The limitations to this work include difficulty in meeting with the fishers both at first contact as well as during subsequent visits due to the nature of their work and poor access routes to some of the fishing communities. The study showed there were various factors that make fishers to be vulnerable to climate change. But some may not be caused directly by climate change. However, the incidence of climate change makes them vulnerability indicators. The fishers prefer that the climate does not continue to change because of its adverse effect on their livelihood. They migrate from one fishing community to another as one of their adaptive strategies. There should be more efforts aimed to provide necessary information which could assist them prepare adequately for future climate change. They also need effective government extension services as well as non-governmental agencies to get them informed about the need for other sources of income and assistance to set up alternative businesses.

References

Adelekan, I., & Fregene, T. (2014). *Vulnerability of artisanal fishing communities to flood risks in Coastal Southwest Nigeria, climate and development*. London-UK: Taylor and Francis. doi:10.1080/17565529.2014

Adelekan, I. O. (2010). Vulnerability of poor urban communities to flooding in Lagos, Nigeria. *Environment and Urbanization, 22*(2), 433–450.

Bates, S. P., et al. (2006). Experience-based design: from redesigning the system around the patient to co-designing services with the patient. *Quality and Safety in Health care, 15*, 307–310.

Boko, M., Niang, I., Nyong, A., Vogel, C., Githeko, A., Medany, M., et al. (2007). Africa. In M. L. Parry, O. F. Canziani, J. P. Palutikof, P. J. Van Der Linden, & C. E. Hanson (Eds.), *Climate change 2007: Impacts, adaptation and vulnerability. Contributions of Working Group II to the fourth assessment report of the Intergovernmental Panel on Climate Change* (pp. 433–467). Cambridge: Cambridge University Press.

Brugere, C. (2015). *Climate change vulnerability in fisheries and aquaculture: A synthesis of six regional studies*. FAO Fisheries Circular No.1104, Rome, FAO, 88 p.

FAO. (2008). Climate change implications for fisheries and aquaculture. *The state of fisheries and aquaculture 2008* (pp. 87–91). Rome, Italy: FAO.

Ficke, A. D., Myrick, C. A., & Hansen L. J. (2007). Potential impacts of global climate change on freshwater fisheries. *Reviews in Fish Biology and Fisheries, 17*(4), 581–613. Springer Science +Business Media B.V. 2007.

French, G. T., Awosika, L. F., & Ibe, C. E. (1995). Sea-level rise and Nigeria: Potential impacts and consequences. *Journal of Coastal Research, 14*, 224–242. http://www.jstor.org/stable/25735710

Halls, A. S. (2009). Addressing fisheries in the climate change and adaptation initiative. *Fish Res Dev Melkong Region, 15*.

IPCC. (2007a). Summary for policymakers. In S. Solomon, D. Qin, M. Manning et al. (Eds.), *Climate Change 2007: The Physical Science Basis. Contribution of the Working Group I to the Fourth Assessment Report of the Intergovernmental Panel on Climate Change*. Cambridge University Press, Cambridge, UK, 22 p. Available at http://www.ipcc.ch/pdf/assessment-report/ar4/wg1/ar4-wg1-spm.pdf. Date accessed December 6, 2010.

IPCC. (2007b). Climate change 2007: Synthesis report. Contribution of working groups I, II and III to the fourth assessment report of the Intergovernmental Panel on Climate Change. In R. K. Pachauri & A. Reisinger (Eds.), *Intergovernmental Panel on Climate Change*. Geneva, Switzerland, 8 p.

Johnston, C. A., Zedler, J. B., Tulbure, M. G., Frieswyk, C. B., Bedford, B. L., & Vaccaro, L. (2009). A unifying approach for evaluating the condition of wetland plant communities and identifying related stressors. *Ecological Applications, 19*, 1739–1757.

Macchi, M., Oviedo, G., Gotheil, S., Cross, K., Boedhihartono, A., Wolfangel, C., et al. (2008). *Indigenous and traditional peoples and climate change*. IUCN Issues paper. Gland, Switzerland.

Medugu, I. N., Majid, M. R., & Leal, W. (2014). Assessing the vulnerability of farmers, fishermen and herdsmen to climate change: A case study from Nigeria. *International Journal of Global Warming, 6*(1).

Nicholls, R. J., Wong, P. P., Burkett, V. R., Codignotto, J. O., Hay, J. E., Maclean, R. E., … Woodroffe, C. D. (2007). Coastal systems and low-lying areas. Climate change, 2007. Impacts, adaptation and vulnerability. In M. L. Parry, O. F. Canziani, J. P. Palutikof, P. J. Vander Linden, & C. E. Hanson (Eds.), *Contribution of Working Group II to the fourth assessment report of the Intergovernmental Panel On Climate Change* (pp. 315–356). Cambridge: Cambridge University Press.

Nwilo P. C. (1997). *Managing the impacts of storms surges on Victoria Island, Lagos* (IAHS Publications No. 239, pp. 325–330). Available from iahs.info.

Osei, A. J., Merem, E. C., & Twumasi Y. A. (2006). Use of GIS and remote sensing technology as a decision support tool in land administration: The case of Lagos, Nigeria. In Promoting Land Administration and Good Governance 5th FIG Regional Conference Accra, Ghana, March 8–11, 2006.

Salick, J., & Byg, A. (Eds.). (2007). *Indigenous people and climate change*. Oxford, UK: Tyndall Centre.

Stanley, I. (2012). *Nigeria and climate change adaptation*. ISSP, Insight.

Turner, R. K., Doktor, P., & Adger, W. N. (1995). Assessing the costs of sea level rise. *Environment and Planning, 27,* 1777–17796.

Williams, L., & Rota, A. (2012). Impact of climate change of fisheries and aquaculture in the developing world and opportunities for adaptation. IFAD Fisheries Thematic Paper: Tool for Project Design. http://www.ifad.org/lrkm/pub/fisheries.pdf

World Fish Center. (2007). *Fisheries and aquaculture can provide solutions to cope with climate change*. Issues brief. Penang, Malaysia: World Fish Center.

World Fish Center. (2009). *Climate change and fisheries-vulnerability and adaptation in Cambodia*. Pubs. ICLARM. Net resource-centre. Penang, Malaysia: World Fish Center.

The manufacturer's authorised representative in the EU is Springer Nature Customer Service Centre GmbH, Europaplatz 3, 69115 Heidelberg, Germany. If you have any concerns regarding our products, please contact ProductSafety@springernature.com

Printed and bound by CPI Group (UK) Ltd, Croydon, CR0 4YY
30/09/2026
02225229-0020